Applied Regression Analysis

Applied Regression Analysis

THIRD EDITION

Norman R. Draper

Harry Smith

A Wiley-Interscience Publication

JOHN WILEY & SONS, INC.

New York · Chichester · Weinheim · Brisbane · Singapore · Toronto

This book is printed on acid-free paper. ∞

Copyright © 1998 by John Wiley & Sons, Inc. All rights reserved.

Published simultaneously in Canada.

Library of Congress Cataloging-in-Publication Data:

Draper, Norman Richard.
    Applied regression analysis / N.R. Draper, H. Smith. — 3rd ed.
        p.   cm. — (Wiley series in probability and statistics.   Texts
    and references section)
    "A Wiley-Interscience publication."
    Includes bibliographical references (p.   –   ) and index.
    ISBN 0-471-17082-8 (acid-free paper)
    1. Regression analysis.   I. Smith, Harry, 1923–   . II. Title.
    III. Series.
    QA278.2.D7   1998
    519.5′36—dc21   97-17969
                                                              CIP

Printed in the United States of America.

10 9 8 7

Contents

Preface to the Third Edition

The second edition had 10 chapters; this edition has 26. On the whole (but not entirely) we have chosen to use smaller chapters, and so distinguish more between different types of material. The tabulation below shows the major relationships between second edition and third edition sections and chapters.

Material dropped consists mainly of second edition Sections 6.8 to 6.13 and 6.15, Sections 7.1 to 7.6, and Chapter 8. New to this edition are Chapters 16 on multicollinearity, 18 on generalized linear models, 19 on mixture ingredients, 20 and 21 on the geometry of least squares, 25 on robust regression, and 26 on resampling procedures. Small revisions have been made even in sections where the text is basically unchanged. Less prominence has been given to printouts, which nowadays can easily be generated due to the excellent software available, and to references and bibliography, which are now freely available (either in book or computer form) via the annual updates in *Current Index to Statistics*. References are mostly given in brief either in situ or close by, at the end of a section or chapter. Full references are in a bibliography but some references are also given in full in sections or within the text or in exercises, whenever this was felt to be the appropriate thing to do. There is no precise rule for doing this, merely the authors' predilection. Exercises have been grouped as seemed appropriate. They are intended as an expansion to the text and so most exercises have full or partial solutions; there are a very few exceptions. One hundred and one true/false questions have also been provided; all of these are in "true" form to prevent readers remembering erroneous material. Instructors can reword them to create "false" questions easily enough. Sections 24.5 and 24.6 have some duplication with work in Chapter 20, but we decided not to eliminate this because the sections contain some differences and have different emphases. Other smaller duplications occur; in general, we feel that duplication is a good feature, and so we do not avoid it.

Our viewpoint in putting this book together is that it is desirable for students of regression to work through the straight line fit case using a pocket calculator and then to proceed quickly to analyzing larger models on the computer. We are aware that many instructors like to get on to the computer right away. Our personal experience is that this can be unwise and, over the years, we have met many students who enrolled for our courses saying "I know how to put a regression on the computer but I don't understand what I am doing." We have tried to keep such participants constantly in mind.

We have made no effort to explain any of the dozens of available computing systems. Most of our specific references to these were removed after we received reviews of an earlier draft. Reviewers suggested we delete certain specifics and replace them by others. Unfortunately, the reviewers disagreed on the specifics! In addition, many specific program versions quickly become obsolete as new versions are issued. Quite often students point out to us in class that "the new version of BLANK does (or doesn't!) do that now." For these reasons we have tried to stay away from advocating any particular way to handle computations. A few mild references to MINITAB (used in our University of Wisconsin classes) have been retained but readers will find it easy to ignore these, if they wish.

We are grateful for help from a number of people, many of these connected with N. R. Draper at the University of Wisconsin. Teaching assistants contributed in many ways, by working new assignments, providing class notes of lectures spoken but not recorded, and discussing specific problems. Former University of Wisconsin student Dennis K. J. Lin, now a faculty member at Pennsylvania State University, contributed most in this regard. More generally, we profited from teaching for many years from the excellent Wiley book *Linear Regression Analysis,* by George A. F. Seber, whose detailed algebraic treatment has clearly influenced the geometrical presentations of Chapters 20 and 21.

N. R. Draper is grateful to the University of Wisconsin and to his colleagues there for a timely sabbatical leave, and to Professor Friedrich Pukelsheim of the University of Augsburg, Germany, for inviting him to spend the leave there, providing full technical facilities and many unexpected kindnesses as well. Support from the German Alexander von Humboldt Stiftung is also gratefully acknowledged. N. R. Draper is also thankful to present and former faculty and staff at the University of Southampton, particularly Fred (T. M. F.) Smith, Nye (J. A.) John (now at Waikato University, New Zealand), Sue Lewis, Phil Prescott, and Daphne Turner, all of whom have made him most welcome on annual visits for many years. The enduring influence of R. C. Bose (1901–1987) is also gratefully acknowledged.

The staff at the Statistics Department, Mary Esser (staff supervisor, retired), Candy Smith, Mary Ann Clark (retired), Wanda Gray (retired), and Gloria Scalissi, have all contributed over the years. Our special thanks go to Gloria Scalissi who typed much of a difficult and intricate manuscript.

For John Wiley & Sons, the effects of Bea Shube's help and wisdom linger on, supplemented more recently by those of Kate Roach, Jessica Downey, and Steve Quigley. We also thank Alison Bory on the editorial side and Production Editor Lisa Van Horn for their patience and skills in the final stages.

We are grateful to all of our reviewers, including David Belsley and Richard (Rick) Chappell and several anonymous ones. The reviews were all very helpful and we followed up most of the suggestions made, but not all. We ourselves have often profited by reading varying presentations in different places and so we sometimes resisted changing our presentation to conform to presentations elsewhere.

Many others contributed with correspondence or conversation over the years. We do not have a complete list, but some of them were Cuthbert Daniel, Jim Durbin, Xiaoyin (Frank) Fan, Conrad Fung, Stratis Gavaris, Michael Haber, Brian Joiner, Jane Kawasaki, Russell Langley, A. G. C. Morris, Ella Munro, Vedula N. Murty, Alvin P. Rainosek, J. Harold Ranck, Guangheng (Sharon) Shen, Jake Sredni, Daniel Weiner, William J. Welch, Yonghong (Fred) Yang, Yuyun (Jessie) Yang, and Lisa Ying. Others are mentioned within the text, where appropriate. We are grateful to them all.

To notify us of errors or misprints, please e-mail to draper@stat.wisc.edu. An updated list of such discrepancies will be returned e-mail, if requested. For a hardcopy of the list, please send a stamped addressed envelope to N. R. Draper, University of Wisconsin Statistics Department, 1210 West Dayton Street, Madison, WI 53706, U.S.A.

NORMAN R. DRAPER
HARRY SMITH

Relationships of Second Edition and Third Edition Text Material

Topic	Second Edition	Third Edition	Topic	Second Edition	Third Edition
Straight line fit	1.0–1.4	1.0–1.5	Polynomial models	5.1, 5.2	12.1
Pure error	1.5	2.1–2.2	Transformations	5.3	13
Correlation	1.6	1.6	Dummy variables	5.4	14
Inverse regression	1.7	3.2	Centering and scaling	5.5	16.2, 16.3
Practical implications	1.8	3.3	Orthogonal polynomials	5.6	22.2
			Orthogonalizing $\mathbf{X}$	5.7	16A
Straight line, matrices	2.0–2.5	4	Summary data	5.8	22.3
General regression	2.6	5			
Extra SS	2.7–2.9	6.1, 6.2, 6A	Selection procedures	6.0–6.6, 6.12	15
General linear hypothesis	2.10	9.1	Ridge regression	6.7	17
Weighted least			Ridge, canonical form	6A	17A
squares	2.11	9.2, 9.3, 9.4	Press	6.8	—
Restricted least			Principal components	6.9	—
squares	2.13	9.5	Latent root regression	6.10	—
Inverse regression	2.15	9.6	Stagewise regression	6.13	—
Errors in multiple X's	—	9.7	Robust regression	6.14	25
Bias in estimates	2.12	10			
Errors in X and Y	2.14	3.4	Data example	7.0–7.6	—
Inverse regression	2.15	9.6	Polynomial example	7.7	12.2
Matrix results	2A	5A			
E (Extra SS)	2B	10.4	Model building talk	8	—
How significant?	2C	11			
Lagrange's multipliers	2D	9A	ANOVA models	9	23
			Nonlinear estimation	10	24
Residuals plots	3.1–3.8	2	Multicollinearity	—	16
Serial correlation	3.9–3.11	7	GLIM	—	18
Influential observations	3.12	8.3, 8.4	Mixtures models	—	19
Normal plots	3A	2A	Geometry of LS	—	20
Two X's example	4.0, 4.2	6.3	More geometry	—	21.1–21.6
Geometry	4.1	21.7	Robust regression	—	25
			Resampling methods	—	26

About the Software

The diskette that accompanies the book includes data files for the examples used in the chapters and for the exercises. These files can be used as input for standard statistical analysis programs. When writing program scripts, please note that descriptive text lines are included above data sections in the files.

The data files are included in the REGRESS directory on the diskette, which can be placed on your hard drive by your computer operating system's usual copying methods. You can also use the installation program on the diskette to copy the files by doing the following.

1. Type a:install at the Run selection of the File menu in a Windows 3.1 system or access the floppy drive directory through a Windows file manager and double click on the INSTALL.EXE file.
2. After skipping through the introductory screens, select a path for installing the files. The default directory for the file installation is C:\REGRESS. You may edit this selection to choose a different drive or directory. Press Enter when done.
3. The files will be installed to the selected directory.

Applied Regression Analysis

CHAPTER 0

Basic Prerequisite Knowledge

Readers need some of the knowledge contained in a basic course in statistics to tackle regression. We summarize some of the main requirements very briefly in this chapter. Also useful is a pocket calculator capable of getting sums of squares and sums of products easily. Excellent calculators of this type cost about $25–50 in the United States. Buy the most versatile you can afford.

0.1. DISTRIBUTIONS: NORMAL, *t*, AND *F*

Normal Distribution

The normal distribution occurs frequently in the natural world, either for data "as they come" or for transformed data. The heights of a large group of people selected randomly will look normal in general, for example. The distribution is symmetric about its mean μ and has a standard deviation σ, which is such that practically all of the distribution (99.73%) lies inside the range $\mu - 3\sigma \le x \le \mu + 3\sigma$. The frequency function is

$$f(x) = \frac{1}{\sigma(2\pi)^{1/2}} \exp\left(\frac{(x - \mu)^2}{-2\sigma^2}\right), \qquad -\infty \le x \le \infty. \qquad (0.1.1)$$

We usually write that $x \sim N(\mu, \sigma^2)$, read as "x is normally distributed with mean μ and variance σ^2." Most manipulations are done in terms of the *standard normal* or *unit normal* distribution, $N(0, 1)$, for which $\mu = 0$ and $\sigma = 1$. To move from a general normal variable x to a standard normal variable z, we set

$$z = (x - \mu)/\sigma. \qquad (0.1.2)$$

A standard normal distribution is shown in Figure 0.1 along with some properties useful in certain regression contexts. All the information shown is obtainable from the normal table in the Tables section. Check that you understand how this is done. Remember to use the fact that the total area under each curve is 1.

Gamma Function

The gamma function $\Gamma(q)$, which occurs in Eqs. (0.1.3) and (0.1.4), is defined as an integral in general:

$$\Gamma(q) = \int_0^\infty e^{-x} x^{q-1} \, dx.$$

1

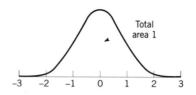

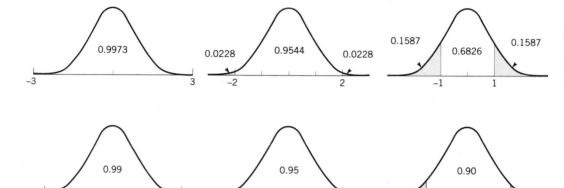

Figure 0.1. The standard (or unit) normal distribution $N(0, 1)$ and some of its properties.

However, it is easier to think of it as a generalized factorial with the basic property that, for any q,

$$\Gamma(q) = (q - 1)\Gamma(q - 1)$$
$$= (q - 1)(q - 2)\Gamma(q - 2),$$

and so on. Moreover,

$$\Gamma(\tfrac{1}{2}) = \pi^{1/2} \quad \text{and} \quad \Gamma(1) = 1.$$

So, for the applications of Eqs. (0.1.3) and (0.1.4), where ν, m, and n are integers, the gamma functions are either simple factorials or simple products ending in $\pi^{1/2}$.

Example 1

$$\Gamma(5) = 4 \times \Gamma(4) = 4 \times 3 \times \Gamma(3) = 4 \times 3 \times 2 \times \Gamma(2)$$
$$= 4 \times 3 \times 2 \times 1 \times \Gamma(1) = 24.$$

Example 2

$$\Gamma(\tfrac{5}{2}) = \tfrac{3}{2} \times \Gamma(\tfrac{3}{2}) = \tfrac{3}{2} \times \tfrac{1}{2} \times \Gamma(\tfrac{1}{2}) = 3\pi^{1/2}/4.$$

t-Distribution

There are many t-distributions, because the form of the curve, defined by

$$f_\nu(t) = \frac{\Gamma\left(\dfrac{\nu + 1}{2}\right)}{(\nu\pi)^{1/2}\Gamma\left(\dfrac{\nu}{2}\right)} \left(1 + \frac{t^2}{\nu}\right)^{-(\nu + 1)/2} \qquad (-\infty \leq t \leq \infty), \tag{0.1.3}$$

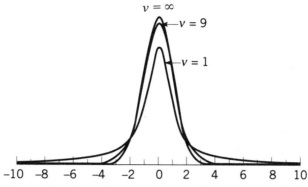

Figure 0.2. The *t*-distributions for $\nu = 1, 9, \infty$; $t(\infty) = N(0, 1)$.

depends on ν, the number of degrees of freedom. In general, the $t(\nu)$ distribution looks somewhat like a standard (unit) normal but is "heavier in the tails," and so lower in the middle, because the total area under the curve is 1. As ν increases, the distribution becomes "more normal." In fact, $t(\infty)$ *is* the $N(0, 1)$ distribution, and, when ν exceeds about 30, there is so little difference between $t(\nu)$ and $N(0, 1)$ that it has become conventional (but not mandatory) to use the $N(0, 1)$ instead. Figure 0.2 illustrates the situation. A two-tailed table of percentage points is given in the Tables section.

F-Distribution

The *F*-distribution depends on two separate degrees of freedom m and n, say. Its curve is defined by

$$f_{m,n}(F) = \frac{\Gamma\left(\dfrac{m + n}{2}\right)\left(\dfrac{m}{n}\right)^{m/2}}{\Gamma\left(\dfrac{m}{2}\right)\Gamma\left(\dfrac{n}{2}\right)} \frac{F^{m/2-1}}{(1 + mF/n)^{(m+n)/2}} \qquad (F \geq 0). \qquad (0.1.4)$$

The distribution rises from zero, sometimes quite steeply for certain m and n, and reaches a peak, falling off very skewed to the right. See Figure 0.3. Percentage points for the upper tail levels of 10%, 5%, and 1% are in the Tables section.

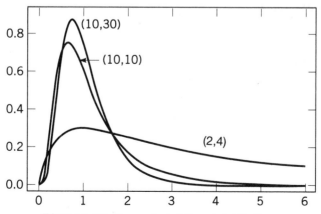

Figure 0.3. Some selected $f(m, n)$ distributions.

The F-distribution is usually introduced in the context of testing to see whether two variances are equal, that is, the null hypothesis that H_0: $\sigma_1^2/\sigma_2^2 = 1$, versus the alternative hypothesis that H_1: $\sigma_1^2/\sigma_1^2 \neq 1$. The test uses the statistic $F = s_1^2/s_2^2, s_1^2$ and s_2^2 being statistically independent estimates of σ_1^2 and σ_2^2, with ν_1 and ν_2 degrees of freedom (df), respectively, and depends on the fact that, if the two samples that give rise to s_1^2 and s_2^2 are independent and normal, then $(s_1^2/s_2^2)/(\sigma_1^2/\sigma_2^2)$ follows the $F(\nu_1, \nu_2)$ distribution. Thus *if* $\sigma_1^2 = \sigma_2^2$, $F = s_1^2/s_2^2$ follows $F(\nu_1, \nu_2)$. When given in basic statistics courses, this is usually described as a two-tailed test, which it usually is. In regression applications, it is typically a one-tailed, upper-tailed test. This is because regression tests always involve putting the "s^2 that could be too big, but cannot be too small" at the top and the "s^2 that we think estimates the true σ^2 well" at the bottom of the F-statistic. In other words, we are in the situation where we test H_0: $\sigma_1^2 = \sigma_2^2$ versus H_1: $\sigma_1^2 > \sigma_2^2$.

0.2. CONFIDENCE INTERVALS (OR BANDS) AND *t*-TESTS

Let θ be a parameter (or "thing") that we want to estimate. Let $\hat{\theta}$ be an estimate of θ ("estimate of thing"). Typically, $\hat{\theta}$ will follow a normal distribution, either exactly because of the normality of the observations in $\hat{\theta}$, or approximately due to the effect of the Central Limit Theorem. Let $\sigma_{\hat{\theta}}$ be the standard deviation of $\hat{\theta}$ and let $se(\hat{\theta})$ be the standard error, that is, the estimated standard deviation, of $\hat{\theta}$ ("standard error of thing"), based on ν degrees of freedom. Typically we get $se(\hat{\theta})$ by substituting an estimate (based on ν degrees of freedom) of an unknown standard deviation into the formula for $\sigma_{\hat{\theta}}$.

1. A $100(1 - \alpha)\%$ confidence interval (CI) for the parameter θ is given by

$$\hat{\theta} \pm t(\nu, 1 - \alpha/2)se(\hat{\theta}) \tag{0.2.1}$$

where $t_\nu(1 - \alpha/2)$ is the percentage point of a t-variable with ν degrees of freedom (df) that leaves a probability $\alpha/2$ in the upper tail, and so $1 - \alpha/2$ in the lower tail. A two-tailed table where these percentage points are listed under the heading of $2(\alpha/2) = \alpha$ is given in the Tables section. Equation (0.2.1) in words is

$$\left\{\begin{array}{l}\text{Estimate}\\ \text{of thing}\end{array}\right\} \pm \left\{\begin{array}{l}\text{A } t \text{ percentage point}\\ \text{leaving } \alpha/2 \text{ in the}\\ \text{upper tail, based on}\\ \nu \text{ degrees of freedom}\end{array}\right\} \left\{\begin{array}{l}\text{Standard error}\\ \text{of estimate}\\ \text{of thing}\end{array}\right\}. \tag{0.2.2}$$

2. To test $\theta = \theta_0$, where θ_0 is some specified value of θ that is presumed to be valid (often $\theta_0 = 0$ in tests of regression coefficients) we evaluate the statistic

$$t = \frac{\hat{\theta} - \theta_0}{se(\hat{\theta})} \tag{0.2.3}$$

or, in words,

$$t = \frac{\left\{\begin{array}{l}\text{Estimate}\\ \text{of thing}\end{array}\right\} - \left\{\begin{array}{l}\text{Postulated or test}\\ \text{value of thing}\end{array}\right\}}{\left\{\begin{array}{l}\text{Standard error of}\\ \text{estimate of thing}\end{array}\right\}}. \tag{0.2.4}$$

(a)

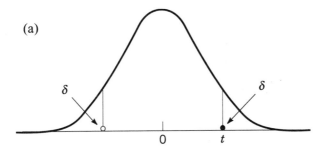

(b)

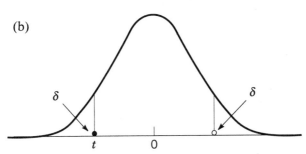

Figure 0.4. Two cases for a *t*-test. (*a*) The observed *t* is positive (black dot) and the upper tail area is δ. A two-tailed test considers that this value could just as well have been negative (open "phantom" dot) and quotes "a two-tailed *t*-probability of 2δ." (*b*) The observed *t* is negative; similar argument, with tails reversed.

This "observed value of *t*" (our "dot") is then placed on a diagram of the $t(\nu)$ distribution. [Recall that ν is the number of degrees of freedom on which $\mathrm{se}(\hat{\theta})$ is based and that is the number of df in the estimate of σ^2 that was used.] The tail probability beyond the dot is evaluated and doubled for a two-tail test. See Figure 0.4 for the probability 2δ. It is conventional to ask if the 2δ value is "significant" or not by concluding that, if $2\delta < 0.05$, *t* is significant and the idea (or hypothesis) that $\theta = \theta_0$ is unlikely and so "rejected," whereas if $2\delta > 0.05$, *t* is nonsignificant and we "do not reject" the hypothesis $\theta = \theta_0$. The alternative hypothesis here is $\theta \neq \theta_0$, a two-sided alternative. Note that the value 0.05 is not handed down in holy writings, although we sometimes talk as though it is. Using an "alpha level" of $\alpha = 0.05$ simply means we are prepared to risk a 1 in 20 chance of making the wrong decision. If we wish to go to $\alpha = 0.10$ (1 in 10) or $\alpha = 0.01$ (1 in 100), that is up to us. Whatever we decide, we should remain consistent about this level throughout our testing.

However, it is pointless to agonize too much about α. A journal editor who will publish a paper describing an experiment if $2\delta = 0.049$, but will not publish it if $2\delta = 0.051$ is placing a purely arbitrary standard on the work. (Of course, it is the editor's right to do that.) Such an attitude should not necessarily be imposed by experimenters on themselves, because it is too restrictive a posture in general. Promis-

T A B L E 0.1. Example Applications for Formulas (0.2.1)–(0.2.4)

Situation	θ	$\hat{\theta}$	$se(\hat{\theta})$
Straight line fit $Y = \beta_0 + \beta_1 X + \varepsilon$	β_1	b_1	$\dfrac{s}{S_{XX}^{1/2}}, S_{XX} = \Sigma(X_i - \overline{X})^2$
	β_0	b_0	$s\left\{\dfrac{\Sigma X_i^2}{nS_{XX}}\right\}^{1/2}$
Predicted response $\hat{Y}_0 = b_0 + b_1 X_0$ at $X = X_0$	$E(Y)$ at X_0	$\hat{Y}_0$	$s\left\{\dfrac{1}{n} + \dfrac{(X_0 - \overline{X})^2}{S_{XX}}\right\}^{1/2}$

ing experimental leads need to be followed up, even if the arbitrary α standard has not been attained. For example, an $\alpha = 0.05$ person might be perfectly justified in following up a $2\delta = 0.06$ experiment by performing more experimental runs to further elucidate the results attained by the first set of runs. To give up an avenue of investigation merely because one experiment did not provide a significant result may be a mistake. The α value should be thought of as a guidepost, not a boundary.

In every application of formulas (0.2.1)–(0.2.4), we have to ask what θ, $\hat{\theta}$, θ_0, $se(\hat{\theta})$, and the t percentage point are. The actual formulas we use are always the same. Table 0.1 contains some examples of θ, $\hat{\theta}$, and $se(\hat{\theta})$ we shall subsequently meet. (The symbol s replaces the σ of the corresponding standard deviation formulas.)

0.3. ELEMENTS OF MATRIX ALGEBRA

Matrix, Vector, Scalar

A $p \times q$ matrix **M** is a rectangular array of numbers containing p rows and q columns written

$$\mathbf{M} = \begin{bmatrix} m_{11} & m_{12} & \cdots & m_{1q} \\ m_{21} & m_{22} & \cdots & m_{2q} \\ \vdots & \vdots & \ddots & \vdots \\ m_{p1} & m_{p2} & \cdots & m_{pq} \end{bmatrix}.$$

For example,

$$\mathbf{A} = \begin{bmatrix} 4 & 1 & 3 & 7 \\ -1 & 0 & 2 & 2 \\ 6 & 5 & -2 & 1 \end{bmatrix}$$

is a 3 × 4 matrix. The plural of *matrix* is *matrices*. A "matrix" with only one row is called a *row vector*: a "matrix" with only one column is called a *column vector*. For example, if

$$\mathbf{a}' = [1, 6, 3, 2, 1] \quad \text{and} \quad \mathbf{b} = \begin{bmatrix} -1 \\ 0 \\ 1 \end{bmatrix},$$

then $\mathbf{a}'$ is a row vector of length five and $\mathbf{b}$ is a column vector of length three. A 1×1 "vector" is an ordinary number or *scalar*.

Equality

Two matrices are equal if and only if their dimensions are identical and they have exactly the same entries in the same positions. Thus a matrix equality implies as many individual equalities as there are terms in the matrices set equal.

Sum and Difference

The sum (or difference) of two matrices is the matrix each of whose elements is the sum (or difference) of the corresponding elements of the matrices added (or subtracted). For example,

$$\begin{bmatrix} 7 & 6 & 9 \\ 4 & 2 & 1 \\ 6 & 5 & 3 \\ 2 & 1 & 4 \end{bmatrix} - \begin{bmatrix} 1 & 2 & 4 \\ -1 & 3 & -2 \\ 6 & 2 & 1 \\ 7 & 0 & 2 \end{bmatrix} = \begin{bmatrix} 6 & 4 & 5 \\ 5 & -1 & 3 \\ 0 & 3 & 2 \\ -5 & 1 & 2 \end{bmatrix}.$$

The matrices must be of exactly the same dimensions for addition or subtraction to be carried out. Otherwise the operations are not defined.

Transpose

The transpose of a matrix $\mathbf{M}$ is a matrix $\mathbf{M}'$ whose rows are the columns of $\mathbf{M}$ and whose columns are the rows of $\mathbf{M}$ in the same original order. Thus for $\mathbf{M}$ and $\mathbf{A}$ as defined above,

$$\mathbf{M}' = \begin{bmatrix} m_{11} & m_{21} & \cdots & m_{p1} \\ m_{12} & m_{22} & \cdots & m_{p2} \\ \vdots & \vdots & \ddots & \vdots \\ m_{1q} & m_{2q} & \cdots & m_{pq} \end{bmatrix},$$

$$\mathbf{A}' = \begin{bmatrix} 4 & -1 & 6 \\ 1 & 0 & 5 \\ 3 & 2 & -2 \\ 7 & 2 & 1 \end{bmatrix}.$$

Note that the transpose notation enables us to write, for example,

$$\mathbf{b}' = (-1, 0, 1) \quad \text{or alternatively} \quad \mathbf{b} = (-1, 0, 1)'.$$

Note: The parentheses around a matrix or vector can be square-ended or curved. Often, capital letters are used to denote matrices and lowercase letters to denote vectors. Boldface print is often used, but this is not universal.

Symmetry

A matrix $\mathbf{M}$ is said to be *symmetric* if $\mathbf{M}' = \mathbf{M}$.

Multiplication

Suppose we have two matrices, $\mathbf{A}$, which is $p \times q$, and $\mathbf{B}$, which is $r \times s$. They are *conformable* for the product $\mathbf{C} = \mathbf{AB}$ only if $q = r$. The resulting product is then a $p \times s$ matrix, the multiplication procedure being defined as follows: If

$$\mathbf{A} = \begin{bmatrix} a_{11} & a_{12} & \cdots & a_{1q} \\ a_{21} & a_{22} & \cdots & a_{2q} \\ \cdot & \cdot & \cdot & \cdot \\ \cdot & \cdot & \cdot & \cdot \\ a_{p1} & a_{p2} & \cdots & a_{pq} \end{bmatrix}, \quad \mathbf{B} = \begin{bmatrix} b_{11} & b_{12} & \cdots & b_{1s} \\ b_{21} & b_{22} & \cdots & b_{2s} \\ \cdot & \cdot & \cdot & \cdot \\ \cdot & \cdot & \cdot & \cdot \\ b_{q1} & b_{q2} & \cdots & b_{qs} \end{bmatrix},$$

$$p \times q \qquad\qquad q \times s$$

then the product

$$\mathbf{AB} = \mathbf{C} = \begin{bmatrix} c_{11} & c_{12} & \cdots & c_{1s} \\ c_{21} & c_{22} & \cdots & c_{2s} \\ \cdot & \cdot & \cdot & \cdot \\ \cdot & \cdot & \cdot & \cdot \\ c_{p1} & c_{p2} & \cdots & c_{ps} \end{bmatrix}$$

$$p \times s$$

is such that

$$c_{ij} = \sum_{l=1}^{q} a_{il} b_{lj};$$

that is, the entry in the ith row and jth column of $\mathbf{C}$ is the *inner product* (the element by element cross-product) of the ith row of $\mathbf{A}$ with the jth column of $\mathbf{B}$. For example,

$$\begin{bmatrix} 1 & 2 & 1 \\ -1 & 3 & 0 \end{bmatrix} \begin{bmatrix} 1 & 2 & 3 \\ 4 & 0 & -1 \\ -2 & 1 & 3 \end{bmatrix}$$

$$2 \times 3 \qquad\qquad 3 \times 3$$

$$= \begin{bmatrix} 1(1) + 2(4) + 1(-2) & 1(2) + 2(0) + 1(1) & 1(3) + 2(-1) + 1(3) \\ -1(1) + 3(4) + 0(-2) & -1(2) + 3(0) + 0(1) & -1(3) + 3(-1) + 0(3) \end{bmatrix}$$

$$= \begin{bmatrix} 7 & 3 & 4 \\ 11 & -2 & -6 \end{bmatrix}.$$

$$2 \times 3$$

We say that, in the product **AB,** we have *premultiplied* **B** by **A** or we have *postmultiplied* **A** by **B.** Note that, in general, **AB** and **BA,** even if both products are permissible (conformable), do not lead to the same result. In a matrix multiplication, the order in which the matrices are arranged is crucially important, whereas the order of the numbers in a scalar product is irrelevant.

When several matrices and/or vectors are multiplied together, the product should be carried out in the way that leads to the least work. For example, the product

$$\begin{matrix} \mathbf{W} & \mathbf{Z}' & \mathbf{y} \\ p \times p & p \times n & n \times 1 \end{matrix}$$

could be carried out as $(\mathbf{WZ}')\mathbf{y}$, or as $\mathbf{W}(\mathbf{Z}'\mathbf{y})$, where the parenthesized product is evaluated first. In the first case we would have to carry out pn p-length cross-products and p n-length cross-products; in the second case p p-length and p n-length, clearly a saving in effort.

Special Matrices and Vectors

We define

$$\mathbf{I}_n = \begin{bmatrix} 1 & 0 & 0 & \cdots & 0 \\ 0 & 1 & 0 & \cdots & 0 \\ 0 & 0 & 1 & \cdots & 0 \\ \vdots & \vdots & \vdots & \ddots & \vdots \\ 0 & 0 & 0 & \cdots & 1 \end{bmatrix},$$

a square $n \times n$ matrix with 1's on the diagonal and 0's elsewhere, as the *unit matrix* or *identity matrix*. This fulfills the same role as the number 1 in ordinary arithmetic. If the size of $\mathbf{I}_n$ is clear from the context, the subscript n is often omitted. We further use **0** to denote a vector

$$\mathbf{0} = (0, 0, \ldots, 0)'$$

or a matrix

$$\mathbf{0} = \begin{bmatrix} 0 & 0 & \cdots & 0 \\ 0 & 0 & \cdots & 0 \\ \vdots & \vdots & \ddots & \vdots \\ 0 & 0 & \cdots & 0 \end{bmatrix},$$

all of whose values are zeros; the actual size of **0** is usually clear from the context. We also define

$$\mathbf{1} = (1, 1, \ldots, 1)'$$

a vector of all 1's; the size of **1** is either specified or is clear in context. Note that $\mathbf{1}'\mathbf{1} =$ the squared length of the vector **1**, but that $\mathbf{11}'$ is a square matrix, each entry of which is 1, with the number of rows and columns each equal to the length of **1**.

Orthogonality

A vector $\mathbf{a} = (a_1, a_2, \ldots, a_n)'$ is said to be *orthogonal* to a vector $\mathbf{b} = (b_1, b_2, \ldots, b_n)'$ if the sum of products of their elements is zero, that is, if

$$\sum_{i=1}^{n} a_i b_i = \mathbf{a}'\mathbf{b} = \mathbf{b}'\mathbf{a} = 0.$$

Inverse Matrix

The inverse $\mathbf{M}^{-1}$ of a square matrix $\mathbf{M}$ is the unique matrix such that

$$\mathbf{M}^{-1}\mathbf{M} = \mathbf{I} = \mathbf{M}\mathbf{M}^{-1}.$$

The columns $\mathbf{m}_1, \mathbf{m}_2, \ldots, \mathbf{m}_n$ of an $n \times n$ matrix are *linearly dependent* if there exist constants $c_1, c_2, \ldots, c_n$, not all zero, such that

$$c_1\mathbf{m}_1 + c_2\mathbf{m}_2 + \cdots + c_n\mathbf{m}_n = \mathbf{0}$$

and similarly for rows. A square matrix, some of whose rows (or some of whose columns) are linearly dependent, is said to be *singular* and does not possess an inverse. A square matrix that is not singular is said to be *nonsingular* and can be inverted.
If $\mathbf{M}$ is symmetric, so is $\mathbf{M}^{-1}$.

Obtaining an Inverse

The process of matrix inversion is a relatively complicated one and is best appreciated by considering an example. Suppose we wish to obtain the inverse $\mathbf{M}^{-1}$ of the matrix

$$\mathbf{M} = \begin{bmatrix} 3 & 4 & 5 \\ 1 & 2 & 6 \\ 7 & 1 & 9 \end{bmatrix}.$$

Let

$$\mathbf{M}^{-1} = \begin{bmatrix} a & b & c \\ d & e & f \\ g & h & k \end{bmatrix}.$$

Then we must find $(a, b, c, \ldots, h, k)$ so that

$$\begin{bmatrix} a & b & c \\ d & e & f \\ g & h & k \end{bmatrix} \begin{bmatrix} 3 & 4 & 5 \\ 1 & 2 & 6 \\ 7 & 1 & 9 \end{bmatrix} = \begin{bmatrix} 1 & 0 & 0 \\ 0 & 1 & 0 \\ 0 & 0 & 1 \end{bmatrix},$$

that is, so that

$$3a + b + 7c = 1, \quad 3d + e + 7f = 0, \quad 3g + h + 7k = 0,$$
$$4a + 2b + c = 0, \quad 4d + 2e + f = 1, \quad 4g + 2h + k = 0,$$
$$5a + 6b + 9c = 0, \quad 5d + 6e + 9f = 0, \quad 5g + 6h + 9k = 1.$$

Solving these three sets of three linear simultaneous equations yields

$$\mathbf{M}^{-1} = \begin{bmatrix} \frac{12}{103} & -\frac{31}{103} & \frac{14}{103} \\ \frac{33}{103} & -\frac{8}{103} & -\frac{13}{103} \\ -\frac{13}{103} & \frac{25}{103} & \frac{2}{103} \end{bmatrix} = \frac{1}{103} \begin{bmatrix} 12 & -31 & 14 \\ 33 & -8 & -13 \\ -13 & 25 & 2 \end{bmatrix}.$$

(Note the removal of a common factor, explained below.) In general, for an $n \times n$ matrix there will be n sets of n simultaneous linear equations. Accelerated methods for inverting matrices adapted specifically for use with electronic computers permit inverses to be obtained with great speed, even for large matrices. Working out inverse matrices "by hand" is obsolete, nowadays, except in simple cases (see below).

Determinants

An important quantity associated with a square matrix is its *determinant*. Determinants occur naturally in the solution of linear simultaneous equations and in the inversion of matrices. For a 2×2 matrix

$$\mathbf{M} = \begin{bmatrix} a & b \\ c & d \end{bmatrix},$$

the determinant is defined as

$$\det \mathbf{M} = \begin{vmatrix} a & b \\ c & d \end{vmatrix} = ad - bc.$$

For a 3×3 matrix

$$\begin{bmatrix} a & b & c \\ d & e & f \\ g & h & k \end{bmatrix},$$

it is

$$a\begin{vmatrix} e & f \\ h & k \end{vmatrix} - b\begin{vmatrix} d & f \\ g & k \end{vmatrix} + c\begin{vmatrix} d & e \\ g & h \end{vmatrix} = aek - afh - bdk + bfg + cdh - ceg.$$

Note that we expand by the first row, multiplying a by the determinant of the matrix left when we cross out the row and column containing a, multiplying b by the determinant of the matrix left when we cross out the row and column containing b, multiplying c by the determinant of the matrix left when we cross out the row and column containing c. We also attach alternate signs $+, -, +$ to these three terms, counting from the top left-hand corner element: $+$ to a, $-$ to b, $+$ to c, and so on, alternately, if there were more elements in the first row.

In fact, the determinant can be written down as an expansion of *any* row or column by the same technique. The signs to be attached are counted $+ - + -$, and so on, from the top left corner element alternating either along row or column (but *not* diagonally). In other words the signs

$$\begin{bmatrix} + & - & + \\ - & + & - \\ + & - & + \end{bmatrix}$$

are attached and any row or column is used to write down the determinant. For example, using the second row we have

$$-d\begin{vmatrix} b & c \\ h & k \end{vmatrix} + e\begin{vmatrix} a & c \\ g & k \end{vmatrix} - f\begin{vmatrix} a & b \\ g & h \end{vmatrix}$$

to obtain the same result as before.

The same principle is used to get the determinant of any matrix. Any row or column is used for the expansion and we multiply each element of the row or column by:

1. Its appropriate sign, counted as above.
2. The determinant of the submatrix obtained by deletion of the row and column in which the element of the original matrix stands.

Determinants arise in the inversion of matrices as follows. The inverse M^{-1} may be obtained by first replacing each element m_{ij} of the original matrix M by an element calculated as follows:

1. Find the determinant of the submatrix obtained by crossing out the row and column of M in which m_{ij} stands.
2. Attach a sign from the $+ - + -$ count, as above.
3. Divide by the determinant of M.

When all elements of M have been replaced, *transpose the resulting matrix.* The transpose will be M^{-1}.

The reader might like to check these rules by showing that

$$M^{-1} = \begin{bmatrix} a & b \\ c & d \end{bmatrix}^{-1} = \begin{bmatrix} d/D & -b/D \\ -c/D & a/D \end{bmatrix},$$

where $D = ad - bc$ is the determinant of M; and that

$$Q^{-1} = \begin{bmatrix} a & b & c \\ d & e & f \\ g & h & k \end{bmatrix}^{-1} = \begin{bmatrix} A & B & C \\ D & E & F \\ G & H & K \end{bmatrix},$$

where

$$A = (ek - fh)/Z, \quad B = -(bk - ch)/Z, \quad C = (bf - ce)/Z,$$
$$D = -(dk - fg)/Z, \quad E = (ak - cg)/Z, \quad F = -(af - cd)/Z,$$
$$G = (dh - eg)/Z, \quad H = -(ah - bg)/Z, \quad K = (ae - bd)/Z,$$

and where

$$Z = aek + bfg + cdh - afh - bdk - ceg$$

is the determinant of Q. Note that, if M is symmetric (so that $b = c$), M^{-1} is also symmetric. Also, if Q is symmetric (so that $b = d$, $c = g$, $f = h$), then Q^{-1} is also symmetric because then $B = D$, $C = G$, and $F = H$.

The determinant is essentially a measure of the volume contained in a parallelepiped defined by the vectors in the rows (or columns) of the square matrix. See, for example, Schneider and Barker (1973, pp. 161–169).

If the square matrix is of the form $X'X$, dimension $p \times p$, say, the equation

$$\mathbf{u'X'Xu} = \text{constant}$$

defines an ellipsoid in the space of the variables $(u_1, u_2, \ldots, u_p) = \mathbf{u'}$ and $\det(\mathbf{X'X}) = |\mathbf{X'X}|$ is proportional to the volume contained by the ellipse. The exact area depends on the choice of the constant. This result has application in the construction of joint confidence regions for regression parameters in Section 5.4.

Common Factors

If *every* element of a matrix has a common factor, it can be taken outside the matrix. Conversely, if a matrix is multiplied by a constant c, every element of the matrix is multiplied by c. For example,

$$\begin{bmatrix} 4 & 6 & -2 \\ 8 & 6 & 2 \end{bmatrix} = 2 \begin{bmatrix} 2 & 3 & -1 \\ 4 & 3 & 1 \end{bmatrix}.$$

Note that, if a matrix is square and of size $p \times p$, and if c is a common factor, then the determinant of the matrix has a factor c^p, not just c. For example,

$$\begin{vmatrix} 4 & 6 \\ 8 & 6 \end{vmatrix} = 2^2 \begin{vmatrix} 2 & 3 \\ 4 & 3 \end{vmatrix} = 2^2(6 - 12) = -24.$$

Additional information on matrices is given where needed in the text. Also see Appendix 5A.

CHAPTER 1

Fitting a Straight Line by Least Squares

1.0. INTRODUCTION: THE NEED FOR STATISTICAL ANALYSIS

In today's industrial processes, there is no shortage of "information." No matter how small or how straightforward a process may be, measuring instruments abound. They tell us such things as input temperature, concentration of reactant, percent catalyst, steam temperature, consumption rate, pressure, and so on, depending on the characteristics of the process being studied. Some of these readings are available at regular intervals, every five minutes perhaps or every half hour; others are observed continuously. Still other readings are available with a little extra time and effort. Samples of the end product may be taken at intervals and, after analysis, may provide measurements of such things as purity, percent yield, glossiness, breaking strength, color, or whatever other properties of the end product are important to the manufacturer or user.

In research laboratories, experiments are being performed daily. These are usually small, carefully planned studies and result in sets of data of modest size. The objective is often a quick yet accurate analysis, enabling the experimenter to move on to "better" experimental conditions, which will produce a product with desirable characteristics. Additional data can easily be obtained if needed, however, if the decision is initially unclear.

A Ph.D. researcher may travel into an African jungle for a one-year period of intensive data-gathering on plants or animals. She will return with the raw material for her thesis and will put much effort into analyzing the data she has, searching for the messages that they contain. It will not be easy to obtain more data once her trip is completed, so she must carefully analyze every aspect of what data she has.

Regression analysis is a technique that can be used in any of these situations. Our purpose in this book is to explain in some detail something of the technique of extracting, from data of the types just mentioned, the main features of the relationships hidden or implied in the tabulated figures. (Nevertheless, the study of regression analysis techniques will also provide certain insights into how to plan the collection of data, when the opportunity arises. See, for example, Section 3.3.)

In any system in which variable quantities change, it is of interest to examine the effects that some variables exert (or appear to exert) on others. There may in fact be a simple functional relationship between variables; in most physical processes this is the exception rather than the rule. Often there exists a functional relationship that is too complicated to grasp or to describe in simple terms. In this case we may wish to approximate to this functional relationship by some simple mathematical function, such as a polynomial, which contains the appropriate variables and which graduates or approximates to the true function over some limited ranges of the variables involved. By examining such a graduating function we may be able to learn more about the underlying true relationship and to appreciate the separate and joint effects produced by changes in certain important variables.

Even where no sensible physical relationship exists between variables, we may wish to relate them by some sort of mathematical equation. While the equation might be physically meaningless, it may nevertheless be extremely valuable for predicting the values of some variables from knowledge of other variables, perhaps under certain stated restrictions.

In this book we shall use one particular method of obtaining a mathematical relationship. This involves the initial assumption that a certain type of relationship, linear in unknown parameters (except in Chapter 24, where nonlinear models are considered), holds. The unknown parameters are estimated under certain other assumptions with the help of available data, and a fitted equation is obtained. The value of the fitted equation can be gauged, and checks can be made on the underlying assumptions to see if any of these assumptions appears to be erroneous. The simplest example of this process involves the construction of a fitted straight line when pairs of observations (X_1, Y_1), (X_2, Y_2), ... , (X_n, Y_n) are available. We shall deal with this in a simple algebraic way in Chapters 1–3. To handle problems involving large numbers of variables, however, matrix methods are essential. These are introduced in the context of fitting a straight line in Chapter 4. Matrix algebra allows us to discuss concepts in a larger linear least squares regression context in Chapters 5–16 and 19–23. Some non-least-squares topics are discussed in Chapters 17 (ridge regression), 18 (generalized linear models), 24 (nonlinear estimation), 25 (robust regression), and 26 (resampling procedures).

We assume that anyone who uses this book has had a first course in statistics and understands certain basic ideas. These include the ideas of parameters, estimates, distributions (especially normal), mean and variance of a random variable, covariance between two variables, and simple hypothesis testing involving one- and two-sided t-tests and the F-test. We believe, however, that a reader whose knowledge of these topics is rusty or incomplete will nevertheless be able to make good progress after a review of Chapter 0.

We do not intend this as a comprehensive textbook on all aspects of regression analysis. Our intention is to provide a sound basic course plus material necessary to the solution of some practical regression problems. We also add some excursions into related topics.

We now take an early opportunity to introduce the reader to the data in Appendix 1A. Here we see 25 observations taken at intervals from a steam plant at a large industrial concern. Ten variables, some of them in coded form, were recorded as follows:

1. Pounds of steam used monthly, in coded form.
2. Pounds of real fatty acid in storage per month.
3. Pounds of crude glycerin made.
4. Average wind velocity (in mph).
5. Calendar days per month.
6. Operating days per month.
7. Days below 32°F.
8. Average atmospheric temperature (°F).

9. Average wind velocity squared.

10. Number of start-ups.

We can distinguish two main types of variable at this stage. We shall usually call these *predictor variables* and *response variables*. (For alternative terms, see below.) By predictor variables we shall usually mean variables that can either be set to a desired value (e.g., input temperature or catalyst feed rate) or else take values that can be observed but not controlled (e.g., the outdoor humidity). As a result of changes that are deliberately made, or simply take place in the predictor variables, an effect is transmitted to other variables, the response variables (e.g., the final color or the purity of a chemical product). In general, we shall be interested in finding out how changes in the predictor variables affect the values of the response variables. If we can discover a simple relationship or dependence of a response variable on just one or a few predictor variables we shall, of course, be pleased. The distinction between predictor and response variables is not always completely clear-cut and depends sometimes on our objectives. What may be considered a response variable at the midstage of a process may also be regarded as a predictor variable in relation to (say) the final color of the product. In practice, however, the roles of variables are usually easily distinguished.

Other names frequently seen are the following:

$$\text{Predictor variables} = \text{input variables} = \text{inputs}$$

$$= X\text{-variables} = \text{regressors}$$

$$= \text{independent variables.}$$

(We shall try to avoid using the last of these names, because it is often misleading. In a particular body of data, two or more "independent" variables may vary together in some definite way due, perhaps, to the method in which an experiment is conducted. This is not usually desirable—for one thing it restricts the information on the separate roles of the variables—but it may often be unavoidable.)

$$\text{Response variables} = \text{output variables} = \text{outputs}$$

$$= Y\text{-variables}$$

$$= \text{dependent variables.}$$

Returning to the data in Appendix 1A, which we shall refer to as the *steam data*, we examine the 25 sets of observations on the variables, one set for each of 25 different months. Our primary interest here is in the monthly amount of steam produced and how it changes due to variations in the other variables. Thus we shall regard variable X_1 as a dependent or response variable, Y, in what follows, and the others as predictor variables, $X_2, X_3, \ldots, X_{10}$.

We shall see how the method of analysis called the *method of least squares* can be used to examine data and to draw meaningful conclusions about dependency relationships that may exist. This method of analysis is often called *regression analysis*. (For historical remarks, see Section 1.8.)

Throughout this book we shall be most often concerned with relationships of the form

$$\text{Response variable} = \text{Model function} + \text{Random error.}$$

The model function will usually be "known" and of specified form and will involve the predictor variables as well as *parameters* to be estimated from data. The distribution of the random errors is often assumed to be a normal distribution with mean zero, and errors are usually assumed to be independent. All assumptions are usually checked after the model has been fitted and many of these checks will be described.

(*Note:* Many engineers and others call the parameters *constants* and the predictors *parameters*. Watch out for this possible difficulty in cross-discipline conversations!)

We shall present the least squares method in the context of the simplest application, fitting the "best" straight line to given data in order to relate two variables X and Y, and will discuss how it can be extended to cases where more variables are involved.

1.1. STRAIGHT LINE RELATIONSHIP BETWEEN TWO VARIABLES

In much experimental work we wish to investigate how the changes in one variable affect another variable. Sometimes two variables are linked by an exact straight line relationship. For example, if the resistance R of a simple circuit is kept constant, the current I varies directly with the voltage V applied, for, by Ohm's law, $I = V/R$. If we were not aware of Ohm's law, we might obtain this relationship empirically by making changes in V and observing I, while keeping R fixed and then observing that the plot of I against V more or less gave a straight line through the origin. We say "more or less" because, although the relationship actually is exact, our measurements may be subject to slight errors and thus the plotted points would probably not fall exactly on the line but would vary randomly about it. For purposes of predicting I for a particular V (with R fixed), however, we should use the straight line through the origin. Sometimes a straight line relationship is not exact (even apart from error) yet can be meaningful nevertheless. For example, suppose we consider the height and weight of adult males for some given population. If we plot the pair $(Y_1, Y_2) = $ (height, weight), a diagram something like Figure 1.1 will result. (Such a presentation is conventionally called a *scatter diagram*.)

Note that for any given height there is a range of observed weights, and vice versa. This variation will be partially due to measurement errors but primarily due to variation between individuals. Thus no unique relationship between actual height and weight

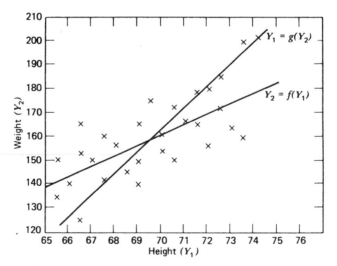

Figure 1.1. Heights and weights of 30 American males.

can be expected. But we can note that the average observed weight for a given observed height increases as height increases. This locus of average observed weight for given observed height (as height varies) is called the *regression curve* of weight on height. Let us denote it by $Y_2 = f(Y_1)$. There also exists a regression curve of height on weight, similarly defined, which we can denote by $Y_1 = g(Y_2)$. Let us assume that these two "curves" are both straight lines (which in general they may not be). In general, these two curves are *not* the same, as indicated by the two lines in the figure.

Suppose we now found we had recorded an individual's height but not his weight and we wished to estimate this weight. What could we do? From the regression line of weight on height we could find an average observed weight of individuals of the given height and use this as an estimate of the weight that we did not record.

A pair of random variables such as (height, weight) follows some sort of bivariate probability distribution. When we are concerned with the dependence of a random variable Y on a quantity X that is variable but *not* randomly variable, an equation that relates Y to X is usually called a *regression equation*. Although the name is, strictly speaking, incorrect, it is well established and conventional. In nearly all of this book we assume that the predictor variables are *not* subject to random variation, but that the response variable is. From a practical point of view, this is seldom fully true but, if it is not, a much more complicated fitting procedure is needed. (See Sections 3.4 and 9.7.) To avoid this, we use the least squares procedure only in situations where we can assume that any random variation in any of the predictor variables is so small compared with the *range* of that predictor variable observed that we can effectively ignore the random variation. This assumption is rarely stated, but it is implicit in all least squares work in which the predictors are assumed "fixed." (The word "fixed" means "not random variables" in such a context; it does not mean that the predictors cannot take a variety of values or levels.) For additional comments see Section 3.4.

We can see that whether a relationship is exactly a straight line or a line only insofar as mean values are concerned, knowledge of the relationship will be useful. (The relationship might, of course, be more complicated than a straight line but we shall consider this later.)

A straight line relationship may also be a valuable one even when we *know* that such a relationship cannot be true. Consider the response relationship shown in Figure 1.2. It is obviously not a straight line over the range $0 \le X \le 100$. However, if we were interested primarily in the range $0 \le X \le 45$, a straight line relationship evaluated from observations in this range might provide a perfectly adequate representation of the function *in this range*. The relationship thus fitted would, of course, not apply to values of X outside this restricted range and could not be used for predictive purposes outside this range.

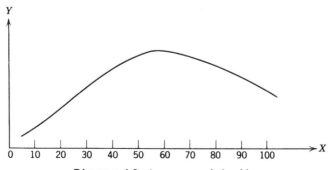

Figure 1.2. A response relationship.

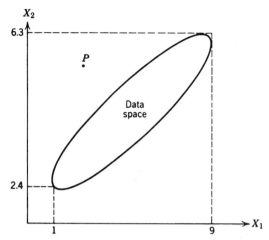

Figure 1.3. A point P outside the data space, whose coordinates nevertheless lie within the ranges of the predictor variables observed.

Similar remarks can be made when more than one predictor variable is involved. Suppose we wish to examine the way in which a response Y depends on variables X_1, $X_2, \ldots, X_k$. We determine a regression equation from data that "cover" certain regions of the "X-space." Suppose the point $\mathbf{X}_0' = (X_{10}, X_{20}, \ldots, X_{k0})$ lies *outside* the regions covered by the original data. While we can mathematically obtain a predicted value $\hat{Y}(\mathbf{X}_0')$ for the response at the point $\mathbf{X}_0'$, we must realize that reliance on such a prediction is extremely dangerous and becomes more dangerous the further $\mathbf{X}_0'$ lies from the original regions, unless some additional knowledge is available that the regression equation is valid in a wider region of the X-space. Note that it is sometimes difficult to realize at first that a suggested point lies outside a region in a multidimensional space. To take a simple example, consider the region indicated by the ellipse in Figure 1.3, within which all the data points (X_1, X_2) lie; the corresponding Y values, plotted vertically up from the page, are not shown. We see that there are points in the region for which $1 \leq X_1 \leq 9$ and for which $2.4 \leq X_2 \leq 6.3$. Although the X_1 and X_2 coordinates of P lie individually within these ranges, P itself lies outside the region. A simple review of the printed data would often not detect this. When more dimensions are involved, misunderstandings of this sort easily arise.

1.2. LINEAR REGRESSION: FITTING A STRAIGHT LINE BY LEAST SQUARES

We have mentioned that in many situations a straight line relationship can be valuable in summarizing the observed dependence of one variable on another. We now show how the equation of such a straight line can be obtained by the method of least squares when data are available. Consider, in Appendix 1A, the 25 observations of variable 1 (pounds of steam used per month) and variable 8 (average atmospheric temperature in degrees Fahrenheit). The corresponding pairs of observations are given in Table 1.1 and are plotted in Figure 1.4.

Let us tentatively assume that the regression line of variable 1, which we shall denote by Y, on variable 8 (X) has the form $\beta_0 + \beta_1 X$. Then we can write the linear, first-order model

$$Y = \beta_0 + \beta_1 X + \epsilon; \tag{1.2.1}$$

TABLE 1.1. Twenty-five Observations of Variables 1 and 8

Observation Number	Variable Number	
	1 (Y)	8 (X)
1	10.98	35.3
2	11.13	29.7
3	12.51	30.8
4	8.40	58.8
5	9.27	61.4
6	8.73	71.3
7	6.36	74.4
8	8.50	76.7
9	7.82	70.7
10	9.14	57.5
11	8.24	46.4
12	12.19	28.9
13	11.88	28.1
14	9.57	39.1
15	10.94	46.8
16	9.58	48.5
17	10.09	59.3
18	8.11	70.0
19	6.83	70.0
20	8.88	74.5
21	7.68	72.1
22	8.47	58.1
23	8.86	44.6
24	10.36	33.4
25	11.08	28.6

that is, for a given X, a corresponding observation Y consists of the value $\beta_0 + \beta_1 X$ plus an amount ϵ, the increment by which any individual Y may fall off the regression line. Equation (1.2.1) is the *model* of what we believe. $\beta_0 + \beta_1 X$ is the *model function* here and β_0 and β_1 are called the *parameters* of the model. We begin by assuming that the model holds; but we shall have to inquire at a later stage if indeed it does. In many aspects of statistics it is necessary to assume a mathematical model to make progress. It might be well to emphasize that what we are usually doing is to *consider* or *tentatively entertain* our model. The model must always be critically examined somewhere along the line. It is our "opinion" of the situation at one stage of the investigation and our "opinion" must be changed if we find, at a later stage, that the facts are against it.

Meaning of Linear Model

When we say that a model is linear or nonlinear, we are referring to linearity or nonlinearity *in the parameters*. The value of the highest power of a predictor variable in the model is called the *order* of the model. For example,

$$Y = \beta_0 + \beta_1 X + \beta_{11} X^2 + \epsilon$$

is a second-order (in X) linear (in the β's) regression model. Unless a model is specifically called nonlinear it can be taken that it is linear in the parameters, and the

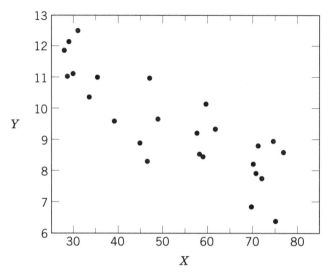

Figure 1.4. Plot of the steam data for variables 1 (Y) and 8 (X).

word linear is usually omitted and understood. The order of the model could be of any size. Notation of the form β_{11} is often used in polynomial models; β_1 is the parameter that goes with X while β_{11} is the parameter that goes with $X^2 = XX$. The natural extension of this sort of notation appears, for example, in Chapter 12, where β_{12} is the parameter associated with X_1X_2 and so on.

Least Squares Estimation

Now β_0, β_1, and ϵ are unknown in Eq. (1.2.1), and in fact ϵ would be difficult to discover since it changes for each observation Y. However, β_0 and β_1 remain fixed and, although we cannot find them exactly without examining all possible occurrences of Y and X, we can use the information provided by the 25 observations in Table 1.1 to give us *estimates* b_0 and b_1 of β_0 and β_1; thus we can write

$$\hat{Y} = b_0 + b_1X, \tag{1.2.2}$$

where $\hat{Y}$, read "Y hat," denotes the *predicted* value of Y for a given X, when b_0 and b_1 are determined. Equation (1.2.2) could then be used as a predictive equation; substitution for a value of X would provide a prediction of the true mean value of Y for that X.

The use of small roman letters b_0 and b_1 to denote estimates of the parameters given by Greek letters β_0 and β_1 is standard. However, the notation $\hat{\beta}_0$ and $\hat{\beta}_1$ for the estimates is also frequently seen. We use the latter type of notation ourselves in Chapter 24, for example.

Our estimation procedure will be that of least squares.

Under certain assumptions to be discussed in Chapter 5, the method of least squares has certain properties. For the moment we state it as our chosen method of estimating the parameters without a specific justification. Suppose we have available n sets of observations (X_1, Y_1), (X_2, Y_2), ... , (X_n, Y_n). (In our steam data example $n = 25$.) Then by Eq. (1.2.1) we can write

$$Y_i = \beta_0 + \beta_1X_i + \epsilon_i, \tag{1.2.3}$$

for $i = 1, 2, \ldots, n$, so that the sum of squares of deviation from the true line is

$$S = \sum_{i=1}^{n} \epsilon_i^2 = \sum_{i=1}^{n} (Y_i - \beta_0 - \beta_1 X_i)^2. \tag{1.2.4}$$

S is also called the *sum of squares function*. We shall choose our estimates b_0 and b_1 to be the values that, when substituted for β_0 and β_1 in Eq. (1.2.4), produce the least possible value of S; see Figure 1.5. [Note that, in (1.2.4), X_i, Y_i are the fixed numbers that we have observed.] We can determine b_0 and b_1 by differentiating Eq. (1.2.4) first with respect to β_0 and then with respect to β_1 and setting the results equal to zero. Now

$$\frac{\partial S}{\partial \beta_0} = -2 \sum_{i=1}^{n} (Y_i - \beta_0 - \beta_1 X_i),$$

$$\frac{\partial S}{\partial \beta_1} = -2 \sum_{i=1}^{n} X_i(Y_i - \beta_0 - \beta_1 X_i), \tag{1.2.5}$$

so that the estimates b_0 and b_1 are solutions of the two equations

$$\sum_{i=1}^{n} (Y_i - b_0 - b_1 X_i) = 0,$$

$$\sum_{i=1}^{n} X_i(Y_i - b_0 - b_1 X_i) = 0, \tag{1.2.6}$$

where we substitute (b_0, b_1) for (β_0, β_1), when we equate Eq. (1.2.5) to zero. From Eq. (1.2.6) we have

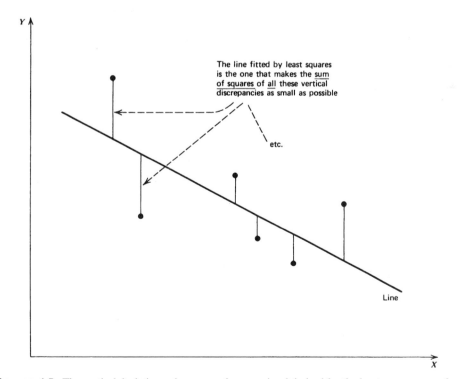

The line fitted by least squares is the one that makes the sum of squares of all these vertical discrepancies as small as possible

etc.

Line

Figure 1.5. The vertical deviations whose sum of squares is minimized for the least squares procedure.

$$\sum_{i=1}^{n} Y_i - nb_0 - b_1 \sum_{i=1}^{n} X_i = 0$$

$$\sum_{i=1}^{n} X_i Y_i - b_0 \sum_{i=1}^{n} X_i - b_1 \sum_{i=1}^{n} X_i^2 = 0$$

(1.2.7)

or

$$b_0 n + b_1 \sum_{i=1}^{n} X_i = \sum_{i=1}^{n} Y_i$$

$$b_0 \sum_{i=1}^{n} X_i + b_1 \sum_{i=1}^{n} X_i^2 = \sum_{i=1}^{n} X_i Y_i.$$

(1.2.8)

These equations are called the *normal equations*. (Normal here means perpendicular, or orthogonal, a geometrical property explained in Chapter 20. The normal equations can also be obtained via a geometrical argument.)

The solution of Eq. (1.2.8) for b_1, the slope of the fitted straight line, is

$$b_1 = \frac{\sum X_i Y_i - [(\sum X_i)(\sum Y_i)]/n}{\sum X_i^2 - (\sum X_i)^2/n} = \frac{\sum (X_i - \overline{X})(Y_i - \overline{Y})}{\sum (X_i - \overline{X})^2},$$

(1.2.9)

where all summations are from $i = 1$ to n and the two expressions for b_1 are just slightly different forms of the same quantity. For, defining

$$\overline{X} = (X_1 + X_2 + \cdots + X_n)/n = \sum X_i/n,$$

$$\overline{Y} = (Y_1 + Y_2 + \cdots + Y_n)/n = \sum Y_i/n,$$

we have that

$$\Sigma(X_i - \overline{X})(Y_i - \overline{Y}) = \sum X_i Y_i - \overline{X} \sum Y_i - \overline{Y} \sum X_i + n\overline{X}\,\overline{Y}$$

$$= \sum X_i Y_i - n\overline{X}\,\overline{Y}$$

$$= \sum X_i Y_i - (\sum X_i)(\sum Y_i)/n.$$

This shows the equivalence of the numerators in (1.2.9), and a parallel calculation, in which Y is replaced by X, shows the equivalence of the denominators. The quantity $\sum X_i^2$ is called the *uncorrected sum of squares of the X's* and $(\sum X_i)^2/n$ is the *correction for the mean of the X's*. The difference is called the *corrected sum of squares of the X's*. Similarly, $\sum X_i Y_i$ is called the *uncorrected sum of products*, and $(\sum X_i)(\sum Y_i)/n$ is the *correction for the means*. The difference is called the *corrected sum of products of X and Y*.

Pocket-Calculator Form

The first form in Eq. (1.2.9) is normally used for pocket-calculator evaluation of b_1, because it is easier to work with and does not involve the tedious adjustment of each X_i and Y_i to $(X_i - \overline{X})$ and $(Y_i - \overline{Y})$, respectively. To avoid rounding error, however, it is best to carry as many significant figures as possible in this computation. (Such advice is good in general; rounding is best done at the "reporting stage" of a calculation, not at intermediate stages.) Most digital computers obtain more accurate answers using the second form in Eq. (1.2.9); this is because of their round-off characteristics and the form in which most regression programs are written.

A convenient notation, now and later, is to write

$$S_{XY} = \Sigma(X_i - \overline{X})(Y_i - \overline{Y})$$
$$= \Sigma(X_i - \overline{X})Y_i$$
$$= \Sigma X_i(Y_i - \overline{Y})$$
$$= \Sigma X_iY_i - (\Sigma X_i)(\Sigma Y_i)/n$$
$$= \Sigma X_iY_i - n\overline{X}\overline{Y}.$$

Note that all these forms are equivalent. Similarly, we can write

$$S_{XX} = \Sigma(X_i - \overline{X})^2$$
$$= \Sigma(X_i - \overline{X})X_i$$
$$= \Sigma X_i^2 - (\Sigma X_i)^2/n$$
$$= \Sigma X_i^2 - n\overline{X}^2$$

and

$$S_{YY} = \Sigma(Y_i - \overline{Y})^2$$
$$= \Sigma(Y_i - \overline{Y})Y_i$$
$$= \Sigma Y_i^2 - (\Sigma Y_i)^2/n$$
$$= \Sigma Y_i^2 - n\overline{Y}^2.$$

The easily remembered formula for b_1 is then

$$b_1 = S_{XY}/S_{XX}. \tag{1.2.9a}$$

The solution of Eqs. (1.2.8) for b_0, the intercept at $X = 0$ of the fitted straight line, is

$$b_0 = \overline{Y} - b_1\overline{X}. \tag{1.2.10}$$

The predicted or fitted equation is $\hat{Y} = b_0 + b_1X$ as in (1.2.2), we recall. Substituting Eq. (1.2.10) into Eq. (1.2.2) gives the estimated regression equation in the alternative form

$$\hat{Y} = \overline{Y} + b_1(X - \overline{X}), \tag{1.2.11}$$

where b_1 is given by Eq. (1.2.9). From this we see immediately that if we set $X = \overline{X}$ in (1.2.11), then $\hat{Y} = \overline{Y}$. This means that the point $(\overline{X}, \overline{Y})$ lies on the fitted line. In other words, this least squares line contains the center of gravity of the data.

Calculations for the Steam Data

Let us now perform these calculations on the selected steam data given in Table 1.1. We find the following:

$$n = 25,$$
$$\Sigma Y_i = 10.98 + 11.13 + \cdots + 11.08 = 235.60,$$
$$\overline{Y} = 235.60/25 = 9.424,$$
$$\Sigma X_i = 35.3 + 29.7 + \cdots + 28.6 = 1315,$$
$$\overline{X} = 1315/25 = 52.60,$$

$$\Sigma\, X_i Y_i = (10.98)(35.3) + (11.13)(29.7) + \cdots + (11.08)(28.6)$$

$$= 11821.4320,$$

$$\Sigma\, X_i^2 = (35.3)^2 + (29.7)^2 + \cdots + (28.6)^2 = 76323.42,$$

$$b_1 = \frac{\Sigma\, X_i Y_i - (\Sigma\, X_i)(\Sigma\, Y_i)/n}{\Sigma\, X_i^2 - (\Sigma\, X_i)^2/n} = \frac{S_{XY}}{S_{XX}}$$

$$= \frac{11821.4320 - (1315)(235.60)/25}{76323.42 - (1315)^2/25} = \frac{-571.1280}{7154.42}$$

$$= -0.079829.$$

The fitted equation is thus

$$\hat{Y} = \overline{Y} + b_1(X - \overline{X})$$

$$= 9.4240 - 0.079829(X - 52.60)$$

$$= 13.623005 - 0.079829X.$$

The foregoing form of $\hat{Y}$ shows that $b_0 = 13.623005$. The fitted regression line is plotted in Figure 1.6. We can tabulate for each of the 25 values X_i, at which a Y_i observation is available, the fitted value $\hat{Y}_i$ and the *residual* $Y_i - \hat{Y}_i$ as in Table 1.2. The residuals are given to the same number of places as the original data. They are our "estimates of the errors ϵ_i" and we write $e_i = Y_i - \hat{Y}_i$ in a parallel notation.

Note that since $\hat{Y}_i = \overline{Y} + b_1(X_i - \overline{X})$,

$$Y_i - \hat{Y}_i = (Y_i - \overline{Y}) - b_1(X_i - \overline{X}),$$

which we can sum to give

$$\sum_{i=1}^{n} (Y_i - \hat{Y}_i) = \sum_{i=1}^{n} (Y_i - \overline{Y}) - b_1 \sum_{i=1}^{n} (X_i - \overline{X}) = 0.$$

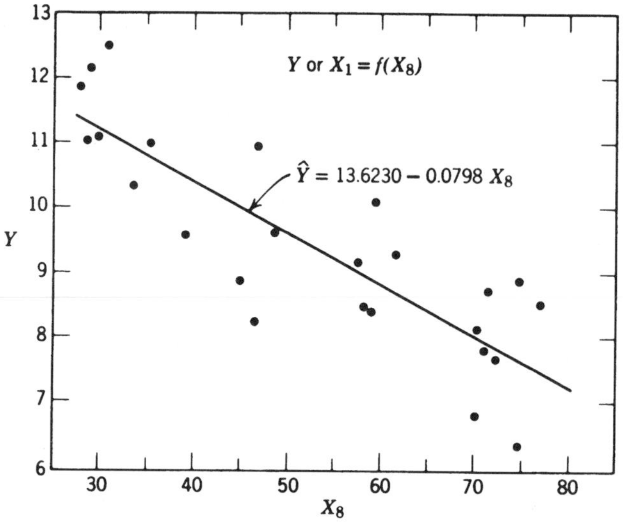

Figure 1.6. Plot of the steam data—variables 1 (Y) and 8 (X)—and the least squares line.

T A B L E 1.2. Observations, Fitted Values, and Residuals

Observation Number	Y_i	$\hat{Y}_i$	$Y_i - \hat{Y}_i$
1	10.98	10.81	0.17
2	11.13	11.25	−0.12
3	12.51	11.17	1.34
4	8.40	8.93	−0.53
5	9.27	8.72	0.55
6	8.73	7.93	0.80
7	6.36	7.68	−1.32
8	8.50	7.50	1.00
9	7.82	7.98	−0.16
10	9.14	9.03	0.11
11	8.24	9.92	−1.68
12	12.19	11.32	0.87
13	11.88	11.38	0.50
14	9.57	10.50	−0.93
15	10.94	9.89	1.05
16	9.58	9.75	−0.17
17	10.09	8.89	1.20
18	8.11	8.03	0.08
19	6.83	8.03	−1.20
20	8.88	7.68	1.20
21	7.68	7.87	−0.19
22	8.47	8.98	−0.51
23	8.86	10.06	−1.20
24	10.36	10.96	−0.60
25	11.08	11.34	−0.26

This piece of algebra tells us that the residuals sum to zero, in theory. In practice, the sum may not be exactly zero due to rounding. The sum of residuals in any regression problem is always zero when there is a β_0 term in the model as a consequence of the first normal equation. The omission of β_0 from a model implies that the response is zero when all the predictor variables are zero. This is a very strong assumption, which is usually unjustified. In a straight line model $Y = \beta_0 + \beta_1 X + \epsilon$, omission of β_0 implies that the line passes through $X = 0$, $Y = 0$; that is, the line has a zero *intercept* $\beta_0 = 0$ at $X = 0$.

Centering the Data

We note here, before the more general discussion in Section 16.2, that physical removal of β_0 from the model is always possible by "centering" the data, but this is quite different from setting $\beta_0 = 0$. For example, if we write Eq. (1.2.1) in the form

$$Y - \overline{Y} = (\beta_0 + \beta_1 \overline{X} - \overline{Y}) + \beta_1(X - \overline{X}) + \epsilon$$

or

$$y = \beta_0' + \beta_1 x + \epsilon,$$

say, where $y = Y - \overline{Y}$, $\beta_0' = \beta_0 + \beta_1 \overline{X} - \overline{Y}$, and $x = X - \overline{X}$, then the least squares estimates of β_0' and β_1 are given as follows:

$$b_1 = \frac{\Sigma(x_i - \bar{x})(y_i - \bar{y})}{\Sigma(x_i - \bar{x})^2} = \frac{\Sigma(X_i - \bar{X})(Y_i - \bar{Y})}{\Sigma(X_i - \bar{X})^2},$$

identical to Eq. (1.2.9), while

$$b_0' = \bar{y} - b_1\bar{x} = 0, \qquad \text{since } \bar{x} = \bar{y} = 0,$$

whatever the value of b_1. Because this always happens, we can write and fit the centered model as

$$Y - \bar{Y} = \beta_1(X - \bar{X}) + \epsilon,$$

omitting the β_0' (intercept) term entirely. We have lost one parameter but there is a corresponding loss in the data since the quantities $Y_i - \bar{Y}, i = 1, 2, \ldots, n$, represent only $(n - 1)$ separate pieces of information due to the fact that their sum is zero, whereas $Y_1, Y_2, \ldots, Y_n$ represent n separate pieces of information. Effectively the "lost" piece of information has been used to enable the proper adjustments to be made to the model so that the intercept term can be removed. The model fit is exactly the same as before but is written in a slightly different manner, pivoted around $(\bar{X}, \bar{Y})$.

1.3. THE ANALYSIS OF VARIANCE

We now tackle the question of how much of the variation in the data has been explained by the regression line. Consider the following identity:

$$Y_i - \hat{Y}_i = Y_i - \bar{Y} - (\hat{Y}_i - \bar{Y}). \tag{1.3.1}$$

What this means geometrically for the fitted straight line is illustrated in Figure 1.7. The residual $e_i = Y_i - \hat{Y}_i$ is the difference between two quantities: (1) the deviation of the observed Y_i from the overall mean $\bar{Y}$ and (2) the deviation of the fitted $\hat{Y}_i$ from the overall mean $\bar{Y}$. Note that the average of the $\hat{Y}_i$, namely,

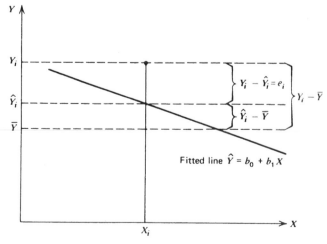

Figure 1.7. Geometrical meaning of the identity (1.3.1).

$$\Sigma \, \hat{Y}_i/n = \Sigma (b_0 + b_1 X_i)/n$$
$$= (nb_0 + b_1 n \overline{X})/n$$
$$= b_0 + b_1 \overline{X} \tag{1.3.2}$$
$$= \overline{Y},$$

is the same as the average of the Y_i. This fact also reconfirms that $\Sigma \, e_i = \Sigma (Y_i - \hat{Y}_i) = n\overline{Y} - n\overline{Y} = 0$, as stated previously.

We can also rewrite Eq. (1.3.1) as

$$(Y_i - \overline{Y}) = (\hat{Y}_i - \overline{Y}) + (Y_i - \hat{Y}_i).$$

If we square both sides of this and sum from $i = 1, 2, \ldots, n$, we obtain

$$\Sigma (Y_i - \overline{Y})^2 = \Sigma (\hat{Y}_i - \overline{Y})^2 + \Sigma (Y_i - \hat{Y}_i)^2. \tag{1.3.3}$$

The cross-product term, $\mathrm{CPT} = 2 \, \Sigma (\hat{Y}_i - \overline{Y})(Y_i - \hat{Y}_i)$, can be shown to vanish by applying Eq. (1.2.11) with subscript i, so that

$$\hat{Y}_i - \overline{Y} = b_1 (X_i - \overline{X})$$
$$Y_i - \hat{Y}_i = Y_i - \overline{Y} - b_1 (X_i - \overline{X}). \tag{1.3.4}$$

It follows that the cross-product term is

$$\mathrm{CPT} = 2 \, \Sigma \, b_1 (X_i - \overline{X})\{(Y_i - \overline{Y}) - b_1 (X_i - \overline{X})\}$$
$$= 2b_1\{S_{XY} - b_1 S_{XX}\} \tag{1.3.5}$$
$$= 0$$

by Eq. (1.2.9a). It is also clear that

$$\Sigma (\hat{Y}_i - \overline{Y})^2 = b_1^2 S_{XX}$$
$$= b_1 S_{XY} \tag{1.3.6}$$
$$= S_{XY}^2/S_{XX}.$$

This provides a simple pocket calculator way of computing this quantity when calculating b_1.

Sums of Squares

We now return to a discussion of Eq. (1.3.3). The quantity $(Y_i - \overline{Y})$ is the deviation of the ith observation from the overall mean and so the left-hand side of Eq. (1.3.3) is the sum of squares of deviations of the observations from the mean; this is shortened to *SS about the mean* and is also the *corrected sum of squares of the Y's*, namely, S_{YY}. Since $\hat{Y}_i - \overline{Y}$ is the deviation of the predicted value of the ith observation from the mean, and $Y_i - \hat{Y}_i$ is the deviation of the ith observation from its predicted or fitted value (i.e., the ith *residual*), we can express Eq. (1.3.3) in words as follows:

$$\left(\begin{array}{c}\text{Sum of squares} \\ \text{about the mean}\end{array}\right) = \left(\begin{array}{c}\text{Sum of squares} \\ \text{due to regression}\end{array}\right) + \left(\begin{array}{c}\text{Sum of squares} \\ \text{about regression}\end{array}\right). \tag{1.3.7}$$

This shows that, of the variation in the Y's about their mean, some of the variation can be ascribed to the regression line and some, $\Sigma (Y_i - \hat{Y}_i)^2$, to the fact that the actual observations do not all lie on the regression line: if they all did, the sum of squares

T A B L E 1.3. Analysis of Variance (ANOVA) Table: The Basic Split of S_{YY}

Source of Variation	Degrees of Freedom (df)	Sum of Squares (SS)	Mean Square (MS)
Due to regression	1	$\sum_{i=1}^{n} (\hat{Y}_i - \overline{Y})^2$	MS_{Reg}
About regression (residual)	$n - 2$	$\sum_{i=1}^{n} (Y_i - \hat{Y}_i)^2$	$s^2 = \dfrac{SS}{(n-2)}$ [a]
Total, corrected for mean $\overline{Y}$	$n - 1$	$\sum_{i=1}^{n} (Y_i - \overline{Y})^2$	

[a] Some older regression programs have documentation that labels the quantity $\Sigma(Y_i - \overline{Y})^2/(n-1) = S_{YY}/(n-1)$ as s^2. For us, this would be true *only* if the model fitted were $Y = \beta + \epsilon$. In this case, the regression sum of squares due to b_0 would be (as it is in general—e.g., Table 1.4) $n\overline{Y}^2 = (\Sigma Y_i)^2/n$ and S_{YY} would be the appropriate residual sum of squares for the corresponding fitted model $\hat{Y} = \overline{Y}$.

about the regression (soon to be called the *residual sum of squares*) would be zero! From this procedure we can see that a sensible first way of assessing how useful the regression line will be as a predictor is to see how much of the SS about the mean has fallen into the SS due to regression and how much into the SS about the regression. We shall be pleased if the SS due to regression is much greater than the SS about regression, or what amounts to the same thing, if the ratio $R^2 = $ (SS due to regression)/(SS about mean) is not too far from unity.

Degrees of Freedom (df)

Any sum of squares has associated with it a number called its degrees of freedom. This number indicates how many independent pieces of information involving the n independent numbers $Y_1, Y_2, \ldots, Y_n$ are needed to compile the sum of squares. For example, the SS about the mean needs $(n-1)$ independent pieces [of the numbers $Y_1 - \overline{Y}, Y_2 - \overline{Y}, \ldots, Y_n - \overline{Y}$, only $(n-1)$ are independent since all n numbers sum to zero by definition of the mean]. We can compute the SS due to regression from a single function of $Y_1, Y_2, \ldots, Y_n$, namely, b_1 [since $\Sigma(\hat{Y}_i - \overline{Y})^2 = b_1^2 \Sigma(X_i - \overline{X})^2$ as in Eq. (1.3.6)], and so this sum of squares has one degree of freedom. By subtraction, the SS about regression, which we shall in future call the residual sum of squares (it is, in fact, the sum of squares of the residuals $Y_i - \hat{Y}_i$) has $(n-2)$ degrees of freedom (df). This reflects the fact that the present residuals are from a fitted straight line model, which required estimation of *two* parameters. In general, the residual sum of squares is based on (number of observations − number of parameters estimated) degrees of freedom. Thus corresponding to Eq. (1.3.3), we can show the split of degrees of freedom as

$$n - 1 = 1 + (n - 2). \qquad (1.3.8)$$

Analysis of Variance Table

From Eqs. (1.3.3) and (1.3.8) we can construct an *analysis of variance* table in the form of Table 1.3. The "Mean Square" column is obtained by dividing each sum of squares entry by its corresponding degrees of freedom.

A more general form of the analysis of variance table, which we do not need here but which is useful for comparison purposes later, is obtained by incorporating the

T A B L E 1.4. Analysis of Variance (ANOVA) Table Incorporating SS(b_0)

Source	df	SS	MS = SS/df
Due to $b_1 \mid b_0$	1	$SS(b_1 \mid b_0) = \sum_{i=1}^{n} (\hat{Y}_i - \bar{Y})^2$	MS_{Reg}
Residual	$n - 2$	$\sum_{i=1}^{n} (Y_i - \hat{Y}_i)^2$	s^2
Total, corrected	$n - 1$	$\sum_{i=1}^{n} (Y_i - \bar{Y})^2$	
Correction factor (due to b_0)	1	$SS(b_0) = \left(\sum_{i=1}^{n} Y_i\right)^2 / n = n\bar{Y}^2$	
Total	n	$\sum_{i=1}^{n} Y_i^2$	

correction factor for the mean of the Y's into the table, where, for reasons explained below, it is called SS(b_0). The table takes the form of Table 1.4. (Note the abbreviated headings.) An alternative way of presenting Table 1.4 is to drop the line labeled "Total, corrected" and the rule above it. The "Total" line is then the sum of the remaining three entries. If we rewrite the entries in the order of Table 1.5, we can see a logical development. The first SS entry "Due to b_0" is the amount of variation $n\bar{Y}^2$ explained by a horizontal straight line $\hat{Y} = \bar{Y}$. In fact, if we fit the model $Y = \beta_0 + \epsilon$ via least squares, the fitted model is $\hat{Y} = \bar{Y}$. If we subsequently fit the "with slope also" model $Y = \beta_0 + \beta_1 X + \epsilon$, the "Due to $b_1 \mid b_0$" SS entry S_{XY}^2/S_{XX} is the *extra* variation picked up by the slope term *over and above* that picked up by the intercept alone. This is a special case of the "extra sum of squares" principle to be explained in Chapter 6. Note, however, that most computer programs produce an analysis of variance table in a form that omits SS(b_0), that is, a table like Table 1.4 down to the "Total, corrected" line. Often the word "corrected" is omitted in a printout, but if the "Total" df is only $n - 1$, the source should be labeled as "Total, corrected."

When the calculations for Tables 1.3–1.5 are actually carried out on a pocket calculator, the residual SS is rarely calculated directly as shown, but is usually obtained by subtracting "SS($b_1 \mid b_0$)" from the "total, corrected, SS." As already mentioned in (1.3.6), the sum of squares due to regression SS($b_1 \mid b_0$) can be calculated a number of ways as follows (all summations are over $i = 1, 2, \ldots, n$).

T A B L E 1.5. Analysis of Variance Table in Extra SS Order

Source	df	SS	MS = SS/df
b_0	1	$n\bar{Y}^2$	—
$b_1 \mid b_0$	1	S_{XY}^2/S_{XX}	MS_{Reg}
Residual	$n - 2$	By subtraction	s^2
Total	n	$\sum_{i=1}^{n} Y_i^2$	

T A B L E 1.6. Analysis of Variance Table for the Example

Source	df	SS	MS = SS/df	Calculated F-Value
Regression	1	45.5924	45.5924	$57.54 = \frac{45.5924}{0.7923}$
Residual	23	18.2234	$s^2 = 0.7923$	
Total, corrected	24	63.8158		

$$SS(b_1|b_0) = \Sigma(\hat{Y}_i - \overline{Y})^2 = b_1\{\Sigma(X_i - \overline{X})(Y_i - \overline{Y})\} = b_1 S_{XY} \tag{1.3.9}$$

$$= \frac{\{\Sigma(X_i - \overline{X})(Y_i - \overline{Y})\}^2}{\Sigma(X_i - \overline{X})^2} = \frac{S_{XY}^2}{S_{XX}} \tag{1.3.10}$$

$$= \frac{\{\Sigma X_i Y_i - (\Sigma X_i)(\Sigma Y_i)/n\}^2}{\Sigma X_i^2 - (\Sigma X_i)^2/n} = \frac{S_{XY}^2}{S_{XX}} \tag{1.3.11}$$

$$= \frac{\{\Sigma(X_i - \overline{X})Y_i\}^2}{\Sigma(X_i - \overline{X})^2}. \tag{1.3.12}$$

We leave it to the reader to verify the algebraic equivalence of these formulas, which follows from the algebra previously given. Of these forms, Eq. (1.3.9) is perhaps the easiest to use on a pocket calculator because the two pieces have already been calculated to fit the straight line. However, rounding off b_1 can cause inaccuracies, so Eq. (1.3.11) with division performed last is the formula we recommend for calculator evaluation.

Note that the total corrected SS, $\Sigma(Y_i - \overline{Y})^2$, can be written and evaluated either as

$$S_{YY} = \Sigma Y_i^2 - (\Sigma Y_i)^2/n \tag{1.3.13}$$

or as

$$S_{YY} = \Sigma Y_i^2 - n\overline{Y}^2. \tag{1.3.14}$$

The notation $SS(b_1|b_0)$ is read "the sum of squares for b_1 after allowance has been made for b_0" or, more simply, as "the sum of squares for b_1 given b_0." This notation is further expanded in Section 6.1.

The mean square about regression, s^2, will provide an estimate *based on* $(n - 2)$ *degrees of freedom* of the *variance about the regression*, a quantity we shall call $\sigma_{Y \cdot X}^2$. If the regression equation were estimated from an indefinitely large number of observations, the variance about the regression would represent a measure of the error with which any observed value of Y could be predicted from a given value of X using the determined equation (see note 1 of Section 1.4).

Steam Data Calculations

We now carry out and display in Table 1.6 the calculations of this section for our steam data example and then discuss a number of ways in which the regression equation can be examined. The SS due to regression given b_0 is, using (1.3.11),

$$SS(b_1|b_0) = \frac{\{\Sigma X_i Y_i - (\Sigma X_i)(\Sigma Y_i)/n\}^2}{\{\Sigma X_i^2 - (\Sigma X_i)^2/n\}} = \frac{S_{XY}^2}{S_{XX}}$$

$$= \frac{(-571.1280)^2}{7154.42}$$

$$= 45.5924.$$

The Total (corrected) SS is

$$\Sigma Y_i^2 - (\Sigma Y_i)^2/n$$

$$= 2284.1102 - (235.60)^2/25$$

$$= 63.8158.$$

Our estimate of $\sigma_{Y \cdot X}^2$ is $s^2 = 0.7923$ based on 23 degrees of freedom. The F-value will be explained shortly.

Skeleton Analysis of Variance Table

A *skeleton* analysis of variance table consists of the "source" and "df" columns only. In many situations, for example, as in Section 3.3 when comparing several possible arrangements of experimental runs not yet performed, it is useful to compare the corresponding skeleton analysis of variance tables to see which might be most desirable.

R^2 Statistic

A useful statistic to check is the R^2 value of a regression fit. We define this as

$$R^2 = \frac{(\text{SS due to regression given } b_0)}{(\text{Total SS, corrected for the mean } \overline{Y})} \tag{1.3.15}$$

$$= \frac{\Sigma(\hat{Y}_i - \overline{Y})^2}{\Sigma(Y_i - \overline{Y})^2},$$

where both summations are over $i = 1, 2, \ldots, n$. (This is a general definition.) Then R^2 measures the "proportion of total variation about the mean $\overline{Y}$ explained by the regression." It is often expressed as a percentage by multiplying by 100. In fact, R is the correlation [see Eq. (1.6.5)] between Y and $\hat{Y}$ and is usually called the multiple correlation coefficient. R^2 is then "the square of the multiple correlation coefficient." For a straight line fit

$$R^2 = SS(b_1|b_0)/S_{YY}$$

$$= S_{XY}^2/(S_{XX}S_{YY}).$$

Example (*Continued*). From Table 1.6,

$$R^2 = \frac{45.5924}{63.8158} = 0.7144.$$

Thus the fitted regression equation $\hat{Y} = 13.623 - 0.0798X$ explains 71.44% of the total variation in the data about the average $\overline{Y}$. This is quite a large proportion.

R^2 can take values as high as 1 (or 100%) when all the X values are different. When

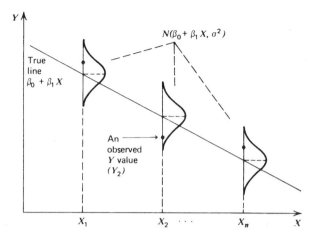

Figure 1.8. Each response observation is assumed to come from a normal distribution centered vertically at the level implied by the assumed model. The variance of each normal distribution is assumed to be the same, σ^2.

repeat runs exist in the data, however, as described in Section 2.1, the value of R^2 cannot attain 1 no matter how well the model fits. This is because no model, however good, can explain the variation in the data due to pure error. An algebraic proof of this fact is given in the solution to Exercise M in "Exercises for Chapters 1–3."

A number of criticisms have been leveled at R^2 in the literature and we discuss them in Section 11.2. In spite of these criticisms, we continue to like R^2 as a "useful first thing to look at" in a regression printout.

1.4. CONFIDENCE INTERVALS AND TESTS FOR β_0 AND β_1

Up to this point we have made no assumptions at all that involve probability distributions. A number of specified algebraic calculations have been made and that is all. We now make the basic assumptions that, in the model $Y_i = \beta_0 + \beta_1 X_i + \epsilon_i$, $i = 1, 2, \ldots, n$:

1. ϵ_i is a random variable with mean zero and variance σ^2 (unknown); that is, $E(\epsilon_i) = 0$, $V(\epsilon_i) = \sigma^2$.

2. ϵ_i and ϵ_j are uncorrelated, $i \neq j$, so that $\text{cov}(\epsilon_i, \epsilon_j) = 0$. Thus

$$E(Y_i) = \beta_0 + \beta_1 X_i, \qquad V(Y_i) = \sigma^2, \qquad (1.4.1)$$

and Y_i and Y_j, $i \neq j$, are uncorrelated.

A further assumption, which is not immediately necessary and will be recalled when used, is that:

3. ϵ_i is a normally distributed random variable, with mean zero and variance σ^2 by assumption 1; that is,

$$\epsilon_i \sim N(0, \sigma^2).$$

Under this additional assumption, ϵ_i, ϵ_j are not only uncorrelated but necessarily independent.

The situation is illustrated in Figure 1.8.

Notes

1. σ^2 may or may not be equal to $\sigma^2_{Y \cdot X}$, the variance about the regression mentioned earlier. If the postulated model is the true model, then $\sigma^2 = \sigma^2_{Y \cdot X}$. If the postulated model is not the true model, then $\sigma^2 < \sigma^2_{Y \cdot X}$. If follows that s^2, the residual mean square that estimates $\sigma^2_{Y \cdot X}$ in any case, is an estimate of σ^2 if the model is correct, but not otherwise. If $\sigma^2_{Y \cdot X} > \sigma^2$ we shall say that the postulated model is incorrect or *suffers from lack of fit*. Ways of deciding this will be discussed later.

2. There is a tendency for errors that occur in many real situations to be normally distributed due to the Central Limit Theorem. If an error term such as ϵ is a sum of errors from several sources, and no source dominates, then no matter what the probability distribution of the separate errors may be, their sum ϵ will have a distribution that will tend more and more to the normal distribution as the number of components increases, by the Central Limit Theorem. An experimental error, in practice, may be a composite of a meter error, an error due to a small leak in the system, an error in measuring the amount of catalyst used, and so on. Thus the assumption of normality is not unreasonable in most cases. In any case we shall later check the assumption by examining residuals (see Chapters 2 and 8).

We now use these assumptions in examining the regression coefficients.

Standard Deviation of the Slope b_1; Confidence Interval for β_1

We know that

$$b_1 = \Sigma(X_i - \overline{X})(Y_i - \overline{Y})/\Sigma(X_i - \overline{X})^2$$
$$= \Sigma(X_i - \overline{X})Y_i/\Sigma(X_i - \overline{X})^2$$

[since the other term removed[1] from the numerator is $\Sigma(X_i - \overline{X})\overline{Y} = \overline{Y} \Sigma(X_i - \overline{X}) = 0$]

$$b_1 = \{(X_1 - \overline{X})Y_1 + \cdots + (X_n - \overline{X})Y_n\}/\Sigma(X_i - \overline{X})^2. \qquad (1.4.2)$$

Now the variance of a function

$$a = a_1 Y_1 + a_2 Y_2 + \cdots + a_n Y_n$$

is

$$V(a) = a_1^2 V(Y_1) + a_2^2 V(Y_2) + \cdots + a_n^2 V(Y_n), \qquad (1.4.3)$$

if the Y_i are pairwise uncorrelated and the a_i are constants; furthermore, if $V(Y_i) = \sigma^2$,

$$V(a) = (a_1^2 + a_2^2 + \cdots + a_n^2)\sigma^2$$
$$= (\Sigma a_i^2)\sigma^2. \qquad (1.4.4)$$

In the expression for b_1, $a_i = (X_i - \overline{X})/\Sigma(X_i - \overline{X})^2$, since the X_i can be regarded as constants. Hence after reduction

$$V(b_1) = \frac{\sigma^2}{\Sigma(X_i - \overline{X})^2} = \frac{\sigma^2}{S_{XX}}. \qquad (1.4.5)$$

---

[1]This term could be left off altogether, but it is conventional to write the numerator of b_1 in a symmetrical form. See the definition of S_{XY} above Eq. (1.2.9a).

The standard deviation of b_1 is the square root of the variance, that is,

$$\mathrm{sd}(b_1) = \frac{\sigma}{\{\Sigma(X_i - \overline{X})^2\}^{1/2}} = \frac{\sigma}{S_{XX}^{1/2}} \tag{1.4.6}$$

or, if σ is unknown and we use the estimate s in its place, assuming the model is correct, the *estimated* standard deviation of b_1 is given by

$$\mathrm{est.\ sd}(b_1) = \frac{s}{\{\Sigma(X_i - \overline{X})^2\}^{1/2}} = \frac{s}{S_{XX}^{1/2}}. \tag{1.4.7}$$

An alternative terminology for the estimated standard deviation is the *standard error*, se. We shall use mostly this alternative terminology.

Confidence Interval for β_1

If we assume that the variations of the observations about the line are normal—that is, the errors ϵ_i are all from the same normal distribution, $N(0, \sigma^2)$—it can be shown that we can assign $100(1 - \alpha)\%$ confidence limits for β_1 by calculating

$$b_1 \pm \frac{t(n - 2, 1 - \tfrac{1}{2}\alpha)s}{\{\Sigma(X_i - \overline{X})^2\}^{1/2}} \tag{1.4.8}$$

where $t(n - 2, 1 - \tfrac{1}{2}\alpha)$ is the $100(1 - \tfrac{1}{2}\alpha)$ percentage point of a t-distribution, with $(n - 2)$ degrees of freedom (the number of degrees of freedom on which the estimate s^2 is based).

Test for $H_0\!: \beta_1 = \beta_{10}$ Versus $H_1\!: \beta_1 \neq \beta_{10}$

On the other hand, if a test is appropriate, we can test the null hypothesis that β_1 is equal to β_{10}, where β_{10} is a specified value that could be zero, against the alternative that β_1 is different from β_{10} (usually stated "$H_0\!: \beta_1 = \beta_{10}$ versus $H_1\!: \beta_1 \neq \beta_{10}$") by calculating

$$t = \frac{(b_1 - \beta_{10})}{\mathrm{se}(b_1)}$$

$$= \frac{(b_1 - \beta_{10})\{\Sigma(X_i - \overline{X})^2\}^{1/2}}{s} \tag{1.4.9}$$

and comparing $|t|$ with $t(n - 2, 1 - \tfrac{1}{2}\alpha)$ from a t-table with $(n - 2)$ degrees of freedom—the number on which s^2 is based. The test will be a two-sided test conducted at the $100\alpha\%$ level in this form. Calculations for our example follow.

Example (*Continued*). Confidence interval for β_1.

$$V(b_1) = \sigma^2/\Sigma(X_i - \overline{X})^2$$

$$= \sigma^2/7154.42$$

$$\mathrm{est.\ } V(b_1) = s^2/7154.42$$

$$= 0.7923/7154.42$$

$$= 0.00011074$$

$$\mathrm{se}(b_1) = \sqrt{\mathrm{est.\ } V(b_1)} = 0.0105.$$

Suppose $\alpha = 0.05$, so that $t(23, 0.975) = 2.069$. Then 95% confidence limits for β_1 are

$$b_1 \pm t(23, 0.975) \cdot s/\{\Sigma(X_i - \overline{X})^2\}^{1/2}, \text{ or}$$

$$-0.0798 \pm (2.069)(0.0105),$$

providing the interval

$$-0.1015 \le \beta_1 \le -0.0581.$$

In words, the true value β_1 lies in the interval (-0.1015 to -0.0581), and this statement is made with 95% confidence.

Example (*Continued*). Test of $H_0 : \beta_1 = 0$.
   We shall also test the null hypothesis that the true value β_1 is zero, or that there is no straight line sloping relationship between atmospheric temperature (X) and the amount of steam used (Y). As noted above, we write (using $\beta_{10} = 0$)

$$H_0 : \beta_1 = 0, \qquad H_1 : \beta_1 \ne 0$$

and evaluate

$$t = (b_1 - 0)/\text{se}(b_1)$$

$$= -0.0798/0.0105$$

$$= -7.60.$$

   Since $|t| = 7.60$ exceeds the appropriate critical value of $t(23, 0.975) = 2.069$, $H_0 : \beta_1 = 0$ is rejected. [Actually, 7.60 also exceeds $t(23, 0.9995)$; we chose a two-sided 95% level test here, however, so that the confidence interval and the t-test would both make use of the same probability level. In this case we can effectively make the test by examining the confidence interval to see if it includes zero, as described below.] The data we have seen cause us to reject the idea that a linear relationship between Y and X might not exist.

Reject or Do Not Reject

If it had happened that the observed $|t|$ value had been smaller than the critical value, we would have said that we *could not reject* the hypothesis. Note carefully that we do not use the word "accept," since we normally cannot accept a hypothesis. The most we can say is that on the basis of certain observed data we cannot reject it. It may well happen, however, that in another set of data we can find evidence that is contrary to our hypothesis and so reject it.
   For example, if we see a man who is poorly dressed we may hypothesize, H_0: "This man is poor." If the man walks to save bus fare or avoids lunch to save lunch money, we have no reason to reject this hypothesis. Further observations of this kind may make us feel H_0 is true, but we still cannot accept it unless we know all the true facts about the man. However, a single observation against H_0, such as finding that the man owns a bank account containing $500,000, will be sufficient to reject the hypothesis.

Confidence Interval Represents a Set of Tests

Once we have the confidence interval for β_1 we do not actually have to compute the $|t|$ value for a particular two-sided t-test at the same α-level. It is simplest to examine the confidence interval for β_1 and see if it contains the value β_{10}. If it does, then the hypothesis $\beta_1 = \beta_{10}$ cannot be rejected; if it does not, the hypothesis is rejected. This can be seen from Eq. (1.4.9), for $H_0: \beta_1 = \beta_{10}$ is rejected at the α-level if $|t| > t(n - 2, 1 - \frac{1}{2}\alpha)$, which implies that

$$|b_1 - \beta_{10}| > t(n - 2, 1 - \tfrac{1}{2}\alpha) \cdot s/\{\Sigma(X_i - \overline{X})^2\}^{1/2};$$

that is, β_{10} lies outside the limits of Eq. (1.4.8).

Standard Deviation of the Intercept; Confidence Interval for β_0

A confidence interval for β_0 and a test of whether or not β_0 is equal to some specified value can be constructed in a way similar to that just described for β_1. We can show that

$$\text{sd}(b_0) = \left\{ \frac{\Sigma X_i^2}{n \Sigma(X_i - \overline{X})^2} \right\}^{1/2} \sigma. \tag{1.4.10}$$

Replacement of σ by s provides the estimated $\text{sd}(b_0)$, that is, $\text{se}(b_0)$. Thus $100(1 - \alpha)\%$ confidence limits for β_0 are given by

$$b_0 \pm t(n - 2, 1 - \tfrac{1}{2}\alpha) \left\{ \frac{\Sigma X_i^2}{n \Sigma(X_i - \overline{X})^2} \right\}^{1/2} s. \tag{1.4.11}$$

A t-test for the null hypothesis $H_0: \beta_0 = \beta_{00}$ against the alternative $H_1: \beta_0 \neq \beta_{00}$, where β_{00} is a specified value, will be rejected at the α-level if β_{00} falls outside the confidence interval, or will not be rejected if β_{00} falls inside, or may be conducted separately by finding the quantity

$$t = \frac{(b_0 - \beta_{00})}{\left\{ \dfrac{\Sigma X_i^2}{n \Sigma(X_i - \overline{X})^2} \right\}^{1/2} s} \tag{1.4.12}$$

and comparing it with percentage points $t(n - 2, 1 - \frac{1}{2}\alpha)$ since $n - 2$ is the number of degrees of freedom on which s^2, the estimate of σ^2, is based. Testing the value of the intercept is seldom of practical interest.

Note: It is also possible to get a *joint confidence region* for β_0 and β_1 simultaneously. See Exercise M in "Exercises for Chapters 5 and 6."

1.5. *F*-TEST FOR SIGNIFICANCE OF REGRESSION

Since the Y_i are random variables, any function of them is also a random variable; two particular functions are MS_{Reg}, the mean square due to regression, and s^2, the mean square due to residual variation, which arise in the analysis of variance tables, Tables 1.3–1.6. These functions then have their own distribution, mean, variance, and moments. It can be shown that the mean values are as follows:

$$E(\text{MS}_{\text{Reg}}) = \sigma^2 + \beta_1^2 \Sigma(X_i - \overline{X})^2$$

$$E(s^2) = \sigma^2, \tag{1.5.1}$$

where, if Z is a random variable, $E(Z)$ denotes its mean value or expected value. Suppose that the errors ϵ_i are independent $N(0, \sigma^2)$ variables. Then it can be shown that if $\beta_1 = 0$, the variable MS_{Reg} multiplied by its degrees of freedom (here, one) and divided by σ^2 follows a χ^2 distribution with the same (one) number of degrees of freedom. In addition, $(n-2)s^2/\sigma^2$ follows a χ^2 distribution with $(n-2)$ degrees of freedom. Also, since these two variables are independent, a statistical theorem tells us that the ratio

$$F = \frac{MS_{Reg}}{s^2} \qquad (1.5.2)$$

follows an F-distribution with (here) 1 and $(n-2)$ degrees of freedom provided (recall) that $\beta_1 = 0$. This fact can thus be used as a test of $H_0 : \beta_1 = 0$ versus $H_1 : \beta_1 \neq 0$. We compare the ratio $F = MS_{Reg}/s^2$ with the $100(1-\alpha)\%$ point of the tabulated $F(1, n-2)$ distribution in order to determine whether β_1 can be considered nonzero on the basis of the data we have seen.

Example (*Continued*). From Table 1.6, we see that the required ratio is $F = 45.5924/0.7923 = 57.54$. If we look up percentage points of the $F(1, 23)$ distribution, we see that the 95% point $F(1, 23, 0.95) = 4.28$. Since the calculated F exceeds the critical F-value in the table—that is, $F = 57.54 > 4.28$—we reject the hypothesis $H_0 : \beta_1 = 0$, running a risk of less than 5% of being wrong.

p-Values for F-Statistics

Many computer printouts give the tail area beyond the observed F-value, typically to three or four decimal places. For our example, this is $Prob\{F(1, 23) > 57.54\} = 0.0000$. (This simply means that the tail area, when rounded, is smaller than 0.0001 or 0.01%.) This can then be judged in relation to the risk level adopted by the person making the test. Thus a 5% person (one prepared to run the risk of rejecting H_0: $\beta_1 = 0$ versus $H_1 : \beta_1 \neq 0$ wrongly once in every 20 tests, on average) and a 10% person (... once in every 10 ...) can look at the same printout and make decisions appropriate to their own percentage level. (In this example, both would come to the same conclusion.)

F = t²

The reader will have noticed that we have had two tests for the test of $H_0 : \beta_1 = 0$ versus $H_1 : \beta_1 \neq 0$, a t-test and an F-test. In fact, the two tests are equivalent and mathematically related here, due to the theoretical fact that $F(1, v) = \{t(v)\}^2$; that is, the square of a t-variable with v df is an F-variable with 1 and v df. (*Note:* This only happens when the first df of F is 1.) For the test statistic we have

$$F = \frac{MS_{Reg}}{s^2} = \frac{b_1\{\Sigma(X_i - \overline{X})(Y_i - \overline{Y})\}}{s^2}$$

$$= \frac{b_1^2 \Sigma(X_i - \overline{X})^2}{s^2} \qquad \text{(by the definition of } b_1) \qquad (1.5.3)$$

$$= \left[\frac{b_1\{\Sigma(X_i - \overline{X})^2\}^{1/2}}{s} \right]^2$$

$$= t^2$$

from Eq. (1.4.9) with $\beta_{10} = 0$. Since the variable $F(1, n-2)$ is the square of the $t(n-2)$ variable, and this carries over to the percentage points (upper α tail of the

F and two-tailed t, total of α), we find exactly the same test results. When there are more regression coefficients the overall F-test for regression, which is the extension of the one given here, does not correspond to the t-test of a coefficient. This is why we have to know about both t- and F-tests, in general. However, tests for *individual coefficients* can be made either in t or $t^2 = F$ form by a similar argument. The t form is often seen in computer programs.

In the example we had an observed F-value of 57.54 and an observed t-value of -7.60. Note that $(-7.6)^2 = 57.76$. Were it not for round-off error this would be equal to the observed value of F. In all such comparisons, the effects of round-off must be taken into consideration.

p-Values for t-Statistics

As in the case of the F-statistic discussed above, many computer programs print out a tail area for the observed t-statistic. This is typically the two-tailed probability value, that is, the area outside the t-value observed and outside minus the t-value observed. Each user can then decide on the message he or she reads from this, depending on the user's chosen α-level. In regression contexts where $F(1, v) = t^2(v)$, the one-sided F-level corresponds to the two-sided t-level.

1.6. THE CORRELATION BETWEEN X AND Y

When we fit a postulated straight line model $Y = \beta_0 + \beta_1 X + \epsilon$, we are tentatively entertaining the idea that Y can be expressed, apart from error, as a first-order function of X. In such a relationship, X is usually assumed to be "fixed," that is, does not have a probability distribution, while Y is usually assumed to be a random variable that follows a probability distribution with mean $\beta_0 + \beta_1 X$ and variance $V(\epsilon)$. (Even if this is not true exactly for X, in many practical circumstances we can behave as though it is, as discussed in Section 1.1.)

More generally for the moment, let us consider two random variables, U and W, say, which follow some continuous joint bivariate probability distribution $f(U, W)$. Then we define the correlation coefficient between U and W as

$$\rho_{UW} = \frac{\text{cov}(U, W)}{\{V(U)V(W)\}^{1/2}},\tag{1.6.1}$$

where

$$\text{cov}(U, W) = \int_{-\infty}^{\infty}\int_{-\infty}^{\infty}\{U - E(U)\}\{W - E(W)\}f(U, W)\,dU\,dW,\tag{1.6.2}$$

and

$$V(U) = \int_{-\infty}^{\infty}\int_{-\infty}^{\infty}\{U - E(U)\}^2 f(U, W)\,dU\,dW,\tag{1.6.3}$$

where

$$E(U) = \int_{-\infty}^{\infty}\int_{-\infty}^{\infty} Uf(U, W)\,dU\,dW.\tag{1.6.4}$$

$V(W)$ and $E(W)$ are similarly defined in terms of W. (If the distributions are discrete, summations replace integrals in the usual way.)

It can be shown that $-1 \leq \rho_{UW} \leq 1$. The quantity ρ_{UW} is a measure of the linear association between the random variables U and W. For example, if $\rho_{UW} = 1$, U and W are perfectly positively correlated and the possible values of U and W all lie on a straight line with positive slope in the (U, W) plane. If $\rho_{UW} = 0$, the variables are said to be uncorrelated, that is, linearly unassociated with each other. This does *not* mean that U and W are statistically independent, as most elementary textbooks emphasize. If $\rho_{UW} = -1$, U and W are perfectly negatively correlated and the possible values of U and W again all lie on a straight line, this time with negative slope, in the (U, W) plane.

If a sample of size n, (U_1, W_1), (U_1, W_2), ..., (U_n, W_n), is available from the joint distribution, the quantity

$$r_{UW} = \frac{\sum_{i=1}^{n}(U_i - \overline{U})(W_i - \overline{W})}{\{\sum_{i=1}^{n}(U_i - \overline{U})^2\}^{1/2}\{\sum_{i=1}^{n}(W_i - \overline{W})^2\}^{1/2}}, \tag{1.6.5}$$

where $n\overline{U} = \Sigma U_i$ and $n\overline{W} = \Sigma W_i$, called the *sample correlation coefficient between U and W*, is an estimate of ρ_{UW} and provides an empirical measure of the *linear association* between U and W. [If factors $1/(n-1)$ are placed before all summations, then r_{UW} has the form of ρ_{UW} with the covariance and variances replaced by sample values.] Like ρ_{UW}, $-1 \leq r_{UW} \leq 1$.

When the U_i and W_i, $i = 1, 2, ..., n$, are all constants, rather than sample values from some distribution, r_{UW} can still be used as a measure of linear association. Because the set of values (U_i, W_i), $i = 1, 2, ..., n$, can be thought of as a complete finite distribution, r_{UW} is, effectively, a population rather than a sample value; that is, $r_{UW} = \rho_{UW}$ in this case. [If factors of $1/n$ are inserted before all summations in Eq. (1.6.5), we obtain Eq. (1.6.1) for the discrete case.]

When we are concerned with the situation where $X_1, X_2, ..., X_n$ represent the values of a finite X-distribution and corresponding observations $Y_1, Y_2, ..., Y_n$ are observed values of random variables whose mean values depend on the corresponding X's (as in this chapter), the correlation coefficient ρ_{XY} can still be defined by Eq. (1.6.1), provided that all integrations with respect to X in expressions like Eqs. (1.6.2)–(1.6.4) are properly replaced by summations over the discrete values $X_1, X_2, ..., X_n$. Equation (1.6.5), with X and Y replacing U and W, of course, can still be used to estimate ρ_{XY} by r_{XY} if a sample of observations $Y_1, Y_2, ..., Y_n$ at the n X-values X_1, $X_2, ..., X_n$, respectively, is available.

In this book we shall make use of expressions of the form of r_{UW} in Eq. (1.6.5); their actual names and roles will depend on whether the quantities that stand in place of U and W are to be considered sample or population values. We shall call all such quantities r_{UW} the *correlation (coefficient) between U and W* and use them as appropriate measures of linear association between various quantities of interest. The distinction made above as to whether they are actually population or sample values is not necessary for our purpose and will be ignored throughout.

When the correlation r_{XY} is nonzero, this means that there exists a linear association between the specific values of the X_i and the Y_i in the data set, $i = 1, 2, ..., n$. In our regression situation, we assume that the X_i are values not subject to random error (to a satisfactory approximation at least; such assumptions are rarely strictly true, as discussed in Section 1.1) and the Y_i are random about mean values specified by the model. Later, when we get involved with more than one predictor variable, we shall use the correlation coefficient [e.g., Eq. (1.6.5) with X_1 and X_2 replacing U and W, which we can then call r_{12}] to measure the linear association between the specific values (X_{1i}, X_{2i}) that occur in a data set. In neither of these cases are we sampling a bivariate distribution.

One final and extremely important point is the following. The value of a correlation r_{XY} shows only the extent to which X and Y are linearly associated. It does not by itself imply that any sort of causal relationship exists between X and Y. Such a false assumption has led to erroneous conclusions on many occasions. (For some examples of such conclusions, including "Lice make a man healthy," see Chapter 8 of *How to Lie with Statistics* by Darrell Huff, W.W. Norton, New York, 1954. Outside North America this book is available as a Pelican paperback.)

Correlation and Regression

Suppose that data $(X_1, Y_1), (X_2, Y_2), \ldots , (X_n, Y_n)$ are available. We can obtain $r_{XY} = r_{YX}$ by applying Eq. (1.6.5) and, if we postulate a model $Y = \beta_0 + \beta_1 X + \epsilon$, we can also obtain an estimated regression coefficient b_1 given by Eq. (1.2.9). Our emphasis in the foregoing on the fact that r_{XY} is a measure of linear association between X and Y invites the question of how r_{XY} and b_1 are connected. We first note that the formula in Eq. (1.6.5) is unaffected by changes in the origins or the scales of U and W. Comparing Eq. (1.6.5), with X and Y replacing U and W, with Eq. (1.2.9) we see that

$$b_1 = \left\{ \frac{\Sigma(Y_i - \overline{Y})^2}{\Sigma(X_i - \overline{X})^2} \right\}^{1/2} r_{XY},$$
(1.6.6)

where summations are over $i = 1, 2, \ldots , n$. In other words, b_1 is a scaled version of r_{XY}, scaled by the ratio of the "spread" of the Y_i divided by the spread of the X_i. If we write

$$(n - 1)s_Y^2 = \Sigma(Y_i - \overline{Y})^2,$$

$$(n - 1)s_X^2 = \Sigma(X_i - \overline{X})^2,$$

then

$$b_1 = \frac{s_Y}{s_X} r_{XY}.$$
(1.6.7)

Thus b_1 and r_{XY} are closely related but provide different interpretations. The unit-free and scale-free correlation r_{XY} measures linear association between X and Y, while b_1 measures the size of the change in Y which can be predicted when a unit change is made in X. Scale changes in the data will affect b_1 but *not* r_{XY}. In more general regression problems, the regression coefficients are also related to simple correlations of the type of Eq. (1.6.5) but in a more complicated manner. These relationships are rarely of practical interest.

r_{XY} and R Connections

Two additional relationships should be noted:

$$r_{XY} = (\text{sign of } b_1)R = (\text{sign of } b_1)(R^2)^{1/2}$$
(1.6.8)

for a straight line fit only, where R is the positive multiple correlation coefficient whose square is

$$R^2 = \frac{\Sigma(\hat{Y}_i - \overline{Y})^2}{\Sigma(Y_i - \overline{Y})^2} \tag{1.6.9}$$

defined in Section 1.3. Also,

$$r_{Y\hat{Y}} = R; \tag{1.6.10}$$

that is, R is equal to the correlation between the given observations Y_i and the predicted values $\hat{Y}_i$. Equation (1.6.10) is true for any linear regression with any number of predictors [whereas Eq. (1.6.8) holds only for the straight line case]. The reader can confirm these relationships by some straightforward algebra (see the solution to Exercise P in "Exercises for Chapters 1–3.").

(In one unusual application, it was desired to compare the Y_i values also with prespecified W_i values, as well as with the $\hat{Y}_i$. This was done by computing r_{YW} in addition to $r_{Y\hat{Y}}$.)

Testing a Single Correlation

Suppose we have evaluated a single correlation coefficient r_{XY} (which we shall call r without subscripts in what follows), which is an estimate of a true (but unknown) parameter ρ. We can obtain a confidence interval for ρ, or test the null hypothesis $H_0 : \rho = \rho_0$, where ρ_0 is the specified value (perhaps zero) versus any of the alternative hypotheses $H_1 : \rho \neq \rho_0$ or $\rho > \rho_0$, or $\rho < \rho_0$, using the approximation known as Fisher's[2] z-transformation. This is

$$z' = \tfrac{1}{2}\ln\left(\frac{1+r}{1-r}\right) \equiv \tanh^{-1}(r) \sim N\left(\tanh^{-1}\rho, \frac{1}{n-3}\right) \tag{1.6.11}$$

approximately. Thus an approximate $100(1 - \alpha)\%$ confidence interval for ρ is given by solving

$$\tfrac{1}{2}\ln\left(\frac{1+r}{1-r}\right) \pm z\left(1 - \frac{\alpha}{2}\right)\left\{\frac{1}{n-3}\right\}^{1/2} = \tfrac{1}{2}\ln\left(\frac{1+\rho}{1-\rho}\right), \tag{1.6.12}$$

where $z(1 - \alpha/2)$ is the upper $\alpha/2$ percentage point of the $N(0, 1)$ distribution, for the two values of ρ that satisfy the $\pm$ alternatives on the left. A test statistic for testing H_0 is

$$z = (n - 3)^{1/2}\left\{\tfrac{1}{2}\ln\left(\frac{1+r}{1-r}\right) - \tfrac{1}{2}\ln\left(\frac{1+\rho_0}{1-\rho_0}\right)\right\}, \tag{1.6.13}$$

which is compared to preselected percentage points of the $N(0, 1)$ distribution. The three alternative hypotheses require a two-tailed test for $H_1 : \rho \neq \rho_0$, an upper-tailed test for $H_1 : \rho > \rho_0$, and a lower-tailed test for $\rho < \rho_0$ respectively. Because a t-distribution with infinite degrees of freedom is a unit normal distribution, a selection of percentage points is given in the bottom row of the t-table. For other percentage points, use the table of the normal distribution itself.

---

[2]For more about R. A. Fisher, see *R. A. Fisher: The Life of a Scientist*, by Joan Fisher Box. Wiley, New York, 1978.

Example. Suppose $n = 103$, $r = 0.5$. Choose $\alpha = 0.05$. Then Eq. (1.6.12) reduces to

$$\tfrac{1}{2}\ln 3 \pm 0.196 = \tfrac{1}{2}\ln \{(1 + \rho)/(1 - \rho)\}$$

and the 95% confidence interval for ρ is 0.339–0.632. Any value ρ_0 of ρ outside this interval would thus be rejected in a 5% level two-sided test of $H_0: \rho = \rho_0$ versus $H_1: \rho \neq \rho_0$, due to the parallel arithmetic involved.

Suppose we wish to test $H_0: \rho = 0.6$ versus $H_1: \rho < 0.6$ at the 1% level. The percentage point needed is -2.326 from the 0.02 column (since we need only a lower-tail test) of the t-table with infinite degrees of freedom. From Eq. (1.6.13) we find the test statistic

$$z = 10\{\tfrac{1}{2}\ln 3 - \tfrac{1}{2}\ln (1.6/0.4)\} = -1.438,$$

which falls above the percentage point -2.326; we do not reject H_0 at the one-sided 1% level.

A good reference for this material is *Biometrika Tables for Statisticians*, Vol. I, by E. S. Pearson and H. O. Hartley, Cambridge University Press, 1958. See pp. 28–32 and 139.

1.7. SUMMARY OF THE STRAIGHT LINE FIT COMPUTATIONS

Data: $(X_1, Y_1), (X_2, Y_2), \ldots, (X_n, Y_n)$.
Model: $Y = \beta_0 + \beta_1 X + \epsilon$.

Pocket-Calculator Computations

Evaluate
$$S_{XX} = \Sigma X_i^2 - (\Sigma X_i)^2/n,$$
$$S_{XY} = \Sigma X_i Y_i - (\Sigma X_i)(\Sigma Y_i)/n,$$
$$S_{YY} = \Sigma Y_i^2 - (\Sigma Y_i)^2/n,$$
$$b_1 = S_{XY}/S_{XX},$$
$$b_0 = \bar{Y} - b_1 \bar{X},$$
$$\mathrm{SS}(b_1|b_0) = S_{XY}^2/S_{XX},$$
$$\text{Residual SS} = S_{YY} - S_{XY}^2/S_{XX} = (n - 2)s^2.$$

Get Table 1.5 or 1.3. Then:

$$R^2 = S_{XY}^2/(S_{XX}S_{YY}),$$
$$F = \{S_{XY}^2/S_{XX}\}/s^2 \quad \text{with } (1, n - 2) \text{ df},$$
$$t = F^{1/2}.$$

(Either F or t can be used to test $H_0: \beta_1 = 0$ versus $H_1: \beta_1 \neq 0$.)

1.8. HISTORICAL REMARKS

It appears that Sir Francis Galton (1822–1911), a well-known British anthropologist and meteorologist, was responsible for the introduction of the word "regression." Originally he used the term "reversion" in an unpublished address "Typical laws of heredity in man" to the Royal Institution on February 9, 1877. The later term "regression" appears in his Presidential address made before Section H of the British Association at Aberdeen, 1885, printed in *Nature*, September 1885, pp. 507–510, and also in a paper "Regression towards mediocrity in hereditary stature," *Journal of the Anthropological Institute*, **15**, 1885, 246–263. In the latter, Galton reports on his initial discovery (p. 246) that the offspring of seeds "did *not* tend to resemble their parent seeds in size, but to be always more mediocre [i.e., more average] than they—to be smaller than the parents, if the parents were large; to be larger than the parents if the parents were very small.... The experiments showed further that the mean filial regression towards mediocrity was directly proportional to the parental deviation from it." Galton then describes how the same features were observed in the records of "heights of 930 adult children and of their respective parentages, 205 in number." Essentially, he shows that, if $Y =$ child's height and $X =$ "parents height" (actually a weighted average of the mother's and father's heights; see the original paper for the details), a regression equation something like $\hat{Y} = \overline{Y} + \frac{2}{3}(X - \overline{X})$ is appropriate, although he does not phrase it in this manner. (The notation is explained in Section 1.1.) Galton's paper makes fascinating reading, as does the account in *The History of Statistics*, by S. M. Stigler, 1986, pp. 294–299, Belknap Press of Harvard University. Galton's analysis would be called a "correlation analysis" today, a term for which he is also responsible. The term "regression" soon came to be applied to relationships in situations other than the one from which it originally arose, including situations where the predictor variables were *not* random, and its use has persisted to this day. In most model-fitting situations today, there is no element of "regression" in the original sense. Nevertheless, the word is so established that we continue to use it.

There has been a dispute about who first discovered the method of least squares. It appears that it was discovered independently by Carl Friedrich Gauss (1777–1855) and Adrien Marie Legendre (1752–1833), that Gauss started using it before 1803 (he claimed in about 1795, but there is no corroboration of this earlier date), and that the first account was published by Legendre in 1805. When Gauss wrote in 1809 that he had used the method earlier than the date of Legendre's publication, controversy concerning the priority began. The facts are carefully sifted and discussed by R. L. Plackett in "Studies in the history of probability and statistics. XXIX. The discovery of the method of least squares," *Biometrika*, **59**, 1972, 239–251, a paper we enthusiastically recommend. Also recommended are accounts by C. Eisenhart, "The meaning of 'least' in least squares," *Journal of the Washington Academy of Sciences*, **54**, 1964, 24–33 (reprinted in *Precision Measurement and Calibration*, edited by H. H. Ku, National Bureau of Standards Special Publication 300, Vol. 1, 1969), and "Gauss, Carl Friedrich," in the *International Encyclopedia of the Social Sciences*, Vol. 6, 1968, pp. 74–81, Macmillan Co., Free Press Division, New York, and the *Encyclopedia of Statistical Sciences*, Vol. 3, 1983, pp. 305–309, Wiley, New York; and a related account by S. M. Stigler, "Gergonne's 1815 paper on the design and analysis of polynomial regression experiments," *Historia Mathematica*, **1**, 1974, 431–447 (see p. 433). See also Stigler's book *The History of Statistics*, 1986, Belknap Press of Harvard University.

APPENDIX 1A. STEAM PLANT DATA

The response is X1, the predictors X2–X10.

1	10.98	5.20	0.61	7.4	31	20	22	35.3	54.8	4
2	11.13	5.12	0.64	8.0	29	20	25	29.7	64.0	5
3	12.51	6.19	0.78	7.4	31	23	17	30.8	54.8	4
4	8.40	3.89	0.49	7.5	30	20	22	58.8	56.3	4
5	9.27	6.28	0.84	5.5	31	21	0	61.4	30.3	5
6	8.73	5.76	0.74	8.9	30	22	0	71.3	79.2	4
7	6.36	3.45	0.42	4.1	31	11	0	74.4	16.8	2
8	8.50	6.57	0.87	4.1	31	23	0	76.7	16.8	5
9	7.82	5.69	0.75	4.1	30	21	0	70.7	16.8	4
10	9.14	6.14	0.76	4.5	31	20	0	57.5	20.3	5
11	8.24	4.84	0.65	10.3	30	20	11	46.4	106.1	4
12	12.19	4.88	0.62	6.9	31	21	12	28.9	47.6	4
13	11.88	6.03	0.79	6.6	31	21	25	28.1	43.6	5
14	9.57	4.55	0.60	7.3	28	19	18	39.1	53.3	5
15	10.94	5.71	0.70	8.1	31	23	5	46.8	65.6	4
16	9.58	5.67	0.74	8.4	30	20	7	48.5	70.6	4
17	10.09	6.72	0.85	6.1	31	22	0	59.3	37.2	6
18	8.11	4.95	0.67	4.9	30	22	0	70.0	24.0	4
19	6.83	4.62	0.45	4.6	31	11	0	70.0	21.2	3
20	8.88	6.60	0.95	3.7	31	23	0	74.5	13.7	4
21	7.68	5.01	0.64	4.7	30	20	0	72.1	22.1	4
22	8.47	5.68	0.75	5.3	31	21	1	58.1	28.1	6
23	8.86	5.28	0.70	6.2	30	20	14	44.6	38.4	4
24	10.36	5.36	0.67	6.8	31	20	22	33.4	46.2	4
25	11.08	5.87	0.70	7.5	31	22	28	28.6	56.3	5

EXERCISES

Exercises for Chapter 1 are located in the section "Exercises for Chapters 1–3", at the end of Chapter 3.

CHAPTER 2

Checking the Straight Line Fit

We discuss basic methods of checking a fitted regression model. Although we talk about these in terms of fitting a straight line, the basic methods apply generally whenever a linear model is fitted, no matter how many predictors there are. Other techniques too advanced for our current context are given in Chapter 8. Here we examine the following:

1. The lack of fit F-test when the data contain *repeat observations*, that is, when *pure error* is available (Sections 2.1 and 2.2).
2. Basic visual checks that can be made on the residuals $e_i = Y_i - \hat{Y}_i$ (Sections 2.3–2.6).
3. The Durbin–Watson test for checking serial correlation (Section 2.7).

2.1. LACK OF FIT AND PURE ERROR

General Discussion of Variance and Bias

We have already remarked that the fitted regression line is a calculated line based on a certain model or assumption, an assumption we should not blindly accept but should *tentatively entertain*. In certain circumstances we can check whether or not the model is correct. First, we can examine the consequences of an incorrect model. Let us recall that $e_i = Y_i - \hat{Y}_i$ is the *residual* at X_i. This is the amount by which the actual observed value Y_i differs from the fitted value $\hat{Y}_i$. As shown in Section 1.2, $\Sigma\, e_i = 0$. The residuals contain all available information on the way in which the model fitted fails to properly explain the observed variation in the dependent variable Y. Let $\eta_i = E(Y_i)$ denote the value given by the true model, whatever it is, at $X = X_i$. Then we can write

$$Y_i - \hat{Y}_i = (Y_i - \hat{Y}_i) - E(Y_i - \hat{Y}_i) + E(Y_i - \hat{Y}_i)$$

$$= \{(Y_i - \hat{Y}_i) - (\eta_i - E(\hat{Y}_i))\} + (\eta_i - E(\hat{Y}_i))$$

$$= q_i + B_i,$$

say, where

$$q_i = \{(Y_i - \hat{Y}_i) - (\eta_i - E(\hat{Y}_i))\}, \qquad B_i = \eta_i - E(\hat{Y}_i).$$

The quantity B_i is the bias error at $X = X_i$. If the model is correct, then $E(\hat{Y}_i) = \eta_i$ and B_i is zero. If the model is not correct, $E(\hat{Y}_i) \neq \eta_i$ and B_i is not zero but has a value that depends on the true model and the value of X_i. The quantity q_i is a random variable that has zero mean since $E(q_i) = E(Y_i - \hat{Y}_i) - (\eta_i - E(\hat{Y}_i)) = \eta_i - E(\hat{Y}_i) - (\eta_i - E(\hat{Y}_i)) = 0$, and this is true whether the model is correct or not, that is, whether $E(\hat{Y}_i) = \eta_i$ or not.

The q_i, it can be shown, are correlated, and the quantity $q_1^2 + q_2^2 + \cdots + q_n^2$ has expected or mean value $(n - 2)\sigma^2$, where $V(Y_i) = V(\epsilon_i) = \sigma^2$ is the error variance. From this it can be shown further that the residual mean square value

$$\frac{1}{n-2}\left\{\sum_{i=1}^{n}(Y_i - \hat{Y}_i)^2\right\} \tag{2.1.1}$$

has expected or mean value σ^2 if the postulated model is of the correct form, or $\sigma^2 + \Sigma B_i^2/(n - 2)$ if the model is not correct. If the model is correct, that is, if $B_i = 0$, then the residuals are (correlated) random deviations q_i and the residual mean square can be used as an estimate of the error variance σ^2.

However, if the model is not correct, that is, if $B_i \neq 0$, then the residuals contain both random (q_i) and systematic (B_i) components. We can refer to these as the variance error and bias error components of the residuals, respectively. Also, the residual mean square will tend to be inflated and will no longer provide a satisfactory measure of the random variation present in the observations. (Since, however, the mean square is a random variable it may, by chance, not have a large value even when bias does exist. For some similar work on the general regression case see Section 10.2.)

How Big is σ^2?

In the simple case of fitting a straight line, bias error can usually be detected merely by examining a plot of the data. When the model is more complicated and/or involves more variables this may not be possible. If a prior estimate of σ^2 is available (by "prior estimate" we mean one obtained from previous experience of the variation in the situation being studied) we can see (or test by an F-test) whether or not the residual mean square is significantly greater than this prior estimate. If it is significantly greater we say that there is lack of fit and we would reconsider the model, which would be inadequate in its present form. If no prior estimate of σ^2 is available, but repeat measurements of Y (i.e., two or more measurements) have been made at the same value of X, we can use these repeats to obtain an estimate of σ^2. Such an estimate is said to represent "pure error" because, if the setting of X is identical for two observations, only the random variation can influence the results and provide differences between them. Such differences will usually then provide an estimate of σ^2 which is much more reliable than we can obtain from any other source. For this reason, it is sensible when designing experiments to arrange for repeat observations.

Genuine Repeats Are Needed

It is important to understand that repeated runs must be genuine repeats and not just repetitions of the same reading. For example, suppose we were attempting to relate, by regression methods, Y = intelligence quotient to X = height of person. A genuine repeat point would be obtained if we measured the separate IQs of two people of exactly the same height. If, however, we measure the IQ of one person of some specified height *twice*, this would not be a genuine repeat point in our context but merely a "reconfirmed" single point. The latter would, it is true, supply information on the variation of the testing method, which is part of the variation σ^2, but it would *not* provide information on the variation in IQ between people of the same height, which is the σ^2 of our problem. In chemical experiments, a succession of readings made during steady-state running does not provide genuine repeat points. However, if a certain set of conditions was reset anew, after intermediate runs at other X-levels had been made, and provided drifts in the response level had not occurred, genuine repeat runs would be obtained. With this in mind, repeat runs that show remarkable agreement which is contrary to expectation should always be regarded cautiously and subjected to additional investigation.

Calculation of Pure Error and Lack of Fit Mean Squares

When there are repeat runs in the data, we need additional notation to take care of the multiple observations on Y at the same value of X. Suppose we have m different values of X and, at the jth of these m particular values, X_j, where $j = 1, 2, \ldots, m$, there are n_j observations; we say that:

$Y_{11}, Y_{12}, \ldots, Y_{1n_1}$ are n_1 repeat observations at X_1;
$Y_{21}, Y_{22}, \ldots, Y_{2n_2}$ are n_2 repeat observations at X_2;
$\cdots$

Y_{ju} is the uth observation ($u = 1, 2, \ldots, n_j$) at X_j;
$\cdots$

$Y_{m1}, Y_{m2}, \ldots, Y_{mn_m}$, are n_m repeat observations at X_m.

Altogether, there are

$$n = \sum_{j=1}^{m} \sum_{u=1}^{n_j} 1 = \sum_{j=1}^{m} n_j$$

observations. The contribution to the pure error sum of squares from the n_1 observations at X_1 is the internal sum of squares of the Y_{1u} about their average $\overline{Y}_1$; that is,

$$\sum_{u=1}^{n_1} (Y_{1u} - \overline{Y}_1)^2 = \sum_{u=1}^{n_1} Y_{1u}^2 - n_1 \overline{Y}_1^2$$

$$= \sum_{u=1}^{n_1} Y_{1u}^2 - \frac{1}{n_1} \left(\sum_{u=1}^{n_1} Y_{1u} \right)^2. \tag{2.1.2}$$

Provided we are sure that the pure error variation is of the same order of magnitude throughout the data (see Sections 2.2 and 2.3) we next pool the internal sums of squares from all the sites with repeat runs to obtain the overall pure error SS as

$$\sum_{j=1}^{m} \sum_{u=1}^{n_j} (Y_{ju} - \overline{Y}_j)^2 \tag{2.1.3}$$

with degrees of freedom

$$n_e = \sum_{j=1}^{m} (n_j - 1) = \sum_{j=1}^{m} n_j - m. \tag{2.1.4}$$

Thus the pure error mean square is

$$s_e^2 = \frac{\sum_{j=1}^{m} \sum_{u=1}^{n_j} (Y_{ju} - \overline{Y}_j)^2}{\sum_{j=1}^{m} n_j - m} \tag{2.1.5}$$

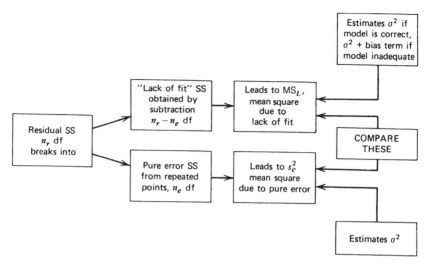

Figure 2.1. Breakup of residual sum of squares into lack of fit and pure error sums of squares.

and is an estimate of σ^2 irrespective of whether the model being fitted is correct or not. In words this quantity is the total of the "within repeats" sums of squares divided by the total of the corresponding degrees of freedom.

Special Formula when $n_j = 2$

If there are only two observations Y_{j1}, Y_{j2} at the point X_j, then

$$\sum_{u=1}^{2} (Y_{ju} - \overline{Y}_j)^2 = \tfrac{1}{2}(Y_{j1} - Y_{j2})^2 \qquad (2.1.6)$$

and this is an easier form to compute. This SS has one degree of freedom.

Split of the Residual SS

Now the pure error sum of squares is actually part of the residual sum of squares as we now show. We can write the residual for the uth observation at X_j as

$$Y_{ju} - \hat{Y}_j = (Y_{ju} - \overline{Y}_j) - (\hat{Y}_j - \overline{Y}_j), \qquad (2.1.7)$$

using the fact that all the repeat points at any X_j will have the *same* predicted value $\hat{Y}_j$. If we square both sides and sum over both u and j, we obtain

$$\sum_{j=1}^{m}\sum_{u=1}^{n_j} (Y_{ju} - \hat{Y}_j)^2 = \sum_{j=1}^{m}\sum_{u=1}^{n_j} (Y_{ju} - \overline{Y}_j)^2 + \sum_{j=1}^{m} n_j(\hat{Y}_j - \overline{Y}_j)^2, \qquad (2.1.8)$$

the cross-product vanishing in the summation over u for each j. The left-hand side of Eq. (2.1.8) is the residual sum of squares; the first term on the right-hand side is the pure error sum of squares. The remainder we call the lack of fit sum of squares. It follows that the pure error sum of squares can be introduced into the analysis of variance table as shown in Figure 2.1. The usual procedure is then to compare the ratio $F = \text{MS}_L/s_e^2$ with the $100(1 - \alpha)\%$ point of an F-distribution with $(n_r - n_e)$ and n_e degrees of freedom. If the ratio is:

T A B L E 2.1. Twenty-three Observations with Same Repeat Runs

Time Order	Y	X	Time Order	Y	X	Time Order	Y	X
12	2.3	1.3	19	1.7	3.7	3	3.5	5.3
23	1.8	1.3	20	2.8	4.0	6	2.8	5.3
7	2.8	2.0	5	2.8	4.0	10	2.1	5.3
8	1.5	2.0	2	2.2	4.0	4	3.4	5.7
17	2.2	2.7	21	3.2	4.7	9	3.2	6.0
22	3.8	3.3	15	1.9	4.7	13	3.0	6.0
1	1.8	3.3	18	1.8	5.0	14	3.0	6.3
11	3.7	3.7				16	5.9	6.7

1. *Significant.* This indicates that the model appears to be inadequate. Attempts would be made to discover where and how the inadequacy occurs. (See comments on the various residuals plots discussed in Sections 2.3–2.6. Note, however, that the plotting of residuals is a standard technique for all regression analyses, not only those in which lack of fit can be demonstrated by this particular test.)

2. *Not Significant.* This indicates that there appears to be no reason *on the basis of this test* to doubt the adequacy of the model and both pure error and lack of fit mean squares can be used as estimates of σ^2. A pooled estimate of σ^2 can be obtained by recombining the pure error and lack of fit sums of squares into the residual sum of squares and dividing by the residual degrees of freedom n_r to give $s^2 =$ (Residual SS)$/n_r$. (Note that the residuals should *still* be examined because there are other aspects of the residuals to be checked.)

We discussed earlier the fact that repeat runs must be genuine repeats. If they are not genuine repeats, s_e^2 will tend to underestimate σ^2, and the lack of fit F-test will tend to wrongly "detect" nonexistent lack of fit.

Example. Since our previous example, which involved data taken from Appendix 1A, did not contain repeat observations, we shall employ a specially constructed example (see Table 2.1) to illustrate the lack of fit and pure error calculations. A regression line $Y = 1.426 + 0.316X$ was estimated from the data in Table 2.1. The analysis of variance table is shown in Table 2.2. Note that the F-value for regression is not checked *at this stage* because we do not yet know if the model suffers from lack of fit or not.

We now find the pure error, and hence the lack of fit.

1. Pure error SS from repeats at $X = 1.3$ is $\frac{1}{2}(2.3 - 1.8)^2 = 0.125$, with 1 degree of freedom.

2. Pure error SS from repeats at $X = 4.0$ is

T A B L E 2.2. ANOVA Table for the Data of Table 2.1

Source	df	SS	MS	F-Ratio
Regression	1	5.499	5.499	7.56 significant at $\alpha = 0.05$ level if no lack of fit
Residual	21	15.278	$0.728 = s^2$	
Total, corrected	22	20.777		

$$(2.8)^2 + (2.8)^2 + (2.2)^2 - 3\{(2.8 + 2.8 + 2.2)/3\}^2$$

$$= 20.52 - (7.8)^2/3$$

$$= 20.52 - 20.28$$

$$= 0.24, \quad \text{with 2 degrees of freedom.}$$

Similar calculations provide the following quantities;

Level of X	$\Sigma_{u=1}^{n_j} (Y_{ju} - \overline{Y}_j)^2$	df
1.3	0.125	1
2.0	0.845	1
3.3	2.000	1
3.7	2.000	1
4.0	0.240	2
4.7	0.845	1
5.3	0.980	2
6.0	0.020	1
Totals	7.055	10

We can thus rewrite the analysis of variance as shown in Table 2.3. The F-ratio $MS_L/ s_e^2 = 1.061$ is not significant. Thus, on the basis of this test at least, we have no reason to doubt the adequacy of our model and can use $s^2 = 0.728$ as an estimate of σ^2, in order to carry out an F-test for significance of the overall regression. This latter F-test is valid only if no lack of fit is exhibited by the model and if no other violation of the regression assumptions is apparent. To emphasize this point we summarize the steps to be taken when our data contain repeat observations:

1. Fit the model, write down the usual analysis of variance table with regression and residual entries. Do not perform an F-test for overall regression yet.

2. Work out the pure error sum of squares and break up the residual as in Figure 2.1. (If there is no pure error, lack of fit has to be checked via residuals plots instead; see Sections 2.3–2.6 and Chapter 8.)

3. Perform the F-test for lack of fit. If significant lack of fit is exhibited, go to step 4a. If the lack of fit test is not significant, so that there is no reason to doubt the adequacy of the model, go to step 4b.

4a. Significant lack of fit. Stop the analysis of the model fitted and seek ways to improve the model by examining residuals. Do *not* carry out the F-test for overall regression, and do not attempt to obtain confidence intervals. The assumptions on which these calculations are based are not true if there is lack of fit in the model fitted. (See Section 10.2.)

T A B L E 2.3. ANOVA (Showing Lack of Fit Calculation)

Source	df	SS	MS	F-Ratio
Regression	1	5.499	5.499	7.56 significant at $\alpha = 0.05$
Residual	21	15.278	$0.728 = s^2$	
Lack of fit	11	8.233	$0.748 = MS_L$	1.061 (not significant)
Pure error	10	7.055	$0.706 = s_e^2$	
Total, corrected	22	20.777		

4b. Lack of fit test *not* significant. Examine the residuals to see if any other violations of assumptions come to light. If not, recombine the pure error and lack of fit sums of squares into the residual sum of squares, use the residual mean square s^2 as an estimate of $V(Y) = \sigma^2$, carry out an F-test for overall regression, obtain confidence bands for the true mean value of Y, evaluate R^2, and so on.

Note that the fact that the model passes the lack of fit test does not mean that it is *the* correct model—merely that it is a plausible one that has not been found inadequate by the data so far. If lack of fit had been found, a different model would have been necessary—perhaps (here) the quadratic one $Y = \beta_0 + \beta_1 X + \beta_{11} X^2 + \epsilon$. Even though the model in our example does not exhibit lack of fit, and has a statistically significant F for overall regression, it is nevertheless not very useful. The R^2 value is only $R^2 = 5.4992/20.7774 = 0.2647$, so that only a small proportion of the variation around $\overline{Y}$ is explained. However, even this pessimistic view of R^2 has to be modified slightly as we now describe.

Effect of Repeat Runs on R^2

As we have already remarked in Section 1.3, it is impossible for R^2 to attain 1 when repeat runs exist, no matter how many terms are used in the model. (A trivial exception is when $s_e^2 = 0$, which rarely happens in practice when there are repeat runs.) No model can pick up the pure error variation (see the solution to Exercise M in "Exercises for Chapters 1–3.")

To illustrate this in our most recent example, we note that the pure error sum of squares is 7.055 with 10 degrees of freedom. No matter what model is fitted to these data, this 7.055 will remain unchanged and unexplained. Thus the maximum R^2 attainable with these data is

$$\text{Max } R^2 = \frac{\text{Total SS, corrected} - \text{Pure error SS}}{\text{Total SS, corrected}}$$

$$= \frac{20.777 - 7.055}{20.777}$$

$$= 0.6604.$$

The value of R^2 actually attained by the fitted model, however, is 0.2674. In other words, we have explained $0.2674/0.6604 = 0.4049$, or about 40% of the amount that can be explained. This figure, while still not too encouraging, looks slightly better. Such a calculation often gives a better sense of what the model actually is achieving in terms of what can be achieved.

Looking at the Data and Fitted Model

The data and the fitted model $\hat{Y} = 1.426 + 0.316X$ are shown in Figure 2.2. We see clearly that, overall, the variation of the points off the line is comparable to the variation within sets of repeats, as already shown by our test for lack of fit with F-ratio slightly over 1. We notice, however, a possible defect not picked up by the lack of fit test. The last observation $(X, Y) = (6.7, 5.9)$ looks a bit remote both from the data and from the line. Clearly other checks are needed to be able to detect this, particularly in larger regressions with several predictors (X's), where a simple plot is not feasible. We shall get to this, and other possible defects, in Chapters 7 and 8. We first finish off our discussion of pure error.

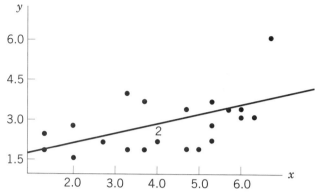

Figure 2.2. Plot of Table 2.1 data and fitted line.

Pure Error in the Many Predictors Case

The formulas given above in the single predictor context apply generally no matter how many predictor variables, $X_1, X_2, \ldots$, are in the data. The only point to watch is that a set of repeat runs must all have the same X_1 value, the same X_2 value, and so on. For example, the four responses at the four points

$$(X_1, X_2, X_3, X_4) = (4, 2, 17, 1), (4, 2, 17, 1), (4, 2, 17, 1), (4, 2, 17, 1)$$

provide repeat runs; however, the four responses at the four points

$$(X_1, X_2, X_3, X_4) = (4, 2, 7, 1), (4, 2, 16, 1), (4, 2, 18, 1), (4, 2, 29, 1)$$

do not, for example, because their X_3 coordinates are all different.

Adding (or Dropping) X's Can Affect Maximum R^2

Note that, if additional predictor variables are added to the model, the maximum R^2 value may increase. This is because observations that were repeats before may not be repeats when the additional predictor(s) are introduced. For comments on the treatment of pure error when predictors are *dropped* from the model, see Section 12.2. Dropping predictors can create (pseudo) pure error. An eye needs to be kept on the effects, on the pure error calculation, of all changes of these types.

Approximate Repeats

Some sets of data have no, or very few, repeat runs but do have *approximate repeats*, that is, sets of runs that are close together in the X-space compared with the general spread of the points in the X-space. In such cases, we can often use these pseudo-repeats as though they were repeat runs and evaluate an approximate pure error sum of squares from them. This is then incorporated in the analysis in the usual way. For an example of such a use, see Exercise L, in "Exercises for Chapters 1–3." The major problem here is in deciding what the words "close together" mean.

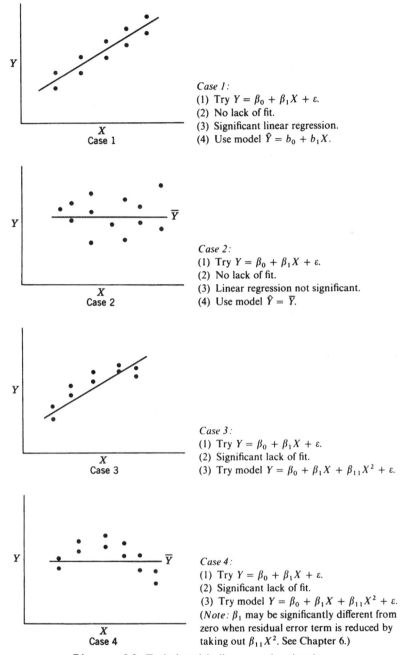

Figure 2.3. Typical straight line regression situations.

Case 1:
(1) Try $Y = \beta_0 + \beta_1 X + \varepsilon$.
(2) No lack of fit.
(3) Significant linear regression.
(4) Use model $\hat{Y} = b_0 + b_1 X$.

Case 2:
(1) Try $Y = \beta_0 + \beta_1 X + \varepsilon$.
(2) No lack of fit.
(3) Linear regression not significant.
(4) Use model $\hat{Y} = \bar{Y}$.

Case 3:
(1) Try $Y = \beta_0 + \beta_1 X + \varepsilon$.
(2) Significant lack of fit.
(3) Try model $Y = \beta_0 + \beta_1 X + \beta_{11} X^2 + \varepsilon$.

Case 4:
(1) Try $Y = \beta_0 + \beta_1 X + \varepsilon$.
(2) Significant lack of fit.
(3) Try model $Y = \beta_0 + \beta_1 X + \beta_{11} X^2 + \varepsilon$.
(*Note:* β_1 may be significantly different from zero when residual error term is reduced by taking out $\beta_{11} X^2$. See Chapter 6.)

Generic Pure Error Situations Illustrated Via Straight Line Fits

The diagrams in Figure 2.3 illustrate some situations that may arise when a straight line is fitted to data and the consequent action to be taken. All of these situations are obvious in the context of a straight line fit, but they illustrate situations that occur in more general regressions and our comments should be viewed in that light.

Case 1. The model we try shows no lack of fit and we need to use all of the model.

Case 2. The model we try shows no lack of fit but we do not need all of it.

Case 3. The model we try shows lack of fit and a higher-order (or a different) model must be formulated.

Case 4. The model we try shows lack of fit and, moreover, some of the terms in it seem to be too small to be useful. (A test is not valid here because of the lack of fit.) We must formulate a higher order (or a different) model and must not jump to premature conclusions about terms currently in the model. (For more on such difficulties see Chapter 12.)

(The words "a different model" also include the possibility of transforming Y or X, for example, by using ln Y as a response.)

2.2. TESTING HOMOGENEITY OF PURE ERROR

In practice, we most often look at a plot of the spreads of the repeat runs and decide by eye whether or not they look a lot different from one another. Formal tests exist, if really needed, but all have drawbacks. Cochran's test and Hartley's test require the same number of replicates at each site, plus special tables, so we do not discuss these. Bartlett's test is commonly used but is sensitive to non-normality; that is, if the data are non-normal, the validity of the test is greatly affected. Nevertheless, we describe it and a modified version below. We also describe and recommend Levene's test using group medians rather than means, if such a test is desired. (This essentially converts a test of variances into a test of means, which is relatively unaffected by non-normality. The price paid for this is lowered testing power.)

Bartlett's Test

Let $s_1^2, s_2^2, \ldots, s_m^2$ be the estimates of σ^2 from the m groups of repeats with $\nu_1, \nu_2, \ldots, \nu_m$ degrees of freedom, respectively. In terms of previous notation, $\nu_j = n_j - 1$ and

$$s_j^2 = \sum_{u=1}^{n_j} (Y_{ju} - \overline{Y}_j)^2/(n_j - 1). \tag{2.2.1}$$

As before,

$$s_e^2 = (\nu_1 s_1^2 + \nu_2 s_2^2 + \cdots + \nu_m s_m^2)/(\nu_1 + \nu_2 + \cdots + \nu_m), \tag{2.2.2}$$

and we write $\nu = \nu_1 + \nu_2 + \cdots + \nu_m$. The constant C is defined as

$$C = 1 + \{\nu_1^{-1} + \nu_2^{-1} + \cdots + \nu_m^{-1} - \nu^{-1}\}/\{3(m-1)\}, \tag{2.2.3}$$

where m is the number of groups with repeat runs. The test statistic is then

$$B = \left\{ \nu \ln s_e^2 - \sum_{j=1}^{m} \nu_j \ln s_j^2 \right\}/C. \tag{2.2.4}$$

When the variances of the groups are all the same, B is distributed as χ_{m-1}^2 approximately. A significant B value could indicate inhomogeneous variances. It could also indicate non-normality, so it makes sense to actually look at the shapes of plots of the m samples, too.

Example. Consider the data used to illustrate lack of fit in Section 2.1. We have

$$s_e^2 = 7.055/10 = 0.7055,$$

$$C = 1 + \{6(1/1) + 2(1/2) - 1/10\}/\{3(8-1)\} = 1.328571,$$

$$B = \{10 \ln (0.7055) - \ln (0.125) - \ln (0.845) - \cdots$$

$$-2 \ln (0.980) - \ln (0.020)\}/1.328571$$

$$= \{-3.488485 + 7.836646\}/1.328571 = 3.273(7 \text{ df}).$$

The value of the statistic is very small, indicating no reason to doubt homogeneity of variances. (For example, the 0.95 percentage point of χ_7^2 is 14.1.)

Bartlett's Test Modified for Kurtosis

In this variation, the statistic B of Eq. (2.2.4) is multiplied by $d = 2/(\hat{\beta} - 1)$, where

$$\hat{\beta} = N \sum_{j=1}^{m} \sum_{u=1}^{n_j} (Y_{ju} - \overline{Y}_j)^4 / \left\{ \sum_{j=1}^{m} \sum_{u=1}^{n_j} (Y_{ju} - \overline{Y}_j)^2 \right\}^2 \qquad (2.2.5)$$

estimates the (assumed common) kurtosis of the sets of repeats. For normally distributed data the true β would be 3 and d would typically be close to 1. The same χ^2-test as before is used for this statistic; here N is the total number of observations in the (usually reduced) data set used for the test, that is, the total number of observations in all the sets of repeats, ignoring all the single observations in the data.

Example (*Continued*)

$$\hat{\beta} = 18\{[0.25^4 + (-0.25)^4] + \cdots + [0.1^4 + (-0.1)^4]\}/(7.055)^2$$

$$= 18\{5.231\}/(7.055)^2 = 1.891761$$

$$d = 2/(0.891761) = 2.242753$$

$$Bd = 3.273 \times 2.242753 = 7.3405 \quad (7 \text{ df}).$$

The modified test statistic remains nonsignificant compared with $\chi_7^2(0.95) = 14.1$.

Levene's Test Using Means

Consider, in the jth group of repeats, the absolute deviations

$$z_{ju} = |Y_{ju} - \overline{Y}_j|, \qquad u = 1, 2, \ldots, n_j, \qquad (2.2.6)$$

of the Y's from the means of their repeats group. Consider this as a one-way classification and compare the "between groups" mean square with the "within groups" mean square via an F-test. The appropriate F-statistic is then

$$\frac{\sum_{j=1}^{m} n_j(\overline{z}_j - \overline{z})^2/(m-1)}{\sum_{j=1}^{m} \sum_{u=1}^{n_j} (z_{ju} - \overline{z}_j)^2 / \sum_{j=1}^{m} (n_j - 1)}, \qquad (2.2.7)$$

T A B L E 2.4. Details for Levene's Test Using Means for the Data of Table 2.1

X-Level, X_j	z_{ju}	n_j	$\bar{z}_j$
1.3	0.25, 0.25	2	0.25
2.0	0.65, 0.65	2	0.65
3.3	1.00, 1.00	2	1.00
3.7	1.00, 1.00	2	1.00
4.0	0.20, 0.20, 0.40	3	0.2\dot{6}
4.7	0.65, 0.65	2	0.65
5.3	0.00, 0.70, 0.70	3	0.4\dot{6}
6.0	0.10, 0.10	2	0.10

where

$$\bar{z}_j = \sum_{u=1}^{n_j} z_{ju}/n_j, \qquad \bar{z} = \sum_{j=1}^{m} \sum_{u=1}^{n_j} z_{ju} / \sum_{j=1}^{m} n_j. \tag{2.2.8}$$

The F-value is referred to $F\{m - 1, \sum_{j=1}^{m} (n_j - 1)\}$, using only the upper tail.

Example *(Continued).* We have $m = 8$, $\sum_{j=1}^{m} n_j = 18$, $\sum_{j=1}^{m} (n_j - 1) = 10$, $\bar{z} = 9.5/18 = 0.52\dot{7}$. The z_{ju} and the row means $\bar{z}_j$ are shown in Table 2.4, where we use only the repeat runs, ignoring the singles, which do not contribute here.

The numerator is then $1.687783/(8 - 1) = 0.24112$ and the denominator is $0.353333/10 = 0.035333$, whereupon $F = 6.824$, which we can compare to $F(7, 10, 0.95) = 3.14$. This would indicate that there *are* differences between the variances of the various groups. We comment on this further below.

Levene's Test Using Medians

Consider, in the jth group of repeats, the absolute deviations

$$z_{ju} = |Y_{ju} - \tilde{Y}_j|, \qquad u = 1, 2, \ldots, n_j,$$

of the Y's from the *medians* $\tilde{Y}_j$ of their repeats group. Consider these in a one-way classification and compare the "between groups" mean square with the "within groups" mean square via an F-test. The appropriate F-statistic is again (2.2.7), and it is tested in the same way as before. See Carroll and Schneider (1985).

Example *(Continued).* Note that when only two observations are in a group, the mean and median are identical. So only groups with three or more observations can give z_{ju} and $\bar{z}_j$ values different from those in Table 2.4. For $X = 4.0$, the median is 2.8; the mean was 2.6. For $X = 5.3$ the median is 2.8, identical to the mean, as it happens. So the F-statistic changes only slightly *for this example* through the changed calculation for $X = 4.0$; the new z_{ju} values there are 0, 0, and 0.6 with $\bar{z}_j = 0.2$ (replacing 0.2, 0.2, and 0.4 with mean $0.26\dot{6}$). Now $\bar{z} = 9.3/18 = 0.51\dot{6}$. This gives $F = \{1.803333/(8 - 1)\}/\{0.566667/10\} = 4.546$, smaller than in the foregoing test, but still greater than $F(7, 10, 0.95) = 3.14$. So again differences are declared between the variances of the groups.

Some Cautionary Remarks

Our numerical example is (on the one hand) not a particularly good one to illustrate the Levene tests because the denominator of the test statistic is estimated by only two sets of three z_{ju} values; the pairs do not contribute to the within sum of squares. On the other hand, it does alert us to such possible problems! It is also worrying that, although the pairs do not contribute to the within groups numerical value, they are granted a degree of freedom! Alan Miller has suggested a sensible possible adjustment, reducing these df to zero, but this does not seem to solve the problem either. Simulations performed by T.-S. Lim and W.-Y. Loh, some of which are mentioned in Lim and Loh (1996) and some of which were performed privately as a favor to the authors of this book, seem to indicate that the best test is Levene's test using medians. (Our example would indicate that the data should not contain too many *pairs* of repeats, however.) At the same time, it makes practical sense to plot the Y-values and visually to compare the repeat groupings with one another. So that is our somewhat cautious joint recommendation, with the plots always taking preference. (In using these example data again later, we do not make any adjustments for possible unequal variances, since the evidence for this seems weak.)

A Second Example

The groups of data below are values from our Exercise 23D, adapted via $(Y - 1430)/5$.

Group:	1	2	3	4	5
	52	24	39	59	20
	30	3	4	24	23
	63	43	16	0	27
	51	23	—	3	18

The Bartlett test value is, from (2.2.4), $B = 6.83$. Adjusting via (2.2.5) leads to $Bd = 6.41$. Both values are less than $\chi^2_{4,0.95} = 9.49$. The F-statistics from Levene's tests are 1.56 (using means) and 1.16 (using medians). Both are less than $F_{4,14,0.95} = 3.11$. So in this example we have consistent conclusions *not* rejecting homogeneity.

2.3. EXAMINING RESIDUALS: THE BASIC PLOTS

As we have already mentioned, the residuals $e_i = Y_i - \hat{Y}_i$ contain within them information on why the model might not fit the data. So it is well worthwhile to check the behavior of the residuals and allow them to tell us of any peculiarities of the regression fit that might have occurred.

The study of residuals is not new, as the following quotation makes clear.

> Almost all the greatest discoveries in astronomy have resulted from the consideration of what we have elsewhere termed RESIDUAL PHENOMENA, of a quantitative or numerical kind, that is to say, of such portions of the numerical or quantitative results of observation as remain outstanding and unaccounted for after subducting and allowing for all that would result from the strict application of known principles. (Sir John F. W. Herschel, Bart. K. H., 1849, p. 548)

An enormous amount has been written about the study of residuals. There are, in fact, several excellent books (see Section 2.8). In this section we discuss only the basic plots that allow the most useful checks. These are the checks that should be done on a routine basis for every regression. More sophisticated methods are discussed in later chapters for those wishing to explore further.

The work of this section is useful and valid not only for linear regression models but also for nonlinear regression models and analysis of variance models. In fact, this section applies to *any* situation where a model is fitted and measures of unexplained variation (in the form of a set of residuals) are available for examination. Thus, like the pure error calculations in Section 2.1, it is *not* restricted only to the straight line regression case, even though we find it convenient to talk about it here.

How Should the Residuals Behave?

The residuals are defined as the *n* differences $e_i = Y_i - \hat{Y}_i$, $i = 1, 2, \ldots, n$, where Y_i is an observation and $\hat{Y}_i$ is the corresponding fitted value obtained by use of the fitted regression equation.

   Note: Usually, the residuals would be evaluated to the same number of significant figures as appeared in the original response observations. Sometimes one additional significant figure is used, but to go beyond this is generally a waste of effort. Computer printouts typically contain more figures than necessary, of course, but these would be cut back if the data were transcribed for reporting purposes.

   We can see from their definition that the residuals e_i are the differences between what is actually observed, and what is predicted by the regression equation—that is, the amount that the regression equation has not been able to *explain*. Thus we can think of the e_i as the *observed errors if the model is correct*. (There are, however, restrictions on the e_i induced by the normal equations.) Now in performing the regression analysis we have made certain assumptions about the errors; the usual assumptions are that the errors are independent, have zero mean, have a constant variance σ^2, and follow a normal distribution. The last assumption is required for making *F*-tests. Thus if our fitted model is correct, the residuals should exhibit tendencies that tend to confirm the assumptions we have made or, at least, should not exhibit a denial of the assumptions. This latter idea is the one that should be kept in mind when examining the residuals. We should ask: "Do the residuals make it appear that our assumptions are wrong?" After we have examined the residuals we shall be able to conclude either that (1) the assumptions appear to be violated (in a way that can be specified) or (2) the assumptions do not appear to be violated. Note that (2) does not mean that we are concluding that the assumptions are correct; it means merely that on the basis of the data we have seen, we have no reason to say that they are incorrect. The same spirit occurs in making tests of hypotheses when we either *reject* or *do not reject* (rather than *accept*). We now give ways of examining the residuals in order to check the model. These ways are all graphical, are easy to do, and are usually very revealing when the assumptions are violated. The principal ways of plotting the residuals e_i are:

1. To check for non-normality.
2. To check for time effects if the time order of the data is known.
3. To check for nonconstant variance and the possible need for a transformation on *Y*.
4. To check for curvature of higher order than fitted in the *X*'s.

   In addition to these basic plots, the residuals should also be plotted:

5. In any way that is sensible for the particular problem under consideration.

   (We remark before proceeding that the basic plots should always be done and will often pick up any deficiencies present in many sets of residuals. It is also possible for these simple plots to be "fooled" or "defeated," however, if a sophisticated defect, or a combination of defects, occurs. That is why more complicated methods of analyzing residuals have been developed. The methods of this chapter are, however, the first line of defense for detection of an unsuitable model.)

We talk here, and continue to do so in this chapter, of looking at the ordinary residuals, defined as $Y_i - \hat{Y}_i$. Actually there are several types of residuals, any or all of which could be obtained and could be plotted as well, or instead. We discuss the various types in Chapter 8. For most regressions, it is not crucial which set of residuals is plotted. Occasionally it makes a great deal of difference which choice of residuals is made; this would show up when the plots of various residual sets are compared.

2.4. NON-NORMALITY CHECKS ON RESIDUALS

We usually assume that $\epsilon_i \sim N(0, \sigma^2)$ and that all errors are independent of one another. Their estimates, the residuals, cannot be independent. The estimation of the parameters (p of them, say; $p = 2$ for the straight line) means that the n residuals carry only $(n - p)$ df. The p normal equations [for $p = 2$, see Eqs. (1.2.8)] are restrictions on the e_i, essentially. Unless p is large compared with n, this typically has little effect on our non-normality checks. We first note that:

> For any model with a β_0 (intercept) term in it, the least squares residuals must, in theory, add to zero.

This is seen from the first normal equation obtained by differentiating the error sum of squares with respect to β_0. If the model fitted is $E(Y) = \beta_0 + \beta_1 X_1 + \cdots + \beta_k X_k$, the equation can be written

$$-2 \Sigma(Y_i - b_0 - b_1 X_{1i} - \cdots - b_k X_{ki}) = 0,$$

where the summation is taken over $i = 1, 2, \ldots, n$. This reduces to

$$\Sigma(Y_i - \hat{Y}_i) = 0.$$

Thus

$$\Sigma \, e_i = 0.$$

Because the least squares fitting procedure guarantees this, there is no need to check that the mean $\bar{e} = \Sigma e_i / n$ is zero. We have made it so!

Often a simple histogram, or a stem and leaf plot, will be enough. Figure 2.4 shows histograms for the residuals from (a) the steam data fit of Chapter 1 and (b) the straight line fit for the lack of fit test data in Section 2.1. We conclude that these (somewhat crude) plots look "normal enough," sometimes a difficult judgment, except for the highest observation in Figure 2.4b, which looks like an outlier.

Normal Plot of Residuals

An alternative and (we believe) better check is to make a *normal probability plot*. This is not difficult to do "by hand" but is better done in the computer. A full explanation is given in Appendix 2A. Here we merely set out the steps required in the MINITAB system of computing. Find out which column contains the residuals you wish to "normal plot", say, c11. Write

```
nscore c11   c12
plot c12 ,c11
```

Draw (or imagine) a straight line through the main middle bulk of the plot. Ask:

Midpoint	Count	
−1.6	1	*
−1.2	3	***
−0.8	1	*
−0.4	4	****
0.0	7	*******
0.4	2	**
0.8	3	***
1.2	4	****

(*a*) Histogram of steam residuals $n = 25$

Midpoint	Count	
−1.0	4	****
−0.5	6	******
0.0	6	******
0.5	4	****
1.0	1	*
1.5	1	*
2.0	0	
2.5	1	*

(*b*) Histogram of residuals $n = 23$

Figure 2.4. Histograms of residuals from (*a*) steam data fit and (*b*) lack of fit test data.

"Do all the points lie on such a line, more or less?" If the answer is yes, one would conclude that the residuals do not deny the assumption of "normality of errors" made in performing tests and getting confidence intervals. We see from the plots in Figures 2A.4, 2A.5 and 2A.6 that the steam data residuals look alright, but that the pure error example data show signs of an *outlier*, an observation that falls unusually out of the pattern for a normal sample. For *why* it is an outlier, see Appendix 2A.

(*Note carefully:* If the plot is made "the other way around," that is, as
nscore c11 c12
plot c11 c12
the criteria for an outlier to exist are different. It follows that every normal plot must be looked at carefully to see which axis is which, before conclusions are reached. This step is *very* important.)

There are other ways of assessing normality. For information on the Shapiro and Wilk test, a useful starter reference is Royston (1995).

2.5. CHECKS FOR TIME EFFECTS, NONCONSTANT VARIANCE, NEED FOR TRANSFORMATION, AND CURVATURE

We plot the residuals e_i vertically against, in turn:

1. The time order of the data, if known.
2. The corresponding fitted values $\hat{Y}_i$, using the fitted model.
3. The corresponding X_i values if there is only one predictor variable; or, in general, each set of X_{ji}, where $j = 1, 2, \ldots, k$ represent the X's in the regression.

Figure 2.5. A satisfactory residuals plot should give this overall impression.

In all of these cases, a satisfactory plot is one that shows a (more or less) horizontal band of points giving the impression of Figure 2.5. There are many possible unsatisfactory plots. Three typical ones appear in Figure 2.6. The first of these three (the funnel) displays the band of residuals widening to the right showing nonconstant variance. The second is an upward trend and the third is curvature. (All of these defective plots can appear in other directions, of course, for example, a reversed funnel or a downward curve.) It is difficult to be absolutely specific about what to do if these defects are found but Table 2.5 gives some general indications.

Plots for the steam data and the lack of fit example data appear in Figures 2.7 and 2.8. We see that, for the steam data, none of the three plots seems to show any worrying anomaly that would indicate the regression fit is defective. For the lack of fit example data, we must remember that the observation with the largest residual (value 2.36) is a likely outlier. If we ignore this residual in the residuals versus time plot, there is still a hint of a funnel shape, but perhaps too little to act on—opinions would differ on this. The plot of residuals versus $\hat{Y}$ values again shows up the outlier but is otherwise unremarkable. (The apparent slight "downward slope look" caused by ignoring the outlier is essentially "caused by" the presence of the outlier.) The plot of residuals versus X is similar to the foregoing plot, because Y and X rise together. In our two examples, the residuals are equally spaced in time. If they were not, and the correct spacings were known, the residuals would be plotted using those spacings, of course.

Three Questions and Answers

Query 1. Why do we plot the residuals $e_i = Y_i - \hat{Y}_i$ against the $\hat{Y}_i$ and not against the Y_i, for the usual linear model?

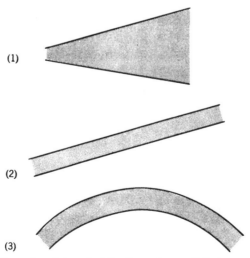

(1)

(2)

(3)

Figure 2.6. Examples of characteristics shown by unsatisfactory residuals behavior.

T A B L E 2.5. Possible Remedies for Unsatisfactory Residuals Plots

Unsatisfactory Plot: See Figure 2.6	Plot of e_i Versus		
	Time Order	Fitted $\hat{Y}_i$	X_{ji} Values
Funnel indicating nonconstant variance	Use weighted[a] least squares	Use weighted[a] least squares or transform[b] the Y_i	Use weighted[a] least squares or transform[b] the Y_i
Ascending or descending band	Consider adding first-order term in time	Error in analysis or wrongful omission of β_0	Error in the calculations; first-order effect of X_j not removed
Curved band	Consider adding first- and second-order terms in time	Consider adding extra terms to the model or transform[b] the Y_i	Consider adding extra terms to the model or transform[b] the Y_i

[a]See Section 9.2.
[b]See Chapter 13.

Answer. Because the e's and the Y's are usually correlated but the e's and the $\hat{Y}$'s are not. One way to see this is to think of plots of the e_i as ordinate against (i) the Y_i and (ii) the $\hat{Y}_i$, and find the slope of a least squares lines through the points. For (i) it will be $1 - R^2$; for (ii) 0. This means that, unless $R^2 = 1$, there will always be a slope of $1 - R^2$ in the e_i versus Y_i plot, even if there is nothing wrong. However, a slope in the e_i versus $\hat{Y}_i$ plot *indicates* that something is wrong. See Exercise X in "Exercises for Chapters 5 and 6."

Query 2. Why does the plot of residuals $e_i = Y_i - \hat{Y}_i$ versus $\hat{Y}_i$ exhibit a series of straight lines, with slopes of -1?

Answer. This feature is in fact *always* present but is usually not obvious. A line is formed by any set of plotted points with the same Y value. Suppose, for example, we have m points with the same value of Y, $Y = a$, say. Then we plot this subset of residuals

$$a - \hat{Y}_1, \quad a - \hat{Y}_2, \ldots, \quad a - \hat{Y}_m,$$

versus $\quad \hat{Y}_1, \qquad \hat{Y}_2, \ldots, \qquad \hat{Y}_m, \qquad$ respectively.

These m points are all on a line with slope -1 through the points $(\hat{Y}, e) = (a, 0)$ and $(0, a)$. For, if $\tilde{Y}$ is the average $\tilde{Y} = (\hat{Y}_1 + \hat{Y}_2 + \cdots + \hat{Y}_m)/m$, the slope of the line, via least squares, is

$$\frac{S_{e\hat{Y}}}{S_{\hat{Y}\hat{Y}}} = \frac{\Sigma(a - \hat{Y}_i - (a - \tilde{Y}))(\hat{Y}_i - \tilde{Y})}{\Sigma(\hat{Y}_i - \tilde{Y})^2} = -1,$$

after cancellation of the a's. The intercept is $(a - \tilde{Y}) - (-1)\tilde{Y} = a$.

In data sets where only a limited number of Y's are recorded (e.g., color levels of a dyestuffs product, percentages of pests present on a plant leaf) this feature may become very obvious. Searle (1988, p. 211) who drew attention to this feature also points out that:

1. The lines always exist. When no Y's are repeated, there is only one point on each line.

2. The lines occur no matter what model is fitted, and whether linear, nonlinear, or generalized linear model estimation techniques have been used.

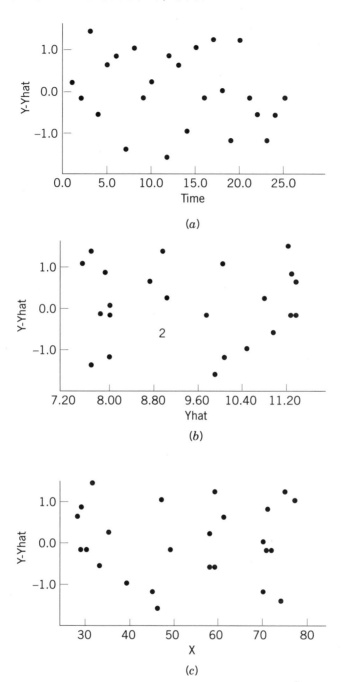

Figure 2.7. Plots of steam data residuals versus (a) order, (b) $\hat{Y}$, and (c) X.

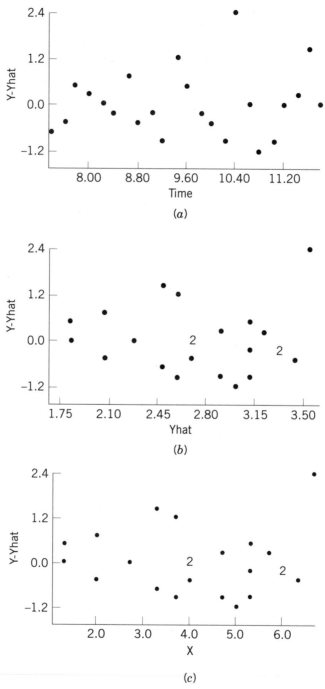

F i g u r e 2.8. Plots of lack of fit data residuals versus (a) order, (b) $\hat{Y}$, and (c) X.

Example. The data in Table 2.1 have four data points with $Y = 2.8$ with residuals from a straight line fit $\hat{Y} = 1.426 + 0.316X$ of:

Number	e	$\hat{Y}$
(7)	$2.8 - 2.05 = 0.75$ plotted against 2.05	
(20)	$2.8 - 2.68 = 0.12$ plotted against 2.68	
(5)	$2.8 - 2.68 = 0.12$ plotted against 2.68	
(6)	$2.8 - 3.10 = -0.30$ plotted against 3.10	

It is easily confirmed that the four points lie on a line with slope -1 and through points $(0, 2.8)$, $(2.8, 0)$, that is, on the line $e + \hat{Y} = 2.8$.

Query 3. Is it possible to work out some test statistics instead of looking at the diagrams?

Answer. It is possible to evaluate test statistics, but it is often difficult to know if they are sufficiently deviant to require action. In practical regression situations, a detailed examination of the corresponding residuals plots is usually far more informative, and the plots will almost certainly reveal any violations of assumptions serious enough to require corrective action.

Consider the plot of e_i against $\hat{Y}_i$ described above. Three particular types of discrepancies were mentioned and related to the diagrams of Figure 2.6. We can measure each of these defects with appropriate statistics as follows. Define

$$T_{pq} = \Sigma_{i=1}^n e_i^p \hat{Y}_i^q. \tag{2.5.1}$$

Then:

1. $T_{21} = \Sigma_{i=1}^n e_i^2 \hat{Y}_i$ provides a measure for the type of defect shown in Figure 2.6(1).

2. $T_{11} = \Sigma_{i=1}^n e_i \hat{Y}_i$. This should always be zero. This provides a measure for the defect shown in Figure 2.6(2). Evaluation of this statistic could be done as a routine check, if desired.

3. $T_{12} = \Sigma_{i=1}^n e_i \hat{Y}_i^2$ provides a measure for the type of defect shown in Figure 2.6(3). It is related to Tukey's "one degree of freedom for nonadditivity" statistic. (See also Exercise O in "Exercises for Chapters 5 and 6.")

Other types of statistics are also available. Readers who would like to learn more about them should consult the texts listed in Section 2.8.

Comment

The plots we have discussed are very basic ones and can be criticized in a number of ways because they may not show up defects of specific types. A vast literature has grown up, and many more sophisticated methods have been suggested. We deal with some of these in Chapter 8, and these provide further references to which interested readers can turn after that.

2.6. OTHER RESIDUALS PLOTS

Specialist knowledge of the problem under study often suggests that other types of residuals plots should be examined. For example, suppose it were known that the 23 observations that led to the 23 residuals from the lack of fit test example came

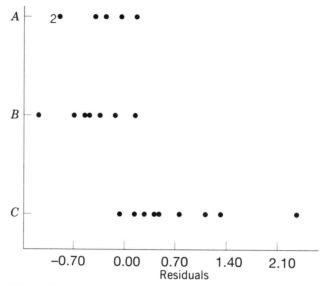

Figure 2.9. Residuals plot indicating block effects not incorporated in the fitted model.

from three separate machines A, B, and C, so that the residuals when grouped by machines were

A: -0.08, -0.89, 0.11, -1.01, -0.30, -1.00, -0.42

B: -0.56, -0.67, 0.11, -0.49, -1.20, -0.12, -0.32

C: 0.46, -0.04, 0.74, 1.33, 1.11, 0.29, 0.40, 0.17, 2.36

(The machine order for the observations in Table 2.1 is thus *CCCBA   CBCAA BBCAB   CAACB   BAC.*)

Figure 2.9 shows a plot against machines. This would suggest that there is a basic difference in level of response Y of machine C compared with A and B. Such a difference could be incorporated into the model by the introduction of a dummy variable; this is discussed in Chapter 14.

Another example of "other residual plots" occurs when a possible new variable comes into consideration. Suppose it is suspected that the ambient temperature is affecting the contents of a large vessel. Although vessel temperature has been recorded at a selected, protected, measuring point, the temperature at the other side of the vessel may possibly be affected by exposure to the outside air. If the ambient temperatures are recorded for the period during which data were collected, the residuals could now be plotted against the temperatures observed, to see if any dependency of response on ambient temperature is revealed. If it is, new terms of appropriate kinds can be added to the model to take account of the dependency.

These are two examples of what "other residuals plots" might be used. In general, residuals should be plotted in *any* reasonable way that occurs to the experimenter or statistician, based on specialist knowledge of the problem under study. The plots already described are, however, the basic ones and should always be performed for a full analysis.

Dependencies Between Residuals

As we have remarked, the residuals, unlike the errors they estimate, are not independent. Does this affect the plots? Yes. Does it invalidate the plots? In most situations, no. Anscombe and Tukey (1963, p. 144) remark on this point. In discussing the two-way analysis of variance (where there are several constraints on the residuals) they remark that, although correlations and constraints affect distributions of functions of the residuals, the "corresponding effects on the graphical procedures ... can usually be neglected. This is mainly because of the way in which graphical appearances arise from residuals, though in part because of the absence of precisely defined significance levels. (This is also true for most other balanced designs.)" In a later sentence Anscombe and Tukey state that in a two-way table with four or more rows and four or more columns, "the effect of correlation upon graphic procedures is usually negligible. ...." It would appear that in general regression situations the effect of correlations between residuals need not be considered when plots are made, except when the ratio $(n - p)/n$—that is, (number of degrees of freedom in residuals)/(number of residuals)—is quite small.

In Chapter 8 we shall see how to evaluate the pairwise correlations between the residuals. If these correlations are relatively small, there is usually little effect on the residuals plots.

2.7. DURBIN-WATSON TEST

We later (Chapter 7) explain the Durbin-Watson test in some detail for multiple predictors. Here we merely sketch its application to the residuals obtained from fitting a straight line $\hat{Y} = b_0 + b_1 X$. It is assumed here that the observations, and so the residuals, have a natural order such as a time order or space order, here indicated by the order $Y_1, Y_2, \ldots, Y_n$. In practice, the given data might have to be recast to obtain the proper ordering. The residuals $e_1, e_2, \ldots, e_n$ are estimates for errors assumed to be independent. If they are not independent, the residuals may reveal it. The Durbin-Watson test checks for a sequential dependence in which each error (and so residual) is correlated with those before and after it in the sequence. The test focuses specifically on the differences between successive residuals in the following way. Consider the Durbin-Watson statistic

$$d = \sum_{u=2}^{n} (e_u - e_{u-1})^2 / \sum_{u=1}^{n} e_u^2. \qquad (2.7.1)$$

It can be shown that:

1. $0 \leq d \leq 4$ always.

2. If successive residuals are positively serially correlated, that is, positively correlated in their sequence, d will be near 0.

3. If successive residuals are negatively correlated, d will be near 4, so that $4 - d$ will be near 0.

4. The distribution of d is symmetric about 2.

Because of (4), a $d < 2$ should be used as is; a $d > 2$ should be tested as $4 - d$ and point (3) should be kept in mind. The test is conducted as follows. Compare d (or $4 - d$, whichever is closer to zero) with d_L and d_U in Table 2.6. If $d < d_L$, conclude that positive serial correlation is a possibility; if $d > d_U$, conclude that no serial correlation is indicated. (If $4 - d < d_L$, conclude that negative serial correlation is a possibility; if $4 - d > d_U$, conclude that no serial correlation is indicated.) If the d (or $4 - d$) value lies between d_L and d_U, the test is inconclusive. An indication of

T A B L E 2.6. Significance Points of d_L and d_U for a Straight Line Fit

n^a	1%		2.5%		5%	
	d_L	d_U	d_L	d_U	d_L	d_U
15	0.81	1.07	0.95	1.23	1.08	1.36
20	0.95	1.15	1.08	1.28	1.20	1.41
25	1.05	1.21	1.18	1.34	1.29	1.45
30	1.13	1.26	1.25	1.38	1.35	1.49
40	1.25	1.34	1.35	1.45	1.44	1.54
50	1.32	1.40	1.42	1.50	1.50	1.59
70	1.43	1.49	1.51	1.57	1.58	1.64
100	1.52	1.56	1.59	1.63	1.65	1.69
150	1.61	1.64	—	—	1.72	1.75
200	1.66	1.68	—	—	1.76	1.78

[a]Interpolate linearly for intermediate n-values.

positive or negative serial correlation would be cause for the model to be reexamined. Weighted, or generalized, least squares (Section 9.2) becomes a possibility.

2.8. REFERENCE BOOKS FOR ANALYSIS OF RESIDUALS

Full reference details of the following are in the bibliography: Atkinson (1985); Barnett and Lewis (1994); Belsley (1991); Belsley, Kuh, and Welsch (1980); Chatterjee and Hadi (1988); Cook and Weisberg (1982); Hawkins (1980); and Rousseeuw and Leroy (1987).

APPENDIX 2A. NORMAL PLOTS

The area under an $N(0, 1)$ distribution from $-\infty$ to some point x is given by

$$y = \int_{-\infty}^{x} \frac{1}{\sqrt{2\pi}} \exp\left(-\frac{1}{2}t^2\right) dt. \tag{2A.1}$$

If we plot $100y$ as ordinate, against x as abscissa, we obtain the "S-shaped" curve, called the cumulative probability curve of the $N(0, 1)$ distribution. Some points on this curve are, for example, $(x, y) = (-1.96, 2.5)$, $(0, 50)$, and $(1.96, 97.5)$, all of which are easily obtained from tables of the cumulative $N(0, 1)$ distribution. (See Figures 2A.1 and 2A.2.)

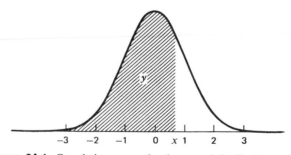

Figure 2A.1. Cumulative area under the normal distribution to point x.

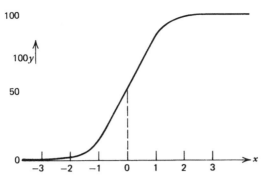

Figure 2A.2. Cumulative normal curve.

Normal probability paper is a specially constructed type of graph paper that is available at most technical bookstores. Although the unnumbered horizontal axis is marked by equal divisions in the usual way, the vertical axis has a special scale. The vertical scale goes from 0.01 to 99.99 but the spacing of the divisions becomes wider as we move up from the 50 point to the 99.99 point and down from the 50 point to the 0.01 point, with symmetry about the horizontal 50 line. The scaling is such that if 100 times the value of y in (2A.1) is plotted against x, the resulting "curve" will be a straight line. Thus the vertical scaling, determined from the inverse functions of Eq. (2A.1), $x = F^{-1}(y)$, "straightens out" the top and bottom of the S-shaped curve in Figure 2A.2. Note that since the points $(-\infty, 0)$ and $(\infty, 100)$ are on the straight line plot, the values 0 and 100 cannot be shown on the scale since the horizontal scale is of limited length and cannot go from $-\infty$ to ∞. A further point on the straight line mentioned above is $(1, 84.13)$. We shall find this point useful in a moment.

If points from the cumulative $N(\mu, \theta^2)$ distribution are plotted on normal probability paper [rather than points from the $N(0, 1)$ distribution], then a straight line will pass through such points as $(x, y) = (\mu - 1.96\theta, 2.5), (\mu, 50), (\mu + \theta, 84.13), (\mu + 1.96\theta, 97.5)$, and so on. This fact is very useful if we have a sample $x_1, x_2, \ldots, x_m$ and wish to decide if it could have come from a normal distribution, and if so, to obtain a quick estimate of the standard deviation θ. First, the sample is arranged in ascending order, due regard being given to sign. Let us assume this has been done already so that $x_1, x_2, \ldots, x_m$ is the correct order. We now plot x_i against the ordinate[1] with value

$$100(i - \tfrac{1}{2})/m. \qquad (2A.2)$$

The rationale behind this is that, if we divide the unit area under the normal curve into m equal areas, we might "expect" that one observation lies in each section so marked out. Thus the ith observation in order, x_i, is plotted against the cumulative area to the middle of the ith section, which is $(i - \tfrac{1}{2})/m$. The factor 100 adapts this to the vertical scale given on the normal probability paper. (See Figure 2A.3.)

If the sample is a normal sample it will be found that a well-fitting straight line can be drawn (by eye) through the bulk of the points plotted, although none of the points may necessarily fall right on the line. We can then use the best-fitting straight line to estimate θ as follows. Find x_{50} and $x_{84.13}$, the values of x for which the line crosses

---

[1]For possible alternatives to $100(i - \tfrac{1}{2})/m$, see Barnett (1975); note especially the last paragraph of p. 101 and the first paragraph of p. 104. The BMDP programs use $100(3i - 1)/(3m + 1) = 100(i - \tfrac{1}{3})(m + \tfrac{1}{3})$ and also produce a "detrended normal probability plot" from which the slope has been removed. MINITAB uses $100(i - \tfrac{3}{8})/(m + \tfrac{1}{4})$ and converts this to a normal score. The differences between these different systems are typically unimportant in practical use.

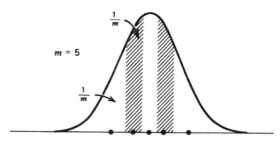

Figure 2A.3. Splitting the area under the normal curve into *m* equal pieces; we might "expect" one observation in each piece at a location that divides the area of the piece into two equal portions.

horizontal lines drawn at 50 and 84.13 ordinate levels. Then the difference $x_{84.13} - x_{50}$ is an estimate of $[(\mu + \theta) - \mu] = \theta$. (See Figure 2A.4 write-up below.)

An instructive way to gain experience to make decisions on these types of plots is to look up samples of various sizes from a table of random normal deviates and to plot them on normal probability paper. This will give an idea of the variation from linearity that *can* occur and that is *not* abnormal. Plots of this type are given by Daniel and Wood (1980, Appendix 3A).

Normal Scores

Most normal plots are done on the computer, and there the vertical axis is often converted to a *normal score*, that is, the normal deviate value that would correspond to the plotted probability level. (For example, 2.28% would be converted to -2, 2.5% to -1.96, 50% to zero, and 99.865% to 3. See the normal probability table.) In the MINITAB system, this is particularly simple to achieve, If the residuals are in column C6, we write

```
nscore c6   c7
plot c7   c6
```

and the plot is made. Note that the plot instruction must be written in that way to get the diagram to look similar to our discussion and examples below. Use of

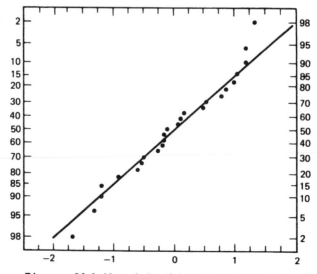

Figure 2A.4. Normal plot of the residuals of Table 1.2.

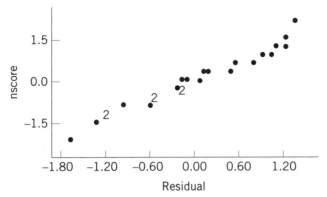

Figure 2A.5. Normal plot of the residuals of Table 1.2.

```
nscore c6   c7
plot c6   c7
```
would reverse the axes and change all the connected explanations. The point is a trivial one, but the consequences can be enormous. So watch it!

Example 1. Consider the $m = 25$ residuals given in Table 1.2. We first arrange these in ascending order, giving due regard to sign:

$$-1.68, -1.32, -1.20, -1.20, -0.93, -0.60, -0.53, -0.51, -0.26, -0.19, -0.17, -0.16,$$
$$-0.12, 0.08, 0.11, 0.17, 0.50, 0.55, 0.80, 0.87, 1.00, 1.05, 1.20, 1.20, 1.34.$$

To obtain a full normal plot of these we set $m = 25$ in Eq. (2A.2) and successively set $i = 1, 2, \ldots, m$ to give the ordinate values:

$$2, 6, 10, 14, 18, 22, 26, 30, 34, 38, 42, 46, 50, 54, 58, 62, 66, 70, 74, 78, 82, 86, 90, 94, 98.$$

These ordinate values are associated with the ordered residuals. Thus the bottom point in Figure 2A.4 is at abscissa -1.68, ordinate $= 2$, that is, $(-1.68, 2)$. Readers will note that to plot this, it is necessary to use the bottom and *right-hand* scale on the probability paper. The left-hand scale shows this point as $(-1.68, 98)$, the left-hand scale being $(100 - \text{right-hand scale})$; this is a peculiarity of probability paper that seems destined to persist. The second point plotted is at $(-1.32, 6)$ and so on. The line shown is drawn by eye and represents an attempted rough fit to the majority of the points with somewhat more weight given to the central points. Usually the

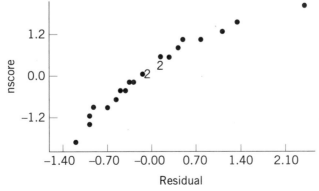

Figure 2A.6. Normal plot of the residuals from Section 2.1.

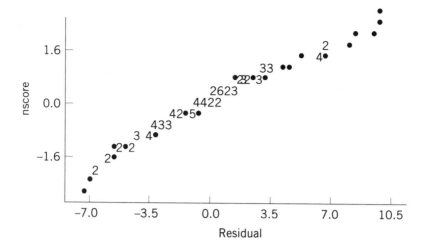

(a) Normal sample exhibiting lighter-tail characteristics

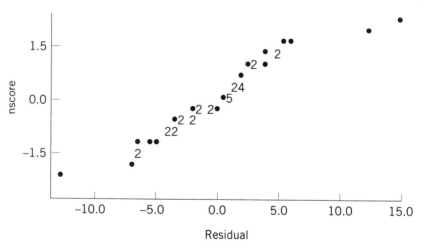

(b) Sample with one negative and two positive outliers

F i g u r e 2A.7. Some characteristics of normal plots.

abscissa at which the line cuts the 50 ordinate would provide an estimate of the mean
of the sample plotted but, in fact, we have drawn the line through (0, 50) here because
the residuals sum to zero in theory. (In practice, rounding errors may occur, as we
have noted.) An estimate of the standard deviation is $x_{84.13} - x_{50} = 0.97 - 0 = 0.97$
approximately. This compares well with $s = (0.7923)^{1/2} = 0.89$ from Table 1.6. The
normal plot is not atypical of plots of normal samples of this size. The lowest two
values and the highest two values are both "pulled in," to a minor extent, but this is
not uncommon with least squares residuals. Certainly there are no outliers, which
would show up as points well out to the left of the lower part of the line and well out
to the right of the upper part of the line.

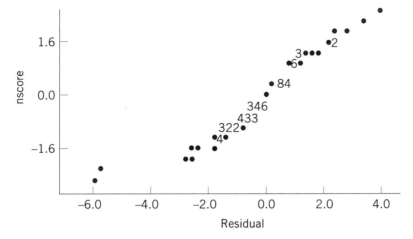

(c) Distribution "heavier-tailed" than the normal (e.g., t)

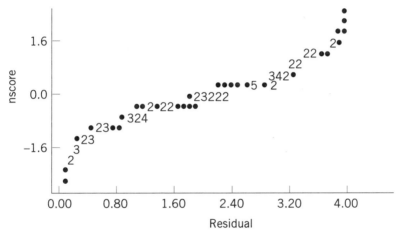

(d) Distribution "lighter-tailed" than the normal (e.g., uniform)

Figure 2A.7. (*Continued*)

Example 2. Figure 2A.5 shows a MINITAB computed plot of residuals versus nscores for the steam data. The plot is essentially identical to that of Figure 2A.4 except that the nscores were derived from MINITAB's $100(i - \frac{3}{8})/(m + \frac{1}{4})$ rather than our suggested $100(i - \frac{1}{2})/m$.

Example 3. Figure 2A.6 shows a MINITAB constructed plot of the residuals from the pure error example of Section 2.1. The outlying observation previously remarked upon appears off and to the right. It is too large to fall nicely on a line through the central bulk of the points.

Outliers

An outlier among residuals is one that is far greater than the rest in absolute value and perhaps lies three or four standard deviations or further from the mean of the residuals. The outlier is a peculiarity and indicates a data point that is not at all typical of the rest of the data. It follows that an outlier should be submitted to particularly careful examination to see if the reason for its peculiarity can be determined.

Rules have been proposed for rejecting outliers [i.e., for deciding to remove the corresponding observation(s) from the data, after which the data are reanalyzed without these observations]. Automatic rejection of outliers is not always a very wise procedure. Sometimes the outlier is providing information that other data points cannot due to the fact that it arises from an unusual combination of circumstances, which may be of vital interest and requires further investigation rather than rejection. As a general rule, outliers should be rejected out of hand only if they can be traced to causes such as errors in recording the observations or in setting up the apparatus. Otherwise careful investigation is in order.

Some General Characteristics of Normal Plots

The diagrams in Figure 2A.7 show characteristics that can occur when residuals are displayed in a probability plot. In all diagrams, the probability or normal score is on the vertical axis and the residuals values are on the horizontal axis.

Normal plots are also used to examine effects (contrasts) from factorial experiments. In that context, the plot sometimes exhibits the look of two parallel lines. This typically indicates that (at least) one of the observations is suspect. See Box and Draper (1987, p. 132).

Making Your Own Probability Paper

The books of our childhood often had "projects for a rainy afternoon." Here is such a project. Probability paper on which the cumulative distribution curve becomes a straight line can be constructed for *any* continuous distribution as follows. Draw the cumulative distribution function. Draw horizontal lines at equal intervals of the vertical probability scale 0 to 1. At the points where the horizontal lines hit the curve, drop perpendiculars onto any horizontal line l, labeling the foot of the perpendicular 100 times the vertical probability scale reading from which it arose. The scale on the horizontal line l then provides the new spacings that should be employed on the vertical scale of the probability paper. Effectively, we have applied the inverse transformation $x = F^{-1}(y)$, where $y = F(x)$ is the cumulative probability function, to equal intervals of y. In labeling the new vertical axis, we multiply by 100 for convenience.

APPENDIX 2B. MINITAB INSTRUCTIONS

The MINITAB program below will obtain many of the details discussed in Chapters 1 and 2.

```
oh=0
set c2
2.3 1.8 2.8 1.5 2.2 3.8 1.8 3.7 1.7 2.8 &
2.8 2.2 3.2 1.9 1.8 3.5 2.8 2.1 3.4 3.2 &
3 3 5.9
set c1
1.3 1.3 2 2 2.7 3.3 3.3 3.7 3.7 4 &
4 4 4.7 4.7 5 5.3 5.3 5.3 5.7 6 &
6 6.3 6.7
set c20
12 23 7 8 17 22 1 11 19 20 &
```

```
5 2 21 15 18 3 6 10 4 9 13 14 16
end of data
corr c1 c2 m1

regress c2 1 c1 c11 c12;
resi c14;
pure.

plot c1 c2
print m1
print c11 c12
histogram c14
plot c14 c12
plot c14 c1
plot c14 c20
nscore c14 c41
plot c41 c14
end
stop
-------------------------
Comments:
The data are from Table 2.1.
c2 contains the Y's, c1 the X's.
c20 contains the time order.
c11 contains internally studentized residuals (Section 8.1).
c12 contains fitted values.
c14 contains residuals.
c41 contains nscores for the residual.
```

EXERCISES

Exercises for Chapter 2 are located in the section "Exercises for Chapters 1–3" at the end of Chapter 3.

CHAPTER 3

Fitting Straight Lines: Special Topics

3.0. SUMMARY AND PRELIMINARIES

This chapter, which can be omitted in a first reading, deals with some special topics relating to straight line fits. These are:

(3.1) Predictions, and confidence statements, for the true mean value of Y at a given X.

(3.2) Inverse regression. We want to predict an X, given a Y-value, after the fit has been made.

(3.3) How we can pick a good design for fitting a straight line using results in Chapters 1 and 2.

(3.4) A brief discussion of, and a suggested method for dealing with, the situation where the errors in X cannot be ignored, as is usually done.

For Section 3.1 we need a preliminary result about the covariance of two linear combinations of $Y_1, Y_2, \ldots, Y_n$.

Covariance of Two Linear Functions

Let a_i and c_i be constants, let Y_i be random, and suppose that

$$a = a_1 Y_1 + a_2 Y_2 + \cdots + a_n Y_n,$$

$$c = c_1 Y_1 + c_2 Y_2 + \cdots + c_n Y_n.$$

Suppose that all the Y_i have the same variance $\sigma^2 = V(Y_i)$ and that the Y's are pairwise uncorrelated, that is, cov $(Y_i, Y_j) = 0, i \neq j$. It then follows that

$$\text{cov}(a, c) = (a_1 c_1 + a_2 c_2 + \cdots + a_n c_n)\sigma^2. \qquad (3.0.1)$$

[*Note:* If $\text{cov}(Y_i, Y_j) = \rho_{ij}\sigma^2$ for $i \neq j$, then all possible terms of form $(a_i c_j + a_j c_i)\rho_{ij}$ would be added inside the parentheses on the right of (3.0.1). However, we do not need this for our purposes here.]

Example. Let $a = \bar{Y}$, and $c = b_1$, the slope from a straight line fit. Then

$$a_i = 1/n \qquad (3.0.2)$$

and

$$c_i = \frac{X_i - \overline{X}}{\Sigma(X_i - \overline{X})^2} = \frac{X_i - \overline{X}}{S_{XX}} \tag{3.0.3}$$

from the algebra just above (1.4.2). It follows that

$$a_1 c_1 + a_2 c_2 + \cdots + a_n c_n = (n S_{XX})^{-1} \Sigma(X_i - \overline{X}) = 0,$$

and so

$$\text{cov}(\overline{Y}, b_1) = 0, \tag{3.0.4}$$

that is, $\overline{Y}$ and b_1 are uncorrelated random variables.

3.1. STANDARD ERROR OF $\hat{Y}$

The fitted equation of a straight line, least squares fit can be written $\hat{Y} = b_0 + b_1 X$ or, with $b_0 = \overline{Y} - b_1\overline{X}$, as

$$\hat{Y} = \overline{Y} + b_1(X - \overline{X}) \tag{3.1.1}$$

where both $\overline{Y}$ and b_1 are subject to error, which will affect $\hat{Y}$. At a specified value of X, say, X_0, we predict

$$\hat{Y}_0 - \overline{Y} + b_1(X_0 - \overline{X}) \tag{3.1.2}$$

for the mean value of Y at X_0. Because X_0 and $\overline{X}$ are fixed, and because cov $(\overline{Y}, b_1) = 0$, see above, we obtain

$$V(\hat{Y}_0) = V(\overline{Y}) + (X_0 - \overline{X})^2 V(b_1)$$

$$= \frac{\sigma^2}{n} + \frac{(X_0 - \overline{X})^2 \sigma^2}{\Sigma(X_i - \overline{X})^2}. \tag{3.1.3}$$

Hence

$$\text{se}(\hat{Y}_0) = \text{est. sd}(\hat{Y}_0) = s\left\{\frac{1}{n} + \frac{(X_0 - \overline{X})^2}{\Sigma(X_i - \overline{X})^2}\right\}^{1/2}. \tag{3.1.4}$$

This is a minimum when $X_0 = \overline{X}$ and increases as we move X_0 away from $\overline{X}$ in either direction. In other words, the greater distance an X_0 is (in either direction) from $\overline{X}$, the larger is the error we may expect to make when predicting, from the regression line, the mean value of Y at X_0. This is intuitively very reasonable. To state the matter loosely, we might expect to make our "best" predictions in the "middle" of our observed range of X and would expect our predictions to be less good away from the "middle." For values of X outside our experience—that is, outside the range observed—we should expect our predictions to be less good, becoming worse as we moved away from the range of observed X-values.

We now apply these results to the steam data of Chapter 1.

Example

$$n = 25, \qquad \Sigma(X_i - \overline{X})^2 = 7154.42;$$

$$s^2 = 0.7923, \qquad \overline{X} = 52.60;$$

$$\text{est. } V(\hat{Y}_0) = s^2 \left\{ \frac{1}{n} + \frac{(X_0 - \overline{X})^2}{\Sigma(X_i - \overline{X})^2} \right\}$$

$$= 0.7923 \left\{ \frac{1}{25} + \frac{(X_0 - 52.60)^2}{7154.42} \right\}.$$

For example, if $X_0 = \overline{X}$, then $\hat{Y}_0 = \overline{Y}$ and

$$\text{est. } V(\hat{Y}_0) = 0.7923 \left\{ \frac{1}{25} \right\} = 0.031692,$$

that is, $\text{se}(\hat{Y}_0) = \sqrt{0.031692} = 0.1780$.
   If $X_0 = 28.6$,

$$\text{est. } V(\hat{Y}_0) = 0.7923 \left\{ \frac{1}{25} + \frac{(28.60 - 52.60)^2}{7154.42} \right\} = 0.095480,$$

that is, $\text{se}(\hat{Y}_0) = \sqrt{0.095480} = 0.3090$.
   Correspondingly, $\text{se}(\hat{Y}_0)$ is also 0.3090 when $X_0 = 76.60$.
   The 95% confidence limits for the true mean value of Y for a given X_0 are then given by $\hat{Y}_0 \pm (2.069) \text{se}(\hat{Y}_0)$. The t-value for $\nu = 23$ df is used because the $s^2 = 0.7923$ estimate is based on 23 df. The limits are thus:

$$\text{At } X_0 = \overline{X} = 52.60, \quad \overline{Y} \pm 2.069(0.1780) = 9.424 \pm 0.368$$

$$= (9.056, 9.792).$$

$$\text{At } X_0 = 28.6, \quad (13.623 - 0.079829(28.6)) \pm 2.069(0.3090)$$

$$= 11.340 \pm 0.639 = (10.701, 11.979).$$

$$\text{At } X_0 = 76.6, \quad (13.623 - 0.079829(76.6)) \pm 0.639$$

$$= 7.508 \pm 0.639 = (6.869, 8.147).$$

Note that because the X_0 values 28.6 and 76.6 are the same distance from $\overline{X} = 52.6$, the *widths* of the respective intervals are the same but their positions differ vertically.
   If we joined up all the lower end points and all the upper end points of such intervals as X_0 changes, we would get Figure 3.1. The two curves are hyperbolas.
   These limits can be interpreted as follows. Suppose repeated samples of Y_i are taken of the same size and at the same fixed values of X as were used to determine the fitted line above. Then of all the 95% confidence intervals constructed for the mean value of Y for a given value of X, say X_0, 95% of these intervals will contain the true mean value of Y at X_0.
   If only one prediction $\hat{Y}_0$ is made, say, for $X = X_0$, then we act as though the probability that the calculated interval at this point will contain the true mean is 0.95.

Intervals for Individual Observations and Means of q Observations

The variance and standard deviation formulas above apply to the predicted *mean value* of Y for a given X_0. Since the actual observed value of Y varies about the true mean value with variance σ^2 [independent of the $V(\hat{Y})$], a predicted value of an *individual* observation will still be given by $\hat{Y}$ but will have variance

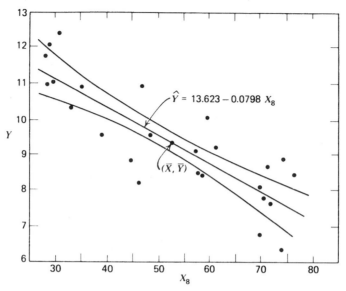

Figure 3.1. The loci of 95% confidence bands for the true mean value of Y.

$$\sigma^2 \left\{ 1 + \frac{1}{n} + \frac{(X_0 - \overline{X})^2}{\Sigma(X_i - \overline{X})^2} \right\} \qquad (3.1.5)$$

with corresponding estimated value obtained by inserting s^2 for σ^2. Confidence values can be obtained in the same way as before; that is, we can calculate a 95% confidence interval for a new observation, which will be centered on $\hat{Y}_0$ and whose length will depend on an estimate of this new variance from

$$\hat{Y}_0 \pm t(\nu, 0.975) \left\{ 1 + \frac{1}{n} + \frac{(X_0 - \overline{X})^2}{\Sigma(X_i - \overline{X})^2} \right\}^{1/2} s, \qquad (3.1.6)$$

where ν is the number of degrees of freedom on which s^2 is based (and equals $n - 2$ here). A confidence interval for the average of q new observations about $\hat{Y}_0$ is obtained similarly as follows:

Let $\overline{Y}_0$ be the mean of q future observations at X_0 (where q could equal 1 to give the case above). Then

$$\overline{Y}_0 \sim N(\beta_0 + \beta_1 X_0, \sigma_0^2/q),$$

$$\hat{Y}_0 \sim N(\beta_0 + \beta_1 X_0, V(\hat{Y}_0)),$$

so that

$$\overline{Y}_0 - \hat{Y}_0 \sim N(0, \sigma_0^2/q + V(\hat{Y}_0)),$$

and $[(\overline{Y}_0 - \hat{Y}_0)/\text{se}(\overline{Y}_0 - \hat{Y}_0)]$ is distributed as a $t(\nu)$ variable, where ν is the number of degrees of freedom on which s^2, the estimate of σ^2, is based. Thus

$$\text{Prob}\left\{ |\overline{Y}_0 - \hat{Y}_0| \leq t(\nu, 0.975) \left[s^2 \left(\frac{1}{q} + \frac{1}{n} + \frac{(X_0 - \overline{X})^2}{\Sigma(X_i - \overline{X})^2} \right) \right]^{1/2} \right\} = 0.95$$

from which we can obtain 95% confidence limits for $\overline{Y}_0$ about $\hat{Y}_0$ of

$$\hat{Y}_0 \pm t(v, 0.975) \left[\frac{1}{q} + \frac{1}{n} + \frac{(X_0 - \overline{X})^2}{\Sigma(X_i - \overline{X})^2} \right]^{1/2} s. \tag{3.1.7}$$

These limits are of course wider than those for the mean value of Y for given X_0, since these limits are the ones within which 95% of future observations at X_0 (for $q = 1$) or future means of q observations at X_0 ($q > 1$) are expected to lie.

Note: To obtain simultaneous confidence curves appropriate for the whole regression function over its entire range, it would be necessary to replace $t(v, 1 - \tfrac{1}{2}\alpha)$ by $\{2F(2, n - 2, 1 - \alpha)\}^{1/2}$. See, for example, pp. 110–116 of *Simultaneous Statistical Inference*, 2nd ed., by R. G. Miller, published by Springer-Verlag, New York, in 1981.

Confidence bands are rarely drawn in practice. However, the idea is important to understand and a suitable confidence interval around any $\hat{Y}$ value can always be evaluated numerically by applying a general algebraic formula, no matter how many X's there are. This latter aspect is a valuable one.

3.2. INVERSE REGRESSION (STRAIGHT LINE CASE)

Suppose we have fitted a straight line $\hat{Y} = b_0 + b_1 X$ to a set of data (X_i, Y_i), $i = 1$, $2, \ldots, n$, and now, for a specified value of Y, say Y_0, we wish to obtain a predicted value $\hat{X}_0$, the corresponding value of X, as well as some sort of interval confidence statement for X surrounding $\hat{X}_0$. A practical example of such a problem is the following: X_i is an age estimate obtained from counting tree rings, while Y_i is a corresponding age estimate obtained from a carbon-dating process. The fitted straight line provides a "calibration curve" for the carbon-dating method, related to the more accurate tree ring data. Application of the carbon-dating method to an object now gives a reading Y_0. What statements can we make about the object's true age? This problem is called the inverse regression problem. (In other examples, Y_0 might be a true mean value or the average of q observations.)

There are several alternative ways of obtaining the (same) solution to this type of problem. First, let us assume Y_0 is a true mean value, not a single observation or average of q observations. Intuitively reasonable is the following. Draw the fitted straight line and curves that give the end points of the $100(1 - \alpha)\%$ confidence intervals for the true mean value of Y given X. (See Figure 3.2.) Draw a horizontal line parallel to the X axis at a height Y_0. Where this straight line cuts the confidence interval curves, drop perpendiculars onto the X-axis to give lower and upper $100(1 - \alpha)\%$ "fiducial limits," labeled X_L and X_U in Figure 3.2. The perpendicular from the point of intersection of the two straight lines onto the X-axis gives the inverse estimate of X, defined by solving $Y_0 = b_0 + b_1 \hat{X}_0$ for $\hat{X}_0$, namely,

$$\hat{X}_0 = (Y_0 - b_0)/b_1.$$

To obtain the values of X_L and X_U we can proceed as follows. In the figure, X_L is the X-coordinate of the point of intersection of the line

$$Y = Y_0 \qquad (\text{i.e., } Y = b_0 + b_1 \hat{X}_0) \tag{3.2.1}$$

and the curve

$$Y = Y_{X_L} - ts \left\{ \frac{1}{n} + \frac{(X_L - \overline{X})^2}{S_{XX}} \right\}^{1/2}, \tag{3.2.2}$$

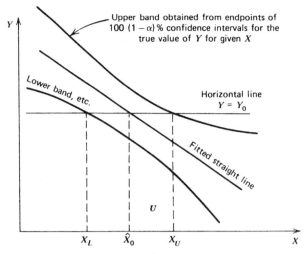

Figure 3.2. Inverse regression: estimating X by $\hat{X}_0$ from a given Y_0 value, and obtaining a $100(1 - \alpha)\%$ "fiducial interval" for X.

where $S_{XX} = \Sigma(X_i - \overline{X})^2$, $Y_{X_L} = b_0 + b_1 X_L$, $t = t(\nu, 1 - \alpha/2)$ is the usual t percentage point, and ν is the number of degrees of freedom of s^2. Setting Eqs. (3.2.1) and (3.2.2) equal, canceling a b_0, rearranging to leave the square root alone on one side of the equation, and squaring both sides to get rid of the square root leads to a quadratic equation

$$PX_L^2 + 2QX_L + R = 0, \tag{3.2.3}$$

in X_L, where

$$P = b_1^2 - t^2 s^2 / S_{XX},$$

$$Q = t^2 s^2 \overline{X} / S_{XX} - b_1^2 \hat{X}_0, \tag{3.2.4}$$

$$R = b_1^2 \hat{X}_0^2 - t^2 s^2 / n - t^2 s^2 \overline{X}^2 / S_{XX}.$$

We get exactly the same equation for X_U, so that X_L, X_U are the roots of (3.2.3). These, after some manipulation, are found to be

$$\left.\begin{matrix} X_U \\ X_L \end{matrix}\right\} = \overline{X} + \frac{b_1(Y_0 - \overline{Y}) \pm ts\{[(Y_0 - \overline{Y})^2/S_{XX}] + (b_1^2/n) - (t^2 s^2/nS_{XX})\}^{1/2}}{b_1^2 - (t^2 s^2/S_{XX})} \tag{3.2.5}$$

or, in an alternative form,

$$= \hat{X}_0 + \frac{(\hat{X}_0 - \overline{X})g \pm (ts/b_1)\{[\hat{X}_0 - \overline{X})^2 S_{XX}] + (1 - g)/n\}^{1/2}}{1 - g}, \tag{3.2.6}$$

where $g = t^2 s^2 / (b_1^2/S_{XX})$. When g is "small," say 0.05 or smaller, it is convenient to set $g = 0$ for an approximate answer. Note that we can write

$$g = t^2/\{b_1/(s^2 S_{XX})^{1/2}\}^2$$

$$= \left\{\frac{t(\nu, 1 - \alpha/2) \text{ percentage point}}{t\text{-statistic derived from } b_1 \text{ divided by its standard error}}\right\}^2. \tag{3.2.7}$$

Thus the "more significant" b_1 is, the larger will be the denominator of g and the

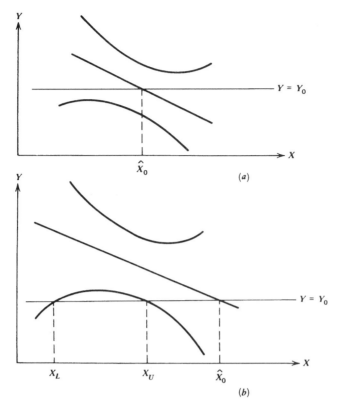

Figure 3.3. Inverse regression peculiarities: (a) complex roots and (b) real roots but both on the same side of the regression line. Inverse regression would not be of much practical value in these sorts of circumstances. The respective intervals are (a) $(-\infty, \infty)$ and (b) (X_U, ∞).

smaller will be g. Clearly g will tend to be large if $|b_1|$ is small or badly determined, in which case s^2 will tend to be large or S_{XX} will tend to be small or both. Inverse estimation is, typically, not of much practical value unless the regression is well determined, that is, b_1 is significant, which implies that g should be smaller than (say) about 0.20. (A t-statistic value of 2.236 would achieve this, for example.)

When the regression line is not well determined, peculiarities can arise. For example, the roots X_L and X_U may be complex; or they may both be real but both on one side of the regression line. Drawing a figure will usually make it quite obvious why the peculiarity has arisen. When the line is not well estimated, the hyperbolas defined by the end points of the confidence intervals usually flare out badly, or turn back down (or up as the case may be) sharply. Figure 3.3 shows two examples.

An alternative way of writing the quadratic equation (3.2.3), which enables the generalization to the multiple predictor case to be made more easily, is

$$\{-Y_0 + b_0 + b_1 X\}^2 = t^2 s^2 \left\{ \frac{1}{n} + \frac{(X - \overline{X})^2}{S_{XX}} \right\}, \tag{3.2.8}$$

where the X represents the X_L in (3.2.2), or X_U.

Note: The calculations above are true mean value calculations. To obtain a more general formula in which Y_0 is regarded not as a true mean value but as the mean of q observations, replace each $1/n$ by $1/q + 1/n$ in Eqs. (3.2.2), (3.2.4), (3.2.5), (3.2.6), and (3.2.8), as described above. Then $q = 1$ provides an individual observation formula, applicable to a single new reading, as in the carbon-dating example that introduced this section. When $q = \infty$ we obtain the true mean value formulas shown above.

We have adopted the term "fiducial limits" for (X_L, X_U) from Williams (1959) whose account of this topic in his Chapter 6 is excellent. Rather than becoming involved in the theoretical arguments behind the use of this term, we ask the reader simply to regard such intervals as inverse confidence limits for X given Y_0.

3.3. SOME PRACTICAL DESIGN OF EXPERIMENT IMPLICATIONS OF REGRESSION

We have dealt with the fitting of a straight line model $Y = \beta_0 + \beta_1 X + \epsilon$ to a set of data (X_i, Y_i), $i = 1, 2, \ldots, n$. We have also carefully considered a detailed analysis of how well the line fits, and whether or not repeated observations indicated whether or not there was any evidence in the data to indicate that an alternative model should be used. When we have only one predictor variable X and when the postulated model is a straight line, the alternatives we would consider would often be higher-order polynomials in X; for example, the quadratic $Y = \beta_0 + \beta_1 X + \beta_{11} X^2 + \epsilon$, or the cubic, and so on. We now put all this information into a practical perspective by considering the problem of choosing an experimental strategy for the one-predictor-variable case.

Experimental Strategy Decisions

Suppose an experimenter wants to collect data on a response variable Y at n selected values of a controllable predictor variable in order to determine an empirical relationship between Y and the predictor variable. We assume that the latter is (at least to a satisfactory approximation) not subject to random error, but that Y is, and that the n values of the predictor are not necessarily all distinct, that is, repeat runs are permitted. The experimenter first has to ask, and answer, a number of questions.

1. What range of values of the predictor variable is she currently interested in? This is often difficult to decide. The range must be wide enough to permit useful inference, yet narrow enough to permit representation by as simple a model function as possible. Once the decision is made, the interval can be coded to $(-1, 1)$ without loss of generality. For example, if a temperature range of $140°F \leq T \leq 200°F$ is selected, the coding

$$X = (T - 170)/30$$

will convert it to the interval $-1 \leq X \leq 1$. In general, the transformation is given by

$$X = \frac{\text{Original variable} - \text{Midpoint of original interval}}{\text{Half the original range}}.$$

2. What kind of relationship does the experimenter anticipate will hold over the selected range? Is it first order (i.e., straight line), second order (i.e., quadratic), or what? To decide this, she will not only bring to bear her own knowledge but will usually seek the expertise of others as well. To fix ideas, let us assume the experimenter believes the relationship is probably first order but is not absolutely sure.

3. If the relationship tentatively decided upon in (2) is wrong, what alternative does the experimenter expect? For example, if she believes the true model is a straight line, she is most likely to expect that, if she is wrong, the model will be somewhat

curved in a quadratic manner. A more remote possibility is that the true model may be cubic. Typically, she will decide that she *may be* one order too low, if anything. Otherwise, she would probably postulate a higher-order model to begin with.

4. What is the inherent variation in the response? That is, what is $V(Y) = \sigma^2$? The experimenter may have a great deal of experience with similar data and may "know" what σ^2 is. More typically, she may wish to incorporate repeat runs into the experiments so that σ^2 can be estimated at the same time as the relationship between Y and X, and also so that the usual assumptions about the constancy of σ^2 throughout the chosen range of values of the predictor can be checked.

5. How many experimental runs are possible? The experimenter has only limited dollars, staff, facilities, and time. How many runs are justified by the importance of the problem and the costs involved?

6. How many sites (i.e., different X-values) should be chosen? How many repeat runs should be performed at each of the chosen sites?

Let us now continue our discussion in terms of a specific example.

An Example

Suppose our experimenter decides that a straight line relationship is most likely over the range $-1 \leq X \leq 1$ of her coded predictor, that she most fears a quadratic alternative, that she does not know σ^2, and that 14 runs are possible. At what values of X (i.e., what sites) should she perform experimental runs, how many where, and with what justification?

Figure 3.4 shows some of the possibilities. (Each dot represents a run; a pile of dots represents repeated runs.) Let us see how each matches the requirements we have set out.

Each design has 14 degrees of freedom to begin with. Two of these are taken up by the parameter estimates b_0 and b_1 leaving 12 residual degrees of freedom to be allocated between lack of fit and pure error. Lines (1) and (2) of Table 3.1 show how these residual degrees of freedom split up for the various designs. Line (3) gives the value of

$$S_{XX}^{-1/2} = \{\Sigma(X_i - \overline{X})^2\}^{-1/2},$$

which, by Eq. (1.4.6), is proportional to the standard deviation of b_1 when a straight line is fitted. Line (4) shows the number of parameters that it is *possible* to fit to the design data. We can fit a polynomial of order $p - 1$ (with p parameters including β_0) to a design with p sites. A second reason that this is shown is that p is proportional (when n and σ^2 are fixed) to $p\sigma^2/n$, and the latter is the average size of $V(\hat{Y}(X))$ averaged over all the points of the design when the polynomial of order $p - 1$ is fitted. In other words

$$\sum_{i=1}^{n} V\{\hat{Y}(X_i)\}/n = p\sigma^2/n.$$

This result is true in general for any linear model. For the straight line case, when $p = 2$, it can be deduced by replacing the subscript 0 by i in Eq. (3.1.3), summing from $i = 1, 2, \ldots, n$, and dividing by n. For a general proof using matrices see Exercise R in "Exercises for Chapters 5 and 6."

Note that the lack of fit degrees of freedom is equal to the number of distinct X-sites in the data minus the number of parameters in the postulated model. In fact,

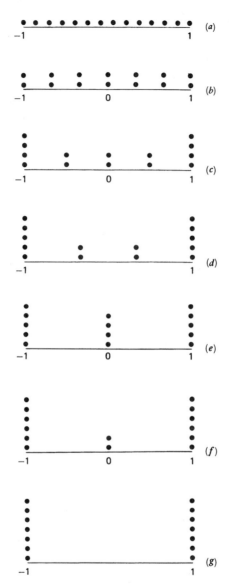

Figure 3.4. Some possible experimental arrangements for obtaining data for fitting a straight line: (*a*) 14 sites, (*b*) 7 sites, (*c*) 5 sites, (*d*) 4 sites, (*e*) 3 sites, (*f*) 3 sites, and (*g*) 2 sites. Which are good, which are bad, in the circumstances described in the text? The sites are equally spaced in cases (*a*)–(*f*).

T A B L E 3.1. Characteristics of Various Strategies Depicted in Figure 3.4

	(*a*)	(*b*)	(*c*)	(*d*)	(*e*)	(*f*)	(*g*)
(1) Lack of fit df:	12	5	3	2	1	1	0
(2) Pure error df:	0	7	9	10	11	11	12
(3) sd $(b_1)/\sigma$:	0.43	0.40	0.33	0.31	0.32	0.29	0.27
(4) p sites:	14	7	5	4	3	3	2

because our example has two parameters to be estimated, β_0 and β_1, line (4) minus line (1) equals 2 throughout Table 3.1.

Comments on Table 3.1

The requirement in our example that σ^2 needs to be estimated via pure error makes (*a*) a poor strategy here. If we are to be able to check lack of fit, (*g*) is automatically eliminated, too.

Next consider (*b*). Is it really sensible to use seven different levels if our major alternative is a quadratic model? Not really, because we do not need that many levels to check our alternative. Moreover, of designs remaining, it has the highest sd(b_1)/σ. So we drop (*b*) from consideration.

Our best choice clearly lies with one of (*c*), (*d*), (*e*), and (*f*); exactly which would be selected depends on the experimenter's preferences. Only three levels (sites) are strictly necessary to achieve a lack of fit test versus a quadratic alternative but there is only one degree of freedom for lack of fit in (*e*) and (*f*). The latter is a better choice on the basis of the sd(b_1)/σ values. Design (*d*) allows two df for lack of fit; design (*c*) perhaps goes too far in number of sites. So the final choice lies between (*f*) and (*d*) with (*f*) slightly preferred perhaps if the quadratic alternative is all that is anticipated.

Perhaps the most important aspect of this discussion is not so much a specific choice of design, but the immediate elimination of designs that might in some other contexts be regarded as reasonable. For example, design (*a*) would be a very poor choice—who needs 14 levels to estimate a straight line? Again, design (*g*) provides the smallest variance for the slope b_1 but is of no use at all if we want to be able to check possible lack of fit against a quadratic (or indeed any) alternative. When a design must be chosen from a list of alternatives, we recommend consideration of details like those in Table 3.1; such a display can be both helpful and revealing.

3.4. STRAIGHT LINE REGRESSION WHEN BOTH VARIABLES ARE SUBJECT TO ERROR[1]

Whenever we fit the model

$$Y_i = \beta_0 + \beta_1 X_i + \epsilon_i, \quad i = 1, 2, \ldots, n, \quad (3.4.1)$$

by least squares to a set of n data values (X_i, Y_i), we usually take it for granted that Y_i is subject to the error ϵ_i and X_i is not subject to error. If this is true, and if the errors $\epsilon_1, \epsilon_2, \ldots, \epsilon_n$ are independently and normally distributed, $N(0, \sigma^2)$, maximum likelihood estimation, and least squares estimation, namely,

$$\underset{\beta_0, \beta_1}{\text{Minimize}} \sum_{i=1}^{n} (Y_i - \beta_0 - \beta_1 X_i)^2$$

provide the same estimates (b_0, b_1) of (β_0, β_1).

What if both X and Y are subject to error? We can write

$$Y_i = \eta_i + \epsilon_i, \quad (3.4.2)$$

$$X_i = \xi_i + \delta_i. \quad (3.4.3)$$

---

[1]This section is a condensed version of Draper (1992).

We assume that a straight line relationship

$$\eta_i = \beta_0 + \beta_1 \xi_i \qquad (3.4.4)$$

holds between the true but unobserved values η_i and the n unknown parameters ξ_i. Substituting (3.4.4) into (3.4.2) and then substituting for ξ_i from (3.4.3) gives

$$Y_i = \beta_0 + \beta_1 X_i + (\epsilon_i - \beta_1 \delta_i). \qquad (3.4.5)$$

Let us assume that $\epsilon_i \sim N(0, \sigma^2)$, with the ϵ_i uncorrelated, and $\delta_i \sim N(0, \sigma_\delta^2)$, with the δ_i uncorrelated, with ϵ_i and δ_i uncorrelated, and define

$$\sigma_\xi^2 = \sum_{i=1}^{n} (\xi_i - \bar{\xi})^2/n, \qquad (3.4.6)$$

$$\sigma_{\xi\delta} = \text{Covariance } (\xi, \delta), \qquad (3.4.7)$$

$$\rho = \sigma_{\xi\delta}/(\sigma_\xi \sigma_\delta), \qquad (3.4.8)$$

$$r = \sigma_\delta/\sigma_\xi. \qquad (3.4.9)$$

In (3.4.7), $\sigma_{\xi\delta}$ would typically be zero; however, see case (2) below. If, mistakenly, we fit (3.4.1) by least squares, b_1 will be biased. In fact,

$$E(b_1) = \beta_1 - \frac{\beta_1 r(\rho + r)}{1 + 2\rho r + r^2}. \qquad (3.4.10)$$

The bias is negative if $\sigma_\xi^2 + \sigma_{\xi\delta} > 0$, that is, if $\rho + r > 0$. The bias arises from the fact that X_i is not independent of the error in (3.4.5), in general. In fact,

$$\text{Covariance } [X_i, (\epsilon_i - \beta_1 \delta_i)] = -\beta_1(\rho + r)\sigma_\xi \sigma_\delta. \qquad (3.4.11)$$

We thus see that there are cases where fitting (3.4.1) by least squares will provide little or no bias. These are:

1. If σ_δ^2 is small compared with σ_ξ^2, the errors in the X's are small compared with the spread in the ξ's (and so in the X's) and r will be small. The bias in (3.4.10) is then small. This is what is often assumed in practice, when least squares is used.

2. If the X's are fixed and determined by the experimenter (see Berkson, 1950), then $\sigma_{\xi\delta} = \text{Covariance}(X_i - \delta, \delta) = -\sigma_\delta^2$, which means that $\sigma_{\xi\delta} + \sigma_\delta^2 = 0$, or $\rho + r = 0$, implying zero bias in (3.4.10).

3. We wish to fit $Y_i = \eta_i + \epsilon_i$, where $\eta_i = \beta_0 + \beta_1 X_i$ (the observed X_i, note), and not as in (3.4.4).

These formats will not fit all practical cases. One case that occurred at the University of Wisconsin, in connection with a study on wild birds, required the observation of X_i = "the distance the bird was from a path." The student doing the study pointed out that, as she approached a bird, it flew away before she got close enough to see precisely where it had perched. Thus error in recording X was unavoidable.

If we attempt to obtain the maximum likelihood estimates of β_0 and β_1 under the distributional assumptions made in connection with (3.4.5), we find that there is an identifiability problem. The estimation cannot be carried through without some additional information being added, for example, knowledge of the ratio $\lambda = \sigma^2/\sigma_\delta^2$ (Barnett, 1967; Wong, 1989). This is Case III of Sprent and Dolby (1980), discussed below. Various authors have suggested alternative analyses. The literature is too vast to discuss in full detail here, and we provide a selective, perhaps biased, discussion.

Sprent and Dolby (1980) distinguish four cases:

I. (X, Y) are bivariate normal variables and $E(Y|X) = \beta_0 + \beta_1 X$.

II. $Y \sim N(\beta_0 + \beta_1 X, \sigma^2)$. The observed X-values are fixed on realizations of a random variable with any (reasonable) distribution.

In both I and II, estimates via maximum likelihood are the usual least squares estimates $b_1 = S_{XY}/S_{XX}$ and $b_0 = \overline{Y} - b_1 \overline{X}$, where $S_{XY} = \Sigma_{i=1}^n (X_i - \overline{X})(Y_i - \overline{Y})$ and $S_{XX} = \Sigma_{i=1}^n (X_i - \overline{X})^2$.

III. The case of Eqs. (3.4.1) through (3.4.5). If λ were known, maximum likelihood leads to estimates

$$\hat{\beta}_1 = [S_{YY} - \lambda S_{XX} + \{(S_{YY} - \lambda S_{XX})^2 + 4\lambda S_{XY}^2\}^{1/2}]/(2S_{XY})$$

$$\hat{\beta}_0 = \overline{Y} - \hat{\beta}_1 \overline{X}, \tag{3.4.12}$$

where $S_{YY} = \Sigma_{i=1}^n (Y_i - \overline{Y})^2$. Note that, if $\lambda = S_{YY}/S_{XX}$, $\hat{\beta}_1 = (S_{YY}/S_{XX})^{1/2}$, which is the geometric mean functional relationship, after attachment of the sign of S_{XY}. This case is often called the functional relationship model. Note also that, when $\lambda = 1$, the solution (3.4.12) defines the line that minimizes the sum of squares of *perpendicular* deviations from the line. Many people find such a solution intuitively satisfying, but it is appropriate only when $\sigma^2 = \sigma_\delta^2$, that is, when $\lambda = 1$. This solution was first given by Adcock (1878).

IV. Similar to III but with ξ_i a normal random variable, independent of δ, so that because of (3.4.4), (ξ_i, η_i) follow a joint degenerate bivariate normal distribution. This is the so-called structural relationship model and again the case III solution applies if λ is known.

Sprent and Dolby "do not recommend *ad hoc* use of the geometric mean functional relationship when there are errors in both variables," arguing that other *ad hoc* estimates could equally be used. Nevertheless, we shall recommend it below for reasons to be stated.

Riggs, Guarnieri, and Addelman (1978) study, partially through simulations, a variety of 34 different methods of fitting (X, Y) data. While they favor (3.4.12), they warn that a reasonably accurate estimate of λ is desirable. They also point out that the geometric mean functional relationship occupies a "central position" in compromises between the two least squares solutions, Y on X and X on Y, an appealing characteristic (see their Figure 8, p. 1338).

Practical Advice

Although many of us try to avoid the issue of errors in both X and Y by advising "take data where the X-range is large compared with the X-error," this cannot always be done, and one must often suggest something specific. If λ is known (or can reasonably be estimated) use of the maximum likelihood solution (3.4.12) is probably best.

A simple alternative initially suggested by Wald (1940), using two groups, and amended by Bartlett (1949) to three groups is the following: Divide the data into three equal (or as equal as possible) groups with: (1) the smaller, or most negative, X-values; let $P_1 \equiv (\overline{X}_1, \overline{Y}_1)$ be the center of gravity of these. (2) The larger, or least negative, X-values; let $P_3 \equiv (\overline{X}_3, \overline{Y}_3)$ be their center of gravity. (3) The remainder, which are used only in estimating the overall center of gravity, $(\overline{X}, \overline{Y})$. Use the line passing through $(\overline{X}, \overline{Y})$ with slope $(\overline{Y}_3 - \overline{Y}_1)/(\overline{X}_3 - \overline{X}_1)$, that is, parallel to $P_1 P_3$. For reasoning, see Wald (1940) and Barlett (1949). Later studies by Gibson and Jowett (1957) indicate that maximum efficiency is achieved by a division of observations closer to the ratio $1:2:1$, but the exact split is not crucial.

Geometric Mean Functional Relationship

Our own preference is to suggest the geometric mean functional relationship for which the estimators are

$$\hat{\beta}_1 = (S_{YY}/S_{XX})^{1/2}, \qquad \hat{\beta}_0 = \overline{Y} - \hat{\beta}_1 \overline{X}. \tag{3.4.13}$$

This does assume, it is true, that we are using (3.4.12) with $\lambda = S_{YY}/S_{XX}$. However, it is appealing otherwise, if this assumption is not unreasonable. The estimator $\hat{\beta}_1$ is the geometric mean of the quantities

$$b_1 = S_{XY}/S_{XX}, \qquad a_1^{-1} = (S_{XY}/S_{YY})^{-1},$$

where b_1 and a_1 are, respectively, the slopes in least squares fits of Y versus X ($\hat{Y} = b_0 + b_1 X$) and of X versus Y ($\hat{X} = a_0 + a_1 Y$). Inverting the latter relationship leads to

$$Y = -a_0/a_1 + a_1^{-1}\hat{X}$$

so the geometric mean $\hat{\beta}_1 = (b_1 a_1^{-1})^{1/2}$ is a compromise value lying in between the two "Y on X equation" slopes. Note that, if the roles of X and Y are reversed, exactly the same line emerges, that is, the fitted line

$$Y = \hat{\beta}_0 + \hat{\beta}_1 X \tag{3.4.14}$$

with coefficients from (3.4.13) is uniquely defined. This natural symmetry is most appealing. The attractiveness of the geometrical mean functional relationship has greatly been enhanced by the independent discoveries of Teissier (1948) and Barker, Soh, and Evans (1988) that this solution is an optimum solution to a specific problem. (See also Harvey and Mace, 1982.) That is, the geometric mean functional relationship minimizes the sum of the areas obtained by drawing horizontal (parallel to the X-axis) and vertical (parallel to the Y-axis) lines from each data point (see Figure 3.5). The symmetry of the solution is again obvious; interchange of the X and Y axes leaves the areas unchanged. One disadvantage of the geometric mean functional relationship is that no easy calculations are available for conducting tests on the parameters or constructing confidence intervals for them. (For the complications involved, see, for example, Creasy, 1956.) It is possible that applying nonlinear estimation and defining the loss function as the sum of the areas might offer some help here. The geometric

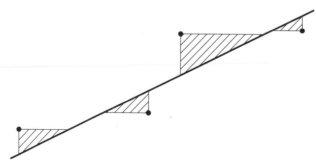

Figure 3.5. The geometrical mean functional relationship line minimizes the sum of the shaded areas. (The dots are data points.)

**T A B L E 3.2. Star Data from Dressler (1984) and
Jefferys (1990)**[a]

Coma Sample ($z = 1$)		Virgo Sample ($z = 0$)	
V_{26}	log σ	V_{26}	log σ
12.60	2.449	11.39	2.242
13.12	2.394	12.53*	1.716*
14.23	2.285	9.98	2.412
14.86	2.166	9.37	2.480
15.88*	1.863*	12.24	2.059
13.92	2.286	9.20	2.355
15.45*	1.761*	12.17	2.009
14.36	2.209	12.01	1.949
15.07	2.113	12.50	2.079
14.07	2.301	8.56	2.474
14.53	2.243	10.28	2.268
15.60	2.169	11.28	2.170
12.27	2.383	8.79	2.528
14.36	2.311	12.02*	1.778*
14.50	2.339	11.92	2.021
15.52	2.251	9.95	2.391
13.46	2.361	11.30	2.185
11.85	2.584	9.88	2.338
15.31	2.007	9.82	2.303
13.98	2.180	8.90	2.514
15.28	2.099	10.97	2.262
14.26	2.275	9.28	2.276
14.11	2.320	11.37	2.207
14.87	2.191		
14.82	2.247		
15.37	2.059		
13.49	2.394		
15.04	2.154		
13.67	2.274		
12.88	2.383		

[a]In the example, the four observations marked with an asterisk will be ignored.

mean functional relationship has been condemned as being inconsistent, that is, the estimates do not tend to their true values as n tends to infinity. However, other estimators are biased, and what happens for large n is often not of concern to those with practical problems and small data sets. While it is true that maximum likelihood methods can make appeal to asymptotic results at this point, such results do not seem to apply too well when n is small, judging by the comments of various authors.

Examples

Example 1. The data in Table 3.2 were used by Jefferys (1990) and taken from Dressler (1984). "They consist of the integrated V magnitudes V_{26}, and log of the central velocity dispersion, log σ, of a sample of 53 galaxies from two galaxy clusters, the Coma and Virgo cluster" (Jefferys, 1990, p. 602). The model

$$\log \sigma = \beta_0 + \beta_1 V_{26}$$

is deemed appropriate with a common β_1 and a different β_0 for each cluster. Four outliers are present (marked by asterisks in Table 3.2), which we ignore. (This bypasses some of the points made by Jefferys, which are not our concern here.) We adopt a dummy or indicator variable z; $z = 1$ for the Coma sample and $z = 0$ for the Virgo sample. Two least squares fits using the models

$$\log \sigma = \beta_0 + \beta_1 V_{26} + \beta_2 z + \epsilon$$

and

$$V_{26} = \alpha_0 + \alpha_1 \log \sigma + \sigma_2 z + \epsilon$$

provide, respectively, fitted equations

$$\hat{\log} \sigma = 3.4795 - 0.116334 V_{26} + 0.44097 z$$

and

$$\hat{V}_{26} = 25.329 - 6.5641 \log \sigma + 3.7685 z.$$

The slope of the geometric mean functional relationship is thus

$$-\{-0.116334/(-6.5641)\}^{1/2} = -0.133127.$$

Putting parallel straight lines with this slope through the individual centers of gravity of the two sets of data provides fitted equations

$$\hat{\log} \sigma = 4.159 - 0.133 V_{26} \quad \text{(Coma sample)}$$

and

$$\hat{\log} \sigma = 3.656 - 0.133 V_{26} \quad \text{(Virgo sample)}.$$

These are very close to the reference solution of Jefferys (1990), which was "an errors-in-variables least squares fit" to the same data. (The method is not further explained.) They are virtually identical to Dressler's (1984) values obtained via a sensible ad hoc procedure (Jefferys's: 4.14, 3.65, −0.132; Dressler's: 4.156, 3.656, −0.1333).

Example 2. The data in Table 3.3, from Kelly (1984), were taken from Miller (1980). Kelly uses the data to illustrate points she is making about (i) estimating the variance of the classical estimators of (3.4.12) and (ii) detecting influential observations. We analyze them using the geometric mean functional relationship estimator.

The two fits to all the data (X = heelstick, Y = catheter) are $\hat{Y} = 2.786 + 0.8805X$, and $\hat{X} = 4.210 + 0.7870Y$, which we can invert to the form $Y = -5.349 + 1.2706\hat{X}$. The geometric mean functional relationship is thus $Y = -0.91 + 1.058X$. Both individual regressions indicate that the second observation is influential, however, and a plot of the data indicates we might consider dropping it. The two fits to the remaining 19 observations give

$$\hat{Y} = -1.628 + 1.1147X$$

and

$$\hat{X} = 5.482 + 0.70462Y,$$

which we can invert to the form

$$Y = -7.780 + 1.4192\hat{X}.$$

T A B L E 3.3. Serum Kanamycin Levels in Blood Samples
Drawn Simultaneously from an Umbilical Catheter and a Heel
Venipuncture in 20 Babies

Baby	Heelstick (X)	Catheter (Y)
1	23.0	25.2
2	33.2	26.0
3	16.6	16.3
4	26.3	27.2
5	20.0	23.2
6	20.0	18.1
7	20.6	22.2
8	18.9	17.2
9	17.8	18.8
10	20.0	16.4
11	26.4	24.8
12	21.8	26.8
13	14.9	15.4
14	17.4	14.9
15	20.0	18.1
16	13.2	16.3
17	28.4	31.3
18	25.9	31.2
19	18.9	18.0
20	13.8	15.6

Then $\hat{\beta} = (1.1147/0.70462)^{1/2} = 1.258$ and the geometric mean functional relationship is

$$\hat{Y} = -4.52 + 1.258X.$$

(If the second observation is not deleted, the parallel result would be $\hat{Y} = -0.91 + 1.058X$, where the 1.058 is the geometric mean of the slopes 0.8805 and 1.2706.) Kelly (1984) obtains two 95% confidence intervals for the slope using all the data, getting (0.76, 1.38) via a bootstrap method, and (0.76, 1.52) via a method based on normal assumptions, given by Kendall and Stuart (1961, pp. 388–390). She concludes that these support the hypothesis that $\beta_0 = 0$, $\beta_1 = 1$, which implies that the methods of measurement that gave rise to Table 3.3 are equivalent. She then points out that removal of the second observation takes the estimated point for (β_0, β_1) "to approximately the edge of a 60% confidence region around" her original estimates based on a maximum likelihood analysis assuming $\lambda = 1$. We interpret that to mean that the hypothesis $\beta_0 = 0$, $\beta_1 = 1$ is no longer supported.

The geometric mean functional relationship does not provide confidence intervals, but we can get a rough feel for the situation by looking at the estimates when all equations are written in Y on X form. When observation 2 is included, the two slopes are 0.8805 and 1.2706 and their geometric mean is 1.0577; the two intercept values are 2.786 and -5.349 and the intercept of the geometric mean functional relationship is -0.91. One feels that the hypothesis—intercept = 0, slope = 1—is not unreasonable. Now remove the second observation. The slopes are now 1.1147 and 1.4192 with a geometric mean of 1.258 (all > 1) and the two intercepts are -1.628 and -7.780 (both < 0) with an intercept of -4.52 from the geometric mean functional relationship. The impression we get is that the hypothesis is not valid. Thus the situation turns on the one influential data point. Can we regard the two lines that lead to the geometric mean functional relationship as confidence limits of some sort? No properties of them are known, it seems, but using them appears to be common sense.

References

Adcock (1878); Barker, Soh, and Evans (1988); Barnett (1967); Barlett (1949); Creasy (1956); Draper (1992); Dressler (1984); Gibson and Jowett (1957); Harvey and Mace (1982); Jefferys (1990); Kelly (1984); Kendall and Stuart (1961); Miller (1980); Riggs, Guarnieri and Addelman (1978); Sprent and Dolby (1980); Teissier (1948); Wald (1940); Wong (1989). (We are grateful to Dr. Penny Reynolds for supplying some of these references.)

EXERCISES FOR CHAPTERS 1–3

Readers who have not yet read all the material needed in Chapters 1–3 should temporarily skip over parts of questions that refer to unread portions. The residuals from all fits should be calculated and examined, whether or not it is specified in the question.

A. A study was made on the effect of temperature on the yield of a chemical process. The following data (in coded form) were collected:

X	Y
−5	1
−4	5
−3	4
−2	7
−1	10
0	8
1	9
2	13
3	14
4	13
5	18

  1. Assuming a model, $Y = \beta_0 + \beta_1 X + \epsilon$, what are the least squares estimates of β_0 and β_1? What is the prediction equation?
  2. Construct the analysis of variance table and test the hypothesis $H_0: \beta_1 = 0$ with an α risk of 0.05.
  3. What are the confidence limits ($\alpha = 0.05$) for β_1?
  4. What are the confidence limits ($\alpha = 0.05$) for the true mean value of Y when $X = 3$?
  5. What are the confidence limits ($\alpha = 0.05$) for the difference between the true mean value of Y when $X_1 = 3$ and the true mean value of Y when $X_2 = -2$?
  6. Are there any indications that a better model should be tried?
  7. Comment on the number of levels of temperature investigated with respect to the estimate of β_1 in the assumed model.

B. A test is to be run on a given process for the purpose of determining the effect of an independent variable X (such as process temperature) on a certain characteristic property of the finished product Y (such as density). Four observations are to be taken at each of five settings of the independent variable X.
  1. In what order would you take the 20 observations required in this test?
  2. When the test was actually run, the following results were obtained:

$$\overline{X} = 5.0 \qquad \Sigma(X_i - \overline{X})^2 = 160.0 \qquad \Sigma(X_i - \overline{X})(Y_i - \overline{Y}) = 80.0$$
$$\overline{Y} = 3.0 \qquad \Sigma(Y_i - \overline{Y})^2 = 83.2.$$

Assume a model of the type $Y = \beta_0 + \beta_1 X + \epsilon$.
a. Calculate the fitted regression equation.
b. Prepare the analysis of variance table.
c. Determine 95% confidence limits for the true mean value of Y when
   (1) $X = 5.0$;
   (2) $X = 9.0$.
3. Suppose that the actual data plot is as shown in Figure B1 and that the sum of squares due to replication (pure error) is 42. Answer the following questions on the basis of this additional information.

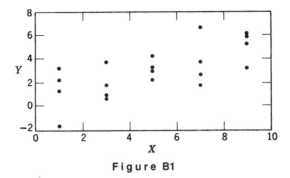

Figure B1

a. Does the regression equation you derived in 2a adequately represent the data? Give reasons and include the result of a test for lack of fit.
b. Are the confidence limits you calculated in 2c applicable? If not, state your reasons.
c. If the model used in part 2 does not seem appropriate, suggest a possible alternate.
4. Suppose that the actual data plot is as shown in Figure B2 and that the sum of squares due to replication is 42.0. Answer the questions 3a, 3b, and 3c on the basis of this information.

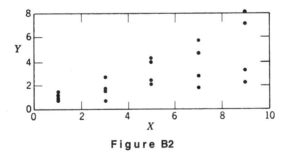

Figure B2

5. Suppose that the actual data plot is as shown in Figure B3, and the sum of squares due to replication is 23.2. Answer the questions 3a, 3b, and 3c on the basis of this information.

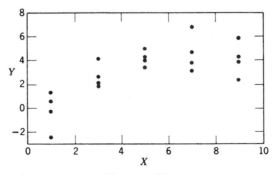

Figure B3

C. Thirteen specimens of 90/10 Cu–Ni alloys, each with a specific iron content, were tested in a corrosion-wheel setup. The wheel was rotated in salt seawater at 30 ft/s for 60 days. The corrosion was measured in weight loss in milligrams/square decimeter/day, MDD. The following data were collected:

X (Fe)	Y (loss in MDD)
0.01	127.6
0.48	124.0
0.71	110.8
0.95	103.9
1.19	101.5
0.01	130.1
0.48	122.0
1.44	92.3
0.71	113.1
1.96	83.7
0.01	128.0
1.44	91.4
1.96	86.2

Requirement. Determine if the effect of iron content on the corrosion resistance of 90/10 Cu–Ni alloys in seawater can justifiably be represented by a straight line model. Assume $\alpha = 0.05$.

D. (*Source:* "Leverage and regression through the origin," by G. Casella, *The American Statistician*, **37**, 1983, 147–152.) There are very few occasions where it makes sense to fit a model without an intercept β_0. If there *were* occasion to fit the model $Y = \beta X + \epsilon$ to a set of data $(X_1, Y_1), (X_2, Y_2), \ldots, (X_n, Y_n)$, the least squares estimate of β would be

$$b = \Sigma X_i Y_i / \Sigma X_i^2.$$

Suppose you have a programmed calculator that will fit only the intercept model

$$Y = \beta_0 + \beta_1 X + \epsilon,$$

but you want to fit the no-intercept model. By adding one more fake data point $(m\bar{X}, m\bar{Y})$ to the data above, where

$$m = n/\{(n + 1)^{1/2} - 1\} = n/a,$$

say, and letting the calculator fit $Y = \beta_0 + \beta_1 X + \epsilon$, you can estimate β by using the b_1 estimate but ignoring b_0, which is not zero in general. Try to show this algebraically.

   Check this out with the very small data set $(X, Y) = (1, 0.5), (3, 1.0), (5, 0.9)$ for which $(m\bar{X}, m\bar{Y}) = (9, 2.4)$.

   (In MINITAB, the subcommand "noconstant" will get you the fit through the origin.)

E. The data below consist of seven pairs of values of

   $X =$ price of alcohol relative to take-home pay,

   $Y_1 =$ consumption of absolute alcohol in liters per head of population per year,

   $Y_2 =$ cirrhosis deaths per 100,000 of populations,

for seven European countries. Plot Y_1 versus X and Y_2 versus X (Y as ordinate, X as abscissa) and judge whether, in your opinion, a straight line would fit each set of data reasonably well.

u	Country	X	Y_1	Y_2
1	France	0.016	24.66	51.7
2	Italy	0.027	18.00	30.5
3	West Germany	0.026	13.63	29.0

u	Country	X	Y_1	Y_2
4	Belgium	0.022	8.42	14.2
5	United Kingdom	0.057	7.66	4.1
6	Ireland	0.092	7.64	5.0
7	Denmark	0.096	7.50	11.6

1. For the (X, Y_2) only, evaluate the fitted least squares line and draw it on your data plot.
2. Evaluate the residuals to two decimal places. Check that $\Sigma e_u = 0$, within rounding error.
3. Obtain the analysis of variance table.
4. Find the standard errors of b_0 and b_1.
5. Find the formula for the standard error of $\hat{Y}$, and construct 95% confidence bands for the true mean value of Y. (Plot about half a dozen points over the X-range and join them up smoothly.)
6. Test the overall regression via an F-test, and find how much of the variation about the mean $\bar{Y}$ is explained by the fitted line.
7. The data have no actual repeats. One practical trick that is often useful, however, is to consider points that are "sufficiently close together" as being, approximately, repeat runs, and to use this as a basis for working out an approximate pure error sum of squares. After this has been evaluated, the usual steps are followed. The major difficulty in doing such a thing is in deciding exactly what the words "sufficiently close together" mean.
    For our data, it might be sensible to regard the following sets of runs as approximate repeats.
   a. $X = 0.027, 0.026, 0.022$.
   b. $X = 0.092, 0.096$.
   Evaluate an approximate pure error sum of squares using these sets, and proceed with the appropriate approximate analysis. State your conclusions.
8. Reviewing the whole problem, comment on the apparent value and usefulness of the straight line you have fitted. It has been suggested by some that, if the price of alcohol were raised, fewer cirrhosis deaths would result. On the basis of these data, what do *you* think? What possible flaws might exist in your conclusion? (Discuss the situation, briefly.)

F. The moisture of the wet mix of a product is considered to have an effect on the finished product density. The moisture of the mix was controlled and finished product densities were measured as shown in the following data:

Mix Moisture (Coded) X	Density (Coded) Y
4.7	3
5.0	3
5.2	4
5.2	5
5.9	10
4.7	2
5.9	9
5.2	3
5.3	7
5.9	6
5.6	6
5.0	4
$\Sigma X = 63.6$	$\Sigma Y = 62$
$\Sigma X^2 = 339.18$	$\Sigma Y^2 = 390$
$\Sigma x^2 = 2.10$	$\Sigma y^2 = 69.67$
$\bar{X} = 5.3$	$\bar{Y} = 5.17$
$\Sigma XY = 339.1$	
$\Sigma xy = 10.5$	

Requirements

   1. Fit the model $Y = \beta_0 + \beta_1 X + \epsilon$ to the data.
   2. Place 95% confidence limits on β_1.
   3. Is there any evidence in the data that a more complex model should be tried? (Use $\alpha = 0.05$.)

G. The cost of the maintenance of shipping tractors seems to increase with the age of the tractor. The following data were collected.

Age (yr) X	6 Months Cost ($) Y
4.5	619
4.5	1049
4.5	1033
4.0	495
4.0	723
4.0	681
5.0	890
5.0	1522
5.5	987
5.0	1194
0.5	163
0.5	182
6.0	764
6.0	1373
1.0	978
1.0	466
1.0	549

Requirements

   1. Determine if a straight line relationship is sensible. (Use $\alpha = 0.10$.)
   2. Can a better model be selected?

H. It has been proposed in a manufacturing organization that a cup loss figure performed on the line supersede a bottle loss analysis, which is a costly, time-consuming laboratory procedure. The cup loss analysis would yield better control of the process because of the gain in time. If it can be shown that cup loss is a function of bottle loss, it would be a reasonable decision to make. Given the following data, what is your conclusion? (Use $\alpha = 0.05$.)

Bottle Loss (%) X	Cup Loss (%) Y
3.0	3.1
3.1	3.9
3.0	3.4
3.6	4.0
3.8	3.6
2.7	3.6
3.1	3.1
2.7	3.6
2.7	2.9
3.3	3.6
3.2	4.1
2.1	2.6
3.0	3.1
2.6	2.8

I. It is thought that the number of cans damaged in a boxcar shipment of cans is a function of the speed of the boxcar at impact. Thirteen boxcars selected at random were used to examine whether this appeared to be true. The data collected were as follows:

Speed of Car at Impact X	Number of Cans Damaged Y
4	27
3	54
5	86
8	136
4	65
3	109
3	28
4	75
3	53
5	33
7	168
3	47
8	52

What are your conclusions? (Use $\alpha = 0.05$.)

J. The effect of the temperature of the deodorizing process on the color of the finished product was determined experimentally. The data collected were as follows:

Temperature X	Color Y
460	0.3
450	0.3
440	0.4
430	0.4
420	0.6
410	0.5
450	0.5
440	0.6
430	0.6
420	0.6
410	0.7
400	0.6
420	0.6
410	0.6
400	0.6

1. Fit the model $Y = \beta_0 + \beta_1 X + \epsilon$.
2. Is this model sensible? (Use $\alpha = 0.05$.)
3. Obtain a 95% confidence interval for the true mean value of Y at any given value of X, say, X_0.

K. The data below (provided by Tom B. Whitaker) show 34 pairs of values of

$X =$ average level of aflatoxin in a mini-lot sample
    of 120 pounds of peanuts, ppb,

$Y =$ percentage of noncontaminated peanuts in the batch $-$ 99.

Y	X	Y	X	Y	X
0.971	3.0	0.942	18.8	0.863	46.8
0.979	4.7	0.932	18.9	0.811	46.8
0.982	8.3	0.908	21.7	0.877	58.1
0.971	9.3	0.970	21.9	0.798	62.3
0.957	9.9	0.985	22.8	0.855	70.6
0.961	11.0	0.933	24.2	0.788	71.1
0.956	12.3	0.858	25.8	0.821	71.3
0.972	12.5	0.987	30.6	0.830	83.2
0.889	12.6	0.958	36.2	0.718	83.6
0.961	15.9	0.909	39.8	0.642	99.5
0.982	16.7	0.859	44.3	0.658	111.2
0.975	18.8				

1. Plot the data (Y as ordinate, X as abscissa) and, "by eye" only, draw what appears to you to be a "well-fitting straight line 'through' the points." Keep this figure; you will need it for comparison purposes later. Would you say your line is a "good fit"?
2. Also, evaluate ΣX_i, ΣY_i, ΣX_i^2, ΣY_i^2, $X_i Y_i$, all summations being from $i = 1$ through 34.
3. Fit the model $Y = \beta_0 + \beta_1 X + \epsilon$ by least squares. Draw the fitted line on your plot and check how good your "eye-fit" was.
4. Evaluate the residuals to three decimal places. Check that $\Sigma e_i = 0$, within rounding error. Examine plots of the residuals. Give conclusions.
5. Obtain the analysis of variance table in the form given in Table 1.4.
6. Find the standard errors of b_0 and b_1.
7. Find the formula for the standard error of $\hat{Y}$, and construct 95% confidence bands for the true mean value of Y. (Plot about half a dozen points over the X-range and join them up smoothly.)
8. Test the overall regression via an F-test, and find how much of the variation about the mean $\overline{Y}$ is explained by the fitted line.

L. The data given in Exercise K have only two pairs of actual repeats, in fact, at $X = 18.8$ and at $X = 46.8$. One practical trick that is often useful, however, is to consider points that are "sufficiently close together" as being, approximately, repeat runs, and to use this as a basis for working out an approximate pure error sum of squares. After this has been evaluated, the usual steps are followed. The major difficulty in doing such a thing is in deciding exactly what the words "sufficiently close together" mean.

For our data here, it would be sensible to regard the following seven sets of runs as approximate repeats:

$X = 9.3, 9.9$
$X = 12.3, 12.5, 12.6$
$X = 18.8, 18.8, 18.9$
$X = 21.7, 21.9$
$X = 46.8, 46.8$ (these are exact repeats, of course!)
$X = 70.6, 71.1, 71.3$
$X = 83.2, 83.6$

Evaluate an approximate pure error sum of squares using these sets, and proceed with the appropriate approximate analysis, following the details in Section 2.1. State your conclusions.

Reviewing the whole problem in Exercises K and L, comment on the apparent value and usefulness of the straight line you have fitted.

M. For fitting a straight line we have defined

$$R^2 = SS(b_1|b_0)/\Sigma(Y_i - \overline{Y})^2.$$

What is the maximum possible value of R^2 if:

1. There are no repeat runs at all in the data?

2. There *are* proper repeat runs in the data?

Are your conclusions extendable to general regression situations, do you think?

N. The straight line model $\hat{Y} = 1.692 - 0.0546X$ is fitted to the data below, and the analysis of variance table is as shown. Continue the analysis.

X	Y	$\hat{Y}$	e	X	Y	$\hat{Y}$	e
10	−2	1.1	−3.1	40	0	−0.5	0.5
10	−4	1.1	−5.1	40	1	−0.5	1.5
20	1	0.6	0.4	40	2	−0.5	2.5
20	3	0.6	2.4	50	−2	−1.0	−1.0
30	2	0.1	1.9	50	−3	−1.0	−2.0
30	5	0.1	4.9	50	−4	−1.0	−3.0

ANOVA

Source	df	SS	MS	F
b_0	1	0.083		
$b_1\|b_0$	1	7.240	7.240	0.845
Residual	10	85.677	8.568	
Total	12	93.000		

O. In a certain Federal Power Commission hearing some years ago, witnesses presented the following data (here rounded):

X	Y
13.3	3.5
16.9	5.1
19.9	4.8
23.2	6.7
26.3	6.0
30.1	9.5
42.6	8.1

The variables are

Predictor X = Percentage of liquids in gas production output;

Response Y = Unit cost in cents of processing output.

Fit a straight line to these data and find the residual variation via an analysis of variance table.

The witnesses extrapolated the straight line to the points $X = 0$ and $X = 100$ to provide unit costs for situations in which production consisted of no liquids and all liquids. Find the values of $\hat{Y}(0)$ and $\hat{Y}(100)$ and determine 95% confidence limits for the true mean value of Y at $X = 0$ and $X = 100$. What do you conclude?

P. Show that, for a straight line fit, $r_{XY}^2 = R^2 = r_{Y\hat{Y}}^2$. (The second equality is true in general.)

Q. Consider the hypothesis that the titles of papers in a journal tend to be longer for short papers and shorter for long papers. Check this hypothesis in the following way. Look at recent issues of any journal in your own field and make a list, for a number of papers (the more the better, but let's say at least 25), of the following data:

X = Number of pages in a paper or note;

Y = Number of words in the paper or note title (hyphenated
words count as one word).

Plot these data and fit a straight line $Y = \beta_0 + \beta_1 X + \epsilon$ to them via least squares.

Check for lack of fit, if repeat points (or near repeats) exist. Then, provided that there is no significant lack of fit or, if there is no pure error, provided that the residuals do not exhibit any obvious irregularity that would indicate lack of fit, test $H_0: \beta_1 = 0$ versus $H_1: \beta_1 < 0$ via a one-sided t test. What is your conclusion?

R. (*Source:* Ice crystal growth data from B. F. Ryan, E. R. Wishart, and D. E. Shaw, Commonwealth Scientific and Industrial Research Organisation (C.S.I.R.O.), Australia. See "The growth rates and densities of ice crystals between $-3°C$ and $-21°C$," *Journal of the Atmospheric Sciences,* **33,** 1976, 842–850.)

Ice crystals are introduced into a chamber, the interior of which is maintained at a fixed temperature $(-5°C)$ and a fixed level of saturation of air with water. The growth of the crystals with time is observed. The 43 sets of measurements presented here are of axial length of the crystals (A) in micrometers for times (T) of 50 seconds to 180 seconds from the introduction of the crystals. Each measurement represents a single complete experiment; the experiments were conducted over a number of days, and were randomized as to observation time. (The actual order in which they were conducted is not available.) It was desired to learn whether or not a straight line model $A = \beta_0 + \beta_1 T + \epsilon$ provided an adequate representation of the growth with time of the axial length of the ice crystal. Perform a full analysis and state your conclusions.

T	A	T	A
50	19	125	28
60	20, 21	130	31, 32
70	17, 22	135	34, 25
80	25, 28	140	26, 33
90	21, 25, 31	145	31
95	25	150	36, 33
100	30, 29, 33	155	41, 33
105	35, 32	160	40, 30, 37
110	30, 28, 30	165	32
115	31, 36, 30	170	35
120	36, 25, 28	180	38

S. Refer to the steam data in Table 1.1, and its subsequent analysis.
1. Suppose we specify a true mean value of $Y_0 = 10$. Use the methods of inverse estimation to provide an estimate $\hat{X}_0$ of the corresponding X value and 95% fiducial limits (X_L, X_U).
2. State the value of g you found in (1). Obtain the approximate "$g = 0$" (X_L, X_U) values and compare them with those in (1).
3. If Y_0 were not a true mean value but the result of a single new observation, what would $\hat{X}_0$ and (X_L, X_U) be?
4. State the value of g you found in (3). Obtain the approximate "$g = 0$" (X_L, X_U) values and compare them with those in (3).
   Use accurate figures and carry enough places to minimize round-off errorrs.

T. Recall that

$$V(\hat{Y}_0) = \sigma^2 \left[\frac{1}{n} + \frac{(X_0 - \bar{X})^2}{\Sigma(X_i - \bar{X})^2} \right].$$

An experimenter tells you that he has drawn the loci (loci = plural of locus) of end points of 95% confidence bands for the true mean value of Y given X and that "for all practical purposes over the range I am interested in, $150 \leq X \leq 170$, these loci are parallel straight lines." Would you anticipate that the X-range of his data is
1. Much smaller than 20 X-units?
2. Roughly equal to 20 X-units? or
3. Much greater than 20 X-units?
Why?

U. 1. You are given some data (X_i, Y_i), $i = 1, 2, \ldots, n$ and asked to fit a straight line $\hat{Y} = b_0 + b_1X$ to it by least squares, and to perform a "full analysis." As you finish, the experimenter comes in and says "I've just found out all my X's are biased. Each X value I gave you is only 90% of the true X. Would you please redo the analysis." Using $1 - q$ instead of 90% (so you can handle the problem more generally) explain what parts of the "full analysis" are changed by this new information, and in what manner.
   (*Hint:* New $S_{XX} \equiv S_{XX}^* = S_{XX}/(1 - q)^2$, etc., so $b_1^* = S_{XY}^*/S_{XX}^* = b_1(1 - q)$, and so on.)
   2. What would happen if the proportion differed for each observation, for example, was $(1 - q_i)$ for $i = 1, 2, \ldots, n$?
   3. If in (1) or (2) the q or q_i were small, so that q^2 or q_i^2 could be ignored, how would the usual full analysis be changed?

V. (*Source:* "Life expectancy" by M. E. Wilson and L. E. Mather, Letter to the Editor, *Journal of the American Medical Association,* **229,** No. 11, 1974, 1421–1422.) The 50 pairs of observations below arose during a study carried out by Dr. L. E. Mather and Dr. M. E. Wilson. The variables are

$$X = \text{Age of person at death (to nearest year)};$$

$$Y = \text{Length of lifeline on left hand in centimeters (to nearest 0.15 cm)}.$$

Many people believe that the length of one's life is linearly related to the length of one's lifeline. What light do these data throw on such a belief? You may assume that:

$$\Sigma X = 3333, \qquad \Sigma X^2 = 231{,}933, \qquad \Sigma XY = 30{,}549.75,$$

$$\Sigma Y = 459.9, \qquad \Sigma Y^2 = 4308.57.$$

X = Age (yr)	Y = Length (cm)	X = Age (yr)	Y = Length (cm)
19	9.75	68	9.00
40	9.00	69	7.80
42	9.60	69	10.05
42	9.75	70	10.50
47	11.25	71	9.15
49	9.45	71	9.45
50	11.25	71	9.45
54	9.00	72	9.45
56	7.95	73	8.10
56	12.00	74	8.85
57	8.10	74	9.60
57	10.20	75	6.45
58	8.55	75	9.75
61	7.20	75	10.20
62	7.95	76	6.00
62	8.85	77	8.85
65	8.25	80	9.00
65	8.85	82	9.75
65	9.75	82	10.65
66	8.85	82	13.20
66	9.15	83	7.95
66	10.20	86	7.95
67	9.15	88	9.15
68	7.95	88	9.75
68	8.85	94	9.00

W. In *The Chicago Maroon* for Friday, November 10, 1972, The Party Mart advertised per bottle prices for vintage port as given in the accompanying table.

1. Plot the data and examine them. Would it be sensible to fit a regression of the response "price" on the predictor "year"? What disadvantages can you see?
2. What transformation of the predictor "year" would be sensible? (*Hint:* Pretend it is 1972, and think, for example, how you describe your own "year," typically.) Plot price versus your new predictor, and examine the plot. What type of transformation on price would seem sensible here to make the data look "more straight-line-ish"?
3. Plot the data $Y = \ln(\text{price})$ versus $Z = $ age of bottle. Fit a straight line through the data by least squares, evaluate the residuals, and produce the analysis of variance table.
4. What do you conclude about the price of vintage port as exhibited by this set of data and your analysis? To the nearest cent, at what per-year rate would you expect the price of a bottle of vintage port to rise *if* a similar price pattern continued into the future?
5. A subsequent advertisement three years later on Tuesday, November 25, 1975, offered 1937 vintage port at $20.00 per bottle. If it can be assumed that a straight line relationship is preserved, and applies also to this new data point, how much per bottle per year does it appear prices have risen in the intervening three years? Are your answers here and in (4) consistent, or does it appear that per year price rises have accelerated?

Year	Price($)	Year	Price ($)
1890	50.00	1941	10.00
1900	35.00	1944	5.99
1920	25.00	1948	8.98
1931	11.98	1950	6.98
1934	15.00	1952	4.99
1935	13.00	1955	5.98
1940	6.98	1960	4.98

Source: Data via Steve Stigler, 1976.

X. (*Source:* "Graphs in statistical analysis," by F. J. Anscombe, The *American Statistician,* **27,** 1973, 17–21.) Fit a straight line model $Y = \beta_0 + \beta_1 X + \epsilon$ to each of the four sets of data below and show that *for each set,* $n = 11$, $\bar{X} = 9$, $\bar{Y} = 7.5$, $\hat{Y} = 3 + 0.5X$, $S_{XX} = 110$, Regression SS $= S_{XY}^2/S_{XX} = 27.5$ (1 df), Residual SS $= S_{YY} - S_{XY}^2/S_{XX} = 13.75$ (9 df), $se(b_1) = 0.118$, $R^2 = 0.667$.

Plot all four sets of data and explain how the sets of data differ and what their main characteristics are.

(Note that data sets 1–3 all have the same X values but different Y's.)

Data Set No.:	1–3	1	2	3	4	4
Variable:	X	Y	Y	Y	X	Y
Obs. No.: 1	10	8.04	9.14	7.46	8	6.58
2	8	6.95	8.14	6.77	8	5.76
3	13	7.58	8.74	12.74	8	7.71
4	9	8.81	8.77	7.11	8	8.84
5	11	8.33	9.26	7.81	8	8.47
6	14	9.96	8.10	8.84	8	7.04
7	6	7.24	6.13	6.08	8	5.25
8	4	4.26	3.10	5.39	8	5.56
9	12	10.84	9.13	8.15	8	7.91
10	7	4.82	7.26	6.42	8	6.89
11	5	5.68	4.74	5.73	19	12.50

Y. Suppose you were asked to do a simulation as follows:

1. Choose a "true" straight line $\eta = \beta_0 + \beta_1 X$.
2. Select n values of X, X_1, X_2, ... , X_n.
3. Generate n random errors ϵ_i, $i = 1, 2, \ldots , n$, from an $N(0, \sigma^2)$ and so obtain n "observations," $Y_i = \beta_0 + \beta_1 X_i + \epsilon_i$.
4. Fit a straight line $\hat{Y} = b_0 + b_1 X$ to the "observations."
5. Obtain a set of residuals $e_i = Y_i - \hat{Y}_i$.
6. Repeat steps 3–5 a total of N times, using the following (for example) parameter values: $n = 11$, $X_i = -1 + (i - 1)0.2$, $\sigma^2 = 1$, $N = 1000$.

Question: Why have no values been specified for β_0 and β_1?

(*Hint:* Show that the residuals e_i do not depend on β_0 and β_1. Thus any values can be used; the simplest choice is $\beta_0 = \beta_1 = 0$.)

Z. The data in Table Z, published by the *Chicago Tribune* on December 7, 1993, show consumption of candy (Y) and population (X) for 17 countries in 1991.

1. Plot Y versus X.
2. Fit $Y = \beta_0 + \beta_1 X + \epsilon$ by least squares and plot the fitted line on your diagram.
3. Evaluate the residuals to two decimal places. Check that $\Sigma e_u = 0$, within rounding error.
4. Obtain an analysis of variance table.
5. Find out how much of the variation about the mean $\overline{Y}$ is explained by the fitted line.
6. There are no exact repeat runs. As an approximation, however, treat the Y values at $X = 3.5$ and 4.3 as repeats; also those at $X = 5.0$ and 5.1; and finally those at $X = 56.9$, 57.7, and 57.8. (One could argue about the first and second of these groupings, of course, and you might wish to do a second, alternative calculation of your choice, to see what difference it makes.) Use the pseudo repeat runs to test (approximately) for lack of fit and state your conclusion. If you find lack of fit, omit parts (7), (8), (9), and (12).
7. Is it appropriate to test for overall regression via the F-test? If so, do it.
8. Find standard errors for b_0 and b_1.
9. Find the formula for the standard error of $\hat{Y}$ and, assuming that s^2 is an appropriate estimate of σ^2, construct 95% confidence bands for the true mean value of Y. (Plot about half a dozen points over the X-range and join them up smoothly.)

T A B L E   Z. Candy Consumption for Selected Countries in 1991

Country	Consumption (in millions of pounds)	Population (in millions)
1 Australia	327.4	17.3
2 Austria	179.5	7.7
3 Belgium/Luxembourg	279.4	10.4
4 Denmark	139.1	5.1
5 Finland	92.5	5.0
6 France	926.7	56.9
7 Germany	2186.3	79.7
8 Ireland	96.8	3.5
9 Italy	523.9	57.8
10 Japan	935.9	124.0
11 Netherlands	444.2	15.1
12 Norway	119.7	4.3
13 Spain	300.7	39.0
14 Sweden	201.9	8.7
15 Switzerland[a]	194.7	6.9
16 United Kingdom	1592.9	57.7
17 United States[a]	5142.2	252.7

[a]Both Switzerland and the United States include sugar-free candy in their statistics. The U.S. consumption excludes chewing gum.

Source: International Statistics Committee; *Chicago Tribune.*

10. There is a hint in the table footnote that the Swiss and U.S. data might not be consistent with the rest. If they were not, how much effect would the inconsistencies have on the fitted regression, do you think? A lot? Not much?

11. Plot the residuals versus $\hat{Y}$. Does this plot seem to confirm, or deny, the basic assumption of constant error variance. If it does deny it, suggest what could be done.

12. Fit the model $X = \alpha_0 + \alpha_1 Y + \epsilon$ by least squares. Find the fitted line that passes through $(\overline{X}, \overline{Y})$ and has slope

$$c_1 = (b_1/a_1)^{1/2},$$

where b_1 is the least squares estimate of β_1 in the $Y = \beta_0 + \beta_1 X + \epsilon$ fit (2) and a_1 is the least squares estimate of α_1 in the model of this part. (This fitted line is the "geometric mean functional relationship"; see Section 3.4.) State exactly what this third line is in the form $\hat{Y} = c_0 + c_1 X$, and plot all three lines on a new diagram.

AA. The height of soap suds in the dishpan is of importance to soap manufacturers. An experiment was performed by varying the amount of soap and measuring the height of the suds in a standard dishpan after a given amount of agitation. The data are as follows:

Grams of Product, X	Suds Height, Y
4.0	33
4.5	42
5.0	45
5.5	51
6.0	53
6.5	61
7.0	62

Assume that a model of the form $Y = \beta_0 + \beta_1 X + \epsilon$ is reasonable.

Requirements
1. Determine the best-fitting equation.
2. Test it for statistical significance.
3. Calculate the residuals and see if there is any evidence suggesting that a more complicated model would be more suitable.

BB. In an experiment similar to that given in Exercise AA, the experimenter stated that the model $Y = \beta_0 + \beta_1 X + \epsilon$ was "a ridiculous model unless $\beta_0 = 0$, for anyone knows that if you don't put any soap in the dishpan there will be no suds." Thus he insists on using the model $Y = \beta_1 X + \epsilon$. His data are shown as follows:

Grams of Product, X	Suds Height, Y
3.5	24.4
4.0	32.1
4.5	37.1
5.0	40.4
5.5	43.3
6.0	51.4
6.5	61.9
7.0	66.1
7.5	77.2
8.0	79.2

Requirements
1. Accepting the experimenter's model, determine the best estimate of β_1.
2. Using this equation, estimate $\hat{Y}$ for each X.

3. Examine the residuals.
4. Draw conclusions and make recommendations to the experimenter.

CC. The following data indicate the relationship between the amount of β-erythroidine in an aqueous solution and the colorimeter reading of the turbidity:

X Concentration (mg/mL)	Y Colorimeter Reading
40	69
50	175
60	272
70	335
80	490
90	415
40	72
60	265
80	492
50	180

Requirements. Fit the equation $Y = \beta_0 + \beta_1 X + \epsilon$, obtain the residuals, examine them, and comment on the adequacy of the model.

DD. A synthetic fiber, which because of its hairlike appearance has been found suitable in the manufacture of wigs, must necessarily be preshrunk prior to manufacture. This is accomplished in two steps:

Step 1. The fiber is soaked in a dilute solution of chemical A, which is necessary to preserve the luster of the fiber during step 2.

Step 2. The fiber is baked in large ovens at a very high temperature for 1 hour.

It is suggested that the temperature at which the fiber is baked may influence the effectiveness of the preshrinking process. An experiment is performed in which the baking temperature T is varied for various batches of fiber. The finished fiber is then soaked in rainwater for a suitable length of time and put out in the sun to dry. The amount of further shrinkage Y (in percent) resulting from the rainwater test is recorded along with the value of T for each batch:

Batch No.	T	Y
1	280	2.1
2	250	3.0
3	300	3.2
4	320	1.4
5	310	2.6
6	280	3.9
7	320	1.3
8	300	3.4
9	320	2.8

1. Fit a regression line $\hat{Y} = b_0 + b_1 T$ to the data by least squares.
   (*Note:* Coding the variable T may simplify the calculations, but remember in the end to express the fitted equation in terms of the original variable T.)
2. Perform an analysis of variance and test:
   a. The lack of fit.
   b. The significance of the regression.
   What is the percentage variation explained by the regression equation?
3. What is the standard error of b_1? Give a 95% confidence interval for the true regression coefficient β_1.
4. Give the fitted value $\hat{Y}_i$ and the residual $Y_i - \hat{Y}_i$ corresponding to each run (batch).

5. For $T_0 = 315$, find an interval about the predicted value $\hat{Y}_0$ within which a single future observation Y will fall with probability 0.95.
6. Could we use the fitted equation to predict a value of Y at $T = 360$? Give reasons for your answer.
   (See Exercise L in "Exercises for Chapters 5 and 6" for a continuation.)

EE. The tree data below, obtained by Tamra J. Burcar and used with her permission, consist of 23 observations of X = tree diameter in inches measured at breast height (DBH) and Y = height in feet.

X	Y	X	Y	X	Y
5.5	58	8.3	68	10.6	80
5.7	60	8.6	65	10.8	82
6.5	64	9.5	70	11.3	70
6.6	60	10.0	63	11.3	74
6.7	65	10.1	75	11.6	68
6.9	56	10.2	72	11.6	68
7.0	57	10.4	78	13.0	82
7.3	70	10.6	65		

1. Plot Y versus X.
2. Fit $Y = \beta_0 + \beta_1 X + \epsilon$ by least squares and plot the fitted line on your diagram.
3. Evaluate the residuals to two decimal places. Check that $\Sigma e_i = 0$, within rounding error.
4. Obtain the analysis of variance table.
5. Find out how much of the variation about the mean $\overline{Y}$ is explained by the fitted line.
6. Use the repeat runs to test for lack of fit and state your conclusions.
7. Is it appropriate to test for overall regression via the F-test? If so, do it.
8. Find standard errors for b_0 and b_1, using s_e^2 or s^2 as seems appropriate from part 6.
9. Find the formula for the standard error of $\hat{Y}$ and, assuming that s^2 is an appropriate estimate of σ^2 for this part, construct 95% confidence bands for the true mean value of Y. (Plot about half a dozen points over the X-range and join them up smoothly.)

In fact two more data points were obtained by Ms. Burcar but were deleted above. They were:

X	Y
5.8	42
18.0	88

The small tree was clearly stunted; the large tree may have been damaged by winds because it protruded above the tree line.

10. Show the new trees on your plot and comment briefly on what you see.

FF. Reconsider the data of Exercise EE. Fit the model $X = \alpha_0 + \alpha_1 Y + \epsilon$ by least squares. Find the fitted line that passes through $(\overline{X}, \overline{Y})$ and has slope

$$c_1 = (b_1/a_1)^{1/2},$$

where b_1 is the least squares estimate of β_1 in the $Y = \beta_0 + \beta_1 X + \epsilon$ fit of the previous exercise and a_1 is the least squares estimate of α_1. (This fitted line is the geometric mean functional relationship, as described in Section 3.4.) State exactly what this third line is in the form $\hat{Y} = c_0 + c_1 X$, and plot all three lines on a new diagram.

(The following comment is not connected specifically to the data of Exercises EE and FF in any way, but since it concerns data of the same type, we reproduce it here for the benefit of those taking similar data.

Experience suggests certain cautions. First, there is a tendency to select vigorously growing trees of good form for the dimensional analysis, unless this tendency is consciously counteracted. The preference for "good" sample trees implies overestimation of productivity when regressions from these trees are applied to field quadrat data. Second, the largest errors result from applying regressions to the largest trees in the samples. . . . If, from the population of large trees in the stand, many of them senescent or with partly broken crowns, a particularly "good" individual has been chosen, the slope of the regression as it extends to larger tree sizes is biased by this individual. The production estimates for the few large trees in the sample quadrat will be overestimates for most of these trees. It is therefore important that errors of estimation for large trees be controlled by some means. . . ."

This excellent advice, which can be summarized by saying "Make sure your data are representative of your population," was written by R. H. Whittaker and P. L. Marks in "Methods of assessing terrestial productivity," Chapter 4, p. 85 of *Primary Productivity of the Biosphere*, edited by H. Lieth and R. H. Whittaker, and published by Springer-Verlag, New York, in 1975.)

GG. The *New York Times* of Thursday, May 24, 1990, showed, on page B5, a chart containing data obtained in the National Child Care Staffing Study released in Fall 1989 by the Child Care Employee Project in Oakland, California. These data are displayed below. Plot Y = turnover versus X = hourly wage. The article stated that one city had a high unemployment rate. Which data point do you think represents that city? Omit the Detroit data point and fit a straight line $Y = \beta_0 + \beta_1 X + \epsilon$ to the remaining four data points. Estimate the true mean turnover if the wage were $6.00, and put 95% confidence bands around the estimate.

Place	Hourly Wage (in $) of Staff Who Care for Children	Average Annual Staff Turnover (in %)
Atlanta	4.96	57
Boston	7.28	29
Detroit	4.88	27
Phoenix	4.45	64
Seattle	5.21	41

HH. (*Source:* R. Peter Hypher, Ottawa, Ontario, Canada.) Below are 29 observations on

$$X = \text{Monthly Protestant receipts in Canadian dollars,}$$

$$Y = \text{Monthly attendance at Catholic mass,}$$

for a certain Canadian town some years ago. (The figures, read off from a plot, vary slightly from the originals, which are not available.)

Plot the data, and fit the model $Y = \beta_0 + \beta_1 X + \epsilon$ by least squares. Assuming conditions remain the same, can we validly predict attendance at Catholic mass from the Protestant monthly receipts? Estimate the Catholic mass attendance when $X = 0$ and provide 95% confidence limits on the true mean value of Y at $X = 0$.

X	Y	X	Y	X	Y
100	2000	290	2875	390	2500
100	2475	290	2875	400	2875
100	2825	300	2900	400	3800
155	2000	300	3000	440	3825
160	1650	330	3100	440	3900
205	2475	350	2875	450	3100
230	2725	350	3225	460	3050
250	2125	350	3400	500	3475
250	3000	350	3450	550	3200
		360	2950	600	3675

II. The data below consist of

$$X = \text{Course number,}$$

$$Y = \text{Questionnaire score on "question 11,"}$$

for 32 teaching assistants in the Statistics Department, University of Wisconsin, in the Spring Semester of 1978. Plot the data, fit a straight line model $Y = \beta_0 + \beta_1 X + \epsilon$, perform all the usual analyses, and state the practical conclusions that arise from your analysis.

Row	Y	X	Row	Y	X	Row	Y	X
1	5.00	349	12	4.41	224	23	4.00	301
2	4.86	333	13	4.38	333	24	3.92	201
3	4.75	301	14	4.29	314	25	3.88	201
4	4.64	301	15	4.25	201	26	3.86	312
5	4.66	302	16	4.22	424	27	3.74	224
6	4.62	824	17	4.21	424	28	3.69	224
7	4.55	710	18	4.15	301	29	3.57	224
8	4.47	301	19	4.07	201	30	3.27	301
9	4.43	301	20	4.08	301	31	2.96	224
10	4.44	310	21	4.00	224	32	2.67	224
11	4.43	702	22	4.00	301			

JJ. (*Source: USA Today* for Monday, December 6, 1993, page 5B.) The data in the accompanying Table JJ consist of average Sunday circulation (Y) and average daily circulation (X) for 48 U.S. newspapers. All values are in thousands, rounded to the nearest thousand. Two of the "top 50" were omitted; the second and 20th were removed because they had no Sunday edition. The figures are for the 6 months ended September 30, 1993 and include bulk sales, defined as lower price sales to, for example, hotels and airlines, who give them free to customers.

1. Plot Y versus X.
2. Fit $Y = \beta_0 + \beta_1 X + \epsilon$ by least squares and plot the fitted line on your diagram.

T A B L E JJ. Average Sunday (Y) and Daily (X) Circulations in Thousands for 48 of the Top 50 Newspapers in the United States for the Period March–September 1993

Newspaper	Sunday (Y)	Daily (X)	Newspaper	Sunday (Y)	Daily (X)
1	2313	1886	25	559	342
2	1766	1155	26	483	338
3	1497	1097	27	441	333
4	1140	816	28	428	324
5	928	764	29	214	322
6	826	748	30	378	302
7	1107	694	31	431	291
8	1187	558	32	440	285
9	708	549	33	321	284
10	524	536	34	342	282
11	814	527	35	338	270
12	814	508	36	401	264
13	944	487	37	320	264
14	704	474	38	375	263
15	709	442	39	346	255
16	607	413	40	347	252
17	696	411	41	456	250
18	543	396	42	340	246
19	509	387	43	326	236
20	455	384	44	301	233
21	1186	367	45	496	232
22	548	348	46	411	232
23	408	344	47	321	230
24	453	343	48	504	229

Source: USA Today, Monday, December 6, 1993, p. 5B.

3. Evaluate the residuals to one decimal place. Check that $\Sigma e_u = 0$, within rounding error.

4. Obtain an analysis of variance table.

5. Find out how much of the variation about the mean $\overline{Y}$ is explained by the fitted line.

6. There are only two pairs of exact repeat runs in the rounded data quoted. As an approximation, however, treat the Y values at $X = 229, 230, 232, 232, 233$, and 236 as repeats; also, those at $X = 246, 250, 252$, and 255; those at $X = 263, 264$, and 264; those at $X = 282, 284$, and 285; those at $X = 333$ and 338; those at $X = 342, 343, 344$, and 348; and finally those at $X = 411$ and 413. (One could argue about these groupings, of course, and you might wish to do a second, alternative calculation using only the two pairs, to see what difference it makes.) Use the pseudo repeat runs to test (approximately) for lack of fit and state your conclusion. If lack of fit is shown, omit parts (7), (8), and (9).

7. Is it appropriate to test for overall regression via the F-test? If so, do it.

8. Find standard errors for b_0 and b_1.

9. Find the formula for the standard error of $\hat{Y}$ and, assuming that s^2 is an appropriate estimate of σ^2 for this part, construct 95% confidence bands for the true mean value of Y. (Plot about half a dozen points over the X-range and join them up smoothly.)

10. Plot the residuals versus $\hat{Y}$. Does this plot seem to confirm, or deny, the basic assumption of constant error variance. If it does deny it, suggest what could be done.

11. Below are the average daily bulk sales for the first nine newspapers in the data list. Adjust the data for these numbers (subtract them) and repeat the analysis. State your conclusions overall.

Newspaper	Sunday (Y)	Daily X
1	408	391
2	10	14
3	9	7
4	1	2
5	1	0
6	1	1
7	5	3
8	1	2
9	6	5

KK. (*Source:* Audit Bureau of Circulations, as given in the *New York Times,* October 31, 1995, page C7.) The average daily circulations of the largest 12 U.S. newspapers (apart from the *Detroit Free Press,* whose union workers were on strike in 1995) for the 6 months ending September 30 are listed in the table for 1994 (X) and 1995 (Y). Circulation figures for papers 2 and 10 do not include Fridays. Fit a straight line model $Y = \beta_0 + \beta_1 X + \epsilon$, evaluate the residuals, check the fit, and amend it as appropriate. State conclusions.

Paper	1994	1995
1. *Wall Street Journal*	1,780,422	1,763,140
2. *USA Today*	1,465,936	1,523,610
3. *New York Times*	1,114,168	1,081,541
4. *Los Angeles Times*	1,062,199	1,012,189
5. *Washington Post*	810,675	793,660
6. *Daily News*	753,024	738,091
7. *Chicago Tribune*	678,081	684,366
8. *Newsday*	693,556	634,627
9. *Houston Chronicle*	409,007	541,478
10. *Dallas Morning News*	491,480	500,358
11. *Boston Globe*	506,543	498,853
12. *San Francisco Chronicle*	509,548	489,238

LL. The manager of a small mail order house hires additional personnel every time there is a peak demand that exceeds the work load of his normal three employees. To check the effectiveness of this idea, he records the daily output of his total crew on various days during various periods, both peak and otherwise. These data are given below. Fit a straight line model $Y = \beta_0 + \beta_1 X + \epsilon$ to the data via least squares, check for lack of fit, and (if there is no lack of fit) test for overall regression. Examine the residuals, in any case, and state the conclusions you draw from the study.

Number of Parcels Dispatched Y	Number of Employees X	Number of men Z^a
50	1^b	
110	2^b	
90	2^b	
150	3	
140	3	
180	3	
190	4	
310	6	
330	6	
340	7	
360	8	
380	10	
360	10	

[a]Needed for Exercise H in "Exercises for Chapters 5 and 6." Ignore for now.
[b]Regular employee(s) sick or on vacation.

Useful Facts. $n = 13$

$$\Sigma X_i = 65, \qquad \Sigma Y_i = 2990, \qquad \Sigma X_i Y_i = 19{,}120,$$

$$\Sigma X_i^2 = 437, \qquad \Sigma Y_i^2 = 857{,}500, \qquad \Sigma Y_i^2 - (\Sigma Y_i)^2/n = 169{,}800.$$

MM. (Source: Steven LeMire.) An important factor in the performance of an air filter is its resistance. Measuring this accurately on a large section of filter material is expensive. An inexpensive alternative device was suggested and was used to obtain the data X below. The Y values were obtained from the more expensive method. (All values have been multiplied by 1000.) Fit straight line models $Y = \beta_0 + \beta_1 X + \epsilon$ and $X = \alpha_0 + \alpha_1 Y + \epsilon$ to these data, and plot each set of residuals versus its corresponding set of fitted values, $\hat{Y}$ or $\hat{X}$, respectively. Also fit the geometric mean functional relationship line which has slope $(b/a)^{1/2}$ and which passes through $(\overline{X}, \overline{Y}) = (96.25, 89.1875)$, where b and a are the estimates of β_1 and α_1. Plot the two sets of residuals, horizontal and vertical, from this line versus their corresponding fitted values. Does each set of residuals sum to zero? Now review all the results and comment.

X	Y	X	Y
0	0	79	72
22	22	93	87
26	25	133	118
41	40	122	120
62	59	142	125
67	64	165	150
70	67	190	182
78	70	250	226

CHAPTER 4

Regression in Matrix Terms: Straight Line Case

We shall now present the steam data example given earlier in terms of matrix algebra. The use of matrices has many advantages, not the least of these being that once the problem is written and solved in matrix terms the solution can be applied to any regression problem no matter how many terms there are in the regression equation. Although there is a matrix algebra section in Chapter 0, some explanations are deliberately duplicated in this section.

Matrices

A matrix (plural matrices) is a rectangular array of symbols or numbers and is usually denoted by a single letter in **boldface** type, for example, $\mathbf{Q}$ or $\mathbf{q}$. There are several rules for manipulation of such arrays. Quite complicated expressions or equations can often be represented very simply by just a few letters properly defined and grouped.

We shall not introduce matrices formally but will use them in the context of the example. The reader with sound knowledge of matrices might wish to omit this chapter.

4.1. FITTING A STRAIGHT LINE IN MATRIX TERMS

We define $\mathbf{Y}$ to be the *vector of observations* Y_i, $\mathbf{X}$ to be the *matrix of predictor variables*, $\boldsymbol{\beta}$ to be the *vector of parameters to be estimated*, $\boldsymbol{\epsilon}$ to be a *vector of errors*, and $\mathbf{1}$ to be a vector of ones. In terms of the steam data in Table 1.1, and Eq. (1.2.3), we thus define

$$
\mathbf{Y} = \begin{bmatrix} 10.98 \\ 11.13 \\ 12.51 \\ 8.40 \\ \vdots \\ 10.36 \\ 11.08 \end{bmatrix} \quad
\mathbf{X} = \begin{bmatrix} 1 & 35.3 \\ 1 & 29.7 \\ 1 & 30.8 \\ 1 & 58.8 \\ \vdots & \vdots \\ 1 & 33.4 \\ 1 & 28.6 \end{bmatrix} \quad
\boldsymbol{\beta} = \begin{bmatrix} \beta_0 \\ \beta_1 \end{bmatrix} \quad
\boldsymbol{\epsilon} = \begin{bmatrix} \epsilon_1 \\ \epsilon_2 \\ \epsilon_3 \\ \epsilon_4 \\ \vdots \\ \epsilon_{24} \\ \epsilon_{25} \end{bmatrix} \quad
\mathbf{1} = \begin{bmatrix} 1 \\ 1 \\ 1 \\ 1 \\ \vdots \\ 1 \\ 1 \end{bmatrix}.
$$

$$(4.1.1)$$

Note that

Y is a 25 × 1 vector.
X is a 25 × 2 matrix.
$\boldsymbol{\beta}$ is a 2 × 1 vector.
$\boldsymbol{\epsilon}$ is a 25 × 1 vector.
1 is a 25 × 1 vector.

(Any matrix with one column is called a column vector; any matrix with one row is called a row vector. A 1 × 1 "matrix" is just an ordinary number or scalar.)

The dots in the matrices and vectors represent data not reproduced to save space in making the definition—a conventional procedure in matrix work. The column **1** of ones is not strictly needed at this stage, but it is convenient to define it here; it is extremely useful in matrix manipulations. Note that **X** is formed of two column vectors. The first is simply **1**, the second is an (unnamed) vector of X-values in the data usually called the "X-column." Many writers call the column of ones in **X** the "X_0-column," pretending that there is a predictor variable X_0 that is always set at the value one. A variable chosen in some such arbitrary way is usually called a *dummy variable* and useful extended applications of such devices will be given in Chapters 14 and 23.

Manipulating Matrices

The rules of multiplication for matrices and vectors insist that two matrices must be *conformable*. For example, if **A** is an $n \times p$ matrix we can:

1. *Postmultiply* it by a $p \times q$ matrix to give as a result an $n \times p \times p \times q = n \times q$ matrix.
2. *Premultiply* it by an $m \times n$ matrix to give as a result an $m \times n \times n \times p = m \times p$ matrix.

Thus, for example, the multiplication $\boldsymbol{\beta}\mathbf{X}$ is not possible since $\boldsymbol{\beta}$ is 2 × 1 and **X** is 25 × 2. But $\mathbf{X}\boldsymbol{\beta}$ is possible as follows:

$$\mathbf{X}\boldsymbol{\beta} = \begin{bmatrix} 1 & 35.3 \\ 1 & 29.7 \\ \vdots & \vdots \\ 1 & 28.6 \end{bmatrix} \begin{bmatrix} \beta_0 \\ \beta_1 \end{bmatrix} = \begin{bmatrix} \beta_0 + 35.3\beta_1 \\ \beta_0 + 29.7\beta_1 \\ \vdots \\ \beta_0 + 28.6\beta_1 \end{bmatrix}. \tag{4.1.2}$$

$$25 \times 2 \quad 2 \times 1 \qquad 25 \times 1$$

As a more general example consider the product

$$\overset{\mathbf{A}}{\begin{bmatrix} 1 & 2 & 4 \\ -1 & 0 & 1 \\ 2 & 3 & 1 \end{bmatrix}} \overset{\mathbf{B}}{\begin{bmatrix} 1 & -1 \\ 2 & 1 \\ 3 & 5 \end{bmatrix}} = \overset{\mathbf{C}}{\begin{bmatrix} 17 & 21 \\ 2 & 6 \\ 11 & 6 \end{bmatrix}}.$$

$$3 \times 3 \qquad 3 \times 2 \qquad 3 \times 2$$

To find the element in row i and column j of **C**, we take row i of **A** and column j of **B**, find the cross-product of corresponding elements, and add. For example,

Row 2 of $\mathbf{A}$ is $\quad\quad -1 \quad 0 \quad 1$

Column 1 of $\mathbf{B}$ is $\quad\quad 1 \quad 2 \quad 3$

Thus the element in row 2, column 1 of $\mathbf{C}$ is

$$-1(1) + 0(2) + 1(3) = 2.$$

Orthogonality

Definition. If the sum of the cross-products of corresponding elements of row i and column j is zero, then row i is said to be *orthogonal* to column j. (The same definition applies to row and row or column and column.)

The Model in Matrix Form

The sum of two matrices or vectors is just the matrix whose elements are the sums of corresponding elements in the separate matrices or vectors. For example,

$$\mathbf{X\beta} + \mathbf{\epsilon} = \begin{bmatrix} \beta_0 + 35.3\beta_1 \\ \beta_0 + 29.7\beta_1 \\ \vdots \\ \beta_0 + 28.6\beta_1 \end{bmatrix} + \begin{bmatrix} \epsilon_1 \\ \epsilon_2 \\ \vdots \\ \epsilon_{25} \end{bmatrix} = \begin{bmatrix} \beta_0 + 35.3\beta_1 + \epsilon_1 \\ \beta_0 + 29.7\beta_1 + \epsilon_2 \\ \vdots \\ \beta_0 + 28.6\beta_1 + \epsilon_{25} \end{bmatrix}. \tag{4.1.3}$$

The two matrices or vectors must have the same dimensions for this to be possible. (The difference between two matrices is similarly defined with differences instead of sums.) If two matrices or vectors are equal, corresponding elements are equal. Thus writing the matrix equation

$$\mathbf{Y} = \mathbf{X\beta} + \mathbf{\epsilon} \tag{4.1.4}$$

implies that

$$10.98 = \beta_0 + 35.3\beta_1 + \epsilon_1$$
$$\vdots \tag{4.1.5}$$
$$11.08 = \beta_0 + 28.6\beta_1 + \epsilon_{25}$$

or

$$Y_i = \beta_0 + \beta_1 X_i + \epsilon_i \quad\quad (i = 1, \ldots, 25) \tag{4.1.6}$$

for each of the 25 observations. Thus the matrix equation, Eq. (4.1.4), and Eq. (4.1.6) express the same model. Equation (4.1.6) is identical to Eq. (1.2.3).

Setup for a Quadratic Model

In setting up the model in matrix form, only the choice of $\mathbf{X}$ usually presents any difficulty to the beginner. The simplest way of obtaining $\mathbf{X}$ is first to write down all of the parameters shown in the model as a vector $\mathbf{\beta}$, and then to see what corresponding X-columns would be needed to reproduce the model in its given algebraic form from the $\mathbf{X\beta}$ product. For example, if the model is $Y = \beta_0 + \beta_1 X + \beta_{11} X^2 + \epsilon$, the vector $\mathbf{\beta}$ will be a column of three elements β_0, β_1, and β_{11} and the corresponding X-columns must necessarily be 1 (or X_0 if we use that notation), X, and X^2. Thus the ith row of $\mathbf{X}$ will consist of $(1, X_i, X_i^2)$, where X_i is the ith of the n observations. Note that a reordering of the elements of $\boldsymbol{\beta}$ requires a reordering of the columns of $\mathbf{X}$ to correspond.

Transpose

We now define the *transpose* of a matrix. It is the matrix obtained by writing all rows as columns in the order in which they occur so that the columns all become rows. The transpose of a matrix $\mathbf{M}$ is written $\mathbf{M}'$, for example,

$$\mathbf{M} = \begin{bmatrix} 3 & 2 \\ 1 & 4 \\ 7 & 0 \end{bmatrix}, \qquad \mathbf{M}' = \begin{bmatrix} 3 & 1 & 7 \\ 2 & 4 & 0 \end{bmatrix}.$$

$$3 \times 2 \qquad\qquad\qquad 2 \times 3$$

Thus, for example,

$$\boldsymbol{\epsilon}' = (\epsilon_1, \epsilon_2, \ldots, \epsilon_n).$$

Note that we can then write

$$\epsilon_1^2 + \epsilon_2^2 + \cdots + \epsilon_n^2 = \boldsymbol{\epsilon}'\boldsymbol{\epsilon},$$

$$Y_1^2 + Y_2^2 + \cdots + Y_n^2 = \mathbf{Y}'\mathbf{Y},$$

$$n\overline{Y} = Y_1 + Y_2 + \cdots + Y_n = \mathbf{1}'\mathbf{Y},$$

$$n\overline{Y}^2 = (\Sigma\, Y_i)^2/n = \mathbf{Y}'\mathbf{1}\mathbf{1}'\mathbf{Y}/n.$$

Furthermore

$$\mathbf{X}'\mathbf{X} = \begin{bmatrix} 1 & 1 & \cdots & 1 \\ 35.3 & 29.7 & \cdots & 28.6 \end{bmatrix} \begin{bmatrix} 1 & 35.3 \\ 1 & 29.7 \\ \vdots & \vdots \\ 1 & 28.6 \end{bmatrix} = \begin{bmatrix} 25 & 1315 \\ 1315 & 76323.42 \end{bmatrix}.$$

In general, for a straight line model, we see that

$$\mathbf{X}'\mathbf{X} = \begin{bmatrix} 1 & 1 & \cdots & 1 \\ X_1 & X_2 & \cdots & X_n \end{bmatrix} \begin{bmatrix} 1 & X_1 \\ 1 & X_2 \\ \vdots & \vdots \\ 1 & X_n \end{bmatrix} = \begin{bmatrix} n & \Sigma\, X_i \\ \Sigma\, X_i & \Sigma\, X_i^2 \end{bmatrix}. \qquad (4.1.7)$$

In addition,

$$\mathbf{X}'\mathbf{Y} = \begin{bmatrix} 1 & 1 & \cdots & 1 \\ 35.3 & 29.7 & \cdots & 28.6 \end{bmatrix} \begin{bmatrix} 10.98 \\ 11.13 \\ \vdots \\ 11.08 \end{bmatrix} = \begin{bmatrix} 235.60 \\ 11821.4320 \end{bmatrix}$$

so that, generally, for a straight line fit,

$$\mathbf{X'Y} = \begin{bmatrix} 1 & 1 & \cdots & 1 \\ X_1 & X_2 & \cdots & X_n \end{bmatrix} \begin{bmatrix} Y_1 \\ Y_2 \\ \vdots \\ Y_n \end{bmatrix} = \begin{bmatrix} \Sigma Y_i \\ \Sigma X_i Y_i \end{bmatrix}. \tag{4.1.8}$$

This means that the normal equations (1.2.8) can be written

$$\mathbf{X'Xb} = \mathbf{X'Y}, \tag{4.1.9}$$

where $\mathbf{b'} = (b_0, b_1)$, and these equations, when solved, provide the least squares estimates (b_0, b_1) of (β_0, β_1). How do we solve these equations in matrix form? To do so we define the *inverse* of a matrix. This exists only when a matrix is square and when the determinant of the matrix (a quantity that we shall not define here but of which we shall provide some examples) is nonzero. This latter condition is usually stated as *when the matrix is nonsingular*. This will be true in our applications unless otherwise stated. In regression work we wish to invert the $\mathbf{X'X}$ matrix. If it is *singular*, and so does not have an inverse, this will be reflected in the fact that some of the normal equations will be linear combinations of others; see, for example, Eq. (4.2.3). In this case there will be fewer equations than there are unknowns for which to solve. In such a case unique estimates are not possible unless some additional conditions on the parameters apply. (See Chapter 23 for additional comments on this point.)

Inverse of a Matrix

Suppose now that $\mathbf{M}$ is a nonsingular $p \times p$ matrix. The inverse of $\mathbf{M}$ is written $\mathbf{M}^{-1}$, is $p \times p$, and is such that

$$\mathbf{M}^{-1}\mathbf{M} = \mathbf{M}\mathbf{M}^{-1} = \mathbf{I}_p,$$

where $\mathbf{I}_p$ is the *unit matrix of order p*, which consists of unities (i.e., ones) in every position of the main diagonal (i.e., the diagonal running from the upper left corner to the lower right corner) and zeros elsewhere; for example,

$$\mathbf{I}_4 = \begin{bmatrix} 1 & 0 & 0 & 0 \\ 0 & 1 & 0 & 0 \\ 0 & 0 & 1 & 0 \\ 0 & 0 & 0 & 1 \end{bmatrix}.$$

(When the size of the unit matrix is obvious, the subscript is often omitted.) The unit matrix plays the same role in matrix multiplication that 1 does in ordinary multiplication—it leaves the multiplicand unchanged. The inverse of a matrix is unique.

Inverses of Small Matrices

The formulas for inverting matrices of sizes two and three are as follows:

$$\mathbf{M}^{-1} = \begin{bmatrix} a & b \\ c & d \end{bmatrix}^{-1} = \begin{bmatrix} d/D & -b/D \\ -c/D & a/D \end{bmatrix}, \tag{4.1.10}$$

where $D = ad - bc$ is the *determinant* of the 2×2 matrix $\mathbf{M}$.

$$\mathbf{Q}^{-1} = \begin{bmatrix} a & b & c \\ d & e & f \\ g & h & k \end{bmatrix}^{-1} = \begin{bmatrix} A & B & C \\ D & E & F \\ G & H & K \end{bmatrix}, \tag{4.1.11}$$

where

$$A = (ek - fh)/Z \qquad B = -(bk - ch)/Z \qquad C = (bf - ce)/Z$$
$$D = -(dk - fg)/Z \qquad E = (ak - cg)/Z \qquad F = -(af - cd)/Z$$
$$G = (dh - eg)/Z \qquad H = -(ah - bg)/Z \qquad K = (ae - bd)/Z$$

and where

$$Z = a(ek - fh) - b(dk - fg) + c(dh - eg)$$
$$= aek + bfg + cdh - ahf - dbk - gec$$

is the determinant of $\mathbf{Q}$.

Matrix Symmetry for Square Matrices

Matrices of the form $\mathbf{X}'\mathbf{X}$ met in regression work are always symmetric, that is, the element in the ith row and jth column is the same as the element in the jth row and ith column. Thus the transpose of a symmetric matrix is the matrix itself. This is easy to see if we apply the general rule $(\mathbf{AB})' = \mathbf{B}'\mathbf{A}'$ for transposes of a product. Because $(\mathbf{A}')' = \mathbf{A}$ itself, we can write $(\mathbf{X}'\mathbf{X})' = \mathbf{X}'\mathbf{X}$. (Working with a few simple numerical cases will clarify this point.) If the matrix $\mathbf{M}$ of size two above is symmetric, $b = c$ and the inverse also becomes symmetric. If the matrix $\mathbf{Q}$ above is symmetric, $b = d$, $c = g$, $f = h$. Then, relabeling the matrix $\mathbf{S}$, we obtain the symmetric inverse

$$\mathbf{S}^{-1} = \begin{bmatrix} a & b & c \\ b & e & f \\ c & f & k \end{bmatrix}^{-1} = \begin{bmatrix} A & B & C \\ B & E & F \\ C & F & K \end{bmatrix}, \tag{4.1.12}$$

where

$$A = (ek - f^2)/Y \qquad B = -(bk - cf)/Y \qquad C = (bf - ce)/Y$$
$$E = (ak - c^2)/Y \qquad F = -(af - bc)/Y$$
$$K = (ae - b^2)/Y$$

and where

$$Y = a(ek - f^2) - b(bk - cf) + c(bf - ce)$$
$$= aek + 2bcf - af^2 - b^2k - c^2e$$

is the determinant of $\mathbf{S}$. The inverse of any symmetric matrix is, itself, a symmetric matrix.

Diagonal Matrices

Matrices of sizes greater than three are usually cumbersome to invert unless they have a special form. One matrix that is easy to invert, no matter what its size, is a *diagonal matrix*, which consists of nonzero elements in the main upper-left to lower-right diagonal, and zeros elsewhere. The inverse is obtained by inverting all nonzero elements where they stand. For example,

$$
\begin{bmatrix} a_1 & & & \mathbf{0} \\ & a_2 & & \\ & & \ddots & \\ \mathbf{0} & & & a_n \end{bmatrix}^{-1} = \begin{bmatrix} 1/a_1 & & & \mathbf{0} \\ & 1/a_2 & & \\ & & \ddots & \\ \mathbf{0} & & & 1/a_n \end{bmatrix}. \tag{4.1.13}
$$

(Note, in this special case, the use of **0** to denote a large triangular block of zeros. This is often seen.)

Inverting Partitioned Matrices with Blocks of Zeros

Another type of simplification sometimes occurs when some columns of the **X** matrix are orthogonal to *all* other columns. The **X'X** matrix then takes the *partitioned* form

$$
\begin{bmatrix} \mathbf{P} & \mathbf{0} \\ \mathbf{0} & \mathbf{R} \end{bmatrix},
$$

where, for example, **P** might be $p \times p$, **R** might be $r \times r$, and the symbol **0** is used to denote two differently shaped blocks of zeros, a $p \times r$ one in the top right-hand corner and an $r \times p$ one in the lower left-hand corner. The inverse of this matrix is then

$$
\begin{bmatrix} \mathbf{P} & \mathbf{0} \\ \mathbf{0} & \mathbf{R} \end{bmatrix}^{-1} = \begin{bmatrix} \mathbf{P}^{-1} & \mathbf{0} \\ \mathbf{0} & \mathbf{R}^{-1} \end{bmatrix}. \tag{4.1.14}
$$

For example, if

$$
\mathbf{P} = \begin{bmatrix} 1 & 3 \\ 2 & 8 \end{bmatrix}, \quad \mathbf{P}^{-1} = \begin{bmatrix} 4 & -\frac{3}{2} \\ -1 & \frac{1}{2} \end{bmatrix},
$$

$$
\mathbf{R} = \begin{bmatrix} 1 & 0 & 1 \\ 2 & 3 & 2 \\ 4 & 1 & 1 \end{bmatrix}, \quad \mathbf{R}^{-1} = \begin{bmatrix} -\frac{1}{9} & -\frac{1}{9} & \frac{1}{3} \\ -\frac{2}{3} & \frac{1}{3} & 0 \\ \frac{10}{9} & \frac{1}{9} & -\frac{1}{3} \end{bmatrix},
$$

then

$$
\begin{bmatrix} 1 & 3 & 0 & 0 & 0 \\ 2 & 8 & 0 & 0 & 0 \\ 0 & 0 & 1 & 0 & 1 \\ 0 & 0 & 2 & 3 & 2 \\ 0 & 0 & 4 & 1 & 1 \end{bmatrix}^{-1} = \begin{bmatrix} 4 & -\frac{3}{2} & 0 & 0 & 0 \\ -1 & \frac{1}{2} & 0 & 0 & 0 \\ 0 & 0 & -\frac{1}{9} & -\frac{1}{9} & \frac{1}{3} \\ 0 & 0 & -\frac{2}{3} & \frac{1}{3} & 0 \\ 0 & 0 & \frac{10}{9} & \frac{1}{9} & -\frac{1}{3} \end{bmatrix}.
$$

When there are more than two nonzero blocks, the obvious extension holds. It is important to note that the blocks must be on the main diagonal, and the off-diagonal blocks must consist entirely of zeros for the extension to apply.

Less Obvious Partitioning

The inverse formula (4.1.14) also applies even when the rows and columns containing nonzero elements are intermingled, provided that the matrix can be divided in such a way that the portions, such as $\mathbf{P}$ and $\mathbf{R}$ above, are completely separated from each other by zeros. For example, using the same numbers as above, the matrix

$$\begin{bmatrix} 1 & 0 & 0 & 0 & 1 \\ 0 & 1 & 0 & 3 & 0 \\ 2 & 0 & 3 & 0 & 2 \\ 0 & 2 & 0 & 8 & 0 \\ 4 & 0 & 1 & 0 & 1 \end{bmatrix}$$

can be partitioned and the separate portions inverted separately. Note that the second and fourth rows *and* columns are completely isolated, or insulated, from the first, third, and fifth rows and columns by zeros. Thus the nonzero elements in the second and fourth rows and columns comprise a 2×2 matrix, which can be separately inverted, whereas the other nonzero elements form a completely separate 3×3 matrix, which also can be separately inverted. Thus the inverse has the form

$$\begin{bmatrix} -\frac{1}{9} & 0 & -\frac{1}{9} & 0 & \frac{1}{3} \\ 0 & 4 & 0 & -\frac{3}{2} & 0 \\ -\frac{2}{3} & 0 & \frac{1}{3} & 0 & 0 \\ 0 & -1 & 0 & \frac{1}{2} & 0 \\ \frac{10}{9} & 0 & \frac{1}{9} & 0 & -\frac{1}{3} \end{bmatrix}.$$

Situations like this often occur when carefully designed experiments are analyzed using regression analysis.

The correctness of all these inverses can be confirmed by actually multiplying the inverse by the original, both before and behind. The result is an $\mathbf{I}$ matrix of appropriate size in every case. In practical situations, when the size of a matrix exceeds 3×3, and no simplified form is possible, finding the inverse can be a lengthy procedure. The work would usually be performed within an electronic computer.

Back to the Straight Line Case

We wish now to invert the $\mathbf{X'X}$ matrix of our example. This is of size 2×2 and of the general form of Eq. (4.1.7). Using Eq. (4.1.10) with $b = c$, we obtain the inverse as

$$(\mathbf{X'X})^{-1} = \begin{bmatrix} \dfrac{\Sigma X_i^2}{n\,\Sigma(X_i - \overline{X})^2} & \dfrac{-\overline{X}}{\Sigma(X_i - \overline{X})^2} \\ \dfrac{-\overline{X}}{\Sigma(X_i - \overline{X})^2} & \dfrac{1}{\Sigma(X_i - \overline{X})^2} \end{bmatrix}. \tag{4.1.15}$$

If *every* element of a matrix has a common factor it can be taken outside the matrix. (Conversely, if a matrix is multiplied by a constant C, every element of the matrix must be multiplied by C.) Thus an alternative form is

$$(\mathbf{X'X})^{-1} = \frac{1}{n\,\Sigma(X_i - \overline{X})^2} \begin{bmatrix} \Sigma X_i^2 & -\Sigma X_i \\ -\Sigma X_i & n \end{bmatrix}. \tag{4.1.16}$$

Since $\mathbf{X'X}$ is symmetric, so is its inverse $(\mathbf{X'X})^{-1}$ as mentioned earlier. The quantity taken outside the matrix is the determinant of $\mathbf{X'X}$, written det $(\mathbf{X'X})$ or $|\mathbf{X'X}|$. Using the form of Eq. (4.1.15) on the steam data example we find that

$$(\mathbf{X'X})^{-1} = \begin{bmatrix} 0.4267941 & -0.0073535 \\ -0.0073535 & 0.0001398 \end{bmatrix}.$$

Solving the Normal Equations

If we premultiply Eq. (4.1.9) by $(\mathbf{X'X})^{-1}$, we obtain

$$(\mathbf{X'X})^{-1}(\mathbf{X'X})\mathbf{b} = (\mathbf{X'X})^{-1}\mathbf{X'Y},$$

that is,

$$\mathbf{b} = (\mathbf{X'X})^{-1}\mathbf{X'Y} \tag{4.1.17}$$

since $(\mathbf{X'X})^{-1}\mathbf{X'X} = \mathbf{I}$. This is an important result to remember since the solution of linear regression normal equations can *always* be written in this form, provided $\mathbf{X'X}$ is nonsingular and the regression problem is properly expressed.

Using the data of our example we find that

$$\mathbf{b} = \begin{bmatrix} 0.4267941 & -0.0073535 \\ -0.0073535 & 0.0001398 \end{bmatrix} \begin{bmatrix} 235.60 \\ 11,821.4320 \end{bmatrix}$$

$$= \begin{bmatrix} 13.623790 \\ -0.079848 \end{bmatrix}.$$

A Small Sermon on Rounding Errors

Note that the results are not identical, to six places of decimals, to the values obtained in Section 1.2. Such discrepancies frequently occur because of the rounding off of numbers used in the calculation, and carelessness in such matters can cause serious errors, depending on the numbers involved. Here the numerical discrepancies are slight from a practical point of view, but they emphasize the fact that in general as many figures as possible should be carried in regression calculations. Sometimes, due to the magnitudes of the numbers in the calculation, the entire significance of the results can be lost through careless rounding.

Certain ways of performing the calculations (especially when they are done "by hand," i.e., on a pocket calculator) are better than others since they are less affected by round-off error. In particular, it is wise to postpone divisions to as late a stage as possible. For example, if we had employed the form of Eq. (4.1.16) instead of Eq. (4.1.15) to obtain $(\mathbf{X'X})^{-1}$ we would have obtained

$$(\mathbf{X'X})^{-1} = \frac{1}{178{,}860.5} \begin{bmatrix} 76{,}323.42 & -1315 \\ -1315 & 25 \end{bmatrix}.$$

Then we could have obtained **b** from

$$\mathbf{b} = \frac{1}{178{,}860.5} \begin{bmatrix} 76{,}323.42 & -1315 \\ -1315 & 25 \end{bmatrix} \begin{bmatrix} 235.60 \\ 11{,}821.432 \end{bmatrix}$$

$$= \frac{1}{178{,}860.5} \begin{bmatrix} 2{,}436{,}614.672 \\ -14{,}278.2 \end{bmatrix}$$

$$= \begin{bmatrix} 13.622989 \\ -0.079829 \end{bmatrix},$$

the division being performed last of all.

Doing the calculations the three separate ways gives the following answers:

	Formulas (Section 1.2)	Inverse Matrix	Inverse Matrix (Division Last)
b_0	13.623005	13.623790	13.622989
b_1	−0.079829	−0.079848	−0.079829

As we have said, these differences are of slight consequence in this example. The third method is actually the most accurate. To see what the consequences of rounding can be, we suggest the reader make use of the inverse matrix in the second method, and round the elements in several ways—for example, rounding to 6, 5, 4, or 3 places of decimals or the same numbers of significant figures. Rounding errors provide a major share of disagreements when several people work the same regression problem using pocket calculators.

Modern computer programs vary somewhat in accuracy, but the variation is typically well below the horizon needed by the average user.

Section Summary

If we express the straight line model to be fitted to the data of our example in the form

$$\mathbf{Y} = \mathbf{X}\boldsymbol{\beta} + \boldsymbol{\epsilon}$$

as in Eq. (4.1.4), then the least squares estimates of (β_0, β_1), that is, of

$$\boldsymbol{\beta} = \begin{bmatrix} \beta_0 \\ \beta_1 \end{bmatrix},$$

are given by the normal equations

$$\mathbf{X'Xb} = \mathbf{X'Y},$$

which have the solution

$$\begin{bmatrix} b_0 \\ b_1 \end{bmatrix} = \mathbf{b} = (\mathbf{X'X})^{-1}\mathbf{X'Y}.$$

This result is of great importance and should be memorized. Note that the fitted values $\hat{\mathbf{Y}}$ are obtained by evaluating

$$\hat{\mathbf{Y}} = \mathbf{Xb}.$$

4.2. SINGULARITY: WHAT HAPPENS IN REGRESSION TO MAKE X'X SINGULAR? AN EXAMPLE

In presenting the inverse matrix in Section 4.1, we said it need not exist. We think about this a little more via a simple example. We recall that, because terms in an inverse $(\mathbf{X'X})^{-1}$ are always divided by the determinant of $\mathbf{X'X}$, written $|\mathbf{X'X}|$, the inverse "blows up" if the determinant is zero. We then say that $\mathbf{X'X}$ is *singular* and that $(\mathbf{X'X})^{-1}$ does not exist. The next question is how, in a regression situation, the determinant of $\mathbf{X'X}$ can be zero. We illustrate this for a straight line case. Figure 4.1 shows several data points, all at the same value X_* of X. (There could be n points at X_* without affecting the argument below, so we use n in what follows.)

Suppose we are asked to fit a straight line to these data. Common sense tells us that we need data at two or more X-sites to determine a straight line "properly," that is, uniquely. So our first reaction might be that the fit cannot be made; but, of course, it can. *Any* straight line $b_0 + b_1 X$ through the point $(X_*, \overline{Y})$ will minimize the sum of squares of deviations

$$S = \sum_{i=1}^{n} (Y_i - \beta_0 - \beta_1 X_i)^2$$

because, for our data, $X_i = X_* = \overline{X}$, so that

$$S = \sum_{i=1}^{n} (Y_i - \overline{Y} + \overline{Y} - \beta_0 - \beta_1 \overline{X})^2$$

$$= \sum_{i=1}^{n} (Y_i - \overline{Y})^2 + \sum_{i=1}^{n} (\overline{Y} - \beta_0 - \beta_1 \overline{X})^2 + (\text{zero cross-product})$$

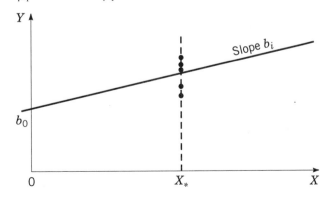

Figure 4.1.

and the second term vanishes when $\overline{Y} - \beta_0 - \beta_1 \overline{X} = 0$. Thus any straight line $Y = b_0 + b_1 X$, which passes through or contains $(\overline{X}, \overline{Y})$, is a least squares line. The solution exists but is not unique. There is an infinity of solutions.

Let us follow this case through via matrix algebra. We see that

$$\mathbf{X}' = \begin{bmatrix} 1 & 1 & \cdots & 1 \\ X_* & X_* & \cdots & X_* \end{bmatrix}, \qquad \mathbf{X} = \begin{bmatrix} 1 & X_* \\ 1 & X_* \\ \vdots & \vdots \\ 1 & X_* \end{bmatrix} \qquad (4.2.1)$$

Note that the second column of $\mathbf{X}$ (i.e., the second row of $\mathbf{X}'$) is X_* times the first column, namely, a linear combination of it. Furthermore,

$$\mathbf{X}'\mathbf{X} = \begin{bmatrix} n & nX_* \\ nX_* & nX_*^2 \end{bmatrix} \qquad (4.2.2)$$

with determinant $n(nX_*^2) - (nX_*)^2 = 0$. Thus we cannot evaluate $(\mathbf{X}'\mathbf{X})^{-1}$ nor carry out the formal computation $\mathbf{b} = (\mathbf{X}'\mathbf{X})^{-1} \mathbf{X}'\mathbf{Y}$ because $\mathbf{X}'\mathbf{X}$ is singular. Note that the dependence in the columns of $\mathbf{X}$ has also been transmitted to the columns (and rows) of $\mathbf{X}'\mathbf{X}$; the second column of $\mathbf{X}'\mathbf{X}$ is X_* times the first column (and similarly with rows).

The implication of the singularity of $\mathbf{X}'\mathbf{X}$ is that we cannot estimate β_0 and β_1 *uniquely*. The normal equations can still be written down as $\mathbf{X}'\mathbf{X}\mathbf{b} = \mathbf{X}\mathbf{Y}$, however. They just cannot be solved uniquely. For our simple example, the normal equations are

$$nb_0 + nX_*b_1 = n\overline{Y},$$
$$nX_*b_0 + nX_*^2 b_1 = nX_*\overline{Y}, \qquad (4.2.3)$$

and we see immediately that the second equation is X_* times the first, so that there is really only one distinct normal equation, not two. This single equation implies that $\overline{Y} = b_0 + b_1 X_*$ and leads to an infinity of solutions expressible as

$$(b_0, b_1) = \{b_0, (\overline{Y} - b_0)/X_*\} \qquad (4.2.4)$$

for any b_0. Essentially, we can pick any intercept value b_0 we please, and our nonunique least squares line so selected is created by joining the point $(0, b_0)$ to the point $(X_*, \overline{Y})$, giving a slope $b_1 = (\overline{Y} - b_0)/X_*$. (Alternatively, we could select a slope b_1 and determine b_0 via $b_0 = \overline{Y} - b_1 X_*$.)

The basic computational difficulty inherent in our simple example occurs in more complex forms for multiple-X regression problems but, nevertheless, the overall principle is the same.

Singularity in the General Linear Regression Context

If, for a given set of data and a given model, the $\mathbf{X}$ matrix is such that any of its columns can be expressed as a linear combination of other columns, this dependency will be transferred to $\mathbf{X}'\mathbf{X}$ and so $\mathbf{X}'\mathbf{X}$ will have a zero determinant and be singular. This means that $(\mathbf{X}'\mathbf{X})^{-1}$ cannot be computed, and the least squares procedure will not give unique estimates but many alternative solutions. This arises because the data are inadequate for fitting the model or, what is the same thing, the model is too complex for the available data. One needs either more data, or a simpler model for the available data.

Computer programs are written so as to detect these problems. MINITAB, for example, leaves out of the regression fit any variable whose X-column is a linear combination of previous columns. Because all computations have rounding errors in them, columns that are not fully dependent on others, but almost so, are sometimes omitted. Recoding variables sometimes helps to avoid that possibility.

4.3. THE ANALYSIS OF VARIANCE IN MATRIX TERMS

We recall from Section 1.3 that, in a more general form of the analysis of variance table, we wrote

$$SS(b_1|b_0) = b_1\left[\Sigma X_i Y_i - \frac{(\Sigma X_i)(\Sigma Y_i)}{n}\right] = b_1[\Sigma X_i Y_i - n\overline{XY}],$$

$$SS(b_0) = \text{Correction for mean} = \frac{(\Sigma Y_i)^2}{n} = n\overline{Y}^2.$$

Each of these sums of squares has one degree of freedom. Now, adding these together,

$$\begin{aligned} SS(b_1|b_0) + SS(b_0) &= b_1 \Sigma X_i Y_i - b_1 n\,\overline{XY} + n\overline{Y}^2 \\ &= b_1 \Sigma X_i Y_i + n\overline{Y}(\overline{Y} - b_1\overline{X}) \\ &= b_1 \Sigma X_i Y_i + b_0 \Sigma Y_i \qquad\qquad (4.3.1) \\ &= (b_0, b_1)\begin{bmatrix} \Sigma Y_i \\ \Sigma X_i Y_i \end{bmatrix} \\ &= \mathbf{b'X'Y} \end{aligned}$$

in matrix terms, with two degrees of freedom. Thus we can write the analysis of variance table in matrix form as follows:

Source	df	SS	MS
$\mathbf{b'} = (b_0, b_1)$	2	$\mathbf{b'X'Y}$	
Residual	$n - 2$	$\mathbf{Y'Y} - \mathbf{b'X'Y}$	s^2
Total (uncorrected)	n	$\mathbf{Y'Y}$	

In this way we can split the total variation $\mathbf{Y'Y}$ into two portions, one due to the straight line we have estimated, namely, $\mathbf{b'X'Y}$, and a residual that shows the remaining variation of the points about the regression line. In order to find what portion of the total variation can be attributed to the addition of the term $\beta_1 X_i$ to the simpler model $Y_i = \beta_0 + \epsilon_i$, we would just subtract the correction factor $n\overline{Y}^2$ from the sum of squares $\mathbf{b'X'Y}$ in order to obtain $SS(b_1|b_0)$ as before. The quantity $n\overline{Y}^2$ would be $SS(b_0)$ if the model $Y_i = \beta_0 + \epsilon_i$ were fitted. The remainder of $\mathbf{b'X'Y}$ thus measures the *extra sum of squares removed by* b_1 when the model $Y_i = \beta_0 + \beta_1 X_i + \epsilon_i$ is used. If an estimate of pure error from repeat points is available, it is subtracted from the residual sum of squares to provide the same breakup and the same tests as described in Section 2.1.

Example. For our steam data example we had

$$\mathbf{b} = \begin{bmatrix} 13.623790 \\ -0.079848 \end{bmatrix}, \qquad \mathbf{X'Y} = \begin{bmatrix} 235.60 \\ 11{,}821.4320 \end{bmatrix}.$$

Hence

$$SS(\mathbf{b}) = \mathbf{b'X'Y} = 2265.8472,$$

$$SS(b_0) = (\Sigma Y_i)^2/n = 2220.2944,$$

$$SS(b_1|b_0) = SS(b_1, \text{after allowance for } b_0)$$

$$= \mathbf{b'X'Y} - (\Sigma Y_i)^2/n = 45.5528.$$

We see that this result differs from the value in Table 1.6 due to rounding error in the second decimal place.

Note that $SS(b_1|b_0)$ can be written in matrix form as

$$SS(b_1|b_0) = \mathbf{b'X'Y} - \mathbf{Y'11'Y}/n$$

$$= \mathbf{Y'X(X'X)^{-1}X'Y} - \mathbf{Y'11'Y}/n$$

$$= \mathbf{Y'(H - 11'}/n)\mathbf{Y}$$

where $\mathbf{H} = \mathbf{X(X'X)^{-1}X'}$ is a useful symmetric, idempotent ($\mathbf{H}^2 = \mathbf{H}$) matrix that occurs repeatedly in regression work. In replacing $\mathbf{b'}$ by $\mathbf{Y'X(X'X)^{-1}}$ above, we have made use of the important rule that the transpose of a product of matrices is the product of the transposes *in the reverse order*. In symbols, for example,

$$(\mathbf{ABC})' = \mathbf{C'B'A'}.$$

If we apply this rule to $\mathbf{b} = \mathbf{(X'X)^{-1}X'Y}$, identification of $\mathbf{A} = \mathbf{(X'X)^{-1}}$, $\mathbf{B} = \mathbf{X'}$, and $\mathbf{C} = \mathbf{Y}$, plus the realizations that $\mathbf{(X'X)^{-1}}$ is symmetric so that it is its own transpose, and

$$(\mathbf{X'})' = \mathbf{X}$$

(i.e., if we transpose a transpose, we are back to our starting point) bring the quoted result. (The transpose rule applies to a product of any size, by the way.)

Some other matrix regression results on which the reader might wish to test his or her matrix manipulative skills are the following:

$$\mathbf{e} = \mathbf{(I - H)Y},$$

$$\mathbf{e'1} = \mathbf{1'e} = 0,$$

$$\mathbf{e'\hat{Y}} = \mathbf{\hat{Y}'e} = 0.$$

4.4. THE VARIANCES AND COVARIANCE OF b_0 AND b_1 FROM THE MATRIX CALCULATION

We recall that $V(b_1) = \sigma^2/\Sigma(X_i - \overline{X})^2$. Also,

$$V(b_0) = V(\overline{Y} - b_1\overline{X}) = \sigma^2 \left[\frac{1}{n} + \frac{\overline{X}^2}{\Sigma(X_i - \overline{X})^2} \right] = \frac{\sigma^2 \Sigma X_i^2}{n\Sigma(X_i - \overline{X})^2} \qquad (4.4.1)$$

since, as we showed earlier, $\overline{Y}$ and b_1 have zero covariance, and the X's are regarded as constants. In addition,

$$\text{cov}(b_0, b_1) = \text{cov}[(\overline{Y} - b_1\overline{X}), b_1]$$
$$= -\overline{X} V(b_1) \tag{4.4.2}$$
$$= -\overline{X}\sigma^2/\Sigma(X_i - \overline{X})^2.$$

Thus we can write the *variance–covariance* matrix of the vector **b** as follows:

$$\mathbf{V(b)} = \mathbf{V}\begin{pmatrix} b_0 \\ b_1 \end{pmatrix} = \begin{bmatrix} V(b_0) & \text{cov}(b_0, b_1) \\ \text{cov}(b_0, b_1) & V(b_1) \end{bmatrix}$$

$$= \begin{bmatrix} \dfrac{\sigma^2\Sigma X_i^2}{n\,\Sigma(X_i - \overline{X})^2} & -\dfrac{\overline{X}\sigma^2}{\Sigma(X_i - \overline{X})^2} \\ -\dfrac{\overline{X}\sigma^2}{\Sigma(X_i - \overline{X})^2} & \dfrac{\sigma^2}{\Sigma(X_i - \overline{X})^2} \end{bmatrix}. \tag{4.4.3}$$

Now if every element of a matrix has a common factor, we can remove it and set it outside the matrix, so that we can remove σ^2. The matrix that remains is seen to be $(\mathbf{X'X})^{-1}$ from Eq. (4.1.15). Thus

$$\mathbf{V(b)} = (\mathbf{X'X})^{-1}\sigma^2. \tag{4.4.4}$$

This is an important result and should be remembered. When σ^2 is unknown we use, instead, s^2, the estimate of σ^2 obtained from the analysis of variance table, if there is no lack of fit, or s_e^2, the pure error mean square if lack of fit is shown. This provides us with the *estimated variance–covariance matrix of* **b**. The standard errors of the regression coefficients are the square roots of the diagonal entries.

Correlation Between b_0 and b_1

The correlation between b_0 and b_1 can be obtained directly from the $(\mathbf{X'X})^{-1}\sigma^2$ matrix, or the $(\mathbf{X'X})^{-1}s^2$ matrix or even just the $(\mathbf{X'X})^{-1}$ matrix because any common factor cancels anyway. We obtain

$$\text{Correlation}(b_0, b_1) = \frac{\text{cov}(b_0, b_1)}{\{V(b_0)V(b_1)\}^{1/2}} \tag{4.4.5}$$

$$= \frac{-\overline{X}\sigma^2/S_{XX}}{\left\{\dfrac{\sigma^2\Sigma X_i^2}{nS_{XX}}\dfrac{\sigma^2}{S_{XX}}\right\}^{1/2}} \tag{4.4.6}$$

$$= -\overline{X}(n/\Sigma X_i^2)^{1/2}. \tag{4.4.7}$$

For the steam data, $\overline{X} = 52.6$, $n = 25$, $\Sigma X_i^2 = 76{,}323.42$, and the correlation is -0.952. Or we can use directly the numbers below (4.1.16) to give

$$\text{Correlation}(b_0, b_1) = -0.0073535/\{0.4267941)(0.0001398)\}^{1/2} = -0.952. \tag{4.4.8}$$

This is a relatively high value. This high negative correlation shows up in the relative position of the joint confidence region for (β_0, β_1) compared with the rectangular formed by the individual (marginal) confidence intervals. The negative sign implies the upper-left to lower-right slant of the larger axis of the ellipse while the high numerical value of 0.952 implies that it will run essentially from corner to corner of the rectangle. An accurate diagram is shown in Figure 5.3.

4.5. VARIANCE OF Ŷ USING THE MATRIX DEVELOPMENT

Let X_0 be a selected value of X. The predicted mean value of Y for this value of X is

$$\hat{Y}_0 = b_0 + b_1 X_0.$$

Let us define the vector $\mathbf{X}_0$ as

$$\mathbf{X}_0' = (1, X_0).$$

We can then write

$$\hat{Y}_0 = (1, X_0) \begin{bmatrix} b_0 \\ b_1 \end{bmatrix} = \mathbf{X}_0'\mathbf{b} = \mathbf{b}'\mathbf{X}_0.$$

Since $\hat{Y}_0$ is a linear combination of the random variables b_0 and b_1, it follows that

$$V(\hat{Y}_0) = V(b_0) + 2X_0 \operatorname{cov}(b_0, b_1) + X_0^2 V(b_1).$$

As can be verified by working out the indicated matrix and vector products, the above quantity can be expressed in the alternative form

$$V(\hat{Y}_0) = [1, X_0] \begin{bmatrix} V(b_0) & \operatorname{cov}(b_0, b_1) \\ \operatorname{cov}(b_0, b_1) & V(b_1) \end{bmatrix} \begin{bmatrix} 1 \\ X_0 \end{bmatrix}$$

$$= \mathbf{X}_0'(\mathbf{X}'\mathbf{X})^{-1}\sigma^2\mathbf{X}_0$$

$$= \mathbf{X}_0'(\mathbf{X}'\mathbf{X})^{-1}\mathbf{X}_0\sigma^2.$$

Although now given in a different form, this is identical in value to Eq. (3.1.3). This important matrix result should be remembered. With suitable redefinition of $\mathbf{X}_0$ and $\mathbf{X}$, it is applicable to the general linear regression situation. A estimated variance is obtained when σ^2 is replaced by an estimate s^2.

4.6. SUMMARY OF MATRIX APPROACH TO FITTING A STRAIGHT LINE (NONSINGULAR CASE)

1. Set down the model in the form $\mathbf{Y} = \mathbf{X}\boldsymbol{\beta} + \boldsymbol{\epsilon}$.
2. Find $\mathbf{b} = (\mathbf{X}'\mathbf{X})^{-1}\mathbf{X}'\mathbf{Y}$ to obtain the least squares estimate $\mathbf{b}$ of $\boldsymbol{\beta}$ provided by the data. (This solves the normal equations $\mathbf{X}'\mathbf{X}\mathbf{b} = \mathbf{X}'\mathbf{Y}$.)
3. Construct $\mathbf{b}'\mathbf{X}'\mathbf{Y}$ the sum of squares due to coefficients and hence obtain the basic analysis of variance as follows:

Source	df	SS	MS
Regression	2	$\mathbf{b}'\mathbf{X}'\mathbf{Y}$	
Residual	$n - 2$	$\mathbf{Y}'\mathbf{Y} - \mathbf{b}'\mathbf{X}'\mathbf{Y}$	s^2 (estimates σ^2 if the model is correct)
Total	n	$\mathbf{Y}'\mathbf{Y}$	

Additional subdivision of the sum of squares is achieved by finding $SS(b_1|b_0)$, the extra sum of squares due to b_1, and introducing pure error. The more detailed analysis of variance table will take the following form:

Source		df	SS	MS	
SS(**b**)	$SS(b_0)$	1	$n\bar{Y}^2$		
	$SS(b_1\|b_0)$	1	$\mathbf{b'X'Y} - n\bar{Y}^2$		
Residual	Lack of fit	$n - 2 - n_e$	$\mathbf{Y'Y} - \mathbf{b'X'Y} - SS(pe)$	MS_L	s^2
	Pure error	n_e	$SS(pe)$	s_e^2	
Total		n	$\mathbf{Y'Y}$		

The second table is often rewritten with the *corrected* total sum of squares at the bottom, omitting the sum of squares due to the mean $n\bar{Y}^2$. [Incidentally, as we noted previously, we can write $n\bar{Y}^2$ in matrix form as $\mathbf{Y'11'Y}/n$ if we wish, although this is not usually done. This computation is, in fact, less subject to round-off error if performed as $(\Sigma Y_i)^2/n$.] The abbreviated table takes the following form:

Source		df	SS	MS	
$SS(b_1\|b_0)$		1	$\mathbf{b'X'Y} - n\bar{Y}^2$		
Residual	Lack of fit	$n - 2 - n_e$	$\mathbf{Y'Y} - \mathbf{b'X'Y} - SS(pe)$	MS_L	s^2
	Pure error	n_e	$SS(pe)$	s_e^2	
Total, corrected		$n - 1$	$\mathbf{Y'Y} - n\bar{Y}^2$		

The tests for lack of fit and (if there is no lack of fit) for $H_0 : \beta_1 = 0$ versus $H_1 : \beta_1 \neq 0$ are performed as described in Chapters 1 and 2. An additional measure of the regression is provided by the ratio

$$R^2 = \frac{(\mathbf{b'X'Y} - n\bar{Y}^2)}{(\mathbf{Y'Y} - n\bar{Y}^2)}$$

4. If no lack of fit is shown, so that s^2 can be used as an estimate of σ^2, $(\mathbf{X'X})^{-1}s^2$ will provide estimates of $V(b_0)$, $V(b_1)$, and $\text{cov}(b_0, b_1)$ and enable individual coefficients to be tested or other calculations made as in Chapter 1.

5. The following quantities can be found:

$$\text{The vector of fitted values:} \quad \hat{\mathbf{Y}} = \mathbf{Xb};$$

$$\text{A prediction of } Y \text{ at } X_0 : \quad \hat{Y}_0 = \mathbf{X}_0'\mathbf{b} = \mathbf{b'X}_0;$$

$$\text{with variance:} \quad V(\hat{Y}_0) = \mathbf{X}_0'(\mathbf{X'X})^{-1}\mathbf{X}_0\sigma^2;$$

where $\mathbf{X}_0' = (1, X_0)$.

4.7. THE GENERAL REGRESSION SITUATION

We have seen how the problem of fitting a straight line by least squares can be handled through the use of matrices. This approach is important for the following reason. If we wish to fit *any* model linear in parameters $\beta_0, \beta_1, \beta_2, \ldots$, by least squares, the calculations necessary are of exactly the same form (in matrix terms) as those for the straight line involving only two parameters β_0 and β_1. Only the sizes of the matrices and the numbers of certain degrees of freedom change. The mechanics of calculation, however, increase sharply with the number of parameters. Thus while the formulas are easy to remember, the use of a computer is essential. Even when few parameters are involved, or when the data arise from a designed experiment that provides an **X'X** matrix of simple or patterned form, computer evaluation is preferable.

We deal with the general regression situation in Chapter 5.

EXERCISES FOR CHAPTER 4

A. In this question we define

$$A = \begin{bmatrix} 4 & 1 \\ 3 & -2 \\ 1 & 7 \end{bmatrix}, \quad B = \begin{bmatrix} 2 & 1 \\ 1 & 2 \end{bmatrix}, \quad C = \begin{bmatrix} 3 & -1 & 2 \\ -2 & 3 & 1 \\ 1 & 4 & 1 \end{bmatrix}.$$

Are the statements below true or false?

1. $B + C = \begin{bmatrix} 5 & 0 & 2 \\ -1 & 5 & 1 \\ 1 & 4 & 1 \end{bmatrix}$.

2. $AC = \begin{bmatrix} 7 & 9 & 12 \\ 14 & 21 & 7 \end{bmatrix}$.

3. $AB = \begin{bmatrix} 9 & 6 \\ 4 & -1 \\ 9 & 15 \end{bmatrix}$.

4. $B^{-1} = \dfrac{1}{3}\begin{bmatrix} 2 & -1 \\ -1 & 2 \end{bmatrix}$.

5. $A^{-1} = \dfrac{1}{11}\begin{bmatrix} 2 & 1 & 0 \\ 3 & -4 & 0 \end{bmatrix}$.

6. $(A'A)^{-1}A'ABB^{-1} = \begin{bmatrix} 1 & 0 & 0 \\ 0 & 1 & 0 \\ 0 & 0 & 1 \end{bmatrix}$.

B. We define

$$A = \begin{bmatrix} 4 & 0 & 3 \\ 0 & 4 & 0 \\ 3 & 0 & 2 \end{bmatrix}, \quad B = \begin{bmatrix} -1 & 1 \\ 2 & 3 \\ 3 & 2 \end{bmatrix}, \quad C = \begin{bmatrix} 4 & 3 \\ 3 & 2 \end{bmatrix}.$$

Calculate the matrices below, or say that it is impossible to do so, if it is impossible.
   1. $B + C$.
   2. BB'.
   3. $A + B'B$.
   4. BC.

5. AA^{-1}BC.
6. CB$'$.
7. CAB.

8. BC^{-1}, where $\mathbf{C}^{-1} = \begin{bmatrix} -2 & 3 \\ 3 & -4 \end{bmatrix}$.

9. A^{-1}.
10. A$'$A(A$'$)$^{-1}$A^{-1}.

C. We define

$$\mathbf{A} = \begin{bmatrix} 1 & 1 & 1 \\ 1 & 2 & 3 \end{bmatrix}, \quad \mathbf{b} = \begin{bmatrix} -1 \\ 1 \end{bmatrix}, \quad \mathbf{C} = \begin{bmatrix} 1 & 0 \\ 2 & 3 \end{bmatrix}, \quad \mathbf{D} = \begin{bmatrix} 1 & 1 & 1 \\ 1 & 2 & 0 \\ -1 & 1 & 0 \end{bmatrix}.$$

Are the results below true or false? If false, say (briefly) why.
1. Ab $= [0, 1, 2]$.

2. A$'$C $= \begin{bmatrix} 3 & 3 \\ 5 & 6 \\ 7 & 9 \end{bmatrix}$.

3. AD $= \begin{bmatrix} 1 & 4 & 1 \\ 6 & 8 & 1 \end{bmatrix}$.

4. C^{-1} $= \begin{bmatrix} 1 & 0 \\ -\frac{2}{3} & \frac{1}{3} \end{bmatrix}$.

5. bC $= \begin{bmatrix} 1 \\ 3 \end{bmatrix}$.

6. A $-$ **D** $= \begin{bmatrix} 0 & 0 & 0 \\ 0 & 0 & 3 \\ -1 & 1 & 0 \end{bmatrix}$.

7. C^{-1}CCC^{-1}C^{-1}CCC^{-1}C $=$ **C**.

8. D^{-1} $= \dfrac{1}{3}\begin{bmatrix} 0 & 1 & -2 \\ 0 & 1 & 1 \\ 3 & -2 & 1 \end{bmatrix}$.

9. b$'$Cb $= 2$.

D. Using the matrix development throughout, fit to the data below, the model $Y = \beta_0 + \beta_1 X + \epsilon$, and so obtain b_0 and b_1. Plot the data and the fitted line. Find the fitted values and the residuals correct to one decimal place. Evaluate the analysis of variance table, and test for lack of fit. Find the variance–covariance matrix of the estimated parameters, and the matrix expression for $V(\hat{Y})$. Hence find $V(\hat{Y})$ when $X = 65$, and construct a 95% confidence interval for $E(Y|X = 65)$.

X:	30	40	50	80	30	40	60	70	70	70	30	80	70	70
Y:	13	17	20	29	12	15	22	25	23	27	15	27	24	26

E. Using the matrix development, fit a straight line $Y = \beta_0 + \beta_1 X + \epsilon$ to the data below, and provide a complete analysis (i.e., all relevant details of Chapter 4).

X	Y
1	4.2
1	3.8
2	3.0
3	2.3
4	1.8
4	2.0
4	2.2
5	2.0
6	2.5
6	2.7

F. Using the data below, go through the following steps:
1. Fit a straight line model $Y = \beta_0 + \beta_1 X + \epsilon$, $\epsilon \sim N(0, \mathbf{I}\sigma^2)$.
2. Plot the data and your fitted line.
3. Obtain the basic ANOVA table.
4. Add additional details on the ANOVA table.
5. Test for lack of fit.
6. Carry out the usual F-test for overall regression. Is this test valid?
7. Find, via the usual calculations, a 95% confidence interval for the true mean value of Y at $X_0 = \sqrt{122} + 6$. Is it valid?
8. Evaluate the fitted values and residuals. Plot each e_i versus its corresponding $\hat{Y}_i$.
9. Write down the form of the variance–covariance matrix of the b's in terms of σ^2.
10. Evaluate R^2.
11. Evaluate r_{XY}^2. What is its relationship to R^2, for this model?
12. What are your overall conclusions?

	X	Y	XY
	0	−2	0
	2	0	0
	2	2	4
	5	1	5
	5	3	15
	9	1	9
	9	0	0
	9	0	0
	9	1	9
	10	−1	−10
Sum	60	5	32
Sum of Squares	482	21	528

Note: When using a calculator, it is best to work with integers and simple fractions as long as you can. Convert to decimals only at the last possible moment.

CHAPTER 5

The General Regression Situation

In presenting the general regression situation we state many results without proving them. For proofs, the reader could consult, for example, Plackett (1960), Seber (1977), or Rao (1973).

5.1. GENERAL LINEAR REGRESSION

Suppose we have a model under consideration, which can be written in the form

$$\mathbf{Y} = \mathbf{X}\boldsymbol{\beta} + \boldsymbol{\epsilon}, \tag{5.1.1}$$

where

$\mathbf{Y}$ is an $(n \times 1)$ vector of observations,
$\mathbf{X}$ is an $(n \times p)$ matrix of known form,
$\boldsymbol{\beta}$ is a $(p \times 1)$ vector of parameters,
$\boldsymbol{\epsilon}$ is an $(n \times 1)$ vector of errors,

and where $E(\boldsymbol{\epsilon}) = \mathbf{0}$, $V(\boldsymbol{\epsilon}) = \mathbf{I}\sigma^2$, so the elements of $\boldsymbol{\epsilon}$ are uncorrelated.
Since $E(\boldsymbol{\epsilon}) = \mathbf{0}$, an alternative way of writing the model is

$$E(\mathbf{Y}) = \mathbf{X}\boldsymbol{\beta}. \tag{5.1.1a}$$

The error sum of squares is then

$$\begin{aligned}
\boldsymbol{\epsilon}'\boldsymbol{\epsilon} &= (\mathbf{Y} - \mathbf{X}\boldsymbol{\beta})'(\mathbf{Y} - \mathbf{X}\boldsymbol{\beta}) \\
&= \mathbf{Y}'\mathbf{Y} - \boldsymbol{\beta}'\mathbf{X}'\mathbf{Y} - \mathbf{Y}'\mathbf{X}\boldsymbol{\beta} + \boldsymbol{\beta}'\mathbf{X}'\mathbf{X}\boldsymbol{\beta} \\
&= \mathbf{Y}'\mathbf{Y} - 2\boldsymbol{\beta}'\mathbf{X}'\mathbf{Y} + \boldsymbol{\beta}'\mathbf{X}'\mathbf{X}\boldsymbol{\beta}.
\end{aligned} \tag{5.1.2}$$

[This follows due to the fact that $\boldsymbol{\beta}'\mathbf{X}'\mathbf{Y}$ is a 1×1 matrix, or a scalar, whose transpose $(\boldsymbol{\beta}'\mathbf{X}'\mathbf{Y})' = \mathbf{Y}'\mathbf{X}\boldsymbol{\beta}$ must have the same value.]

The least squares estimate of $\boldsymbol{\beta}$ is the value $\mathbf{b}$, which, when substituted in Eq. (5.1.2), minimizes $\boldsymbol{\epsilon}'\boldsymbol{\epsilon}$. It can be determined by differentiating Eq. (5.1.2) with respect to $\boldsymbol{\beta}$ and setting the resultant matrix equation equal to zero, at the same time replacing $\boldsymbol{\beta}$ by $\mathbf{b}$. (Differentiating $\boldsymbol{\epsilon}'\boldsymbol{\epsilon}$ with respect to a vector quantity $\boldsymbol{\beta}$ is equivalent to differentiating $\boldsymbol{\epsilon}'\boldsymbol{\epsilon}$ separately with respect to each element of $\boldsymbol{\beta}$ in order, writing down the resulting derivatives one below the other, and rearranging the whole into matrix form.) This provides the *normal equations*

$$(\mathbf{X}'\mathbf{X})\mathbf{b} = \mathbf{X}'\mathbf{Y}. \tag{5.1.3}$$

Two main cases arise: either Eq. (5.1.3) consists of p independent equations in p unknowns, or some equations depend on others so that there are fewer than p independent equations in the p unknowns (the p unknowns are the elements of **b**). If some of the normal equations depend on others, $\mathbf{X'X}$ is singular, so that $(\mathbf{X'X})^{-1}$ does not exist. Then either the model should be expressed in terms of fewer parameters or else additional restrictions on the parameters must be given or assumed. Some examples of this situation are given in Chapter 23. If the p normal equations are independent, $\mathbf{X'X}$ is nonsingular, and its inverse exists. In this case the solution of the normal equations can be written

$$\mathbf{b} = (\mathbf{X'X})^{-1}\mathbf{X'Y}. \tag{5.1.4}$$

This solution **b** has the following properties:

1. It is an estimate of $\boldsymbol{\beta}$ that minimizes the error sum of squares $\boldsymbol{\epsilon'\epsilon}$, *irrespective* of any distribution properties of the errors.
Note: An assumption that the errors $\boldsymbol{\epsilon}$ are normally distributed is *not* required in order to obtain the estimate **b** but it *is* required later in order to make tests that depend on the assumption of normality, such as t- or F-tests, or for obtaining confidence intervals based on the t- and F-distributions.

2. The elements of **b** are linear functions of the observations $Y_1, Y_2, \ldots, Y_n$ and provide unbiased estimates of the elements of $\boldsymbol{\beta}$ which have the minimum variance (of *any* linear functions of the Y's that provide unbiased estimates), irrespective of distribution properties of the errors.
Note: Suppose we have an expression $T = l_1 Y_1 + l_2 Y_2 + \cdots + l_n Y_n$, which is a linear function of observations $Y_1, Y_2, \ldots, Y_n$, and which we use as an estimate of a parameter θ. Then T is a random variable whose probability distribution will depend on the distribution from which the Y's arise. If we repeatedly take samples of Y's and evaluate the corresponding T's, we shall generate the distribution of T empirically. Whether we do this or not, the distribution of T will have some definite mean value that we can write as $E(T)$ and a variance that we can write as $V(T)$. If it happens that the mean of the distribution of T is equal to the parameter θ we are estimating by T—that is, if $E(T) = \theta$—then we say that T is an unbiased estimator of θ. The word *estimator* is normally used when referring to the theoretical expression for T in terms of a sample of Y's. A specific numerical value of T would be called an unbiased *estimate* of θ. This distinction, though correct, is not always maintained in statistical writings. If we have all possible linear functions $T_1, T_2, \ldots$, say, of n observations $Y_1, Y_2, \ldots, Y_n$, and if the T's satisfy

$$\theta = E(T_1) = E(T_2) \cdots,$$

that is, they are all unbiased estimators of θ, then the one with the smallest value of $V(T_j), j = 1, 2, \ldots$, is the *minimum variance unbiased estimator* of θ. [The result (2) is *Gauss's Theorem* or the *Gauss–Markov* Theorem. See Jaske (1994).]

A Justification for Using Least Squares

3. If the errors are independent and $\epsilon_i \sim N(0, \sigma^2)$, then **b** is the maximum likelihood estimate of $\boldsymbol{\beta}$. (In vector terms we can write $\boldsymbol{\epsilon} \sim N(\mathbf{0}, \mathbf{I}\sigma^2)$, meaning that $\boldsymbol{\epsilon}$ follows an n-dimensional multivariate normal distribution with $E(\boldsymbol{\epsilon}) = \mathbf{0}$ (where $\mathbf{0}$ denotes a vector consisting entirely of zeros and of the same length as $\boldsymbol{\epsilon}$) and $\mathbf{V}(\boldsymbol{\epsilon}) = \mathbf{I}\sigma^2$; that is, $\boldsymbol{\epsilon}$ has a variance–covariance matrix whose diagonal elements, $V(\epsilon_i), i = 1, 2, \ldots,$

n, are all σ^2 and whose off-diagonal elements, covariance (ϵ_i, ϵ_j), $i \neq j = 1, \ldots, n$, are all zero. The likelihood function for the sample $Y_1, Y_2, \ldots, Y_n$ is defined in this case as the product

$$\prod_{i=1}^{n} \frac{1}{\sigma(2\pi)^{1/2}} e^{-\epsilon_i^2/(2\sigma^2)} = \frac{1}{\sigma^n(2\pi)^{n/2}} e^{-\epsilon'\epsilon/2\sigma^2}. \tag{5.1.5}$$

Thus for a fixed value of σ, maximizing the likelihood function is equivalent to minimizing the quantity $\epsilon'\epsilon$. Note that this fact can be used to provide a justification for the least squares procedure (i.e., for minimizing the sum of *squares* of errors), because in many physical situations the assumption that errors are normally distributed is quite sensible. We shall, in any case, find out if this assumption appears to be violated by examining the residuals from the regression analysis.

If any definite *a priori* knowledge is available about the error distribution, perhaps from theoretical considerations or from sound prior knowledge of the process under study, the maximum likelihood argument could be used to obtain estimates based on a criterion other than least squares. For example, suppose the errors ϵ_i, $i = 1, 2, \ldots$, n, were independent and followed the double exponential distribution:

$$f(\epsilon_i) = (2\sigma)^{-1} e^{-|\epsilon_i|/\sigma} \qquad (-\infty \leq \epsilon_i \leq \infty) \tag{5.1.6}$$

rather than the normal distribution:

$$f(\epsilon_i) = \frac{1}{\sigma(2\pi)^{1/2}} e^{-\epsilon_i^2/2\sigma^2} \tag{5.1.7}$$

as is usually assumed. The double exponential frequency function has a pointed peak of height $1/2\sigma$ at $\epsilon_i = 0$, and tails off to zero as ϵ_i goes to both plus and minus infinity. Then application of the maximum likelihood principle for estimating β, assuming σ fixed, would involve minimization of

$$\sum_{i=1}^{n} |\epsilon_i|,$$

the sum of absolute errors, and not the minimization of

$$\sum_{i=1}^{n} \epsilon_i^2,$$

the sum of *squares* of errors.

5.2. LEAST SQUARES PROPERTIES

Assuming that $E(\epsilon) = 0$, $V(\epsilon) = I\sigma^2$, we can proceed with the following steps whether the errors are normally distributed or not.

1. The fitted values are obtained from $\hat{\mathbf{Y}} = \mathbf{Xb}$.
2. The vector of residuals is given by $\mathbf{e} = \mathbf{Y} - \hat{\mathbf{Y}}$.
    It is true that $\sum_{i=1}^{n} e_i \hat{Y}_i = 0$, whatever the linear model. This can be seen by multiplying the jth normal equation by the jth b and adding the results. If there is a β_0 term in the model, it is also true that $\sum_{i=1}^{n} e_i = 0$. (The e_i and $\hat{Y}_i$, $i = 1, 2, \ldots, n$, are the ith elements of the vectors $\mathbf{e}$ and $\hat{\mathbf{Y}}$, respectively. Thus $\mathbf{e}'\hat{\mathbf{Y}} = 0 = \hat{\mathbf{Y}}'\mathbf{e}$ always, and $\mathbf{e}'\mathbf{1} = 0 = \mathbf{1}'\mathbf{e}$ when the model contains β_0.)

3. $V(\mathbf{b}) = (\mathbf{X}'\mathbf{X})^{-1}\sigma^2$ provides the variances (diagonal terms) and covariances (off-diagonal terms) of the estimates. (An estimate of σ^2 is obtained as described below.)

4. Suppose $\mathbf{X}_0'$ is a specified $1 \times p$ vector whose elements are of the same form as a row of $\mathbf{X}$ so that $\hat{Y}_0 = \mathbf{X}_0'\mathbf{b} = \mathbf{b}'\mathbf{X}_0$ is *the fitted value at a specified location defined by* $\mathbf{X}_0$. For example, if the model were $Y = \beta_0 + \beta_1 X + \beta_{11} X^2 + \epsilon$, then $\mathbf{X}_0' = (1, X_0, X_0^2)$ for a given value X_0. Then $\hat{Y}_0$ is the value *predicted at* $\mathbf{X}_0$ *by the regression equation* and has variance

$$V(\hat{Y}_0) = \mathbf{X}_0'V(\mathbf{b})\mathbf{X}_0 = \mathbf{X}_0'(\mathbf{X}'\mathbf{X})^{-1}\mathbf{X}_0\sigma^2. \tag{5.2.1}$$

5. A basic analysis of variance table can be constructed as follows:

Source	df	SS	MS
Regression	p	$\mathbf{b}'\mathbf{X}'\mathbf{Y}$	$MS_{Regression}$
Residual	$n - p$	$\mathbf{Y}'\mathbf{Y} - \mathbf{b}'\mathbf{X}'\mathbf{Y}$	$MS_{Residual}$
Total	n	$\mathbf{Y}'\mathbf{Y}$	

A further subdivision of the parts of this table can be carried out as follows.

5a. If a β_0 term is in the model we can subdivide the regression sum of squares into

$$SS(b_0) = \frac{(\Sigma Y_i)^2}{n} = n\bar{Y}^2 \tag{5.2.2}$$

$$SS(\text{Regression}|b_0) = SS(\text{Reg}|b_0) = \mathbf{b}'\mathbf{X}'\mathbf{Y} - \frac{(\Sigma Y_i)^2}{n}. \tag{5.2.3}$$

These sums of squares are based on 1 and $p - 1$ degrees of freedom, respectively.

5b. If repeat observations are available we can split the residual SS into SS(pure error) with n_e degrees of freedom, which estimates $n_e\sigma^2$ and SS(lack of fit) with $(n - p - n_e)$ degrees of freedom.

"Repeats" now must be repeats in *all* coordinates $X_1, X_2, \ldots, X_k$ of the predictor variables (though approximate use of "very close" points is sometimes seen in practice). This provides an analysis of variance table as follows. (Note: *lof* = lack of fit, *pe* = pure error).

Source	df	SS	MS	
b_0	1	$SS(b_0)$		
Regression$\|b_0$	$p - 1$	$SS(\text{Reg}\|b_0)$	$MS(\text{Reg}\|b_0)$ } $MS_{Regression}$	
Lack of fit	$n - p - n_e$	$SS(lof)$	$MS(lof)$ } $MS_{Residual}$	
Pure error	n_e	$SS(pe)$	$MS(pe)$	
Total	n	$\mathbf{Y}'\mathbf{Y}$		

The R^2 Statistic

The ratio

$$R^2 = \frac{SS(\text{Reg}|b_0)}{\mathbf{Y}'\mathbf{Y} - SS(b_0)} = \frac{\Sigma(\hat{Y}_i - \bar{Y})^2}{\Sigma(Y_i - \bar{Y})^2} \tag{5.2.4}$$

is an extension of the quantity defined for the straight line regression and is the square of the *multiple correlation coefficient*. Another name for R^2 is the *coefficient of multiple determination*.

R^2 is the square of the correlation between $\mathbf{Y}$ and $\hat{\mathbf{Y}}$ and $0 \le R^2 \le 1$. If pure error exists, it is impossible for R^2 to actually attain 1; see the remarks in Section 2.1. A perfect fit to the data for which $\hat{Y}_i = Y_i$, an unlikely event in practice, would give $R^2 = 1$.

If $\hat{Y}_i = \overline{Y}$, that is, if $b_1 = b_2 = \cdots b_{p-1} = 0$ (or if a model $Y = \beta_0 + \epsilon$ alone has been fitted), then $R^2 = 0$. Thus R^2 is a measure of the usefulness of the terms, other than β_0, in the model.

R^2 Can Be Deceptive

It is important to realize that, if there is no pure error, R^2 can be made unity simply by employing n properly selected coefficients in the model, including β_0, since a model can then be chosen that fits the data exactly. (For example, if we have an observation of Y at four different values of X, a cubic polynomial

$$Y = \beta_0 + \beta_1 X + \beta_2 X^2 + \beta_3 X^3$$

passes exactly through all four points.) Since R^2 is often used as a convenient measure of the success of the regression equation in explaining the variation in the data, we must be sure that an improvement in R^2 due to adding a new term to the model has some real significance and is not due to the fact that the number of parameters in the model is getting close to saturation point—that is, the number of distinct X-sites. This is an *especial* danger when there are *repeat* observations.

For example, if we have 100 observations that occur in five groups each of 20 repeats, we have effectively five pieces of information, represented by five mean values, and 95 degrees of freedom for pure error, 19 at each repeat point. Thus a five-parameter model will provide a perfect fit to the five means and may give a very large value of R^2, especially if the experimental error is small compared with the spread of the five means. In this case the fact that 100 observations can be well predicted by a model with only five parameters is not surprising since there are really only five distinct data sites and not 100 as it first seemed. When there are no exact repeats, but the points in the X-space (at which observations Y are available) *are* close together, this type of situation can occur and yet be well concealed within the data. Plots of the data, and the residuals, will usually reveal such "clusters" of points.

Adjusted R^2 Statistic

Suppose p is the total number of parameters in a fitted model (including β_0) and RSS_{n-p} is the corresponding residual sum of squares. We have defined the R^2 statistic, a measure of the amount of variation about the mean explained by the fitted equation, as

$$R^2 = \frac{\mathbf{b}'\mathbf{X}'\mathbf{Y} - n\overline{Y}^2}{\mathbf{Y}'\mathbf{Y} - n\overline{Y}^2} = 1 - \frac{\text{RSS}_{n-p}}{\text{CTSS}} \tag{5.2.5}$$

where CTSS denotes the corrected total sum of squares $\mathbf{Y}'\mathbf{Y} - n\overline{Y}^2$, and where n is the total number of observations.

A related statistic, which is preferred by some workers, is the *adjusted R^2* defined, in our context, as

$$R_a^2 = 1 - \frac{(\text{RSS}_{n-p})/(n-p)}{(\text{CTSS})/(n-1)} = 1 - (1 - R^2)\left(\frac{n-1}{n-p}\right). \tag{5.2.6}$$

An "adjustment" has been made for the corresponding degrees of freedom of the two quantities RSS_{n-p} and CTSS, the idea being that the statistic R_a^2 can be used to compare equations fitted not only to a specific set of data but also to two or more entirely different sets of data. (The value of this statistic for the latter purpose is, in our opinion, not high; R_a^2 might be useful as an initial gross indicator, but this is all.)

As pointed out by Kennard (1971), adjusted R^2 is closely related to the C_p statistic, a statistic used in one type of regression selection procedure. We discuss the use of C_p in Chapter 15. Apart from this, we do not use adjusted R^2 in this book.

The equivalence of the numerators in Eqs. (5.2.4) and (5.2.5) may be established as follows:

$$\Sigma(\hat{Y}_i - \overline{Y})^2 = \Sigma \hat{Y}_i^2 - (\Sigma Y_i)^2/n$$

and

$$\Sigma \hat{Y}_i^2 = \hat{Y}'\hat{Y} = (\mathbf{Xb})'(\mathbf{Xb})$$

$$= \mathbf{b}'\mathbf{X}'\mathbf{Xb}$$

$$= \mathbf{b}'\mathbf{X}'\mathbf{Y}$$

because $\mathbf{X}'\mathbf{Xb} = \mathbf{X}'\mathbf{Y}$ from the normal equations.

5.3. LEAST SQUARES PROPERTIES WHEN $\epsilon \sim N(0, \mathbf{I}\sigma^2)$

The analysis of variance breakup is an algebraic equality (or a geometric one, depending on one's viewpoint—see Chapter 20) only and does not depend on distributive properties of the errors. However, if we assume additionally that $\epsilon_i \sim N(0, \sigma^2)$ and that the ϵ_i are independent—that is, $\epsilon \sim N(0, \mathbf{I}\sigma^2)$—we can do the following.

1. Test lack of fit by treating the ratio

$$\left[\frac{\text{SS(lack of fit)}/(n - p - n_e)}{\text{SS(pure error)}/n_e}\right] \tag{5.3.1}$$

as an $F[(n - p - n_e), n_e]$ variate and by comparing its value with $F[(n - p - n_e), n_e, 1 - \alpha]$. If there is no lack of fit, $\text{SS(residual)}/(n - p) = \text{MS}_E$, usually called s^2, is an unbiased estimate of σ^2. If lack of fit cannot be tested, use of s^2 as an estimate of σ^2 *implies* an assumption that the model is correct. (If it is not, s^2 will usually be too large since it is a random variable with a mean *greater* than σ^2. Note carefully, however, that due to sampling fluctuation—since it *is* a random variable—it could also be too small.)

2. Test the overall regression equation (more specifically, test $H_0: \beta_1 = \beta_2 = \cdots = \beta_{p-1} = 0$ against H_1: not all $\beta_i = 0$) by treating the mean square ratio

$$\frac{[\text{SS}(\text{Reg}|b_0)/(p-1)]}{s^2} \tag{5.3.2}$$

as an $F(p - 1, \nu)$ variate, where $\nu = n - p$.

Just Significant Regressions May Not Predict Well

Suppose we decide on a specified risk level α. The fact that the observed mean square ratio exceeds $F(p - 1, \nu, 1 - \alpha)$ means that a "statistically significant" regression has been obtained; in other words, the proportion of the variation in the data which has been accounted for by the fitted equation is deemed greater than would be expected by chance in similar sets of data with the same values of n and $\mathbf{X}$. This does not necessarily mean that the equation is useful for predictive purposes. Unless the range of values predicted by the fitted equation is considerably greater than the size of the random error, prediction will often be of no value even though a "significant" F-value has been obtained, since the equation will be "fitted to the errors" only. For more on this, see Section 11.1.

The Distribution of R^2

We see that

$$R^2 = \frac{\mathrm{SS(Regression}|b_0)}{\sum_{i=1}^{n}(Y_i - \overline{Y})^2}$$

$$= \frac{\mathrm{SS(Regression}|b_0)}{\mathrm{SS(Regression}|b_0) + \text{Residual SS}} \tag{5.3.3}$$

$$= \frac{\nu_1 F}{\nu_1 F + \nu_2},$$

where the quantity

$$F = \frac{\mathrm{SS(Regression}|b_0)/\nu_1}{\text{Residual SS}/\nu_2}$$

is our usual F-statistic for testing overall regression given b_0, that is, for testing the null hypothesis H_0: that all the β's (excluding β_0) are zero against the alternative hypothesis H_1: that at least one of the β's (excluding β_0) is not zero. The value of β_0 is irrelevant to the test. To correspond to Eq. (5.3.2) we can set $\nu_1 = p - 1$, $\nu_2 = n - p$. Under H_0, F is distributed as an $F(\nu_1, \nu_2)$ variable. A statistical theorem tells us that R^2 follows a $\beta(\frac{1}{2}\nu_1, \frac{1}{2}\nu_2)$ distribution, called the beta-distribution and (here) degrees of freedom $\frac{1}{2}\nu_1$ and $\frac{1}{2}\nu_2$. We shall not discuss the beta-distribution at all but, clearly, if we had appropriate tables we could test H_0 against H_1 using R^2. The result would be *exactly* equivalent to that of our standard F-test, the significance point for R^2 being obtained from Eq. (5.3.3) with $F(p - 1, n - p, 1 - \alpha)$ substituted for F. For this reason, and because tables of the beta-distribution are not as universally available as those of F, a test on R^2 is rarely done.

Properties, Continued

3. If we use an estimate s_ν^2 for σ^2, $100(1 - \alpha)\%$ confidence limits for the true mean value of Y at $\mathbf{X}_0$ are obtained from

$$\hat{Y}_0 \pm t(\nu, 1 - \tfrac{1}{2}\alpha)s_\nu \sqrt{\mathbf{X}_0'(\mathbf{X}'\mathbf{X})^{-1}\mathbf{X}_0}. \tag{5.3.4}$$

4. State that

$$\mathbf{b} \sim N(\boldsymbol{\beta}, (\mathbf{X'X})^{-1}\sigma^2). \tag{5.3.5}$$

5. Obtain individual $100(1 - \alpha)\%$ confidence intervals for the various parameters separately from the formula

$$b_i \pm t(\nu, 1 - \alpha/2)\text{se}(b_i) \tag{5.3.6}$$

where the "$\text{se}(b_i)$" is the square root of the ith diagonal term of the matrix $(\mathbf{X'X})^{-1}s^2$. These intervals can be used to define a rectangular block in the space of the β's. This block is *not* a proper joint confidence region for the β's, however; see (6) instead.

6. Obtain a joint $100(1 - \alpha)\%$ confidence region for *all* the parameters $\boldsymbol{\beta}$ from the equation

$$(\boldsymbol{\beta} - \mathbf{b})'\mathbf{X'X}(\boldsymbol{\beta} - \mathbf{b}) = ps^2 F(p, \nu, 1 - \alpha), \tag{5.3.7}$$

where $F(p, \nu, 1 - \alpha)$ is the $1 - \alpha$ point ("upper α-point") of the $F(p, \nu)$ distribution and where s^2 has the same meaning as in (1) above and the model is assumed correct. This equality is the equation of the boundary of an "elliptically shaped" (or "ellipsoidally shaped" more generally) contour in a space that has as many dimensions, p, as there are parameters in $\boldsymbol{\beta}$. (Such regions can also be constructed for a subset of the parameters.)

Comparisons between (5) and (6) are discussed in Sections 5.4 and 5.5.

Bonferroni Limits

A more conservative (wider) set of intervals of the form of (5.3.6) is obtained if we replace $t(\nu, 1 - \alpha/2)$ by $t(\nu, 1 - \alpha/(2p))$. It can be shown that these intervals jointly have a confidence coefficient of at least $1 - \alpha$. These, and other sets of rectangular limits, are described in Nickerson (1994).

5.4. CONFIDENCE INTERVALS VERSUS REGIONS

Figure 5.1 illustrates a possible situation that may arise when two parameters are considered. The joint 95% confidence region for the true parameters, β_1 and β_2, is

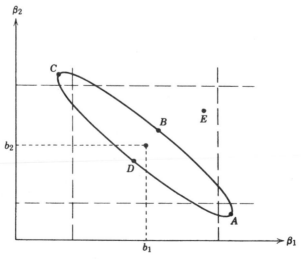

Figure 5.1. Joint and individual confidence statements. The point (b_1, b_2) defined by the least squares estimates is at the center of both ellipse and rectangle.

shown as a long thin ellipse and encloses values (β_1, β_2), which the data regard as *jointly* reasonable for the parameters. It takes into account the correlation between the estimates b_1 and b_2. The individual 95% confidence intervals for β_1 and β_2 separately are appropriate for specifying ranges for the individual parameters irrespective of the value of the other parameter. If an attempt is made to interpret these intervals simultaneously—that is (wrongly) regard the rectangle that they define as a joint confidence region—then, for example, it may be thought that the coordinates of the point E provide reasonable values for (β_1, β_2). The joint confidence region, however, clearly indicates that such a point is not reasonable. When only two parameters are involved, construction of the confidence ellipse is not difficult. In practice, even for two parameters, it is rarely drawn.

If some knowledge of the ellipsoidal region were desired, it would be possible to find the coordinates of the points at the ends of the major axes of the region. (In Figure 5.1 these would be the points A, B, C, and D.) This would involve obtaining the confidence contour and reducing it to canonical form. This also is not difficult, but we do not discuss it, because it is rarely done. The major point to be made here is that the "joint' message of individual confidence intervals should be regarded with caution, and attention should be paid both to the relative sizes of the $V(b_i)$ and to the sizes of the covariances of b_i and b_j. When b_i and b_j have variances of different sizes and the correlation between b_i and b_j, namely,

$$\rho_{ij} = \frac{\text{cov}(b_i, b_j)}{[V(b_i)V(b_j)]^{1/2}}$$

is not small, the situation illustrated in Figure 5.1 occurs. If ρ_{ij} is close to zero then the rectangular region defined by individual confidence intervals will approximate to the correct joint confidence region, though the joint region is correct. The elongation of the region will depend on the relative sizes of $V(b_i)$ and $V(b_j)$. Some examples are shown in Figure 5.2.

Note: If the model is written originally, and fitted, in the alternative form

$$E(Y - \overline{Y}) = \beta_1(X_1 - \overline{X}_1) + \beta_2(X_2 - \overline{X}_2) + \cdots + \beta_k(X_k - \overline{X}_k),$$

where $\overline{Y}, \overline{X}_1, \overline{X}_2, \ldots, \overline{X}_k$ are the observed means of the actual data, then joint confidence intervals can be obtained that do not involve β_0, which sometimes is of little interest.

See Exercise M in "Exercises for Chapters 5 and 6."

Moral

We shall nearly always look at individual confidence intervals that form a "rectangular brick" in the number of dimensions defined by the number of parameters. This brick is not a correct joint confidence region, which, in general, is a "difficult to see and appreciate" ellipsoidal shape. Knowledge of the correlations between the parameter estimates could be helpful in relating the brick and the ellipse, if the effort were thought to be worthwhile.

5.5. MORE ON CONFIDENCE INTERVALS VERSUS REGIONS

We now discuss a one-number calculation that can be useful in comparing a confidence interval block with an ellipsoidal joint confidence region; see Figures 5.1 and 5.2. In

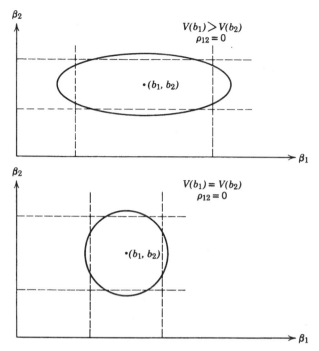

F i g u r e 5.2. Examples of situations where individual confidence intervals combine well to approximate a joint confidence region for two parameters.

this section, we number the model parameters as $\beta_1, \beta_2, \ldots, \beta_p$ (rather than $\beta_0, \beta_1,$ $\ldots, \beta_{p-1}$) to simplify the notation slightly. We first rewrite (5.3.6) and (5.3.7) in the forms

$$b_i \pm t(\nu, 1 - \alpha/2)(V_{ii}s^2)^{1/2}, \qquad i = 1, 2, \ldots, p, \tag{5.5.1}$$

where

$$(V_{ij}) = \mathbf{V} = (\mathbf{X'X})^{-1}, \qquad i, j = 1, 2, \ldots, p,$$

and

$$(\boldsymbol{\beta} - \mathbf{b})'\mathbf{X'X}(\boldsymbol{\beta} - \mathbf{b}) = ps^2 F(p, \nu, 1 - \theta). \tag{5.5.2}$$

In our discussion in Section 5.3, we took $\theta = \alpha$, but this choice is not necessary. In Exercise M in "Exercises for Chapters 5 and 6," we argue that $\alpha = 0.05$ and $\theta = 1 - (1 - \alpha)^2 = 0.10$, approximately, might be appropriate. Such a choice might be sensible if there were no correlations between the estimates b_i. This is unlikely unless the regression follows up a carefully designed experiment. If all the b_i were mutually uncorrelated, however, $\mathbf{X'X}$ would be a diagonal matrix, and the major axes of the ellipse would be parallel to the sides of the rectangular block. Suppose α is given; it is often chosen as 0.05. Then we could choose θ in such a way that, *when all the b_i are uncorrelated*, the rectangular block and the ellipsoid are of the same size. In general, the volume of the rectangular region is

$$R \equiv 2^p t^p s^p (V_{11} V_{22} \cdots V_{pp})^{1/2}, \tag{5.5.3}$$

where $t = t(\nu, 1 - \alpha/2)$. The volume of the ellipsoidal region is given by a constant (depending on the dimension p) times the product of the semi-axial lengths. It can be shown that this volume is

$$E \equiv \frac{\pi^{p/2}}{\Gamma(p/2+1)} (ps^2F)^{p/2}c_p^{1/2}(V_{11}V_{22}\cdots V_{pp})^{1/2}, \tag{5.5.4}$$

where $\pi = 3.14159$, and where the gamma functions required satisfy $\Gamma(u) = (u - 1)\Gamma(u-1)$, $\Gamma(1) = 1$, $\Gamma(\frac{1}{2}) = \pi^{1/2}$. The quantity c_p is defined as the determinant of a normalized form of $(\mathbf{X'X})^{-1}$, namely, of $\{V_{ij}/(V_{ii}V_{jj})^{1/2}\}$. Thus c_p is simply the determinant of the correlation matrix of $b_1, b_2, \ldots, b_p$.

The ratio of the volumes of the two regions is, in general,

$$\frac{E}{R} = \frac{\pi^{p/2}}{\Gamma(p/2+1)} \frac{p^{p/2}F^{p/2}c_p^{1/2}}{2^p t^p}. \tag{5.5.5}$$

We now link α and θ in the following manner. We specify that when $c_p = 1$, that is, when the b_i are uncorrelated, $E = R$, and so their ratio is 1. this implies that, in the case where the major axes of the ellipsoid are parallel to the axes of the β's, we would wish to link α and θ so that the ellipsoid and its approximating rectangular block have equal volume. This requires that

$$F(p, \nu, \theta) = \left\{ \frac{\Gamma(p/2+1)}{\pi^{p/2}} \frac{2^p t^p}{p^{p/2}} \right\}^{2/p}$$

$$= \frac{4\{\Gamma(p/2+1)\}^{2/p}}{\pi p} \{t(\nu, 1-\alpha/2)\}^2 \tag{5.5.6}$$

$$= \frac{4\{\Gamma(p/2+1)\}^{2/p}}{\pi p} \{F(1, \nu, \alpha)\}.$$

Note, as a check, that when $p = 1$, we get the obvious $\theta = \alpha$, for the one-parameter case. When $n \geq 10$, and $\alpha = 0.05$, the θ values vary little for given p. For $p = 2$, for example, the case most often depicted, and for uncorrelated b_i values, we obtain an elliptical confidence region of size equal to the rectangular region based on separate 95% confidence intervals if we use a 91.3% confidence ellipse (approximately). Similar approximate results for some other cases are 88.4% for $p = 3$, 86.0% for $p = 4$, 83.9% for $p = 5$, and 82.1% for $p = 6$.

The above calculations lead us to an easy way to assess how well the rectangular block can represent the correct ellipsoidal region in a regression for any value of p. Using any selected linked values α and θ that satisfy (5.5.6) we see that the right-hand side of (5.5.5) reduces to $c_p^{1/2}$. This value gives the ratio of the volume of the ellipsoidal confidence region compared to the volume of the rectangular block. Note that $0 \leq c_p^{1/2} \leq 1$, the zero corresponding to linear dependence in the $\mathbf{X}$-columns and the 1 to an orthogonal set of $\mathbf{X}$-columns. A relative volume calculation can be made from (5.5.5) even if an ellipsoid other than the one satisfying (5.5.6) is selected, of course. Note that our calculations can also be applied to the slightly different but closely related suggestions for confidence regions made by Weisberg (1985, pp. 97–99, Eqs. (4.2.6) and (4.2.7)). Further information can also be found, if needed, by a canonical analysis of (5.5.2) using the value of θ linked to α.

Example, $p = 2$. Consider the straight line steam-data fit in Chapter 1. Take $\alpha = 0.05$. For this fit, $p = 2$, $\nu = 23$, the right-hand side of (5.5.6) is 2.7241, $1 - \theta = 0.9132$, which we round to 0.913. The correlation between intercept and slope is $r = V_{12}/(V_{11}V_{22})^{1/2} = (-0.0073535)/[(0.4267941)(0.0001398)]^{1/2} = -(0.90628)^{1/2} = -0.952$. Thus $c_p^{1/2} = (1 - r^2)^{1/2} = 0.306$ and so the 91.3% ellipse covers about 30.6% of the area of the rectangle. Moreover, the high negative correlation indicates a diagonal upper-left-

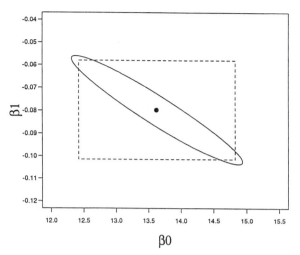

Figure 5.3. Individual 95% confidence bands and a 91.3% joint confidence region for the steam data (Appendix 1A, Y and X_8).

to-lower-right-lying ellipse. Figure 5.3 shows that these simple calculations describe the situation well. (If the ellipse in Figure 5.3 were replaced by the 95% ellipse, which surrounds the 91.3% ellipse and juts out somewhat more at the upper-left and lower-right extremes, the area covered would increase to about 38.4% of the area of the rectangle.)

Example, p = 4. Consider (see Appendix 1A) the steam data again, in particular, the planar fit of the response variable onto predictors X_5, X_6, and X_8 and an intercept. We have $p = 4$, $\nu = 21$ and the sides of the rectangular t-block for $\alpha = 0.05$ are $-13.02 \le \beta_0 \le -7.08$, $0.075 \le \beta_5 \le 0.729$, $0.114 \le \beta_6 \le 0.284$, and $-0.089 \le \beta_8 \le -0.074$. The predictors are very highly correlated, however, and $c_p^{1/2} = 0.000977$. Thus the ellipsoid defined by $\theta = 0.14$ $(1 - \theta = 0.86)$ has a volume of only about 0.1% of the rectangular block. The latter thus gives a totally misleading impression; the ellipsoid is an extremely long thin one. (Even if variable X_5 is dropped, the ellipsoid still represents only about 4.3% of the three-dimensional rectangular t-block.)

Conclusion. The value of $c_p^{1/2}$, where c_p is the determinant of the correlation matrix of the b_i, is a useful calculation to display in regression problems. It provides the ratio of the volume of the (correct) ellipsoidal joint confidence region for the β's to the (wrong but easily obtained) rectangular t-block region, for linked α and θ values that achieve equal volumes in the uncorrelated case. Supplementary information can also be obtained from a canonical reduction of the ellipse's equation. (See, for example, Box and Draper, 1987, pp. 332–372.) The eigenvectors will give the exact orientations of the major axes of the ellipsoid with respect to the β-axes. These orientations depend on the correlations between the various pairs of b's. In particular, if $c_p = 1$, the axes of the ellipse are aligned exactly with the β-axes. Diagrams such as Figures 5.1–5.3 are not needed, once their nature is understood.

When *F*-Test and *t*-Tests Conflict

Occasionally one finds a practical regression problem where an overall F-test for regression given b_0 is significant, but all the t-tests for individual hypotheses $H_0: \beta_i = 0$

are not significant. Largey and Spencer (1996) discuss how such occurrences are related to diagrams of the form of Figure 5.3. (The reverse case, a nonsignificant F but significant t-values, is possible but even rarer.)

References

Box and Draper (1987); Draper and Guttman (1995); Largey and Spencer (1996); Weisberg (1985); Willan and Watts (1978).

APPENDIX 5A. SELECTED USEFUL MATRIX RESULTS

For a more comprehensive list of results see, for example, Graybill (1961) or Rao (1973).

1. $(\mathbf{AB})' = \mathbf{B}'\mathbf{A}'$, $(\mathbf{ABC})' = \mathbf{C}'\mathbf{B}'\mathbf{A}'$, etc. If $\mathbf{M}' = \mathbf{M}$, both are symmetric.

2. $(\mathbf{AB})^{-1} = \mathbf{B}^{-1}\mathbf{A}^{-1}$.

3. A square matrix $\mathbf{C}$ is said to be *orthogonal* if $\mathbf{C}'\mathbf{C} = \mathbf{I}$. Then $\mathbf{C}' = \mathbf{C}^{-1}$.

4. A square matrix $\mathbf{M}$ is said to be idempotent if $\mathbf{MM} = \mathbf{M}$. This can be written as $\mathbf{M}^2 = \mathbf{M}$, also.

5. If $\mathbf{M}$ is symmetric and idempotent,

$$(\mathbf{I} - 2\mathbf{M})'(\mathbf{I} - 2\mathbf{M}) = \mathbf{I}.$$

Thus any matrix of the form $\mathbf{I} - 2\mathbf{M}$, where M is symmetric and idempotent, is orthogonal.

6. Trace $(\mathbf{AB})$ = trace $(\mathbf{BA})$, where trace denotes the sum of the diagonal elements of a square matrix. ($\mathbf{A} = p \times q$, $\mathbf{B} = q \times p$, say.)

7. If

$$\mathbf{M} = \begin{bmatrix} \mathbf{A} & \mathbf{B} \\ \mathbf{C} & \mathbf{D} \end{bmatrix} \quad \text{and if} \quad \begin{cases} \mathbf{P} = \mathbf{A} - \mathbf{BD}^{-1}\mathbf{C}, \\ \mathbf{Q} = \mathbf{D} - \mathbf{CA}^{-1}\mathbf{B}, \end{cases}$$

then

$$\mathbf{M}^{-1} = \begin{bmatrix} \mathbf{P}^{-1} & -\mathbf{A}^{-1}\mathbf{BQ}^{-1} \\ -\mathbf{D}^{-1}\mathbf{CP}^{-1} & \mathbf{Q}^{-1} \end{bmatrix}$$

assuming all matrices shown inverted are nonsingular. Alternatively,

$$\mathbf{M}^{-1} = \begin{bmatrix} \mathbf{A}^{-1} + \mathbf{A}^{-1}\mathbf{BQ}^{-1}\mathbf{CA}^{-1} & -\mathbf{A}^{-1}\mathbf{BQ}^{-1} \\ -\mathbf{Q}^{-1}\mathbf{CA}^{-1} & \mathbf{Q}^{-1} \end{bmatrix}.$$

If $\mathbf{M}$ is symmetric, set $\mathbf{C} = \mathbf{B}'$.

8. If $\mathbf{E}$ is $n \times p$, and $\mathbf{F}$ is $p \times n$, then

$$(\mathbf{I}_n + \mathbf{EF})^{-1} = \mathbf{I}_n - \mathbf{E}(\mathbf{I}_p + \mathbf{FE})^{-1}\mathbf{F}.$$

This is especially useful when p is much smaller than n.

Special Case 1. If $\mathbf{X}$ is $n \times p$

$$(\mathbf{I}_n + \mathbf{X}(\mathbf{X}'\mathbf{X})^{-1}\mathbf{X}')^{-1} = \mathbf{I}_n - \mathbf{X}(\mathbf{X}'\mathbf{X})^{-1}[\mathbf{I}_p + \mathbf{X}'\mathbf{X}(\mathbf{X}'\mathbf{X})^{-1}]^{-1}\mathbf{X}'$$

$$= \mathbf{I}_n - \tfrac{1}{2}\mathbf{X}(\mathbf{X}'\mathbf{X})^{-1}\mathbf{X}'.$$

Thus $(\mathbf{I}_n + \mathbf{H})^{-1} = \mathbf{I}_n - \tfrac{1}{2}\mathbf{H}$, where $\mathbf{H} = \mathbf{X}(\mathbf{X}'\mathbf{X})^{-1}\mathbf{X}'$ is the hat matrix.

Special Case 2. If $\mathbf{A}$ is $n \times n$, and $\mathbf{u}$, $\mathbf{v}$ are $n \times 1$ vectors, then

$$(\mathbf{A} + \mathbf{uv}')^{-1} = (\mathbf{I} + \mathbf{A}^{-1}\mathbf{uv}')^{-1}\,\mathbf{A}^{-1} = \mathbf{A}^{-1} - (\mathbf{A}^{-1} - (\mathbf{A}^{-1}\mathbf{u})(\mathbf{v}'\mathbf{A}^{-1})/\{1 + \mathbf{v}'\mathbf{A}^{-1}\mathbf{u}\}.$$

This enables inversion of $\mathbf{A} + \mathbf{uv}$ from knowledge of $\mathbf{A}^{-1}$. (Set $\mathbf{E} = \mathbf{A}^{-1}\mathbf{u}$, $\mathbf{F} = \mathbf{v}'$.)

9. If $\mathbf{A}$ is $p \times p$, $\mathbf{B}$ is $p \times q$, $\mathbf{C}$ is $q \times p$, and $\mathbf{D}$ is $q \times q$, then

$$\begin{vmatrix} \mathbf{A} & \mathbf{B} \\ \mathbf{C} & \mathbf{D} \end{vmatrix} = |\mathbf{A}\|\mathbf{D} - \mathbf{C}\mathbf{A}^{-1}\mathbf{B}| = |\mathbf{A} - \mathbf{B}\mathbf{D}^{-1}\mathbf{C}\|\mathbf{D}|$$

Proof. Premultiply the original matrix by

$$\begin{bmatrix} \mathbf{I}_p & \mathbf{0} \\ -\mathbf{C}\mathbf{A}^{-1} & \mathbf{I}_q \end{bmatrix}$$

to give a matrix equation; then take determinants of both sides.

Special Case 1. Set $\mathbf{C} = -\mathbf{B}'$, $\mathbf{D} = \mathbf{I}$ and we obtain the result

$$|\mathbf{A}\|\mathbf{I} + \mathbf{B}'\mathbf{A}^{-1}\mathbf{B}| = |\mathbf{A} + \mathbf{B}\mathbf{B}'|.$$

Special Case 2. Set $\mathbf{C} = \mathbf{B}'$ if the partitioned matrix is symmetric.

A useful reference for some special inverse matrices is Roy and Sarhan (1956).

EXERCISES

Exercises for Chapter 5 are located in the Section "Exercises for Chapters 5 and 6" at the end of Chapter 6.

CHAPTER 6

Extra Sums of Squares and Tests for Several Parameters Being Zero

The ideas connected with the extra sum of squares principle are extremely important and must be understood fully by regression practitioners.

6.1. THE "EXTRA SUM OF SQUARES" PRINCIPLE

In regression work, the question often arises as to whether or not it was worthwhile to include certain terms in the model. This question can be investigated by considering the extra portion of the regression sum of squares which arises due to the fact that the terms under consideration *were* in the model. The mean square derived from this extra sum of squares can then be compared with the estimate, s^2, of σ^2 to see if it appears significantly large. If it does, the terms should have been included; if it does not, the terms would be judged unnecessary and could be removed.

We have already seen an example of this in the case of fitting a straight line where SS $(b_1|b_0)$ represented the extra sum of squares due to including the term $\beta_1 X$ in the model. We now state the procedure more generally. Suppose the functions Z_1, Z_2, $\ldots$, Z_{p-1} are known functions of the basic variables X_1, X_2, $\ldots$, and suppose that values of the X's and the corresponding response Y are available. Consider the two models below.

1. $Y = \beta_0 + \beta_1 Z_1 + \beta_2 Z_2 + \cdots + \beta_{p-1} Z_{p-1} + \epsilon$.

Suppose we obtain the following least squares estimates: $b_0(1), b_1(1), b_2(1), \ldots, b_{p-1}(1)$ and suppose that $SS(b_0(1), b_1(1), b_2(1), \ldots, b_{p-1}(1)) = S_1$, and there is no lack of fit. Let the estimate of σ^2 be s^2, obtained from the residual of Model 1.

2. $Y = \beta_0 + \beta_1 Z_1 + \beta_2 Z_2 + \cdots + \beta_{q-1} Z_{q-1} + \epsilon (q < p)$.

The Z's in this Model 2 are the same functions as in Model 1 when subscripts are the same. There are, however, fewer terms in this second model.

Suppose we now obtain the following least squares estimates: $b_0(2), b_1(2), b_2(2), \ldots, b_{q-1}(2)$. *Note:* These may or may not be the same as $b_0(1), b_1(1), \ldots, b_{q-1}(1)$ above. If they are identical then $b_i(1)$ and $b_j(1)$ are orthogonal linear functions for $1 \leq i \leq q - 1, q \leq j \leq p - 1$. This happens when, in Model 1, the first q columns of the $\mathbf{X}$ matrix are all orthogonal to the last $p - q$ columns. This can happen in planned experiments. It rarely happens otherwise. See Appendix 6A.

Suppose that $SS(b_0(2), b_1(2), b_2(2), \ldots, b_{q-1}(2)) = S_2$, for this second model. Then $S_1 - S_2$ is the *extra sum of squares*, due to the inclusion of the terms $\beta_q Z_q + \cdots + \beta_{p-1} Z_{p-1}$ in Model 1. Since S_1 has p degrees of freedom and S_2 has q degrees of freedom, $S_1 - S_2$ has $(p - q)$ degrees of freedom. It can be shown that, if $\beta_q = \beta_{q+1} = \cdots = \beta_{p-1} = 0$, then $E\{(S_1 - S_2)/(p - q)\} = \sigma^2$. In addition, if the errors are normally distributed, $(S_1 - S_2)$ will then be distributed as $\sigma^2 \chi^2_{p-q}$ independently of s^2. This means we can compare $(S_1 - S_2)/(p - q)$ with s^2 by an $F(p - q, \nu)$ test, where ν is the number of degrees of freedom on which s^2 is based, to test the hypothesis $H_0: \beta_q = \beta_{q+1} = \cdots = \beta_{p-1} = 0$.

We can write $S_1 - S_2$ conveniently as $SS(b_q, \ldots, b_{p-1}|b_0, b_1, \ldots, b_{q-1})$ where we must keep in mind that two models are actually involved since the notation does not show it. This is read as *the sum of squares of $b_q, \ldots, b_{p-1}$ given $b_0, b_1 \ldots, b_{q-1}$.* By continued application of this principle we can obtain, successively, for any regression model, $SS(b_0)$, $SS(b_1|b_0)$, $SS(b_2|b_0, b_1)$, $\ldots$, $SS(b_{p-1}|b_0, b_1, \ldots, b_{p-2})$, if we wish. All these sums of squares are distributed independently of s^2 and equal their mean squares since each has one degree of freedom. The mean squares can be compared with s^2 by a series of F-tests. This is useful when the terms of the model have a logical "order of entry," as would be the case, for example, if $Z_j = X^j$. A judgment can then be made about how many terms should be in the model.

Polynomial Models

When the terms in the model occur in natural groupings, such as happens, for example, in polynomial models with (1) β_0, (2) first-order terms, and (3) second-order terms, we can construct alternative extra sums of squares, for example, $SS(b_0)$, SS(first-order b's$|b_0$), SS(second-order b's$|b_0$, first-order b's), and compare *these* with s^2. The extra sum of squares principle can be used in many ways therefore to achieve whatever breakup of the regression sum of squares seems reasonable for the problem at hand.

Other Points

The number of degrees of freedom for each sum of squares will be the number of parameters before the vertical division line (except when the estimates are linearly dependent; this happens when $\mathbf{X'X}$ is singular and the normal equations are linearly dependent and will not usually concern us. The number of degrees of freedom is then the maximum number of linearly independent estimates in the set being considered). These extra SS are distributed independently of s^2. The corresponding mean squares, which equal (sum of squares)/(degrees of freedom), can be divided by s^2 to provide an F-ratio for testing the hypothesis that the true values of the coefficients whose estimates gave rise to the extra sum of squares are zero.

The expected value of an extra sum of squares is evaluated in Appendix 6B.

The extra sum of squares principle is actually a special case of testing a general linear hypothesis. In the more general treatment the extra sum of squares is calculated from the residual sums of squares and not the regression sum of squares. Since the total sum of squares $\mathbf{Y'Y}$ is the same for both regression calculations, we would obtain the same result numerically whether we used the difference of regression or residual sums of squares. See Eq. (6.1.8).

Two Alternative Forms of the Extra SS

We can decide to remove the correction factor $n\overline{Y}^2$ or not remove it, before we take the difference between two sums of squares to get an extra sum of squares. For example, suppose our initial model (Model 1) is

$$Y = \beta_0 + \beta_1 X_1 + \beta_2 X_2 + \beta_3 X_3 + \beta_4 X_4 + \beta_5 X_5 + \epsilon, \tag{6.1.1}$$

and we want the extra SS for b_3, b_4, and b_5 given b_0, b_1, and b_2. The reduced model (Model 2) is

$$Y = \beta_0 + \beta_1 X_1 + \beta_2 X_2 + \epsilon. \tag{6.1.2}$$

For Model 1, the regression SS is

$$SS(b_0, b_1, b_2, b_3, b_4, b_5) = S_1, \tag{6.1.3}$$

the correction factor is $n\overline{Y}^2$, and the total SS is $\mathbf{Y'Y}$. Thus the residual SS is $\mathbf{Y'Y} - S_1$.
For model 2, the regression SS is

$$SS(b_0, b_1, b_2) = S_2, \tag{6.1.4}$$

where we *no longer show* that the Model 2 b's could be different from the Model 1 b's of same subscript, though in general they are. (One has to get used to this notation and to realize that it conceals a possible confusion!) The correction factor is $n\overline{Y}^2$ and the total sum of squares is $\mathbf{Y'Y}$. Thus the residual SS is $\mathbf{Y'Y} - S_2$.
We now require

$$SS(b_3, b_4, b_5 | b_0, b_1, b_2) = S_1 - S_2. \tag{6.1.5}$$

[*Note:* A reordering of b's before the vertical bar and/or a reordering after the vertical bar does not change the meaning of (6.1.5).] We can rewrite this as

$$S_1 - S_2 = (S_1 - n\overline{Y}^2) - (S_2 - n\overline{Y}^2) \tag{6.1.6}$$

when it becomes a difference between sums of squares corrected for b_0, that is,

$$SS(b_1, b_2, b_3, b_4, b_5 | b_0) - SS(b_1, b_2 | b_0). \tag{6.1.7}$$

There is yet a third way to get this extra sum of squares. We can rewrite the $S_1 - S_2$ as

$$S_1 - S_2 = (\mathbf{Y'Y} - S_2) - (\mathbf{Y'Y} - S_1) \tag{6.1.8}$$

when it becomes a difference of residual SS but in reversed order, because the regression with the larger regression SS (S_1) must have the smaller residual SS; and vice versa for S_2. Of the three calculations, the best is the one you prefer! (For a specific matrix formula for the extra sum of squares in general, see Section 10.4.)

Sequential Sums of Squares

When we call the regression option in a programming system, we tell the computer a certain order for our X's. Sometimes this order has a meaning for us, sometimes it is just the order in which we wrote down the data. Let us suppose we fitted model

(6.1.1) and loaded the X's in the order shown there. Then we would (in some programs, e.g., MINITAB) or could (in others) see a printout of extra SS of form

$$SS_1 = SS(b_1|b_0),$$
$$SS_2 = SS(b_2|b_1, b_0),$$
$$SS_3 = SS(b_3|b_2, b_1, b_0), \tag{6.1.9}$$
$$SS_4 = SS(b_4|b_3, b_2, b_1, b_0),$$
$$SS_5 = SS(b_5|b_4, b_3, b_2, b_1, b_0),$$

often called the sequential sum of squares printout. If, as in our example above, we wanted the extra SS (6.1.5) or (6.1.6) or (6.1.8) we could get it by summing $SS_5 + SS_4 + SS_3$ in (6.1.9). In any printout like this where subscripts 1 and 2 come first and second (in either order 12 or 21) and subscripts 3, 4, and 5 come third, fourth, and fifth (in order 345 or 354 or 435 or 453 or 534 or 543), a similar calculation will give the correct answer. We cannot get $SS(b_1, b_4, b_5|b_0, b_2, b_3)$ from (6.1.9), however. The breakdown given does not allow this, except in special cases where there is enough pairwise orthogonality among the columns of the $\mathbf{X}$ matrix to make the calculation correct. In general, it will not work and there is little point discussing the exceptions in much more detail than is given in Appendix 6A. It is usually easier to rerun the regression with another ordering of the predictor variables.

We follow our example one more step. Suppose we wish to test $H_0: \beta_3 = \beta_4 = \beta_5 = 0$ in (6.1.1) versus H_1: not so. (There are many ways H_0 would not be true, so this is the easiest way to state the alternative hypothesis.) Our F-test would be carried out on

$$F = \{(S_1 - S_2)/(6 - 3)\}/s^2,$$

where s^2 is the residual mean square from the larger of the two models, namely, from the fit of (6.1.1). The degrees of freedom would be $6 - 3 = 3$ for the numerator and $(n - 6)$ for the denominator, where n is the number of observations.

Special Problems with Polynomial Models

In models where the X's are individual predictor variables, it may not make any difference which β's are set equal to zero in H_0 and so tested via an extra SS test. In the case of a polynomial model, however, certain tests do not make practical sense. For example, suppose that in (6.1.1), we had

$$X_3 = X_1^2, \qquad X_4 = X_2^2, \qquad X_5 = X_1 X_2. \tag{6.1.10}$$

Testing $H_0: \beta_3 = \beta_4 = \beta_5 = 0$ *is* sensible because it answers the question: "Do we need the quadratic curvature in the model?" The question of whether $H_0: \beta_1 = \beta_2 = 0$ is true is, in general, *not* a good one, as it is asking if the stationary point of the surface lies at the origin, a very rare event that depends on the coding of the factors as well as the shape of the surface. We discuss this issue in more detail in Chapter 12 and there suggest some rules that may be useful.

Partial Sums of Squares

We have seen how to obtain extra sums of squares for one or more estimated coefficients given other coefficients by considering two models, one of which includes the coefficients in question and one of which does not.

If we have several terms in a regression model we can think of them as "entering" the equation in any desired sequence. If we find

$$SS(b_i | b_0, b_1, \ldots, b_{i-1}, b_{i+1}, \ldots, b_k), \qquad i = 1, 2, \ldots, k, \qquad (6.1.11)$$

we shall have a one degree of freedom sum of squares, which measures the contribution to the regression sum of squares of each coefficient b_i given that all the terms that did not involve β_i were already in the model. In other words, we shall have a measure of the value of *adding a β_i term to the model* that originally did not include such a term. Another way of saying that is that we have a measure of the value of β_i *as though it were added to the model last*. The corresponding mean square, equal to the sum of squares since it has one degree of freedom, can be compared by an F-test to s^2 as described. This particular type of F-test is often called a *partial F-test* for β_i. If the extra term under consideration is $\beta_t X_t$, say, we can talk (loosely) about a partial F-test on the variable X_t, even though we are aware that the test actually is on the coefficient β_t.

When a suitable model is being "built" the partial F-test is a useful criterion for adding or removing terms from the model. The effect of an X-variable (X_q, say) in determining a response may be large when the regression equation includes only X_q. However, when the same variable is entered into the equation after other variables, it may affect the response very little, due to the fact that X_q is highly correlated with variables already in the regression equation. The partial F-test can be made for all regression coefficients as though each corresponding variable were the last to enter the equation—to see the relative effects of each variable in excess of the others. This information can be combined with other information if a choice of variables need be made. Suppose, for example, either X_1 or X_2 alone could be used to provide a regression equation for a response Y. Suppose use of X_1 provided smaller predictive errors than use of X_2. Then if predictive accuracy were desired, X_1 would probably be used in future work. If, however, X_2 were a variable through which the response level could be controlled (whereas X_1 was a measured but noncontrolling variable) and if control were important rather than prediction, then it might be preferable to use X_2 rather than X_1 as a predictor variable for future work.

When $t = F^{1/2}$

The partial F-statistic with 1 and ν degrees of freedom for testing $H_0: \beta_j = 0$ versus $H_1: \beta_j \neq 0$ is exactly equal to the square of the t-statistic with ν degrees of freedom obtained via $t = b_j / \{se(b_j)\}$, where $se(b_j)$, the standard error of b_j, is the square root of the appropriate diagonal term of $(\mathbf{X}'\mathbf{X})^{-1} s^2$ and s^2 is based on ν df. (This is a distributional fact that we do not prove.) The test can be made in either F or t form with the same results. Examination of the tables of percentage points will show that $F(1, \nu, 1 - \alpha) = \{t(\nu, 1 - \alpha/2)\}^2$ for any values of ν and α. (As always, round-off errors will sometimes prevent the numerical relationship from being exact to the number of figures quoted.)

6.2. TWO PREDICTOR VARIABLES: EXAMPLE

We now look at a two-predictor example using part of the steam data. This enables us to illustrate some of the details mentioned in Section 6.1, as well as the matrix algebra of Chapter 5. Consider the first-order linear model of form

$$Y = \beta_0 + \beta_1 X_1 + \beta_2 X_2 + \epsilon. \tag{6.2.1}$$

We shall continue with the example used in Chapter 1 (the data for which are given in Appendix 1A) and will now add variable number 6 to the problem. So that we are clear about which variables are being considered in the model, we shall use the original variable subscripts. Thus our model will be written

$$Y = \beta_0 X_0 + \beta_8 X_8 + \beta_6 X_6 + \epsilon, \tag{6.2.2}$$

where Y = response or number of pounds of steam used per month, coded,
$\quad X_0$ = dummy variable, whose value is always unity,
$\quad X_8$ = average atmospheric temperature in the month (in °F),
$\quad X_6$ = number of operating days in the month.

The following matrices can be constructed. (The complete figures for the vector **Y** and the second and third columns of matrix **X** appear in Appendix 1A and are also given in Table 6.1.)

$$
\mathbf{Y} = \begin{bmatrix} 10.98 \\ 11.13 \\ 12.51 \\ 8.4 \\ \vdots \\ 10.36 \\ 11.08 \end{bmatrix}, \quad
\mathbf{X} = \begin{matrix} X_0 & X_8 & X_6 \\ \begin{bmatrix} 1 & 35.3 & 20 \\ 1 & 29.7 & 20 \\ 1 & 30.8 & 23 \\ 1 & 58.8 & 20 \\ \vdots & \vdots & \vdots \\ 1 & 33.4 & 20 \\ 1 & 28.6 & 22 \end{bmatrix} \end{matrix}, \quad
\boldsymbol{\beta} = \begin{bmatrix} \beta_0 \\ \beta_8 \\ \beta_6 \end{bmatrix}, \quad
\boldsymbol{\epsilon} = \begin{bmatrix} \epsilon_1 \\ \epsilon_2 \\ \epsilon_3 \\ \epsilon_4 \\ \vdots \\ \epsilon_{24} \\ \epsilon_{25} \end{bmatrix}
$$

where **Y** is a (25×1) vector,
$\quad$ **X** is a (25×3) matrix,
$\quad$ **β** is a (3×1) vector,
$\quad$ **ε** is a (25×1) vector.

Using the results of Chapter 5, the least squares estimates of β_0, β_8, and β_6 are given by

$$\mathbf{b} = (\mathbf{X'X})^{-1}\mathbf{X'Y},$$

where **b** is the vector of estimates of the elements of β, provided that $\mathbf{X'X}$ is nonsingular. Thus

$$
\mathbf{b} = \begin{bmatrix} b_0 \\ b_8 \\ b_6 \end{bmatrix} = \left\{ \begin{bmatrix} 1 & 1 & 1 & \cdots & 1 \\ 35.3 & 29.7 & 30.8 & \cdots & 28.6 \\ 20 & 20 & 23 & \cdots & 22 \end{bmatrix} \begin{bmatrix} 1 & 35.3 & 20 \\ 1 & 29.7 & 20 \\ 1 & 30.8 & 23 \\ \vdots & \vdots & \vdots \\ 1 & 28.6 & 22 \end{bmatrix}^{-1} \right\}
$$

$$
\times \begin{bmatrix} 1 & 1 & 1 & \cdots & 1 \\ 35.3 & 29.7 & 30.8 & \cdots & 28.6 \\ 20 & 20 & 23 & \cdots & 22 \end{bmatrix} \begin{bmatrix} 10.98 \\ 11.13 \\ 12.51 \\ \vdots \\ 11.08 \end{bmatrix} .
$$

Note the sizes of the matrices in the above statement:

$$[3 \times 1] = \{[3 \times 25][25 \times 3]\}^{-1}[3 \times 25][25 \times 1].$$

Multiplying the matrices within the large braces, we have

$$
\begin{array}{cc}
[3 \times 1] & [3 \times 3]^{-1} \\
\begin{bmatrix} b_0 \\ b_8 \\ b_6 \end{bmatrix} = \begin{bmatrix} 25.00 & 1315.00 & 506.00 \\ 1315.00 & 76323.42 & 26353.30 \\ 506.00 & 26353.30 & 10460.00 \end{bmatrix}^{-1}
\end{array}
$$

$$
\begin{array}{cc}
[3 \times 25] & [25 \times 1] \\
\times \begin{bmatrix} 1 & 1 & \cdots & 1 \\ 35.3 & 29.7 & \cdots & 28.6 \\ 20 & 20 & \cdots & 22 \end{bmatrix} & \begin{bmatrix} 10.98 \\ 11.13 \\ 12.51 \\ \vdots \\ 11.08 \end{bmatrix} .
\end{array}
$$

Then,

$$
\begin{array}{ccc}
[3 \times 1] & [3 \times 3]^{-1} & [3 \times 1] \\
\begin{bmatrix} b_0 \\ b_8 \\ b_6 \end{bmatrix} = \begin{bmatrix} 25.00 & 1315.00 & 506.00 \\ 1315.00 & 76323.42 & 26353.30 \\ 506.00 & 26353.30 & 10460.00 \end{bmatrix}^{-1} & \begin{bmatrix} 235.6000 \\ 11821.4320 \\ 4831.8600 \end{bmatrix} .
\end{array}
$$

Next, the inverse of the $[3 \times 3]$ matrix is obtained to give

$$
\begin{array}{cc}
[3 \times 1] & [3 \times 3] \\
\begin{bmatrix} b_0 \\ b_8 \\ b_6 \end{bmatrix} = & \begin{bmatrix} 2.778747 & -0.011242 & -0.106098 \\ & 0.146207 \times 10^{-3} & 0.175467 \times 10^{-3} \\ \text{(Symmetric)} & & 0.478599 \times 10^{-2} \end{bmatrix}
\end{array}
$$

$$
\times \begin{array}{c} [3 \times 1] \\ \begin{bmatrix} 235.6000 \\ 11821.4320 \\ 4831.8600 \end{bmatrix} \end{array}.
$$

The inverse calculation can be checked by multiplying $(\mathbf{X'X})^{-1}$ by the original $(\mathbf{X'X})$ to give a 3×3 unit matrix. Note that, since the inverse (like the original matrix) is symmetric, only an upper triangular portion of it is recorded. Performing the matrix multiplication gives

$$
\begin{array}{cc}
[3 \times 1] & [3 \times 1] \\
\begin{bmatrix} b_0 \\ b_8 \\ b_6 \end{bmatrix} = & \begin{bmatrix} 9.1266 \\ -0.0724 \\ 0.2029 \end{bmatrix}.
\end{array}
$$

Thus the fitted least squares equation is

$$\hat{Y} = 9.1266 - 0.0724 X_8 + 0.2029 X_6.$$

Actually, when these matrix calculations are performed by a computer routine, they are not carried through in precisely this way. One reason for this is that large rounding errors may occur when this sequence is followed.

Substitution into the fitted equation of the data values of X_8 and X_6 leads to the fitted values $\hat{Y}_i$ and residuals $Y_i - \hat{Y}_i$ given in Table 6.1. A plot of the observations Y_i and the fitted values $\hat{Y}_i$ is shown in Figure 6.1.

How Useful Is the Fitted Equation?

The analysis of variance table takes the following form:

ANOVA

Source of Variation	df	SS	MS	F
Regression $|b_0$	2	54.1871	27.0936	61.8999
Residual	22	9.6287	0.4377	
Total (corrected)	24	63.8158		
Mean (b_0)	1	2220.2944		
Total (uncorrected)	25	2284.1102		

Provided that further examination of the model and residuals shows no flaw, the least squares equation

T A B L E 6.1. Steam Data, Fitted Values and Residuals

Observation Number	X_8	X_6	Y	$\hat{Y}$	Residual
1	35.3	20	10.98	10.63	0.35
2	29.7	20	11.13	11.03	0.10
3	30.8	23	12.51	11.56	0.95
4	58.8	20	8.40	8.93	−0.53
5	61.4	21	9.27	8.94	0.33
6	71.3	22	8.73	8.43	0.30
7	74.4	11	6.36	5.97	0.39
8	76.7	23	8.50	8.24	0.26
9	70.7	21	7.82	8.27	−0.45
10	57.5	20	9.14	9.02	0.12
11	46.4	20	8.24	9.82	−1.58
12	28.9	21	12.19	11.29	0.90
13	28.1	21	11.88	11.35	0.53
14	39.1	19	9.57	10.15	−0.58
15	46.8	23	10.94	10.40	0.54
16	48.5	20	9.58	9.67	−0.09
17	59.3	22	10.09	9.30	0.79
18	70.0	22	8.11	8.52	−0.41
19	70.0	11	6.83	6.29	0.54
20	74.5	23	8.88	8.40	0.48
21	72.1	20	7.68	7.96	−0.28
22	58.1	21	8.47	9.18	−0.71
23	44.6	20	8.86	9.96	−1.10
24	33.4	20	10.36	10.77	−0.41
25	28.6	22	11.08	11.52	−0.44
			235.60		$\Sigma(Y_i - \hat{Y}_i) = 0$
			$\overline{Y} = 9.424$		$\Sigma(Y_i - \hat{Y}_i)^2 = 9.6432$

$$\hat{Y} = 9.1266 - 0.0724 X_8 + 0.2029 X_6$$

is a significant explanation of the data. The calculated $F = 61.90$ exceeds the tabulated $F(2, 22, 0.95) = 3.44$ by a healthy margin. In fact, the tail area beyond 61.90 is only $p = 0.00007$ for the $F(2, 22)$ distribution.

What Has Been Accomplished by the Addition of a Second Predictor Variable (Namely, X_6)?

There are several useful criteria that can be applied to answer this question, and we now discuss them.

R^2

The square of the multiple correlation coefficient R^2 is defined as

$$R^2 = \frac{\text{Sum of squares due to regression } |b_0}{\text{Total (corrected) sum of squares}}.$$

It is often stated as a percentage, $100R^2$. The larger it is, the better the fitted equation explains the variation in the data. We can compare the value of R^2 at each stage of the regression problem:

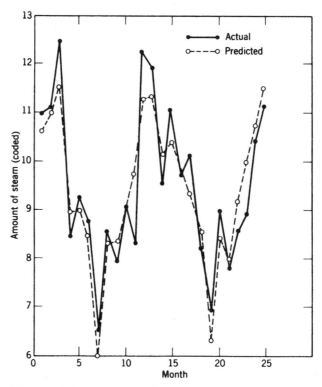

Figure 6.1. Plot of Y_i and $\hat{Y}_i$ values by month, for steam data.

Step 1. $Y = f(X_8)$.

Regression equation $100\,R^2$
$\hat{Y} = 13.6230 - 0.0798X_8$ 71.44% (see Section 1.3)

Step 2. $Y = f(X_8, X_6)$.

Regression equation $100\,R^2$
$\hat{Y} = 9.1266 - 0.0724X_8 + 0.2029X_6$ 84.89%

Thus we see a substantial increase in R^2.

The addition of a new predictor variable to a regression will generally increase R^2. (More exactly, it cannot decrease it and will leave it the same only if the new predictor is a linear combination of the predictors already in the equation.) Moreover, the addition of more and more predictors will give the highest feasible value of R^2 when the number of data sites equals the number of parameters. The pure error can never be explained by any fitted model, however, as already mentioned.

The increase in R^2 from one equation to the other could be tested but it is pointless to do so, because the R^2 statistic is related to the F-test for regression given b_0, while the increase in R^2 is related to an extra SS F-test. Thus all desired R^2 tests are conducted via F-tests.

The Standard Error *s*

The residual mean square s^2 is an estimate of $\sigma_{Y.X}^2$, the variance about the regression. Before and after adding a variable to the model, we can check

$$s = \sqrt{\text{Residual mean square}}.$$

Examination of this statistic indicates that the smaller it is the better, that is, the more precise will be the predictions. Of course, s can be reduced to the pure error value (or to zero if there is no pure error) by including as many parameters as there are data sites. Apart from an approach to such an extreme, reduction of s is desirable. In our example at Step 1,

$$s = \sqrt{0.7926} = 0.89.$$

At Step 2,

$$s = \sqrt{0.4377} = 0.66.$$

Thus the addition of X_6 has decreased s and improved the precision of estimation.

The value of s is not always decreased by adding a predictor variable. This is because the reduction in the residual sum of squares may be less than the original residual mean square. Since one degree of freedom is removed from the residual degrees of freedom as well, the resulting mean square may get larger.

$s/\overline{Y}$

A useful way of looking at the decrease in s is to consider it in relation to the response. In our example, at Step 1, s as a percentage of mean $\overline{Y}$ is

$$0.89/9.424 = 9.44\%.$$

At Step 2, s as a percentage of mean $\overline{Y}$ is

$$0.66/9.424 = 7.00\%$$

The addition of X_6 has reduced the standard error of estimate down to about 7% of the mean response. Whether this level of precision is satisfactory or not is a matter for the experimenter to decide, on the basis of prior knowledge and personal feelings.

Extra SS F-Test Criterion

This method consists of breaking down the sum of squares due to regression given b_0 into two sequential pieces as follows:

ANOVA

Source of Variation	df	SS	MS	F
Regression $\vert b_0$	2	54.1871	27.0936	61.8999
Due to $b_8\vert b_0$	1	45.5924	45.5924	104.1636
Due to $b_6\vert b_8, b_0$	1	8.5947	8.5947	19.6361
Residual	22	9.6287	0.4377	
Total (corrected)	24	63.8158		

The F-value of 19.64 exceeds $F(1, 22, 0.95) = 4.30$ by a factor of more than four, indicating a statistically significant contribution by the addition of X_6 to the equation.

We can, of course, also consider what would have been the effect of adding the variables in the reverse order, X_6 followed by X_8. It is still the same amount $SS(b_8, b_6\vert b_0) = 54.1871$, which is split up, but the split is now different (and will be different in general except when the conditions of Appendix 6A prevail). We obtain:

ANOVA

Source of Variation	df	SS	MS	F
Regression $\|b_0$	2	54.1871	27.0936	61.8999
Due to $b_6\|b_0$	1	18.3424	18.3424	41.9063
Due to $b_8\|b_6, b_0$	1	35.8447	35.8447	81.8933
Residual	22	9.6287	0.4377	
Total (corrected)	24	63.8158		

To get the entry for SS$(b_6\|b_0)$ we have to fit $\hat{Y} = 3.561 + 0.2897X_6$ and evaluate the expression $S_{6Y}^2/S_{66} = 18.3424$. Comparing the sums of squares, we find:

Contribution of	X_8 in First		X_6 in First
X_8	45.59		35.84
X_6	8.59		18.34
Totals	54.18	=	54.18

In this example, each variable picks up more of the variation when it gets into the equation first than it does when it gets in second. This is also reflected in the corresponding F-values. However, X_8 is still the more important variable in both cases, since its contribution in reducing the residual sum of squares is the larger, regardless of the order of introduction of the variables. Behavior like this is common, but is not guaranteed. See Appendix 6B.

Standard Error of b_i

Using the result given in Section 5.2, the variance–covariance matrix of $\mathbf{b}$ is $(\mathbf{X'X})^{-1}\sigma^2$.

Thus variance of $b_i = V(b_i) = c_{ii}\sigma^2$, where c_{ii} is the diagonal element in $(\mathbf{X'X})^{-1}$ corresponding to the ith variable.

The covariance of b_i, $b_j = c_{ij}\sigma^2$, where c_{ij} is the off-diagonal element in $(\mathbf{X'X})^{-1}$ corresponding to the intersection of the ith row and jth column, or jth row and ith column, since $(\mathbf{X'X})^{-1}$ is symmetric.

Thus the standard deviation of b_i is $\sigma \sqrt{c_{ii}}$ and we replace σ by s to obtain the standard error of b_i. For example, the standard error of b_8 is obtained as follows:

$$\text{est. var}(b_8) = s^2 c_{88}$$

$$= (0.4377)(0.146207 \times 10^{-3})$$

$$= 0.639948 \times 10^{-4}.$$

Then $\text{se}(b_8) = \sqrt{\text{est. var}(b_8)} = \sqrt{0.639948 \times 10^{-4}} = 0.008000$.

Note that the t-statistic, $t = b_8/\text{se}(b_8) = -0.0724/0.008 = -9.05$ so that $t^2 = 81.9025$. In theory, this is *identical* to the partial F-value $F_{8|6,0} = 81.8933$ in the foregoing table. As usual, rounding errors have crept into the calculations. The parallel calculation for b_6 is $t = 0.2029/\{(0.4377)(0.478599 \times 10^{-2})\}^{1/2} = 0.2029/0.0458 = 4.433 = (19.6515)^{1/2}$, whereas $F_{6|8,0} = 19.6361$.

Correlations Between Parameter Estimates

We can convert the $(\mathbf{X}'\mathbf{X})^{-1}s^2$ matrix or, more simply, the $(\mathbf{X}'\mathbf{X})^{-1}$ matrix, because s^2 cancels out, into a correlation matrix by dividing each row *and* each column by the square root of the appropriate diagonal entry. Using the $(\mathbf{X}'\mathbf{X})^{-1}$ matrix of the steam data, for example, we divide the first row *and* the first column by the square root of the first diagonal entry, that is by, $(2.778747)^{1/2}$ and so on, moving down the diagonal. We obtain, to three decimal places,

$$
\begin{array}{cccc}
 & 0 & 8 & 6 \\
0 & \begin{bmatrix} 1.000 & -0.558 & -0.920 \\ & 1.000 & 0.210 \\ \text{(Symmetric)} & & 1.000 \end{bmatrix}
\end{array}
$$

The correlations between estimates are $\text{Corr}(b_0, b_8) = -0.558$, $\text{Corr}(b_0, b_6) = -0.920$, and $\text{Corr}(b_8, b_6) = 0.210$.

Confidence Limits for the True Mean Value of Y, Given a Specific Set of Xs

The predicted value $\hat{Y} = b_0 + b_1 X_1 + \cdots + b_{p-1} X_{p-1}$ is an estimate of

$$E(Y) = \beta_0 + \beta_1 X_1 + \cdots + \beta_{p-1} X_{p-1}.$$

The variance of $\hat{Y}$, $V[b_0 + b_1 X_1 + \cdots + b_{p-1} X_{p-1}]$, is

$$V(b_0) + X_1^2 V(b_1) + \cdots + X_{p-1}^2 V(b_{p-1})$$

$$+ 2 X_1 \text{cov}(b_0, b_1) + \cdots + 2 X_{p-2} X_{p-1} \text{cov}(b_{p-2}, b_{p-1}).$$

This expression can be written very conveniently in matrix notation as follows, where $\mathbf{C} = (\mathbf{X}'\mathbf{X})^{-1}$.

$$V(\hat{Y}) = \sigma^2(\mathbf{X}_0'\mathbf{C}\mathbf{X}_0)$$

$$= \sigma^2 \begin{bmatrix} 1 & X_1 & \cdots & X_{p-1} \end{bmatrix} \begin{bmatrix} c_{00} & c_{01} & \cdots & c_{0,p-1} \\ c_{10} & c_{11} & \cdots & c_{1,p-1} \\ \vdots & & \ddots & \vdots \\ c_{p-1,0} & & & c_{p-1,p-1} \end{bmatrix} \begin{bmatrix} 1 \\ X_1 \\ \vdots \\ X_{p-1} \end{bmatrix}.$$

Thus the $1 - \alpha$ confidence limits on the true mean value of Y at $\mathbf{X}_0$ are given by

$$\hat{Y} \pm t\{(n - p), 1 - \tfrac{1}{2}\alpha\} \cdot s \sqrt{\mathbf{X}_0'\mathbf{C}\mathbf{X}_0}.$$

For example, the variance of $\hat{Y}$ for the point in the X-space ($X_8 = 32$, $X_6 = 22$) is obtained as follows:

$$\text{est. var}(\hat{Y}) = s^2(\mathbf{X}_0'\mathbf{C}\mathbf{X}_0)$$

$$= (0.4377)(1, 32, 22)$$

$$\times \begin{bmatrix} 2.778747 & -0.011242 & -0.106098 \\ -0.011242 & 0.146207 \times 10^{-3} & 0.175467 \times 10^{-3} \\ -0.106098 & 0.175467 \times 10^{-3} & 0.478599 \times 10^{-2} \end{bmatrix} \begin{bmatrix} 1 \\ 32 \\ 22 \end{bmatrix}$$

$$= (0.4377)(0.104140) = 0.045582.$$

The 95% confidence limits on the true mean value of Y at $X_8 = 32$, $X_6 = 22$ are given by

$$\hat{Y} \pm t(22, 0.975) \cdot s \sqrt{X_0' C X_0} = 11.2736 \pm (2.074)(0.213499)$$

$$= 11.2736 \pm 0.4418$$

$$= 10.8318, 11.7154.$$

These limits are interpreted as follows. Suppose repeated samples of Y's are taken of the same size each time and at the same fixed values of (X_8, X_6) as were used to determine the fitted equation obtained above. Then of all the 95% confidence intervals constructed for the mean value of Y for $X_8 = 32$, $X_6 = 22$, 95% of these intervals will contain the true mean value of Y at $X_8 = 32$, $X_6 = 22$. From a practical point of view we can say that there is a 0.95 probability that the statement, the true mean value of Y at $X_8 = 32$, $X_6 = 22$ lies between 10.8318 and 11.7154, is correct.

Confidence Limits for the Mean of g Observations Given a Specific Set of X's

These limits are calculated from

$$\hat{Y} \pm t(\nu, 1 - \tfrac{1}{2}\alpha) \cdot s \sqrt{1/g + X_0' C X_0}.$$

For example, the 95% confidence limits for an individual observation for the point $(X_8 = 32, X_6 = 22)$ are

$$\hat{Y} \pm t(22, 0.975) \cdot s \sqrt{1 + X_0' C X_0} = 11.2736 \pm (2.074)(0.661589)\sqrt{1 + 0.10413981}$$

$$= 11.2736 \pm (2.074)(0.661589)(1.050781)$$

$$= 11.2736 \pm 1.4418$$

$$= 9.8318, 12.7154.$$

Note: To obtain simultaneous confidence surfaces appropriate for the whole regression function over its entire range, it would be necessary to replace $t(\nu, 1 - \tfrac{1}{2}\alpha)$ by $\{pF(p, n - p, 1 - \alpha)\}^{1/2}$, where p is the total number of parameters in the model including β_0. (Currently, $\nu = n - p$. In the example, $n = 25$, $p = 3$.) See, for example, Miller (1981).

Examining the Residuals

The residuals shown in Table 6.1 could be examined to see if they provide any indication that the model is inadequate. We leave this as an exercise, except for the following comments:

1. Residual versus $\hat{Y}$ plot (Figure 6.2). No unusual behavior is indicated.
2. The runs test and the Durbin-Watson test indicated no evidence of time-dependent nonrandomness. (See also Exercise A, in "Exercises for Chapter 7.")

6.3. SUM OF SQUARES OF A SET OF LINEAR FUNCTIONS OF Y's

In some applications, for example, two-level factorial designs, the items of interest are *contrasts*, that is, linear combinations of Y's whose coefficients add to zero, so

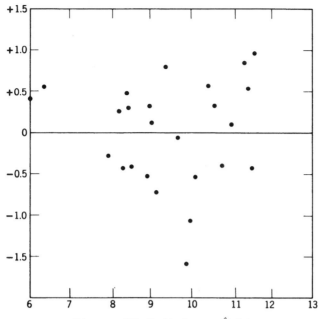

Figure 6.2. Residual versus $\hat{Y}$ plot.

that part of the data is "contrasted" with another part. It is often puzzling in such circumstances to know what is the sum of squares attributable to such a contrast. In regression work, the parameter estimates are also linear combinations of the observations, although not contrasts. We give here a rule that is foolproof for getting an appropriate sum of squares of any set of linear functions $\mathbf{C'Y}$, say, where $\mathbf{C'}$ is an $m \times n$ matrix. The correct answer emerges even if the set contains duplicated linear functions or linear functions that are linear combinations of other linear functions!

Let $\mathbf{C'Y}$ be a set of linear functions of the observations $\mathbf{Y}$. Then the sum of squares due to $\mathbf{C'Y}$ is defined as follows:

$$\text{SS}(\mathbf{C'Y}) = \mathbf{z'C'Y} = \mathbf{Y'Cz}, \qquad (6.3.1)$$

where $\mathbf{z}$ is *any* solution of the equations

$$\mathbf{C'Cz} = \mathbf{C'Y}. \qquad (6.3.2)$$

Sometimes $\mathbf{z}$ is unique, sometimes not. Nevertheless, the resulting sum of squares is always unique. In the above, $\mathbf{C'Y}$ is $m \times 1$, $\mathbf{C'}$ is $m \times n$, $\mathbf{Y}$ is $n \times 1$, and $\mathbf{z}$ is $m \times 1$.

Special Case m = 1. Let $\mathbf{C'} = \mathbf{c'}$, a $1 \times n$ row vector. Then

$$\text{SS}(\mathbf{c'Y}) = (\mathbf{c'Y})^2/\mathbf{c'c}. \qquad (6.3.3)$$

General Nonadditivity of SS. Suppose that

$$\mathbf{C'} = \begin{bmatrix} \mathbf{C'_1} \\ \mathbf{C'_2} \end{bmatrix} \qquad (6.3.4)$$

where $\mathbf{C'_1}$ is $m_1 \times n$ and $\mathbf{C'_2}$ is $m_2 \times n$, where $m_1 + m_2 = m$. Then

$$\text{SS}(\mathbf{C'Y}) = \text{SS}(\mathbf{C'_1Y}) + \text{SS}(\mathbf{C'_2Y}) \qquad (6.3.5)$$

if and only if $C_1'C_2 = 0$, that is, if and only if all the rows of C_1' are orthogonal to all the rows of C_2'. (This can also be expressed as "all the columns of C_1 are orthogonal to all the columns of C_2.")

Example 1. Find the sum of squares due to the least squares estimates $b = (X'X)^{-1}X'Y$ in the nonsingular case.

Here $C' = (X'X)^{-1}X'$, so that z is any solution to

$$(X'X)^{-1}X'X(X'X)^{-1}z = (X'X)^{-1}X'Y,$$

which is clearly given (uniquely) by $z = X'Y$. Thus

$$SS(b) = (Y'C)z = b'X'Y,$$

the familiar formula.

Example 2. Find $SS(\overline{Y})$. We see that $\overline{Y} = \Sigma Y_i/n = c'Y$, where

$$c' = \left(\frac{1}{n}, \frac{1}{n}, \ldots, \frac{1}{n}\right).$$

Thus $c'c = 1/n$ and $SS(\overline{Y}) = \overline{Y}^2/(1/n) = n\overline{Y}^2$.

Example 3. For a straight line fit, find $SS(b_1)$. We write $b_1 = c'Y$, where the ith element of c' is $(X_i - \overline{X})/S_{XX}$. Thus $c'c = 1/S_{XX}$ and $SS(b_1) = S_{XX}b_1^2 = S_{XY}^2/S_{XX}$.

Example 4. For a straight line fit, find $SS(b_0)$. Now $b_0 = \overline{Y} - b_1\overline{X}$ so that the ith element of c' is

$$c_i = \frac{1}{n} - \frac{\overline{X}(X_i - \overline{X})}{S_{XX}}$$

and

$$c'c = \sum_{i=1}^{n} \left\{\frac{1}{n^2} - \frac{2\overline{X}(X_i - \overline{X})}{nS_{XX}} + \frac{\overline{X}^2(X_i - \overline{X})^2}{S_{XX}^2}\right\},$$

$$= \frac{1}{n} + \frac{\overline{X}^2}{S_{XX}}$$

and so

$$SS(b_0) = b_0^2 \left/ \left\{\frac{1}{n} + \frac{\overline{X}^2}{S_{XX}}\right\}\right.$$

Note: This is not the usual $SS(b_0)$ in regression tables, because the usual $SS(b_0) = n\overline{Y}^2$ is really $SS(\overline{Y})$, that is, the SS of the b_0 we would get from the model $Y = \beta_0 + \epsilon$. The $SS(b_0)$ of this example is

$$SS(b_0|b_1) \text{ for the model } Y = \beta_0 + \beta_1 X + \epsilon.$$

Examples 2, 3, and 4 provide important clues as to what formula (6.3.1) produces. It is always the *extra sum of squares*. For the general case, we write

$$\mathbf{Y} = (\mathbf{X}_1, \mathbf{X}_2) \begin{pmatrix} \beta_1 \\ \beta_2 \end{pmatrix} + \boldsymbol{\epsilon},$$

$$\mathbf{X}'\mathbf{X} = \begin{bmatrix} \mathbf{X}_1' \\ \mathbf{X}_2' \end{bmatrix} (\mathbf{X}_1, \mathbf{X}_2) = \begin{bmatrix} \mathbf{X}_1'\mathbf{X}_1 & \mathbf{X}_1'\mathbf{X}_2 \\ \mathbf{X}_2'\mathbf{X}_1 & \mathbf{X}_2'\mathbf{X}_2 \end{bmatrix} = \begin{bmatrix} \mathbf{C}_{11} & \mathbf{C}_{12} \\ \mathbf{C}_{21} & \mathbf{C}_{22} \end{bmatrix},$$

say. We write the inverse as

$$\begin{bmatrix} \mathbf{C}^{11} & \mathbf{C}^{12} \\ \mathbf{C}^{21} & \mathbf{C}^{22} \end{bmatrix}$$

so that

$$\begin{bmatrix} \mathbf{b}_1 \\ \mathbf{b}_2 \end{bmatrix} = \begin{bmatrix} \mathbf{C}^{11}\mathbf{X}_1' + \mathbf{C}^{12}\mathbf{X}_2' \\ \mathbf{C}^{21}\mathbf{X}_1' + \mathbf{C}^{22}\mathbf{X}_2' \end{bmatrix} \mathbf{Y} = \begin{bmatrix} \mathbf{D}_1'\mathbf{Y} \\ \mathbf{D}_2'\mathbf{Y} \end{bmatrix}, \text{ say.}$$

If formulas (6.3.1) and (6.3.2) are applied to $\mathbf{b}_2$ the result will be the extra $SS(\mathbf{b}_2|\mathbf{b}_1)$.

We omit the proof, which is intricate but not difficult. It requires use of the second inverse in Result 7, Appendix 5A, and involves showing that $\mathbf{D}_2'\mathbf{D}_2 = \mathbf{Q}^{-1}$, where $\mathbf{Q} = \mathbf{X}_2'(\mathbf{I} - \mathbf{X}_1(\mathbf{X}_1'\mathbf{X}_1)^{-1}\mathbf{X}_1')\mathbf{X}_2 = \mathbf{X}_2'\mathbf{Z}$, say, where $\mathbf{Z}$ is the residual matrix when $\mathbf{X}_2$ is regressed on $\mathbf{X}_1$. Thus $\mathbf{Z}$ is "the part of $\mathbf{X}_2$ orthogonal to $\mathbf{X}_1$," and it is easy to see that $\mathbf{X}_1'\mathbf{Z} = 0$. It then follows that $\mathbf{z} = \mathbf{Q}\mathbf{D}_2'\mathbf{Y}$ and that $\mathbf{z}'\mathbf{D}_2\mathbf{Y} = \mathbf{Y}'\mathbf{Z}(\mathbf{Z}'\mathbf{Z})^{-1}\mathbf{Z}'\mathbf{Y}$. This is one form of the extra sum of squares.

APPENDIX 6A. ORTHOGONAL COLUMNS IN THE X MATRIX

Suppose we have a regression problem involving parameters β_0, β_1, and β_2. Using the extra sum of squares principle we can calculate a number of quantities such as:

$SS(b_2)$	from the model	$Y = \beta_2 X_2 + \epsilon$
$SS(b_2|b_0)$	from the model	$Y = \beta_0 + \beta_2 X_2 + \epsilon$
$SS(b_2|b_0, b_1)$	from the model	$Y = \beta_0 + \beta_1 X_1 + \beta_2 X_2 + \epsilon$

These will usually have completely different numerical values except when the "β_2" column of the $\mathbf{X}$ matrix is orthogonal to the "β_0" and the "β_1" columns. When this happens we can unambiguously talk about "$SS(b_2)$." We now examine this situation in more detail.

Suppose in the model $\mathbf{Y} = \mathbf{X}\boldsymbol{\beta} + \boldsymbol{\epsilon}$ we divide the matrix $\mathbf{X}$ up into t sets of columns denoted in matrix form by

$$\mathbf{X} = \{\mathbf{X}_1, \mathbf{X}_2, \dots, \mathbf{X}_t\}.$$

A corresponding division can be made in $\boldsymbol{\beta}$ so that

$$\boldsymbol{\beta} = \begin{bmatrix} \beta_1 \\ \beta_2 \\ \vdots \\ \beta_t \end{bmatrix},$$

where the number of columns in $\mathbf{X}_i$ is equal to the number of rows in $\boldsymbol{\beta}_i$, $i = 1, 2,$ $\ldots, t$. The model can then be written

$$E(\mathbf{Y}) = \mathbf{X}\boldsymbol{\beta} = \mathbf{X}_1\boldsymbol{\beta}_1 + \mathbf{X}_2\boldsymbol{\beta}_2 + \cdots + \mathbf{X}_t\boldsymbol{\beta}_t.$$

Suppose that

$$\mathbf{b} = \begin{bmatrix} \mathbf{b}_1 \\ \mathbf{b}_2 \\ \vdots \\ \mathbf{b}_t \end{bmatrix}$$

is the vector estimate of $\boldsymbol{\beta}$ for this model (and given data) obtained from the normal equations

$$\mathbf{X'Xb} = \mathbf{X'Y}.$$

Result. If the columns of $\mathbf{X}_i$ are orthogonal to the columns of $\mathbf{X}_j$ for all $i, j = 1, 2,$ $\ldots, t (i \neq j)$, that is, if $\mathbf{X}_i'\mathbf{X}_j = \mathbf{0}$, it is true that

$$\mathrm{SS}(\mathbf{b}) = \mathrm{SS}(\mathbf{b}_1) + \mathrm{SS}(\mathbf{b}_2) + \cdots + \mathrm{SS}(\mathbf{b}_t)$$

$$= \mathbf{b}_1'\mathbf{X}_1'\mathbf{Y} + \mathbf{b}_2'\mathbf{X}_2'\mathbf{Y} + \cdots + \mathbf{b}_t'\mathbf{X}_t'\mathbf{Y}$$

and $\mathbf{b}_i$ is the least square estimate of $\boldsymbol{\beta}_i$, and $\mathrm{SS}(\mathbf{b}_i) = \mathbf{b}_i'\mathbf{X}_i'\mathbf{Y}$ *whether any of the other terms are in the model or not.* Thus

$$\mathrm{SS}(\mathbf{b}_i) = \mathrm{SS}(\mathbf{b}_i|\text{any set of } \mathbf{b}_j, j \neq i).$$

(Note that it is *not* necessary for the columns of $\mathbf{X}_i$ to be orthogonal to *each other*—only for the $\mathbf{X}_i$ columns all to be orthogonal to all other columns of $\mathbf{X}$.)

We consider the case $t = 2$. Here

$$\mathbf{X} = (\mathbf{X}_1, \mathbf{X}_2),$$

where $\mathbf{X}_1'\mathbf{X}_2 = \mathbf{X}_2'\mathbf{X}_1 = \mathbf{0}$. (This means that all the columns in $\mathbf{X}_1$ are orthogonal to all the columns in $\mathbf{X}_2$.) We can write the model as

$$\mathbf{Y} = \mathbf{X}\boldsymbol{\beta} + \boldsymbol{\epsilon} = \mathbf{X}_1\boldsymbol{\beta}_1 + \mathbf{X}_2\boldsymbol{\beta}_2 + \boldsymbol{\epsilon},$$

where $\boldsymbol{\beta}' = (\boldsymbol{\beta}_1', \boldsymbol{\beta}_2')$ is split into the two sets of coefficients, which correspond to the $\mathbf{X}_1$ and $\mathbf{X}_2$ sets of columns. The normal equations are $\mathbf{X'Xb} = \mathbf{X'Y}$; that is,

$$\begin{bmatrix} \mathbf{X}_1'\mathbf{X}_1 & \mathbf{X}_1'\mathbf{X}_2 \\ \mathbf{X}_2'\mathbf{X}_1 & \mathbf{X}_2'\mathbf{X}_2 \end{bmatrix} \begin{bmatrix} \mathbf{b}_1 \\ \mathbf{b}_2 \end{bmatrix} = \begin{bmatrix} \mathbf{X}_1'\mathbf{Y} \\ \mathbf{X}_2'\mathbf{Y} \end{bmatrix},$$

where a split in $\mathbf{b}$ corresponding to that in $\boldsymbol{\beta}$ has been made. Since the off-diagonal terms $\mathbf{X}_1'\mathbf{X}_2 = \mathbf{0}$, $\mathbf{X}_2'\mathbf{X}_1 = \mathbf{0}$, the normal equations can be split into the two sets of equations

$$\mathbf{X}_1'\mathbf{X}_1\mathbf{b}_1 = \mathbf{X}_1'\mathbf{Y}; \qquad \mathbf{X}_2'\mathbf{X}_2\mathbf{b}_2 = \mathbf{X}_2'\mathbf{Y}$$

with solutions

$$\mathbf{b}_1 = (\mathbf{X}_1'\mathbf{X}_1)^{-1}\mathbf{X}_1'\mathbf{Y}; \qquad \mathbf{b}_2 = (\mathbf{X}_2'\mathbf{X}_2)^{-1}\mathbf{X}_2'Y,$$

assuming that the matrices shown inverted are nonsingular. Thus $\mathbf{b}_1$ is the least squares estimate of $\boldsymbol{\beta}_1$ whether $\boldsymbol{\beta}_2$ is in the model or not, and vice versa. Now

$$SS(\mathbf{b}_1) = \mathbf{b}_1'\mathbf{X}_1'\mathbf{Y} \quad \text{and} \quad SS(\mathbf{b}_2) = \mathbf{b}_2'\mathbf{X}_2'\mathbf{Y}.$$

Thus

$$SS(\mathbf{b}_1, \mathbf{b}_2) = \mathbf{b}'\mathbf{X}'\mathbf{Y}$$

$$= (\mathbf{b}_1', \mathbf{b}_2')(\mathbf{X}_1, \mathbf{X}_2)'\mathbf{Y}$$

$$= (\mathbf{b}_1', \mathbf{b}_2')\begin{pmatrix} \mathbf{X}_1'\mathbf{Y} \\ \mathbf{X}_2'\mathbf{Y} \end{pmatrix}$$

$$= \mathbf{b}_1'\mathbf{X}_1'\mathbf{Y} + \mathbf{b}_2'\mathbf{X}_2'\mathbf{Y}$$

$$= SS(\mathbf{b}_1) + SS(\mathbf{b}_2).$$

It follows that

$$SS(\mathbf{b}_1|\mathbf{b}_2) = SS(\mathbf{b}_1, \mathbf{b}_2) - SS(\mathbf{b}_2) = SS(\mathbf{b}_1).$$

Similarly,

$$SS(\mathbf{b}_2|\mathbf{b}_1) = SS(\mathbf{b}_2)$$

and this depends only on the orthogonality of $\mathbf{X}_1$ and $\mathbf{X}_2$. The extension to cases where $t > 2$ is immediate.

APPENDIX 6B. TWO PREDICTORS: SEQUENTIAL SUMS OF SQUARES

We saw in the example of Section 6.2 that the $SS(b_8, b_6|b_0)$ splits differently according to the sequence of entry.

SS Contribution of Predictor Below When:	First Predictor in Is	
	X_8	X_6
X_8	45.59	35.84
X_6	8.59	18.34
Totals	54.18	54.18

This type of split-up where one variable (here X_8) has the larger SS whether it enters first or second happens quite frequently but other cases can occur, some of them quite strange. Schey (1993) has discussed and illustrated this point using seven "contrived examples."

Table 6B.1 shows the seven data sets and Table 6B.2 shows the disposition of the regression sums of squares using the same formation as above. (Our recomputations vary slightly from Schey's numbers, but the point is not affected.) The model fitted is

$$Y = \beta_0 + \beta_1 X_1 + \beta_2 X_2 + \epsilon,$$

T A B L E 6B.1. Seven Example Sets of Data Devised by Schey (1993)

Set	X_1	X_2	Y	Set	X_1	X_2	Y
1	1.80	12.80	3.71	5	4.66	0.56	6.23
	8.90	12.21	12.29		8.05	−3.94	5.66
	4.76	15.46	−0.79		9.61	−2.86	4.16
	1.86	8.93	0.90		8.59	−1.63	3.43
	2.69	3.92	3.04		3.47	0.22	−2.46
	1.36	1.64	−4.00		4.04	−1.13	−0.41
	7.84	16.07	10.33		0.63	2.25	0.15
	2.79	9.56	4.80		5.91	−0.90	1.18
	3.94	15.55	6.03		8.18	−1.87	8.44
	4.31	3.14	0.86		4.38	−1.10	1.55
2	1.83	−4.20	−0.53	6	0.23	2.12	1.01
	7.17	3.55	7.20		0.15	5.41	−3.73
	6.18	1.57	5.55		5.03	1.90	11.00
	9.44	−2.64	3.02		7.99	−3.36	11.17
	0.86	12.27	−0.56		1.08	2.68	0.32
	0.34	−5.94	−6.36		3.24	4.03	6.65
	5.02	2.35	8.64		9.41	−1.84	7.52
	9.98	5.47	12.24		6.34	−1.62	0.86
	2.00	12.21	5.07		8.17	−4.15	1.55
	7.92	8.14	9.86		5.00	1.48	6.24
3	9.09	−3.97	31.60	7	0.26	2.36	−1.81
	4.09	37.43	21.21		3.74	0.45	1.72
	0.97	−17.97	−4.36		8.46	−1.43	8.13
	1.45	24.94	10.18		9.27	−4.54	1.62
	0.73	7.61	−0.54		0.67	2.45	−1.06
	3.31	−6.93	9.53		9.51	−5.07	7.89
	7.97	19.45	28.53		8.91	−5.49	1.97
	9.81	−41.36	16.77		3.97	3.31	6.69
	0.79	−8.44	−5.94		5.77	−2.69	1.02
	2.15	3.56	2.03		3.50	3.95	8.43
4	3.82	13.36	11.77				
	8.06	20.41	10.79				
	0.01	−1.42	1.27				
	3.27	3.92	1.55				
	5.10	7.49	3.19				
	8.22	12.69	−1.79				
	5.49	15.16	7.86				
	2.98	15.02	9.52				
	8.97	19.46	3.75				
	9.48	12.14	−3.53				

which can be fitted in the form

$$Y - \overline{Y} = \beta_1(X_1 - \overline{X}_1) + \beta_2(X_2 - \overline{X}_2) + \epsilon$$

with identical results. If the columns $X_1 - \overline{X}_1$ and $X_2 - \overline{X}_2$ were orthogonal, the entry order would not matter, as in Appendix 6A.

We see the following:

T A B L E 6B.2. Sequential Sums of Squares in Seven Examples Devised by Schey (1993)

Data Set Number	SS Contribution of Predictor Below When:	First Predictor in Is	
		X_1	X_2
1	X_1	134.10	71.89
	X_2	9.62	71.83
2	X_1	169.12	169.09
	X_2	56.34	56.37
3	X_1	1162.89	1446.14
	X_2	387.22	103.14
4	X_1	14.70	164.41
	X_2	204.51	54.80
5	X_1	43.94	23.60
	X_2	3.16	23.50
6	X_1	73.87	74.21
	X_2	24.85	24.51
7	X_1	28.83	115.16
	X_2	86.33	0.00

1. Case 1 shows X_1 very important when it enters first, and X_1 and X_2 are equally important in the reverse order.
2. Case 2 has the vectors of $(X_1 - \overline{X}_1)$ and $(X_2 - \overline{X}_2)$ orthogonal, so the entry order is irrelevant, as in Appendix 6A.
3. Each variable contributes more when it comes in second than it does when it comes in first! When $SS(b_2|b_1, b_0) > SS(b_2|b_0)$, X_1 is said to be a suppressor variable. Here, both X_1 and X_2 are suppressor variables.
4. This case is similar to the third, but it differs in certain geometrical aspects discussed in the reference, aspects which we have omitted.
5. Similar to 1, but different geometry.
6. A remarkable case because the vectors of $(X_1 - \overline{X}_1)$ and $(X_2 - \overline{X}_2)$ are not orthogonal, but the sums of squares are essentially unchanged. So orthogonality is sufficient, but not necessary, for this SS behavior.
7. An extreme case that is unlikely to arise from real data. Variable X_2 contributes nothing when it goes in first, but accounts for $SS(b_2|b_1, b_0) = 86.33$ when it goes in second!

References

Freund (1988); Hamilton (1987a, b); Mitra (1988); Schey (1993).

EXERCISES FOR CHAPTERS 5 AND 6

A. Consider the data in the following table:

X_0	X_1	X_2	Y
1	1	8	6
1	4	2	8
1	9	-8	1
1	11	-10	0
1	3	6	5
1	8	-6	3
1	5	0	2
1	10	-12	-4
1	2	4	10
1	7	-2	-3
1	6	-4	5

Requirements

1. Using least squares procedures, estimate the β's in the model:

$$Y = \beta_0 X_0 + \beta_1 X_1 + \beta_2 X_2 + \epsilon.$$

2. Write out the analysis of variance table.

3. Using $\alpha = 0.05$, test to determine if the overall regression is statistically significant.

4. Calculate the square of the multiple correlation coefficient, namely, R^2. What portion of the total variation about $\overline{Y}$ is explained by the two variables?

5. The inverse of the $\mathbf{X'X}$ matrix for this problem is as follows:

$$\begin{bmatrix} 4.3705 & -0.8495 & -0.4086 \\ -0.8495 & 0.1690 & 0.0822 \\ -0.4086 & 0.0822 & 0.0422 \end{bmatrix}.$$

Using the results of the analysis of variance table with this matrix, calculate estimates of the following:

a. Variance of b_1.

b. Variance of b_2.

c. The variance of the predicted value of Y for the point $X_1 = 3$, $X_2 = 5$.

6. How useful is the regression using X_1 alone? What does X_2 contribute, given that X_1 is already in the regression?

7. How useful is the regression using X_2 alone? What does X_1 contribute, given that X_2 is already in the regression?

8. What are your conclusions?

B. The table below gives 12 sets of observations on three variables X, Y, and Z. Find the regression plane of X on Y and Z—that is, the linear combination of Y and Z that best predicts the value of X when only Y and Z are given. By constructing an analysis of variance table for X, or otherwise, test whether it is advantageous to include both Y and Z in the prediction formula.

X	Y	Z
1.52	98	77
1.41	76	139
1.16	58	179
1.45	94	95
1.24	73	142
1.21	57	186
1.63	97	82
1.38	91	100
1.37	79	125

X	Y	Z
1.36	92	96
1.40	92	99
1.03	54	190

Source: Cambridge Diploma, 1949.

C. The data below are selected from a much larger body of data referring to candidates for the General Certificate of Education who were being considered for a special award. Here, Y denotes the candidate's total mark, out of 1000, in the G.C.E. examination. Of this mark the subjects selected by the candidate account for a maximum of 800; the remainder, with a maximum of 200, is the mark in the compulsory papers—"General" and "Use of English"—this mark is shown as X_1. X_2 denotes the candidate's mark, out of 100, in the compulsory School Certificate English Language paper taken on a previous occasion.

Compute the multiple regression of Y on X_1 and X_2, and make the necessary tests to enable you to comment intelligently on the extent to which current performance in the compulsory papers may be used to predict aggregate performance in the G.C.E. examination, and on whether previous performance in School Certificate English Language has any predictive value independently of what has already emerged from the current performance in the compulsory papers.

Candidate	Y	X_1	X_2	Candidate	Y	X_1	X_2
1	476	111	68	9	645	117	59
2	457	92	46	10	556	94	97
3	540	90	50	11	634	130	57
4	551	107	59	12	637	118	51
5	575	98	50	13	390	91	44
6	698	150	66	14	562	118	61
7	545	118	54	15	560	109	66
8	574	110	51				

Source: Cambridge Diploma, 1953. (Exercises B and C are published with permission of Cambridge University Press.)

D. Eight runs were made at various conditions of saturation (X_1) and transisomers (X_2). The response, SCI, is listed below as Y for the corresponding levels of X_1 and X_2.

Y	X_1	X_2
66.0	38	47.5
43.0	41	21.3
36.0	34	36.5
23.0	35	18.0
22.0	31	29.5
14.0	34	14.2
12.0	29	21.0
7.6	32	10.0

1. Fit the model $Y = \beta_0 + \beta_1 X_1 + \beta_2 X_2 + \epsilon$.
2. Is the overall regression significant? (Use $\alpha = 0.05$.)
3. How much of the variation in Y about $\overline{Y}$ is explained by X_1 and X_2?

E. The effect of sealer plate temperature and sealer plate clearance in a soap wrapping machine affects the percentage of wrapped bars that pass inspection. Some data on these variables were collected and are shown as follows:

Sealer Plate Clearance, X_1	Sealer Plate Temperature, X_2	% Sealed Properly, Y
130	190	35.0
174	176	81.7
134	205	42.5
191	210	98.3
165	230	52.7
194	192	82.0
143	220	34.5
186	235	95.4
139	240	56.7
188	230	84.4
175	200	94.3
156	218	44.3
190	220	83.3
178	210	91.4
132	208	43.5
148	225	51.7

Requirements

1. Assume a linear model $Y = \beta_0 + \beta_1 X_1 + \beta_2 X_2 + \epsilon$ and determine least squares estimates of β_0, β_1, and β_2.
2. Is the overall regression significant? (Use $\alpha = 0.05$.)
3. Is one of the two variables more useful than the other in predicting the percentage sealed properly?
4. What recommendations would you make concerning the operation of the wrapping machine?

F. Using the 17 observations given below:
1. Fit the model $Y = \beta_0 + \beta_1 X_1 + \beta_2 X_2 + \epsilon$.
2. Test for lack of fit, using pure error.
3. Examine the residuals.
4. Assess the value of including each of the variables X_1 and X_2 in the regression model.

X_1	X_2	Y
17	42	90
19	45	71,76
20	29	63, 63, 80, 80
21	93	80, 64, 82, 66
25	34	75, 82
27	98	99
28	9	73
30	73	67, 74

G. Fit the model $Y = \beta_0 + \beta_1 X_1 + \beta_2 X_2 + \epsilon$ to the data below. Check for lack of fit via both pure error and the examination of residuals. Assess the value of including each of the predictors X_1 and X_2 in the regression model.

X_1	X_2	Y
2.6	3.9	83
2.8	4.2	64
2.8	4.2	69
2.9	2.6	56
2.9	2.6	56
2.9	2.6	73

X_1	X_2	Y
2.9	2.6	73
3.0	9.0	57
3.0	9.0	59
3.0	9.0	73
3.0	9.0	75
3.4	3.1	68
3.4	3.1	75
3.6	9.5	92
3.7	0.6	66
3.9	7.0	60
3.9	7.0	67

H. (Please refer to Exercise LL in "Exercises for Chapters 1–3" first.) After he had analyzed the original data, the manager rechecked the records to try to find additional information that would improve his model. He drew out the facts that the numbers of men working at any one time were, for the listing of days shown originally, as follows:

$$Z = 0, 0, 0, 0, 0, 0, 1, 0, 0, 1, 3, 6, 6.$$

Fit a planar model $Y = \beta_0 + \beta_1 X + \beta_2 Z + \epsilon$ to the entire data set via least squares, check for lack of fit, and (if there is no lack of fit) test for overall regression. Also test $H_0: \beta_2 = 0$ versus $H_1: \beta_2 \neq 0$ using the extra sum of squares principle. What conclusions do you draw from your analysis?

Useful Facts.

$$(\mathbf{X'X})^{-1} = \begin{bmatrix} 13 & 65 & 17 \\ 65 & 437 & 155 \\ 17 & 155 & 83 \end{bmatrix}^{-1} = \frac{1}{24{,}780} \begin{bmatrix} 12{,}246 & -2{,}760 & 2{,}646 \\ -2{,}760 & 790 & -910 \\ 2{,}646 & -910 & 1{,}456 \end{bmatrix},$$

$$\mathbf{X'Y} = \begin{bmatrix} 2{,}990 \\ 19{,}120 \\ 6{,}050 \end{bmatrix}, \quad \mathbf{b} = \frac{1}{24{,}780} \begin{bmatrix} -147{,}360 \\ 1{,}346{,}900 \\ -678{,}860 \end{bmatrix} = \begin{bmatrix} -5.947 \\ 54.354 \\ -27.396 \end{bmatrix}.$$

I. Fit the model $Y = \beta_0 + \beta_1 X_1 + \beta_2 X_2 + \epsilon$ to the data below, provide an analysis of variance table, and perform the partial F-tests to test $H_0: \beta_i = 0$ versus $H_1: \beta_i \neq 0$ for $i = 1, 2$, given that the other variable is already in the model. Comment on the relative contributions of the variables X_1 and X_2 depending on whether they enter the model first or second.

X_1	X_2	Y
−5	5	11
−4	4	11
−1	1	8
2	−3	2
2	−2	5
3	−2	5
3	−3	4

J. Fit the model $Y = \beta_0 + \beta_1 X_1 + \beta_2 X_2 + \epsilon$ to the data below. After testing for lack of fit, find the appropriate extra SS F-statistic for testing $H_0: \beta_2 = 0$ versus $H_1: \beta_2 \neq 0$, and find its degrees of freedom. Relate this F-statistic numerically to the t-statistic typically used to test the same hypothesis.

X_1	X_2	Y
-1	-1	8
-1	1	13
-1	1	12
-1	1	11
1	-1	9
1	-1	8
1	-1	7
1	1	13
0	0	11
0	0	13

K. The questions below relate to fitting the model $Y = \beta_0 + \beta_1 X_1 + \beta_2 X_2 + \epsilon$ to the following data:

	X_1	X_2	Y
	-1	-1	7.2
	-1	0	8.1
	0	0	9.8
	1	0	12.3
	1	1	12.9
Sum	0	0	50.3
Sum of squares	4	2	531.19

1. Write down the normal equation $(\mathbf{X'X})\mathbf{b} = \mathbf{X'Y}$ in matrix format.
2. Obtain the solution $\mathbf{b} = (\mathbf{X'X})^{-1}\mathbf{X'Y}$ using matrix manipulations.
3. Find SS(b_0, b_1, b_2) via matrix manipulations.
4. Find the residual sum of squares, and obtain s^2.
5. Evaluate se(b_i), $i = 0, 1, 2$.
6. Find $\hat{Y}_0$ at the point $(X_{10}, X_{20}) = (0.5, 0)$.
7. Obtain se$(\hat{Y}_0)$.
8. Find SS$(b_2|b_1, b_0)$.
9. You are now told that the 9.8 value at $(0, 0)$ in the data is the average of four observations so that the variance of this Y is $\sigma^2/4$ and not σ^2. However, all observations are still independent. Your informant says that you should have used weighted least squares $\mathbf{b} = (\mathbf{X'V}^{-1}\mathbf{X})^{-1}\mathbf{X'V}^{-1}\mathbf{Y}$, where $\mathbf{V}$ is a diagonal matrix here, to get your estimates. Provide the new (weighted least squares) values of b_0, b_1, and b_2. (See Section 9.2.)

L. For the experimental situation described in Exercise DD of "Exercises for Chapters 1–3," suppose that data for the concentratioin of chemical A had been recorded for each run. It is suggested that the variation found in this factor might be causing the large variation in response discovered previously. The readings of the concentration factor C (in percent) are:

Batch	1	2	3	4	5	6	7	8	9
C	6	6	8	7	9	8	5	9	11

where the batch numbers are the same as in Exercise DD.
1. Plot the residuals obtained in Exercise DD versus C. Notice anything?

2.* Fit the model $Y = \beta_0 + \beta_1 X_1 + \beta_2 X_2 + \epsilon$ to the data, where $X_1 = (T - 300)/10$ and $X_2 = C - 8$.

3. Perform an analysis of variance and test:

  a. The lack of fit.

  b. The significance of the effect of including β_1 and β_2 in the model, rather than only β_0.

  c. The significance of including β_2 in the model, rather than just β_0 and β_1.

4. What percentage of the total variation (corrected) has been "taken up" by including the concentration effect in our model?

5. What is the standard error of $\tilde{b}_1$? Of $\tilde{b}_2$? ($\tilde{b}_1$ and $\tilde{b}_2$ are the regression coefficients in the expression for $\hat{Y}$ in terms of the original variables T and C.)

6. Write the fitted value and the residual for each batch. Notice anything?

7. What is the estimated variance of the predicted value $\hat{Y}$ at the point $T = 315$, $C = 8$?

M. Using the data given in Table 1.1, find a joint 90% confidence region for (β_0, β_1). $[F(2, 23, 0.90) = 2.55.]$

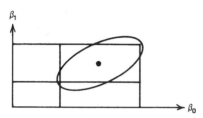

Figure M.1.

On an accurate figure (Figure M1 shows the appropriate format) draw the following:

1. The estimated point (b_0, b_1).

2. The 90% confidence contour for (β_0, β_1).

3. The 95% confidence intervals for β_0 and β_1, separately, and the rectangle these jointly indicate.

Comment briefly on your results.

Hints: (a) Equation (5.3.7), written out, is a quadratic equation in β_0 and β_1. To get the ellipse, set a value of β_0 and solve the resulting quadratic for two values of β_1 to get an upper and lower point on the ellipse (see Figure M2). Imaginary roots mean that your selected β_0 value is outside the ellipse (see Figure M3). Repeat for several values of β_0 and join up the points. It is easiest to use the computer for the calculations.

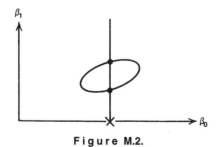

Figure M.2.

*If you employ the coding system $X_1 = (T - 300)/10$, $X_2 = C - 8$, you may use the following hint:

$$\begin{bmatrix} 9 & -2 & -3 \\ -2 & 46 & 13 \\ -3 & 13 & 29 \end{bmatrix}^{-1} = \frac{1}{10111} \begin{bmatrix} 1165 & 19 & 112 \\ 19 & 252 & -111 \\ 112 & -111 & 410 \end{bmatrix}.$$

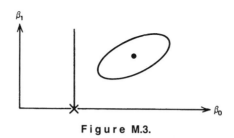

Figure M.3.

(b) Because the two intervals are 95% ones, the rectangle is a sort of $(0.95)(0.95) = 0.9025$ or $90\frac{1}{4}\%$ joint region—an incorrect one, we know, but that is the probability level. Thus we compare it with a 90% true region, the nearest we can easily look up from the F-tables.

N. Show that the square of the multiple correlation coefficient R^2 is equal to the square of the correlation between $\mathbf{Y}$ and $\hat{\mathbf{Y}}$

O. Consider the formal regression of the residuals e_i onto a quadratic function $\alpha_0 + \alpha_1 \hat{Y}_i + \alpha_2 \hat{Y}_i^2$ of the fitted values $\hat{Y}_i$, by least squares. Show that all three estimated coefficients depend on $T_{12} = \Sigma e_i \hat{Y}_i^2$. What does this imply?

P. We fit a straight line model to a set of data using the formulas $\mathbf{b} = (\mathbf{X'X})^{-1}\mathbf{X'Y}$, $\hat{\mathbf{Y}} = \mathbf{Xb}$ with the usual definitions. We define $\mathbf{H} = \mathbf{X(X'X)}^{-1}\mathbf{X'}$. Show that

$$\text{SS(due to regression)} = \mathbf{Y'HY}$$

$$= \hat{\mathbf{Y}}'\hat{\mathbf{Y}}$$

$$= \hat{\mathbf{Y}}'\mathbf{H}^3\mathbf{Y}.$$

Q. Show that $\mathbf{X'e} = \mathbf{0}$.

R. Show that, for any linear model

$$\sum_{i=1}^{n} V(\hat{Y}_i)/n = \text{trace}\{\mathbf{X(X'X)}^{-1}\mathbf{X'}\}\sigma^2/n = p\sigma^2/n.$$

S. See Exercise Y in "Exercises for Chapters 1–3." Would that result extend if there were more X's? (Yes.)

T. Suppose $\mathbf{Y} = \mathbf{X\boldsymbol{\beta}} + \boldsymbol{\epsilon}$ is a regression model containing a β_0 term in the first position, and $\mathbf{1} = (1, 1, \ldots, 1)'$ is an $n \times 1$ vector of ones. Show that $(\mathbf{X'X})^{-1}\mathbf{X'1} = (1, 0, \ldots, 0)'$ and hence that $\mathbf{1'X(X'X)}^{-1}\mathbf{X'1} = n$. (*Hint:* $\mathbf{X'1}$ is the first column of $\mathbf{X'X}$.) These results can be useful in regression matrix manipulations. For connected reading, see letters in The *American Statistician*, April 1972, 47–48.

U. By noting that $\mathbf{X}_0 = (1, \overline{X}_1, \overline{X}_2, \ldots)'$ can be written as $\mathbf{X'1}/n$, and applying the result in Exercise T above, show that $V(\hat{Y})$ at the point $(\overline{X}_1, \overline{X}_2, \ldots, \overline{X}_n)$ is σ^2/n.

V. Look again at the (X, Y_2) data of Exercise E in "Exercises for Chapters 1–3." Fit the quadratic model

$$Y = \beta_0 + \beta_1 X + \beta_{11} X^2 + \epsilon$$

to these data and provide the usual subsidiary analyses. Draw the fitted curve on a plot of the data and estimate the abscissa value at the minimum point of the curve. [*Hint:* It is at $-b_1/(2b_{11})$.] What does your conclusion mean?

[*Note:* You may have to code the data if your intended regression is frustrated by the computer. If that happens, why does it happen? A suggested coding is $U = (X - 0.048)/0.048$. You do not have to use this if you do not wish to!]

W. Look at Appendix 6B. Perform at least one of the regressions yourself and check the results against those given.

X. Show that, in the general linear regression situation with a β_0 term in the model:
1. The correlation between the vectors $\mathbf{e}$ and $\mathbf{Y}$ is $(1 - R^2)^{1/2}$. The implication of this

result is that it is a mistake to attempt to find defective regressions by a plot of residuals e_i versus observations Y_i as this will always show a slope.

2. Show that this slope is $1 - R^2$.

3. Show, further, that the correlation between **e** and $\hat{\mathbf{Y}}$ is zero.

Y. Four levels, coded as -3, -1, 1, and 3, were chosen for each of two variables X_1 and X_2, to provide a total of 16 experimental conditions when all possible combinations (X_1, X_2) were taken. It was decided to use the resulting 16 observations to fit a regression equation including a constant term, all possible first-order, second-order, third-order, and fourth-order terms in X_1 and X_2. The data were fed into a computer routine, which usually obtains a vector estimate

$$\mathbf{b} = (\mathbf{X'X})^{-1}\mathbf{X'Y}.$$

The computer refused to obtain the estimates. Why?

The experimeter, who had meanwhile examined the data, decided at this stage to ignore the levels of variable X_2 and fit a fourth-order model in X_1 only to the *same* observations. The computer again refused to obtain the estimates. Why?

Z. The cloud point of a liquid is a measure of the degree of crystallization in a stock that can be measured by the refractive index. It has been suggested that the percentage of I-8 in the base stock is an excellent predictor of cloud point using the second-order model:

$$Y = \beta_0 + \beta_1 X + \beta_{11}X^2 + \epsilon.$$

The following data were collected on stocks with known percentage of I-8.

% I-8, X	Cloud Point, Y	% I-8, X	Cloud Point, Y
0	22.1	2	26.1
1	24.5	4	28.5
2	26.0	6	30.3
3	26.8	8	31.5
4	28.2	10	33.1
5	28.9	0	22.8
6	30.0	3	27.3
7	30.4	6	29.8
8	31.4	9	31.8
0	21.9		

Requirements

1. Determine the best-fitting second-order model.

2. Using $\alpha = 0.05$, check the overall regression.

3. Test for lack of fit.

4. Would the first-order model, $Y = \beta_0 + \beta_1 X + \epsilon$, have been sufficient? Use the residuals from this simpler model to support your conclusions.

5. Comment on the use of the fitted second-order model as a predictive equation.

AA. A certain experiment gives observations $(Y_1, Y_2, Y_3, Y_4) = (4, 2, 1, 5)$. What is the sum of squares of the set of linear functions $L_1 = Y_1 + 2Y_2 + 2Y_3 + Y_4$ and $L_2 = Y_1 - Y_2 - Y_3 + Y_4$? What is the sum of squares of the set of linear functions L_1, L_2, and L_3, where $L_3 = -3Y_1 + 3Y_4$.

CHAPTER 7

Serial Correlation in the Residuals and the Durbin–Watson Test

In what follows it is assumed that the time order (or some other type of order) of the observations is known, and that observations are equally spaced in that ordering.

7.1. SERIAL CORRELATION IN RESIDUALS

In regression work, we typically assume that the observational errors are pairwise uncorrelated. If this assumption were substantially untrue, we would expect that the plot of residuals in time order, or some other sensible order defined by the practical circumstances, would help us to detect it. There are, of course, many ways in which the errors may be correlated. A common way is that they may be *serially correlated*, that is, the correlations between errors s steps apart are always the same. We shall use the notation ρ_s for this correlation, $s = 1, 2, \ldots$.

More specifically, if residuals exhibit local positive serial correlation, successive residuals in a time sequence tend to be more alike than otherwise, and a time plot of them will have the general characteristics of Figure 7.1a, rising and falling but with close points more alike than otherwise. The correlation between residuals one (or two, or three, ...) step(s) apart is called the lag-1 (or 2, or 3, ...) serial correlation. The empirical lag-1 serial correlation can be examined by plotting each residual except the first against the one preceding it. The positive lag-1 serial correlation present in the data of Figure 7.1a reveals itself in the "lower-left to upper-right" tendency of such a plot, shown in Figure 7.1b. To view correlations for higher lags we can make similar plots for residuals two steps apart, three steps apart, and so on.

Negative serial correlation between successive residuals can also arise. One cause is a phenomenon known as carryover, which occurs in batch processes. This can happen as follows. Suppose that, for a particular batch of product in a process, incomplete recovery occurs because some of the product is left in the pipelines and pumps of the reactor system. The recorded yield for this batch will be unusually low. In the next batch, however, there would be a tendency for the material left behind to be recovered, thus giving an unusually high batch yield. A pattern of residuals may result like those plotted in Figure 7.2a, in which a positive value tends to be followed by a negative one, and vice versa. The existence of negative lag-1 serial correlation for these data is shown by the "lower-right to upper-left" pattern of Figure 7.2b.

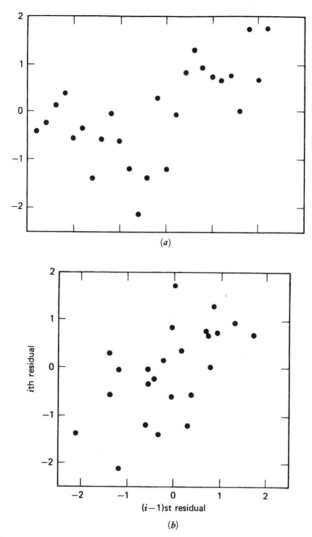

Figure 7.1. (*a*) A series of residuals exhibiting local positive serial correlation. (*b*) Lag-1 serial plot for this series.

Figure 7.3*a* shows a random series of residuals and Figure 7.3*b* shows the corresponding lag-1 serial plot that exhibits no tendency of trend at all.

We can characterize the behavior shown in the three figures as *attraction* (successive residuals are "like" one another but the plot wanders around), *repulsion* (successive residuals repel or are "unlike" one another), and unrelated (successive residuals are "almost independent"; they are of course related by the normal equations.)

The study of serial correlation patterns is one of the techniques used in time-series analysis. Such special analysis of correlated data can often be rewarding. The interested reader should look at texts such as Box, Jenkins and Reinsel (1994). Here we are concerned only with the detection of serial correlation in regression residuals.

A well-known way of checking for serial correlation patterns in an equally spaced sequence of residuals is via the Durbin–Watson test, which we now describe. A simpler, less sophisticated runs test is described in Section 7.3.

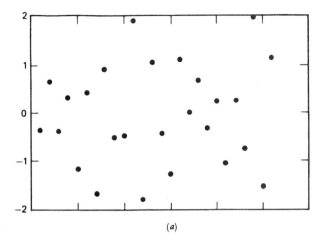

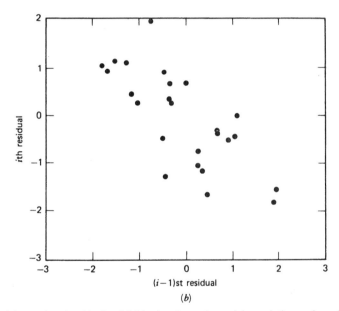

Figure 7.2. (*a*) A series of residuals exhibiting local negative serial correlation perhaps due to carryover. (*b*) Lag-1 serial plot for this series.

7.2. THE DURBIN–WATSON TEST FOR A CERTAIN TYPE OF SERIAL CORRELATION

A popular test for detecting a certain type of serial correlation is the *Durbin–Watson test*. (This is named after the two authors who discussed its use for testing regression residuals and provided suitable testing tables in 1951. It was originally put foward by Von Neumann for nonregression problems in 1941. Selected sources are listed at the end of the chapter.)

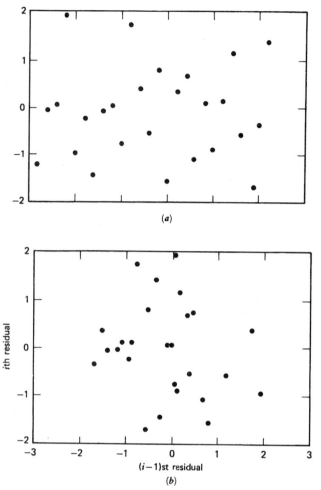

Figure 7.3. (a) An uncorrelated series of residuals. (b) Lag-1 serial plot for this series.

Suppose we wish to fit a postulated linear model

$$Y_u = \beta_0 + \sum_{i=1}^{k} \beta_i X_{iu} + \epsilon_u \tag{7.2.1}$$

by least squares to observations $(Y_u, X_{1u}, X_{2u}, \ldots X_{ku})$, $u = 1, 2, \ldots, n$. We would usually assume that the errors ϵ_u are independent $N(0, \sigma^2)$ variables, so that all serial correlations $\rho_s = 0$. We want to see if this assumption is justified by checking the residuals. We shall test this null hypothesis H_0: all $\rho_s = 0$ via the Durbin–Watson test against the alternative

$$H_1: \rho_s = \rho^s \tag{7.2.2}$$

$(\rho \neq 0$ and $|\rho| < 1)$, an alternative that arises from the assumption that the errors ϵ_u are such that

$$\epsilon_u = \rho\epsilon_{u-1} + z_u, \tag{7.2.3}$$

where $z_u \sim N(0, \sigma^2)$ and is independent of $\epsilon_{u-1}, \epsilon_{u-2}, \ldots$, and of $z_{u-1}, z_{u-2}, \ldots$. It is also assumed that both the mean and variance of ϵ_u are constant, independent of u, whereupon it follows, necessarily, that $\epsilon_u \sim N\{0, \sigma^2/(1 - \rho^2)\}$. Note that, under the null hypothesis $H_0: \rho = 0$, this reduces to $\epsilon_u \sim N(0, \sigma^2)$, our usual assumptions for all $u = 1, 2, \ldots, n$.

Although the test is fully appropriate only against the specific alternative (7.2.2), it is typically applied in a general way without much thought of alternatives. This means that it will often lose power compared to when it is properly employed.

To test H_0 against H_1 we fit the model from Eq. (7.2.1) and find the residuals e_1, $e_2, \ldots, e_n$. We then form the Durbin–Watson statistic

$$d = \sum_{u=2}^{n} (e_u - e_{u-1})^2 / \sum_{u=1}^{n} e_u^2 \qquad (7.2.4)$$

and determine whether or not to reject the null hypothesis H_0 on the basis of the value of d. The distribution of d depends on the X-data and is not independent of them (as, for example, is the t-distribution, apart from degrees of freedom, in other regression contexts). The distribution of d lies between 0 and 4 and is symmetric about 2. Percentage points also depend on the X-data and would have to be calculated for each application to perform the test properly. Because of the difficulty of doing this routinely, the test is usually performed using tabled bounds (d_L, d_U), where L = lower and U = upper, on the percentage points. Thus instead of looking up a single critical value, we have to look up two critical values. Moreover, d is used only for a lower-tailed test against alternatives $\rho > 0$. To test against the alternative $\rho < 0$, we theoretically need an upper-tailed test; fortunately, this can simply be handled as a lower-tailed test using the statistic $(4 - d)$.

Note that the extremes 0 and 4 are attainable only for very large samples. The minimum attainable values depend on the sample size n in the following way:

n	15	30	50	100	200	300	500
Minimum d	0.0437	0.0110	0.0039	0.0010	0.0002	0.0001	0.0000

The corresponding maximum d values are $(4 - $ minimum $d)$. This display and several portions of tables below are quoted and adapted from Savin and White (1977), fully referenced as (2) below.

Primary Test, Tables of d_L and d_U

Tables 7.1, 7.2, and 7.3 show pairs of 1%, 2.5%, and 5% lower-tail significance points, that is, critical values for probability levels $\alpha = 0.01, 0.025$, and 0.05, respectively, called (d_L, d_U). These are given for various numbers of observations n, and for $k = 1, 2, \ldots, 20$ predictor variables X_i [see k in Eq. (7.2.1)]. The sources of these tables are two papers:

1. J. Durbin and G. S. Watson (1951). Testing for serial correlation in least squares regression, II. *Biometrika*, **38**, 159–178 (DW51).
2. N. E. Savin and K. J. White (1977). The Durbin–Watson test for serial correlation with extreme sample sizes or many regressors. *Econometrica*, **45**, 1989–1996 (SW77).

In our amended versions of tables reproduced from SW77, figures are rounded to two decimal places and some combinations of (n, k) values that result in relatively few residual degrees of freedom have been omitted. Also, we have omitted all d_U values that exceed 1.99. If the observed d statistic equaled or exceeded the distribution mean 2, there would be little point doing a lower-tailed test.

T A B L E   7.1. Significance Points of d_L and d_U:1%

	k = 1		k = 2		k = 3		k = 4		k = 5	
n	d_L	d_U	d_L	d_U	d_L	d_U	d_L	d_U	d_L	d_U
15	0.81	1.07	0.70	1.25	0.59	1.46	0.49	1.70	0.39	1.96
16	0.84	1.09	0.74	1.25	0.63	1.44	0.53	1.66	0.44	1.90
17	0.87	1.10	0.77	1.25	0.67	1.43	0.57	1.63	0.48	1.85
18	0.90	1.12	0.80	1.26	0.71	1.42	0.61	1.60	0.52	1.80
19	0.93	1.13	0.83	1.26	0.74	1.41	0.65	1.58	0.56	1.77
20	0.95	1.15	0.86	1.27	0.77	1.41	0.68	1.57	0.60	1.74
21	0.97	1.16	0.89	1.27	0.80	1.41	0.72	1.55	0.63	1.71
22	1.00	1.17	0.91	1.28	0.83	1.40	0.75	1.54	0.66	1.69
23	1.02	1.19	0.94	1.29	0.86	1.40	0.77	1.53	0.70	1.67
24	1.04	1.20	0.96	1.30	0.88	1.41	0.80	1.53	0.72	1.66
25	1.05	1.21	0.98	1.30	0.90	1.41	0.83	1.52	0.75	1.65
26	1.07	1.22	1.00	1.31	0.93	1.41	0.85	1.52	0.78	1.64
27	1.09	1.23	1.02	1.32	0.95	1.41	0.88	1.51	0.81	1.63
28	1.10	1.24	1.04	1.32	0.97	1.41	0.90	1.51	0.83	1.62
29	1.12	1.25	1.05	1.33	0.99	1.42	0.92	1.51	0.85	1.61
30	1.13	1.26	1.07	1.34	1.01	1.42	0.94	1.51	0.88	1.61
31	1.15	1.27	1.08	1.34	1.02	1.42	0.96	1.51	0.90	1.60
32	1.16	1.28	1.10	1.35	1.04	1.43	0.98	1.51	0.92	1.60
33	1.17	1.29	1.11	1.36	1.05	1.43	1.00	1.51	0.94	1.59
34	1.18	1.30	1.13	1.36	1.07	1.43	1.01	1.51	0.95	1.59
35	1.19	1.31	1.14	1.37	1.08	1.44	1.03	1.51	0.97	1.59
36	1.21	1.32	1.15	1.38	1.10	1.44	1.04	1.51	0.99	1.59
37	1.22	1.32	1.16	1.38	1.11	1.45	1.06	1.51	1.00	1.59
38	1.23	1.33	1.18	1.39	1.12	1.45	1.07	1.52	1.02	1.58
39	1.24	1.34	1.19	1.39	1.14	1.45	1.09	1.52	1.03	1.58
40	1.25	1.34	1.20	1.40	1.15	1.46	1.10	1.52	1.05	1.58
45	1.29	1.38	1.24	1.42	1.20	1.48	1.16	1.53	1.11	1.58
50	1.32	1.40	1.28	1.45	1.24	1.49	1.20	1.54	1.16	1.59
55	1.36	1.43	1.32	1.47	1.28	1.51	1.25	1.55	1.21	1.59
60	1.38	1.45	1.35	1.48	1.32	1.52	1.28	1.56	1.25	1.60
65	1.41	1.47	1.38	1.50	1.35	1.53	1.31	1.57	1.28	1.61
70	1.43	1.49	1.40	1.52	1.37	1.55	1.34	1.58	1.31	1.61
75	1.45	1.50	1.42	1.53	1.39	1.56	1.37	1.59	1.34	1.62
80	1.47	1.52	1.44	1.54	1.42	1.57	1.39	1.60	1.36	1.62
85	1.48	1.53	1.46	1.55	1.43	1.58	1.41	1.60	1.39	1.63
90	1.50	1.54	1.47	1.56	1.45	1.59	1.43	1.61	1.41	1.64
95	1.51	1.55	1.49	1.57	1.47	1.60	1.45	1.62	1.42	1.64
100	1.52	1.56	1.50	1.58	1.48	1.60	1.46	1.63	1.44	1.65
150	1.61	1.64	1.60	1.65	1.58	1.67	1.57	1.68	1.56	1.69
200	1.66	1.68	1.65	1.69	1.64	1.70	1.63	1.72	1.62	1.72

Source: DW51 for $n \leq 100$ and SW77 for $n = 150,200$ (see text).

The testing procedures are as follows:

1. *One-sided test against alternatives $\rho > 0$.*

   If $d < d_L$, conclude d is significant, reject H_0, at level α.
   If $d > d_U$, conclude d is not significant, do not reject H_0.
   If $d_L \leq d \leq d_U$, the test is said to be inconclusive.

T A B L E 7.1. Significance Points of d_L and d_U:1% (*Continued*)

n	k = 6		k = 7		k = 8		k = 9		k = 10	
	d_L	d_U	d_L	d_U	d_L	d_U	d_L	d_U	d_L	d_U
20	0.52	1.92	0.44		0.36		0.29		0.23	
21	0.55	1.88	0.47		0.40		0.33		0.27	
22	0.59	1.85	0.51		0.44		0.37		0.30	
23	0.62	1.82	0.55	1.98	0.47		0.40		0.34	
24	0.65	1.80	0.58	1.94	0.51		0.44		0.38	
25	0.68	1.78	0.61	1.92	0.54		0.47		0.41	
26	0.71	1.76	0.64	1.89	0.57		0.51		0.44	
27	0.74	1.74	0.67	1.88	0.60		0.54		0.47	
28	0.76	1.73	0.70	1.85	0.63	1.97	0.57		0.50	
29	0.79	1.72	0.72	1.83	0.66	1.95	0.60		0.53	
30	0.81	1.71	0.75	1.81	0.68	1.93	0.62		0.56	
31	0.83	1.70	0.77	1.80	0.71	1.91	0.65		0.59	
32	0.86	1.69	0.79	1.79	0.73	1.89	0.67		0.61	
33	0.88	1.68	0.82	1.78	0.76	1.87	0.70	1.98	0.64	
34	0.90	1.68	0.84	1.77	0.78	1.86	0.72	1.96	0.67	
35	0.91	1.67	0.86	1.76	0.80	1.85	0.74	1.94	0.69	
36	0.93	1.67	0.88	1.75	0.82	1.84	0.77	1.93	0.71	
37	0.95	1.66	0.90	1.74	0.84	1.83	0.79	1.91	0.73	
38	0.97	1.66	0.91	1.74	0.86	1.82	0.81	1.90	0.75	1.99
39	0.98	1.66	0.93	1.73	0.88	1.81	0.83	1.89	0.77	1.97
40	1.00	1.65	0.95	1.72	0.90	1.80	0.84	1.88	0.79	1.96
45	1.07	1.64	1.02	1.70	0.97	1.77	0.97	1.83	0.88	1.90
50	1.12	1.64	1.08	1.69	1.04	1.75	1.00	1.81	0.96	1.86
55	1.17	1.64	1.13	1.69	1.10	1.73	1.06	1.79	1.02	1.84
60	1.21	1.64	1.18	1.68	1.14	1.73	1.11	1.77	1.07	1.82
65	1.25	1.64	1.22	1.68	1.19	1.72	1.15	1.76	1.12	1.80
70	1.28	1.65	1.25	1.68	1.22	1.72	1.19	1.75	1.16	1.79
75	1.31	1.65	1.28	1.68	1.26	1.72	1.23	1.75	1.20	1.79
80	1.34	1.65	1.31	1.68	1.29	1.71	1.26	1.75	1.23	1.78
85	1.36	1.66	1.34	1.69	1.31	1.71	1.29	1.74	1.26	1.77
90	1.38	1.66	1.36	1.69	1.34	1.71	1.31	1.74	1.29	1.77
95	1.40	1.67	1.38	1.69	1.36	1.72	1.34	1.74	1.31	1.77
100	1.42	1.67	1.40	1.69	1.38	1.72	1.36	1.74	1.34	1.77
150	1.54	1.71	1.53	1.72	1.52	1.74	1.50	1.75	1.49	1.77
200	1.61	1.74	1.60	1.75	1.59	1.76	1.58	1.77	1.57	1.80

Source: SW77 (see text). Copyright The Econometric Society.

2. *One-sided test against alternative $\rho < 0$. Repeat (1) using $(4 - d)$ in place of d.*
3. *Two-sided equal-tailed test against alternatives $\rho \neq 0$.*

If $d < d_L$ or $4 - d < d_L$, conclude d is significant, reject H_0 at level 2α.
If $d > d_U$ and $4 - d > d_U$, conclude d is not significant, do not reject H_0 at level 2α. Otherwise, the test is said to be inconclusive.

A Simplified Test

The inconclusive feature of the tests above is not attractive, but the problem is a difficult one. In later work, procedures for deciding inconclusive cases were formulated,

T A B L E 7.1. Significance Points of d_L and d_U:1% (Continued)

	k = 11		k = 12		k = 13		k = 14		k = 15	
n	d_L	d_U	d_L	d_U	d_L	d_U	d_L	d_U	d_L	d_U
25	0.35		0.29		0.24		0.19		0.15	
26	0.38		0.32		0.27		0.22		0.18	
27	0.41		0.36		0.30		0.25		0.21	
28	0.44		0.39		0.33		0.28		0.24	
29	0.47		0.42		0.36		0.31		0.27	
30	0.50		0.45		0.39		0.34		0.29	
31	0.53		0.48		0.42		0.37		0.32	
32	0.56		0.50		0.45		0.40		0.35	
33	0.59		0.53		0.48		0.43		0.38	
34	0.61		0.56		0.50		0.45		0.40	
35	0.63		0.58		0.53		0.48		0.43	
36	0.66		0.61		0.55		0.50		0.46	
37	0.68		0.63		0.58		0.53		0.48	
38	0.70		0.65		0.60		0.55		0.50	
39	0.72		0.67		0.62		0.58		0.53	
40	0.74		0.69		0.65		0.60		0.55	
45	0.84	1.97	0.79		0.74		0.70		0.66	
50	0.91	1.93	0.87	1.99	0.83		0.79		0.75	
55	0.98	1.89	0.94	1.95	0.90		0.86		0.83	
60	1.04	1.87	1.00	1.91	0.97	1.96	0.93		0.89	
65	1.09	1.85	1.05	1.89	1.02	1.93	0.99	1.98	0.95	
70	1.13	1.83	1.10	1.87	1.07	1.91	1.04	1.95	1.01	
75	1.17	1.82	1.14	1.86	1.11	1.89	1.08	1.93	1.05	1.97
80	1.21	1.81	1.18	1.84	1.15	1.88	1.12	1.91	1.09	1.95
85	1.24	1.80	1.21	1.83	1.18	1.87	1.16	1.90	1.13	1.93
90	1.26	1.80	1.24	1.83	1.22	1.86	1.19	1.89	1.17	1.92
95	1.29	1.79	1.27	1.82	1.24	1.85	1.22	1.88	1.20	1.91
100	1.31	1.79	1.29	1.82	1.27	1.84	1.25	1.87	1.23	1.90
150	1.47	1.78	1.46	1.80	1.44	1.81	1.43	1.83	1.41	1.85
200	1.56	1.79	1.55	1.80	1.54	1.81	1.53	1.82	1.52	1.84

Source: SW77 (see text). Copyright The Econometric Society

but they are more complicated, and we shall not discuss them here. It has been discovered, however, that, in many situations, treating the test as though d_L did not exist and d_U were the appropriate single critical value is a very good approximation[1] to the truth. Thus a simplified, approximate test procedure is the following

1S. *Simplified one-sided test against alternatives $\rho > 0$. If $d < d_U$, reject H_0 at level α, otherwise do not reject.*

2S. *Simplified one-sided test against alternatives $\rho < 0$. If $4 - d < d_U$, reject H_0 at level α, otherwise do not reject.*

3S. *Simplified two-sided test against alternatives $\rho \neq 0$. If $d < d_U$ or $4 - d < d_U$, reject H_0 at level 2α.*

There is no simple way to immediately determine if this simplified test is valid. For practical purposes at this level of complication, we suggest first applying the (d_L, d_U)

---

[1]For a discussion of the accuracy of the d_U approximation, and alternatives, see Durbin and Watson (1971).

T A B L E 7.1. Significance Points of d_L and d_U:1% (Continued)

	$k = 16$		$k = 17$		$k = 18$		$k = 19$		$k = 20$	
n	d_L	d_U	d_L	d_U	d_L	d_U	d_L	d_U	d_L	d_U
30	0.25		0.21		0.17		0.14		0.11	
31	0.28		0.23		0.20		0.16		0.13	
32	0.30		0.26		0.22		0.18		0.15	
33	0.33		0.29		0.25		0.21		0.17	
34	0.36		0.31		0.27		0.23		0.20	
35	0.38		0.34		0.30		0.26		0.22	
36	0.41		0.36		0.32		0.28		0.24	
37	0.43		0.39		0.35		0.31		0.27	
38	0.46		0.41		0.37		0.33		0.29	
39	0.48		0.44		0.40		0.36		0.32	
40	0.51		0.46		0.42		0.38		0.34	
45	0.61		0.57		0.53		0.49		0.45	
50	0.71		0.67		0.63		0.59		0.55	
55	0.79		0.75		0.71		0.67		0.64	
60	0.86		0.82		0.79		0.75		0.72	
65	0.92		0.87		0.85		0.82		0.79	
70	0.97		0.94		0.91		0.88		0.85	
75	1.02		0.99		0.96		0.93		0.91	
80	1.07	1.98	1.04		1.01		0.98		0.96	
85	1.11	1.97	1.08		1.05		1.03		1.00	
90	1.14	1.95	1.12	1.98	1.09		1.07		1.04	
95	1.17	1.93	1.50	1.96	1.13	1.99	1.10		1.08	
100	1.20	1.92	1.18	1.95	1.16	1.98	1.14		1.11	
150	1.40	1.86	1.39	1.88	1.37	1.90	1.36	1.91	1.34	1.93
200	1.51	1.85	1.50	1.86	1.48	1.87	1.47	1.88	1.46	1.90

Source: SW77 (see text). Copyright The Econometric Society.

test to see if a clear decision is reached in that manner. Inconclusive results from this test will, of course, be judged significant by the simplified test, but this second-level decision can be regarded as having either a tentative "warning flag" attached to it or perhaps a slightly higher α-risk than the one indicated by the simplified test. Example 2 below illustrates this sort of judgment.

Example 1. The residuals from a straight line fit to $n = 50$ pairs of values of (X, Y) gave rise to a d-statistic of value $d = 0.625$. For a two-sided test of $H_1 : \rho = 0$ against the two-sided alternative $\rho \neq 0$ we first compare d and $4 - d = 3.375$ against suitable d_L and d_U values from Tables 7.1–7.3. From the $\alpha = 0.01$ table, with $k = 1$ and $n = 50$, we find

$$d = 0.625 < d_L = 1.32.$$

It follows, from applying procedure (3), that we reject H_0 at the $2\alpha = 0.02$ level and conclude that there *does* appear to be serial correlation, of the type tested against, present in the data. Doubt is thus cast on the fitted model and the data should be reconsidered in the light of this new information.

Example 2. For a set of 70 residuals from a linear model involving $k = 4$ predictor variables, we find a d-statistic of value $d = 1.51$. Test $H_0 : \rho_s = 0$ against the one-sided alternative $H_1 : \rho_s = \rho^s$, where $\rho > 0$.

T A B L E 7.2. Significance Points of d_L and d_U: 2.5%

	k = 1		k = 2		k = 3		k = 4		k = 5	
n	d_L	d_U	d_L	d_U	d_L	d_U	d_L	d_U	d_L	d_U
15	0.95	1.23	0.83	1.40	0.71	1.61	0.59	1.84	0.48	2.09
16	0.98	1.24	0.86	1.40	0.75	1.59	0.64	1.80	0.53	2.03
17	1.01	1.25	0.90	1.40	0.79	1.58	0.68	1.77	0.57	1.98
18	1.03	1.26	0.93	1.40	0.82	1.56	0.72	1.74	0.62	1.93
19	1.06	1.28	0.96	1.41	0.86	1.55	0.76	1.72	0.66	1.90
20	1.08	1.28	0.99	1.41	0.89	1.55	0.79	1.70	0.70	1.87
21	1.10	1.30	1.01	1.41	0.92	1.54	0.83	1.69	0.73	1.84
22	1.12	1.31	1.04	1.42	0.95	1.54	0.86	1.68	0.77	1.82
23	1.14	1.32	1.06	1.42	0.97	1.54	0.89	1.67	0.80	1.80
24	1.16	1.33	1.08	1.43	1.00	1.54	0.91	1.66	0.83	1.79
25	1.18	1.34	1.10	1.43	1.02	1.54	0.94	1.65	0.86	1.77
26	1.19	1.35	1.12	1.44	1.04	1.54	0.96	1.65	0.88	1.76
27	1.21	1.36	1.13	1.44	1.06	1.54	0.99	1.64	0.91	1.75
28	1.22	1.37	1.15	1.45	1.08	1.54	1.01	1.64	0.93	1.74
29	1.24	1.38	1.17	1.45	1.10	1.54	1.03	1.63	0.96	1.73
30	1.25	1.38	1.18	1.46	1.12	1.54	1.05	1.63	0.98	1.73
31	1.26	1.39	1.20	1.47	1.13	1.55	1.07	1.63	1.00	1.72
32	1.27	1.40	1.21	1.47	1.15	1.55	1.08	1.63	1.02	1.71
33	1.28	1.41	1.22	1.48	1.16	1.55	1.10	1.63	1.04	1.71
34	1.29	1.41	1.24	1.48	1.17	1.55	1.12	1.63	1.06	1.70
35	1.30	1.42	1.25	1.48	1.19	1.55	1.13	1.63	1.07	1.70
36	1.31	1.43	1.26	1.49	1.20	1.56	1.15	1.63	1.09	1.70
37	1.32	1.43	1.27	1.49	1.21	1.56	1.16	1.62	1.10	1.70
38	1.33	1.44	1.28	1.50	1.23	1.56	1.17	1.62	1.12	1.70
39	1.34	1.44	1.29	1.50	1.24	1.56	1.19	1.63	1.13	1.69
40	1.35	1.45	1.30	1.51	1.25	1.57	1.20	1.63	1.15	1.69
45	1.39	1.48	1.34	1.53	1.30	1.58	1.25	1.63	1.21	1.69
50	1.42	1.50	1.38	1.54	1.34	1.59	1.30	1.64	1.26	1.69
55	1.45	1.52	1.41	1.56	1.37	1.60	1.33	1.64	1.30	1.69
60	1.47	1.54	1.44	1.57	1.40	1.61	1.37	1.65	1.33	1.69
65	1.49	1.55	1.46	1.59	1.43	1.62	1.40	1.66	1.36	1.69
70	1.51	1.57	1.48	1.60	1.45	1.63	1.42	1.66	1.39	1.70
75	1.53	1.58	1.50	1.61	1.47	1.64	1.45	1.67	1.42	1.70
80	1.54	1.59	1.52	1.62	1.49	1.65	1.47	1.67	1.44	1.70
85	1.56	1.60	1.53	1.63	1.51	1.65	1.49	1.68	1.46	1.71
90	1.57	1.61	1.55	1.64	1.53	1.66	1.50	1.69	1.48	1.71
95	1.58	1.62	1.56	1.65	1.54	1.67	1.52	1.69	1.50	1.71
100	1.59	1.63	1.57	1.65	1.55	1.67	1.53	1.70	1.51	1.72

Source: Durbin and Watson (1951).

We obtain the following significance points from Tables 7.1–7.3:

	d_L	d_U
$\alpha = 0.05$	1.49	1.74
$\alpha = 0.025$	1.42	1.66
$\alpha = 0.01$	1.34	1.58

We see that, using procedure (1), the primary test is inconclusive at all levels. Applying

T A B L E 7.3. Significance Points of d_l and d_U:5%

n	$k = 1$		$k = 2$		$k = 3$		$k = 4$		$k = 5$	
	d_L	d_U	d_L	d_U	d_L	d_U	d_L	d_U	d_L	d_U
15	1.08	1.36	0.95	1.54	0.82	1.75	0.69	1.97	0.56	2.21
16	1.10	1.37	0.98	1.54	0.86	1.73	0.74	1.93	0.62	2.15
17	1.13	1.38	1.02	1.54	0.90	1.71	0.78	1.90	0.67	2.10
18	1.16	1.39	1.05	1.53	0.93	1.69	0.82	1.87	0.71	2.06
19	1.18	1.40	1.08	1.53	0.97	1.68	0.86	1.85	0.75	2.02
20	1.20	1.41	1.10	1.54	1.00	1.68	0.90	1.83	0.79	1.99
21	1.22	1.42	1.13	1.54	1.03	1.67	0.93	1.81	0.83	1.96
22	1.24	1.43	1.15	1.54	1.05	1.66	0.96	1.80	0.86	1.94
23	1.26	1.44	1.17	1.54	1.08	1.66	0.99	1.79	0.90	1.92
24	1.27	1.45	1.19	1.55	1.10	1.66	1.01	1.78	0.93	1.90
25	1.29	1.45	1.21	1.55	1.12	1.66	1.04	1.77	0.95	1.89
26	1.30	1.46	1.22	1.55	1.14	1.65	1.06	1.76	0.98	1.88
27	1.32	1.47	1.24	1.56	1.16	1.65	1.08	1.76	1.01	1.86
28	1.33	1.48	1.26	1.56	1.18	1.65	1.10	1.75	1.03	1.85
29	1.34	1.48	1.27	1.56	1.20	1.65	1.12	1.74	1.05	1.84
30	1.35	1.49	1.28	1.57	1.21	1.65	1.14	1.74	1.07	1.83
31	1.36	1.50	1.30	1.57	1.23	1.65	1.16	1.74	1.09	1.83
32	1.37	1.50	1.31	1.57	1.24	1.65	1.18	1.73	1.11	1.82
33	1.38	1.51	1.32	1.58	1.26	1.65	1.19	1.73	1.13	1.81
34	1.39	1.51	1.33	1.58	1.27	1.65	1.21	1.73	1.15	1.81
35	1.40	1.52	1.34	1.58	1.28	1.65	1.22	1.73	1.16	1.80
36	1.41	1.52	1.35	1.59	1.29	1.65	1.24	1.73	1.18	1.80
37	1.42	1.53	1.36	1.59	1.31	1.66	1.25	1.72	1.19	1.80
38	1.43	1.54	1.37	1.59	1.32	1.66	1.26	1.72	1.21	1.79
39	1.43	1.54	1.38	1.60	1.33	1.66	1.27	1.72	1.22	1.79
40	1.44	1.54	1.39	1.60	1.34	1.66	1.29	1.72	1.23	1.79
45	1.48	1.57	1.43	1.62	1.38	1.67	1.34	1.72	1.29	1.78
50	1.50	1.59	1.46	1.63	1.42	1.67	1.38	1.72	1.34	1.77
55	1.53	1.60	1.49	1.64	1.45	1.68	1.41	1.72	1.38	1.77
60	1.55	1.62	1.51	1.65	1.48	1.69	1.44	1.73	1.41	1.77
65	1.57	1.63	1.54	1.66	1.50	1.70	1.47	1.73	1.44	1.77
70	1.58	1.64	1.55	1.67	1.52	1.70	1.49	1.74	1.46	1.77
75	1.60	1.65	1.57	1.68	1.54	1.71	1.51	1.74	1.49	1.77
80	1.61	1.66	1.59	1.69	1.56	1.72	1.53	1.74	1.51	1.77
85	1.62	1.67	1.60	1.70	1.57	1.72	1.55	1.75	1.52	1.77
90	1.63	1.68	1.61	1.70	1.59	1.73	1.57	1.75	1.54	1.78
95	1.64	1.69	1.62	1.71	1.60	1.73	1.58	1.75	1.56	1.78
100	1.65	1.69	1.63	1.72	1.61	1.74	1.59	1.76	1.57	1.78
150	1.72	1.75	1.71	1.76	1.69	1.77	1.68	1.79	1.67	1.80
200	1.76	1.78	1.75	1.79	1.74	1.80	1.73	1.81	1.72	1.82

Source: DW 51 for $n \leq 100$ and SW77 for $n = 150,200$ (see text).

procedure (1S) we then come to the secondary conclusion that H_0 should be rejected because it falls below d_U at the $\alpha = 0.01$ level. Our true rejection level is perhaps not as low as $\alpha = 0.01$ because we are using the simplified test. However, we also note that we would *almost* reject at the $\alpha = 0.05$ level using an ordinary test because 1.51 is close to $d_L = 1.49$. Thus we can think, with reasonable safety, of the rejection of

T A B L E 7.3. Significance Points of d_L and d_U:5% (Continued)

	k = 6		k = 7		k = 8		k = 9		k = 10	
n	d_L	d_U	d_L	d_U	d_L	d_U	d_L	d_U	d_L	d_U
20	0.69		0.60		0.50		0.42		0.34	
21	0.73		0.64		0.55		0.46		0.38	
22	0.77		0.68		0.59		0.50		0.42	
23	0.80		0.72		0.63		0.55		0.47	
24	0.84		0.75		0.67		0.58		0.51	
25	0.87		0.78		0.70		0.62		0.54	
26	0.90	1.99	0.82		0.74		0.66		0.58	
27	0.93	1.97	0.85		0.77		0.69		0.62	
28	0.95	1.96	0.87		0.80		0.72		0.65	
29	0.98	1.94	0.90		0.83		0.75		0.68	
30	1.00	1.93	0.93		0.85		0.78		0.71	
31	1.02	1.92	0.95		0.88		0.81		0.74	
32	1.04	1.91	0.97		0.90		0.84		0.77	
33	1.06	1.90	0.99	1.99	0.93		0.86		0.80	
34	1.08	1.89	1.02	1.98	0.95		0.89		0.82	
35	1.10	1.88	1.03	1.97	0.97		0.91		0.85	
36	1.11	1.88	1.05	1.96	0.99		0.93		0.87	
37	1.13	1.87	1.07	1.95	1.01		0.95		0.89	
38	1.15	1.86	1.09	1.94	1.03		0.97		0.91	
39	1.16	1.86	1.10	1.93	1.05		0.99		0.93	
40	1.18	1.85	1.12	1.92	1.06		1.01		0.95	
45	1.24	1.84	1.19	1.90	1.14	1.96	1.09		1.04	
50	1.29	1.82	1.25	1.88	1.20	1.93	1.16	1.99	1.11	
55	1.33	1.81	1.29	1.86	1.25	1.91	1.21	1.96	1.17	
60	1.37	1.81	1.34	1.85	1.30	1.89	1.26	1.94	1.22	1.98
65	1.40	1.81	1.37	1.84	1.34	1.88	1.30	1.92	1.27	1.96
70	1.43	1.80	1.40	1.84	1.37	1.87	1.34	1.91	1.31	1.95
75	1.46	1.80	1.43	1.83	1.40	1.87	1.37	1.90	1.34	1.94
80	1.48	1.80	1.45	1.83	1.43	1.86	1.40	1.89	1.37	1.93
85	1.50	1.80	1.47	1.83	1.45	1.86	1.42	1.89	1.40	1.92
90	1.52	1.80	1.49	1.83	1.47	1.85	1.44	1.88	1.42	1.91
95	1.54	1.80	1.51	1.83	1.49	1.85	1.47	1.88	1.44	1.90
100	1.55	1.80	1.53	1.83	1.51	1.85	1.48	1.87	1.46	1.90
150	1.65	1.82	1.64	1.83	1.62	1.85	1.61	1.86	1.59	1.88
200	1.71	1.83	1.70	1.84	1.69	1.85	1.68	1.86	1.67	1.87

Source: SW77 (see text). Copyright The Econometric Society.

H_0 being at a level of somewhat between $\alpha = 0.05$ and $\alpha = 0.01$.[2] Doubt is cast on the fitted model and a reanalysis of the data, taking account of the indicated serial correlation, is appropriate.

Width of the Primary Test Inconclusive Region

Figure 7.4 shows a plot of the 5% values of d_L and d_U versus the number of observations n, for $15 \le n \le 100$ and $1 \le k \le 5$, joined up by smooth lines to clarify the plot. (See Durbin and Watson, 1951.) This plot could be extended using Table 7.3, of course.

[2]An alternative interpretation is that the primary test would be significant at about the 0.06 level.

T A B L E 7.3. Significance Points of d_L and d_U: 5% (Continued)

	$k = 11$		$k = 12$		$k = 13$		$k = 14$		$k = 15$	
n	d_L	d_U	d_L	d_U	d_L	d_U	d_L	d_U	d_L	d_U
25	0.47		0.40		0.34		0.28		0.22	
26	0.51		0.44		0.37		0.31		0.26	
27	0.54		0.48		0.41		0.35		0.29	
28	0.58		0.51		0.45		0.38		0.33	
29	0.61		0.54		0.48		0.42		0.36	
30	0.64		0.58		0.51		0.45		0.39	
31	0.67		0.61		0.55		0.48		0.43	
32	0.70		0.64		0.58		0.52		0.46	
33	0.73		0.67		0.61		0.55		0.49	
34	0.76		0.70		0.63		0.58		0.52	
35	0.78		0.72		0.66		0.60		0.55	
36	0.81		0.75		0.69		0.63		0.58	
37	0.83		0.77		0.71		0.66		0.60	
38	0.85		0.80		0.74		0.68		0.63	
39	0.88		0.82		0.76		0.71		0.65	
40	0.90		0.84		0.79		0.73		0.68	
45	0.99		0.94		0.89		0.84		0.79	
50	1.06		1.02		0.97		0.93		0.88	
55	1.13		1.09		1.05		1.00		0.96	
60	1.18		1.15		1.11		1.07		1.03	
65	1.23		1.20		1.16		1.12		1.09	
70	1.27	1.99	1.24		1.21		1.17		1.14	
75	1.31	1.97	1.28		1.25		1.22		1.18	
80	1.34	1.96	1.31	1.99	1.28		1.25		1.22	
85	1.37	1.95	1.34	1.98	1.32		1.29		1.26	
90	1.40	1.94	1.37	1.97	1.34		1.32		1.29	
95	1.42	1.93	1.39	1.96	1.37	1.98	1.35		1.32	
100	1.44	1.92	1.42	1.95	1.39	1.97	1.37		1.35	
150	1.58	1.89	1.56	1.91	1.55	1.92	1.54	1.94	1.52	1.96
200	1.65	1.89	1.64	1.90	1.63	1.91	1.62	1.92	1.61	1.93

Source: SW77 (see text). Copyright The Econometric Society.

Note that the vertical distance between pairs of correspondingly numbered curves is the region of indecision involved in the standard test, and that the width of this region becomes smaller as n increases. The moral is obvious: the more observations we have, the more likely it is that we shall be able to make a definite decision via the Durbin–Watson test. Workers in time series have a rule of thumb that $n \geq 50$ observations are needed in order for their analyses to produce worthwhile conclusions. As Figure 7.4 shows, such a rule of thumb would not be out of place for application of the Durbin–Watson test.

Mean Square Successive Difference

Readers viewing Figure 7.4 may be curious about the dashed line lying between the various upper and lower pairs of curves. This is a join of the 5% points for testing for serial correlation in a model $Y = \beta_0 + \epsilon$. In other words, it is the Durbin–Watson test for $k = 0$ X's in the model. Table 7.4 shows selected lower-tail percentage points

T A B L E 7.3. Significance Points of d_L and d_U:5% (Continued)

	$k = 16$		$k = 17$		$k = 18$		$k = 19$		$k = 20$	
n	d_L	d_U	d_L	d_U	d_L	d_U	d_L	d_U	d_L	d_U
30	0.34		0.29		0.24		0.20		0.16	
31	0.37		0.32		0.27		0.22		0.18	
32	0.40		0.35		0.30		0.25		0.21	
33	0.43		0.38		0.33		0.28		0.24	
34	0.46		0.41		0.36		0.31		0.27	
35	0.49		0.44		0.39		0.34		0.30	
36	0.52		0.47		0.42		0.37		0.32	
37	0.55		0.50		0.45		0.40		0.35	
38	0.58		0.52		0.47		0.42		0.38	
39	0.60		0.55		0.50		0.45		0.40	
40	0.63		0.58		0.53		0.48		0.43	
45	0.74		0.69		0.64		0.60		0.55	
50	0.84		0.79		0.75		0.70		0.66	
55	0.92		0.88		0.84		0.80		0.75	
60	0.99		0.95		0.91		0.87		0.84	
65	1.05		1.02		0.98		0.94		0.91	
70	1.11		1.07		1.04		1.01		0.97	
75	1.15		1.12		1.09		1.06		1.03	
80	1.20		1.17		1.14		1.11		1.08	
85	1.23		1.21		1.18		1.15		1.12	
90	1.27		1.24		1.21		1.19		1.16	
95	1.30		1.27		1.25		1.22		1.20	
100	1.32		1.30		1.28		1.25		1.23	
150	1.50	1.97	1.49	1.99	1.47		1.46		1.44	
200	1.60	1.94	1.59	1.96	1.58	1.97	1.57	1.98	1.55	1.99

Source: SW77 (see text). Copyright The Econometric Society.

given by Nelson (1980), here rounded to two decimal places. Nelson's figures allow the dashed line to be extended for higher n. We have omitted Nelson's one-tailed 10% values, which do not match with any of the other tables shown here. The blank 2.5% column in Table 7.4 indicates the absence of these values, not given by Nelson.

It has been argued that, when the model contains a lagged response variable as well as predictors, the Durbin–Watson test is inappropriate. An example of such a model is

$$Y_{i+1} = \beta_0 + \beta Y_i + \beta_1 X_{1i} + \beta_2 X_{2i} + \epsilon_i.$$

Rayner (1994) concluded that, in spite of the fact that the Durbin–Watson test has a bias toward nonrejection for such models, it may well be better than competitors. Details and related references are given in the quoted paper.

7.3. EXAMINING RUNS IN THE TIME SEQUENCE PLOT OF RESIDUALS: RUNS TEST

This test provides a quick but approximate alternative to the Durbin–Watson test. Since it ignores the actual sizes of the residuals and uses only their signs in time sequence, it throws away a lot of information, but it is easy to apply.

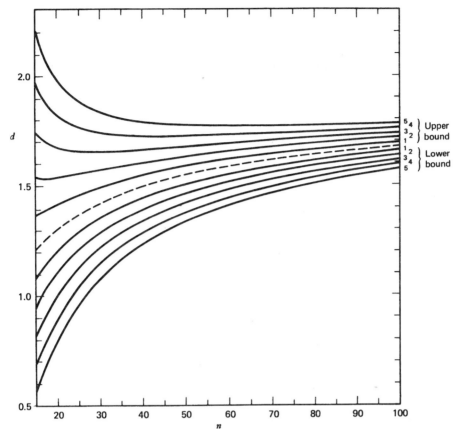

Figure 7.4. Graphs of 5% values of d_U and d_L against n for $k = 1, 2, 3, 4, 5$.

When the time sequence of a set of residuals is known, it is sometimes noticeable that groups of positive or negative residuals occur in what might be an unusual pattern. To take an extreme case, if 30 residuals in time sequence consisted of 16 negative followed by 14 positive residuals, we might first suspect that an unconsidered variable had changed levels between the 16th and 17th runs. Such behavior can also arise from a positive serial correlation between successive residuals, however. Similarly, a very large number of sign switches in a sequence might arise from negative serial correlation. When there is a sequence of such runs it is useful to have a method that will enable a decision to be made on whether the run pattern is "unusual" or not. We now explore how this might be done.

Runs

Suppose we have a sequence of signs such as

$$+ + - + - - - - + + - + + + +.$$

These may be the signs of residuals in time sequence (which will be our application) but equally well the "plus and minus" signs might denote "male and female," "head and tail," "better and worse," "treatment A and treatment B," or the two levels of any other dichotomous classification. Suppose there are n signs in all, n_1 plus signs

T A B L E 7.4. **Selected Lower Significance Points for the Mean Square Successive Difference Test (Durbin–Watson Test with $k = 0$ Predictor Variables): The Corresponding Upper Points Are $4 - $ (Those Shown)**

n	1%	2.5%	5%
10	0.75		1.06
20	1.04		1.30
30	1.20		1.42
40	1.29		1.49
50	1.36		1.54
60	1.42		1.58
70	1.46		1.61
80	1.49		1.64
90	1.52		1.66
100	1.54		1.67
150	1.62		1.73
200	1.67		1.77
300	1.73		1.81
400	1.77		1.84
500	1.79		1.85
600	1.81		1.87
800	1.84		1.88
1000	1.85		1.90

and n_2 minus signs, and there are r runs. In the above example, $n_1 = 9$, $n_2 = 6$, and there are $r = 7$ runs indicated by the parentheses below

$$(+ +)(-)(+)(- - - -)(+ +)(-)(+ + + +).$$

We can ask if the particular arrangement of signs we observe is an "extreme" arrangement or not.

We first examine a case where there are only 15 possible sign sequences. Suppose there are six signs, two of which are plus. The following sign arrangements are possible:

Arrangement	Number of Runs, u
+ + − − − −	2
+ − + − − −	4
+ − − + − −	4
+ − − − + −	4
+ − − − − +	3
− + + − − −	3
− + − + − −	5
− + − − + −	5
− + − − − +	4
− − + + − −	3
− − + − + −	5
− − + − − +	4
− − − + + −	3
− − − + − +	4
− − − − + +	2

The distribution of runs is as follows:

$$r = 2 \qquad 3 \qquad 4 \qquad 5$$
$$\text{Frequency} = 2 \qquad 4 \qquad 6 \qquad 3 \qquad (\text{Total} = 15)$$
$$\text{Cumulative Probability} = 0.133 \quad 0.400 \quad 0.800 \quad 1.000$$

Thus five runs would occur in $\frac{3}{15}$ or 20% of the possible cases, that is, with a probability of 0.2. Alternatively, two runs would occur in $\frac{2}{15}$ or 13.3% of the possible cases, that is, with a probability of 0.133. A low number of runs in the residuals sequence might indicate positive serial correlation, while a high number might arise from negative serial correlation. We discuss the situation in the "too few runs" context first. If we observed only $r = 2$ runs in a set of six residuals of which two were positive, we would have observed an event that occurs with a probability of 0.133. Obviously, nothing significant will occur in such a small example. For any given sequence of signs we can find the probability that the observed value of r (or a lesser value) will occur. (*Example:* When $n_1 = 2$, $n_2 = 4$, and the observed number of runs is 3, Prob($r \leq 3$) = $(2 + 4)/15 = 0.4$, a not unusual event occurring in 40% of cases.) On the basis of such a probability level we can decide whether or not we believe that a random arrangement of signs has occurred. [We might, for example, compare the probability with a preassigned value, say, $\alpha = 0.05$, and reject the idea of a random arrangement if Prob($u \leq$ observed u) ≤ 0.05.] For a "too many runs" test we would cumulate the frequencies downward from the high-r end.

Tables for Modest n_1 and n_2

Tables 7.5 and 7.6 show, respectively, lower-tail and upper-tail cumulative probabilities for selected n_1, n_2, with $3 \leq n_1 \leq n_2 \leq 10$. Only probability values less than or equal to 0.10 are shown. Typically, n_1 and n_2 would roughly be equal in a set of residuals. (When $n_1 > n_2$, interchange n_1 and n_2.) These distributions were given originally by Swed and Eisenhart (1943). In that paper more decimal places are given and the arrangement of cases is different. Related tables also appear in Lindley and Scott (1984, see pp. 60–62).

Example. Twenty residuals—half positive, half negative—show five runs of signs in their time sequence. Is this an unusually small number? We see that $p(r \leq 5) = 0.004$ from the (10, 10) line, so the answer is yes. (At the $\alpha = 0.10$ level, seven or fewer runs would be a small number.) We can now reexamine the residuals and data and evaluate the lag-1 serial correlation. In the upper tail for this case, 15 or more runs would be a large number at the $\alpha = 0.10$ level.

Note that $(\frac{1}{2}n, \frac{1}{2}n)$ cases have a symmetric distribution, that is, $p(r \leq m) = p(r \geq n + 2 - m)$. Also, the distribution mean is at $(\frac{1}{2}n + 1)$; see below.

Larger n_1 and n_2 Values

Outside the range of small n_1, n_2 values in the tables, it is convenient to use a normal approximation to the actual distribution. Let

$$\mu = \frac{2n_1 n_2}{n_1 + n_2} + 1, \tag{7.3.1}$$

$$\sigma^2 = \frac{2n_1 n_2 (2n_1 n_2 - n_1 - n_2)}{(n_1 + n_2)^2 (n_1 + n_2 - 1)}. \tag{7.3.2}$$

T A B L E 7.5. Cumulative Lower-Tail Areas in the Distribution of the Total Number of Runs r in Samples of Size (n_1, n_2) for $3 \le n_1 \le n_2 \le 10$, $n_1 + n_2 \ge 10$, and Tail Areas ≤ 0.10 only (for $n_1 \ge n_2$, Simply Interchange n_1 and n_2)

(n_1, n_2) $r =$	2	3	4	5	6	7
(3, 7)	0.017	0.083				
(3, 8)	0.012	0.067				
(3, 9)	0.009	0.055				
(3,10)	0.007	0.045				
(4, 6)	0.010	0.048				
(4, 7)	0.006	0.033				
(4, 8)	0.004	0.024				
(4, 9)	0.003	0.018	0.085			
(4, 10)	0.002	0.014	0.068			
(5, 5)	0.008	0.040				
(5, 6)	0.004	0.024				
(5, 7)	0.003	0.015	0.076			
(5, 8)	0.002	0.010	0.054			
(5, 9)	0.001	0.007	0.039			
(5, 10)	0.001	0.005	0.029	0.095		
(6, 6)	0.002	0.013	0.067			
(6, 7)	0.001	0.008	0.043			
(6, 8)	0.001	0.005	0.028	0.086		
(6, 9)	0.000	0.003	0.019	0.063		
(6, 10)	0.000	0.002	0.013	0.047		
(7, 7)	0.001	0.004	0.025	0.078		
(7, 8)	0.000	0.002	0.015	0.051		
(7, 9)	0.000	0.001	0.010	0.035		
(7, 10)	0.000	0.001	0.006	0.024	0.080	
(8, 8)	0.000	0.001	0.009	0.032	0.100	
(8, 9)	0.000	0.001	0.005	0.020	0.069	
(8, 10)	0.000	0.000	0.003	0.013	0.048	
(9, 9)	0.000	0.000	0.003	0.012	0.044	
(9, 10)	0.000	0.000	0.002	0.008	0.029	0.077
(10, 10)	0.000	0.000	0.001	0.004	0.019	0.051

Source: Adapted from Swed and Eisenhart (1943).

It can be shown that these are the actual mean and variance of the discrete distribution of r. Then approximately, for a lower-tail test,

$$z = \frac{(r - \mu + \frac{1}{2})}{\sigma} \tag{7.3.3}$$

is a unit normal deviate where the $\frac{1}{2}$ is the usual *continuity correction*, which helps compensate for the fact that a continuous distribution is being used to approximate

T A B L E 7.6. Cumulative Upper-Tail Areas in the Distribution of the Total Number of Runs r in Samples of Size (n_1, n_2) for $3 \leq n_1 \leq n_2 \leq 10$, $n_1 + n_2 \geq 10$, and Tail Areas ≤ 0.10 only (for $n_1 \geq n_2$, simply interchange n_1 and n_2)

(n_1, n_2) $r =$	9	10	11	12	13	14	15	16	17	18	19	20
(4, 6)	0.024											
(4, 7)	0.046											
(4, 8)	0.071											
(4, 9)	0.098											
(4, 10)												
(5, 5)	0.040	0.008										
(5, 6)	0.089	0.024	0.002									
(5, 7)		0.045	0.008									
(5, 8)		0.071	0.016									
(5, 9)		0.098	0.028									
(5, 10)			0.042									
(6, 6)		0.067	0.013	0.002								
(6, 7)			0.034	0.008	0.001							
(6, 8)			0.063	0.016	0.002							
(6, 9)			0.098	0.028	0.006							
(6, 10)				0.042	0.010							
(7, 7)			0.078	0.025	0.004	0.001						
(7, 8)				0.051	0.012	0.002	0.000					
(7, 9)				0.084	0.025	0.006	0.001					
(7, 10)					0.043	0.010	0.002					
(8, 8)				0.100	0.032	0.009	0.001	0.000				
(8, 9)					0.061	0.020	0.004	0.001	0.000			
(8, 10)					0.097	0.036	0.010	0.002	0.000			
(9, 9)						0.044	0.012	0.003	0.000	0.000		
(9, 10)						0.077	0.026	0.008	0.001	0.000	0.000	
(10, 10)							0.051	0.019	0.004	0.001	0.000	0.000

Source: Adapted from Swed and Eisenhart (1943).

to a discrete distribution in the lower tail. For a "too many runs" upper-tail test, the continuity correction is $-\frac{1}{2}$ so that we use instead

$$z = \frac{r - \mu - \frac{1}{2}}{\sigma} \qquad (7.3.4)$$

and look up the upper tail of the $N(0, 1)$ distribution. How do we know which tail we want? We first evaluate μ. If $r > \mu$ we use the upper-tail test, and if $r < \mu$ the lower-tail test.

What if there are zeros in the residuals? Do they receive a minus sign or a plus sign? Exact zeros are unlikely in a regression fit but, if one occurs, the easiest way out is to assume first a plus and then a minus and see if the results are the same. This situation is unlikely to arise except in constructed class examples.

Example. Examination of a set of 27 residuals, 15 of which were of one sign and 12 of which were of the opposite sign, arranged in time sequence, revealed $r = 7$ runs. Does the arrangement of signs appear to have "too few runs"?

Here $n_1 = 15$, $n_2 = 12$, $r = 7$. From Eqs. (7.3.1) and (7.3.2), $\mu = \frac{43}{3}$, $\sigma^2 = \frac{740}{117}$. Thus the observed value of z from Eq. (7.3.3) is

$$z = \frac{(7 - \frac{44}{3} + \frac{1}{2})}{(\frac{740}{117})^{1/2}} = -2.713. \tag{7.3.5}$$

The probability of obtaining a unit normal deviate of value -2.713 or smaller is 0.0033 (or 0.33%) so that an unusually low number of runs appears to have occurred. We should reject the idea that the arrangement of signs is random. The model would be suspect and we would now search for an assignable cause for the pattern of residuals.

Comments

Strictly speaking, the test for runs is applicable only when the occurrences that produce the pattern of runs are independent. In a time sequence of residuals this is not true due to the correlations that exist among the residuals, and the probability level obtained from the procedure will be affected in a way that depends on the particular structure of the data. In most practical regression situations, unless the ratio $(n - p)/n$, that is, (number of degrees of freedom in residuals)/(number of residuals), is quite small, the effect can be ignored.

Time plots of residuals can also be subjected to calculations suggested by Cleveland and Kleiner (1975). Three curves of moving statistics are drawn, involving (1) the *midmean* (the average of all observations between the quartiles of the data to that point in time), (2) the *lower* semi-midmean (the midmean of all observations *below* the median of the data to that point in time), and (3) the *upper* midmean(...*above*...). "These three statistics summarize the location, spread, and skewness of the data." (See p. 449 of the reference cited.)

REFERENCES

Box, Jenkins, and Reinsel (1994); Diggle (1990); Durbin (1969, 1970); Durbin and Watson (1950, 1951, 1971); Savin and White (1977); Wei (1990).

EXERCISES FOR CHAPTER 7

A. Fit, to the appropriate portion of the steam data in Appendix 1A, the model $Y = \beta_0 + \beta_5 X_5 + \beta_6 X_6 + \beta_8 X_8 + \epsilon$. This will give you the following fitted model and analysis of variance table, and also the fitted values and residuals in Table A:

$$\hat{Y} = -2.968 + 0.4020 X_5 + 0.19892 X_6 - 0.073924 X_8$$

Source	df	SS	MS	F
Regression $|b_0$	3	56.472	18.824	53.83
Residual	21	7.344	0.350	
Total, corrected	24	63.816		

TABLE A. Observations, Fitted Values, and Residuals

Row	Y	$\hat{Y}$	Residual
1	10.98	10.86	0.12
2	11.13	10.47	0.66
3	12.51	11.79	0.72
4	8.40	8.72	−0.32
5	9.27	9.13	0.14
6	8.73	8.20	0.53
7	6.36	6.18	0.18
8	8.50	8.40	0.10
9	7.82	8.04	−0.22
10	9.14	9.22	−0.08
11	8.24	9.64	−1.40
12	12.19	11.54	0.65
13	11.88	11.60	0.28
14	9.57	9.18	0.39
15	10.94	10.61	0.33
16	9.58	9.49	0.09
17	10.09	9.49	0.60
18	8.11	8.29	−0.18
19	6.83	6.51	0.32
20	8.88	8.56	0.32
21	7.68	7.74	−0.06
22	8.47	9.38	−0.91
23	8.86	9.77	−0.91
24	10.36	11.00	−0.64
25	11.08	11.76	−0.68

1. Plot the residuals in a histogram, in a normal probability (nscore) plot, versus order 1–25, and versus $\hat{Y}$.
2. Comment on what you see in the four plots in part 1.
3. Evaluate the Durbin–Watson statistic and use it to test $H_0 : \rho = 0$ versus the two-sided alternative $H_1 : \rho \neq 0$, assuming that $\rho_s = \rho^s$ in the usual notation.
4. Carry out a test for "too few runs" using the runs tests, on the residuals in the sequence order displayed in Table A.

B. (*Source:* "Using an hyperbola as a transition model to fit two-regime straight-line data," by D. G. Watts and D. W. Bacon, *Technometrics,* **16,** 1974, 369–373.) A set of sediment settling data was subjected to three different regression calculations, using three different models. The residuals from these three separate calculations, multiplied by 1000, are shown in Table B, in the time order in which the data occurred; the actual times of observation appear in the first column. Plot the residuals against time and analyze their behavior by applying a two-sided runs test to each set. What are your conclusions?

C. (*Source:* "Car accidents—environmental aspects," by D. F. Andrews, *International Statistical Review,* **41,** 1973, 235–239.) The data in Table C consist of 50 observations of the response variable Y = "driving deaths" and six possible predictor variables $X_1, X_2, \ldots,$ X_6 for 49 states and the District of Columbia. Figure C1 shows the plot of Y versus $X_1 = 1964$ drivers $\times 10^{-4}$, while Figure C2 shows $y = \log Y$ versus $Z_1 = \log X_1$. To the latter data the model

$$\hat{y} = -0.101 + 0.938Z_1$$

T A B L E B. Three Sets of Residuals (Multiplied by 1000) Versus Time t

Time, t	Set 1	Set 2	Set 3
0.5	−19	[a]	[a]
1	−19	0	2
1.5	−18	0	2
2	−28	−10	−8
2.5	−27	0	2
3	−27	0	2
4.5	−45	−20	−15
6	−23	19	25
9	−19	2	10
12	−5	12	20
14	18	23	27
16	−9	−25	−23
18	4	13	16
20	−3	−5	−4
22	−9	−6	−5
24	−6	4	4
26	8	14	13
28	3	−4	−7
30	7	6	2
32	22	17	11
34	27	8	0
36	33	9	−1
40	26	1	−24
42	14	−9	−24
44	3	−9	−25
46	3	1	−16
48	4	2	−16
50	−4	−6	−26
52	−9	−5	−25
54	−11	−3	−23
56	−11	0	−21
58	−17	−6	−27
60	−17	−2	−23
62	−11	5	−17
64	−6	3	−20
66	−12	−6	−31
68	−7	5	−22
70	−9	−2	−32
72	−8	1	−31
74	4	14	−20
76	7	6	−29
78	8	5	−27
80	17	12	−17
82	31	19	−6
90	45	34	−16
106	−11	−17	11
120	−23	−4	47
150	−36	−9	33

[a]The particular analysis used did not provide a first residual.

T A B L E C. **Fifty Observations of Motor Vehicle Death Data Together with Some Possible Explanatory Predictor Variables**

Region	Y, 1964 Deaths	X_1, 1964 Drivers $\times 10^{-4}$	X_2, 1960 Persons /sq. mi.	X_3, 1963 Road (Rural) Mileage $\times 10^{-3}$	X_4, 1960 More Males than Females	X_5, normal January Maximum Temperature	X_6, 1964 Highway Fuel Consumption gallons $\times 10^7$
AL	968	158	64	66	No	62	119
AK	43	11	0.4	5.9	Yes	30	6.2
AZ	588	91	12	33	Y	64	65
AR	640	92	34	73	N	51	74
CA	4743	952	100	118	N	65	105
CO	566	109	17	73	N	42	78
CT	325	167	518	5.1	N	37	95
DE	118	30	226	3.4	N	41	20
DC	115	35	12524	—	N	44	23
FL	1545	298	91	57	N	67	216
GA	1302	203	68	83	N	54	162
ID	262	41	8.1	40	Y	36	29
IL	2207	544	180	102	N	33	350
IN	1410	254	129	89	N	37	196
IA	833	150	49	100	N	30	109
KS	669	136	27	124	N	42	94
KY	911	147	76	65	N	44	104
LA	1037	146	72	40	N	65	109
ME	196	46	31	19	N	30	37
MD	616	157	314	29	N	44	113
MA	766	255	655	17	N	37	166
MI	2120	403	137	95	N	33	306
MN	841	189	43	110	N	22	132
MS	648	85	46	59	N	57	77
MO	1289	234	63	100	N	40	180
MT	259	38	4.6	72	Y	29	31
NE	450	89	18.4	97	N	32	61
NV	215	23	2.6	44	Y	40	24
NH	158	37	67	13	N	32	23
NJ	1071	329	807	21	N	43	231
NM	387	54	7.8	62	Y	46	48
NY	2745	744	350	84	N	31	439
NC	1580	226	93	71	N	51	177
ND	185	38	9.1	102	Y	20	24
OH	2096	530	237	84	N	41	358
OK	785	137	34	94	N	46	107
OR	575	108	18	73	N	45	81
PA	1889	570	252	89	N	39	353
RI	100	46	812	1.3	N	38	27
SC	870	122	79	52	N	61	86
SD	270	40	9	87	Y	23	28
TN	1059	177	85	67	N	49	135
TX	3006	515	37	196	N	50	448
UT	295	57	10.8	32	N	37	38
VT	131	20	42	13	N	25	15
VA	1050	208	100	50	N	50	150
WA	730	160	43	59	Y	46	109
WV	467	88	77	32	N	43	54
WI	1059	207	72	87	N	26	141
WY	148	22	3.4	67	Y	37	20

is fitted, giving rise to the residuals in Figure C3. By examining the data corresponding to the residuals marked with the names of states, suggest what variables appear to have influence on the data. Which would be the most logical candidate for entry into regression next?

D. A set of 56 residuals equally spaced in time order contains 26 positive residuals and 30 negative residuals. There are 38 runs. Is this an "unusually large" number of runs, do you think?

E. A set of 25 residuals equally spaced in time order has 12 positive values and 13 negative values and exhibits five runs. Is that an unusually small number in your opinion?

F. A regression fit $\hat{Y} = b_0 + b_1 X_1 + b_2 X_2 + b_3 X_3$ on 85 observations equally spaced in time produces a Durbin–Watson statistic of $d = 2.33$. Might this indicate serial correlation? Test at a two-tailed $\alpha = 0.05$ level.

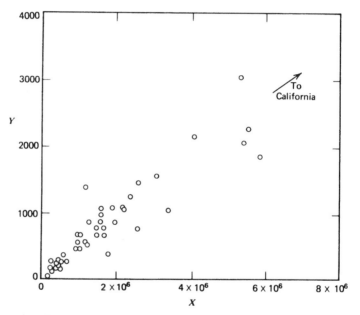

Figure C1. Motor vehicle deaths and drivers by state.

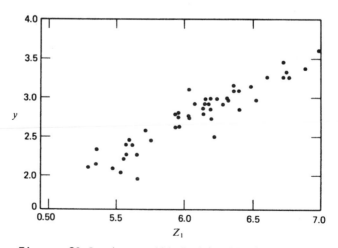

Figure C2. Log (motor vehicle deaths) and log (drivers by state).

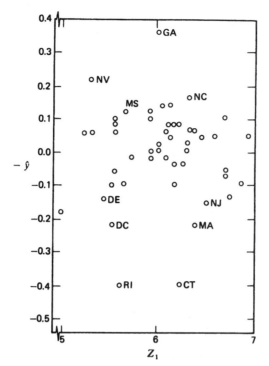

Figure C3. Residuals with some identifiers added.

G. An experimenter tells you: "I have looked at my 51 residuals in time order. Each residual differs from the previous one by an amount that is always within the range $(-1 \leq d_i \leq 1)$. Also, the sum of squares of the residuals is 50. My regression equation involves five predictors. What would you advise me to do?"

H. The following 24 residuals from a straight line fit are equally spaced in time and are given in time sequential order. Is there any evidence of lag-1 serial correlation, do you think? (Use a two-sided test at level $\alpha = 0.05$.)

$$8, -5, 7, 1, -3, -6, 1, -2, 10, 1, -1, 8, -6, 1, -6, -8, 10, -6, 9, -3, 3, -5, 1, -9$$

CHAPTER 8

More on Checking Fitted Models

We have already discussed the basic residuals plots in Chapter 2, and the Durbin–Watson test for serial correlation in Chapter 7. In many regression problems, these checks are perfectly adequate. In this chapter we discuss further techniques that can be of value. Readers should form their own opinions about which methods are likely to be useful in their own regression applications.

8.1. THE HAT MATRIX H AND THE VARIOUS TYPES OF RESIDUALS

Up to this point we have featured only the "ordinary residuals," that is, the $e_i = Y_i - \hat{Y}_i$ values. These are perfectly adequate for residuals checks on most regressions. It is also possible to look at other types of residual quantities. Some workers would look at these other types routinely; some only when they wanted to submit the regression fit to more study. These other types of residuals have been created to overcome problems that are occasionally concealed by the ordinary $e_i = Y_i - \hat{Y}_i$ residuals. We first remind the reader of the notation: the model is $\mathbf{Y} = \mathbf{X}\boldsymbol{\beta} + \boldsymbol{\epsilon}$, the normal equations are $\mathbf{X}'\mathbf{X}\mathbf{b} = \mathbf{X}'\mathbf{Y}$ with solution $\mathbf{b} = (\mathbf{X}'\mathbf{X})^{-1}\mathbf{X}'\mathbf{Y}$, if $\mathbf{X}'\mathbf{X}$ is nonsingular. The fitted values are $\hat{\mathbf{Y}} = \mathbf{X}\mathbf{b} = \mathbf{X}(\mathbf{X}'\mathbf{X})^{-1}\mathbf{X}'\mathbf{Y} = \mathbf{H}\mathbf{Y}$, say, where

$$\mathbf{H} = \mathbf{X}(\mathbf{X}'\mathbf{X})^{-1}\mathbf{X}'. \tag{8.1.1}$$

This matrix $\mathbf{H}$ occurs repeatedly in regression work. It is usually called the "hat matrix," as mentioned below, above (8.1.11).

The elements of $\mathbf{H}$ will be denoted by h_{ij}, namely,

$$\mathbf{H} = \begin{bmatrix} h_{11} & h_{12} & \cdots & h_{1n} \\ h_{21} & h_{22} & \cdots & h_{2n} \\ \vdots & \vdots & & \vdots \\ h_{n1} & h_{n2} & \cdots & h_{nn} \end{bmatrix} \tag{8.1.2}$$

$\mathbf{H}$ is symmetric, that is, $\mathbf{H}' = \mathbf{H}$, so that $h_{ij} = h_{ji}$. $\mathbf{H}$ is also idempotent, that is,

$$\begin{aligned} \mathbf{H}^2 = \mathbf{H}\mathbf{H} &= \{\mathbf{X}(\mathbf{X}'\mathbf{X})^{-1}\mathbf{X}'\}\{\mathbf{X}(\mathbf{X}'\mathbf{X})^{-1}\mathbf{X}'\} \\ &= \mathbf{X}(\mathbf{X}'\mathbf{X})^{-1}\mathbf{X}'\mathbf{X}(\mathbf{X}'\mathbf{X})^{-1}\mathbf{X}' \\ &= \mathbf{X}(\mathbf{X}'\mathbf{X})^{-1}\mathbf{X}' \\ &= \mathbf{H}. \end{aligned} \tag{8.1.3}$$

(In fact, $\mathbf{H}^p = \mathbf{H}$, where p is any power.)

The residuals can be expressed as

$$\mathbf{e} = \mathbf{Y} - \hat{\mathbf{Y}} = \mathbf{Y} - \mathbf{HY} = (\mathbf{I} - \mathbf{H})\mathbf{Y}. \tag{8.1.4}$$

(The matrix $\mathbf{I} - \mathbf{H}$ is also symmetric and idempotent, by the way. We show this below.)

Variance–Covariance Matrix of e

Since $E(\mathbf{Y}) = \mathbf{X}\boldsymbol{\beta}$, and because $(\mathbf{I} - \mathbf{H})\mathbf{X} = \mathbf{0}$, it follows that

$$\mathbf{e} - E(\mathbf{e}) = (\mathbf{I} - \mathbf{H})(\mathbf{Y} - \mathbf{X}\boldsymbol{\beta}) = (\mathbf{I} - \mathbf{H})\boldsymbol{\epsilon} \tag{8.1.5}$$

and the variance–covariance matrix of $\mathbf{e}$ is defined as

$$\mathbf{V}(\mathbf{e}) = E\{[\mathbf{e} - E(\mathbf{e})][\mathbf{e} - E(\mathbf{e})]'\} = (\mathbf{I} - \mathbf{H})E(\boldsymbol{\epsilon}\boldsymbol{\epsilon}')(\mathbf{I} - \mathbf{H})'. \tag{8.1.6}$$

Now $E(\boldsymbol{\epsilon}\boldsymbol{\epsilon}') = \mathbf{V}(\boldsymbol{\epsilon}) = \mathbf{I}\sigma^2$ if $E(\boldsymbol{\epsilon}) = \mathbf{0}$, as we usually assume, and when unweighted least squares are used. Furthermore, $(\mathbf{I} - \mathbf{H})' = (\mathbf{I} - \mathbf{H}') = \mathbf{I} - [\mathbf{X}(\mathbf{X}'\mathbf{X})^{-1}\mathbf{X}']' = \mathbf{I} - \mathbf{X}(\mathbf{X}'\mathbf{X})^{-1}\mathbf{X}' = \mathbf{I} - \mathbf{H}$. Thus the matrix $\mathbf{I} - \mathbf{H}$ is symmetric, and

$$\begin{aligned}
\mathbf{V}(\mathbf{e}) &= (\mathbf{I} - \mathbf{H})\mathbf{I}\sigma^2(\mathbf{I} - \mathbf{H})' \\
&= (\mathbf{I} - \mathbf{H})(\mathbf{I} - \mathbf{H})\sigma^2 \\
&= (\mathbf{I} - \mathbf{H} - \mathbf{H} + \mathbf{H}\mathbf{H})\sigma^2 \\
&= (\mathbf{I} - \mathbf{H})\sigma^2
\end{aligned} \tag{8.1.7}$$

since $\mathbf{HH} = \mathbf{H}^2 = \mathbf{H}$. Thus $V(e_i)$ is given by the ith diagonal element $1 - h_{ii}$, and $\text{cov}(e_i, e_j)$ is given by the (i, j)th element $-h_{ij}$ of the matrix $(\mathbf{I} - \mathbf{H})\sigma^2$. The correlation between e_i and e_j is given by

$$\rho_{ij} = \frac{\text{cov}(e_i, e_j)}{\{V(e_i) \cdot V(e_j)\}^{1/2}} = \frac{-h_{ij}}{\{(1 - h_{ii})(1 - h_{jj})\}^{1/2}}. \tag{8.1.8}$$

The values of these correlations thus depend entirely on the elements of the matrix $\mathbf{X}$, since σ^2 cancels. [In situations where we "design our experiment," that is, choose our $\mathbf{X}$ matrix, we thus have the opportunity to affect these correlations. We cannot get all zero correlations, of course, because the n residuals carry only $(n - p)$ degrees of freedom and are linked by the normal equations.]

Other Facts About H

1. $\text{SS (all parameters)} = \text{SS}(\mathbf{b}) = \mathbf{b}'\mathbf{X}'\mathbf{Y}$

$$\begin{aligned}
&= \hat{\mathbf{Y}}'\mathbf{Y} \\
&= \mathbf{Y}'\mathbf{H}'\mathbf{Y} \\
&= \mathbf{Y}'\mathbf{HY} \\
&= \mathbf{Y}'\mathbf{H}^2\mathbf{Y} \\
&= \hat{\mathbf{Y}}'\hat{\mathbf{Y}}.
\end{aligned} \tag{8.1.9}$$

2. The average $V(\hat{Y}_i)$ over all the data points is

$$\sum_{i=1}^{n} V(\hat{Y}_i)/n = \text{trace } (\mathbf{H}\sigma^2)/n = p\sigma^2/n, \tag{8.1.10}$$

where p is the number of parameters. (See Exercise R in "Exercises for Chapters 5 and 6.")

3. $\mathbf{H1} = \mathbf{1}$ when the model contains a β_0 term. This means that every row of $\mathbf{H}$ adds up to 1. So does every column, since $\mathbf{H}$ is symmetric. Note that $\mathbf{1}' = \mathbf{1}'\mathbf{H}' = \mathbf{1}'\mathbf{H}$ and $\mathbf{1}'\mathbf{H1} = n$.

4. Because $\hat{\mathbf{Y}} = \mathbf{HY}$, $\mathbf{H}$ is often called the "hat matrix," that is, the matrix that converts Y's into $\hat{Y}$'s. The diagonal elements are often called the leverages, since examination of

$$\hat{Y}_i = h_{ii}Y_i + \sum_{j \neq i} h_{ij}Y_j \tag{8.1.11}$$

indicates via h_{ii} how heavily Y_i contributes to $\hat{Y}_i$. The messages from the leverages are not clear-cut. Cook and Weisberg (1982, p. 15) say that "for any $h_{ii} > 0$, $\hat{Y}_i$ will be dominated by $h_{ii}Y_i$ if Y_i is sufficiently different from the other elements of $\mathbf{Y}$ (that is, an outlier)." Hadi (1992, p. 5) says that "because high-leverage points are outliers in the X-space, some authors define leverage in terms of outlyingness in the X-space. However, leverage and outlyingness in the X-space are two different concepts . . . . High-leverage points are outliers in the X-space but the converse is not necessarily true." We shall deemphasize use of the leverages.

Internally Studentized Residuals[1]

It is clear from (8.1.7) that $V(e_i) = (1 - h_{ii})\sigma^2$, and these may well vary considerably. Usually σ^2 would be estimated by s^2, the residual mean square, that is, by

$$s^2 = \mathbf{e}'\mathbf{e}/(n - p) = \Sigma e_i^2/(n - p). \tag{8.1.12}$$

We can thus *studentize* the residuals by defining

$$s_i = \frac{e_i}{s(1 - h_{ii})^{1/2}}, \tag{8.1.13}$$

namely, by dividing each residual by its standard error $\{\hat{V}(e_i)\}^{1/2} = \{(1 - h_{ii})s^2\}^{1/2}$. These studentized residuals are said to be *internally studentized* because the s has, within it, e_i itself. So e_i is both on top and (concealed) underneath.

Extra Sum of Squares Attributable to e_i

We recall that, since $\mathbf{e} = (\mathbf{I} - \mathbf{H})\mathbf{Y}$,

$$e_i = -h_{i1}Y_1 - h_{i2}Y_2 - \cdots + (1 - h_{ii})Y_i \cdots - h_{in}Y_n$$
$$= \mathbf{c}'\mathbf{Y}, \tag{8.1.14}$$

say, where $\mathbf{c}' = (-h_{i1}, -h_{i2}, \ldots, (1 - h_{ii}), \ldots - h_{in})$. Now

---

[1]Some versions of MINITAB call these standardized residuals. They are obtained by allocating a column for them in the regression command.

$$\mathbf{c'c} = \sum_{j=1}^{n} h_{ij}^2 + (1 - 2h_{ii})$$

$$(8.1.15)$$

$$= (1 - h_{ii}).$$

This follows because the summation is the ith row of $\mathbf{H}$ multiplied by the ith column of $\mathbf{H}$, which gives h_{ii} by the idempotency of $\mathbf{H}$. So, by (6.3.3) applied to (8.1.14), the 1 df extra SS for e_i is

$$\mathrm{SS}(e_i) = e_i^2/(1 - h_{ii}).$$

$$(8.1.16)$$

Thus

$$s^2(i) = \frac{(n - p)s^2 - e_i^2/(1 - h_{ii})}{n - p - 1}$$

$$(8.1.17)$$

provides an estimate of σ^2 after deletion of the contribution of e_i.

Externally Studentized Residuals[2]

Analogously to (8.1.13), we can define

$$t_i = \frac{e_i}{s(i)(1 - h_{ii})^{1/2}}$$

$$(8.1.18)$$

as the *externally studentized* residuals. An advantage of this is that, if e_i is large, it is thrown into emphasis even more by the fact that $s(i)$ has excluded it. The t_i follow a $t(n - p - 1)$ distribution under the usual normality of errors assumption.

Example. Consider again the data of Table 2.1 used to illustrate pure error. We show the calculation details for the influential observation Y_{23}. The ordinary residual is 2.36; $h_{23,23} = 0.158$, $1 - h_{23,23} = 0.842$; $s^2 = 0.7275$,

$$s^2(23) = \{21(0.7275) - (2.36)^2/0.842\}/20 = 0.4336,$$

much smaller than s^2. The internally studentized residual is

$$s_{23} = 2.36/\{0.7275(0.842)\}^{1/2} = 3.01.$$

The externally studentized residual is

$$t_{23} = 2.36/\{0.4336(0.842)\}^{1/2} = 3.90.$$

We see how external studentization has placed the t-residual nearly four standard errors away from zero rather than about three, and has thus given it extra prominence. (Of course, it was already large enough to notice at three standard errors, in this example!)

Calculations like those above are carried out automatically by many programs and the reader does not actually have to go through the details.

The amended residuals can be used in place of the ordinary residuals in any of the plots mentioned in Chapter 2. Comparison of e_i, s_i, and t_i plots is sometimes useful.

---

[2]Some versions of MINITAB call these the t-residuals. Their derivation is also described differently in MINITAB source material, where they are described as $\{Y_i - \hat{Y}(i)\}/\{\mathrm{MSE}_i + V(\hat{Y}(i))\}^{1/2}$, where $\hat{Y}(i)$ is predicted from a regression on data omitting Y_i and $\mathrm{MSE}_i = s^2(i)$. They are, nevertheless, identical to the t_i we describe. A subcommand "tresids C19" gives them in column 19, say.

Other Comments

Gray and Woodall (1994) mention several issues.

1. The internally studentized residuals s_i are such that the marginal distribution of $s_i^2/(n-p)$ is $\beta(\frac{1}{2}, (n-p-1)/2)$, which implies that $|s_i|$ cannot exceed $(n-p)^{1/2}$. This bound is reached only when deleting the ith set of data results in a perfect fit to the data that remain. When the residual degrees of freedom $(n-p)$ are small, no $|s_i|$ residual can be very large as a consequence.

2. Max $|s_i|$ can be tested, if desired; references are given.

3. The externally studentized residuals t_i are not bounded and t_i has a marginal t-distribution with $(n-p-1)$ df.

LaMotte (1994) covers some of the same ground from the viewpoint of when ratio statistics are actually t-variables and when not.

Berk and Booth (1995) indicate that several types of diagnostic plots can be useful to detect curvature omitted in a first-order model, but all of the diagnostics can be misleading in certain circumstances.

8.2. ADDED VARIABLE PLOT AND PARTIAL RESIDUALS

Added Variable Plot

We suggested in Section 2.6 that residuals could also be plotted against variables that are new candidates for entry. For example, if $\mathbf{z} = (z_1, z_2, \ldots, z_n)'$ is a column of values of a variable observed with the rest of the data, we could plot the e_i versus the z_i. We can also first regress $\mathbf{z}$ on the other X's in the model and get its residual vector

$$\mathbf{e}_z = \mathbf{z} - \hat{\mathbf{z}} = (\mathbf{I} - \mathbf{X}(\mathbf{X}'\mathbf{X})^{-1}\mathbf{X})\mathbf{z}.$$

This will remove from $\mathbf{z}$ any effect due to the columns of $\mathbf{X}$. We can then plot $\mathbf{e}$ versus $\mathbf{e}_z$. Such a plot is called an *added variable plot*. Note that the plot is centered around the origin since both variables are residuals from a model with β_0 in it. Assessing the slope of a straight line through the origin in this plot is essentially an assessment of the value of adding z to the group of regression predictors. So this could also be done directly using an extra sum of squares F-test.

Partial Residuals

Partial residuals are residuals that have not been adjusted for a particular X variable. Suppose we focus on X_i and rewrite $\mathbf{X} = (\mathbf{X}_1, \mathbf{x}_i)$ and

$$\hat{\mathbf{Y}} = \mathbf{X}_1\mathbf{b}_1 + \mathbf{x}_i b_i, \tag{8.2.2}$$

where correspondingly

$$\mathbf{b} = (\mathbf{b}_1', b_i)' = \begin{pmatrix} \mathbf{b}_1 \\ b_i \end{pmatrix}.$$

Then the set of partial residuals for X_i would be

$$\mathbf{e}_i^* = \mathbf{Y} - \mathbf{X}_1\mathbf{b}_1 \qquad\qquad (8.2.3)$$

$$= \mathbf{e} + x_i b_i$$

where $\mathbf{e}$ is the usual vector of residuals $\mathbf{Y} - \hat{\mathbf{Y}}$. A plot of e_i^* versus X_i has slope b_i and the vector $\mathbf{e}$ would provide the residuals if a straight line fit were made.

(This has also been called a "component plus residual plot." A variation adds back a quadratic term in X_i to the residuals as well as a first-order term.)

8.3. DETECTION OF INFLUENTIAL OBSERVATIONS: COOK'S STATISTICS

First we consider the (rather extreme) example where we fit a straight line to a set of data consisting of five observations, four at $X = a$, and one at $X = b$. If $V(Y_i) = \sigma^2$ we can show that, at $X = a$, $V(e_i) = 0.75\sigma^2$, $i = 1, 2, 3, 4$, while, at $X = b$, $V(e_5) = 0$. At first sight, a zero variance seems very desirable but it in fact arises because the fitted straight line is determined as the join of the average level of Y at $X = a$ and the single observed level of Y at $X = b$. The residual at $X = b$ is zero whatever the value of the corresponding Y and, in fact, the parameter estimates depend heavily on this particular observation. A large error in this observation is not detectable in the model-fitting process, and examination of the residuals would not reveal it either, if it existed. The observation at $X = b$ is an extremely *influential* one, whether it is correct or not.

The fact that an observation provides a large outlier is not, of course, good, but it does not necessarily mean that the observation is influential in fitting the chosen model. For example, in Figure 8.1, where the data of Table 8.1 are plotted, we see that the observation marked 19 will certainly be an outlier for most simple models fitted through the data. Its possible influence is moderated by the fact that there are

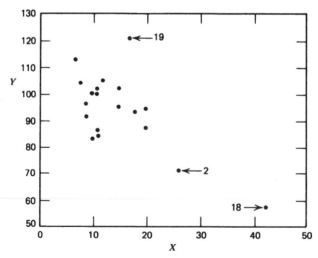

Figure 8.1. A regression with an observation (19) that may not be influential and one (18) that may well be. X represents the age of a child at first word (in months) and Y represents the child's score on an aptitude test. Reproduced by permission from Andrews and Pregibon (1978). The original data were recorded by Dr. L. M. Linde of UCLA and were given by Mickey, Dunn, and Clark (1967). See Table 8.1 for the data.

T A B L E 8.1. Age at First Word (X) and Gesell Adaptive Score (Y)

Case	X	Y
1	15	95
2	26	71
3	10	83
4	9	91
5	15	102
6	20	87
7	18	93
8	11	100
9	8	104
10	20	94
11	7	113
12	9	96
13	10	83
14	11	84
15	11	102
16	10	100
17	12	105
18	42	57
19	17	121
20	11	86
21	10	100

Source: Data from Mickey, Dunn, and Clark (1967) but recorded by L. M. Linde of UCLA.

other observations at neighboring X-values. Observation 18, on the other hand, could well be an influential one. Being alone in its territory, it may have a major influence on the position of the fitted model there. It may or may not have a large residual, depending on the model fitted and the rest of the data.

In any data set where the estimation of one or more parameters depends heavily on a very small number of the observations, problems of interpretation can arise. One way to anticipate such problems is to check whether the deletion of observations greatly affects the fit of the model and the subsequent conclusions. If it does, the conclusions are shaky and more data are probably needed.

Cook (1977) proposed that the influence of the ith data point be measured by the squared scaled distance

$$D_i = (\hat{\mathbf{Y}} - \hat{\mathbf{Y}}(i))'(\hat{\mathbf{Y}} - \hat{\mathbf{Y}}(i))/(ps^2). \qquad (8.3.1)$$

Here, $\hat{\mathbf{Y}} = \mathbf{Xb}$ is the usual vector of predicted values, while $\hat{\mathbf{Y}}(i) = \mathbf{Xb}(i)$ is the vector of predicted values from a least squares fit when the ith data point is deleted, where $\mathbf{b}(i)$ is the corresponding least squares estimator.

When $\mathbf{v}$ is a vector, $\mathbf{v}'\mathbf{v}$ is the squared length of that vector, so D_i is the squared distance between (the ends of) the vectors $\hat{\mathbf{Y}}$ and $\hat{\mathbf{Y}}(i)$, divided by ps^2. If omission of the ith observation makes little difference to the fitted values, D_i will be small. Large D_i indicate observations whose deletion greatly affects the fitted values. Because $\hat{\mathbf{Y}} - \hat{\mathbf{Y}}(i) = \mathbf{X}\{\mathbf{b} - \mathbf{b}(i)\}$, we can also write

$$D_i = \{\mathbf{b} - \mathbf{b}(i)\}'\mathbf{X}'\mathbf{X}\{\mathbf{b} - \mathbf{b}(i)\}/(ps^2). \qquad (8.3.2)$$

A third representation of D_i is the form

$$D_i = \left\{\frac{e_i}{s(1-h_{ii})^{1/2}}\right\}^2 \left\{\frac{h_{ii}}{1-h_{ii}}\right\} \frac{1}{p}, \qquad (8.3.3)$$

where e_i is the ith residual when the full data set is used, s^2 is the estimate of the variance $V(Y_i) = \sigma^2$ provided by the residual mean square when the full data set is used, and h_{ii} is the ith diagonal entry of the matrix $\mathbf{H} = \mathbf{X}(\mathbf{X}'\mathbf{X})^{-1}\mathbf{X}'$. We see that the first factor in Eq. (8.3.3) is the ith internally studentized residual, that is, the residual divided by its standard error (see Section 8.1), while the second term is the ratio (variance of the ith predicted value)/(variance of the ith residual). Note that $0 \leq h_{ii} \leq 1$. D_i can be large if either the first or second factor is large, and these factors measure two separate characteristics of each data point.

For our example above, with four observations at $X = a$, one at $X = b$,

$$D_i = \left\{\frac{e_i^2}{0.75s^2}\right\} \left(\frac{1}{3}\right)\left(\frac{1}{2}\right), \qquad i = 1, 2, 3, 4, \qquad (8.3.4)$$

and

$$D_5 = \text{indeterminate},$$

where, for $i = 1, 2, 3, 4$, each $e_i = Y_i - \overline{Y}_a$, where $\overline{Y}_a = (Y_1 + Y_2 + Y_3 + Y_4)/4$ and $e_5 = 0$. The fifth observation, at $X = b$, is thus "flagged" as being a peculiar one and examination of the circumstances reveals its overwhelming influence.

For the Mickey, Dunn, and Clark data of Table 8.1, the values of the Cook's statistics for omission of observations $1, 2, \ldots, 21$ are 0.01 times, respectively,

$$0, 8, 7, 3, 2, 0, 0, 0, 0, 2, 5, 0, 7, 5, 0, 0, 2, 68, 22, 3, 0.$$

Observation 18 is very influential, and 19 is somewhat influential. There is no formal test for this. We simply compare the sizes of the "big ones" with the base level indicated by the bulk of the numbers. This base level is in the single digits here and the statistics 68 (for observation 18) and 22 (for observation 19) are clearly higher. (Some writings suggest that the Cook's statistics follow an F-distribution, but this is incorrect.) (What happens to the fit when 18 is omitted is shown in Figure 8.2.)

Evaluation of the residuals for these data show that observation 19 appears to be an outlier. Because X is the age (in months) of a child at first word, it is clearly not sensible to plan to collect data to fill in the gap between the lower X-values where most of the data occur, and the X-values of observations 2 and 18, which may not be reliable anyway. If we omit 2 and 18 as atypical X-values and omit 19 as an outlier, the message sent by the data is much reduced. That could well be the appropriate course of action here.

Evaluation of D_i as a routine technique is recommended. It is widely available as a standard option on many regression systems. In MINITAB, the subcommand "cookd C20;" will place the Cook's statistics in column 20, for example.

Higher-Order Cook's Statistics

Cook's statistics can also be evaluated for omitted pairs, omitted triplets, and so on, in an obvious extension of (8.3.1) and (8.3.2). Much more calculation is required in these cases and this is not routine. Here are a few example calculations of Cook's

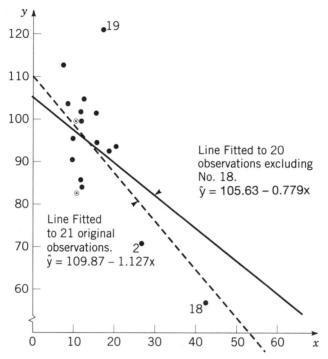

Figure 8.2. Plot of the Mickey, Dunn, and Clark (1967) data together with fitted lines that include and exclude data point 18.

statistics for some omitted pairs of observations from the data of Table 8.1. (These numbers are in their original scaling and have *not* been multiplied by 100 to clear decimals.)

Omitted	Cook's Statistic
18, 2	6.37
18, 3	0.48
18, 11	1.52
18, 19	0.15
19, 2	0.10
19, 3	0.12
19, 11	0.41

Clearly (18, 2) is the most influential pair. The statistic's low value for the pair (18, 19) comes as somewhat of a surprise at first sight. Each exerts a clockwise pull on the line but when both are omitted together, the line does not move very much.

Omitting more than one observation at a time can be useful in cases where two or more points conceal each or one another in a single deletion check. For example, if there were a point (let us call it 22) close to point 18 in the Table 8.1 data, omission of 18 might not give a large Cook's statistic value because 22 would "hold the line close" to its original position. Similarly, omission of 22 alone might have the same result. Omission of both 18 and 22 would, however, reveal them as an influential pair, if that were the case.

Another Worked Example

For the data used to illustrate the lack of fit test in Section 2.1, the Cook's statistics are 0.01 times these numbers:

$$4, 0, 7, 4, 0, 8, 2, 4, 3, 0, 0, 1, 0, 4, 6, 1, 0, 5, 0, 0, 1, 2, 85.$$

The last observation shows up as very influential. We have already mentioned its strangeness in Section 2.1, where it showed up clearly in the plot of Figure 2.2. Evaluation of Cook's statistics usually enables us to pinpoint such observations even when we have not been able to look at a plot, for example, in a case with multiple predictors.

Plots

Some suggestions for plotting the values of Cook's statistics are given by Hines and Hines (1995).

8.4. OTHER STATISTICS MEASURING INFLUENCE

The DFFITS Statistics

These statistics are cousins of Cook's influence measures. They are defined by Belsley, Kuh, and Welsch (1980) as

$$\text{DFFITS}_i = [\{\mathbf{b} - \mathbf{b}(i)\}'\mathbf{X}'\mathbf{X}\{\mathbf{b} - \mathbf{b}(i)\}/s^2(i)]^{1/2} \tag{8.4.1}$$

$$= [D_i p s^2/s^2(i)]^{1/2}, \tag{8.4.2}$$

where D_i is Cook's statistic (8.3.1). [For $s^2(i)$, see Section 8.1. It is an estimate of σ^2 obtained after deletion of the contribution of the ith residual.]

Atkinson's Modified Cook's Statistics

By replacing s^2 by $s^2(i)$, scaling by a factor $(n - p)/p$ instead of $1/p$, and taking the square root, we obtain another set of relatives of Cook's statistics:

$$A_i = [\{\mathbf{b} - \mathbf{b}(i)\}'\mathbf{X}'\mathbf{X}\{\mathbf{b} - \mathbf{b}(i)\}(n - p)/\{ps^2(i)\}]^{1/2} \tag{8.4.3}$$

$$= [D_i(n - p)s^2/s^2(i)]^{1/2} \tag{8.4.4}$$

$$= \text{DFFITS}_i\{(n - p)/n\}^{1/2}. \tag{8.4.5}$$

See Atkinson (1985).

   Cook's, DFFITS, and Atkinson's statistics tell parallel stories for most sets of data. We suggest you go with your preference. While we favor Atkinson's for technical reasons, we usually go with Cook's, which is more widely available.

8.5. REFERENCE BOOKS FOR ANALYSIS OF RESIDUALS

How much effort should one put into examining residuals? It depends on the circumstances. If your regression is one of a series, and tomorrow you move on to other experiments and to other data, perhaps not too much effort beyond the standard plots we have described is needed. On the other hand, if you spent a year in an African jungle and emerged with 61 precious observations, you may wish to spend another year examining all their aspects. In this case, you may need to consult some of the excellent books listed below for sophisticated analyses we have not told you about. The same references appear in Section 2.8.

Atkinson (1985); Barnett and Lewis (1994); Belsley (1991); Belsley, Kuh, and Welsch (1980); Chatterjee and Hadi (1988); Cook and Weisberg (1982); Hawkins (1980); Rousseeuw and Leroy (1987).

EXERCISES FOR CHAPTER 8

A. Find the hat matrix $\mathbf{H} = \mathbf{X}(\mathbf{X}'\mathbf{X})^{-1}\mathbf{X}'$ and the three types of residuals (ordinary, s-residuals, and t-residuals) for any of the regression examples or exercises in this book. Check the characteristics and properties of $\mathbf{H}$ given in Section 8.1. Compare standard plots of the three types of residuals, for example, versus order, overall, or versus $\hat{Y}$. For most (but not all) regressions, it makes little difference which type of residual is used in these plots. When the plots differ materially, the reasons why should be explored. Check for outliers and influential points.

B. If you were asked to fit a straight line to the data

$$(X, Y) = (1, 3), (2, 2.5), (2, 1.2), (3, 1), \text{ and } (6, 4.5),$$

what would you say about it?

C. Find the Cook's statistics for any of the regression examples or exercises in this book. Are any of the observations influential? What can be done if they are?

D. (*Source:* "The hat matrix in regression and anova," by D. C. Hoaglin and R. E. Welsch. *The American Statistician*, **32**, 1978, 17–22. See also p. 146.) We recall that $\hat{\mathbf{Y}} = \mathbf{X}\mathbf{b} = \mathbf{H}\mathbf{Y}$, where $\mathbf{H} = \mathbf{X}(\mathbf{X}'\mathbf{X})^{-1}\mathbf{X}'$ (sometimes called the *hat matrix*) is symmetric and idempotent. Show that each of the diagonal terms h_{uu} (sometimes called the *leverage* of Y_u on $\hat{Y}_u$) for $u = 1, 2, \ldots, n$ lies between 0 and 1, that the n h_{uu} add to p, the number of parameters (note that $\mathbf{X}$ is $n \times p$), and that, if Y_u is replaced by $Y_u^* = Y_u + 1$ in the regression calculation, then $\hat{Y}_u^* = \hat{Y}_u + h_{uu}$.

CHAPTER 9

Multiple Regression: Special Topics

The topics in this chapter are all useful on certain occasions, but most or all of them can probably be passed by on a first reading. Thus it is convenient to group them together at a point in the book where the necessary prerequisites have been established.

9.1. TESTING A GENERAL LINEAR HYPOTHESIS

Experimenters sometimes postulate models that are more general than they hope they need. For example, suppose an experimenter is involved with a response Y and two predictors X_1 and X_2 and has a set of data (Y_i, X_{1i}, X_{2i}), $i = 1, 2, \ldots, n$. She suspects that, although X_1 and X_2 both affect Y, the single predictor of importance is really the difference $X_1 - X_2$. If both X's are needed, she will want to fit the model

$$Y = \beta_0 + \beta_1 X_1 + \beta_2 X_2 + \epsilon, \tag{9.1.1}$$

but, if her suspicion is correct, the model

$$Y = \beta_0 + \beta(X_1 - X_2) + \epsilon \tag{9.1.2}$$

would be good enough. How can she check? Essentially, she has asked the question: "Could it be, in Eq. (9.1.1), that $\beta_1 = -\beta_2 \; (= \beta$, say)?" Or, alternatively, "Is $\beta_1 + \beta_2 = 0$?" She thus will want to test the null hypothesis $H_0 : \beta_1 + \beta_2 = 0$ versus the alternative $H_1 : \beta_1 + \beta_2 \neq 0$. Because H_0 involves a statement about a linear combination of the β's, we call it a *linear hypothesis*.

Linear hypotheses typically arise *from the knowledge of the experimenter and his/ her conjectures about possible models*. They can also arise from a consulting statistician, if he/she is deeply enough involved in the project to understand it at such a level. Ideally, the statistician *should* be that deeply involved, but in practice this does not always happen.

A linear hypothesis can also consist of more than one statement about the β's. We now provide some additional examples of linear hypotheses, explain generally how one is tested, and illustrate the procedure with a simple numerical example, H_1 is always the statement that H_0 is not true in some way, and so is not specifically mentioned in the examples.

(We note that the "extra sum of squares" principle of Section 6.1 is a special case of the work in this section.)

Example 1. Model: $E(Y) = \beta_0 + \beta_1 X_1 + \beta_2 X_2$.

217

$$H_0 : \beta_1 = 0,$$

$$\beta_2 = 0 \qquad \text{(two linear functions, independent).}$$

(By "independent" we mean *linearly* independent, so that one statement cannot be obtained as a linear combination of other statements in the group.)

Example 2. Model: $E(Y) = \beta_0 + \beta_1 X_1 + \beta_2 X_2 + \cdots + \beta_k X_k$.

$$H_0 : \beta_1 = 0,$$

$$\beta_2 = 0,$$

$$\vdots$$

$$\beta_k = 0 \qquad \text{(k linear functions, all independent).}$$

Example 3. Model: $E(Y) = \beta_0 + \beta_1 X_1 + \beta_2 X_2 + \cdots + \beta_k X_k$.

$$H_0 : \beta_1 - \beta_2 = 0,$$

$$\beta_2 - \beta_3 = 0,$$

$$\vdots$$

$$\beta_{k-1} - \beta_k = 0 \qquad \text{($k - 1$ linear functions, independent).}$$

Note that this expresses the hypothesis

$$H_0 : \beta_1 = \beta_2 = \cdots = \beta_k = \beta, \text{ say.}$$

Example 4 (*General Case*). Model: $E(Y) = \beta_0 + \beta_1 X_1 + \beta_2 X_2 + \cdots + \beta_k X_k$.

$$H_0 : c_{10}\beta_0 + c_{11}\beta_1 + c_{12}\beta_2 + \cdots + c_{1k}\beta_k = 0,$$

$$c_{20}\beta_0 + c_{21}\beta_1 + c_{22}\beta_2 + \cdots + c_{2k}\beta_k = 0,$$

$$\vdots$$

$$c_{m0}\beta_0 + c_{m1}\beta_1 + c_{m2}\beta_2 + \cdots + c_{mk}\beta_k = 0.$$

In this hypothesis there are m linear functions of $\beta_0, \beta_1, \beta_2, \ldots, \beta_k$, all of which may not be independent. H_0 can be expressed in matrix form as

$$H_0 : \mathbf{C}\boldsymbol{\beta} = \mathbf{0},$$

where

$$\mathbf{C} = \begin{bmatrix} c_{10} & c_{11} & c_{12} & \cdots & c_{1k} \\ c_{20} & c_{21} & c_{22} & \cdots & c_{2k} \\ \vdots & \vdots & \vdots & & \vdots \\ c_{m0} & c_{m1} & c_{m2} & \cdots & c_{mk} \end{bmatrix}, \qquad \boldsymbol{\beta} = \begin{bmatrix} \beta_0 \\ \beta_1 \\ \beta_2 \\ \vdots \\ \beta_k \end{bmatrix}.$$

We shall suppose in what follows that the m functions are *dependent* and that the last $(m - q)$ of them depend on the first q; that is, if we had these first q independent functions, we could take linear combinations of them to form the other $(m - q)$ linear functions.

We have seen earlier how it is possible to test hypotheses of the forms in Examples 1 and 2. We now explain how more general hypotheses can be tested.

Testing a General Linear Hypothesis $\mathbf{C}\boldsymbol{\beta} = \mathbf{0}$

Suppose that the model under consideration, assumed correct, is

$$E(\mathbf{Y}) = \mathbf{X}\boldsymbol{\beta},$$

where $\mathbf{Y}$ is $(n \times 1)$, $\mathbf{X}$ is $(n \times p)$, and $\boldsymbol{\beta}$ is $(p \times 1)$. If $\mathbf{X}'\mathbf{X}$ is nonsingular we can estimate $\boldsymbol{\beta}$ as

$$\mathbf{b} = (\mathbf{X}'\mathbf{X})^{-1}\mathbf{X}'\mathbf{Y}.$$

The residual sum of squares for this analysis is given, as we have seen, by

$$\text{SSE} = \mathbf{Y}'\mathbf{Y} - \mathbf{b}'\mathbf{X}'\mathbf{Y}.$$

This sum of squares has $(n - p)$ degrees of freedom. The linear hypothesis to be tested, $H_0: \mathbf{C}\boldsymbol{\beta} = \mathbf{0}$, provides q independent conditions on the parameters $\beta_0, \beta_1, \ldots, \beta_k$, on the assumptions (mentioned above) that $\mathbf{C}\boldsymbol{\beta} = \mathbf{0}$ represents m equations, of which only q are independent. We can use the q independent equations to solve for q of the β's in terms of the other $p - q$ of them. Substituting these solutions back into the original model provides us with a reduced model of, say,

$$E(\mathbf{Y}) = \mathbf{Z}\boldsymbol{\alpha},$$

where $\boldsymbol{\alpha}$ is a vector of parameters to be estimated. There will be $p - q$ of these parameters. The right-hand side $\mathbf{Z}\boldsymbol{\alpha}$, where $\mathbf{Z}$ is $n \times (p - q)$ and $\boldsymbol{\alpha}$ is $(p - q) \times 1$, represents the result of substituting into $\mathbf{X}\boldsymbol{\beta}$ for the dependent β's.
We can now estimate the parameter vector $\boldsymbol{\alpha}$ in the new model by

$$\mathbf{a} = (\mathbf{Z}'\mathbf{Z})^{-1}\mathbf{Z}'\mathbf{Y},$$

if $\mathbf{Z}'\mathbf{Z}$ is nonsingular, and can obtain a new residual sum of squares for this regression of

$$\text{SSW} = \mathbf{Y}'\mathbf{Y} - \mathbf{a}'\mathbf{Z}'\mathbf{Y}.$$

This sum of squares has $(n - p + q)$ degrees of freedom.
Since fewer parameters are involved in this second analysis, SSW will always be larger than SSE. The difference $\text{SSW} - \text{SSE}$ is called the *sum of squares due to the hypothesis* $\mathbf{C}\boldsymbol{\beta} = \mathbf{0}$ and has $(n - p + q) - (n - p) = q$ degrees of freedom. A test of the hypothesis $H_0: \mathbf{C}\boldsymbol{\beta} = \mathbf{0}$ is now made by considering the ratio

$$\left(\frac{\text{SSW} - \text{SSE}}{q}\right) \bigg/ \left(\frac{\text{SSE}}{n - p}\right)$$

and referring it to the $F(q, n - p)$ distribution in the usual manner. If the errors are normally distributed and independent, this is an exact test.
The appropriate test for Examples 1 and 2 [already given as Eq. (5.3.2) where $q = k = p - 1$] is a special case of this. The reduced model in both cases consists of

$$E(\mathbf{Y}) = \mathbf{1}\beta_0$$

where $\mathbf{1}' = (1, 1, \ldots, 1)$ is a vector of all ones. Another way of writing this model is

$$E(Y_i) = \beta_0, \qquad i = 1, 2, \ldots, n.$$

Since $b_0 = \bar{Y}$, $\text{SSW} = \mathbf{Y}'\mathbf{Y} - n\bar{Y}^2$ with $(n - 1)$ degrees of freedom, whereas $\text{SSE} =$

$Y'Y - b'X'Y$ with $(n - k - 1)$ degrees of freedom. So the ratio for the test $\beta_1 = \beta_2 = \cdots = \beta_k = 0$ (for Example 2; when $k = 2$, we have Example 1) is simply

$$\left(\frac{b'X'Y - nY^2}{k}\right)\bigg/\left(\frac{Y'Y - b'X'Y}{n - k - 1}\right)$$

and this is referred to the $F(k, n - k - 1)$ distribution. This is exactly the procedure of Eq. (5.3.2) with $k = p - 1$, $v = n - k - 1$, and $s^2 = MS_E = SSE/v$.

We shall now illustrate the use of the procedure in a simple but not so typical case.

Worked Example. Given the model $E(Y) = X\beta$, test the hypothesis $H_0: C\beta = 0$, where

$$Y' = (1, 4, 8, 9, 3, 8, 9),$$

$$\beta' = (\beta_0, \beta_1, \beta_2, \beta_{11}),$$

$$\begin{array}{cccc} 1 & X & X_2 & X_1^2 \end{array}$$

$$X = \begin{bmatrix} 1 & -1 & -1 & 1 \\ 1 & 1 & -1 & 1 \\ 1 & -1 & 1 & 1 \\ 1 & 1 & 1 & 1 \\ 1 & 0 & 0 & 0 \\ 1 & 0 & 1 & 0 \\ 1 & 0 & 2 & 0 \end{bmatrix},$$

and

$$C = \begin{bmatrix} 0 & 0 & 0 & 1 \\ 0 & 1 & -1 & 0 \\ 0 & 1 & -1 & 1 \\ 0 & 2 & -2 & 3 \end{bmatrix}$$

Solution. We first find the residual sum of squares SSE when the original model, of form $E(Y) = \beta_0 + \beta_1 X_1 + \beta_2 X_2 + \beta_{11} X_1^2$, is fitted. We find

$$(X'X)^{-1} = \begin{bmatrix} 7 & 0 & 3 & 4 \\ 0 & 4 & 0 & 0 \\ 3 & 0 & 9 & 0 \\ 4 & 0 & 0 & 4 \end{bmatrix}^{-1} = \begin{bmatrix} \frac{1}{2} & 0 & -\frac{1}{6} & -\frac{1}{2} \\ 0 & \frac{1}{4} & 0 & 0 \\ -\frac{1}{6} & 0 & \frac{1}{6} & \frac{1}{6} \\ -\frac{1}{2} & 0 & \frac{1}{6} & \frac{3}{4} \end{bmatrix},$$

$$X'Y = \begin{bmatrix} 42 \\ 4 \\ 38 \\ 22 \end{bmatrix}, \quad b = (X'X)^{-1}X'Y = \begin{bmatrix} \frac{1}{3} \\ 1 \\ 3 \\ \frac{1}{6} \end{bmatrix}, \quad \begin{array}{l} b'X'Y = 312.33 \\ \\ Y'Y = 316 \end{array}$$

$$SSE = 316 - 312.33 = 3.67.$$

The equations for the null hypothesis $H_0 : C\beta = 0$ are

$$\beta_{11} = 0,$$
$$\beta_1 - \beta_2 = 0,$$
$$\beta_1 - \beta_2 + \beta_{11} = 0,$$
$$2\beta_1 - 2\beta_2 + 3\beta_{11} = 0.$$

The hypothesis can be more simply expressed as $H_0 : \beta_{11} = 0$, $\beta_1 = \beta_2 = \beta$, say, since the third and fourth equations are linear combinations of the first and second equations.

Substituting these conditions in the model gives a reduced model

$$E(Y) = \beta_0 + \beta(X_1 + X_2) = \alpha_0 + \alpha Z,$$

where

$$\alpha_0 = \beta_0, \qquad \alpha = \beta, \qquad Z = X_1 + X_2.$$

Thus

$$\mathbf{Z} = \begin{bmatrix} 1 & (-1-1) \\ 1 & (1-1) \\ 1 & (-1+1) \\ 1 & (1+1) \\ 1 & (0+0) \\ 1 & (0+1) \\ 1 & (0+2) \end{bmatrix} = \begin{bmatrix} 1 & -2 \\ 1 & 0 \\ 1 & 0 \\ 1 & 2 \\ 1 & 0 \\ 1 & 1 \\ 1 & 2 \end{bmatrix}$$

$$\mathbf{Z'Y} = \begin{bmatrix} 42 \\ 42 \end{bmatrix}, \qquad (\mathbf{Z'Z})^{-1} = \begin{bmatrix} 7 & 3 \\ 3 & 13 \end{bmatrix}^{-1} = \frac{1}{82} \begin{bmatrix} 13 & -3 \\ -3 & 7 \end{bmatrix},$$

$$\mathbf{a} = (\mathbf{Z'Z})^{-1}\mathbf{Z'Y} = \frac{21}{41} \begin{bmatrix} 10 \\ 4 \end{bmatrix}, \qquad \mathbf{a'Z'Y} = 301.17,$$

$$\text{SSW} = 316 - 301.17 = 14.83.$$

Now $p = 4$, $n = 7$, $q = 2$, $n - p = 3$, and

$$\text{SSW} - \text{SSE} = 14.83 - 3.67 = 11.16 = \text{SS due to the hypothesis.}$$

The appropriate test statistic for H_0 is thus $(11.16/2) \div 3.67/3 = 4.56$. Since $F(2, 3, 0.95) = 9.55$, we *do not* reject H_0. Since the original model was $E(Y) = \beta_0 + \beta_1 X_1 + \beta_2 X_2 + \beta_{11} X_1^2$ and the hypothesis *not* rejected implies $\beta_{11} = 0$, $\beta_1 = \beta_2 = \beta$, a more plausible model would be $E(Y) = \beta_0 + \beta(X_1 + X_2)$.

9.2. GENERALIZED LEAST SQUARES AND WEIGHTED LEAST SQUARES

It sometimes happens that some of the observations used in a regression analysis are "less reliable" than others. What this usually means is that the variances of the observations are not all equal; in other words the nonsingular matrix $\mathbf{V}(\boldsymbol{\epsilon})$ is not of the form $\mathbf{I}\sigma^2$ but is diagonal with unequal diagonal elements. It may also happen, in some problems, that the off-diagonal elements of $\mathbf{V}(\boldsymbol{\epsilon})$ are not zero, that is, the observations are correlated.

When either or both of these events occur, the ordinary least squares estimation formula $\mathbf{b} = (\mathbf{X}'\mathbf{X})^{-1}\mathbf{X}'\mathbf{Y}$ does not apply and it is necessary to amend the procedures for obtaining estimates. The basic idea is to transform the observations $\mathbf{Y}$ to other variables $\mathbf{Z}$, which *do* appear to satisfy the usual tentative assumptions [that $\mathbf{Z} = \mathbf{Q}\boldsymbol{\beta} + \mathbf{f}$, $E(\mathbf{f}) = \mathbf{0}$, $V(\mathbf{f}) = \mathbf{I}\sigma^2$, and, for F-tests and confidence intervals to be valid, that $\mathbf{f} \sim N(\mathbf{0}, \mathbf{I}\sigma^2)$] and to then apply the usual analysis to the variables so obtained. The estimates can then be reexpressed in terms of the original variables $\mathbf{Y}$. We shall describe how the usual regression procedures are changed. Suppose the model under consideration is

$$\mathbf{Y} = \mathbf{X}\boldsymbol{\beta} + \boldsymbol{\epsilon}, \tag{9.2.1}$$

where

$$E(\boldsymbol{\epsilon}) = \mathbf{0}, \quad \mathbf{V}(\boldsymbol{\epsilon}) = \mathbf{V}\sigma^2, \quad \text{and} \quad \boldsymbol{\epsilon} \sim N(\mathbf{0}, \mathbf{V}\sigma^2). \tag{9.2.2}$$

It can be shown that it is possible to find a nonsingular symmetric matrix $\mathbf{P}$ such that

$$\mathbf{P}'\mathbf{P} = \mathbf{P}\mathbf{P} = \mathbf{P}^2 = \mathbf{V}. \tag{9.2.3}$$

Let us write

$$\mathbf{f} = \mathbf{P}^{-1}\boldsymbol{\epsilon}, \quad \text{so that } E(\mathbf{f}) = \mathbf{0}. \tag{9.2.4}$$

Now it is a fact that, if $\mathbf{f}$ is a vector random variable such that $E(\mathbf{f}) = 0$, then $E(\mathbf{f}\mathbf{f}') = \mathbf{V}(\mathbf{f})$, where the expectation is taken separately for every term in the square $n \times n$ matrix $\mathbf{f}\mathbf{f}'$. Thus

$$\mathbf{V}(\mathbf{f}) = E(\mathbf{f}\mathbf{f}') = E(\mathbf{P}^{-1}\boldsymbol{\epsilon}\boldsymbol{\epsilon}'\mathbf{P}^{-1}), \quad \text{since } (\mathbf{P}^{-1})' = \mathbf{P}^{-1}$$

$$= \mathbf{P}^{-1}E(\boldsymbol{\epsilon}\boldsymbol{\epsilon}')\mathbf{P}^{-1} \tag{9.2.5}$$

$$= \mathbf{P}^{-1}\mathbf{P}\mathbf{P}\mathbf{P}^{-1}\sigma^2$$

$$= \mathbf{I}\sigma^2.$$

It is also true that $\mathbf{f} \sim N(\mathbf{0}, \mathbf{I}\sigma^2)$; that is, $\mathbf{f}$ is normally distributed, since the elements of $\mathbf{f}$ consist of linear combinations of the elements of $\boldsymbol{\epsilon}$, which is itself normally distributed.

Thus if we premultiply Eq. (9.2.1) by $\mathbf{P}^{-1}$ we obtain a new model

$$\mathbf{P}^{-1}\mathbf{Y} = \mathbf{P}^{-1}\mathbf{X}\boldsymbol{\beta} + \mathbf{P}^{-1}\boldsymbol{\epsilon} \tag{9.2.6}$$

or

$$\mathbf{Z} = \mathbf{Q}\boldsymbol{\beta} + \mathbf{f} \tag{9.2.7}$$

with an obvious notation. It is now clear that we can apply the basic least squares theory to Eq. (9.2.7) since $E(\mathbf{f}) = \mathbf{0}$ and $\mathbf{V}(\mathbf{f}) = \mathbf{I}\sigma^2$. The residual sum of squares is

$$\mathbf{f}'\mathbf{f} = \boldsymbol{\epsilon}'\mathbf{V}^{-1}\boldsymbol{\epsilon} = (\mathbf{Y} - \mathbf{X}\boldsymbol{\beta})'\mathbf{V}^{-1}(\mathbf{Y} - \mathbf{X}\boldsymbol{\beta}). \tag{9.2.8}$$

The normal equations $\mathbf{Q}'\mathbf{Q}\mathbf{b} = \mathbf{Q}'\mathbf{Z}$ become

$$\mathbf{X}'\mathbf{V}^{-1}\mathbf{X}\mathbf{b} = \mathbf{X}'\mathbf{V}^{-1}\mathbf{Y} \tag{9.2.9}$$

with solution

$$\mathbf{b} = (\mathbf{X}'\mathbf{V}^{-1}\mathbf{X})^{-1}\mathbf{X}'\mathbf{V}^{-1}\mathbf{Y} \tag{9.2.10}$$

when the matrix just inverted is nonsingular. The regression sum of squares is

$$\mathbf{b'Q'Z} = \mathbf{Y'V^{-1}X(X'V^{-1}X)^{-1}X'V^{-1}Y} \tag{9.2.11}$$

and the total sum of squares is

$$\mathbf{Z'Z} = \mathbf{Y'V^{-1}Y}. \tag{9.2.12}$$

The difference between Eqs. (9.2.12) and (9.2.11) provides the residual sum of squares. The sum of squares due to the mean is $(\Sigma\ Z_i)^2/n$, where Z_i are the n elements of the vector $\mathbf{Z}$. Note that, if we subtract this from Eq. (9.2.11), the remainder is not an extra sum of squares in the usual sense, because the transformed model no longer contains a β_0. Thus an appropriate base sum of squares to subtract here is one due to the first component of Eq. (9.2.7). The variance–covariance matrix of $\mathbf{b}$ is

$$\mathbf{V(b)} = \mathbf{(Q'Q)^{-1}}\sigma^2 = \mathbf{(X'V^{-1}X)^{-1}}\sigma^2. \tag{9.2.13}$$

A joint confidence region for all the parameters can be obtained from

$$(\mathbf{b} - \boldsymbol{\beta})'\mathbf{Q'Q}(\mathbf{b} - \boldsymbol{\beta}) = \left[\frac{p}{(n-p)}\right](\mathbf{Z'Z} - \mathbf{b'Q'Z})F(p, n-p, 1-\alpha) \tag{9.2.14}$$

after substituting from Eqs. (9.2.11) and (9.2.12) and setting $\mathbf{Q} = \mathbf{P^{-1}X}$, if so desired.

Generalized Least Squares Residuals

The residuals that must be checked are the estimates of $\mathbf{f} = \mathbf{P^{-1}\epsilon}$. These residuals are given by

$$\mathbf{P^{-1}(Y - \hat{Y})},$$

where $\hat{\mathbf{Y}} = \mathbf{Xb}$ and $\mathbf{b}$ is taken from Eq. (9.2.11). Thus these residuals are

$$\mathbf{P^{-1}\{I - X(X'V^{-1}X)^{-1}X'V^{-1}\}Y}. \tag{9.2.15}$$

A similar formula applies when $\mathbf{V}$ is estimated.

General Comments

We speak of *generalized least squares* when $\mathbf{V}$ is not a diagonal matrix, and of *weighted least squares* when it is. In the latter case, the observations are independent but have different variances so that

$$\mathbf{V}\sigma^2 = \begin{bmatrix} \sigma_1^2 & & & 0 \\ & \sigma_2^2 & & \\ & & \ddots & \\ 0 & & & \sigma_n^2 \end{bmatrix}$$

where some of the σ_i^2 may be equal.

In practical problems it is often difficult to obtain specific information on the form of $\mathbf{V}$ at first. For this reason it is sometimes necessary to make the (known to be erroneous) assumption $\mathbf{V} = \mathbf{I}$ and then attempt to discover something about the form of $\mathbf{V}$ by examining the residuals from the regression analysis.

If a generalized least squares analysis were called for but an ordinary least squares analysis were performed, the estimates obtained would still be unbiased but would not have minimum variance, since the minimum variance estimates are obtained from the correct generalized least squares analysis.

If standard least squares is used, then the estimates are obtained from $\mathbf{b}_0 = (\mathbf{X}'\mathbf{X})^{-1}\mathbf{X}'\mathbf{Y}$ and

$$E(\mathbf{b}_0) = (\mathbf{X}'\mathbf{X})^{-1}\mathbf{X}'\mathbf{X}\boldsymbol{\beta} = \boldsymbol{\beta}$$

but

$$\mathbf{V}(\mathbf{b}_0) = (\mathbf{X}'\mathbf{X})^{-1}\mathbf{X}'[\mathbf{V}(\mathbf{Y})]\mathbf{X}(\mathbf{X}'\mathbf{X})^{-1}$$
$$= (\mathbf{X}'\mathbf{X})^{-1}\mathbf{X}'\mathbf{V}\mathbf{X}(\mathbf{X}'\mathbf{X})^{-1}\sigma^2.$$

We recall from Eq. (9.2.13) that if the correct analysis is performed,

$$\mathbf{V}(\mathbf{b}) = (\mathbf{X}'\mathbf{V}^{-1}\mathbf{X})^{-1}\sigma^2$$

and, in general, elements of this matrix would provide smaller variances both for individual coefficients and for linear functions of the coefficients.

Application to Serially Correlated Data

The major difficulty in applying generalized least squares methods is in finding $\mathbf{V}$ in Eq. (9.2.2). Suppose we wish to allow for serial correlation, for example. If the observations are listed in time order, the element V_{ij} of $\mathbf{V}$ would be ρ_l, where $l = |i - j|$, with $\rho_0 = 1$. To estimate ρ_l we could lag the observations by l steps and evaluate a correlation coefficient using Eq. (1.6.5), ignoring the unmatched overlap observations. These estimates are substituted into $\mathbf{V}$ to produce a $\hat{\mathbf{V}}$, which is used in formulas such as Eqs. (9.2.10) and (9.2.11). To analyze the residuals from this weighted fit we need the estimates of $\mathbf{f} = \mathbf{P}^{-1}\boldsymbol{\epsilon}$; see Eqs. (9.2.3) and (9.2.4). These estimates are thus

$$\hat{\mathbf{f}} = \hat{\mathbf{P}}^{-1}(\mathbf{Y} - \hat{\mathbf{Y}}), \tag{9.2.16}$$

where

$$\hat{\mathbf{P}}'\hat{\mathbf{P}} = \hat{\mathbf{V}} \tag{9.2.17}$$

and where $\hat{\mathbf{Y}}$ is fitted by generalized least squares, so that

$$\hat{\mathbf{Y}} = \mathbf{X}(\mathbf{X}'\hat{\mathbf{V}}^{-1}\mathbf{X})^{-1}\mathbf{X}'\hat{\mathbf{V}}^{-1}\mathbf{Y}. \tag{9.2.18}$$

In other words, the elements of

$$\hat{\mathbf{f}} = \hat{\mathbf{P}}^{-1}\{\mathbf{I} - \mathbf{X}(\mathbf{X}'\hat{\mathbf{V}}^{-1}\mathbf{X})^{-1}\mathbf{X}'\hat{\mathbf{V}}^{-1}\}\mathbf{Y} \tag{9.2.19}$$

are examined.

(*Note:* This is essentially the same formula as $\mathbf{e} = \{\mathbf{I} - \mathbf{X}(\mathbf{X}'\mathbf{X})^{-1}\mathbf{X}'\}\mathbf{Y}$ for ordinary least squares, but with $\hat{\mathbf{P}}^{-1}\mathbf{X}$ and $\hat{\mathbf{P}}^{-1}\mathbf{Y}$ replacing $\mathbf{X}$ and $\mathbf{Y}$ respectively.)

9.3. AN EXAMPLE OF WEIGHTED LEAST SQUARES

This is an extremely simple example but an interesting one. Suppose we wish to fit the model

$$E(Y) = \beta X.$$

Let us suppose that

$$\mathbf{V}\sigma^2 = \mathbf{V}(\mathbf{Y}) = \begin{bmatrix} 1/w_1 & & & \\ & 1/w_2 & & \mathbf{0} \\ & & \ddots & \\ \mathbf{0} & & & 1/w_n \end{bmatrix} \sigma^2,$$

where the w's are *weights* to be specified. This means that

$$\mathbf{V}^{-1} = \begin{bmatrix} w_1 & & & \\ & w_2 & & \mathbf{0} \\ & & \ddots & \\ \mathbf{0} & & & w_n \end{bmatrix}.$$

Applying the general results above we find, after reduction,

$$b = \frac{\Sigma w_i X_i Y_i}{\Sigma w_i X_i^2},$$

where all summations are from $i = 1, 2, ..., n$.

Case 1. Suppose $\sigma_i^2 = V(Y_i) = kX_i$; that is, the variance of Y_i is proportional to the size of the corresponding X_i. Then $w_i = \sigma^2/kX_i$. Hence

$$b = \frac{\Sigma Y_i}{\Sigma X_i} = \frac{\bar{Y}}{\bar{X}}.$$

Thus if the variance of Y_i is proportional to X_i, the best estimate of the regression coefficient is the mean of the Y_i divided by the mean of the X_i. In addition,

$$V(b) = \frac{\sigma^2}{\Sigma w_i X_i^2} = \frac{k}{\Sigma X_i}.$$

Case 2. Suppose $\sigma_i^2 = V(Y_i) = kX_i^2$; that is, the variance of Y_i is proportional to the square of the corresponding X_i. Then $w_i = \sigma^2/kX_i^2$. Hence

$$b = \frac{\Sigma (Y_i/X_i)}{\Sigma 1}$$

$$= \frac{\Sigma (Y_i/X_i)}{n}.$$

Thus if the variance of the Y_i is proportional to X_i^2, the best estimate of the regression coefficient is the average of the n slopes obtained one from each pair of observations Y_i/X_i. Also,

$$V(b) = \frac{\sigma^2}{\Sigma w_i X_i^2} = \frac{k}{n}.$$

Note: Fitting a straight line through the origin $(X, Y) = (0, 0)$ represents a very strong assumption, which, in general, is not justified. Even when the model is "known" to pass through the origin (as would be the case, for example, if $X =$ speed of car, $Y =$ stopping distance) it does not mean that a straight line fit though the origin is necessarily appropriate. It may be that the available data can be fitted by a straight line not through the origin but that, if more data were available, a higher-order model that did pass through the origin would provide a proper fit. Usually it is best to put an intercept term β_0 in the model and to check on the size of the estimate b_0.

9.4 A NUMERICAL EXAMPLE OF WEIGHTED LEAST SQUARES

The data in Table 9.1, which have been rearranged in an order convenient for purposes of analysis, consist of 35 observations (X_i, Y_i) with a number of sets that are either exact repeats at the same X-value or approximate repeats. These are indicated by the groupings. A fit of the data by (ordinary) least squares produces the fitted model

T A B L E 9.1. Data for Weighted Least Squares Example

X	Y	$\hat{w}_i$
1.15	0.99	1.24028
1.90	0.98	2.18224
3.00	2.60	7.84930
3.00	2.67	7.84930
3.00	2.66	7.84930
3.00	2.78	7.84930
3.00	2.80	7.84930
5.34	5.92	7.43652
5.38	5.35	6.99309
5.40	4.33	6.78574
5.40	4.89	6.78574
5.45	5.21	6.30514
7.70	7.68	0.89204
7.80	9.81	0.84420
7.81	6.52	0.83963
7.85	9.71	0.82171
7.87	9.82	0.81296
7.91	9.81	0.79588
7.94	8.50	0.78342
9.03	9.47	0.47385
9.07	11.45	0.46621
9.11	12.14	0.45878
9.14	11.50	0.45327
9.16	10.65	0.44968
9.37	10.64	0.41435
10.17	9.78	0.31182
10.18	12.39	0.31079
10.22	11.03	0.30672
10.22	8.00	0.30672
10.22	11.90	0.30672
10.18	8.68	0.31079
10.50	7.25	0.28033
10.23	13.46	0.30571
10.03	10.19	0.32680
10.23	9.93	0.30571

Source: Wanda M. Hinshaw.

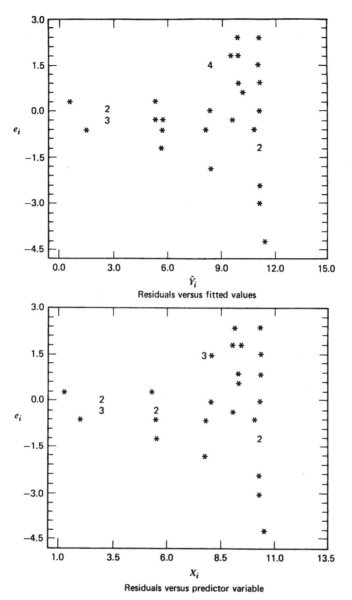

Figure 9.1. Residuals plots, unweighted least squares. (Two indistinguishable points are shown as a 2, and so on.)

$\hat{Y} = -0.5790 + 1.1354X$ and the residuals plots in Figure 9.1. A clear indication that the observations have unequal variances is seen. The overall plot of residuals (not shown) is somewhat skewed toward negative values, also. None of the usual (ordinary) least squares analyses are appropriate and it seems sensible to apply generalized least squares.

We assume (until contrary indications appear) that the Y_i are independent so that **V** has the diagonal pattern with different variances given earlier. We now need to obtain information on the variance pattern. For each of the sets of repeats or near repeats we evaluate the average X-value, X_j, say, and the pure error mean square s_{ej}^2. These are:

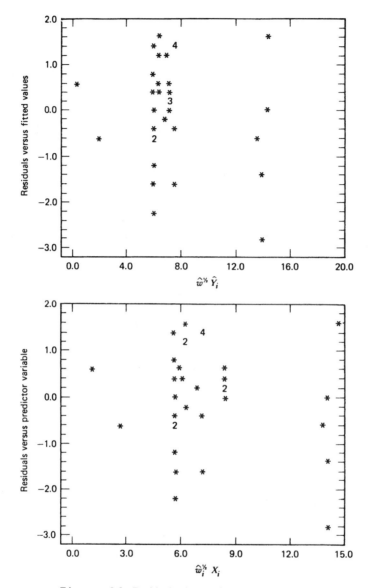

Figure 9.2. Residuals plots, weighted least squares.

$\overline{X}_j$	3.0	5.4	7.8	9.1	10.2
s_{ej}^2	0.0072	0.3440	1.7404	0.8683	3.8964

A plot of these suggests a quadratic relationship, which we estimate by least squares as

$$\hat{s}_e^2 = 1.5329 - 0.7334\overline{X} + 0.0883\overline{X}^2.$$

We can now substitute each individual X_i into this equation, estimate s_{ej}^2, $i = 1, 2,$ ..., 35, and invert these values to give the estimated weights $\hat{w}_i$ shown in the table. The matrix **P** of our text is diagonal with entries $\hat{w}_i^{-1/2}$. Using these weights leads to the weighted least squares prediction equation $\hat{Y} = -0.8891 + 1.1648X$ and an analysis of variance table as follows:

Source	df	SS	MS
$b_1\|b_0$	1	496.96	496.96
Residual	33	42.66	1.29
Total, corrected	34	539.62	

The appropriate "observations" and "fitted values" are now $\hat{w}^{1/2} Y_i$ and $\hat{w}_i^{1/2} \hat{Y}_i$ and the "residuals" to be examined are $\hat{w}_i^{1/2}(Y_i - \hat{Y}_i)$, notice. An overall plot of residuals still shows some skewness but the pattern is slightly better behaved. The residuals plots in Figure 9.2 reveal that the vertical spread of residuals is now roughly the same at the two main levels of the transformed response. (At lower levels there are only two observations so that there is not much of an estimate of the spread there.) The employment of weighted least squares here appears to be justified and useful.

A weighted least squares program exists in most computing systems, but some do not have a generalized least squares program.

9.5 RESTRICTED LEAST SQUARES

For least squares involving restrictions on the parameters see, for example, Waterman (1974) and Judge and Takayama (1966). If the restrictions are of the equality form $\mathbf{C}\boldsymbol{\beta} = \mathbf{d}$, we can use the method of Lagrange's undetermined multipliers (see Appendix 9A) and minimize the Lagrangean function

$$F = (\mathbf{Y} - \mathbf{X}\boldsymbol{\beta})'(\mathbf{Y} - \mathbf{X}\boldsymbol{\beta}) + \boldsymbol{\lambda}'(\mathbf{d} - \mathbf{C}\boldsymbol{\beta}) \qquad (9.5.1)$$

with respect to $\boldsymbol{\beta}$ and $\boldsymbol{\lambda}$. The solution for $\boldsymbol{\beta}$ is

$$\hat{\boldsymbol{\beta}} = \mathbf{b} + (\mathbf{X}'\mathbf{X})^{-1}\mathbf{C}'[\mathbf{C}(\mathbf{X}'\mathbf{X})^{-1}\mathbf{C}']^{-1}(\mathbf{d} - \mathbf{Cb}), \qquad (9.5.2)$$

where $\mathbf{b} = (\mathbf{X}'\mathbf{X})^{-1}\mathbf{X}'\mathbf{Y}$ is the usual unrestricted estimator. See also Chapters 20 and 21 for geometrical aspects.

Note that, if $\mathbf{d} = \mathbf{0}$, so that the restriction is $\mathbf{C}\boldsymbol{\beta} = \mathbf{0}$, we are back in the context of Section 9.1. Even if $\mathbf{d} \neq \mathbf{0}$, we can always substitute back into the model $\mathbf{Y} = \mathbf{X}\boldsymbol{\beta} + \boldsymbol{\epsilon}$ for some parameters using the restrictions $\mathbf{C}\boldsymbol{\beta} = \mathbf{d}$, to obtain a solution in terms of fewer parameters. The solution (9.5.2) is more elegant in that it retains all the original parameters and also ensures that $\mathbf{C}\hat{\boldsymbol{\beta}} = \mathbf{d}$; the latter is obvious from premultiplying Eq. (9.5.2) by $\mathbf{C}$. However, it provides the same predicted values that we would obtain by using the substitution method.

9.6. INVERSE REGRESSION (MULTIPLE PREDICTOR CASE)

Given a fitted regression equation $\hat{Y} = b_0 + b_1 X_1 + b_2 X_2 + \cdots + b_k X_k$ and a true mean value of Y, say, Y_0, we require a "fiducial region" for the point $(X_1, X_2, \ldots, X_k)$. Extending Eq. (3.2.8) we obtain the following equation satisfied by the boundaries of the required region:

$$\{-Y_0 + b_0 + b_1 X_1 + b_2 X_2 + \cdots + b_k X_k\}^2$$

$$= t^2 s^2 \left\{ (1, X_1, X_2, \ldots, X_k)(\mathbf{X'X})^{-1} \begin{bmatrix} 1 \\ X_1 \\ X_2 \\ \vdots \\ X_k \end{bmatrix} \right\}. \quad (9.6.1)$$

This is a hyperbolic surface. Figure 9.3 shows the $k = 2$ case. When Y_0 is the mean of q observations, insert "$1/q\ +$" inside the curly braces on the right-hand side of Eq. (9.6.1). If $\hat{Y}$ is a polynomial, not a plane, the obvious adjustments must be made on both sides of Eq. (9.6.1).

Note: The method indicated above can be applied to other types of problems. For example, the maxima and minima of $\hat{Y} = b_0 = b_1 X + b_2 X^2 + b_3 X^3 + b_4 X^4$ are at the roots of $f \equiv b_1 + 2b_2 X + 3b_3 X^2 + 4b_4 X^3 = 0$. Fiducial limits for the roots can be evaluated from the equation

$$f^2 = t^2 s^2 \{V(f)/\sigma^2\}, \quad (9.6.2)$$

where $V(f)$ denotes the variance of the function f, which has a factor g^2 in it. For a fuller account, including possible problems with imaginary roots, see Williams (1959, pp. 108–109 and 114–116). See also Box and Hunter (1954).

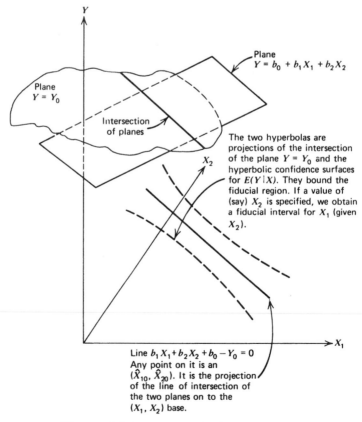

The two hyperbolas are projections of the intersection of the plane $Y = Y_0$ and the hyperbolic confidence surfaces for $E(Y|X)$. They bound the fiducial region. If a value of (say) X_2 is specified, we obtain a fiducial interval for X_1 (given X_2).

Line $b_1 X_1 + b_2 X_2 + b_0 - Y_0 = 0$
Any point on it is an $(\hat{X}_{10}, \hat{X}_{20})$. It is the projection of the line of intersection of the two planes on to the (X_1, X_2) base.

Figure 9.3. Inverse regression for two predictors.

9.7. PLANAR REGRESSION WHEN ALL THE VARIABLES ARE SUBJECT TO ERROR

We briefly describe an extension of the work of Section 3.4 for more than one X. Suppose we have observations $(Y_i, X_{1i}, X_{2i}, \ldots, X_{ki})$, $i = 1, 2, \ldots, n$, with all variables subject to random errors. Consider the quantity

$$D_{Yi}^2 = (\beta_0 + \sum_{j=1}^{k} \beta_j X_{ji} - Y_i)^2, \qquad (9.7.1)$$

the squared deviation of the ith point measured in the Y direction. Similar squared deviations in other directions are defined by, for $\ell = 1, 2, \ldots, k$,

$$D_{\ell i}^2 = (\beta_0 + \sum_{j=1}^{k} \beta_j X_{ji} - Y_i)^2 / \beta_\ell^2. \qquad (9.7.2)$$

The geometric mean of the D^2 values,

$$G_i = (D_{Yi}^2 D_{1i}^2 D_{2i}^2 \cdots D_{ki}^2)^{1/(k+1)}, \qquad (9.7.3)$$

can also be regarded as

$$G_i = V_i^{2/(k+1)}, \qquad (9.7.4)$$

where V_i is the volume created by drawing, from the ith data point $(Y_i, X_{1i}, X_{2i}, \ldots X_{ki})$, lines parallel to the $Y, X_1, X_2, \ldots, X_k$ axes to the plane $Y = \beta_0 + \sum_{j=1}^{k} \beta_j X_j$.

For $k = 1$, V_i is the area of the ith right-angled triangle in Figure 3.5, so that the general G_i is an extension of this concept. The criterion

$$L_G^k = \sum_{i=1}^{n} G_i \qquad (9.7.5)$$

can now be minimized to obtain estimates for the general dimension case. For additional details, see Draper and Yang (1997).

APPENDIX 9A. LAGRANGE'S UNDETERMINED MULTIPLIERS

Notation

Because the method of Lagrange's undetermined multipliers has wide applicability, we have chosen to adopt a fairly "neutral" notation $\theta_1, \theta_2, \ldots, \theta_m$ for the variables involved in the functions f and g_j below. When the method is applied as in Section 9.5, the θ's would be *all* the parameters in the $\boldsymbol{\beta}$ vector. For the ridge regression application in Chapter 17, the θ's would be all the regression β's *except* β_0. In other applications, the θ's might be predictor variables, that is, X's.

Basic Method

Suppose we wish to obtain the stationary or turning values of a function $f(\theta_1, \theta_2, \ldots, \theta_m)$ of m variables $\theta_1, \theta_2, \ldots, \theta_m$, subject to restrictions on the θ_i such as

$$g_j(\theta_1, \theta_2, \ldots, \theta_m) = 0 \qquad (j = 1, 2, \ldots, q).$$

Form the function

$$F = f - \sum_{j=1}^{q} \lambda_j g_j, \tag{9A.1}$$

where $\lambda_1, \lambda_2, \ldots, \lambda_q$ are unknowns. Differentiate Eq. (9A.1) partially with respect to each θ_i and set the results equal to zero. This will provide the m equations

$$\frac{\partial F}{\partial \theta_i} \equiv \frac{\partial f}{\partial \theta_i} - \sum_{j=1}^{q} \lambda_j \frac{\partial g_j}{\partial \theta_i} = 0 \qquad (i = 1, 2, \ldots, m). \tag{9A.2}$$

These m equations, with the additional q equations

$$g_j = 0 \qquad (j = 1, 2, \ldots, q), \tag{9A.3}$$

provide $(q + m)$ equations that can be solved for the $(q + m)$ unknowns $\theta_1, \theta_2, \ldots,$ $\theta_m, \lambda_1, \lambda_2, \ldots, \lambda_q$. Often the quantities λ_j are eliminated and not actually found; for this reason the words "undetermined multipliers" are used to describe them. In some cases, however, the solutions for $\theta_1, \theta_2, \ldots, \theta_m$ are easier to obtain if the λ_j are evaluated first; in other cases, it may be easier to specify values of λ_j in Eqs. (9A.2) and regard other quantities in Eqs. (9A.3) as unknowns, in their place.

Is the Solution a Maximum or Minimum?

Suppose now that $(\theta_1, \theta_2, \ldots, \theta_m) = (a_1, a_2, \ldots, a_m)$ is a solution of Eqs. (9A.2) and (9A.3) after elimination of λ_j. Let

$$\mathbf{M}(\theta) = \mathbf{M}(\theta_1, \theta_2, \ldots, \theta_m) = \begin{bmatrix} \dfrac{\partial^2 F}{\partial \theta_1{}^2} & \dfrac{\partial^2 F}{\partial \theta_1 \partial \theta_2} & \cdots & \dfrac{\partial^2 F}{\partial \theta_1 \partial \theta_m} \\[2mm] \dfrac{\partial^2 F}{\partial \theta_2 \partial \theta_1} & \dfrac{\partial^2 F}{\partial \theta_2{}^2} & \cdots & \dfrac{\partial^2 F}{\partial \theta_2 \partial \theta_m} \\[2mm] \vdots & \vdots & & \vdots \\[2mm] \dfrac{\partial^2 F}{\partial \theta_m \partial \theta_1} & \dfrac{\partial^2 F}{\partial \theta_m \partial \theta_2} & \cdots & \dfrac{\partial^2 F}{\partial \theta_m{}^2} \end{bmatrix} \tag{9A.4}$$

be the matrix of second-order partial derivatives. Let $\mathbf{M}(a_1, a_2, \ldots, a_m) = \mathbf{M}(\mathbf{a})$ be the resulting matrix after the solution $\mathbf{a}' = (a_1, a_2, \ldots, a_m)$ has been substituted into Eq. (9A.4). Then if $\mathbf{M}(\mathbf{a})$ is

1. positive definite, that is, $\mathbf{u}'\mathbf{Mu} > 0$, for all $\mathbf{u}$,
2. negative definite, that is, $\mathbf{u}'\mathbf{Mu} < 0$, for all $\mathbf{u}$,

where $\mathbf{u}' = (u_1, u_2, \ldots, u_m)$ is any $1 \times m$ real vector, the function $f(\theta_1, \theta_2, \ldots, \theta_m)$ achieves

1. a local minimum at $\theta = \mathbf{a}$,
2. a local maximum at $\theta = \mathbf{a}$,

respectively. For, if we expand F about $\mathbf{a}$ as a Taylor series of partial derivatives, remembering that all first partial derivatives of F are zero at $\theta = \mathbf{a}$, we see that

$$F(\mathbf{a} + \mathbf{h}) - F(\mathbf{a}) = \tfrac{1}{2}\mathbf{h}'\mathbf{M}(\mathbf{a})\mathbf{h} + O(h^3),$$

where $\mathbf{h}$ represents a vector of small increments h_i all of the same order and $O(h^3)$

represents a remainder of third order in such increments. Thus, to order h^2, if $\mathbf{M(a)}$ is positive definite, for example, then

$$F(\mathbf{a} + \mathbf{h}) > F(\mathbf{a}), \qquad \text{for } all \text{ small } \mathbf{h}.$$

If $\mathbf{h}$ varies only in such a way that the restrictions are still satisfied, this implies that

$$f(\mathbf{a} + \mathbf{h}) > f(\mathbf{a}),$$

that is, $f(\mathbf{a})$ is, locally, a minimum, subject to the restrictions holding. As we can see from this discussion, it might happen that

$$F(\mathbf{a} + \mathbf{h}) \not> F(\mathbf{a}), \qquad \text{for all small } \mathbf{h}$$

but

$$f(\mathbf{a} + \mathbf{h}) > f(\mathbf{a}),$$

for all $\mathbf{h}$ that satisfy the restrictions. Thus "$\mathbf{M(a)}$ is positive definite" is sufficient, but not necessary, for a local restricted minimum of f at $\theta = \mathbf{a}$. Similar remarks apply to the negative definite case. If $\mathbf{M(a)}$ is indefinite, further investigation of the function near the point $\mathbf{a}$ is required to determine what sort of stationary point has been obtained.

EXERCISES FOR CHAPTER 9

A. Consider the model $Y = \beta_0 + \beta_1 X_1 + \beta_2 X_2 + \beta_3 X_3 + \beta_4 X_4 + \epsilon$. If it is suggested to you that the two variables $Z_1 = X_1 + X_3$ and $Z_2 = X_2 + X_4$ might be adequate to represent the data, what hypothesis, in the form $\mathbf{C}\boldsymbol{\beta} = \mathbf{0}$, would you need to test? (Give the form of $\mathbf{C}$.)

B. For the data $(X_1, X_2, Y) = (-1, -1, 7.2), (-1, 0, 8.1), (0, 0, 9.8), (1, 0, 12.3), (1, 1, 12.9)$, the least squares fit is $\hat{Y} = 10.6 + 2.10X_1 + 0.75X_2$, and the residual sum of squares is 0.107 (2 df). Test the null hypothesis $H_0 : \beta_1 = 2\beta_2$ versus H_1: not so.

C. Look at the Hald data in Appendix 15A. Fit $Y = \beta_0 + \beta_1 X_1 + \beta_2 X_2 + \beta_3 X_3 + \beta_4 X_4 + \epsilon$ (which is done in Appendix 15A) and test $H_0 : \beta_1 = \beta_3, \beta_2 = \beta_4$, versus the alternative not so. Is this a reasonable hypothesis to test?

D. Consider the data of Table 2.1. Suppose you are told that the 23rd observation has variance $4\sigma^2$ rather than σ^2. Refit the equation using weighted least squares with $\mathbf{V}^{-1} = (1, 1, \ldots, 1, 0.25)$.

E. Repeat Exercise D but with $16\sigma^2$ for the variance of the last observation. What changes do you observe?

F. (*Source*: J. A. John.) An experimenter tells you he wishes to fit the model $Y = \beta_0 + \beta_1 X_1 + \beta_2 X_2 + \epsilon$ by least squares, subject to the restriction that $\beta_1 = 1$. He asks specifically if he can just fit $Y - X_1 = \beta_0 + \beta_2 X_2 + \epsilon$ by least squares to get what he wants. Can he? (Yes.)

G. (*Source*: T. J. Mitchell.) An experimenter wishes to fit the quadratic model $Y = \beta_0 + \beta_1 X + \beta_{11} X^2 + \epsilon$. He "knows" (he says) that the response at $X_1 = 1$ is 10 so that, ignoring the error term in the model, $10 = \beta_0 + \beta_1 + \beta_{11}$. He then substitutes for $\beta_0 = 10 - \beta_1 - \beta_{11}$ in the first model to give $Y - 10 = \beta_1 Z_1 + \beta_{11} Z_2 + \epsilon$, where $Z_1 = X - 1$ and $Z_2 = X^2 - 1$. He next fits this second model by least squares to provide b_1 and b_{11}, determines $b_0 = 10 - b_1 - b_{11}$, and announces he has obtained the least squares solution for the first model, subject to the restriction that the response at $X = 1$ is 10. Is he correct? (Yes.)

H. (*Source:* S. C. Piper.) Suppose we wish to fit the model

$$Y = \beta_0 + \beta_1 X_1 + \beta_2 X_2 + \beta_3 X_3 + \beta_4 X_4 + \beta_5 X_5 + \beta_6 X_6 + \beta_7 X_7 + \epsilon$$

by least squares, but it is true that

$$\beta_1 + \beta_2 + \beta_3 + \beta_4 = C_1 \text{ (known constant)}$$

and

$$\beta_5 + \beta_6 + \beta_7 = C_2 \text{ (known constant)}.$$

Suppose we substitute for (say) β_4 and β_7 in the original model using the restrictions and then fit the resulting model

$$Y - C_1 X_4 - C_2 X_7 = \beta_0 + \beta_1(X_1 - X_4) + \beta_2(X_2 - X_4) + \beta_3(X_3 - X_4)$$
$$+ \beta_5(X_5 - X_7) + \beta_6(X_6 - X_7) + \epsilon$$

by least squares. Will this solution be correct? (Yes)

I. (Generalized restricted least squares.) Use the method of Lagrange's undetermined multipliers to show that, for a generalized least squares problem in which Eq. (9.5.1) is replaced by

$$F = (\mathbf{Y} - \mathbf{X}\beta)'\mathbf{V}^{-1}(\mathbf{Y} - \mathbf{X}\beta) + \lambda'(\mathbf{d} - \mathbf{C}\beta),$$

the solution for β replacing (9.5.2) is

$$\hat{\beta} = \mathbf{b}_G + (\mathbf{X}'\mathbf{V}^{-1}\mathbf{X})^{-1}\mathbf{C}'[\mathbf{C}(\mathbf{X}'\mathbf{V}^{-1}\mathbf{X})^{-1}\mathbf{C}']^{-1}(\mathbf{d} - \mathbf{C}\mathbf{b}_G)$$

where $\mathbf{b}_G = (\mathbf{X}'\mathbf{V}^{-1}\mathbf{X})^{-1}\mathbf{X}'\mathbf{V}^{-1}\mathbf{Y}$ is the unrestricted generalized least squares estimator.

CHAPTER 10

Bias in Regression Estimates, and Expected Values of Mean Squares and Sums of Squares

This chapter explores what can be said in situations where we fit one model (e.g., a straight line) but we fear that this model may be somewhat inadequate (e.g., there in fact may be a little quadratic curvature). We can talk in terms of the *fitted model* and the *true model* but it is better to think in terms of the fitted model and the *feared model* alternative. After all, if we knew we were fitting the *wrong* model, why would we do it? We are often interested in what might be wrong with the model fitted *if* some specified alternative were true, however. We first discuss possible biases in the estimates of the parameters of a possibly inadequate model and then see how the consequences of this go through to the analysis of variance table, via the expected values of the various mean squares. Details of how to compute the expected values of mean squares and sums of squares are then given.

10.1. BIAS IN REGRESSION ESTIMATES

We said earlier (Section 5.1) that the least squares estimate $\mathbf{b} = (\mathbf{X'X})^{-1}\mathbf{X'Y}$ of $\boldsymbol{\beta}$ in the model $E(\mathbf{Y}) = \mathbf{X}\boldsymbol{\beta}$ is an unbiased estimate. This means that

$$E(\mathbf{b}) = \boldsymbol{\beta}.$$

That is, if we consider the distribution of $\mathbf{b}$ (obtained by taking repeated samples from the same Y-population keeping $\mathbf{X}$ fixed and estimating $\boldsymbol{\beta}$ for each sample), then the mean value of this distribution is $\boldsymbol{\beta}$.

We now emphasize that this is true *only if the postulated model is the correct model to consider*. If it is *not* the correct model, then the estimates are *biased*; that is, $E(\mathbf{b}) \neq \boldsymbol{\beta}$. The extent of the bias depends, as we shall show, not only on the postulated and the true models but also on the values of the X-variables that enter the regression calculations. When a designed experiment is used, the bias depends on the experimental design, as well as the models.

We shall deal with the general nonsingular regression model from the beginning, since once we have the necessary formulas in matrix terms, they can be applied universally. Special cases can be reworked in their algebraic detail as exercises if desired. Suppose we postulate the model

$$E(\mathbf{Y}) = \mathbf{X}_1\boldsymbol{\beta}_1. \tag{10.1.1}$$

This leads to the least squares estimates:

$$\mathbf{b}_1 = (\mathbf{X}_1'\mathbf{X}_1)^{-1}\mathbf{X}_1'\mathbf{Y}. \tag{10.1.2}$$

If the postulated model is correct, then, since $\mathbf{X}_1$ is a matrix of constants unaffected by expectation, and $\mathbf{b}_1$ and $\mathbf{Y}$ are the random variables,

$$E(\mathbf{b}_1) = (\mathbf{X}_1'\mathbf{X}_1)^{-1}\mathbf{X}_1'E(\mathbf{Y}) = (\mathbf{X}_1'\mathbf{X}_1)^{-1}\mathbf{X}_1'\mathbf{X}_1\boldsymbol{\beta}_1 = \boldsymbol{\beta}_1. \tag{10.1.3}$$

Thus $\mathbf{b}_1$ is an unbiased estimate of $\boldsymbol{\beta}_1$.

Now suppose we once again postulate the model given in Eq. (10.1.1) so that $\mathbf{b}_1$, as defined in Eq. (10.1.2), is still the vector of estimated regression coefficients. Suppose *now,* however, that the true response relationship is in fact not Eq. (10.1.1) but

$$E(\mathbf{Y}) = \mathbf{X}_1\boldsymbol{\beta}_1 + \mathbf{X}_2\boldsymbol{\beta}_2. \tag{10.1.4}$$

That is, there are terms $\mathbf{X}_2\boldsymbol{\beta}_2$ that we did not allow for in our estimation procedure. It now follows that

$$E(\mathbf{b}_1) = (\mathbf{X}_1'\mathbf{X}_1)^{-1}\mathbf{X}_1'E(\mathbf{Y}) = (\mathbf{X}_1'\mathbf{X}_1)^{-1}\mathbf{X}_1'(\mathbf{X}_1\boldsymbol{\beta}_1 + \mathbf{X}_2\boldsymbol{\beta}_2)$$

$$= (\mathbf{X}_1'\mathbf{X}_1)^{-1}\mathbf{X}_1'\mathbf{X}_1\boldsymbol{\beta}_1 + (\mathbf{X}_1'\mathbf{X}_1)^{-1}\mathbf{X}_1'\mathbf{X}_2\boldsymbol{\beta}_2 \tag{10.1.5}$$

$$= \boldsymbol{\beta}_1 + \mathbf{A}\boldsymbol{\beta}_2,$$

where

$$\mathbf{A} = (\mathbf{X}_1'\mathbf{X}_1)^{-1}\mathbf{X}_1'\mathbf{X}_2 \tag{10.1.6}$$

is called the *alias* or *bias* matrix. Note that the bias terms $\mathbf{A}\boldsymbol{\beta}_2$ depend not only on the postulated and the true models but also on the experimental design through the matrices $\mathbf{X}_1$ and $\mathbf{X}_2$. Thus a good choice of design may cause estimates to be less biased than they would otherwise be, even if the wrong model has been postulated and fitted.

Note also that the observations Y_i do not appear in (10.1.5) so that the result can be used to examine potential experimental designs before they are actually performed.

The result (10.1.5) can also be viewed in another way. Look first at (10.1.6). If $\mathbf{X}_1'\mathbf{X}_2 = \mathbf{0}$, there is no bias because $\mathbf{A} = \mathbf{0}$. Suppose that $\mathbf{X}_1$ and $\mathbf{X}_2$ are not orthogonal, however. Then if we regress $\mathbf{X}_2$ on $\mathbf{X}_1$ (i.e., treat each of the columns of $\mathbf{X}_2$ as a "$\mathbf{Y}$ column") to give $\hat{\mathbf{X}}_2 = \mathbf{X}_1(\mathbf{X}_1'\mathbf{X}_1)^{-1}\mathbf{X}_1'\mathbf{X}_2$ for "fitted values," we get the "residuals"

$$\mathbf{X}_2 - \hat{\mathbf{X}}_2 = (\mathbf{I} - \mathbf{X}_1(\mathbf{X}_1'\mathbf{X}_1)^{-1}\mathbf{X}_1')\mathbf{X}_2 = \mathbf{Z}, \tag{10.1.7}$$

say. Note that $\mathbf{X}_1'\mathbf{Z} = \mathbf{0}$. We can thus rewrite the model (10.1.4) as

$$E(\mathbf{Y}) = \mathbf{X}_1\boldsymbol{\beta}_1 + \mathbf{X}_1\mathbf{A}\boldsymbol{\beta}_2 + \mathbf{X}_2\boldsymbol{\beta}_2 - \mathbf{X}_1\mathbf{A}\boldsymbol{\beta}_2$$

$$= \mathbf{X}_1(\boldsymbol{\beta}_1 + \mathbf{A}\boldsymbol{\beta}_2) + (\mathbf{X}_2 - \mathbf{X}_1\mathbf{A})\boldsymbol{\beta}_2 \tag{10.1.8}$$

$$= \mathbf{X}_1\boldsymbol{\beta}^* + \mathbf{Z}\boldsymbol{\beta}_2,$$

say, where $\boldsymbol{\beta}^* = \boldsymbol{\beta}_1 + \mathbf{A}\boldsymbol{\beta}_2$. We have *orthogonalized the model* because $\mathbf{X}_1'\mathbf{Z} = \mathbf{0}$. Note that when we estimate $\boldsymbol{\beta}$, the coefficient of $\mathbf{X}_1$, we obtain an unbiased estimate because $\mathbf{X}_1'\mathbf{Z} = \mathbf{0}$, *but it is an estimate of* $\boldsymbol{\beta}_1 + \mathbf{A}\boldsymbol{\beta}_2$. So our two viewpoints are consistent! This orthogonalization procedure is used in Section 10.4, where $\mathbf{Z}$ is called $\mathbf{X}_{2\cdot1}$, a common and meaningful notation indicating that we have obtained "the portion of $\mathbf{X}_2$ that is orthogonal to $\mathbf{X}_1$."

We now illustrate the application of Eq. (10.1.5) to some simple numerical cases.

Example 1. Suppose we postulate the model

$$E(Y) = \beta_0 + \beta_1 X,$$

but the model

$$E(Y) = \beta_0 + \beta_1 X + \beta_{11} X^2$$

is actually the true response function, unknown to us. If we use observations of Y at $X = -1$, 0, and 1 to estimate β_0 and β_1 in the postulated model, what bias will be introduced? That is, what will the estimates b_0 and b_1 actually estimate? The true model, in terms of the observations, is

$$E(\mathbf{Y}) = E \begin{bmatrix} Y_1 \\ Y_2 \\ Y_3 \end{bmatrix} = \begin{matrix} 1 & \mathbf{X} & \mathbf{X}^2 \\ \begin{bmatrix} 1 & -1 & 1 \\ 1 & 0 & 0 \\ 1 & 1 & 1 \end{bmatrix} \end{matrix} \begin{bmatrix} \beta_0 \\ \beta_1 \\ \beta_{11} \end{bmatrix}$$

$$= \begin{matrix} 1 & \mathbf{X} \\ \begin{bmatrix} 1 & -1 \\ 1 & 0 \\ 1 & 1 \end{bmatrix} \end{matrix} \begin{bmatrix} \beta_0 \\ \beta_1 \end{bmatrix} + \begin{matrix} \mathbf{X}^2 \\ \begin{bmatrix} 1 \\ 0 \\ 1 \end{bmatrix} \end{matrix} \beta_{11}$$

$$= \mathbf{X}_1 \boldsymbol{\beta}_1 + \mathbf{X}_2 \boldsymbol{\beta}_2$$

to achieve the form of Eq. (10.1.4) with Eq. (10.1.1) as the postulated model. It follows that

$$(\mathbf{X}_1'\mathbf{X}_1)^{-1} = \begin{bmatrix} 3 & 0 \\ 0 & 2 \end{bmatrix}^{-1} = \begin{bmatrix} \frac{1}{3} & 0 \\ 0 & \frac{1}{2} \end{bmatrix},$$

$$\mathbf{X}_1'\mathbf{X}_2 = \begin{bmatrix} 1 & 1 & 1 \\ -1 & 0 & 1 \end{bmatrix} \begin{bmatrix} 1 \\ 0 \\ 1 \end{bmatrix} = \begin{bmatrix} 2 \\ 0 \end{bmatrix}.$$

Thus

$$\mathbf{A} = \begin{bmatrix} \frac{1}{3} & 0 \\ 0 & \frac{1}{2} \end{bmatrix} \begin{bmatrix} 2 \\ 0 \end{bmatrix} = \begin{bmatrix} \frac{2}{3} \\ 0 \end{bmatrix}.$$

Applying Eq. (10.1.5) we obtain

$$E \begin{bmatrix} b_0 \\ b_1 \end{bmatrix} = \begin{bmatrix} \beta_0 \\ \beta_1 \end{bmatrix} + \begin{bmatrix} \frac{2}{3} \\ 0 \end{bmatrix} \beta_{11} = \begin{bmatrix} \beta_0 + \frac{2}{3}\beta_{11} \\ \beta_1 \end{bmatrix}$$

or

$$E(b_0) = \beta_0 + \tfrac{2}{3}\beta_{11}, \qquad E(b_1) = \beta_1.$$

Thus b_0 is *biased* by $\frac{2}{3}\beta_{11}$, and b_1 is unbiased.

Example 2. Suppose the postulated model is

$$E(Y) = \beta_0 + \beta_1 X,$$

but the true model is actually

$$E(Y) = \beta_0 + \beta_1 X + \beta_{11} X^2 + \beta_{111} X^3.$$

What biases are induced by taking observations at

$$X = -3, -2, -1, 0, 1, 2, 3?$$

We find

$$\mathbf{X}_1 = \begin{bmatrix} 1 & -3 \\ 1 & -2 \\ 1 & -1 \\ 1 & 0 \\ 1 & 1 \\ 1 & 2 \\ 1 & 3 \end{bmatrix}, \qquad \mathbf{X}_2 = \begin{bmatrix} 9 & -27 \\ 4 & -8 \\ 1 & -1 \\ 0 & 0 \\ 1 & 1 \\ 4 & 8 \\ 9 & 27 \end{bmatrix}$$

$$(\mathbf{X}_1'\mathbf{X}_1)^{-1} = \begin{bmatrix} 7 & 0 \\ 0 & 28 \end{bmatrix}^{-1} = \begin{bmatrix} \frac{1}{7} & 0 \\ 0 & \frac{1}{28} \end{bmatrix},$$

$$\mathbf{X}_1'\mathbf{X}_2 = \begin{bmatrix} 28 & 0 \\ 0 & 196 \end{bmatrix},$$

$$\mathbf{A} = (\mathbf{X}_1'\mathbf{X}_1)^{-1}\mathbf{X}_1'\mathbf{X}_2 = \begin{bmatrix} 4 & 0 \\ 0 & 7 \end{bmatrix}.$$

Thus

$$E(b_0) = \beta_0 + 4\beta_{11},$$

$$E(b_1) = \beta_1 + 7\beta_{111}.$$

By using the general formula, Eq. (10.1.5), we can find the bias in any regression estimates once the postulated model, the feared model, and the design are established. This enables us to find, in specific situations, what effects will be transmitted to our estimates if a particular departure from the assumed model occurs. A sensible procedure in many situations where a polynomial model is postulated is to work on the basis that the postulated model may be wrong because it does not contain terms of one degree higher than those present.

10.2. THE EFFECT OF BIAS ON THE LEAST SQUARES ANALYSIS OF VARIANCE

Note: In this section we shall write $\mathbf{X}$ for the matrix previously called $\mathbf{X}_1$ and $\boldsymbol{\beta}$ for the vector previously called $\boldsymbol{\beta}_1$. The notation $\mathbf{X}_2\boldsymbol{\beta}_2$ will still denote the extra terms of the true models, however. We now summarize the effect bias has on the usual least squares analysis.

Suppose that:

1. The postulated model $E(\mathbf{Y}) = \mathbf{X}\boldsymbol{\beta}$ contains p parameters; $V(\mathbf{Y}) = \mathbf{I}\sigma^2$.
2. The true model is $E(\mathbf{Y}) = \mathbf{X}\boldsymbol{\beta} + \mathbf{X}_2\boldsymbol{\beta}_2$, where $\boldsymbol{\beta}_2$ may be $\mathbf{0}$, in which case the postulated model is correct.
3. The total number of observations taken is n and there are f degrees of freedom

available for lack of fit and e degrees of freedom for pure error, so that $n = p + f + e$. (This means there are $p + f$ distinct points in the design.)

4. The estimates $\mathbf{b} = (\mathbf{X'X})^{-1}\mathbf{X'Y}$ and the fitted values $\hat{\mathbf{Y}} = \mathbf{Xb}$ are obtained as usual.

5. $\mathbf{A} = (\mathbf{X'X})^{-1}\mathbf{X'X}_2$.

Then a number of results are true as given below.

1. The matrix $(\mathbf{X'X})^{-1}\sigma^2$ is always the correct variance–covariance matrix, $V(\mathbf{b})$, of estimated coefficients $\mathbf{b}$.

2. $E(\mathbf{b}) = \boldsymbol{\beta} + \mathbf{A}\boldsymbol{\beta}_2$.

3. $E(\hat{\mathbf{Y}}) = \mathbf{X}\boldsymbol{\beta} + \mathbf{XA}\boldsymbol{\beta}_2$.

4. The analysis of variance table takes the form below:

Source	df	SS	Expected Value of MS
b_0	1	$\mathbf{Y'11'Y}/n$	$\sigma^2 + (\mathbf{X}\boldsymbol{\beta} + \mathbf{X}_2\boldsymbol{\beta}_2)'\mathbf{11'}(\mathbf{X}\boldsymbol{\beta} + \mathbf{X}_2\boldsymbol{\beta}_2)/n$
Other	$p - 1$	$\mathbf{b'X'Y} - \mathbf{Y'11'Y}/n$	$\sigma^2 + (\mathbf{X}\boldsymbol{\beta} + \mathbf{X}_2\boldsymbol{\beta}_2)'\{\mathbf{X}(\mathbf{X'X})^{-1}\mathbf{X'} -$
estimates $\mid b_0$			$\mathbf{11'}/n\}(\mathbf{X}\boldsymbol{\beta} + \mathbf{X}_2\boldsymbol{\beta}_2)/(p - 1)$
Lack of fit	f	By difference	$\sigma^2 + \boldsymbol{\beta}_2'(\mathbf{X}_2 - \mathbf{XA})'(\mathbf{X}_2 - \mathbf{XA})\boldsymbol{\beta}_2/f$
Pure error	e	es_e^2	σ^2
Total	n	$\mathbf{Y'Y}$	

5. When $\boldsymbol{\beta}_2 = \mathbf{0}$, that is, when the postulated model is correct, the results above reduce to the following:

$$E(\mathbf{b}) = \boldsymbol{\beta}, \qquad E(\hat{\mathbf{Y}}) = \mathbf{X}\boldsymbol{\beta},$$

$E(\text{mean square due to other estimates} \mid b_0) = \sigma^2 + \boldsymbol{\beta}'\mathbf{X}'(\mathbf{I} - \mathbf{11'}/n)\mathbf{X}\boldsymbol{\beta}/(p - 1)$. This is why the mean square due to estimates is compared with an estimate of σ^2 to test H_0: all parameters in $\boldsymbol{\beta}$, except β_0, are zero, when the fitted model is not rejected out of hand as a result of a nonsignificant lack of fit test. If the lack of fit test *did* indicate the presence of lack of fit, so that $\boldsymbol{\beta}_2 \neq \mathbf{0}$, then it is useless to carry out a test using the regression mean square, even if we use the pure error mean square s_e^2 to estimate σ^2 rather than the residual mean square. In this case, under H_0, $E(\text{mean square due to other estimates} \mid b_0) = \sigma^2 + \boldsymbol{\beta}_2'\mathbf{X}_2'[\mathbf{X}(\mathbf{X'X})^{-1}\mathbf{X'} - \mathbf{11'}/n]\mathbf{X}_2\boldsymbol{\beta}_2/(p - 1)$, and the F-ratio we would use has a noncentral F-distribution, rather than the ordinary central F-distribution that we assume when we make the (erroneous) test in the usual way.

10.3. FINDING THE EXPECTED VALUES OF MEAN SQUARES

To find the expectation of mean squares, certain special matrix results are useful. Suppose $\mathbf{Q}$ is an $n \times n$ matrix so that $\mathbf{Y'QY}$ is a quadratic form in the elements of $\mathbf{Y}$. Then if E denotes expectation

$$E(\mathbf{Y'QY}) = E(\mathbf{Y})'\mathbf{Q}E(\mathbf{Y}) + \text{trace}(\mathbf{Q\Sigma}),$$

where "trace" means "take the sum of the diagonal elements of the square matrix indicated," and $\boldsymbol{\Sigma} = V(\mathbf{Y})$ is the $n \times n$ variance–covariance matrix of the vector $\mathbf{Y}$. Furthermore, if $\mathbf{M}_1$ and $\mathbf{M}_2$ are any two square matrices of the same size,

$$\text{trace}(\mathbf{M}_1 + \mathbf{M}_2) = \text{trace } \mathbf{M}_1 + \text{trace } \mathbf{M}_2.$$

In addition, if $\mathbf{T}$ is a $t \times s$ matrix and $\mathbf{S}$ is an $s \times t$ matrix so that both products $\mathbf{TS}$ and $\mathbf{ST}$ are feasible, then

$$\text{trace}(\mathbf{TS}) = \text{trace}(\mathbf{ST}).$$

This last result is a remarkably useful result and often leads to quite fantastic simplification. For example, if $\mathbf{X}$ is $n \times p$ and we take $\mathbf{T} = \mathbf{X}(\mathbf{X}'\mathbf{X})^{-1}$, $\mathbf{S} = \mathbf{X}'$, we can quickly evaluate

$$\text{trace}\{\mathbf{X}(\mathbf{X}'\mathbf{X})^{-1}\mathbf{X}'\} = \text{trace}\{\mathbf{X}'\mathbf{X}(\mathbf{X}'\mathbf{X})^{-1}\} = \text{trace}\{\mathbf{I}_p\} = p.$$

We use this particular result in the next example.

Example. Find $E(\mathbf{b}'\mathbf{X}'\mathbf{Y}/p)$, when $E(\mathbf{Y}) = \mathbf{X}\boldsymbol{\beta} + \mathbf{X}_2\boldsymbol{\beta}_2$ and $\mathbf{V}(\mathbf{Y}) = \mathbf{I}\sigma^2$.

$$E(\mathbf{b}'\mathbf{X}'\mathbf{Y}) = E(\mathbf{Y}'\mathbf{X}(\mathbf{X}'\mathbf{X})^{-1}\mathbf{X}'\mathbf{Y})$$

$$= (\mathbf{X}\boldsymbol{\beta} + \mathbf{X}_2\boldsymbol{\beta}_2)'\mathbf{X}(\mathbf{X}'\mathbf{X})^{-1}\mathbf{X}'(\mathbf{X}\boldsymbol{\beta} + \mathbf{X}_2\boldsymbol{\beta}_2)$$
$$+ \text{trace}(\mathbf{X}(\mathbf{X}'\mathbf{X})^{-1}\mathbf{X}'\mathbf{I}\sigma^2)$$

$$= \{(\mathbf{X}'\mathbf{X})^{-1}\mathbf{X}'\mathbf{X}\boldsymbol{\beta} + (\mathbf{X}'\mathbf{X})^{-1}\mathbf{X}'\mathbf{X}_2\boldsymbol{\beta}_2\}'(\mathbf{X}'\mathbf{X})(\mathbf{X}'\mathbf{X})^{-1}\mathbf{X}'\mathbf{X}$$
$$\times \{(\mathbf{X}'\mathbf{X})^{-1}\mathbf{X}'\mathbf{X}\boldsymbol{\beta} + (\mathbf{X}'\mathbf{X})^{-1}\mathbf{X}'\mathbf{X}_2\boldsymbol{\beta}_2\} + p\sigma^2$$

$$= (\boldsymbol{\beta} + \mathbf{A}\boldsymbol{\beta}_2)'\mathbf{X}'\mathbf{X}(\boldsymbol{\beta} + \mathbf{A}\boldsymbol{\beta}_2) + p\sigma^2.$$

Dividing each side by p provides the df-weighted average of the first two entries in the foregoing table. Note, in the manipulations shown, the insertion of the unit matrices $\mathbf{I} = (\mathbf{X}'\mathbf{X})^{-1}\mathbf{X}'\mathbf{X}$ and $\mathbf{I} = \mathbf{X}'\mathbf{X}(\mathbf{X}'\mathbf{X})^{-1}$ to lead to the desired form.

For another example in a different context, see Section 10.4.

10.4. EXPECTED VALUE OF EXTRA SUM OF SQUARES

Write

$$\mathbf{Y} = \mathbf{X}_1\boldsymbol{\beta}_1 + \mathbf{X}_2\boldsymbol{\beta}_2 + \boldsymbol{\epsilon}, \quad \text{for model 1,}$$

$$\mathbf{Y} = \mathbf{X}_1\boldsymbol{\beta}_1 \qquad\quad + \boldsymbol{\epsilon}, \quad \text{for model 2,}$$

where $\mathbf{X}_1$ is $n \times q$, and $\mathbf{X}_2$ is $n \times (p - q)$. Note that the $n \times (p - q)$ matrix

$$\mathbf{X}_{2 \cdot 1} = \mathbf{X}_2 - \mathbf{X}_1(\mathbf{X}_1'\mathbf{X}_1)^{-1}\mathbf{X}_1'\mathbf{X}_2$$

$$= \{\mathbf{I} - \mathbf{X}_1(\mathbf{X}_1'\mathbf{X}_1)^{-1}\mathbf{X}_1'\}\mathbf{X}_2,$$

which is the matrix of "residuals of $\mathbf{X}_2$ regressed on $\mathbf{X}_1$," is orthogonal to $\mathbf{X}_1$. [*Proof.* $\mathbf{X}_1'\mathbf{X}_{2 \cdot 1} = \{\mathbf{X}_1' - \mathbf{X}_1'\mathbf{X}_1(\mathbf{X}_1'\mathbf{X}_1)^{-1}\mathbf{X}_1'\}\mathbf{X}_2 = \{\mathbf{0}\}\mathbf{X}_2 = \mathbf{0}.$] If we define $\mathbf{A} = (\mathbf{X}_1'\mathbf{X}_1)^{-1}\mathbf{X}_1'\mathbf{X}_2$ as the alias or bias matrix, we can also rewrite $\mathbf{X}_{2 \cdot 1}$ as

$$\mathbf{X}_{2 \cdot 1} = \mathbf{X}_2 - \mathbf{X}_1\mathbf{A},$$

whereupon model 1 can be rewritten in the form

$$\mathbf{Y} = \mathbf{X}_1(\boldsymbol{\beta}_1 + \mathbf{A}\boldsymbol{\beta}_2) + \mathbf{X}_{2 \cdot 1}\boldsymbol{\beta}_2 + \boldsymbol{\epsilon},$$

where we have simply added and subtracted $\mathbf{X}_1\mathbf{A}\boldsymbol{\beta}_2$ and regrouped. We can set $\boldsymbol{\alpha}_1 = \boldsymbol{\beta}_1 + \mathbf{A}\boldsymbol{\beta}_2$ and thus rewrite model 1 as

$$\mathbf{Y} = \mathbf{X}_1\boldsymbol{\alpha}_1 + \mathbf{X}_{2 \cdot 1}\boldsymbol{\beta}_2 + \boldsymbol{\epsilon},$$

where the two parts of the model are orthogonal to each other because $\mathbf{X}_1'\mathbf{X}_{2\cdot1} = \mathbf{0}$. Let $\mathbf{a}_1$, $\mathbf{b}_2$ be the least squares estimates of $\boldsymbol{\alpha}_1$, $\boldsymbol{\beta}_2$ in model 1. Then the regression sum of squares S_1 for model 1 is the appropriate "$\mathbf{b}'\mathbf{X}'\mathbf{Y}$," that is,

$$S_1 = (\mathbf{a}_1, \mathbf{b}_2)'[\mathbf{X}_1, \mathbf{X}_{2\cdot1}]'\mathbf{Y}$$

$$= \mathbf{Y}'[\mathbf{X}_1, \mathbf{X}_{2\cdot1}]\begin{bmatrix} \mathbf{X}_1'\mathbf{X}_1 & \mathbf{X}_1'\mathbf{X}_{2\cdot1} \\ \mathbf{X}_{2\cdot1}'\mathbf{X}_1 & \mathbf{X}_{2\cdot1}'\mathbf{X}_{2\cdot1} \end{bmatrix}^{-1}\begin{bmatrix} \mathbf{X}_1' \\ \mathbf{X}_{2\cdot1}' \end{bmatrix}\mathbf{Y}.$$

Because of the orthogonality of $\mathbf{X}_1$ and $\mathbf{X}_{2\cdot1}$, the off-diagonal terms of the inverse matrix vanish, and we can invert the diagonal terms individually to get

$$S_1 = \mathbf{Y}'\mathbf{X}_1(\mathbf{X}_1'\mathbf{X}_1)^{-1}\mathbf{X}_1'\mathbf{Y} + \mathbf{Y}'\mathbf{X}_{2\cdot1}(\mathbf{X}_{2\cdot1}'\mathbf{X}_{2\cdot1})^{-1}\mathbf{X}_{2\cdot1}'\mathbf{Y}$$

$$= S_2 + \mathbf{Y}'\mathbf{Q}\mathbf{Y},$$

say, where S_2 is clearly the appropriate regression sum of squares for model 2, and $\mathbf{Q}$ is defined as implied above. We can thus write the "extra sum of squares for b_2 given b_1" as

$$S_1 - S_2 = \mathbf{Y}'\mathbf{Q}\mathbf{Y}.$$

To get the expectation of this, we apply the general formula

$$E(\mathbf{Y}'\mathbf{Q}\mathbf{Y}) = \{E(\mathbf{Y})\}'\mathbf{Q}\{E(\mathbf{Y})\} + \text{trace}\{\mathbf{Q}\boldsymbol{\Sigma}\},$$

where $\boldsymbol{\Sigma} = V(\mathbf{Y})$. For our situation, $\boldsymbol{\Sigma} = \mathbf{I}\sigma^2$ and $E(\mathbf{Y}) = \mathbf{X}_1\boldsymbol{\beta}_1 + \mathbf{X}_2\boldsymbol{\beta}_2$, so that

$$E(S_1 - S_2) = (\boldsymbol{\beta}_1'\mathbf{X}_1' + \boldsymbol{\beta}_2'\mathbf{X}_2')\{\mathbf{X}_{2\cdot1}(\mathbf{X}_{2\cdot1}'\mathbf{X}_{2\cdot1})^{-1}\mathbf{X}_{2\cdot1}'\}(\mathbf{X}_1\boldsymbol{\beta}_1 + \mathbf{X}_2\boldsymbol{\beta}_2)$$

$$+ \text{trace}\{\mathbf{X}_{2\cdot1}(\mathbf{X}_{2\cdot1}'\mathbf{X}_{2\cdot1})^{-1}\mathbf{X}_{2\cdot1}'\mathbf{I}\sigma^2\}.$$

Now $\mathbf{X}_1'\mathbf{X}_{2\cdot1} = \mathbf{0}$, from above. Write $\mathbf{U} = \mathbf{X}_2'\mathbf{X}_{2\cdot1}(\mathbf{X}_{2\cdot1}'\mathbf{X}_{2\cdot1})^{-1}\mathbf{X}_{2\cdot1}'\mathbf{X}_2$. The trace term can be reduced using the fact that $\text{trace}(\mathbf{ST}) = \text{trace}(\mathbf{TS})$ for two matrices $\mathbf{S}$ and $\mathbf{T}$ that are conformable for both products; with $\mathbf{S} = \mathbf{X}_{2\cdot1}$ and $\mathbf{T} = (\mathbf{X}_{2\cdot1}'\mathbf{X}_{2\cdot1})^{-1}\mathbf{X}_{2\cdot1}'$, the trace term becomes $\sigma^2 \text{trace}(\mathbf{I}_{p-q}) = (p - q)\sigma^2$. Thus overall we obtain

$$E(S_1 - S_2) = \boldsymbol{\beta}_2'\mathbf{U}\boldsymbol{\beta}_2 + (p - q)\sigma^2.$$

It follows then that, *under the null hypothesis $H_0 : \boldsymbol{\beta}_2 = \mathbf{0}$, $E\{(S_1 - S_2)/(p - q)\} = \sigma^2$*.

EXERCISES FOR CHAPTER 10

A. Eight experiments are to be done at the coded levels (±1, ±1) of two predictor variables X_1 and X_2. Two experimenters A and B suggest the following designs.

   A: Take one observation at each of $(X_1, X_2) = (-1, -1)$ and $(1, 1)$ and take three observations at each of $(-1, 1)$ and $(1 - 1)$.

   B: Take two observations at each of the four sites.

1. If a model $Y = \beta_0 + \beta_1 X_1 + \beta_2 X_2 + \epsilon$ is to be fitted by least squares but it is feared there may be some additional quadratic curvature expressed by the extra terms $\beta_{11} X_1^2 + \beta_{22} X_2^2 + \beta_{12} X_1 X_2$, evaluate the anticipated biases in the estimated coefficients b_0, b_1, and b_2 for each design.

2. Suppose n_0 center points were added to each design. Would that affect your results? If yes, how?

Which design is better from the point of view of the variances of the estimated coefficients, $V(b_i)$?

B. Consider again the data of Exercise A. Suppose both experimenters agree that quadratic bias is unlikely and so can be ignored. However, one experimenter wants to omit X_1 from the model and fit only $Y = \beta_0 + \beta_2 X_2 + \epsilon$. What biases would the omission of β_1 cause in the estimates b_0, b_2 for designs A and B?

C. Values of a response Y are observed at six locations of a predictor variable X coded as -5, -3, -1, 1, 3 and 5. A model $Y = \beta_0 + \beta_1 X + \epsilon$ is to be fitted but there is fear that bias in the data, arising from a second-order (quadratic) effect $\beta_2 X^2$, might occur. How would the presence of β_2 bias the estimates b_0 and b_1?

D. Using the data in Exercise C, evaluate all the mean squares shown in the table of Section 10.2.

E. As in Exercises C and D, suppose values of $X = -5, -3, -1, 1, 3$, and 5 were to be employed to fit a straight line $Y = \alpha_0 + \alpha_1 X + \epsilon$. Consider the quadratic alternative $Y = \alpha_0 + \alpha_1 X + \alpha_2\{0.375(X^2 - 35/3)\} + \epsilon$. Are the estimates α_0 and α_1 from the straight line fit biased by α_2? (No.) What is the difference between adding the quadratic term above and the quadratic term $\beta_2 X^2$ in Exercise C?

F. In Section 10.2, show that:
1. E (lack of fit mean square) is as given.
2. E (residual mean square) $= \sigma^2$ if the model is correct, that is, if $\beta_2 = 0$.
(Use the result in Section 10.3.)

G. Suppose we fit, by least squares, the model $E(Y) = \beta_0 + \beta_1 X$, but the model $E(Y) = \beta_0 + \beta_1 X + \beta/X$ is actually the true response function, for $X \geq 1$. If we use five observations of Y at X-values $X = 5, 8, 10, 20, 40$ to estimate β_0 and β_1 in the model actually fitted, what biases will be introduced into the estimates?

C H A P T E R 11

On Worthwhile Regressions, Big F's, and R^2

A message of this chapter is that a regression that is statistically significant is not necessarily a useful one, and that something more is needed. This "something more" is difficult to quantify. A "useful rule of thumb" is offered. Readers who are unhappy with this rule may still profit by reading the chapter and forming their own opinions of how they should feel about their fitted equations.

11.1. IS MY REGRESSION A USEFUL ONE?

Suppose, as in our steam data example, we had 25 observations with no pure error and fitted a straight line. The skeleton analysis of variance table would be as follows:

Source	df	MS	F
b_0	1		
$b_1|b_0$	1	MS_{Reg}	Observed $F = MS_{Reg}/s^2$
Residual	23	s^2	
Total	25		

Assuming no defects were seen in the residuals, our observed F would be an $F(1, 23)$ variable under the null hypothesis of no slope. Suppose this were just significant at the $\alpha = 0.05$ test level, with the observed $F = 4.30 > 4.28 = F(1, 23, 0.95)$. Would we be happy? Momentarily, perhaps, but then we would look at the R^2 statistic. Since F and R^2 are directly connected by the relationship

$$R^2 = \nu_1 F/(\nu_1 F + \nu_2), \tag{11.1.1}$$

where (ν_1, ν_2) are the degrees of freedom of MS_{Reg} and s^2, respectively, we can evaluate $R^2 = 4.30/(4.30 + 23) = 0.1575$ or 15.75%. Not very high. So although our regression is "statistically significant," it is certainly not explaining much of the variation, and it is unlikely that it will predict future observations with much accuracy, or even provide a good summary of the behavior of the data.

One possible viewpoint of a useful or worthwhile model fit is that the range of values predicted by the fitted equation over the X-space should be considerably greater than the size of the errors with which these predictions are made. This viewpoint was investigated by Box and Wetz in 1964. See Box and Wetz (1973). They concluded (details in Appendix 11A) that in order that an equation should be regarded as a satisfactory predictor (in the sense that the range of response values predicted by the equation is substantial compared with the standard error of the response), the observed *F*-ratio of (regression mean square)/(residual mean square) should exceed not merely the selected percentage point of the *F*-distribution, but several times the selected percentage point. How many times depends essentially on how great a ratio (prediction range)/(error of predictions) is specified. We offer the following conservative *Rule of Thumb.* Unless the observed *F* for overall regression exceeds the chosen test percentage point by at least a factor of four, and preferably more, the regression is unlikely to be of practical value for prediction purposes.

(We call this conservative because otherwise we would be tempted to replace "four" by "ten." See Appendix 11A for the reasoning.)

Let us see how this would affect the calculation above. If for $\nu_1 = 1$, $\nu_2 = 23$ we insisted on achieving four times the percentage point, the observed *F* would need to exceed $4(4.28) = 17.12$. The corresponding R^2 would then be, from (11.1.1), $R^2 = 17.12/(17.12 + 23) = 0.4267$ or 42.67%. Adoption of a "ten times" rule would lead to an *F* of at least 42.8, and a corresponding $R^2 = 42.8/(42.8 + 23) = 0.6505$ or 65.05%. Note that the steam data fit of Chapter 1 had $F = 57.54$, which is over 13 times as big as the 5% point 4.28.

A little thought about all this will lead the reader to ask: "Does this mean that if I am a 5% person—one willing to be wrong once in 20 times—I should instead act like a person with a much smaller percentage in general, to ensure that I get a useful model?" Essentially, yes, but it is easier to specify a factor than it is to specify how to change a 5% (or 2.5%, or 1%) test value down to a lower test value. (Also, this argument should be carried out only once, or one argues one's way iteratively down into the tail of the *F*-distribution!)

An Alternative and Simpler Check

A simple but effectively equivalent way of implementing the Box–Wetz type of check is to find the ratio

$$(\text{Max } \hat{Y}_i - \text{Min } \hat{Y}_i)/\{ps^2/n\}^{1/2}. \tag{11.1.2}$$

Provided this is sufficiently large (a rough rule of thumb of about four is suggested as *minimal*) and provided no other defect is seen in the fit, a worthwhile regression interpretation is likely to be possible. [Use of the ratio (11.1.2) in such a manner was suggested by G. E. P. Box. The "four[1] at least" rule of thumb is our suggestion.]

In (11.1.2), the numerator of the ratio is the range of $\hat{Y}$ values that arise *at data points.* The value ps^2/n estimates $p\sigma^2/n$, where p = number of parameters fitted, n = number of observations used, and $\sigma^2 = V(Y_i)$. This quantity $p\sigma^2/n$ is the average variance of the n fitted values, namely,

$$\overline{V}(\hat{Y}) = \sum_{i=1}^{n} V(\hat{Y}_i)/n = p\sigma^2/n. \tag{11.1.3}$$

(The last-mentioned result is proved below.)

---

[1] Some would say ten.

Example 1. For the steam data fit of Y on Y_8 in Chapter 1: Max $\hat{Y}_i = 11.38$, Min $\hat{Y}_i = 7.50$; $n = 25$, $p = 2$, $s^2 = 0.7923$. So the ratio is

$$(11.38 - 7.50)/\{2(0.7923)/25\}^{1/2} = 3.88/0.252 = 15.4.$$

Thus the range of predicted values is over 15 times the average standard error of prediction, clearly satisfactory. This is obviously consistent with the fact that the F-ratio of 57.54 is 13.4 times as great as the 95% point $F(1, 23, 0.95) = 4.28$.

Example 2. For the steam data fit of Y on X_8 and X_6 in Section 6.2, the ratio (11.1.2) is

$$(11.56 - 6.29)/\{3(0.4377)/25\}^{1/2} = 5.27/0.2292 = 23.0.$$

Alternatively, the F ratio of 61.90 is 17.99 times as great as the 95% point $F(2, 22, 0.95) = 3.44$. Both numbers indicate a useful predictive ability for this fit.

Proof of (11.1.3)

$$\begin{aligned}
\mathbf{V(\hat{Y})} &= \mathbf{V(X(X'X)^{-1}X'Y)} && \text{[using rule that} \\
&= \mathbf{X(X'X)^{-1}X'(I\sigma^2)X(X'X)^{-1}X'} && \mathbf{V(AY) = A\{V(Y)\}A'\quad (11.1.4)} \\
&= \mathbf{X(X'X)^{-1}X'}\sigma^2.
\end{aligned}$$

The summation in (11.1.3) is now the sum of the diagonal terms of the above matrix, that is, its trace. In what follows we use result 6 of Appendix 5A, that trace $\mathbf{(AB)} =$ trace $\mathbf{(BA)}$. We have

$$\begin{aligned}
n\overline{V}(\hat{Y}) = \Sigma V(\hat{Y}_i) &= \text{trace } \{\mathbf{X(X'X)^{-1}X'}\sigma^2\} \\
&= \sigma^2 \text{ trace } \{\mathbf{(X'X)^{-1}X'X}\} \\
&= \sigma^2 \text{ trace } \mathbf{I}_p \\
&= p\sigma^2.
\end{aligned}$$

Comment

What we are really getting at, in this section, is that a test using the overall F-statistic in the usual manner is not a very good way to judge the future usefulness of the regression equation. Asking for a larger F-value (four times, or more times, the percentage point) is a crude way to attempt to recalibrate the F-test. Another way is the use of (11.1.2). Or one can decide on a suitable level of R^2. "How big should these statistics be?" you ask. In the end, like many things in life, one has to figure out how to judge regressions by acquiring experience. The statistics provide some guide but are not absolute (as we would wish them to be).

11.2. A CONVERSATION ABOUT R^2

The R^2 statistic is used almost universally in judging regression equations. It is probably the first thing most of us look at, but it is not the only thing. It is an extremely useful indicator even if there are no absolute rules about how big it should be. It does have its drawbacks, however. Suppose we have n observations at m sites and there are n_e df for pure error. Suppose we fit a model with p parameters. Then a skeleton analysis of variance table takes the form:

Source	df	SS	MS
Regression|b_0	$p - 1$	A	a
Lack of fit	$n - p - n_e$	B	b
Pure error	n_e	C	c
Total corrected	$n - 1$	D	d

Then

$$R^2 = A/D = 1 - (B + C)/D \qquad (11.2.1)$$

and we can appreciate that:

1. If there is no pure error, $n_e = 0$, $C = 0$, and $R^2 = 1 - B/D$. B now has $(n - p)$ df. So as we enlarge the model, $n - p$ and B will be reduced, and R^2 may look deceptively good, simply because we are fitting to nearly all, or all, the degrees of freedom.

2. Suppose there is pure error, so that $C > 0$, $c > 0$. Consider the extreme case $B = 0$, $n - p - n_e > 0$, that is, there is no lack of fit but there are more sites of data than there are parameters. Then $R^2 = 1 - C/D$. Even though the model fits well, R^2 is limited by the fact that no model can explain the pure error so its maximum value is $1 - C/D$.

3. Outliers produce Y-values that are inflated or deflated. These either inflate C if they occur in a set of repeat runs, or B if not. In any case they reduce R^2 so R^2 is not resistant to outliers.

4. If the model has no intercept, $D = S_{YY} = \Sigma(Y_i - \overline{Y})^2$ does not occur naturally in the analysis and we recommend that calculation of R^2 should not be made. (Of course, we do not advocate omitting the intercept without careful consideration.)

5. Note that, if generalized least squares is used, the model becomes a no-intercept model, because the **1** vector of the first column is replaced by another vector.

Other problems with R^2 discussed in the literature are that:

6. R^2 is not comparable between models that contain different predictor variables. (If one model contains a subset of the parameters in another model, the comparison *is* valid, however.)

7. R^2 is not comparable between models that involve different transformations of the response vector.

8. Non-least-squares models need specially defined statistics.

What Should One Do for Linear Regression?

Probably the best advice for the moment is: continue to use R^2 (or adjusted R^2, if you prefer that) with caution appropriate to the points made above. For some of the debate, consult the references below.

References

Anderson-Sprecher (1994); Becker and Kennedy (1992); Ding and Bargmann (1991); Draper (1984); Healy (1984); Kvalseth (1985); Magee (1990); Nagelkerke (1991); Schemper (1990); Scott and Wild (1991); Shah (1991); Willett and Singer (1988).

APPENDIX 11A. HOW SIGNIFICANT SHOULD MY REGRESSION BE?

The application to regression situations of work by Box and Wetz is briefly explained. For a "useful" as distinct from a "significant" regression, the observed F-value for regression should exceed the usual percentage point by a multiple. The size of this multiple is arbitrary in the same way that significance levels are arbitrary, but guidelines are given for its choice.

The γ_m Criterion

In regression problems, provided no lack of fit is indicated, a test for regression is usually conducted by looking at the F-ratio of the "regression sum of squares given b_0" to "the residual mean square s^2." This value is then compared with an appropriate upper α-percentage point $F(\nu_m, \nu_r, 1 - \alpha)$, where ν_m and ν_r are, respectively, the degrees of freedom of the numerator and denominator of the F-statistic. When this F-ratio is statistically significant, that is, when $F > F(\nu_m, \nu_r, 1 - \alpha)$, it implies that a significantly large amount of the variation in the data about the mean has been taken up by the regression equation. This does not necessarily mean, however, that the fitted equation is a worthwhile predictor in the sense that the range of predicted values is "substantial" compared with the standard error of the response. The question then arises as to how we can distinguish statistically significant *and* worthwhile prediction equations from statistically significant prediction equations of limited practical value.

Some work that answers this question to a great extent appeared in a 1964 University of Wisconsin Ph.D. thesis "Criteria for judging adequacy of estimation by an approximating response function," by J. Wetz. (See Box and Wetz, 1973). The key finding may be summarized in the following way.

Suppose we fit, by least squares, the model

$$\mathbf{Y} = \boldsymbol{\eta} + \boldsymbol{\varepsilon} = \mathbf{X}\boldsymbol{\beta} + \mathbf{Z}\boldsymbol{\psi} + \boldsymbol{\varepsilon}, \qquad (11\text{A}.1)$$

where $\mathbf{X}\boldsymbol{\beta}$ is the part to be tested in a "test for regression" and $\mathbf{Z}\boldsymbol{\psi}$ represents effects such as the mean, block variables, time trends, and so on, that we wish to eliminate from the variation in the data but in which we otherwise have no interest. We assume $\mathbf{E}(\boldsymbol{\varepsilon}) = \mathbf{0}$, $V(\boldsymbol{\varepsilon}) = \mathbf{I}\sigma^2$. The changes in response values η_i over the n experimental points can conveniently be measured by the quantity

$$\sum_{i=1}^{n} (\eta_i - \bar{\eta}_i)^2 / n, \qquad (11\text{A}.2)$$

where η_i is the true response of the ith observation, and $\bar{\eta}_i$ is the ith element of the vector $\tilde{\boldsymbol{\eta}} = \mathbf{Z}\boldsymbol{\psi}$. When only the intercept term β_0 is eliminated, $\bar{\eta}_i = \overline{\eta}$, the mean of the η_i.

We can compare the quantity in Eq. (11A.2) with the size of the errors we commit in estimating the differences $\eta_i - \bar{\eta}_i$. The least squares estimator of $\eta_i - \bar{\eta}_i$ is $\hat{Y}_i - \tilde{Y}_i$, the ith element of the vector

$$\hat{\mathbf{Y}} - \tilde{\mathbf{Y}} = \mathbf{X}\hat{\boldsymbol{\beta}} = \mathbf{X}(\mathbf{X}'\mathbf{X})^{-1}\mathbf{X}'\mathbf{Y} = \mathbf{H}\mathbf{Y}, \qquad (11\text{A}.3)$$

say, and the variance–covariance matrix of this is given by

$$E\{\mathbf{H}\mathbf{Y} - E(\mathbf{H}\mathbf{Y})\}\{\mathbf{H}\mathbf{Y} - E(\mathbf{H}\mathbf{Y})\}' = \mathbf{H}\sigma^2\mathbf{H}' = \mathbf{H}\sigma^2, \qquad (11\text{A}.4)$$

where $V(\mathbf{Y}) = \mathbf{I}\sigma^2$, due to the fact that $\mathbf{H}$ is symmetric and idempotent. Thus

$V(\hat{Y}_i - \tilde{Y}_i)$ is the ith diagonal element of $\mathbf{H}\sigma^2$, and the average of these variances, which is an overall measure of how well we estimate the quantities $\eta_i - \tilde{\eta}_i$ is given by

$$\sigma^2_{\hat{\mathbf{Y}} - \tilde{\mathbf{Y}}} = \text{trace}(\mathbf{H}\sigma^2)/n$$

$$= \nu_m \sigma^2/n \tag{11A.5}$$

due to the fact that trace $\mathbf{H} = \text{trace } \mathbf{X}\{(\mathbf{X}'\mathbf{X})^{-1}\mathbf{X}'\} = \text{trace}\{(\mathbf{X}'\mathbf{X})^{-1}\mathbf{X}'\}\mathbf{X} = \text{trace}(\mathbf{I}_{\nu_m})$, where ν_m is the number of parameters in $\boldsymbol{\beta}$. It follows that a sensible comparison of the sizes of changes in the $\eta_i - \tilde{\eta}_i$ with their errors of estimate is given by the square root of the ratio of Eq. (11A.2) to Eq. (11A.5), namely,

$$\gamma_m = \left\{ \sum_{i=1}^{n} (\eta_i - \tilde{\eta}_i)^2/(\nu_m \sigma^2) \right\}^{1/2}. \tag{11A.6}$$

Figure 11A.1 shows the situation for a single predictor variable X; γ_m is comparing the spread of the heavy lines (the $\eta_i - \tilde{\eta}_i$) with the average spread of their estimates, whose distributions are shown at the various X_i. How big should γ_m be for a fitted equation to be practically useful as distinct from just statistically significant? This is, to a great extent, arbitrary, in the same sense that a selected statistical significance level is arbitrary. (However, to help fix ideas, we shall soon think in terms of values $\gamma_m = 2, 3$, and 4 so that we can examine the consequences and choose accordingly.) Suppose that γ_0 is the minimally acceptable level of γ_m. Then Box and Wetz show that we shall need to determine a certain value F_0, dependent on γ_0, and, if the usual regression F-ratio exceeds this value F_0, then we shall accept that γ_m is sufficiently large for the regression fit to be a practically useful one. Box and Wetz show further that, approximately, this critical value F_0 is

$$F_0 \simeq (1 + \gamma_0^2)F(\nu_0, \nu_r, 1 - \alpha), \tag{11A.7}$$

where ν_r is the number of residual degrees of freedom and where

$$\nu_0 = \nu_m(1 + \gamma_0^2)^2/(1 + 2\gamma_0^2). \tag{11A.8}$$

In other words, in order for the fit to be practically worthwhile, $F > F_0$. It is, of course, easy to work out F_0 in any specific case, but it is also informative to look at the ratios

$$F_0/F(\nu_m, \nu_r, 1 - \alpha) \tag{11A.9}$$

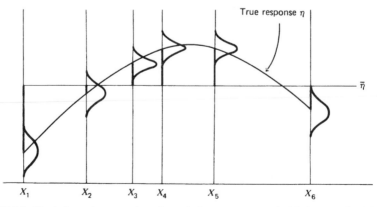

F i g u r e 11A.1. Deviations of true η values about their mean compared with the spreads of the estimates $\hat{Y}_i - \tilde{Y}_i$ for a single X-variable.

T A B L E 11A.1. Ratios $F_0/F(\nu_m, \nu_r, 0.95)$ for $\gamma_0 = 2$

Residual Degrees of Freedom, ν_r	Regression Degrees of Freedom, ν_m								
	1	2	3	4	5	6	10	15	21
3	5	5	5	5	5	5	5	5	5
4	4	4	5	5	5	5	5	5	5
5	4	4	4	5	5	5	5	5	5
10	4	4	4	4	4	4	5	5	5
15	4	4	4	4	4	4	4	5	5
20	4	4	4	4	4	4	4	5	5
30	4	4	4	4	4	4	4	4	5
40 to ∞	3	4	4	4	4	4	4	4	4

for given values of γ_0, and various values of ν_m, ν_r, and α. Tables 11A.1, 11A.2, and 11A.3 show these ratios rounded to the nearest unit for $\gamma_0 = 2$, 3, and 4, respectively, and for $\alpha = 0.05$. We see that, at this probability level, if we are prepared to accept a value of $\gamma_0 = 2$ as being sufficiently informative for our purposes, we need to have an observed F of at least four or five times as large as the usual percentage point to declare the regression practically useful to us. Or, if we are prepared to accept a value of $\gamma_0 = 3$, the F must be at least six to ten times as large as the usual percentage point. Moving to Table 11A.3, we see that, as our selected γ_0 increases, the ratios both increase and vary more, depending on the degrees of freedom involved. (For $\alpha = 0.01$, the general picture is much the same, with ratio values being either the same or one or two units lower in most cases.)

In summary, it is clear that an observed F-ratio must be *at least* four or five times the usual percentage point for the minimum level of proper representation, as Table 11A.1 indicates. In practice, the multiples in Table 11A.2 are probably the ones that should be attained or exceeded in most practical cases, guaranteeing, as they do, a γ_m ratio of 3 or more. However, like the choice of a confidence level, much depends on individual preferences, and the tables are offered as guidelines for personal choice.

The stated results have been given in terms of the F-statistic for overall regression. However, similar results apply for subsets of coefficients in the fitted model, and so the rule may be applied to the F-values arising from such subsets, also, if desired. (See Ellerton, 1978.)

T A B L E 11A.2. Ratios $F_0/F(\nu_m, \nu_r, 0.95)$ for $\gamma_0 = 3$

Residual Degrees of Freedom, ν_r	Regression Degrees of Freedom, ν_m								
	1	2	3	4	5	6	10	15	21
3	9	9	9	9	10	10	10	10	10
4	8	9	9	9	9	9	10	10	10
5	8	8	9	9	9	9	9	10	10
10	7	7	8	8	8	8	9	9	9
15	6	7	7	8	8	8	9	9	9
20	6	6	7	7	8	8	8	9	9
30	6	6	7	7	7	8	8	9	9
40	6	6	7	7	7	7	8	8	9
60	6	6	7	7	7	7	8	8	8
120	6	6	7	7	7	7	8	8	8
∞	6	6	6	7	7	7	7	8	8

T A B L E   11A.3. Ratios $F_0/F(\nu_m, \nu_r, 0.95)$ for $\gamma_0 = 4$

Residual Degrees of Freedom, ν_r	Regression Degrees of Freedom, ν_m								
	1	2	3	4	5	6	10	15	21
3	15	15	16	16	16	16	17	17	17
4	13	14	15	15	16	16	16	16	17
5	12	13	14	15	15	15	16	16	16
10	10	12	13	13	14	14	15	15	16
15	10	11	12	12	13	13	14	15	15
20	9	11	11	12	12	13	14	15	15
30	9	10	11	11	12	12	14	14	15
40	9	10	11	11	12	12	13	14	14
60	9	10	10	11	11	12	13	14	14
120	9	9	10	11	11	11	13	13	14
∞	8	9	10	10	11	11	12	12	12

EXERCISES FOR CHAPTER 11

Note: You will need the formula

$$R^2 = \nu_1 F/(\nu_1 F + \nu_2) \tag{11E}$$

where ν_1 = number of parameters fitted − 1 (for β_0),
ν_2 = residual degrees of freedom.

A. Substitute specific values of ν_1, ν_2 and $F(\nu_1, \nu_2, 1 - \alpha)$ for chosen α (e.g., $\alpha = 0.05$) into the right-hand side of Eq. (11E). The R^2 values you get are those that correspond to "just significant at $100\alpha\%$" F-values. You will be surprised at how low these R^2 values can be, and this will provide additional motivation for reading Appendix 11A. Application of Tables 11A.1, 11A.2, and 11A.3 (in which $\nu_m = \nu_1$, $\nu_r = \nu_2$) will ensure higher R^2 values; you may wish to do a few sample calculations to see what the effects can be.

B. Your friend says he has fitted a plane to $n = 33$ observations on (X_1, X_2, Y) and that his overall regression (given b_0) is just significant at the $\alpha = 0.05$ level. You ask him for his R^2 value but he doesn't know. You work it out for him on the basis of what he has told you.

C. You perform a regression for a colleague. She gives you 46 data points, which include 5 sets of points each with 6 repeat runs, and the model contains 6 parameters including β_0. She tells you: "I hope the R^2 value will exceed 90%." For this to happen, how many times bigger than the 5% tabled F-value would F (for all parameters except b_0) have to be? (Assume that there is no lack of fit.)

D. You are given a regression printout that shows a planar fit to $X_1, X_2, \ldots, X_5$, plus intercept of course, obtained from a set of 50 observations.
  1. The overall F for regression $F(b_1, b_2, \ldots, b_5|b_0)$ is ten times as big as the 5% upper-tail F percentage point. How big is R^2?
  2. Looking at the data, you observe that they consist of six groups of repeats with 8, 8, 8, 8, 8, and 10 observations in them. What would you do now, and why?

C H A P T E R 12

Models Containing Functions of the Predictors, Including Polynomial Models

The general regression methods formulated for linear models in Chapters 5–11 are valid whatever (linear in the parameters) form the model function takes. Thus the model function can in theory be set up in any way one wishes. We discuss this first in a general way. Subsequently, we examine a set of data to which a second-order model is fitted. This example provides motivation for some suggested rules for testing and omitting terms from polynomial models.

12.1. MORE COMPLICATED MODEL FUNCTIONS

Many models fitted in practice involve just the observed predictor variables X_1, $X_2, \ldots, X_k$, say, in their original form; that is, we fit

$$Y = \beta_0 + \beta_1 X_1 + \beta_2 X_2 + \cdots + \beta_k X_k + \epsilon. \tag{12.1.1}$$

Many more general forms are possible. We can write the most general type of linear model in variables $X_1, X_2, \ldots, X_k$ in the form

$$Y = \beta_0 Z_0 + \beta_1 Z_1 + \beta_2 Z_2 + \cdots + \beta_{p-1} Z_{p-1} + \epsilon. \tag{12.1.2}$$

($Z_0 = 1$ is a dummy variable that is always unity and will in general not be shown. However, it is sometimes mathematically convenient to have a Z_0 in the model. For example, if

$$(Z_{1i}, Z_{2i}, \ldots, Z_{p-1}) \qquad i = 1, 2, \ldots, n$$

are n settings of the variables $Z_j, j = 1, 2, \ldots, p - 1$, corresponding to observations $Y_i, i = 1, 2, \ldots, n$, then when $j \neq 0$, and $Z_{0i} = 1$,

$$\sum_{i=1}^{n} Z_{ji} = \sum_{i=1}^{n} Z_{ji} Z_{0i}$$

and thus can be represented by the general cross-product expression

$$\sum_{i=1}^{n} Z_{ji} Z_{li}$$

if the normal equations are written out. Note that $\Sigma_{i=1}^{n} Z_{0i}^2 = n$.)

In (12.1.2), each Z_j, $j = 1, 2, \ldots, p - 1$, is a general function of $X_1, X_2, \ldots, X_k$,

$$Z_j = Z_j(X_1, X_2, \ldots, X_k),$$

and can take any form. In some examples, each Z_j may involve only one X-variable.

Any model that can be written, perhaps after rearrangement or transformation, in the form of Eq. (12.1.2) can be analyzed by the general methods given in Chapters 5–11. We now provide some specific examples of models that can be treated by these methods and relate them to the general form of Eq. (12.1.2).

Polynomial Models of Various Orders in the X_j

First-Order Models

1. If $p = 2$ and $Z_1 = X$ in Eq. (12.1.2), we obtain the simple first-order model with one predictor variable:

$$Y = \beta_0 + \beta_1 X + \epsilon. \tag{12.1.3}$$

2. If $p = k + 1$ and $Z_j = X_j$, we obtain a first-order model with k predictor variables:

$$Y = \beta_0 + \beta_1 X_1 + \beta_2 X_2 + \cdots + \beta_k X_k + \epsilon. \tag{12.1.4}$$

Second-Order Models

1. If $p = 3$, $Z_1 = X$, $Z_2 = X^2$, and $\beta_2 = \beta_{11}$, we obtain a second-order (quadratic) model with one predictor variable:

$$Y = \beta_0 + \beta_1 X + \beta_{11} X^2 + \epsilon. \tag{12.1.5}$$

2. If $p = 6$, $Z_1 = X_1$, $Z_2 = X_2$, $Z_3 = X_1^2$, $Z_4 = X_2^2$, $Z_5 = X_1 X_2$, $\beta_3 = \beta_{11}$, $\beta_4 = \beta_{22}$, and $\beta_5 = \beta_{12}$, we obtain a second-order model with two predictor variables:

$$Y = \beta_0 + \beta_1 X_1 + \beta_2 X_2 + \beta_{11} X_1^2 + \beta_{22} X_2^2 + \beta_{12} X_1 X_2 + \epsilon. \tag{12.1.6}$$

A full second-order model in k variables can be obtained in similar fashion when $p = 1 + k + k + +\frac{1}{2}k(k - 1) = \frac{1}{2}(k + 1)(k + 2)$. Second-order models are used particularly in response surface studies where it is desired to graduate, or approximate to, the characteristics of some unknown response surface by a polynomial of low order. Note that all possible second-order terms are in the model. This is sensible because omission of terms implies possession of definite knowledge that certain types of surface (those that cannot be represented *without* the omitted terms) cannot possibly occur. Knowledge of this sort is not often available. When it is, it would usually enable a more theoretically based study to be made.

An example of a second-order response surface analysis is given in Section 12.2.

Third-Order Models

1. If $p = 4$, $Z_1 = X$, $Z_2 = X^2$, $Z_3 = X^3$, $\beta_2 = \beta_{11}$, and $\beta_3 = \beta_{111}$, we obtain a third-order model with one predictor variable:

$$Y = \beta_0 + \beta_1 X + \beta_{11} X^2 + \beta_{111} X^3 + \epsilon. \tag{12.1.7}$$

2. If $p = 10$ and proper identification of the β_i and Z_i is made (we omit the details now since the examples above should have made the idea clear), the model (12.1.2) can represent a third-order model with two predictor variables given by

$$Y = \beta_0 + \beta_1 X_1 + \beta_2 X_2 + \beta_{11} X_1^2 + \beta_{12} X_1 X_2 + \beta_{22} X_2^2$$
$$+ \beta_{111} X_1^3 + \beta_{112} X_1^2 X_2 + \beta_{122} X_1 X_2^2 + \beta_{222} X_2^3 + \epsilon. \tag{12.1.8}$$

The general third-order model for k factors $X_1, X_2, \ldots, X_k$ can be obtained similarly. Third-order models are also used in response surface work though much less frequently than second-order models. Note the method of labeling the β's. This may seem confusing at first but it is done to enable the coefficients to be readily associated with their corresponding powers of the X's. For example, $X_1 X_2^2 = X_1 X_2 X_2$ has a coefficient β_{122}, and so on. A similar notation is used above for second-order models and is standard in response surface work. (Note that we write β_{122} with lower-valued subscripts first rather than β_{212} or β_{221}.)

Models of *any* desired order can be represented by Eq. (12.1.2) by continuing the process illustrated above.

Transformations

If a second-order model is not adequate, a third-order model may be. However, one should not routinely add higher-order terms. It is often more fruitful to investigate the effects produced by other transformations of the predictor variables, or by transformations of the response variable, or by both. The same remark also applies in the first-order versus second-order decision. For example, a straight-line fit of the response log Y versus X, if appropriate, would usually be preferred to a quadratic fit of Y versus X, assuming the behavior of the residuals showed that either fit was a workable choice.

Models Involving Transformations Other than Integer Powers

The polynomial models above involved powers, and cross-products of powers, of the predictor variables $X_1, X_2, \ldots, X_k$. Here we provide a few examples of other types of transformations that are often useful in forming the model function.

The Reciprocal Transformation. If in Eq. (12.1.2) we take $p = 2$, $Z_1 = 1/X_1$, and $Z_2 = 1/X_2$, we obtain the model

$$Y = \beta_0 + \beta_1 \left(\frac{1}{X_1}\right) + \beta_2 \left(\frac{1}{X_2}\right) + \epsilon. \tag{12.1.9}$$

The Logarithmic Transformation. By taking $p = 2$, $Z_1 = \ln X_1$, and $Z_2 = \ln X_2$, Eq. (12.1.2) can represent

$$Y = \beta_0 + \beta_1 \ln X_1 + \beta_2 \ln X_2 + \epsilon. \tag{12.1.10}$$

The Square Root Transformation. For example,

$$Y = \beta_0 + \beta_1 X_1^{1/2} + \beta_2 X_2^{1/2} + \epsilon. \tag{12.1.11}$$

Clearly, there are many possible transformations, and models can be postulated that contain few or many such terms. *Several different transformations could occur in the same model, of course.* [The examples (12.1.9)–(12.1.11) use the same transformation throughout.] The choice of what, if any, transformation to make is often difficult to decide. The choice would often be made on the basis of previous knowledge of the variables under study. The purpose of making transformations of this type is to be able to use a regression model of simple form in the transformed variables, rather than a more complicated one in the original variables.

Transformations could also involve several X_j variables simultaneously, for example, $Z_1 = X_1^{1/2} \ln X_2$. Transformations of this type are sometimes suggested by the form of the fitted equation in untransformed variables. When the best power of an X to use is not known, a parameter can be substituted. In such cases, nonlinear estimation methods are usually needed.

Plots Can Be Useful

Suitable transformations of the predictor variables are also sometimes suggested by plotting the data in various ways. See, for example, Hoerl (1954). Other references are Dolby (1963) and Tukey (1957).

Our discussion above relates entirely to choosing the model function. (The response Y is untouched.) When, as we assume here, the predictor variables are not subject to error, there are no problems in transforming them.

Transformations on the response are discussed in Chapter 13. For those, one must be especially careful to check that the least squares assumptions [errors independent, $N(0, \sigma^2)$] are not violated by making the transformation. This is usually done by checking the residuals in the *transformed metric,* after the transformation has been chosen and the model fitted.

Using Ratios as Responses and/or Predictors

When ratios with a common component are used in regression analyses, there is a danger that a strong regression relationship will be introduced spuriously by the component. Warnings have appeared in the literature for about a century, but the problems still recur. A useful discussion of possible problems is given by Kronmal (1993). Kronmal's conclusions and recommendations section (his pp. 390–391) begins: "The message of this paper is that ratio variables should only be used in the context of a full linear model in which the variables that make up the ratio are included and the intercept term is also present. The common practice of using ratios for either the [response] or the [predictor] variables in regression analyses can lead to misleading inferences and rarely in any gain." Kronmal then provides some suggestions for responding to researchers who do not wish to "give up their most prized ratio or index."

12.2. WORKED EXAMPLES OF SECOND-ORDER SURFACE FITTING FOR $k = 3$ AND $k = 2$ PREDICTOR VARIABLES

Aia, Goldsmith, and Mooney (1961) reported a pilot plant investigation under the title "Predicting Stoichiometric $CaHPO_4 \cdot 2H_2O$." This section is adapted from that paper with the permission of the American Chemical Society. We omit the chemical details here and also make several minor changes to their original analysis.

In the problem studied, there were seven candidates for predictor variables but four of these were kept fixed throughout the experiment. The three selected for the response surface study and their chosen ranges were as follows:

Variable	Designation	Range Chosen
Mole ratio $NH_3/CaCl_2$ in the calcium chloride solution	r	0.70–1.00
Addition time in minutes of ammoniacal $CaCl_2$ to $NH_4H_2PO_4$	t	10–90
Starting pH of $NH_4H_2PO_4$ solution	pH	2–5

T A B L E 12.1. A Worked Example: The X Matrix and Two Responses

u	1	X_1	X_2	X_3	X_1^2	X_2^2	X_3^2	X_1X_2	X_1X_3	X_2X_3	Y	Y_4
			Design Matrix									
1	1	-1	-1	-1	1	1	1	1	1	1	52.8	6.95
2	1	1	-1	-1	1	1	1	-1	-1	1	67.9	5.90
3	1	-1	1	-1	1	1	1	-1	1	-1	55.4	7.10
4	1	1	1	-1	1	1	1	1	-1	-1	64.2	7.08
5	1	-1	-1	1	1	1	1	1	-1	-1	75.1	5.64
6	1	1	-1	1	1	1	1	-1	1	-1	81.6	5.18
7	1	-1	1	1	1	1	1	-1	-1	1	73.8	6.84
8	1	1	1	1	1	1	1	1	1	1	79.5	5.67
9	1	$-\frac{5}{3}$	0	0	$\frac{25}{9}$	0	0	0	0	0	68.1	6.00
10	1	$\frac{5}{3}$	0	0	$\frac{25}{9}$	0	0	0	0	0	91.2	5.67
11	1	0	$-\frac{5}{3}$	0	0	$\frac{25}{9}$	0	0	0	0	80.6	5.52
12	1	0	$\frac{5}{3}$	0	0	$\frac{25}{9}$	0	0	0	0	77.5	6.47
13	1	0	0	$-\frac{5}{3}$	0	0	$\frac{25}{9}$	0	0	0	36.8	7.17
14	1	0	0	$\frac{5}{3}$	0	0	$\frac{25}{9}$	0	0	0	78.0	5.36
15	1	0	0	0	0	0	0	0	0	0	74.6	6.48
16	1	0	0	0	0	0	0	0	0	0	75.9	5.91
17	1	0	0	0	0	0	0	0	0	0	76.9	6.39
18	1	0	0	0	0	0	0	0	0	0	72.3	5.99
19	1	0	0	0	0	0	0	0	0	0	75.9	5.86
20	1	0	0	0	0	0	0	0	0	0	79.8	5.96

There were seven responses of interest. For each response, the idea was tentatively entertained that it could be graduated by a second-order function of r, t, and pH. We make use here only of the first response (which we call Y) and record the fourth (which we call Y_4) for use in an exercise.

The experimental design chosen was a "cube plus star plus six center points" composite design with $\alpha = \frac{5}{3} = 1.667$. (For "rotatability," α should have the value $2^{3/4} = 1.6818$ so that the selected design is nearly, but not quite, rotatable. The authors ignored this difference and in fact used $\alpha = 1.6818$ in their calculations. For this reason our calculations are slightly different from theirs.)

The design requires five levels of each variable. The experimental variables were coded by the transformations

$$X_1 = (r - 0.85)/0.09, \qquad X_2 = (t - 50)/24, \qquad X_3 = (\text{pH} - 3.5)/0.9. \quad (12.2.1)$$

Thus the equivalence of the design and actual levels can be expressed by the following table.

Coded Levels, X_1 or X_2 or X_3	Actual Levels		
	r	t	pH
$\frac{5}{3}$	1.00	90	5.0
1	0.94	74	4.4
0	0.85	50	3.5
-1	0.76	26	2.6
$-\frac{5}{3}$	0.70	10	2.0

The actual design in coded units is shown as the indicated part of the $\mathbf{X}$ matrix in Table 12.1. In coded units the postulated second-order model can be written

$$E(Y) = \beta_0 + \beta_1 X_1 + \beta_2 X_2 + \beta_3 X_3 + \beta_{11} X_1^2 + \beta_{22} X_2^2$$
$$+ \beta_{33} X_3^2 + \beta_{12} X_1 X_2 + \beta_{13} X_1 X_3 + \beta_{23} X_2 X_3$$

(12.2.2)

for the response Y, and similarly for Y_4.

The runs were performed in random order and the observed values of the seven responses were recorded. Two of these,

$$Y = \text{yield as percentage of theoretical yield},$$

$$Y_4 = \text{bulk density in grams per cubic inch},$$

are given in Table 12.1. The appropriate $\mathbf{X'X}$ matrix is the same no matter which response is being fitted and has the form

$$\mathbf{X'X} = \begin{bmatrix}
N & 0 & 0 & 0 & B & B & B & 0 & 0 & 0 \\
0 & B & 0 & 0 & & & & & & \\
0 & 0 & B & 0 & & \mathbf{0} & & & \mathbf{0} & \\
0 & 0 & 0 & B & & & & & & \\
B & & & & C & D & D & & & \\
B & & \mathbf{0} & & D & C & D & & \mathbf{0} & \\
B & & & & D & D & C & & & \\
0 & & & & & & & D & 0 & 0 \\
0 & & \mathbf{0} & & & \mathbf{0} & & 0 & D & 0 \\
0 & & & & & & & 0 & 0 & D
\end{bmatrix},$$

(12.2.3)

where, for our specific example,

$$N = 20,$$
$$B = 8 + 2\alpha^2 = \tfrac{122}{9},$$
$$C = 8 + 2\alpha^4 = \tfrac{1898}{81},$$
$$D = 8.$$

(12.2.4)

This type of partitioned matrix occurs frequently in planned second-order response surface studies, and its inverse is easily obtained. In the general case, when there are k factors (rather than three) and the same symbols are used in a larger $\mathbf{X'X}$ matrix in the obvious way, the inverse can be written

$$(\mathbf{X}'\mathbf{X})^{-1} = \begin{array}{c} \\ \end{array} \begin{array}{cccccccccccc} 0 & 1 & 2 & & k & 11 & 22 & & kk & 12 & 13 & & (k-1)k \\ \end{array}$$

$$(\mathbf{X}'\mathbf{X})^{-1} = \left[\begin{array}{ccccc|ccccc|cccc} P & 0 & 0 & \cdots & 0 & Q & Q & \cdots & Q & 0 & 0 & \cdots & 0 \\ \hline 0 & 1/B & 0 & \cdots & 0 & & & & & & & & \\ 0 & 0 & 1/B & \cdots & 0 & & & \mathbf{0} & & & & \mathbf{0} & \\ \vdots & \vdots & & & & & & & & & & & \\ 0 & 0 & 0 & \cdots & 1/B & & & & & & & & \\ \hline Q & & & & & R & S & \cdots & S & & & & \\ Q & & & & & S & R & \cdots & S & & & & \\ \vdots & & \mathbf{0} & & & \vdots & & & & & & \mathbf{0} & \\ Q & & & & & S & S & \cdots & R & & & & \\ \hline 0 & & & & & & & & & 1/D & 0 & \cdots & 0 \\ 0 & & & & & & & & & 0 & 1/D & \cdots & 0 \\ \vdots & & \mathbf{0} & & & & & \mathbf{0} & & \vdots & & & \\ 0 & & & & & & & & & 0 & 0 & \cdots & 1/D \end{array}\right] \begin{array}{c} 0 \\ 1 \\ 2 \\ \vdots \\ k \\ 11 \\ 22 \\ \vdots \\ kk \\ 12 \\ 13 \\ \vdots \\ (k-1)k \end{array}$$

$$(12.2.5)$$

The values of P, Q, R, and S are shown in Table 12.2, in the second column marked $C \neq 3D$. [The values in the third column are the simplified forms when $C = 3D$, which happens when the design is "rotatable," that is, when the contours of $V\{\hat{Y}(X)\}$ are spherical ones. In such circumstances, the design can be rotated in the predictor (i.e., the X-) space without affecting the precision of the information obtained.]

In our case, $k = 3$ and $3D = \frac{1944}{81}$ so that $C - 3D$ is small but not zero as it would be for a rotatable design. Consequently, we obtain

$$P = 1597/9614, \qquad Q = -549/9614,$$
$$R = 685.3248/9614, \qquad S = 62.3376/9614. \qquad (12.2.6)$$

Also, $1/B = \frac{9}{122}$, $1/D = \frac{1}{8}$. (The figures are exact; to avoid round-off error later, the final division by 9614 has been postponed until after the subsequent matrix multiplication.) We now need $\mathbf{X}'\mathbf{Y}$. These are given below for the responses Y and Y_4.

T A B L E 12.2. Formulas for Obtaining Elements of $(\mathbf{X}'\mathbf{X})^{-1}$

Symbol	Value when $C \neq 3D$	Value when $C = 3D$ (Rotatable Design)
P	$(C - D)(C + (k - 1)D)/A$	$2(k + 2)D^2/A$
Q	$-(C - D)B/A$	$-2DB/A$
R	$\{N(C + (k - 2)D) - (k - 1)B^2\}/A$	$\{N(k + 1)D - (k - 1)B^2\}/A$
S	$(B^2 - ND)/A$	$(B^2 - ND)/A$
A	$(C - D)\{N(C + (k - 1)D) - kB^2\}$	$2D\{N(k + 2)D - kB^2\}$

Note that A occurs in the formulas for P, Q, R, and S.

$$\mathbf{X'Y} = \tfrac{1}{9} \begin{bmatrix} 12{,}941.1 \\ 671.4 \\ -87.0 \\ 1{,}245.3 \\ 8{,}935.2 \\ 8{,}905.2 \\ 7{,}822.7 \\ -63.9 \\ -105.3 \\ -20.7 \end{bmatrix}, \qquad \mathbf{X'Y_4} = \tfrac{1}{9} \begin{bmatrix} 1{,}108.26 \\ -29.25 \\ 41.43 \\ -60.45 \\ 744.99 \\ 752.99 \\ 766.49 \\ 2.88 \\ -5.04 \\ 3.24 \end{bmatrix}. \tag{12.2.7}$$

Applying the usual formula $\mathbf{b} = (\mathbf{X'X})^{-1}\mathbf{X'Y}$ for the estimates of the regression parameters for the first response Y, we obtain the fitted equation

$$\begin{aligned} \hat{Y} = {}& 76.022 + 5.503X_1 - 0.713X_2 + 10.207X_3 \\ & + 0.712X_1^2 + 0.496X_2^2 - 7.298X_3^2 \\ & - 0.888X_1X_2 - 1.463X_1X_3 - 0.288X_2X_3. \end{aligned} \tag{12.2.8}$$

When the second-order design gives rise to an $\mathbf{X'X}$ matrix of the type shown

T A B L E 12.3. Standard Analysis of Variance Table for Certain Types of Second-Order Designs

Source	df	SS
b_0 (mean)	1	$\left(\sum\limits_{u=1}^{N} Y_u\right)^2 / N$
b_i (first order)	k	$\sum\limits_{i=1}^{k} b_i(iY)$
$b_{ii}\|b_0$ (pure second-order, given b_0)	k	$b_0(0Y) + \sum\limits_{i=1}^{k} b_{ii}(iiY) - \left(\sum\limits_{u=1}^{N} Y_u\right)^2 / N$
b_{ij} (mixed second-order)	$\tfrac{1}{2}k(k-1)$	$\sum\limits_{i=1}^{k}\sum\limits_{j=1}^{k} b_{ij}(ijY)$
Lack of fit	$N - n_e - \tfrac{1}{2}(k+1)(k+2)$	By subtraction
Pure error	n_e	By usual calculation
Total	N	$\sum\limits_{u=1}^{N} Y_u^2$

inverted in (12.2.5) the analysis of variance table is as shown in Table 12.3. Here,

$$(0y) = \sum_{u=1}^{N} Y_u,$$

$$(iy) = \sum_{u=1}^{N} X_{iu} Y_u,$$

$$(iiy) = \sum_{u=1}^{N} X_{iu}^2 Y_u,$$

$$(ijy) = \sum_{u=1}^{N} X_{iu} X_{ju} Y_u,$$

(12.2.9)

all of these expressions being cross-products of columns of the **X** matrix with the column **Y** of observations and so all are elements of the **X'Y** vector. Usually we would combine the $SS(b_{ii}|b_0)$ and $SS(b_{ij})$ to give a

SS(second order terms$|b_0$) with $\frac{1}{2}k(k+1)$ degrees of freedom

but we have displayed them separately in the table to emphasize that the only *extra* sum of squares that arises is $SS(b_{ii}|b_0)$ due to the orthogonality of many pairs of columns in **X,** a feature that is only for specific design choices. We can now proceed, in the usual way, to test lack of fit and the usefulness of the second-order and first-order terms.

It should be noted that many of the special features of this least squares estimation and analysis apply *only* to designs whose **X'X** matrices take the special form given, so that $(\mathbf{X'X})^{-1}$ can be found from the formulas above. Designs without this feature must be subjected to the usual least squares analysis without recourse to special formulas. Nevertheless, the "source" column of the analysis of variance table given above provides a framework to aim at. The pure error sum of squares is obtained as usual, and the successive entries for sums of squares for parameter estimates would all be obtained as extra sums of squares as described in Chapter 6.

For our example, the appropriate analysis of variance table is given as Table 12.4. Since $F(5, 5, 0.95) = 5.05 > 3.04$, no lack of fit is indicated. We can recombine the lack of fit and pure error sums of squares and estimate $V(Y_i) = \sigma^2$ by

$$s^2 = (93.91 + 30.86)/(5 + 5) = 12.477.$$

Dividing this into the first-order mean square gives a ratio $609.93/12.477 = 48.88$, which exceeds $F(3, 10, 0.999) = 12.55$, while from the second-order mean square we obtain the ratio $135.59/12.477 = 10.88$, which exceeds $F(6, 10, 0.999) = 9.93$. Thus both first- and second-order terms appear to be needed in the fitted model.

T A B L E 12.4. Analysis of Variance Table for the Fitted Model

Source	df	SS	MS	F
Mean (b_0)	1	103,377.82		
First-order	3	1,829.80	609.93	
Second-order $|b_0$	6	813.54	135.59	
Lack of fit	5	93.91	18.78	3.04
Pure error	5	30.86	6.17	
Total	20	106,145.93		

T A B L E 12.5. Analysis of Variance for the Reduced Second-Order Model in X_1 and X_3

Source	df	SS	MS	F
First-order	2	1822.91	911.46	89.80
Second-order $\|b_0$	3	803.12	267.71	26.38
Lack of fit	9⎫	111.22⎫	12.36⎫	
	⎬14	⎬142.08	⎬10.15	2.00
Pure error	5⎭	30.86⎭	6.17⎭	
Total (corrected)	19	2768.11		

Do We Need X_2?

In the original paper, the authors noted the small size of all estimated coefficients with a subscript 2 compared with their standard errors and concluded that their model should not contain X_2 at all. When the situation is this clear-cut—all the coefficients being small compared with their standard errors—such a conclusion is unlikely to be wrong. However, the extra sum of squares principle should be applied in such situations in general and we apply it here to illustrate.

Suppose we wish to test the null hypothesis H_0: $\beta_2 = \beta_{22} = \beta_{12} = \beta_{23} = 0$ against the alternative hypothesis H_1 that at least one of these β's is not zero. The regression sum of squares for the full second-order model in X_1, X_2, and X_3 given b_0 is, from the analysis of variance table,

$$S_1 = \text{SS(first-order terms)} + \text{SS(second-order terms}|b_0)$$

$$= 1829.80 + 813.54$$

$$= 2643.34 \qquad \text{(with } 3 + 6 = 9 \text{ df).}$$

Application of the hypothesis H_0 to the original model implies use of the reduced model

$$E(Y) = \beta_0 + \beta_1 X_1 + \beta_3 X_3 + \beta_{11} X_1^2 + \beta_{33} X_3^2 + \beta_{13} X_1 X_3.$$

The appropriate **X** matrix can be obtained from Table 12.1 by deleting the X_2, X_2^2, $X_1 X_2$, and $X_2 X_3$ columns. The **X'X** matrix can be obtained from the previous one by deleting rows and columns corresponding to X_2, X_2^2, $X_1 X_2$, and $X_2 X_3$. The **X'Y** vector is obtained from the previous one by a similar row deletion. The fitted equation is thus

$$\hat{Y} = 76.420 + 5.503 X_1 + 10.207 X_3 + 0.667 X_1^2 - 7.343 X_3^2 - 1.463 X_1 X_3.$$

The regression sum of squares given b_0 for this reduced model is now required. We find it to be $S_2 = 2626.025$ (with 5 df). The extra sum of squares due to b_2, b_{22}, b_{12}, and b_{23} is therefore

$$S_1 - S_2 = 2643.34 - 2626.03$$

$$= 17.31 \qquad \text{(with } 9 - 5 = 4 \text{ df).}$$

This leads to a mean square of $17.31/4 = 4.33$, which can be compared with the residual mean square estimate of σ^2 from the original three-factor regression. The null hypothesis that $\beta_2 = \beta_{22} = \beta_{12} = \beta_{23} = 0$ cannot be rejected. Thus it seems sensible to adopt the reduced model, which does not involve terms in X_2. The analysis of variance table appropriate to this reduced fitted model is shown in Table 12.5. No

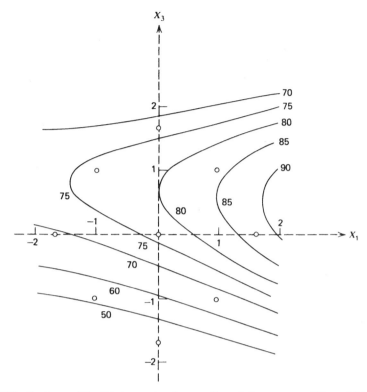

Figure 12.1. Contours of the fitted second-order equation relating response Y to variables X_1 and X_3.

lack of fit is shown and the regression is highly significant for both first- and second-order terms.

In order to examine a fitted second-order response surface, we would usually perform a "canonical analysis" in which the surface is described in terms of coordinates placed along its major axes. Such an analysis is extremely useful and enables the overall situation to be grasped even when many factors are involved. With only two factors, however, as here, we can plot the contours of $\hat{Y}$ directly by writing the fitted equation in the form

$$-7.343X_3^2 + (10.207 - 1.463X_1)X_3 + (0.667X_1^2 + 5.503X_1 + 76.420 - \hat{Y}) = 0.$$

If a value of $\hat{Y}$ is selected, the corresponding contour can be drawn by substituting values of X_1 and solving for X_3. Contours obtained in this way are shown in Figure 12.1. The experimental points are indicated on the diagram by dots. Repeat points are not distinguished, however, and must be obtained by looking at Table 12.6. The contours are those of a rising ridge. Examination of this contour system led the authors to hypothesize on the chemical reactions that could cause such contours. (Quite frequently, response surface investigations provide the initial step in a more fundamental, theoretical, investigation of the system under study.)

These contours may also be viewed in conjunction with the residuals, given in Table 12.6. A "pattern" plot of the residuals in which each residual is placed near its corresponding design point is shown in Figure 12.2. Of the twenty residuals, the six

T A B L E 12.6. The Fitted Values and Residuals Obtained from Fitting a Second-Order Surface $\hat{Y} = f(X_1, X_3)$

u	X_1	X_3	Y	$\hat{Y}$	$e = Y = \hat{Y}$
1	−1	−1	52.8	52.57	0.23
2	1	−1	67.9	66.50	1.40
3	−1	−1	55.4	52.57	2.83
4	1	−1	64.2	66.50	−2.30
5	−1	1	75.1	75.91	−0.81
6	1	1	81.6	83.99	−2.39
7	−1	1	73.8	75.91	−2.11
8	1	1	79.5	83.99	−4.49
9	$-\frac{5}{3}$	0	68.1	69.10	−1.00
10	$\frac{5}{3}$	0	91.2	87.44	3.76
11	0	0	80.6	76.42	4.18
12	0	0	77.5	76.42	1.08
13	0	$-\frac{5}{3}$	36.8	39.01	−2.21
14	0	$\frac{5}{3}$	78.0	73.04	4.97
15	0	0	74.6	76.42	−1.82
16	0	0	75.9	76.42	−0.52
17	0	0	76.9	76.42	0.48
18	0	0	72.3	76.42	−4.12
19	0	0	75.9	76.42	−0.52
20	0	0	79.8	76.42	3.38

with the largest absolute values occur at points $(X_1, X_3) = (0, 0)$ (three); $(\frac{5}{3}, 0)$, $(0, \frac{5}{3})$, and $(1, 1)$ (one each). Thus the model appears to fit least well in the first quadrant of the (X_1, X_3) plane and any conclusions that rely on the validity of the fitted surface in that region could be suspect. (What effect this might have on the authors' original conclusions is a matter for a chemical engineer, rather than a statistician to examine, and we avoid discussion of the point here.) Doubts of this kind can

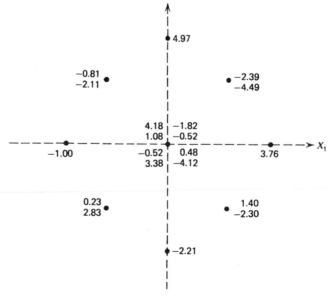

F i g u r e 12.2. Pattern of residuals from the fitted second-order equation relating response Y to variables X_1 and X_3.

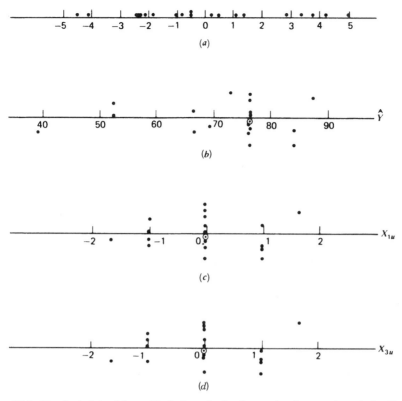

Figure 12.3. Standard plots of the residuals from the fitted second-order equation relating Y to X_1 and X_3: (a) overall, (b) against $\hat{Y}$, (c) against X_1, and (d) against X_3.

sometimes be resolved by doing further experimental runs in the region in which the shape of the fitted surface is suspect and refitting a suitable response function in that more limited region.

We can now examine the residuals in other ways to see if any abnormality is indicated. Figure 12.3 shows the following standard plots of residuals (a) overall, (b) against the fitted values $\hat{Y}_u$, (c) against X_{1u}, and (d) against X_{3u}.

The overall plot does not appear to deny the assumption of normality implicit in the testing of variance ratios in the analysis of variance. The plot against the $\hat{Y}_u$ exhibits "widening" tendencies at first sight but this is deceptive due to the fact that most of the residuals are large and the size of the residuals band is not well established at the lower end of the $\hat{Y}$ scale. Similar behavior occurs in the plots against X_{1u} and X_{3u}, where the size of the residuals band is not well defined at the extremes. Thus one cannot conclude in any of the plots that abnormality is indicated. It does not appear that the basic regression assumptions are unjustified therefore. (Note that, since we do not know the order in which the observations were taken, we are unable to check whether a time trend has affected the response.)

If this investigation were to be continued, additional efforts might involve attempts to account for the large first quadrant residuals by reexamining the original data and their relationships to other predictor variables whose variations have possibly not been considered. In this way, an improvement in the model might be possible. Also, or alternatively, the region in which the fit of the model is questionable could be examined in more detail, as previously suggested.

T A B L E 12.6. **The Fitted Values and Residuals Obtained from Fitting a Second-Order Surface** $\hat{Y} = f(X_1, X_3)$

u	X_1	X_3	Y	$\hat{Y}$	$e = Y = \hat{Y}$
1	−1	−1	52.8	52.57	0.23
2	1	−1	67.9	66.50	1.40
3	−1	−1	55.4	52.57	2.83
4	1	−1	64.2	66.50	−2.30
5	−1	1	75.1	75.91	−0.81
6	1	1	81.6	83.99	−2.39
7	−1	1	73.8	75.91	−2.11
8	1	1	79.5	83.99	−4.49
9	$-\frac{5}{3}$	0	68.1	69.10	−1.00
10	$\frac{5}{3}$	0	91.2	87.44	3.76
11	0	0	80.6	76.42	4.18
12	0	0	77.5	76.42	1.08
13	0	$-\frac{5}{3}$	36.8	39.01	−2.21
14	0	$\frac{5}{3}$	78.0	73.04	4.97
15	0	0	74.6	76.42	−1.82
16	0	0	75.9	76.42	−0.52
17	0	0	76.9	76.42	0.48
18	0	0	72.3	76.42	−4.12
19	0	0	75.9	76.42	−0.52
20	0	0	79.8	76.42	3.38

Treatment of Pure Error When Factors Are Dropped

The foregoing analysis raises a question we have avoided until now. When a factor like X_2 is dropped from a model, should we reassess our treatment of pure error? In Table 12.1, runs 15–20 are the only repeats but, when X_2 is dropped so that the data become as in Table 12.6, the pairs of runs numbered (1, 3), (2, 4), (5, 7), and (6, 8) now apparently form four pairs of repeat runs in variables X_1 and X_3. Also, runs 11 and 12 are now apparently center points. Thus it could be argued that the design should now be considered as a replicated 2^2 factorial (eight points, 1–8) plus a two-factor star (four points, 9, 10, 13, 14) plus eight center points (11, 12, 15, . . . , 20). If this were done, Table 12.5 would have to be revised to show entries for

$$\text{Lack of fit SS} = 78.26 \text{ (3 df)}, \quad \text{MS} = 26.08;$$
$$\text{Pure error SS} = 63.82 \text{ (11 df)}, \quad \text{MS} = 5.80.$$

The consequent F-ratio is $4.50 > F(3, 11, 0.95) = 3.59$, leading to the rather surprising conclusion that there *is* lack of fit. Thus, in this revised analysis, while the variable X_2 appears to be unnecessary in the model, lack of fit is shown if we remove it! However, it is clear that X_2 does little to help explain the variation in the observations. In fact, the size of the pure error mean square is inflated if X_2 *is* used in the model, and, at the same time, the reduction in the degrees of freedom for pure error provides a less sensitive F-test for lack of fit.

Which analysis is correct? One could argue both sides of the issue. On the whole, however, we favor using the pure error as initially calculated before dropping of factors. Presumably, repeats in the original data *are* genuine repeats if so reported, but the same cannot be said of the runs, which look like repeats when a factor is dropped. Also, in many sets of data, the opposite problem may occur, that is, genuine lack of fit will be missed because the "new" repeats will show more variability than the genuine repeats.

A safe way to proceed would be to do the analysis both ways and see if they agree. For many sets of data, they will. If they do not, the data can be subjected to further scrutiny.

How should we proceed in the present example? The model must definitely be placed under suspicion. However, the first- and second-order terms account for a proportion $R^2 = (1822.91 + 803.12)/2768.11 = 0.949$ of the total (corrected for the mean) variation in the data, for the loss of only five degrees of freedom. (When the terms in X_2 are added, the figure rises only to 0.955.) In other words, the model is explaining 95% of the variation about the mean even though, technically, lack of fit is possible. By examining the fitted contours and the residuals together, as we have already done, we can discover where the lack of fit may exist. If the graduation of the true surface appears good over a large region of the X-space, conclusions obtained from the fitted model in that region may still be valid. Further examination of the residuals may also reveal if any of the basic regression assumptions (normality, constant variance, independence of observations) appear to be violated, or may suggest ways to revise the model.

Sometimes, in practical work, the pure error is "too small" simply because the pure error runs have not been randomized (or, at minimum, distributed) over the whole of an experiment. If several pure error runs are done consecutively or close together in time, there is a tendency for the responses to be more alike than they otherwise would be. In other words, the pure error in such a case would not be representative of the range of errors typically found throughout the experiment. This sometimes leads to false signals that lack of fit exists and needs to be investigated carefully. Repeated analyses of the same experimental run also rarely constitute true "repeat runs."

Comment. The example we have just discussed is somewhat unusual in one respect. When, for sound reasons, terms are dropped from a model, lack of fit does not usually appear in the reduced model unless peculiarities exist in the data. We have seen that these peculiarities do not appear to arise from violation of least squares assumptions. Their source remains a matter for speculation.

In a wider sense, this example is not unusual. While the experiment answered some questions, it left others unresolved. These questions become the subject of further conjectures for future work. In this respect, it is typical of much practical experimentation.

Treatment of Pure Error When a Design Is Blocked

The design of Table 12.1 was unblocked. Often, however, response surface designs are performed in blocks in such a way that the blocks are orthogonal to the model. Runs that would be repeat runs in an unblocked design are often divided among the blocks. In such a case, these runs are no longer repeat runs *unless they occur in the same block*, and the pure error must be calculated on that basis. Also, the analysis of variance must contain a sum of squares for blocks. For blocks orthogonal to the model, the appropriate sum of squares for blocks is usually

$$\text{SS(blocks)} = \sum_{w=1}^{m} \frac{B_w^2}{n_w} - \frac{G^2}{N}, \quad \text{with } (m-1) \text{ degrees of freedom,}$$

in the analysis of variance table, where B_w is the total of the n_w observations in the wth block (there are m blocks in all) and G is the grand total of all the observations in all of the m blocks. When blocks are not orthogonal to the model, the extra sum of squares principle applies. (It can, of course, be applied in all cases, orthogonally blocked or not, and produces the answer given above in the former case.)

In some situations, even runs that occur in *different* blocks can be used to measure "almost pure" error, provided an appropriate estimate of the block difference(s) is (are) available. See, for example, Box and Draper (1987, p. 375).

On Dropping Terms

In our second-order polynomial example, we decided to remove all terms in X_2. This raises the more general question of which terms can be considered for removal without damaging the relationships between terms. (Might it have been reasonable to remove, for example, just the term in X_2? We answer this particular question with a resounding no.) For some further conversation and two suggested criteria, please see Section 12.3.

12.3. RETAINING TERMS IN POLYNOMIAL MODELS

We argue in this section that individual terms should not, in general, be dropped from a polynomial model of order two or more, unless the situation is carefully assessed. We also offer two criteria and two consequent rules, which make sense to us and which we recommend. An implication of all this is that to use an equation derived by allowing some mechanical selection procedure to pick a subset of terms from a polynomial model is risky.

Example 1. Quadratic Equation in *X*

Consider the fitted quadratic in a single variable X,

$$\hat{Y} = b_0 + b_1 X + b_{11} X^2. \tag{12.3.1}$$

The maximum or minimum of this quadratic occurs where the first derivative of $\hat{Y}$ with respect to X is zero. This is where $b_1 + 2b_{11}X = 0$, that is, at the location $X = -b_1/(2b_{11})$. If b_{11} is positive, we obtain a minimum; if negative, a maximum.

Let us consider the consequences of dropping just one term. (The equation must be refitted then, of course, to give new coefficients.)

(a) Drop b_0. The intercept at $X = 0$ was $\hat{Y} = b_0$. It will now be forced to be zero, in the refit.

(b) Drop b_1. The maximum or minimum was at $X = -b_1/(2b_{11})$. It will now be forced to be at zero in the refit.

(c) Drop b_{11}. We initially had a quadratic with intercept b_0. We now have a straight line in the refit.

Even if b_0 or b_1 is not statistically significant, actions (a) and (b) will usually produce distortion in the refit. If b_{11} is not significant, however, so that whatever curvature exists is small, the new straight line will tend to do a reasonable job in most cases.

We would thus recommend against dropping either b_0 or b_1 if b_{11} is retained, because this will force the fitting of a quadratic with a built-in restriction and bias. We would argue that dropping b_{11} is reasonable, if the quadratic slope is very slight, because the resulting line will tend to fit well. (Otherwise b_{11} would have been large.) In higher dimensions (more X's), exactly the same sorts of considerations arise.

One way of thinking about the effects of the three different "drops" is to consider what happens to the reduced equations under a shift of origin. Suppose we let $Z = X - a$, corresponding to an origin shift from $X = 0$ to $X = a$. Then we get, by substituting $X = Z + a$,

(a) $b_1 X + b_{11} X^2 = a(b_1 + ab_{11}) + a(b_1 + 2b_{11})Z + b_{11} Z^2$.

(b) $b_0' + b_{11}' X^2 = (b_0' + a^2 b_{11}') + 2ab_{11}' Z + b_{11}' Z^2$.

(c) $b_0'' + b_1'' X = (b_0'' + ab_{11}'') + b_{11}'' Z$.

We see that only the third equation has retained its original form. The other two have reacquired the terms that were deleted! This gives a basis for the first criterion and rule. We also suggest a second criterion and rule, below. In proceeding, we now use the notation for the model function with β's rather than that for a fitted equation with b's, but that does not affect any of our points. So the general question is: Which reduced models are reasonable to adopt, and which are not?

Criterion 1. The Origin Shift Criterion

We shall consider a reduced model to be a sensible one (some say "well formulated") if a shift in the origin of the X-space produces a model of unchanged *form* in the new variables $Z_1, Z_2, \ldots, Z_k$.

Example 2. Second-Order Polynomial in Two X's

Consider the two-X second-order (quadratic) model

$$Y = \beta_0 + \beta_1 X_1 + \beta_2 X_2 + \beta_{11} X_1^2 + \beta_{22} X_2^2 + \beta_{12} X_1 X_2 + \epsilon. \qquad (12.3.2)$$

If we substitute $X_1 = Z_1 + a_1$, $X_2 = Z_2 + a_2$ into the full second-order model (12.3.2), corresponding to an origin shift to the point $(X_1, X_2) = (a_1, a_2)$, we get

$$Y = \beta_0 + \beta_1(Z_1 + a_1) + \beta_2(Z_2 + a_2)$$

$$+ \beta_{11}(Z_1 + a_1)^2 + \beta_{22}(Z_2 + a_2)^2 + \beta_{12}(Z_1 + a_1)(Z_2 + a_2) + \epsilon \qquad (12.3.3)$$

$$= \alpha_0 + \alpha_1 Z_1 + \alpha_2 Z_2 + \alpha_{11} Z_1^2 + \alpha_{22} Z_2^2 + \alpha_{12} Z_1 Z_2 + \epsilon, \qquad (12.3.4)$$

where

$$\alpha_0 = \beta_0 + \beta_1 a_1 + \beta_2 a_2 + \beta_{11} a_1^2 + \beta_{22} a_2^2 + \beta_{12} a_1 a_2, \quad \text{a quadratic in } (a_1, a_2),$$

$$\alpha_1 = \beta_1 + 2\beta_{11} a_1 + \beta_{12} a_2,$$

$$\alpha_2 = \beta_2 + \beta_{12} a_1 + 2\beta_{22} a_2,$$

$$\alpha_{11} = \beta_{11}, \qquad (12.3.5)$$

$$\alpha_{22} = \beta_{22},$$

$$\alpha_{12} = \beta_{12}.$$

What terms can be omitted in (12.3.2) to produce a model of the same (reduced) form in (12.3.4)? Or, equivalently, what β's can be set to zero in (12.3.2) to give a model (12.3.4) of the same form? Obviously, only quadratic coefficients can be dropped. Setting $\beta_{11} = 0$, for example, forces $\alpha_{11} = 0$, but all other coefficients remain in both models. (We ignore the relatively unlikely case where the a's and β's are such that one of the α's vanished identically. This would not occur for all origin moves, nor in most practical cases.)

The general truth, which can be seen via the specific example in (12.3.5), is that only the highest-order coefficients are unaffected by a shift in origin, while all lower-order coefficients become, after the shift, a combination of both lower- and higher-order ones. The consequent rule is the following.

Rule 1

If a model is to be consistent under a shift in origin, only the highest-order terms can be deleted at first and any chosen deletions must keep the model well formulated. Moreover, if any of the highest-order terms are retained, all terms of lower order affected by them in a shift of origin must also be retained, whether or not their estimates are significant in the regression fit.

Note: A model that lacks β_0 cannot be a well-formulated one under origin shift, in any circumstances.

Example 2. Continued

For example, suppose $\beta_{11} \neq 0$, $\beta_{22} = 0$, $\beta_{12} = 0$. Then [see (12.3.5)] $\beta_1 X_1$ must be retained, because α_1 depends on β_{11}; however, $\beta_2 X_2$ is a candidate for possible deletion because $\alpha_2 = \beta_2$ when $\beta_{22} = \beta_{12} = 0$.

Example 3. Third-Order Polynomial in Three Factors

To aid thinking about this in slightly wider contexts, we now give the equivalent of (12.3.5) for second-order (quadratic) and third-order (cubic) models in three factors (X_1, X_2, X_3), with an origin shift to (a_1, a_2, a_3). For the cubic,

$$
\begin{aligned}
Y = {} & \beta_0 + \beta_1 X_1 + \beta_2 X_2 + \beta_3 X_3 \\
& + \beta_{11} X_1^2 + \beta_{22} X_2^2 + \beta_{33} X_3^2 + \beta_{12} X_1 X_2 + \beta_{13} X_1 X_3 + \beta_{23} X_2 X_3 \\
& + \beta_{111} X_1^3 + \beta_{222} X_2^3 + \beta_{333} X_3^3 \\
& + \beta_{122} X_1 X_2^2 + \beta_{133} X_1 X_3^2 + \beta_{112} X_1^2 X_2 + \beta_{233} X_2 X_3^2 + \beta_{113} X_1^2 X_3 \\
& + \beta_{223} X_2^2 X_2 + \beta_{123} X_1 X_2 X_3 + \epsilon,
\end{aligned}
\tag{12.3.6}
$$

we find that:

$$
\begin{aligned}
\alpha_0 = {} & \text{full cubic in } (a_1, a_2, a_3), \\
\alpha_1 = {} & \beta_1 + 2\beta_{11} a_1 + \beta_{12} a_2 + \beta_{13} a_3 + 3\beta_{111} a_1^2 + \beta_{122} a_2^2 + \beta_{133} a_3^2 \\
& + 2\beta_{112} a_1 a_2 + 2\beta_{113} a_1 a_3 + \beta_{123} a_2 a_3, \\
\alpha_2 = {} & \beta_2 + \beta_{12} a_1 + 2\beta_{22} a_2 + \beta_{23} a_3 + \beta_{112} a_1^2 + 3\beta_{222} a_2^2 + \beta_{233} a_3^2 \\
& + 2\beta_{122} a_1 a_2 + 2\beta_{223} a_2 a_3 + \beta_{123} a_1 a_3, \\
\alpha_3 = {} & \beta_3 + \beta_{13} a_1 + \beta_{23} a_2 + 2\beta_{33} a_3 + \beta_{113} a_1^2 + \beta_{223} a_2^2 + 3\beta_{333} a_3^2 \\
& + 2\beta_{133} a_1 a_3 + 2\beta_{233} a_2 a_3 + \beta_{123} a_1 a_2, \\
\alpha_{11} = {} & \beta_{11} + 3\beta_{111} a_1 + \beta_{112} a_2 + \beta_{113} a_3,
\end{aligned}
\tag{12.3.7}
$$

$$\alpha_{22} = \beta_{22} + \beta_{122}a_1 + 3\beta_{222}a_2 + \beta_{223}a_3,$$

$$\alpha_{33} = \beta_{33} + \beta_{133}a_1 + \beta_{233}a_2 + 3\beta_{333}a_3,$$

$$\alpha_{12} = \beta_{12} + 2\beta_{112}a_1 + 2\beta_{122}a_2 + \beta_{123}a_3,$$

$$\alpha_{13} = \beta_{13} + 2\beta_{113}a_1 + \beta_{123}a_2 + 2\beta_{133}a_3,$$

$$\alpha_{23} = \beta_{23} + \beta_{123}a_1 + 2\beta_{223}a_2 + 2\beta_{233}a_3,$$

$$\alpha_{111} = \beta_{111},$$

$$\alpha_{222} = \beta_{222},$$ \hfill (12.3.7)

$$\alpha_{333} = \beta_{333},$$ \hfill continued

$$\alpha_{122} = \beta_{122},$$

$$\alpha_{133} = \beta_{133},$$

$$\alpha_{112} = \beta_{112},$$

$$\alpha_{233} = \beta_{233},$$

$$\alpha_{113} = \beta_{133},$$

$$\alpha_{223} = \beta_{223},$$

$$\alpha_{123} = \beta_{123},$$

The origin-shift parameter situation for the quadratic model in three variables is found by setting all β's with three subscripts equal to zero in (12.3.6) and (12.3.7). In both cases, detailed conclusions developed by extending the ideas of Example 2 are left to the reader.

Example 4

Peixoto (1990) gives the more general polynomial example model

$$Y = \beta_0 + \beta_1 X_1 + \beta_2 X_2 + \beta_{12} X_1 X_2 + \beta_{22} X_2^2 + \beta_{122} X_1 X_2^2$$
$$+ \beta_{222} X_2^3 + \beta_{1222} X_1 X_2^3 + \epsilon$$
\hfill (12.3.8)

as a well-formulated one that satisfies Criterion 1.

[*Exercise:* Substitute $X_1 = Z_1 + a_1$, $X_2 = Z_2 + a_2$ into (12.3.8) and show that no new types of polynomial terms occur in the expansion.]

Criterion 2. The Axes Rotation Criterion

We shall consider a reduced model to be a sensible one if a rotation of the X-axes produces a model of unchanged *form* in the X's.

Example 5. Second-Order Polynomial in Two X's

We consider again the second-order model (12.3.2). An axial rotation from the X's to (say) the W's is such that

$$W_1 = c_1 X_1 + c_2 X_2,$$
$$W_2 = d_1 X_1 + d_2 X_2,$$
\hfill (12.3.9)

or $\mathbf{W} = \mathbf{MX}$, say, where $\mathbf{M}$ is an orthonormal matrix, such that $\mathbf{M'M} = \mathbf{MM'} = \mathbf{MM}^{-1} = \mathbf{I}$. These conditions, which preserve both length and orthogonality of the axes after rotation, imply that $\mathbf{M}$ must take the form

$$\mathbf{M} = \begin{bmatrix} \cos\theta & \sin\theta \\ -\sin\theta & \cos\theta \end{bmatrix} = \begin{bmatrix} c & s \\ -s & c \end{bmatrix}, \tag{12.3.10}$$

say, or a similar form with changes of signs in rows and/or columns.

To substitute for the X's in (12.3.2), we invert via $\mathbf{X} = \mathbf{M}^{-1}\mathbf{W} = \mathbf{M'W}$ to give

$$X_1 = cW_1 - sW_2,$$
$$X_2 = sW_1 + cW_2. \tag{12.3.11}$$

The surface after rotation is then

$$Y = \gamma_0 + \gamma_1 W_1 + \gamma_2 W_2 + \gamma_{11} W_1^2 + \gamma_{22} W_2^2 + \gamma_{12} W_1 W_2 + \epsilon,$$

where

$$\gamma_0 = \beta_0,$$
$$\gamma_1 = c\beta_1 + s\beta_2,$$
$$\gamma_2 = -s\beta_1 + c\beta_2,$$
$$\gamma_{11} = c^2\beta_{11} + s^2\beta_{22} + cs\beta_{12},$$
$$\gamma_{22} = s^2\beta_{11} + c^2\beta_{22} - cs\beta_{12},$$
$$\gamma_{12} = -2cs\beta_{11} + 2cs\beta_{22} + (c^2 - s^2)\beta_{12}. \tag{12.3.12}$$

[The case $\theta = 45°$, $c = s = 2^{-1/2}$ was used by Box and Draper (1987), pp. 447–448.] We see that order is completely preserved, namely, zero-order, first-order, and second-order γ coefficients are linear combinations of only zero-order, first-order, and second-order coefficients, respectively. This example, which extends to more X's, illustrates the more general statement of Rule 2.

Rule 2

If a model is to be consistent under rotation of axes, all terms of a particular order must be considered as a unit. Subsets of terms of a given order cannot be removed.

(For example, it is senseless to consider dropping β_{12} alone, because this does not remove γ_{12}, which is still a combination of β_{11} and β_{22}.)

Application of Rules 1 and 2 Together

If a model is to be consistent both under a shift in origin and under rotation, terms can be dropped only in units of order (e.g., all second-order terms, all first-order terms). Moreover, lower-order terms cannot be deleted if higher-order terms are retained in the model.

In situations where the original axial directions are considered mandatory and untouchable, so that rotation to new axes is out of the question (e.g., from practical considerations in the interpretation of the effects shown by the fitted model), only Rule 1 would apply.

"Do We Need This X?"

Assume now that we do *not* wish to consider rotations of the surface, so that Rule 2 will not come into consideration. We can then ask questions about specific X-variables. Look again at (12.3.5) and consider the question: "Does X_2 contribute to second-order curvature?" This is equivalent to testing $H_0: \beta_{22} = \beta_{12} = 0$ versus H_1: not so. We see that this is equivalent to $\alpha_{22} = \alpha_{12} = 0$, so that Rule 1 is satisfied. We can then test this null hypothesis via an extra sum of squares test in the usual way. It is also possible to test whether X_2 is necessary *at all*, via $H_0: \beta_2 = \beta_{22} = \beta_{12} = 0$, because then $\alpha_2 = \alpha_{22} = \alpha_{12} = 0$. This implies the following:

If rotation of the axes is not an option because of the nature of the predictor variables, subsets of higher-order terms (or all terms) that depend on one (or more) specific predictors can be deleted, but only in combinations that keep the model well formulated, that is, satisfy Rule 1.

For example, in (12.3.7), some possibilities are:

(a) $\beta_{111} = \beta_{122} = \beta_{112} = \beta_{133} = \beta_{113} = \beta_{123} = 0$. (Does X_1 contribute to third-order curvature?)

(b) $\beta_{11} = \beta_{12} = \beta_{13} = 0$, in addition to (a). (Does X_1 contribute to second- and third-order curvature?)

(c) $\beta_1 = 0$, in addition to (b). (Do we need X_1 at all?)

Summary Advice

In deciding what terms can sensibly be deleted from a polynomial model, we suggest:

1. Always apply Rule 1.
2. Consider whether the original predictor variables are either:
    a. not absolute descriptors of the response surface, so that rotation of the axes to give the surface in terms of new variables that are rotational linear combinations of the original variables can be considered; apply Rule 2 in addition to Rule 1; or,
    b. always to be retained (so that descriptions of the response surface are always in these variables); apply only Rule 1, perhaps considering hypotheses involving specific X's, as in the examples above.

Using a Selection Procedure for a Polynomial Fit

Selection procedures (see Chapter 15) do not incorporate rules of the type we have given. Thus their use to get polynomial models is suspect. It might be reasonable to let a selection procedure offer its choice, but the equation should then be reviewed and refined with the criteria given above in mind.

References

Box and Draper (1987); Driscoll and Anderson (1980); Peixoto (1987, 1990).

EXERCISES FOR CHAPTER 12

A. Eighteen observations were obtained on four predictor variables and one response variable in a process. It is suggested that the model

$$Y = \beta_0 X_0 + \beta_1 X_1 + \beta_2 X_2 + \beta_3 X_3 + \beta_4 X_4 + \beta_{11} X_1^2 + \beta_{12} X_1 X_2 + \beta_{22} X_2^2$$
$$+ \beta_{13} X_1 X_3 + \beta_{14} X_1 X_4 + \epsilon$$

is a reasonable one. The data are shown in Table A.

Requirements
  1. Examine the data and the model. Is it possible to fit the proposed model to the data? Why or why not?
  2. Estimate $V(Y_i) = \sigma^2$.

T A B L E A. Data

X_1	X_2	X_3	X_4	Y
20	50	75	15	27
27	55	60	20	23
22	62	68	16	18
27	55	60	20	26
24	75	72	8	23
30	62	73	18	27
32	79	71	11	30
24	75	72	8	23
22	62	68	16	22
27	55	60	20	24
40	90	78	32	16
32	79	71	11	28
50	84	72	12	31
40	90	78	32	22
20	50	75	15	24
50	84	72	12	31
30	62	73	18	29
27	55	60	20	22

B. (*Source:* "Variable shear rate viscosity of SBR-filler-plasticizer systems," by G. C. Derringer, *Rubber Chemistry and Technology,* **47,** September 1974, 825–836.) Fit the model

$$Y = (\alpha_0 + \alpha_1 Z + \alpha_2 Z^2) + (\beta_0 + \beta_1 Z + \beta_{11} Z^2) X_1$$
$$+ (\gamma_0 + \gamma_1 Z + \gamma_{11} Z^2) X_2 + \epsilon,$$

where $Z = \ln(X_3 + 1)$, to the data below, and give a complete analysis. Note that there are six repeat runs.

X_1	X_2	X_3	Y	X_1	X_2	X_3	Y
47.1	33.9	7.5	11.97	47.1	33.9	750	8.46
72.9	33.9	750	8.63	60	21	75	10.65
47.1	8.1	750	8.80	60	21	3000	7.60
60	21	75	10.73	60	21	3	13.06
60	21	75	10.69	39	21	75	10.51
72.9	8.1	7.5	13.12	60	0	75	11.22
47.1	8.1	7.5	12.58	60	21	75	10.67
72.9	33.9	7.5	12.24	60	42	75	10.24
60	21	75	10.64	81	21	75	10.74
72.9	8.1	750	9.09	60	21	75	10.69

C. (*Source:* "A short life test for comparing a sample with previous accelerated test results," by Wayne Nelson, *Technometrics,* **14,** 1972, 175–185.) The data in the table below are accelerated life test results on 24 units of a type of sheathed tabular heater. T is the temperature in °F and Y is the life in hours at that temperature for a single unit. Six units are tested at each temperature. Plot the data and look at them. Fit the model

$$\log_{10} Y = \beta_0 + \beta_1\{1000/(T + 460)\} + \epsilon$$

and perform all the usual analyses. [*Note:* $T + 460$ is the absolute temperature in °F.]

T	Y
1520	1953, 2135, 2471, 4727, 6143, 6314
1620	1190, 1286, 1550, 2125, 2557, 2845
1660	651, 837, 848, 1038, 1361, 1543
1708	511, 651, 651, 652, 688, 729

D. The experiment summarized in the table below was run on a pilot plant to examine the effects of varying the percentage of a certain mix component (X_1), the temperature of the mix(X_2), and the flow-through rate (X_3), on three responses, Y_1, Y_2, and Y_3. The input variables have been coded, but the responses are in their original units. The experimental design shown is a central composite design consisting of eight cube points $(X_1, X_2, X_3) = (\pm 1, \pm 1, \pm 1)$, six axial points $(\pm \alpha, 0, 0)$, $(0, \pm \alpha, 0)$, $(0, 0, \pm \alpha)$, where $\alpha = 1.2154$, and one center point $(0, 0, 0)$. The run order shown is the randomized order in which the design was performed.

X_1	X_2	X_3	Y_1	Y_2	Y_3
−1	−1	1	85.3	72.7	97.1
1	1	−1	72.3	57.6	96.9
0	1.2154	0	71.4	56.5	96.4
0	−1.2154	0	72.0	64.6	96.8
−1	−1	−1	87.0	79.2	97.0
1	1	1	55.6	32.6	96.2
0	0	−1.2154	85.0	75.9	97.2
1.2154	0	0	70.9	53.4	97.9
0	0	0	75.9	59.3	97.4
1	−1	1	76.1	63.2	97.4
−1	1	−1	85.0	75.3	97.2
0	0	1.2154	68.0	57.2	95.5
−1.2154	0	0	89.6	83.6	97.2
−1	1	1	75.0	61.5	96.5
1	−1	−1	74.2	61.0	98.2

Requirements. Is the design rotatable? (See Section 12.2.) Using multiple regression techniques, formulate and fit suitable models of first or second order to Y_1, Y_2, and Y_3 separately. Perform a complete analysis and provide practical conclusions. If larger values of the Y's were more desirable, where in the X-space would it be better to operate?

E. A new product was being considered by the bakery goods research division of a large corporation. Of paramount concern was the maximum peak height obtained from a standard container of mixed dough just prior to baking. Four major ingredients were thought to be important in affecting peak height: percentage of fat, percentage of water, amount of flour in the brew, and the speed of the mixer in rpm. The experiments given in the table were performed, and the values of the maximum peak heights (shown in the body of the table) were recorded for each run. Note that there are four repeat runs at the conditions (12, 50, 20, 130) with responses 492, 523, 530, and 590.

Experimental Data Extensigraph Maximum Peak Height

Percent Fat	Flour in Brew	46 rpm 90	46 rpm 130	46 rpm 170	50 rpm 90	50 rpm 130	50 rpm 170	54 rpm 90	54 rpm 130	54 rpm 170
8	10	833		540				673		493
	20					537				
	30	577		547				660		512
12	10					547				
	20		653		650	492 523 530 590	553		487	
	30					595				
16	10	802		477				710		520
	20					575				
	30	568		401				572		483

Requirements

1. By suitable choice of central levels and scale divisors, code all four predictor variables so that their levels are $(-1, 0, 1)$. Write out the design in the coded variables, and confirm that it is a "cube plus star plus four center points" type design. Is it rotatable?

2. Using multiple regression techniques, construct a suitable model of first or second order for predicting maximum peak heights. In your conclusions, indicate the relative importance of the predictor variables, and make any other comments you find relevant.

F. Fit a full second-order model and perform a complete analysis, using the Y_4 data of Table 12.1.

G. If we believe in the "origin-shift" criterion," is the model

$$Y = \beta_0 + \beta_1 X_1 + \beta_2 X_2 + \beta_{12} X_1 X_2 + \beta_{22} X_2^2 + \beta_{122} X_1 X_2^2 + \epsilon$$

a "well-formulated" one?

H. A proposed model, based on theoretical considerations, is

$$Y = \alpha X_1^\beta X_2^\gamma X_3^\delta \epsilon.$$

Requirements. After transformation, fit the proposed model by least squares. State which predictor variable appears most important and check all coefficients for statistical significance (take $\alpha = 0.05$). Is the model a satisfactory one?

The data shown below, which relate to a study of the quantity of vitamin B_2 in turnip green, are taken from the "Annual progress report on the soils–weather project, 1948," by J. T. Wakeley, University of North Carolina (Raleigh) Institute of Statistics Mimeo Series 19 (1949). The variables are:

X_1 = radiation in relative gram calories per minute during the preceding half day of sunlight (coded by dividing by 100),

X_2 = average soil moisture tension (coded by dividing by 100),

X_3 = air temperature in degrees Fahrenheit (coded by dividing by 10),

Y = milligrams of vitamin B_2 per gram of turnip green.

These data were used by R. L. Anderson and T. A. Bancroft in *Statistical Theory in Research*, McGraw-Hill, New York, 1959, on p. 192, to fit the model

$$Y = \beta_0 + \beta_1 X_1 + \beta_2 X_2 + \beta_3 X_3 + \beta_{12} X_1 X_2 + \epsilon.$$

Requirements. Develop a suitable fitted equation using these data and compare its form with the form of the one fitted by Anderson and Bancroft.

X_1	X_2	X_3	Y	X_1	X_2	X_3	Y
1.76	0.070	7.8	110.4	1.80	0.020	7.3	75.3
1.55	0.070	8.9	102.8	1.80	0.020	6.5	92.0
2.73	0.070	8.9	101.0	1.77	0.020	7.6	82.4
2.73	0.070	7.2	108.4	2.30	0.020	8.2	77.1
2.56	0.070	8.4	100.7	2.03	0.474	7.6	74.0
2.80	0.070	8.7	100.3	1.91	0.474	8.3	65.7
2.80	0.070	7.4	102.0	1.91	0.474	8.2	56.8
1.84	0.070	8.7	93.7	1.91	0.474	6.9	62.1
2.16	0.070	8.8	98.9	0.76	0.474	7.4	61.0
1.98	0.020	7.6	96.6	2.13	0.474	7.6	53.2
0.59	0.020	6.5	99.4	2.13	0.474	6.9	59.4
0.80	0.020	6.7	96.2	1.51	0.474	7.5	58.7
0.80	0.020	6.2	99.0	2.05	0.474	7.6	58.0
1.05	0.020	7.0	88.4				

CHAPTER 13

Transformation of the Response Variable

13.1. INTRODUCTION AND PRELIMINARY REMARKS

Simplifying Models Via Transformation

Suppose we had n data values (f_i, p_i, Y_i), $i = 1, 2, \ldots, n$, that could be well explained by a quadratic equation, say,

$$Y = \beta_0 + \beta_1 f + \beta_2 p + \beta_{11} f^2 + \beta_{22} p^2 + \beta_{12} fp + \epsilon. \tag{13.1.1}$$

We might be perfectly happy with such a fit. Later, however, we might be told that it was customary with this particular type of data to use ln Y, the natural[1] logarithm of Y (logarithm to the base e, or ln) instead of Y. Armed with this knowledge, we might find that a simpler planar equation fit of

$$\ln Y = \beta_0 + \beta_1 f + \beta_2 p + \epsilon \tag{13.1.2}$$

gave a better, as good, or almost as good an explanation of the variation in the data. Essentially we would have used the *response transformation* ln Y to "flatten out" our original six-parameter quadratic surface to be a simpler three-parameter plane. We would also have changed the assumption about the error structure. If the errors in Eq. (13.1.1) were independent $N(0, \sigma^2)$ errors, the errors in Eq. (13.1.2) would not be; and vice versa. So we must give some thought to what is being assumed about the error structure when we transform a response.

Thinking About the Error Structure

It is sometimes reasonable to believe that a model function might be multiplicative rather than additive. Suppose we think that

$$\eta = \alpha X_1^\beta X_2^\gamma X_3^\delta \tag{13.1.3}$$

is a sensible model function for a certain set of data. Let us take natural logarithms in (13.1.3). Then

---

[1] The difference between taking natural logarithms or logarithms to another base (10, say) is a constant multiple. Suppose a given number $A = e^b = 10^c$, say. Then $\ln A = b$ and $\log_{10} A = c$ and their ratio $b/c = \ln 10 = (\log_{10} e)^{-1} = 2.302585$.

$$\ln \eta = \ln \alpha + \beta \ln X_1 + \gamma \ln X_2 + \delta \ln X_3. \qquad (13.1.4)$$

One would then be led to fitting by least squares the model

$$\ln Y = \ln \eta + \epsilon, \qquad (13.1.5)$$

where $\ln \eta$ is given by (13.1.4). In doing this, we would assume that the errors are $\epsilon \sim N(0, I\sigma^2)$. Now let us work backward from (13.1.5). By noting that $\epsilon = \ln (e^\epsilon)$, where e here is the natural logarithm base 2.718282, and exponentiating (13.1.5), we get

$$Y = \alpha X_1^\beta X_2^\gamma X_3^\delta e^\epsilon = \eta e^\epsilon. \qquad (13.1.6)$$

This model does not have *additive* errors (i.e., we do not have $Y = \eta + \text{error}$) but *multiplicative* ones; the model function is *multiplied* by the error. Thus a fit of (13.1.5) is appropriate only if we "believe" that (13.1.6) is a suitable model. Transforming the Y into $\ln Y$ has altered the error structure. If we really believed that the errors were additive and that

$$Y = \alpha X_1^\beta X_2^\gamma X_3^\delta + \text{error}, \qquad (13.1.7)$$

we could not take logarithms and use least squares. We would have to use the methods of nonlinear estimation instead. [However, (13.1.5) could then be fitted to give some *initial estimates*; see Chapter 24.] We say that (13.1.7) is *intrinsically nonlinear*, whereas (13.1.6) is *intrinsically linear*.

To take another example, if we decided to fit the model

$$\frac{1}{Y} = \beta_0 + \beta_1 X_1 + \beta_2 X_2 + \epsilon, \qquad (13.1.8)$$

then we would "believe" that an appropriate model was

$$Y = 1/(\beta_0 + \beta_1 X_1 + \beta_2 X_2 + \epsilon). \qquad (13.1.9)$$

Now, in fact, most people do not think about models in this sequence. In practice, it is simpler to decide on a transformation, fit it, and then examine the residuals *in the metric of the transformed variable* to see if they are reasonably well behaved. If they are, the error specifications in the transformed response space are assumed to be all right. Note that, for (13.1.5), the residuals to examine are of the form $\ln Y_i - (\widehat{\ln Y})_i$, while for (13.1.8) they are $Y_i^{-1} - (\widehat{Y^{-1}})_i$. All tests and confidence statements must be made in the transformed space also.

Predictions in Y-Space

Some scientists are unhappy about working in a transformed Y-space, but a return to the original Y-space can be made after the model has been fitted. Suppose we have fitted a model to $\ln Y$, and we make a prediction $\widehat{\ln Y}$ at a certain set of X's. We can, if we wish, evaluate $\hat{Y} = \exp\{\widehat{\ln Y}\}$ and predict in the original space. Also, a confidence statement on $E(\ln Y)$ with interval (a, b) can be translated into a confidence statement with interval (e^a, e^b) in the Y-space. It will not be symmetric about the predicted value $\hat{Y}$, of course. We can also evaluate residuals $Y_i - \hat{Y}_i$ at the data points, if we wish. These residuals are *not*, however, checked; these are *not* the residuals that should satisfy the residuals checks for normality, and so on.

T A B L E 13.1. Values of Certain Power Functions for Five Benchmark Powers

λ	Y^λ	$W = (Y^\lambda - 1)/\lambda$	$V = (Y^\lambda - 1)/(\lambda \dot{Y}^{\lambda-1})$
1	Y	$Y - 1$	$Y - 1$
$\frac{1}{2}$	$Y^{1/2}$	$2(Y^{1/2} - 1)$	$2\dot{Y}^{1/2}(Y^{1/2} - 1)$
0	$1(?)$	$\ln Y$	$\dot{Y} \ln Y$
$-\frac{1}{2}$	$Y^{-1/2}$	$2(1 - Y^{-1/2})$	$2\dot{Y}^{3/2}(1 - Y^{-1/2})$
-1	Y^{-1}	$1 - Y^{-1}$	$\dot{Y}^2(1 - Y^{-1})$

Preliminary Remarks on the Power Family of Transformations

One extremely useful way of picking a transformation is to assume that a member of the power family will be appropriate, and then to estimate the best power by maximum likelihood. This is often called the "Box–Cox method" in honor of the authors of the seminal paper on this topic, written in 1964. We describe this in the next section. There is a particular difficulty in thinking about powers Y^λ, because as λ approaches zero, Y^λ approaches 1. This would clearly be a senseless transformation! We shall see soon that a zero power is associated with a $\ln Y$ (or $\log Y$) transformation. To make the calculations for choosing the best λ value run smoothly as λ approaches zero, we must perform the Box–Cox calculations using not Y^λ, which gives problems at $\lambda = 0$, but with either $W = (Y^\lambda - 1)/\lambda$, now out of fashion, or (better) $V = W/\dot{Y}^{\lambda-1} = (Y^\lambda - 1)/(\lambda \dot{Y}^{\lambda-1})$. ($\dot{Y}$ is the geometric mean of the Y_i in the data set.) Note that for a power transformation to be applicable, all the Y's must be positive. Table 13.1 will help prepare the reader for the fuller discussion of Section 13.2. It shows what the functions Y^λ, $W = (Y^\lambda - 1)/\lambda$, and $V = W/\dot{Y}^{\lambda-1}$ look like for five benchmark values of λ, namely, $\lambda = 1, \frac{1}{2}, 0, -\frac{1}{2}, -1$. The query next to the one in the second column of Table 13.1 denotes bewilderment at the possibility of a transformation Y^λ when $\lambda = 0$, because all the data would revert to 1's. However, when λ approaches zero, W approaches $\ln Y$ and V thus approaches $\dot{Y} \ln Y$.

Points to Keep in Mind

In general, when we make a transformation, it is impossible to relate the parameters of the model used for the transformed data to the parameters in a model initially intended for the untransformed data. Usually, there is no mathematical equivalence except in an approximate sense via a Taylor series expansion. For example, if instead of fitting $Y = \beta_0 + \beta_1 X + \beta_{11} X^2 + \epsilon$ we fit $Y^\lambda = \alpha_0 + \alpha_1 X + \epsilon$, the relationship of $\beta_0, \beta_1, \beta_{11}$ to $\lambda, \alpha_0, \alpha_1$ is not clear. An attempt to find such a relationship is usually not fruitful.

When several sets of data arise from similar experimental situations, it may not be necessary to carry out complete analyses on all the sets to determine appropriate transformations. Quite often, the same transformation will work for all.

The fact that a general analysis exists for finding transformations does not mean that it should always be used. Often, informal plots of the data will clearly reveal the need for a transformation of an obvious kind (such as $\ln Y$ or $1/Y$). In such a case, the more formal analysis may be viewed as a useful check procedure to hold in reserve.

13.2 POWER FAMILY OF TRANSFORMATIONS ON THE RESPONSE: BOX–COX METHOD

Suppose we have data $(Y_1, Y_2, \ldots, Y_n)$ on a response variable Y that is always positive. (Other cases will be discussed later.) If the ratio of the largest observed Y to the smallest is "considerable," say, 10 or higher, we might consider the possibility of transforming Y. There are many possible types of transformations. A useful idea in many applications is to consider powers, Y^λ, say, and to try to find the best value of λ to use. A snag soon becomes apparent; when $\lambda = 0$, $Y^0 = 1$—making all the data equal! However, if we were to try working with

$$W = \begin{cases} (Y^\lambda - 1)/\lambda, & \text{for } \lambda \neq 0, \\ \ln Y, & \text{for } \lambda = 0, \end{cases} \tag{13.2.1}$$

the problem at $\lambda = 0$ would be overcome, because $\ln Y$ is the appropriate limit, as λ tends to zero, of $(Y^\lambda - 1)/\lambda$, and so the family is now continuous in λ. A disadvantage of (13.2.1) is that, as λ varies, the sizes of the W's can change enormously, leading to minor problems in the analysis and requiring a special program to get the best λ value. For that reason, it is preferable to use the alternative form

$$V = \begin{cases} (Y^\lambda - 1)/(\lambda \, \dot{Y}^{\lambda-1}), & \text{for } \lambda \neq 0, \\ \dot{Y} \ln Y, & \text{for } \lambda = 0, \end{cases} \tag{13.2.2}$$

where the additional divisor $(\dot{Y}^{\lambda-1})$ in (13.2.2), compared with (13.2.1), is the nth power of the appropriate *Jacobian* of the transformation, which converts the set of Y_i into the set of W_i. This ensures that unit volume is preserved in moving from the set of Y_i to the set of V_i in (13.2.2). (To appreciate this remark fully, some knowledge of calculus is needed, but we can proceed without that.)

The quantity $\dot{Y}$ is the geometric mean of the Y_i,

$$\dot{Y} = (Y_1 Y_2 \cdots Y_n)^{1/n}. \tag{13.2.3}$$

$\dot{Y}$ is a constant and it would be evaluated at the beginning of the calculation procedure, usually by antilogging (exponentiating) the formula

$$\ln \dot{Y} = n^{-1} \sum_{i=1}^{n} \ln Y_i. \tag{13.2.4}$$

When formula (13.2.2) is applied to each Y_i we create a vector $\mathbf{V} = (V_1, V_2, \ldots, V_n)'$ and use it to fit a linear model

$$\mathbf{V} = \mathbf{X}\boldsymbol{\beta} + \boldsymbol{\epsilon} \tag{13.2.5}$$

by least squares for any specified value of λ. More generally, we have to estimate λ as well as $\boldsymbol{\beta}$. We do this by invoking the principle of maximum likelihood under the assumption that $\boldsymbol{\epsilon} \sim N(\mathbf{0}, \mathbf{I}\,\sigma^2)$ for the proper choice of λ. This method (and also a Bayesian equivalent of it), applicable to any family of transformations, including the one above, is discussed by Box and Cox (1964). The basic idea is that, if an appropriate λ could be found, an additive model with normally distributed, independent, and homogeneous error structure could be fitted by the maximum likelihood method. We do not have to understand maximum likelihood, Bayesian statistics, or the Jacobian to perform the procedure. The necessary steps are as follows.

Maximum Likelihood Method of Estimating λ

1. Choose a value of λ from a selected range. Usually we look at λ's in the range $(-1, 1)$, or perhaps even $(-2, 2)$, at first, and extend the range later if necessary. We would usually cover the selected range with about 11–21 values of λ. We can always divide up a portion of the interval more finely later if we need the additional detail, but this is often unnecessary—see (3) below.

2. For each chosen λ value, evaluate $\mathbf{V}$ via (13.2.2). Remember to use $V = \dot{Y} \ln Y$ when $\lambda = 0$. Or else avoid using $\lambda = 0$ *exactly*, in covering the selected range of λ. Now fit (13.2.5) and record $S(\lambda, \mathbf{V})$, the residual sum of squares for the regression. Any ordinary least squares regression program can be used for this calculation.

3. Plot $S(\lambda, \mathbf{V})$ versus λ. [Some workers prefer to plot $\ln S(\lambda, \mathbf{V})$ versus λ; make your own choice depending on how big the numbers are.] Draw a smooth curve through the plotted points, and find at what value of λ the lowest point of the curve lies. That value, $\hat{\lambda}$, is the maximum likelihood estimate of λ. Typically, we would not use this precise value of λ in subsequent calculations, but would use instead the nearest convenient value in the sequence, $\ldots$, -2, $-1\frac{1}{2}$, -1, $-\frac{1}{2}$, 0, $\frac{1}{2}$, 1, $1\frac{1}{2}$, 2, $\ldots$ after first checking that such a value lay within a selected confidence interval (see below). For example, if $\hat{\lambda}$ came out to be about 0.11, we would probably use $\lambda = 0$, If $\hat{\lambda}$ were about 0.94, we would use $\lambda = 1$, and so on. (There is, however, considerable leeway for a personal decision in the choice of λ, after the calculations have been examined. In some situations, the values $\frac{1}{3}$, $\frac{2}{3}$ might be appropriate. Some workers prefer to round to the nearest quarter, rather than the nearest half; others feel unhappy with any rounding and proceed using $\hat{\lambda}$ instead.) We then analyze the transformed data—transformed via whatever value of λ was finally selected—and report the results.

Some Conversations on How to Proceed

The last sentence of (3) needs some further explanation. Once we have chosen a λ, how do we actually transform the data? Do we use the form (13.2.2) exactly as it is given but with the selected λ? We can do this if we wish. Alternatively, if a nonzero λ is selected to transform the data, we can carry out our analysis on Y^λ if we wish, rather than on the first line of (13.2.2). Similarly, if $\lambda = 0$ is the value of λ actually chosen to transform the data, we can use either $\ln Y$ (natural logarithms) or $\log Y$ (logarithms to any other base, such as 10, for example). These logarithms differ only by a constant factor, and so only the scale of the numbers involved is affected, not the basic nature of the subsequent analysis. Most people would choose the simplest representation possible (Y^λ or $\ln Y$). We do that in our example. Equation (13.2.2) is then used only for the analysis that determines λ. This form (13.2.2) has several advantages for this purpose. It is conceptually simple (see Box and Cox, 1964, p. 216), it provides better computational accuracy, especially for large λ, and the calculations can be performed using any standard regression program. Also, it allows direct comparison of the residual sums of squares, because the scale factor divisor $\dot{Y}^{\lambda-1}$ essentially reconverts the W_i back to comparable units.

Use of the (13.2.2) form for the final analysis is also acceptable. Only a scale difference and an origin shift are involved and the basic nature of the subsequent analysis is unaffected by these, for a linear model.

Two points relevant to the regression analysis after choice of λ should be noted:

1. The fact that the "best λ" has been selected does not necessarily guarantee an

equation useful in practice. The final equation must be evaluated in the usual ways on its own merits.

2. To allow for the fact that λ has been estimated, some workers remove one df for $\hat{\lambda}$ in the analysis of variance table in the subsequent regression analysis. This reduces the total degrees of freedom from n to $(n - 1)$ and the residual df are adjusted accordingly. This reduction is optional. (Note that no sum of squares is removed. Although we do not make this reduction in our examples below, we are not opposed to it. For large n, it makes little difference, of course.)

Approximate Confidence Interval for λ

The maximum likelihood equations, after simplification, result in an estimation of λ by choosing the λ that minimizes the residual sum of squares function $S(\lambda, \mathbf{V})$. One step back from this point is the equivalent criterion: maximize $L(\lambda)$, where

$$L(\lambda) = -\tfrac{1}{2}n \ln\{S(\lambda, \mathbf{V})/n\}. \tag{13.2.6}$$

Obviously we do not need to plot this form because n is fixed; that is why the simpler plot of $S(\lambda, \mathbf{V})$ versus λ is used. However, an approximate $100(1 - \alpha)\%$ confidence interval for λ consists of those values of λ that satisfy the inequality

$$L(\hat{\lambda}) - L(\lambda) \le \tfrac{1}{2}\chi_1^2(1 - \alpha), \tag{13.2.7}$$

where $\chi_1^2(1 - \alpha)$ is the percentage point of the chi-squared distribution with one degree of freedom, which leaves an area of α in the upper tail of the distribution. Some of these values are as follows:

α	0.10	0.05	0.025	0.01	0.001	
$\chi_1^2(1 - \alpha)$	2.71	3.84	5.02	6.63	10.83	(13.2.8)

To implement Eq. (13.2.7), we could draw, on a plot of $L(\lambda)$ versus λ, a horizontal line at the level

$$L(\hat{\lambda}) - \tfrac{1}{2}\chi_1^2(1 - \alpha) \tag{13.2.9}$$

of the vertical scale. This would cut the curve at two values of λ, and these would be the end points of the approximate confidence interval. Translating this via (13.2.6), we see that we must cut across at heights of

$$S(\lambda, \mathbf{V}) = S(\hat{\lambda}, \mathbf{V})e^{\chi_1^2(1-\alpha)/n} \qquad \text{for } S(\lambda, \mathbf{V}) \tag{13.2.10}$$

or

$$\ln S(\lambda, \mathbf{V}) = \ln S(\hat{\lambda}, \mathbf{V}) + \chi_1^2(1 - \alpha)/n \qquad \text{for } \ln S(\lambda, \mathbf{V}) \tag{13.2.11}$$

according to which plot is used. In both cases we cut across the plot somewhat above the minimum level. $S(\hat{\lambda}, \mathbf{V})$ is the minimum sum of squares value that occurs at $\lambda = \hat{\lambda}$.

The Confidence Statement Has Several Forms

Readers who consult various sources on this matter will perhaps be confused when they read, elsewhere, that instead of the factor $\exp\{\chi_1^2(1 - \alpha)/n\}$ recommended on the right of (13.2.10), they are told to use $1 + t_\nu^2/\nu$, or $1 + z^2/\nu$, or $1 + \chi_1^2(1 - \alpha)/\nu$, or $1 + \chi_1^2(1 - \alpha)/n$, or $1 + z^2/n$, where t_ν and z are the two-tailed percentage points

T A B L E 13.2. Mooney Viscosity MS_4 at 100°C as Function of Filler and Oil Levels in SBR-1500[a]

Naphthenic Oil,[b] phr, p	Filler, phr, f					
	0	12	24	36	48	60
0	26	38	50	76	108	157
10	17	26	37	53	83	124
20	13	20	27	37	57	87
30	—	15	22	27	41	63

[a] Phillips Petroleum Co.
[b] Cirolite Process Oil, Sun Oil Co.

of a t-variable with ν df [the df of the residual SS, $S(\lambda, \mathbf{V})$] and a unit normal variable, respectively. [For (13.2.11), take natural logarithms, ln, of the quantities listed.] All of these alternatives are based on the points that (i) $\exp(x) = 1 + x + x^2/(2!) + x^3/(3!) + \cdots$, which, cut to $1 + x$ as a first approximation, accounts for the "1+" portion; (ii) $\chi_1^2 = z^2 \approx t_\nu^2$ unless ν is small; and (iii) one can argue about whether n df or ν df should be used. For most problems, all these intervals will be more or less the same; thus agonizing about which to use is usually a waste of time. If in doubt, use the most conservative one, the one that gives the largest confidence interval. We recommend (13.2.10) or (13.2.11) in general, however.

Example 1. The data in Table 13.2 are part of a more extensive set given by Derringer (1974). This paper has been adapted with permission of John Wiley & Sons, Inc. We wish to find a transformation of the form $(Y^\lambda - 1)/(\lambda \dot{Y}^{\lambda-1})$ for $\lambda \neq 0$, or $\dot{Y} \ln Y$ for $\lambda = 0$, which will provide a good first-order fit to the data and leave satisfactory residuals. Our model form Eq. (13.2.5) is

$$V = \beta_0 + \beta_1 f + \beta_2 p + \epsilon, \qquad (13.2.12)$$

where f is the filler level and p is the plasticizer level (the latter is indicated in the first column of Table 13.2).

Note that the response data range from 157 to 13, a ratio of $157/13 = 12.1$. When the ratio of the largest response value to the smallest is, or exceeds, about an order of magnitude (i.e., about 10), a transformation on Y is likely to be effective. The geometric mean is $\dot{Y} = 41.5461$.

Table 13.3 shows selected values of $S(\lambda, \mathbf{V})$ for various λ. [An initial set of calcula-

T A B L E 13.3. Values of $S(\lambda, V)$ for Selected Values of λ for the Viscosity Data

λ	$S(\lambda, V)$	λ	$S(\lambda, V)$
−1.0	2456	−0.04	83.5
−0.8	1453	−0.02	85.5
−0.6	779.1	0.00	89.3
−0.4	354.7	0.05	106.7
−0.2	131.7	0.10	135.9
−0.15	104.5	0.2	231.1
−0.10	88.3	0.4	588.0
−0.08	84.9	0.6	1222
−0.06	83.3	0.8	2243
−0.05	83.2	1.0	3821

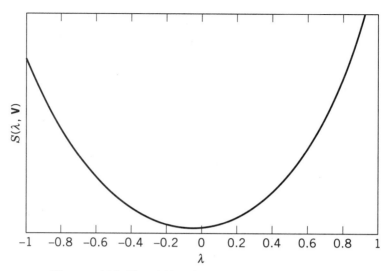

Figure 13.1. Plot of $S(\lambda, \mathbf{V})$ versus λ for the viscosity data.

tions for $\lambda = 2(0.1)2$ was followed by a finer division $\lambda = -0.2(0.01)0.1$ near the bottom of the curve.] A smooth curve through these points is plotted in Figure 13.1. We see that the minimum $(S(\lambda, \mathbf{V})$ occurs at about $\lambda = -0.05$. This is close to zero, suggesting that the transformation

$$V = \dot{Y} \ln Y, \tag{13.2.13}$$

or more simply ln Y, might be a suitable one for this set of data. The approximate 95% confidence interval obtained via Eq. (13.2.10) at a level of $S(\lambda, \mathbf{V}) = 98.3$ is given

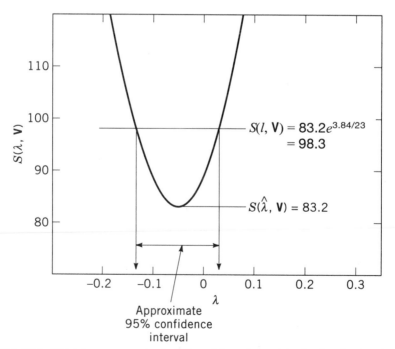

Figure 13.2. Obtaining an approximate 95% confidence interval for λ using the viscosity data.

by $-0.13 \leq \lambda \leq 0.03$. An enlargement of the bottom of the $S(\lambda, \mathbf{V})$ curve is drawn in Figure 13.2 to show this calculation more clearly. We see that the validity of using $\lambda = 0$ is confirmed by this calculation, and that the transformation is well estimated. Values of λ such as $\lambda = 1$ (no transformation at all), $\lambda = \frac{1}{2}$ (the square root transformation), $\lambda = -1$ (the inverse transformation), and many others are completely excluded as possibilities by the data. If the alternative factors for determining approximate 95% confidence intervals are used, the "cuts" across Figure 13.2 would take place at heights of 101.3, or 99.2, or 99.2, or 97.1, or 97.1, respectively. Clearly, whichever is used, there is no practical difference in the conclusion reached.

Note: A wide confidence interval that included two or more of the benchmark levels of $\lambda = -1, -\frac{1}{2}, 0, \frac{1}{2}, 1$ would indicate that λ is not crisply estimated and would imply that it made little difference which of a wide range of possibilities for λ was used. While, in one sense, this seems advantageous, it also may mean that the resulting fitted equation will not predict effectively. If a wide confidence interval includes $\lambda = 1$, the implication is that it may not be worthwhile to transform Y at all.

Application of the natural logarithm transformation to the original data gives us the transformed data of Table 13.4. The best plane, fitted to these transformed data by least squares, is now

$$\widehat{\ln Y} = 3.212 + 0.03088f - 0.03152p. \tag{13.2.14}$$

The corresponding analysis of variance table is shown as Table 13.5. Of the variation about the mean, $100R^2 = 99.51\%$ is explained by the three-parameter model and the F-statistic for overall regression. $F = 2045$ is very significant indeed. Clearly, an excellent fit has been attained.

If we had fitted a first-order model to the *untransformed* data we would have obtained

$$\hat{Y} = 28.184 + 1.55f - 1.717p, \tag{13.2.15}$$

with a $100R^2$ value of 87.93% and an overall $F = 72.9$ (see Table 13.6). This, in itself, is an excellent fit, but the improvement when $\ln Y$ is used in quite dramatic. (In other examples, the initial fit can be quite poor, and the proper transformation enables a significant fit to be achieved; sometimes, the transformation enables a lower degree of polynomial to be fitted than would otherwise be possible. This is true here, too, as we explain below.)

Coding the Predictors. To avoid complicating our example with additional steps, we have used the two predictor variables f and p in the units in which they were given. In a case like the above, where the levels of f and p are equally spaced, the codings

$$x_1 = (f - 30)/6, \qquad x_2 = (p - 15)/5 \tag{13.2.16}$$

would provide coded levels of $x_1 = -5, -3, -1, 1, 3, 5$ and $x_2 = -3, -1, 1, 3$, a slight

T A B L E 13.4. Transformed Values $W = \ln Y$ of the Data in Table 13.2

Naphthenic Oil, phr, p	Filler, phr, f					
	0	12	24	36	48	60
0	3.258	3.638	3.912	4.331	4.682	5.056
10	2.833	3.258	3.611	3.970	4.419	4.820
20	2.565	2.996	3.296	3.611	4.043	4.466
30	—	2.708	3.091	3.296	3.714	4.143

T A B L E 13.5. Analysis of Variance of First-Order Model in f and p Fitted to Logged Viscosity Data

Source	df	SS	MS	F
b_0	1	319.44855	—	
$b_1, b_2 \mid b_0$	2	10.55167	5.27583	2045
Residual	20	0.05171	0.00258	
Total	23	330.05193		

numerical simplification. Note that this sort of simple coding of the predictors has no effect whatsoever on the estimation of λ. However, in some problems, proper coding will simplify the regression calculations. For example, if the $f = 0, p = 30$ observation were not missing in Table 13.2, the coding as shown in Eq. (13.2.16) would make the x_1 and x_2 columns orthogonal to each other, and to the column of 1's in the **X** matrix. (Note, however, that *transformation* of the predictors, say, to $x_1 = f^{\alpha_1}, x_2 = p^{\alpha_2}$ will alter the problem completely and *will* affect the estimation of λ.)

Importance of Checking Residuals

Transformations on the response variable affect the distribution of errors. Our assumption is that, *after* the transformation, the errors in the transformed response will be $N(\mathbf{0}, \mathbf{I}\sigma^2)$. Thus it is important to examine the residuals from the model finally fitted, to see if those assumptions appear to be violated. The residuals from the first-order fit Eq. (13.2.14) are given in Table 13.7. We leave their examination as an exercise for the reader.

13.3. A SECOND METHOD FOR ESTIMATING λ

In the second method of estimation, we choose λ to minimize some quantity that we desire to be small and/or maximize some quantity that we desire to be large. For example, suppose that the original response Y could reasonably be fitted by a second-order model in X_1 and X_2,

$$\beta_0 + \beta_1 X_1 + \beta_2 X_2 + \beta_{11} X_1^2 + \beta_{22} X_2^2 + \beta_{12} X_1 X_2 + \epsilon, \qquad (13.3.1)$$

and that the idea behind transforming from Y to V via Eq. (13.2.2) is to attempt to represent the transformed response by a first-order model $\beta_0 + \beta_1 X_1 + \beta_2 X_2 + \epsilon$. We could fit Eq. (13.3.1) to V by least squares for a selected set of values of λ and choose, as best for our purposes, the value of λ that minimized an appropriate statistic. Possible choices are the F-value connected with the extra sum of squares $SS(b_{11}, b_{22}, b_{12} \mid b_0, b_1, b_2)$ or the ratio of mean squares arising from second- and first-order fitted parame-

T A B L E 13.6. Analysis of Variance of First-Order Model in f and p Fitted to Untransformed Viscosity Data

Source	df	SS	MS	F
$b_1 b_2 \mid b_0$	2	27,842.62	13,921.31	72.9
Residual	20	3,820.60	191.03	
Total, corrected	22	31,663.22		

T A B L E 13.7. Residuals Multiplied by 1000, from First-Order Model Fitted to Logged Viscosity Data

Naphthenic Oil, phr, p	Filler, phr, f					
	0	12	24	36	48	60
0	46	55	−41	7	−13	−9
10	−64	−10	−27	−39	39	70
20	−17	43	−27	−83	−21	31
30	—	71	83	−83	−36	23

ters. For our idea to be successful, second-order terms would have to be nonsignificant for the value of λ finally selected.

Example 2. We again use the viscosity data of Table 13.2. We wish to find a transformation of the form $V = (Y^\lambda - 1)/(\lambda \dot{Y}^{\lambda-1})$ for $\lambda \neq 0$, or $V = \dot{Y} \ln Y$ for $\lambda = 0$, that will allow a good first-order fit without need for second-order terms. We first fit the model

$$V = \beta_0 + \beta_1 f + \beta_2 p + \beta_{11} f^2 + \beta_{22} p^2 + \beta_{12} pf + \epsilon, \qquad (13.3.2)$$

where, as before, f is the filler level and p is the plasticizer level, for a series of chosen values of λ. [Fitting in coded variables such as those given in Eq. (13.2.16) could also be done without affecting the basic results.] For each λ we evaluate

$$MS_1 = \text{mean square arising from SS}(b_1, b_2 \mid b_0)/2,$$

$$MS_2 = \text{mean square arising from SS}(b_{11}, b_{22}, b_{12} \mid b_0, b_1, b_2)/3, \qquad (13.3.3)$$

$$\gamma = MS_2/MS_1,$$

and we plot γ against λ to give Figure 13.3. The numbers needed to create this plot are given in Table 13.8.

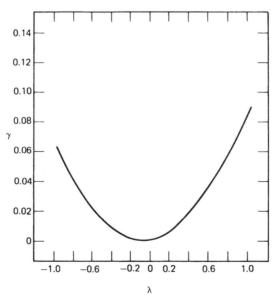

Figure 13.3. Fitting a second-order model to transformed viscosity data: plot of $\gamma = MS_2/MS_1$ versus λ.

T A B L E 13.8. Values of MS$_1$, MS$_2$, and γ = MS$_2$/MS$_1$ for Selected Values of λ for the Viscosity Data

λ	MS$_1$	MS$_2$	γ = MS$_2$/MS$_1$	λ	MS$_1$	MS$_2$	γ = MS$_2$/MS$_1$
-1.0	11162	724	0.0649	0.025	9133	7	0.0008
-0.8	10218	424	0.0415	0.05	9162	11	0.0012
-0.6	9572	218	0.0228	0.1	9233	21	0.0023
-0.4	9183	85	0.0093	0.2	9421	54	0.0057
-0.2	9030	16	0.0018	0.4	10001	173	0.0173
-0.1	9040	3	0.0003	0.6	10891	382	0.0351
-0.05	9066	2	0.0002	0.8	12162	711	0.0584
-0.025	9084	3	0.0003	1.0	13921	1207	0.0867
0.0	9107	4	0.0005				

We see that the minimum γ is attained at about $\lambda = -0.05$, indicating that use of $\lambda = 0$, the logarithmic transformation, is sensible, exactly as we found via the previous method. (A disadvantage[2] of the present procedure, however, is that we cannot easily obtain a confidence interval for λ.) This transformed response leads to the fitted second-order equation

$$\widehat{\ln Y} = 3.231 + 0.02861f - 0.03346p \tag{13.3.4}$$
$$+0.00004416f^2 + 0.00011207p^2 - 0.00003718fp.$$

The corresponding analysis of variance table is shown in Table 13.9. It is clear that the transformation is a successful one, that the full second-order model is not needed, and that the first-order model Eq. (13.2.14) is perfectly adequate. By comparison, the fitted second-order equation for the untransformed data is

$$\hat{Y} = 24.067 + 0.57387f - 0.82628p \tag{13.3.5}$$
$$+0.02639f^2 + 0.02752p^2 - 0.04930fp,$$

with analysis of variance table as in Table 13.10. Thus when no transformation is carried out, significant second-order curvature is present in the data.

Advantages of the Likelihood Method

Of the two methods given for estimating transformation parameters, we would favor the likelihood method in practice for most problems. Through it, we can always obtain

T A B L E 13.9. Analysis of Variance of Second-Order Model in f and p Fitted to Logged Viscosity Data

Source	df	SS	MS	F
$b_1, b_2 \mid b_0$	2	10.55167	5.27583	2037.0
$b_{11}, b_{22}, b_{12} \mid b_0, b_1, b_2$	3	0.00776	0.00259	1.0
Residual	17	0.04395	0.00259	
Total, corrected	22	10.60338		

[2] This disadvantage exists in this example. However, if the method is used on an F-statistic, an approximate confidence interval can be calculated using $F(\nu_1, \nu_2, 1 - \alpha)$ as described in "Transformations: some examples revisited," by N. R. Draper and W. G. Hunter, *Technometrics*, **11**, 1969, 23–40.

T A B L E 13.10. Analysis of Variance of Second-Order Model in *f* and *p* Fitted to Original Viscosity Data

Source	df	SS	MS	F
$b_1, b_2 \mid b_0$	2	27,842.616	13,921.308	1179.7
$b_{11}, b_{22}, b_{12} \mid b_0, b_1, b_2$	3	3,619.987	1,206.662	102.3
Residual	17	200.615	11.801	
Total, corrected	22	31,663.217[a]		

[a] Rounding discrepancy of 0.001.

an approximate confidence interval or region, and we have only to fit the model we are interested in, not a more complicated one, as is usually required in the second method. (In some problems, in fact, the data may be inadequate to fit the desired higher-order alternative.) The second method can be useful, however, when it is desired to examine a variety of criteria. The various plots of the criteria versus λ can be viewed simultaneously, and a compromise value of λ can be selected from these plots.

13.4. RESPONSE TRANSFORMATIONS: OTHER INTERESTING AND SOMETIMES USEFUL PLOTS

The procedures described in Sections 13.2 and 13.3 for estimating a transformation are the basic ones. A number of other plots (not illustrated) provide useful ancillary information when needed.

1. If there are genuine repeat runs in the data set, a pure error sum of squares can be calculated for each λ used and a plot of the lack of fit F-value versus λ can be constructed. A horizontal line can be drawn at the height of some chosen significance level taken from the F-tables, and acceptable values of λ are those for which the curve falls below the significance line. Typically (but not always) the acceptable range of λ's includes $\hat{\lambda}$ and most or all of the confidence interval for λ. When this does not happen, deeper investigation is called for. For examples where this technique can be used, see Exercises 13D and 13E in "Exercises for Chapter 13."

2. The regression coefficients can be plotted versus λ.

3. The t-values for each of the fitted coefficients can be plotted versus λ. Alternatively, t-values regarded as significant can be incorporated into the regression coefficient plot (2) by using, for example, solid lines for those values of the coefficients regarded as significantly different from zero (or from whatever other test value is decided upon), and dashed lines otherwise. (Of course, for those λ-values for which the errors are non-normal, a comparison with the t-distribution is not, strictly speaking, valid. Also, for λ-values at which the variance is nonhomogeneous, the t-values can be misleading.)

4. R^2 may be plotted versus λ to see, for which λ-values, acceptable expanatory levels are to be found. Note that this plot will not be simply a reflection of the $S(\lambda, \mathbf{V})$ plot. It is true that

$$S(\lambda, \mathbf{V}) = (\mathbf{V}'\mathbf{V} - n\overline{V}^2)(1 - R^2),$$
(13.4.1)

so that

$$R^2 = 1 - S(\lambda, \mathbf{V})/(\mathbf{V}'\mathbf{V} - n\bar{V}^2), \tag{13.4.2}$$

but the divisor on the right-hand side is not constant as λ varies. Two such plots could be drawn for the first- and second-order models of Example 2. It is where these two R^2 curves differ the *least* that interesting values of λ lie. (In fact, it would be possible to construct an interval of λ-values for which the difference between the R^2 values does not exceed some desired amount.)

13.5. OTHER TYPES OF RESPONSE TRANSFORMATIONS

A Two-Parameter Family of Response Transformations

By adding an additional parameter, we can extend the range of possible transformations beyond those discussed above. Consider the two-parameter family

$$V = \begin{cases} \dfrac{(Y + \lambda_2)^{\lambda_1} - 1}{\lambda_1 (\dot{Y} + \lambda_2)^{\lambda_1 - 1}}, & \text{for } \lambda_1 \neq 0, \\[2mm] (\dot{Y} + \lambda_2) \ln (Y + \lambda_2), & \text{for } \lambda_1 = 0, \end{cases} \tag{13.5.1}$$

where

$$\dot{Y} + \lambda_2 = \{(Y_1 + \lambda_2)(Y_2 + \lambda_2) \cdots (Y_n + \lambda_2)\}^{1/n} \tag{13.5.2}$$

is the geometric mean of the $(Y_i + \lambda_2)$ terms, and where, necessarily, $Y > -\lambda_2$. The methods used for the one-parameter family (in which $\lambda_2 = 0$) can be extended for this case. We now have $Y + \lambda_2$, where Y was previously, and must search over a two-dimensional (λ_1, λ_2) grid to find the maximum likelihood values $(\hat{\lambda}_1, \hat{\lambda}_2)$. Also, in the calculation to obtain a confidence region for (λ_1, λ_2), we now use χ^2 with *two* degrees of freedom, rather than one, because there are now two transformation parameters. Exactly the same ideas are involved, but the calculations are more complicated. An example is discussed briefly on pp. 225–226 of the 1964 Box and Cox paper. The χ_2^2 $(1 - \alpha)$ percentage points needed for 2 df are these:

α	0.10	0.05	0.025	0.01	0.001
$\chi_2^2(1 - \alpha)$	4.61	5.99	7.38	9.21	13.82

Note that, because $Y > -\lambda_2$, we must avoid searching λ_2 values such that $\lambda_2 \leq -Y_{\min}$. Although a singularity occurs at this λ_2 boundary, it is not relevant; the local maximum satisfying the constraint is what is needed.

A Modulus Family of Response Transformations

When a residuals plot indicates a fairly symmetric but non-normal error distribution, the one-parameter *modulus* power family

$$W = \begin{cases} (\text{sign of } Y)[\{|Y| + 1\}^\lambda - 1]/\lambda, & \lambda \neq 0 \\[2mm] (\text{sign of } Y) \ln \{|Y| + 1\}, & \lambda = 0 \end{cases} \tag{13.5.3}$$

may be useful. For an illustration, see John and Draper (1980).

Transforming Both Sides of the Model

An extremely versatile method of transformation is to transform the model function as well as the response variable, while leaving the error additive. For example, a model

$$Y = f(\mathbf{X}, \boldsymbol{\beta}) + \epsilon \tag{13.5.4}$$

could be transformed and fitted as

$$Y^{\lambda} = \{f(\mathbf{X}, \boldsymbol{\beta})\}^{\lambda} + \epsilon, \tag{13.5.5}$$

where the best λ-value needs to be selected. Alternatively, Y^{λ} and $\{f(\mathbf{X}, \boldsymbol{\beta})\}^{\lambda}$ could be replaced by their respective W or V forms discussed in Section 13.2 This method requires a great deal of computation but promises to be more rewarding in its effects in situations where the model (13.5.4) is reasonable one to fit, but the residuals show signs that they may be non-normal or heteroscedastic. For more on this see, for example, Chapter 4 of Carroll and Ruppert (1988).

A Power Family for Proportions

Exactly the same ideas as in Section 13.2 can be used to estimate the parameter λ in the family

$$P = \{p^{\lambda} - (1 - p)^{\lambda}\}/\lambda, \tag{13.5.6}$$

where p is the observed proportion of times a stated event takes place. The observed values of p would, in general, depend on the values of a number of predictor variables $X_1, X_2, \ldots$, and a model of general form

$$P = f(X_1, X_2, \ldots, \lambda, \boldsymbol{\beta}) + \epsilon \tag{13.5.7}$$

would be postulated where $\boldsymbol{\beta}$ is a vector of parameters. The value of λ would be chosen to give the best fit to these data under the assumption that $\epsilon \sim N(0, \mathbf{I}\sigma^2)$.

The power transformation (13.5.6) was suggested by Tukey. For work on the statistical distribution of P when p has a uniform distribution, good initial reading is Joiner and Rosenblatt (1971); see, also, the references mentioned in the paper.

Two specific transformations for proportions are given in the next section. One is a special case of the above family, and the second is an approximation to a special case.

13.6. RESPONSE TRANSFORMATIONS CHOSEN TO STABILIZE VARIANCE

In situations where the transformed data are to be analyzed by least squares, it is important that the variance of the response to be fitted be independent of its mean value. Where it is known or where it has been found empirically that the standard deviation of the untransformed resonse Y, say, σ_Y, is a particular function $f(\eta)$ of the mean value, $\eta = E(Y)$, we can obtain an appropriate transformation immediately by using the transformed variable $h(y)$, where

$$\frac{\partial h(Y)}{\partial Y} \propto \frac{1}{f(Y)}. \tag{13.6.1}$$

In other words, we obtain $h(Y)$ by integrating $1/f(Y)$ with respect to Y. Some well-

T A B L E 13.11. Appropriate Variance Stabilizing Transformation when $\sigma_Y = f(\eta)$

Nature of Dependence $\sigma_y = f(\eta)$		Variance Stabilizing Transformation[a]
$\sigma_Y \propto \eta^k$ and in particular	$(Y \geq 0)$	Y^{1-k}
$\sigma_Y \propto \eta^{1/2}$ (Poisson)	$(Y \geq 0)$	$Y^{1/2}$
$\sigma_Y \propto \eta$	$(Y \geq 0)$	$\ln Y$
$\sigma_Y \propto \eta^2$	$(Y \geq 0)$	Y^{-1}
$\sigma_Y \propto \eta^{1/2}(1 - \eta)^{1/2}$ (binomial)	$(0 \leq Y \leq 1)$	$\sin^{-1}(Y^{1/2})$
$\sigma_Y \propto (1 - \eta)^{1/2}/\eta$	$(0 \leq Y \leq 1)$	$(1 - Y)^{1/2} - (1 - Y)^{3/2}/3$
$\sigma_Y \propto (1 - \eta^2)^{-2}$	$(-1 \leq Y \leq 1)$	$\ln\{(1 + Y)/(1 - Y)\}$

[a] Modifications for the Poisson and binomial cases have been suggested by Freeman and Tukey (1950). These modifications are discussed in the text.

known transformations that arise in this way are shown in Table 13.11. Note that some of these are members of the power family.

Estimation of k in Table 13.11

If the data contain m sets of replicate runs, it is possible to get an estimate of k fairly quickly by finding the slope of the "best line" through the points (abscissa, ordinate) = $(\ln \overline{Y}_j, \ln s_j)$, $j = 1, \ldots, m$, where $\overline{Y}_j$ and s_j^2 are the sample mean and variance of the jth set of repeat runs. This method is based on the idea that, if σ_Y is proportional to the kth power of η, so that $\sigma_Y \propto \eta^k$, it follows that

$$\ln \sigma_Y = k_0 + k \ln \eta. \tag{13.6.2}$$

The pairs $(\ln s_j, \ln \overline{Y}_j)$ provide us with some data on $(\ln \sigma_Y, \ln \eta)$ from which we can estimate k. This slope estimation is often done simply by eye. A least squares fit calculation would also provide us with "se($\hat{k}$)," which would enable us to judge, at least roughly, how precise our estimation of k appears to be.

When there are no replicate runs, the residuals versus $\hat{Y}$ plot should be examined. At a particular $\hat{Y}_j$ value the width of the band of residuals could be taken as roughly $4s_j$. A few values of $(\ln s_j, \ln \hat{Y}_j)$ can then be plotted and the slope k estimated as in the foregoing paragraph. After transformation, a new residuals versus $\hat{Y}$ plot can be done to see if the transformation was effective or needs readjustment.

Transformations for Responses That Are Proportions

Many types of response data occur as proportions, $0 \leq Y_i \leq 1$, obtained as the number of times a "success" (however that may be defined) happens in a larger number of "trials." For example, six rats of ten complete a given task for a $Y_i = 0.60$. Proportion-type data typically do not have a uniform variance pattern because $V(Y_i) = \pi_i(1 - \pi_i)/m_i$, where $E(Y_i) = \pi_i$ and m_i is the number of trials. Two popular transformations for such data are the following.

1. *The log odds transformation.* (This does *not* stabilize variance.) We set

$$W_i = \ln\{Y_i/(1 - Y_i)\}. \tag{13.6.3}$$

Thus W_i is the natural logarithm of the "odds ratio" $Y_i/(1 - Y_i)$, the ratio of the proportion of successes to the proportion of failures. To fit the model

$$W_i = \beta_0 + \beta_1 X_{1i} + \cdots + \beta_{p-1} X_{p-1,i} + \epsilon_i \tag{13.6.4}$$

to data $(W_i, X_{1i}, \ldots, X_{p-1,i})$, $i = 1, 2, \ldots, n$, we use weighted least squares because

$$V(W_i) = 1/\{\pi_i(1 - \pi_i)m_i\}, \tag{13.6.5}$$

approximately and is not constant. To show this, we use the fact that $\ln (1 + x) = x - x^2/2 + x^3/3 - \cdots \simeq x$ for small x. Then, dropping the subscript i for the moment, we see that (all results are approximate)

$$\ln Y = \ln \pi + (Y - \pi)/\pi \tag{13.6.6}$$

so that

$$E(\ln Y) = \ln \pi \quad \text{and} \quad V(\ln Y) = (1 - \pi)/(\pi m). \tag{13.6.7}$$

Similarly,

$$E\{\ln(1 - Y)\} = \ln(1 - \pi) \quad \text{and} \quad V\{\ln(1 - Y)\} = \pi/\{(1 - \pi)m\} \tag{13.6.8}$$

and

$$\text{cov}\{\ln Y, \ln(1 - Y)\} = -1/m. \tag{13.6.9}$$

It follows that $V(W) = \{(1 - \pi)/\pi + \pi/(1 - \pi) - 2(-1)\}/m = 1/\{\pi(1 - \pi)m\}$. These n variances $V(W_i) = 1/\{\pi_i(1 - \pi_i)m_i\}$ are, of course, unknown but are estimated by the corresponding values $s_i^2 = 1/\{Y_i(1 - Y_i)m_i\}$ for $i = 1, 2, \ldots, n$, and the estimation proceeds as in Section 9.2. In this case, the matrix $\mathbf{V}$ is diagonal with estimated entries $s_1^2, s_2^2, \ldots, s_n^2$. An alternative analysis for the log odds ratio is to use generalized linear models; see Chapter 18.

2. *The arc-sine transformation.* (Use radians or degrees.) As indicated in Table 13.11, the transformation $U = \sin^{-1}Y^{1/2}$ will stablize the variance over a range of Y-values, if the samples that determine the observations Y_i are all of the same size m, say. In fact, $W = 2 \sin^{-1}Y^{1/2}$ is slightly preferred because it has the uniform theoretical variance $1/m$ (for radians; multiply by $\{360/(2\pi)\}^2$ for degrees). Note that, if m is not constant throughout the data, $Z_i = 2m_i^{1/2} \sin^{-1}Y_i^{1/2}$ is needed, where Y_i is determined through m_i trials. Data in the middle range of proportions (say, 0.30–0.70) will not be much affected by these transformations, because of the approximate linearity of the transformation in that range. The range of Y-values over which the arc-sine transformation produces a flat variance function depends on the sample size m. For smaller samples, the flat range does not cover true values of the binomial parameter $E(Y)$ that are close to zero or one. Freeman and Tukey (1950) have suggested the improved transformation

$$U^* = \tfrac{1}{2}[\sin^{-1}\{m_iY_i/(m_i + 1)\}^{1/2} + \sin^{-1}\{(m_iY_i + 1)/(m_i + 1)\}^{1/2}]$$

(instead of U), which extends the range of flatness of the variance function. See Bisgaard and Fuller (1994–95, Figure 2) for examples for $m_i = 20, 50$.

In general, we must keep in mind that there is no guarantee that use of these transformations will necessarily be better than analyzing the proportions directly; much depends on the data. The effectiveness of a transformation is best assessed by trying it on the data and then checking the fit of the model and the pattern of residuals that results.

Basic information on the reasons for some transformations that can be made on the response variable can be found in Bartlett (1947).

Transformations for Responses that Are Poisson Counts

Table 13.11 suggests the square root transformation for Poisson data. Again, a flatter variance profile is achieved by the Freeman and Tukey (1950) suggestion:

$$\tfrac{1}{2}\{Y^{1/2} + (Y + 1)^{1/2}\},$$

which improves the situation for small values of the Poisson parameter $\lambda = E(Y) = V(Y)$. See Bisgaard and Fuller (1994–95, Figure 3).

References

Atkinson (1985); Bartlett (1947); Bisgaard and Fuller (1994–1995); Box and Cox (1964); Carroll and Ruppert (1988); Freeman and Tukey (1950).

EXERCISES FOR CHAPTER 13

A. Consider the following representative data:

Year, X	Speed mph, Y	Means of Attaining Speed
1830	30	Railroad
1905	130	Railroad
1930	400	Airplane
1947	760	Airplane
1952	1,500	Airplane
1969	25,000	Spaceship

1. Plot these (X, Y) data points. Do you feel this plot is an informative one or not? Why?

2. Transform the data by $Z = \log Y$, and plot the (X, Z) points. Is this plot preferable to the previous plot or not? Why?

3. Can you find a reasonably simple transformation $U = f(Y)$ that will produce a (more or less) straight line plot for the points (X, U)?

4. Whatever you conclude in (3), plot the points (X, V), where $V = \log(\log Y)$. Fit a straight line $V = \beta_0 + \beta_1 X + \epsilon$ to these points using least squares. Draw the fitted line on your (X, V) plot. Find the residuals and comment on them.

5. Set down the appropriate analysis of variance table for (4), test for overall regression, and find R^2. Comment appropriately.

6. Use the fitted straight line from (4) to predict when humans will attain the speed of light (186,000 miles per second: note, per *second*).

7. Discuss the reasonability or otherwise of your prediction. On what assumptions does it depend? Whether you feel your prediction is realistic or unrealistic, set out your reasons carefully but succinctly.

B. (*Source:* "An empirical model for viscosity of filled and plasticized elastomer compounds," by G. C. Derringer, *Journal of Applied Polymer Science,* **18,** 1974, 1083–1101.) Two sets of response data (see footnotes c and d) are shown in Table B. For each of the sets, find the best transformation of the form $V = (Y^\lambda - 1)/(\lambda \dot{Y}^{\lambda-1})$ for $\lambda \neq 0$, and $V = \dot{Y} \ln Y$ for $\lambda = 0$, that will allow a useful model of the form $V = \beta_0 + \beta_1 f + \beta_2 p + \epsilon$ to be developed via least squares, where f is the filler level and p is the naphthenic oil level. Perform all the usual regression analyses for your best $\hat{\lambda}$, including examination of residuals. (Note that the coding discussed in Section 13.2 will be especially useful for the second set of data.)

T A B L E B. Mooney Viscosity MS_4 at 100°C as Function of Filter and Oil Levels in SBR-1500[a]

Naphthenic Oil,[b] phr, p	Filler, phr, f					
	0	12	24	36	48	60
0	26[c]	28	30	32	34	37
	25[d]	30	35	40	50	60
10	18	19	20	21	24	24
	18	21	24	28	33	41
20	12	14	14	16	17	17
	13	15	17	20	24	29
30	—	12	12	13	14	14
	11	14	15	17	18	25

[a] SBR 1500, Phillips Petroleum Co.
[b] Circolite Process Oil, Sun Oil Co.
[c] N990, Cabot Corp.
[d] Silica B, Hi-Sil EP, PPG Industries.

C. (*Source*: "An empirical model for viscosity of filled and plasticized elastomer compounds," by G. C. Derringer, *Journal of Applied Polymer Science*, **18**, 1974, 1083–1101.)

T A B L E C. Mooney Viscosities $Y_1(ML_4)$ and $Y_2(MS_4)$ at 100°C for Various Combinations of Coded Variables[a] x_1, x_2, x_3, and x_4

x_1	x_2	x_3	x_4	Y_1	Y_2
−3	−3	−3	−3	51	29
−1	−1	−1	−1	61	34
−1	−1	−1	−1	64	35
−3	−1	−1	1	36	20
−3	1	1	−1	39	21
−1	−3	−1	−1	55	30
1	−3	1	1	50	27
−1	1	−3	1	88	49
1	−1	−3	−1	124	68
−1	−1	1	−3	54	30
1	1	−1	−3	133	74

[a] $x_i = (X_i - 22\frac{1}{2})/7\frac{1}{2}$, $i = 1, 2$, $x_j = (X_j - 15)/5$, $j = 3, 4$, where X_1 = level of Silica A, Hi-Sil 233, PPG Industries, X_2 = level of N330, Cabot Corp., X_3 = Naphthenic oil, Circolite Process Oil, Sun Oil Co., and X_4 = Cumarone indene resin, Camar MH 2½, Allied Chemical Corp.

For each response individually choose the best value of λ in the transformation $V = (Y^\lambda - 1)/(\lambda \dot{Y}^{\lambda-1})$ for $\lambda \neq 0$, $V = \dot{Y} \ln Y$ for $\lambda = 0$, to enable the model $V = \beta_0 + \Sigma \, \beta_i x_i + \epsilon$ to be well fitted to the data by least squares. After choosing the best transformation parameter estimate $\hat{\lambda}$ in each case, carry out all the usual regression analyses including examination of residuals.

D. (*Source*: "Third order rotatable designs in three factors: analysis," by N. R. Draper, *Technometrics*, **4**, 1962, 219–234.) The data in the accompanying table are constructed data, generated from a third-order polynomial model to illustrate use of a third-order sequential design. Find the best value of λ in the transformation $V = (Y^\lambda - 1)/(\lambda \dot{Y}^{\lambda-1})$ for $\lambda \neq 0$, and $V = \dot{Y} \ln Y$ for $\lambda = 0$, to enable a full second-order ten-parameter model in x_1, x_2, x_3 to be fitted. After choosing the best transformation parameter estimate λ, carry out all the usual regression analyses and state your conclusions.

x_1	x_2	x_3	y
-1	-1	-1	34.727
1	-1	-1	38.917
-1	1	-1	44.907
1	1	-1	24.641
-1	-1	1	24.658
1	-1	1	45.636
-1	1	1	33.702
1	1	1	5.374
$2^{1/2}$	0	0	33.414, 34.453
$-2^{1/2}$	0	0	38.540, 39.201
0	$2^{1/2}$	0	40.393, 38.335
0	$-2^{1/2}$	0	40.687, 40.092
0	0	$2^{1/2}$	23.869, 25.823
0	0	$-2^{1/2}$	33.727, 33.068
0	0	0	43.832, 44.562
0	0	0	42.165, 41.187

E. (*Source*: Ice crystal growth data from B. F. Ryan, E. R. Wishart, and D. E. Shaw, Commonwealth Scientific and Industrial Research Organisation (C.S.I.R.O.), Australia. Related journal reference: "The growth rates and densities of ice crystals between $-3°C$ and $-21°C$," *Journal of the Atmospheric Sciences*, **33**, 1976, 842–850.) Ice crystals are introduced into a chamber, the interior of which is maintained at a fixed temperature ($-5°C$) and a fixed level of saturation of air with water. The growth of the crystals with time is observed. The 43 sets of measurement presented here are of the mass of the crystals (M) in nanograms for times (T) of 50–180 seconds from the introduction of the crystals. Each measurement represents a single complete experiment; the experiments were conducted over a number of days and were randomized as to observation time. (The actual order in which they were conducted is not available.) It was desired to connect the response M to the predictor T by a simple fitted relationship. [The possibility that $E(M) = \alpha T^\beta$ was suggested.] Perform the following calculations.

1. Fit the model $V = \gamma + \beta \ln T$, where $V = (M^\lambda - 1)/(\lambda \dot{M}^{\lambda-1})$ for $\lambda \neq 0$, and $V = \dot{M} \ln M$ for $\lambda = 0$, for a suitable set of values of λ, and so pick the best transformation for M, using the methods described in Section 13.2.

2. Use the selected values of λ to carry out all the details of the usual least squares analysis of the data and state your conclusions. In particular, do the residuals bear any suggestion that the variance structure of the errors might not be stable, as assumed?

T	M	T	M
50	11.5	125	47.7
60	8.2, 11.5	130	92.0, 87.2
70	14.1, 17.2	135	58.0, 47.7
80	33.5, 28.8	140	73.2, 58.0
90	15.6, 24.4, 33.5	145	47.7
95	38.8	150	118.9, 58.0
100	47.7, 58.0, 36.1	155	143.9, 87.2
105	47.7, 65.5	160	143.9, 73.2, 73.7
110	58.0, 47.7, 33.5	165	97.0
115	69.5, 69.5, 47.7	170	112.3
120	87.2, 51.0, 33.5	180	113.2

F. It is believed that a response relationship $\eta = \alpha X_1^\beta X_2^\gamma$ is responsible for producing the data below; how experimental errors enter into the situation is not known. Fit the model $\log Y = \log \alpha + \beta \log X_1 + \gamma \log X_2 + \epsilon$ to the data by least squares, examine the resulting fit by whatever methods are available to you, and provide conclusions that will cast some light on the situation. (Use logarithms to the base 10.)

X_1	X_2	Y
10	10	2,040
100	10	7,350
1,000	10	12,210
10,000	10	23,580
10	100	18,200
100	100	10
1,000	100	2,960
10,000	100	108,040
10	1,000	10,370
100	1,000	1,150
1,000	1,000	23,580
10,000	1,000	296,120
10	10,000	9,040
100	10,000	1,960
1,000	10,000	96,980
10,000	10,000	1,004,020

G. The following data were collected on spray congealing:

Values for the Experimental Operating Variables and Average Particle Sizes

Run	Feed Rate per Unit Whetted Wheel Periphery (gm/sec/cm) (X_1)	Peripheral Wheel Velocity (cm/sec) (X_2)	Feed Viscosity (poise) (X_3)	Mean Surface–Volume Particle Size of Product (μ) (Y)
1	0.0174	5300	0.108	25.4
2	0.0630	5400	0.107	31.6
3	0.0622	8300	0.107	25.7
4	0.0118	10800	0.106	17.4
5	0.1040	4600	0.102	38.2
6	0.0118	11300	0.105	18.2
7	0.0122	5800	0.105	26.5
8	0.0122	8000	0.100	19.3
9	0.0408	10000	0.106	22.3
10	0.0408	6600	0.105	26.4
11	0.0630	8700	0.104	25.8
12	0.0408	4400	0.104	32.2
13	0.0415	7600	0.106	25.1
14	0.1010	4800	0.106	39.7
15	0.0170	3100	0.106	35.6
16	0.0412	9300	0.105	23.5
17	0.0170	7700	0.098	22.1
18	0.0170	5300	0.099	26.5
19	0.1010	5700	0.098	39.7
20	0.0622	6200	0.102	31.5
21	0.0622	7700	0.102	26.9
22	0.0170	10200	0.100	18.1
23	0.0118	4800	0.102	28.4
24	0.0408	6600	0.102	27.3
25	0.0622	8300	0.102	25.8
26	0.0170	7700	0.102	23.1
27	0.0408	9000	0.613	23.4
28	0.0170	10100	0.619	18.1
29	0.0408	5300	0.671	30.9
30	0.0622	8000	0.624	25.7
31	0.1010	7300	0.613	29.0
32	0.0118	6400	0.328	22.0

(*continued*)

Run	Feed Rate per Unit Whetted Wheel Periphery (gm/sec/cm) (X_1)	Peripheral Wheel Velocity (cm/sec) (X_2)	Feed Viscosity (poise) (X_3)	Mean Surface–Volume Particle Size of Product (μ) (Y)
33	0.0170	8000	0.341	18.8
34	0.0118	9700	1.845	17.9
35	0.0408	6300	1.940	28.4

Source: "Spray congealing: particle size relationships using a centrifugal wheel atomizer," by M. W. Scott, M. J. Robinson, J. F. Pauls, and R. J. Lantz, *Journal of Pharmaceutical Sciences,* **53**(6), June 1964, 670–675. Reproduced with permission of the copyright owner.

A proposed model, based on theoretical considerations, is

$$Y = \alpha X_1^\beta X_2^\gamma X_3^\delta \epsilon.$$

Requirements. After transformation, fit the proposed model by least squares. State which predictor variable appears most important and check all coefficients for statistical significance (Take $\alpha = 0.05$). Is the model a satisfactory one?

H. Apply the $\sin^{-1}(Y^{1/2})$ transformation to the response observations below, and fit a planar model in all four X's. Compare the root mean square of your analysis, which estimates the standard deviation of the response used, to the theoretical value $(4m)^{-1/2} = 0.05$.

X_1	X_2	X_3	X_4	Fraction Defective
−1	−1	−1	−1	0.16
−1	1	1	1	0.17
−1	−1	1	1	0.12
−1	1	−1	−1	0.06
1	1	−1	1	0.06
1	−1	1	−1	0.68
1	1	1	−1	0.42
1	−1	−1	1	0.26

CHAPTER 14

"Dummy" Variables

What Are "Dummy" Variables?

The variables considered in regression equations usually can take values over some continuous range. Occasionally we must introduce a factor that has two or more distinct levels. For example, data may arise from three machines, or two factories, or six operators. In such a case we cannot set up a continuous scale for the variable "machine" or "factory" or "operator." We must assign to these variables some levels in order to take account of the fact that the various machines or factories or operators may have separate deterministic effects on the response. Variables of this sort are usually called *dummy variables*. They are usually (but not always) unrelated to any physical levels that might exist in the factors themselves.

One example of a dummy variable is found in the attachment of a variable X_0 (whose value is always unity) to the term β_0 in a regression model. The X_0 is unnecessary but provides a notational convenience at times. Other dummy variables are somewhat more than a mere convenience, as we shall see.

An Infinite Number of Choices

The suggestions we make for setting up dummy variable systems are not unique. Typically, there are an infinite number of alternative ways to set up a system to cover any particular type of situation. Given a particular selection of dummy variable vectors that "works," that is, represents the factors as desired, we can derive other sets by taking linear combinations of the vectors in the first set. As long as the second set is chosen so that their vectors are not linearly dependent on one another, all will be well. In general, the most useful dummy variable setups are simple in form, employing levels of 0 and 1, for example, or -1 and 1. Usefulness, however, lies in the eye of the user (as "beauty lies in the eye of the beholder").

14.1. DUMMY VARIABLES TO SEPARATE BLOCKS OF DATA WITH DIFFERENT INTERCEPTS, SAME MODEL

Suppose we wish to introduce into a model the idea that there are two types of machines (types A and B, say) that produce different levels of response, in addition to the variation that occurs due to other predictor variables. One way of doing this is to add to the model a dummy variable Z and regression coefficient α (say) so that an additional term αZ appears in the model. The coefficient α must be estimated at the same time the β's are estimated. Values can be assigned to Z as follows:

$$Z = 0 \text{ if the observation is from machine } A,$$

$$Z = 1 \text{ if the observation is from machine } B.$$

If a is the least squares estimate of α, and if $\hat{f}$ represents the rest of the fitted model, we have

$$\hat{Y} = \hat{f} + aZ \tag{14.1.1}$$

Thus machine A data are estimated by putting $Z = 0$ to get $\hat{Y} = \hat{f}$, while machine B data are predicted by setting $Z = 1$ to give $\hat{Y} = \hat{f} + a$. The value "a" simply estimates the difference in levels between the responses of group B compared to group A and all other factors fitted are represented in $\hat{f}$.

Other Possibilities

Any two distinct values of Z would actually be suitable, though the above is usually best. However, other assignments are sometimes convenient; for example, suppose that of a total of n observations, n_1 come from type A machines and $n_2 = n - n_1$ from type B machines. If we choose levels

$$Z = \frac{-n_2}{\sqrt{n_1 n_2 (n_1 + n_2)}} \quad \text{for machine } A,$$

$$Z = \frac{n_1}{\sqrt{n_1 n_2 (n_1 + n_2)}} \quad \text{for machine } B, \tag{14.1.2}$$

it will be found that the corresponding column of the $\mathbf{X}$ matrix is orthogonal to the "β_0 column" and is "normalized," that is, has sum of squares unity, which may be convenient. (We can also omit the denominators if the normalization of the column is of no consequence.) When the column is normalized, the $\hat{Y}(\text{group } B) - \hat{Y}(\text{group } A)$ difference is $a(n_1 + n_2)^{1/2}/(n_1 n_2)^{1/2}$. Or, if we choose $Z = -1$ for machine A, $Z = 1$ for machine B, the difference $\hat{Y}(\text{group } B) - \hat{Y}(\text{group } A)$ is $2a$. Obviously all of the choices above achieve the same purpose, to provide a difference in levels between the two groups. Thus it is sensible to use a representation that is convenient to the user.

To see how one representation is derived by linear combination from another we *must* count in the dummy $\mathbf{X}_0$ column of the $\mathbf{X}$ matrix. The first representation is covered by the vectors

$$(\mathbf{X}_0, \mathbf{Z}) = \begin{bmatrix} 1 & 0 \\ 1 & 0 \\ \cdots & \\ 1 & 0 \\ 1 & 1 \\ 1 & 1 \\ \cdots & \cdots \\ 1 & 1 \end{bmatrix} \begin{matrix} \\ \\ \\ n_1 \\ \\ \\ \\ n_2 \end{matrix}. \tag{14.1.3}$$

The second representation (14.1.2) has columns $(\mathbf{X}_0, \mathbf{U})$, where

$$\mathbf{U} = \{-n_2/[n_1 n_2(n_1 + n_2)]^{1/2}\}(\mathbf{X}_0 - \mathbf{Z}) + \{n_1/[n_1 n_2(n_1 + n_2)]^{1/2}\}\mathbf{Z}$$
$$= (n_1 + n_2)^{1/2}/(n_1 n_2)^{1/2}\mathbf{Z} - \{n_2/[n_1 n_2(n_1 + n_2)^{1/2}]\}\mathbf{X}_0 \tag{14.1.4}$$

and the third has columns $(\mathbf{X}_0, \mathbf{W})$, where

$$\mathbf{W} = (\mathbf{Z} - \mathbf{X}_0) + \mathbf{Z} = 2\mathbf{Z} - \mathbf{X}_0. \tag{14.1.5}$$

How Many Dummies?

In the above example of two categories (machines A and B) we see we need to construct one dummy column *in addition to* $\mathbf{X}_0$. So two groups require two dummies *including* $\mathbf{X}_0$. This essentially enables us to have the same linear model with two different intercepts. For the first illustration above, the estimated intercepts are, conveniently, b_0 and $(b_0 + a)$. For the third they are $(b_0 - a)$ and $(b_0 + a)$ with a different value of a. (We leave the second to the reader.)

Three Categories, Three Dummies

If we wish to take account of three different categories, two extra dummies (besides $\mathbf{X}_0$) would be needed. The simplest way is to use

$$(Z_1, Z_2) = (1, 0) \quad \text{for machine } A,$$
$$= (0, 1) \quad \text{for machine } B, \tag{14.1.6}$$
$$= (0, 0) \quad \text{for machine } C,$$

and the model would include extra terms $\alpha_1 Z_1 + \alpha_2 Z_2$, with coefficients α_1, α_2 to be estimated. Thus the $\mathbf{X}$ matrix for such data would take the form below, assuming that all the A-data were listed first, then all the B-data, then all the C-data.

$$
\mathbf{X} =
\begin{array}{cccc}
\mathbf{X}_0 & \text{Other } X\text{'s} & \mathbf{Z}_1 & \mathbf{Z}_2 \\
\end{array}
\left[
\begin{array}{cccc}
1 & \cdots & 1 & 0 \\
1 & & 1 & 0 \\
\cdots & & \cdots & \\
1 & & 1 & 0 \\
\hline
1 & \cdots & 0 & 1 \\
1 & & 0 & 1 \\
\cdots & & \cdots & \\
1 & & 0 & 1 \\
\hline
1 & \cdots & 0 & 0 \\
1 & & 0 & 0 \\
\cdots & & \cdots & \\
1 & & 0 & 0 \\
\end{array}
\right]
\begin{array}{l}
\left.\vphantom{\begin{array}{c}1\\1\\1\end{array}}\right\}\text{Group } A \\[1em]
\left.\vphantom{\begin{array}{c}1\\1\\1\end{array}}\right\}\text{Group } B. \\[1em]
\left.\vphantom{\begin{array}{c}1\\1\\1\end{array}}\right\}\text{Group } C
\end{array}
\tag{14.1.7}
$$

Again, many different allocations of levels are possible. If desired, columns that are orthogonal to the $\mathbf{X}_0$ column and that have sum of squares unity can be achieved by setting

$$(Z_1, Z_2) = \left(\frac{-n_3}{\sqrt{n_1 n_3 (n_1 + n_3)}}, \ 0 \right) \qquad \text{for machine } A,$$

$$= \left(0, \ \frac{-n_3}{\sqrt{n_2 n_3 (n_2 + n_3)}} \right) \qquad \text{for machine } B, \qquad (14.1.8)$$

$$= \left(\frac{n_1}{\sqrt{n_1, n_3 (n_1 + n_3)}}, \ \frac{n_2}{\sqrt{n_2 n_3 (n_2 + n_3)}} \right) \qquad \text{for machine } C,$$

where n_1, n_2, and n_3 are, respectively, the numbers of observations from machines A, B, and C. These Z_1, Z_2 columns are not orthogonal to each other but two orthogonal columns could be constructed. Again, all denominators could be dropped in (14.1.8).

r Categories, r Dummies

In general, by an extension of this procedure, we can deal with r levels by the introduction of $(r - 1)$ dummy variables in addition to X_0. The basic allocation pattern is obtained by writing down an $(r - 1) \times (r - 1)$ **I** matrix and adding a row of $(r - 1)$ zeros. The case $r = 6$ is illustrated by the $X_1, X_2, \ldots, X_5$ columns in the third display of Example 2 below.

We now give an example of the use of dummy variables in this manner.

Example 1. The data in Table 14.1 show turkey weights (Y) in pounds, and ages (X) in weeks, of 13 Thanksgiving turkeys. Four of these turkeys were reared in Georgia (G), four in Virginia (V), and five in Wisconsin (W). We would like to relate Y to X via a simple straight line model, but the different origins of the turkeys may cause a problem. If they do, how do we handle it?

Suppose we first regress Y against X to give the fitted equation $\hat{Y} = 1.98 + 0.4167X$. The residuals from this fit are, in order, -0.4, -1.4, -0.2, -0.8, -1.0, -0.8, -0.5, -1.0, 0.8, 1.0, 1.3, 1.5, 1.4. When plotted in sets according to origin, they give rise to

T A B L E 14.1. Turkey Data (X, Y, Origin) and Dummy Variables (Z_1, Z_2)

X	Y	Origin	Z_1	Z_2
28	13.3	G	1	0
20	8.9	G	1	0
32	15.1	G	1	0
22	10.4	G	1	0
29	13.1	V	0	1
27	12.4	V	0	1
28	13.2	V	0	1
26	11.8	V	0	1
21	11.5	W	0	0
27	14.2	W	0	0
29	15.4	W	0	0
23	13.1	W	0	0
25	13.8	W	0	0

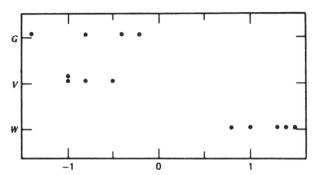

Figure 14.1. Turkey data; residuals from the fitted equation $\hat{Y} = 1.98 + 0.4167X$ plotted against origin of turkey.

Figure 14.1, which clearly signals the need to take account of the different levels of response. To do this, we select the dummy variables Z_1, Z_2 shown in Table 14.1 and fit the model

$$Y = \beta_0 + \beta_1 X + \alpha_1 Z_1 + \alpha_2 Z_2 + \epsilon \tag{14.1.9}$$

by least squares. The fitted equation is

$$\hat{Y} = 1.43 + 0.4868X - 1.92Z_1 - 2.19Z_2. \tag{14.1.10}$$

The estimates $a_1 = -1.92$ and $a_2 = -2.19$ estimate the differences in response levels between (1) sources G and W, and (2) sources V and W, respectively. By substituting for the three sets of values for (Z_1, Z_2) we obtain, for the three different origins,

$$\hat{Y} = -0.49 + 0.4868X, \quad \text{for } G;$$
$$\hat{Y} = -0.76 + 0.4868X, \quad \text{for } V; \tag{14.1.11}$$
$$\hat{Y} = 1.43 + 0.4868X, \quad \text{for } W.$$

The original data and the three fitted straight lines are shown in Figure 14.2. The three lines are all parallel but have different intercepts. The analysis of variance for this fitted model can be written as shown in Table 14.2. Both F-values are highly significant, implying that use of the dummies is clearly worthwhile and that the lines appear to have a definite nonzero slope. Of the variation of the data about the mean, 97.94% has been explained by this equation. (Without the dummies, only 66.47% is explained.)

If desired, t-tests can be constructed to test for differences between the intercepts. For example, the true $W - G$ difference is estimated by $-a_1 = 1.92$ and this, divided by its standard error, namely, the square root of the appropriate diagonal term of the $(\mathbf{X'X})^{-1}s^2$ matrix, gives a t-value whose modulus (positive value) is compared to the percentage point $t(9, 1 - \frac{1}{2}\alpha)$ for a two-sided test of the null hypothesis $H_0: \alpha_1 = 0$ versus $H_1: \alpha_1 \neq 0$. We find, for our data, $t = 1.92/0.201 = 9.55$, which is significant at 0.1%. An alternative and equivalent test is given by using

$$F = \{SS(a_1 | b_0, b_1, a_2)/1\}/s^2 = 8.145/0.090 = 90.50.$$

This is compared with $F(1, 9, 1 - \alpha)$ for a test at the same level. The result is identical, because the F-value is, theoretically, the square of the t-value above; here $t^2 = 91.20$ would be the same as $F = 90.50$ were it not for rounding differences. A test for $H_0: \alpha_2 = 0$, α_2 being the true $V - W$ difference, can be carried out in a similar manner. The t-value is $-2.19/0.21 = -10.43$, which is also significant at 0.1%. The estimated

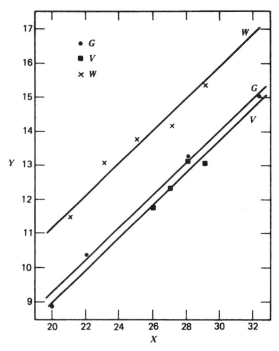

Figure 14.2. Plot of the turkey data and the three fitted straight lines.

$G - V$ difference is given by $a_1 - a_2 = 0.27$, which has an estimated variance of Est. $V(a_1) + $ Est. $V(a_2) - 2$ Est. cov (a_1, a_2), all three terms of which can be obtained from the $(\mathbf{X}'\mathbf{X})^{-1}s^2$ matrix. We find Est. $V(a_1 - a_2) = 0.040369 + 0.044310 - 2(0.018690) = 0.047299 = (0.217)^2$. The t-value is thus $0.27/0.217 = 1.24$, not significant. Thus overall, real differences appear to exist between the G and W levels and between the V and W levels but not between the G and V levels.

An Alternative Analysis of Variance Sequence

The order indicated in the source column of Table 14.2 removes the differences between intercepts first and leaves the remaining variation to $(b_1|b_0, a_1, a_2)$ and residual. It would also be possible to use the order b_0; $(b_1|b_0)$; $(a_1, a_2|b_0, b_1)$; and residual, and so test the hypothesis $H_0: \alpha_1 = \alpha_2 = 0$ (which asks "Do the three sets of data here have different intercepts?") using an extra sum of squares F-test based on $F(2, 9)$. One would then still have to check where those differences were, if a significant result were obtained.

T A B L E 14.2. Analysis of Variance for Turkey Example

Source	df	SS	MS	F
b_0	1	2124.803		
$a_1, a_2 \mid b_0$	2	6.382	3.191	35.46
$b_1 \mid b_0, a_1, a_2$	1	32.224	32.224	358.04
Residual	9	0.811	$s^2 = 0.090$	
Total	13	2164.220		

Will My Selected Dummy Setup Work?

The vectors of (14.1.3) can be described by writing down the components in a two by two matrix as

$$\begin{bmatrix} 1 & 0 \\ 1 & 1 \end{bmatrix},$$

(14.1.12)

where the first row denotes the values of (X_0, Z) for a group A piece of data while the second row applies to a group B piece of data. If this matrix has a nonzero determinant (it is, of course, easy to see that it does) the setup will work. For the turkey data, the corresponding matrix is

$$\begin{bmatrix} 1 & 1 & 0 \\ 1 & 0 & 1 \\ 1 & 0 & 0 \end{bmatrix}$$

(14.1.13)

and the determinant is 1, also. Let us examine the six-group case.

Example 2. Is the dummy variable scheme below a workable one for dealing with possible level differences among six groups?

Group	Z_0	Z_1	Z_2	Z_3	Z_4	Z_5
1	1	1	1	1	1	1
2	1	0	1	1	1	1
3	1	0	0	1	1	1
4	1	0	0	0	1	1
5	1	0	0	0	0	1
6	1	0	0	0	0	0

The determinant value is -1, so the system will work. Evaluating determinants is not difficult, and rules for quick reduction may be found in matrix algebra books. For our purposes, it is usually simplest to use the computer to find the determinant via the so-called eigenvalues, the values $\lambda_1, \lambda_2, \ldots, \lambda_q$, say, which are the roots of the equation det $(\mathbf{M} - \lambda \mathbf{I}) = 0$, where $\mathbf{M}$ is a (square) matrix of interest and det $(\cdot)$ means the determinant of $(\cdot)$. The reason this works is that

$$\det \mathbf{M} = \lambda_1 \lambda_2 \cdots \lambda_q$$

so that, if all the eigenvalues are nonzero, so is the determinant. Problems with "almost-dependent" columns can arise here too, but use of 1's and 0's as dummy levels will usually avoid such problems.

In the MINITAB system, for example, we can write:

```
read 6 6 m1
1 1 1 1 1 1
1 0 1 1 1 1
1 0 0 1 1 1
1 0 0 0 1 1
1 0 0 0 0 1
1 0 0 0 0 0
transpose m1 m2
multiply m2 m1 m3
eigen m3 c3
print c3
end
stop
```

The output from this routine is the column of numbers (which we write as a row here)

$$17.2069 \quad 1.9882 \quad 0.7747 \quad 0.4462 \quad 0.3189 \quad 0.2652;$$

the product of these is 1 not -1. The reason for this is that the MINITAB eigen program requires a symmetric matrix, and we have created one by evaluating $\mathbf{M} = \mathbf{m}_1'\mathbf{m}_1$ before calling for eigenvalues. Because $\mathbf{m}_1$ is square, $\det(\mathbf{m}_1'\mathbf{m}_1) = (\det \mathbf{m}_1')(\det \mathbf{m}_1) = (\det \mathbf{m}_1)^2$. Even though we do not get the sign from our result, we know $\det \mathbf{m}_1$ is nonzero. Evaluation of the eigenvalues of $\mathbf{m}_1\mathbf{m}_1'$ would give identical results because $\mathbf{m}_1$ is square. Note that we do not actually need to take the product of the eigenvalues but can just look at the smallest one to see that it is nonzero (and not almost zero).

Other Verification Methods

A second way of checking a setup like that of Example 2 is to relate it to the basic scheme. Recall that our basic scheme of vectors for this situation, written down with the X_0 column, was as follows:

X_0	X_1	X_2	X_3	X_4	X_5
1	1	0	0	0	0
1	0	1	0	0	0
1	0	0	1	0	0
1	0	0	0	1	0
1	0	0	0	0	1
1	0	0	0	0	0

We see immediately that

$$Z_0 = X_0, \qquad Z_3 = X_1 + X_2 + X_3,$$
$$Z_1 = X_1, \qquad Z_4 = X_1 + X_2 + X_3 + X_4,$$
$$Z_2 = X_1 + X_2, \qquad Z_5 = X_1 + X_2 + X_3 + X_4 + X_5.$$

This establishes the Z's as linear combinations of the X's. Moreover, each Z_j, $j = 2$, 3, 4, 5, in turn introduces an additional X_j so that none of these Z_j can be linearly dependent on prior Z's, and vice versa. It follows that the system will work for separating the intercept levels of six groups of data.

A third, somewhat more tedious, verification is to write down the condition that the columns are dependent and then solve the resulting equations for the constants that form the linear combination. For Example 2, we write that

$$aZ_0 + bZ_1 + cZ_2 + dZ_3 + eZ_4 + fZ_5 = 0,$$

and this must hold row by row, which implies that

$$a + b + c + d + e + f = 0,$$
$$a \quad\quad + c + d + e + f = 0,$$
$$a \quad\quad\quad\quad + d + e + f = 0,$$
$$a \quad\quad\quad\quad\quad\quad + e + f = 0,$$
$$a \quad\quad\quad\quad\quad\quad\quad\quad + f = 0,$$
$$a \quad\quad\quad\quad\quad\quad\quad\quad\quad\quad = 0.$$

Working from the bottom equation up, it is obvious that all the coefficients $a, \ldots, f$ are zero. That means *no* linear combination gives zero, so that the Z's are linearly independent and the system works. If a *nonzero* numerical solution emerges, however, dependence of the Z's is established, making the setup useless.

Example 3. Another workable system in the same context as Example 2 would consist of columns $Z_0 = X_0$, $Z_i = X_0 + X_i$, $i = 1, 2, \ldots, 5$. This would lead to the following scheme:

Group	Z_0	Z_1	Z_2	Z_3	Z_4	Z_5
1	1	2	1	1	1	1
2	1	1	2	1	1	1
3	1	1	1	2	1	1
4	1	1	1	1	2	1
5	1	1	1	1	1	2
6	1	1	1	1	1	1

Note: The dummy variable columns not only must form a linearly independent set themselves but must also form a linearly independent set when they are united with other predictors variable columns in the full regression. In most regressions, the occurrence of such a dependence is unlikely, but it would show up when the full regression is performed, requiring the dummies to be selected differently.

14.2. INTERACTION TERMS INVOLVING DUMMY VARIABLES

The way we used dummy variables in the foregoing section allowed us to fit the same basic model with different intercepts to several sets of data. By an extension of those ideas, we can also allow changes in the way the predictors enter.

Two Sets of Data, Straight Line Models

Suppose A and B denote two data sets, and we are considering fits involving straight lines. There are four possibilities, shown in Figure 14.3.

(*a*) Two distinct lines, $\beta_0 + \beta_1 X$, $\gamma_0 + \gamma_1 X$, four parameters.
(*b*) Two parallel lines, $\beta_0 + \beta_1 X$, $\gamma_0 + \beta_1 X$, three parameters.
(*c*) Two lines with the same intercept $\beta_0 + \beta_1 X$, $\beta_0 + \gamma_1 X$, three parameters.
(*d*) One line, $\beta_0 + \beta_1 X$, two parameters.

We can take care of all four possibilities at once by choosing two dummies (we count the column of 1's associated with the intercept as one of these):

$$
\begin{array}{cc}
X_0 & Z \\
1 & 0 \quad \text{for the } A \text{ data} \\
1 & 1 \quad \text{for the } B \text{ data}
\end{array}
\tag{14.2.1}
$$

and then fitting the model

$$ Y = X_0(\beta_0 + \beta_1 X) + Z(\alpha_0 + \alpha_1 X) + \epsilon $$

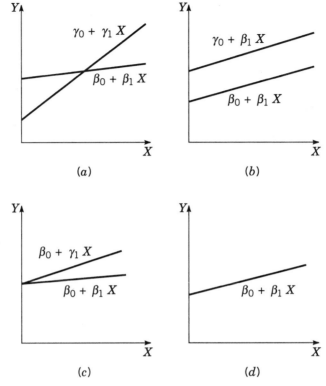

Figure 14.3. Four possibilities for two straight lines.

or

$$Y = \beta_0 + \beta_1 X + \alpha_0 Z + \alpha_1 XZ + \epsilon \tag{14.2.2}$$

when the parentheses are removed. We see that the model contains not only Z but an interaction term involving Z as well. The separate models for the A and B lines are given by setting $Z = 0$ and $Z = 1$ to produce the

A model function: $\beta_0 + \beta_1 X$,

B model function: $(\beta_0 + \alpha_0) + (\beta_1 + \alpha_1)X$.   (14.2.3)

Thus in Figure 14.3a, $\gamma_0 = \beta_0 + \alpha_0$ and $\gamma_1 = \beta_1 + \alpha_1$. The parameters α_0 and α_1 represent changes needed to get from the A model function to the B model function. To test whether the two parallel lines of Figure 14.3b will do, we would fit (14.2.2) and then test $H_0:\alpha_1 = 0$ versus $H_1:\alpha_1 \neq 0$. To test for the appropriateness of the case in Figure 14.3c, the null hypothesis used would be $H_0:\alpha_0 = 0$, while for the case in Figure 14.3d it would be $H_0:\alpha_0 = \alpha_1 = 0$. In all cases, the null hypotheses reduce the two model functions in (14.2.3) to the various reduced forms shown in Figure 14.3, with the identification $\gamma_0 = \beta_0 + \alpha_0$ and $\gamma_1 = \beta_1 + \alpha_1$.

Hierarchical Models

In discussing extra sums of squares in various earlier contexts, for example, Chapter 6, we saw that the terms of the smaller model always formed a part of the larger model. Sequential sums of squares calculations involve a hierarchy of models, a sequence in which each model contains all the terms in the model below it in the hierarchical order (as well as one more term). These hierarchical aspects arise in our current discussion also. Look at Figure 14.3. We can write

$$(a) \quad > \quad (b) \quad > \quad (d)$$
$$\quad\quad \alpha_1 \quad\quad\quad \alpha_0$$

to denote that model (d) is contained in (b) when $\alpha_0 = 0$, which is itself contained in (a) when $\alpha_1 = 0$. Similarly,

$$(a) \quad > \quad (c) \quad > \quad (d)$$
$$\quad\quad \alpha_0 \quad\quad\quad \alpha_1$$

Appropriate sequences of tests on the parameters quickly become clear when one arranges the various models into hierarchical patterns in this way.

Three Sets of Data, Straight Line Models

Before we discuss the general case, let us extend (14.2.2) to three straight lines and illustrate with the turkey data of Table 14.1. To allow the fitting of three separate straight lines, we form the model

$$Y = Z_0(\beta_0 + \beta_1 X) + Z_1(\gamma_0 + \gamma_1 X) + Z_2(\delta_0 + \delta_1 X) + \epsilon, \quad (14.2.4)$$

where $Z_0 = 1$ and Z_1 and Z_2 are two additional dummy variables. We shall choose these as follows:

Z_1	Z_2	Line
1	0	G (Georgia)
0	1	V (Virginia)
0	0	W (Wisconsin)

We rewrite the model as

$$Y = \beta_0 + \beta_1 X + \gamma_0 Z_1 + \gamma_1(Z_1 X) + \delta_0 Z_2 + \delta_1(Z_2 X) + \epsilon \quad (14.2.5)$$

and note that there are two "interaction with dummies" terms, $Z_1 X$ and $Z_2 X$. The fitted equation is

$$\hat{Y} = 2.475 + 0.4450X - 3.454Z_1 - 2.775Z_2$$
$$+ 0.06104(Z_1 X) + 0.02500(Z_2 X). \quad (14.2.6)$$

The three separate straight lines are then

$$\hat{Y} = -0.979 + 0.5060X \quad \text{(setting } Z_1 = 1, Z_2 = 0\text{),}$$
$$\hat{Y} = -0.300 + 0.4700X \quad \text{(setting } Z_1 = 0, Z_2 = 1\text{),} \quad (14.2.7)$$
$$\hat{Y} = 2.475 + 0.4450X \quad \text{(setting } Z_1 = 0, Z_2 = 0\text{).}$$

These lines, which are exactly what one would find if one fitted each subset of data separately, are slightly displaced from the lines shown in Figure 14.2, as the reader can confirm by plotting, or by simply comparing the fitted equations above with those in Eq. (14.1.11). The analysis of variance table for this fit takes the form

ANOVA

Source	df	SS	MS	F
b_0	1	2124.803		
$b_1, c_0, c_1, d_0, d_1 \vert b_0$	5	38.711	7.742	76.6
Residual	7	0.706	0.101	
Total	13	2164.220		

The three fitted lines would be identical if $H_0 : \gamma_0 = \gamma_1 = \delta_0 = \delta_1 = 0$ were true. To test this hypothesis versus $H_1 : H_0$ not true, the extra sum of squares for c_0, c_1, d_0, and d_1, given by

$$SS(b_1, c_0, c_1, d_0, d_1 \vert b_0) - SS(b_1 \vert b_0) = 38.71 - 26.20 = 12.51$$

with 4 df is needed. (The figure 26.20 is the regression sum of squares when a common line is fitted; it was not given earlier.) The appropriate F-statistic is

$$F = (12.51/4)/(0.101) = 30.97,$$

which exceeds $F(4, 7, 0.99) = 7.85$, so that H_0 is rejected. This, of course, is not surprising, as we have already seen when the data were given initially.

We can test the hypothesis that there are three parallel lines, that is, $H_0 : \gamma_1 = \delta_1 = 0$ versus $H_1 : H_0$ not true, by finding the extra sum of squares for c_1 and d_1 via

$$SS(b_1, c_0, c_1, d_0, d_1 \vert b_0) - SS(b_1, c_0, d_0 \vert b_0) = 38.71 - 38.61 = 0.10,$$

the figure 38.61 being the sum of the second and third entries in Table 14.2. This value 0.10 has 2 df and provides the nonsignificant F-ratio $(0.10/2)/0.101 = 0.50$. We do not reject H_0, and so the fit illustrated in Figure 14.2 is clearly a satisfactory one.

As our example shows, the use of interaction terms involving dummy variables makes it easy to formulate appropriate tests and to obtain the right test statistics. This may be the method's greatest virtue.

Two Sets of Data: Quadratic Model

Suppose we have two sets of similar data on a response Y and a predictor X and we have in mind a model of the form

$$Y = \beta_0 + \beta_1 X + \beta_{11} X^2 + \epsilon \tag{14.2.8}$$

for each set. We want to check if we can use the same fitted model for both sets and, if so, what the fitted coefficients should be. We fit, to both sets of data at one time, the model

$$Y = \beta_0 + \beta_1 X + \beta_{11} X^2 + \alpha_0 Z + \alpha_1 XZ + \alpha_{11} X^2 Z + \epsilon, \tag{14.2.9}$$

where Z is a dummy variable with levels 0 for one set of data and 1 for the other. Extra sums of squares tests then enable us to check the various possibilities, as follows, for example:

1. $H_0 : \alpha_0 = \alpha_1 = \alpha_{11} = 0$ versus H_1: not so. If this null hypothesis is rejected, we conclude the models are not the same; if not rejected, we take them to be the same.

2. If H_0 in (1) is rejected, we would look at subsets of the α's. For example, we could test $H_0 : \alpha_1 = \alpha_{11} = 0$ versus H_1: not so. If H_0 were not rejected, we would

conclude that the two sets of data exhibited only a difference in response levels but had the same slope and curvature.

3. If H_0 in (2) is rejected, we could test $H_0 : \alpha_{11} = 0$ versus $H_1 : \alpha_{11} \neq 0$ to see if the models differed only in zero and first-order terms, indicated by nonrejection of H_0.

Other sequences of tests could be used if the were sensible in the context of the problem under study. The sequence chosen above is a natural hierarchical buildup that is often sensible.

General Case: r Sets, Linear Model

In principle, there is no difficulty extending this setup to situations with more sets of data, and other models involving more predictors, $X_1, X_2, \ldots, X_k$. If there were r sets of data, we would specify, in addition to $Z_0 = 1$, $(r - 1)$ dummy variables Z_1, Z_2, $\ldots, Z_{r-1}$ with levels given by writing down an $\mathbf{I}_{r-1}$ matrix with a line of $(r - 1)$ zeros below it. The rows then designate the groups, and the columns the dummies. If the basic model were to be

$$Y = f(\mathbf{X}, \boldsymbol{\beta}) + \epsilon$$

for one set of data, we would fit, to all the data, the model

$$Y = f(\mathbf{X}, \boldsymbol{\beta}) + \sum_{j=1}^{r-1} Z_j f(\mathbf{X}, \boldsymbol{\alpha}_j) + \epsilon, \tag{14.2.10}$$

where $\boldsymbol{\alpha}_j$ is a vector of parameters of the same size as $\boldsymbol{\beta}$, as in the $r = 2$ example above. This can also be written as

$$Y = \sum_{j=0}^{r-1} Z_j f(\mathbf{X}, \boldsymbol{\alpha}_j), \tag{14.2.11}$$

where $\boldsymbol{\alpha}_0$ replaces $\boldsymbol{\beta}$.

We would get the same answers if we fitted the various sets of data individually. For example, if $\mathbf{X}_i$ is the "$\mathbf{X}$ matrix" for the ith set of data and we have two sets, then the model is

$$E(\mathbf{Y}) = \begin{bmatrix} \mathbf{X}_1 & \mathbf{0} \\ \mathbf{X}_2 & \mathbf{X}_2 \end{bmatrix} \begin{bmatrix} \boldsymbol{\beta} \\ \boldsymbol{\alpha} \end{bmatrix}, \tag{14.2.12}$$

which we can regard as

$$E(\mathbf{Y}) = \begin{bmatrix} \mathbf{X}_1 & \mathbf{0} \\ \mathbf{0} & \mathbf{X}_2 \end{bmatrix} \begin{bmatrix} \boldsymbol{\beta} \\ \boldsymbol{\alpha} + \boldsymbol{\beta} \end{bmatrix} = \begin{bmatrix} \mathbf{X}_1 & \mathbf{0} \\ \mathbf{0} & \mathbf{X}_2 \end{bmatrix} \begin{bmatrix} \boldsymbol{\beta} \\ \boldsymbol{\theta} \end{bmatrix}, \tag{14.2.13}$$

say. The advantage of using interaction terms with dummy variables instead is that it enables a single formulation and allows appropriate extra sums of squares tests to be set up in a straightforward way, once the hierarchical aspects of the various submodels have been recognized.

14.3. DUMMY VARIABLES FOR SEGMENTED MODELS

It is sometimes appropriate to fit a model relating a response Y to a single predictor X in terms of segments. (Often, this X is a time variable, or a space variable.) For

T A B L E 14.3. Parity Price (¢) per Pound of Live Weight of Chickens

Date	Parity Price, Y	X or	X'
Jan. 1955	29.1	1	-10
May 1955	29.0	2	-9
Sept. 1955	28.6	3	-8
Jan. 1956	28.1	4	-7
May 1956	28.6	5	-6
Sept. 1956	28.7	6	-5
Jan. 1957	28.2	7	-4
May 1957	28.6	8	-3
Sept. 1957	28.6	9	-2
Jan. 1958	28.1	10	-1
May 1958	28.7	11	0
Sept. 1958	28.6	12	1
Jan. 1959	26.9	13	2
May 1959	27.0	14	3
Sept. 1959	26.8	15	4
Jan. 1960	25.7	16	5
May 1960	25.9	17	6
Sept. 1960	25.6	18	7
Jan. 1961	25.1	19	8
May 1961	25.2	20	9
Sept. 1961	25.1	21	10

example, we might believe that, up to a certain point (rarely known, usually estimated) on the X-scale, a straight line fit is appropriate but, beyond that point, a *different* straight line is needed. Or, we might believe that the initial straight line was followed by a quadratic curve. A variation on the latter case could be that, where the straight line and the quadratic curve intersect, the slope of the model did not change. There could also be several segments rather than two. In addition, there could be other predictor variables, the effects of which were *not* segmented. The main topic of this section will be that of a pair of segmented straight lines in one X with no other predictor variables. This should enable the case of several segments to be tackled using similar ideas. Moreover, adding other predictors typically poses no problems. One simply adds these to the model and **X** matrix in the usual way and fits all terms simultaneously by least squares. We can fit segmented models by using suitably defined dummy variables. As always, there is an infinite choice of ways to select the dummy variable levels. We edge into this topic by first thinking about a single straight line segment in this context.

One Segment

Example. Table 14.3 shows data on the parity price in cents per pound of live weights of chickens at equal intervals of time. Two alternative dummy variables are shown as columns X and X'. Either will do, although the centered column X' may be preferred since it is orthogonal to the column of 1's in the **X** matrix. The appropriate models in the two cases are

$$Y = \beta_0 + \beta_1 X + \text{(other terms with predictor variables)} + \epsilon \qquad (14.3.1)$$

and (since $X_i' = X_i - \overline{X}, i = 1, 2, \ldots, n$)

$$Y = (\beta_0 + \beta_1\overline{X}) + \beta_1 X' + (\text{other terms}) + \epsilon$$

$$= \beta_0' + \beta_1 X' + (\text{other terms}) + \epsilon. \tag{14.3.2}$$

Note: Here, since $n = 21$ is odd, the quantities $X_i' = X_i - \overline{X}$ are all integers. When n is even we can use instead $X_i' = 2(X_i - \overline{X})$ to avoid fractions. For example,

$$X_i = \quad 1 \quad\quad 2 \quad 3 \quad 4 \quad (\overline{X} = 2\tfrac{1}{2})$$

$$X_i - \overline{X} = -1\tfrac{1}{2} \quad -\tfrac{1}{2} \quad \tfrac{1}{2} \quad 1\tfrac{1}{2}$$

$$2(X_i - \overline{X}) = -3 \quad -1 \quad 1 \quad 3$$

(This is a simple orthogonal polynomial. See Chapter 22.)

For a quadratic model, terms $\beta_0 + \beta_1 X + \beta_{11}X^2$ (or $\beta_0' + \beta_1 X' + \beta_{11}X'^2$) would be added, or the trend could be expressed through the first- and second-order orthogonal polynomials described in Chapter 22. Higher-order time trends would be handled in similar fashion with higher-order terms.

The data in Table 14.3 are equally spaced in time, and so the X's are chosen there as equally spaced. If the data were unequally spaced in time, the X's would be chosen accordingly. For example, if the dates were January 1955, February 1955, April 1955, June 1955, $\ldots$, the X's would be 1, 2, 4, 6, $\ldots$, and so on. In such a case, use of the column $X - \overline{X}$ might be inconvenient because it might lead to noninteger values. In that case, either one would use a multiplier to convert the numbers to integers, or, if this were not reasonable, one might use values $X_i - A$, where A was some convenient integer close to $\overline{X}$; however, there is usually little if any advantage to be gained by using $X_i - A$, except that the sizes of the numbers used are reduced somewhat. Because of the unequal spacings, special orthogonal polynomials would have to be evaluated if that kind of approach were desired; consequently, orthogonal polynomials are hardly ever used for unequally spaced data.

Two Segments

When there are two straight line segments, a dummy variable must be set up for each. The problem divides up into two main levels of complication:

1. When it is known which data points lie on which segments, with subcases:
    a. when the abscissa of intersection of the two lines can be assumed to be at a specific value at which one or more observations exists; and
    b. when the abscissa of intersection of the two lines is unknown.
2. When it is not known which data points lie on which segments.

Case 1: When It Is Known Which Points Lie on Which Segments

Example 1a. The equally spaced data plotted in Figure 14.4 fall into case (1a). We suppose that it is known that the first five data points lie (apart from random error) on the first line and that the last five data points lie (apart from random error) on the second line; the fifth point is thus common to both lines. We can (for example) set up two dummy variables X_1 and X_2 for the two lines as follows. Both are set to zero at the known intersection point, namely, the fifth observation, X_1 is stepped back for the first line, X_2 is stepped forward for the second line, and both variables are

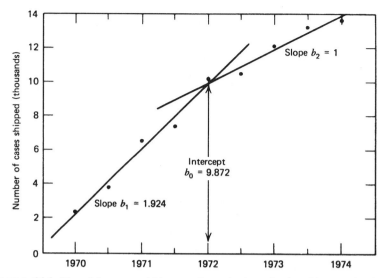

Figure 14.4. Use of dummy variables; two lines, abscissa of point of intersection known.

otherwise zero. (The steps are equally spaced here because the data are equally spaced. If they were not, other appropriate step levels would be chosen instead.) The resulting data matrix, assuming no other predictor variables are involved, is shown in Table 14.4. If we now fit the model

$$Y = \beta_0 + \beta_1 X_1 + \beta_2 X_2 + \epsilon, \tag{14.3.3}$$

the estimates obtained have these roles:

$$b_0 = \text{value of } \hat{Y} \text{ at point of intersection, } X_1 = X_2 = 0,$$

$$b_1 = \text{slope of first trend line,}$$

$$b_2 = \text{slope of second trend line.}$$

For the data shown, the normal equations become

$$9b_0 - 10b_1 + 10b_2 = 79.6,$$

$$-10b_0 + 30b_1 = -41.0, \tag{14.3.4}$$

$$10b_0 + 30b_2 = 128.7,$$

T A B L E 14.4. Dummy Variables for Example of Two Straight Lines Whose Abscissa of Intersection Is Known

Observation Number	Date	X_0	X_1	X_2	Y
1	1970	1	−4	0	2.3
2		1	−3	0	3.8
3	1971	1	−2	0	6.5
4		1	−1	0	7.4
5	1972	1	0	0	10.2
6		1	0	1	10.5
7	1973	1	0	2	12.1
8		1	0	3	13.2
9	1974	1	0	4	13.6

T A B L E 14.5. Alternative Dummy Variable Setup for
Example of Two Straight Lines Whose Point of Intersection
Is Known

Observation Number	Date	X_0	X_1	X_2	Y
1	1970	1	1	0	2.3
2		1	2	0	3.8
3	1971	1	3	0	6.5
4		1	4	0	7.4
5	1972	1	5	0	10.2
6		1	5	1	10.5
7	1973	1	5	2	12.1
8		1	5	3	13.2
9	1974	1	5	4	13.6

with solutions $b_0 = 9.871$, $b_1 = 1.924$, and $b_2 = 1.000$ as drawn in Figure 14.4. If other predictor variables were involved in the problem, appropriate terms would be added on the right-hand side of Eq. (14.3.3).

As with all dummy variable situations, the representation is not unique. For example, an alternative setup is shown in Table 14.5, in which (new X_1) = (previous X_1) + 5. This setup will provide estimates of the slopes as before but the constant term b_0, the value of $\hat{Y}$ when $X_1 = X_2 = 0$, will now be the intercept of the first line at the abscissa $1969\frac{1}{2}$.

Straight Line and Quadratic Curve

Higher-order models would be accommodated by adding higher-order terms. For a straight line followed by a quadratic curve, for example, we would fit

$$Y = \beta_0 + \beta_1 X_1 + \beta_2 X_2 + \beta_{22} X_2^2 + \epsilon. \tag{14.3.5}$$

If, in addition, we wanted the model to have continuous derivative at the join point, we would require $\partial Y / \partial X_1 = \partial Y / \partial X_2$ at the point of intersection, that is, at the fifth data point where $X_2 = 0$,

$$\beta_1 = \beta_2 + 2\beta_{22} X_2. \tag{14.3.6}$$

Thus we would set $\beta_1 = \beta_2 = \beta$, say, in Eq. (14.3.5) and fit the reduced model $Y = \beta_0 + \beta(X_1 + X_2) + \beta_{22} X_2^2 + \epsilon$.

Example 1b. The equally spaced data in Figure 14.5 fall into case (1b). We assume that it is known that the first four data points lie (apart from random error) on one line and that the last five data points lie (apart from random error) on a second line. However, the point of intersection is unknown. A third dummy variable X_3 is needed to take care of the unknown point of intersection. It is set to zero for all points on the first line and then goes to 1 for all points on the second line to allow for a jump (positive or negative) from the first line to the second. The dummies X_1 and X_2 are chosen in either of the ways indicated in Example 1a. Table 14.6 shows an appropriate data matrix. If no other predictor variables are involved we can fit the model

$$Y = \beta_0 + \beta_1 X_1 + \beta_2 X_2 + \beta_3 X_3 + \epsilon. \tag{14.3.7}$$

The parameter β_3 is the step change that comes into effect at the fifth observation

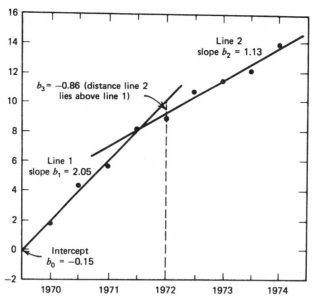

Figure 14.5. Use of dummy variables; two lines, abscissa of point of intersection unknown.

point and is the vertical distance the second line lies *above* the first at this point. (If the second line lies below the first, β_3 is negative.) For the data of Table 14.6, the normal equations are

$$9b_0 + 35b_1 + 10b_2 + 5b_3 = 77.4,$$
$$35b_0 + 155b_1 + 50b_2 + 25b_3 = 347.5,$$
$$10b_0 + 50b_1 + 30b_2 + 10b_3 = 126.3,$$
$$5b_0 + 25b_1 + 10b_2 + 5b_3 = 57.5,$$

(14.3.8)

with solutions

$b_0 = -0.15$ intercept of line 1, when $X_1 = 0$,

$b_1 = 2.05$ (slope of line 1),

$b_2 = 1.13$ (slope of line 2),

$b_3 = -0.86$ (the vertical distance between line 2 and line 1 at the fifth observation point).

The situation is shown graphically in Figure 14.5. The negative sign of b_3 and the fact that $b_1 > b_2$ indicate that the point of intersection of the two lines is to the left of the fifth observation point. In fact, it occurs when $X_1 = 4.065$. This point of intersection can be found by writing both lines in terms of the X_1 scale. The first line is given by

$$\hat{Y} = -0.15 + 2.05X_1,$$

(14.3.9)

and the second by

$$\hat{Y} = -0.15 + 2.05(5) + 1.13X_2 - 0.86,$$

(14.3.10)

that is,

$$\hat{Y} = 9.24 + 1.13X_2.$$

(14.3.11)

T A B L E 14.6. Dummy Variables for Example of Two Straight Lines Whose Abscissa of Intersection Is Unknown

Observation Number	Date	X_0	X_1	X_2	X_3	Y
1	1970	1	1	0	0	1.8
2		1	2	0	0	4.3
3	1971	1	3	0	0	5.6
4		1	4	0	0	8.2
5	1972	1	5	0	1	9.1
6		1	5	1	1	10.7
7	1973	1	5	2	1	11.5
8		1	5	3	1	12.2
9	1974	1	5	4	1	14.0

Looking at the X_1 and X_2 scales in relation to the scale on Figure 14.5, we see that $X_2 = 0$ at $X_1 = 5$ so that we can substitute $X_2 = X_1 - 5$ into the equation of the second line to reduce it to

$$\hat{Y} = 3.59 + 1.13X_1. \qquad (14.3.12)$$

Setting the two right-hand sides of Eqs. (14.3.9) and (14.3.12) equal gives $X_1 = 4.065$ as the point of intersection. These calculations could also be made using X_2 to give $X_2 = -0.935$.

Case 2: When It Is Not Known Which Points Lie on Which Segments

In the previous cases, we estimated the parameters of the composite model by fitting a given model by linear least squares. In the present case, the solution could be obtained by looking at every possible division of the points to the first and second lines, estimating the parameters by linear least squares and evaluating the residual sum of squares for each division, and then picking that division and set of estimates that give rise to the smallest of all the residual sums of squares. (In practice, one would usually not need to look at every division of the points, because a few computations usually show the "ballpark" in which the best division lies and the computations are then confined to those divisions only.) Alternatively, the problem could be cast into the form of a nonlinear estimation problem and solved by the methods discussed in Chapter 24. (Local minima can occur, so some care is needed at times.) Using nonlinear methods is complicated and usually not worthwhile, however.

Remark

In the above illustrations we have assumed that the points belonging to the two segments did not overlap. This would most often be the case. Where overlap occurs, the situation is more complex, but the methods remain valid. Inspection of a plot of the data is usually helpful in determining how to divide the points. When other predictors are present, these can be regressed out first, and the residuals plotted against the variable to be segmented.

EXERCISES FOR CHAPTER 14

A. A response variable is measured on the following dates: November 17, 19, 20, 22, 26, 29, 30, December 1, 2, 3, and 5. It is believed that the response depends on two factors X_1 and

T A B L E B. **Yields Y from Three Tiller Types for Two Varieties of Wheat Subjected to Different Nitrogen Applications**

Y	Tiller Subscript	Waldron/Ciano	N Rate	N per Tiller at Tillering, mg (X)
370	1	W	0	4.2
659	1	W	50	7.2
935	1	W	270	9.8
390	1	C	0	3.6
753	1	C	50	7.6
733	1	C	270	10.3
182	2	W	0	3.1
417	2	W	50	5.2
686	2	W	270	7.8
188	2	C	0	2.8
632	2	C	50	6.0
538	2	C	270	7.7
27	3	W	0	2.0
141	3	W	50	2.8
262	3	W	270	3.6
34	3	C	0	2.7
222	3	C	50	3.1
242	3	C	270	4.4

X_2 whose values have been recorded (but are not given here) and that, in addition, there is a quadratic time trend in the data. How would you take account of it? (If you mention any new variables, you should specify their actual levels.)

B. (*Source:* "Tiller development and yield of standard and semidwarf spring wheat varieties as affected by nitrogen fertilizer," by J. F. Power and J. Alessi, *Journal of Agricultural Science, Cambridge,* **90,** 1978, 97–108. Adapted with the permission of Cambridge University Press.) Table B shows, as Y, grain yields in kg/ha (kilograms per hectare), which resulted from growing two varieties, Waldron and Ciano, of hard red spring wheat, at three different nitrogen rates $N = 0, 50, 270$, representing deficient, adequate, and excessive nitrogen supply. The tiller subscripts refer to tillers (ear-bearing growths from the leaf axils of main stems) developed from the first, second, and third or later leaves of the main stem. (See p. 98 of the source reference.) The last column shows nitrogen content in milligrams (mg) per tiller at tillering, X.

Fit a quadratic model in X to the response Y with additional terms added to the model for tiller subscript, wheat variety, and N rate. Examine your results and provide conclusions.

C. Bars of soap are scored for their appearance in a manufacturing operation. These scores are on a 1–10 scale, and the higher the score the better. The difference between operator performance and the speed of the manufacturing line is believed to measurably affect the quality of the appearance. The following data were collected on this problem:

Operator	Line Speed	Appearance (Sum for 30 Bars)
1	150	255
1	175	246
1	200	249
2	150	260
2	175	223
2	200	231
3	150	265
3	175	247
3	200	256

Requirements
1. Using dummy variables, fit a multiple regression model to these data.
2. Using $\alpha = 0.05$, determine whether operator differences are important in bar appearance. Using the regression model, demonstrate that the average appearance score for operator No. 1 is 250, operator No. 2 is 238, and operator No. 3 is 256.
3. Does line speed affect appearance? (Use $\alpha = 0.05$.)
4. What model would you use to predict bar appearance?

D. An experimenter suggests the following dummy variable scheme to separate possible level differences among six groups. Is it a workable one?

Z_0	Z_1	Z_2	Z_3	Z_4	Z_5
1	1	−1	−1	−1	−1
1	−1	2	−1	−1	−1
1	−1	−1	3	−1	−1
1	−1	−1	−1	4	−1
1	−1	−1	−1	−1	5
1	−1	−1	−1	−1	−1

E. Here is another six-group scheme. Will it work?

Z_1	Z_2	Z_3	Z_4	Z_5
1	1	1	1	1
2	3	3	2	1
3	2	3	3	2
3	3	2	3	3
2	3	3	2	3
1	2	3	3	1

F. 1. An experimenter suggests the following dummy variable scheme to separate possible level differences among six groups. If $u = v = 0$, is it a workable one?

Z_1	Z_2	Z_3	Z_4	Z_5
−1	−1	−1	1	2
1	−1	−1	1	−1
−1	1	−1	1	u
1	1	−1	−1	v
−1	−1	1	−1	−1
1	−1	1	−1	2

2. If you said it *is* workable, are there any nonzero values of u and v that make it unworkable?
3. If you said it is *not* workable, are there any nonzero values of u and v that would make it work?

G. An experimenter says he feels the need to fit two straight lines to ten equally spaced points, the first five of which he believes are on one line, and the second five on another line. He proposes to use dummy system A, below. The statistician on the project suggests system B. Who is right?

	System A				System B		
X_0	X_1	X_2	X_3	X_0	X_1	X_2	X_3
1	1	0	-1	1	1	0	0
1	2	0	-1	1	2	0	0
1	3	0	-1	1	3	0	0
1	4	0	-1	1	4	0	0
1	5	0	-1	1	5	0	0
1	0	0	0	1	5	1	1
1	0	1	0	1	5	2	1
1	0	2	0	1	5	3	1
1	0	3	0	1	5	4	1
1	0	4	0	1	5	5	1

H. An experimenter seeks your help in fitting a "two-piece" relationship to seven observations equally spaced in time. He wants to fit a quadratic followed by a straight line. Four data points lie, apart from error, on the quadratic, three on the line. He does not know where the curve and line will intersect, and different slopes are allowable at the point of intersection. Set up his **X** matrix for him, and write down the model he must fit.

I. The following nine values of Y: 1, 4, 6, 7, 9.5, 11, 11.5, 13, 13.5, are observed at successive equally spaced time intervals. Suppose that one straight line is a suitable model for the first four observations and that a second straight line is suitable to represent the last five observations. Estimate the slopes of the two lines and find where the lines intersect. Is the fit satisfactory?

J. (*Source*: "Nutrition of infants and preschool children in the north central region of the United States of America," by E. S. Eppright, H. M. Fox, B. A. Fryer, G. H. Lamkin, V. M. Vivian, and E. S. Fuller, *World Review of Nutrition and Dietetics*, **14**, 1972, 269–332.) Below are given 72 observations of the response Y = boy's weight/height ratio (W/H), for equally spaced values of the corresponding predictor variable X = age in months. Assume that the observations fall into two groups: (1) the first seven observations and (2) the remaining 65 observations. Assume, further, that the two groups of data can be explained by two straight line time trends. Using the methods of Section 14.3, find the slopes of the two trend lines and their point of intersection. Plot the data and the fitted lines. Also provide an analysis of variance table, and find and analyze the residuals.

W/H	Age	W/H	Age	W/H	Age
0.46	0.5	0.88	24.5	0.92	48.5
0.47	1.5	0.81	25.5	0.96	49.5
0.56	2.5	0.83	26.5	0.92	50.5
0.61	3.5	0.82	27.5	0.91	51.5
0.61	4.5	0.82	28.5	0.95	52.5
0.67	5.5	0.86	29.5	0.93	53.5
0.68	6.5	0.82	30.5	0.93	54.5
0.78	7.5	0.85	31.5	0.98	55.5
0.69	8.5	0.88	32.5	0.95	56.5
0.74	9.5	0.86	33.5	0.97	57.5
0.77	10.5	0.91	34.5	0.97	58.5
0.78	11.5	0.87	35.5	0.96	59.5
0.75	12.5	0.87	36.5	0.97	60.5
0.80	13.5	0.87	37.5	0.94	61.5
0.78	14.5	0.85	38.5	0.96	62.5
0.82	15.5	0.90	39.5	1.03	63.5
0.77	16.5	0.87	40.5	0.99	64.5
0.80	17.5	0.91	41.5	1.01	65.5

W/H	Age	W/H	Age	W/H	Age
0.81	18.5	0.90	42.5	0.99	66.5
0.78	19.5	0.93	43.5	0.99	67.5
0.87	20.5	0.89	44.5	0.97	68.5
0.80	21.5	0.89	45.5	1.01	69.5
0.83	22.5	0.92	46.5	0.99	70.5
0.81	23.5	0.89	47.5	1.04	71.5

K. (This is a shorter version of the foregoing exercise.) Make use of only the first 32 observations. Assume that these 32 observations fall into two groups: (1) the first seven observations and (2) the remaining 25 observations. Assume, further, that the two groups of data can be explained by two straight line time trends. Using the methods of Section 14.3, find the slopes of the two trend lines and their point of intersection. Plot the data and the fitted lines. Also, provide an analysis of variance table, and find and analyze the residuals.

L. "Look at these data," a friend moans. "I don't know whether to fit two straight lines, one straight line, or what." You look at his notes and see that he has two sets of (X, Y) data, given below, which both cover the same X-range. How do you resolve his dilemma? Describe, and give model details, and "things he needs to do."

Set A: X	Y	Set B: X	Y
8	5.3	9	5.1
0	0.9	7	4.4
12	7.1	8	5.2
2	2.4	6	3.8

M. An experimenter has two sets of data, of (X, Y) type, and wishes to fit a quadratic equation to each set. She also wishes (later) to test if the two quadratic fits might be identical in "location" and "curvature" but have different intercept values. Explain how you would set this up for her.

N. You have two sets of data involving values of X and Y, but you are unsure whether to fit the data separately or together. You consider and fit the six-parameter model

$$Y = \beta_0 + \beta_1 X + \beta_{11} X^2 + Z(\alpha_0 + \alpha_1 X + \alpha_{11} X^2) + \epsilon,$$

where Z is a dummy variable whose value is -1 for "set A" and 1 for "set B."
1. What hypothesis would you test to answer the question: "Will a single quadratic model fit all the data?"
2. What hypothesis would you test to answer the question: "Will a single straight line model fit all the data?"
3. How would you obtain *separate* quadratic fits to the two data sets?
4. If a data point in set A and a data point in set B had the same X-value, would those two points be "repeat points" in the fit of the full model written out above?

O. Data on a response Y and k predictor variables $X_1, X_2, \ldots, X_k$ arise from two factories A and B. It is desired to fit a model of the form

$$\hat{Y}_Q = b_{0Q} + b_{1Q} X_1 + b_2 X_2 + \cdots + b_k X_k,$$

where $Q = A$ or B denotes the factory in which the prediction will be made. In other words, the effects of $X_2, \ldots, X_k$ are the same in both factories, but the intercept, and the slope with respect to X_1, are different for each factory. Show that this problem can be handled by using one new dummy variable Z, which takes the value 1 for A, and 0 for B, and then fitting the model

$$Y = \beta_0 + \beta_z Z + \beta_{1z} X_1 Z + \beta_1 X_1 + \cdots + \beta_k X_k + \epsilon$$

and that, in this case,

$$b_{0A} = b_0 + b_z, \qquad b_{1A} = b_1 + b_{1z},$$

$$b_{0B} = b_0, \qquad b_{1B} = b_1.$$

The same idea can be extended to other Xs by adding other $\beta_{jz} Z X_j$ cross-product terms. Show that, if the idea is extended to *all* the X's, we are essentially fitting separate models to the A data and to the B data.

P. (*Source*: "Application of the principle of least squares to families of straight lines," by S. Ergun, *Industrial and Engineering Chemistry*, **48**, November 1956, 2063–2068.)

1. Show that the least squares estimates $a_1, a_2, \ldots, a_m, b$, of the parameters $\alpha_1, \alpha_2, \ldots, \alpha_m, \beta$ in the family of straight lines

$$E(Y_i) = \alpha_i + \beta X_i, \qquad i = 1, 2, \ldots, m,$$

are given by

$$b = \frac{\sum_{i=1}^m \sum_{u=1}^{n_i} (X_{iu} - \overline{X}_i)(Y_{iu} - \overline{Y}_i)}{\sum_{i=1}^m \sum_{u=1}^{n_i} (X_{iu} - \overline{X}_i)^2},$$

$$a_i = \overline{Y}_i - b\overline{X}_i,$$

where

$$X_{i1}, X_{i2}, \ldots, X_{iu}, \ldots, X_{in_i},$$

$$Y_{i1}, Y_{i2}, \ldots, Y_{iu}, \ldots, Y_{in_i}$$

denote the observed values of X_i and Y_i, which relate to the ith line, $i = 1, 2, \ldots, m$. Show also that the residual sum of squares is

$$S^2 = \sum_{i=1}^m \sum_{u=1}^{n_i} (Y_{iu} - \overline{Y}_i)^2 - b^2 \sum_{i=1}^m \sum_{u=1}^{n_i} (X_{iu} - \overline{X}_i)^2$$

with $(\sum_{i=1}^m n_i - m - 1)$ degrees of freedom, that

$$\sigma_b^2 = \frac{\sigma^2}{\sum_{i=1}^m \sum_{u=1}^{n_i} (X_{iu} - \overline{X}_i)^2},$$

and that

$$\sigma_{a_i}^2 = \frac{\sigma^2}{n_i} \left\{ 1 + \frac{n_i \overline{X}_i^2}{\sum_{i=1}^m \sum_{u=1}^{n_i} (X_{iu} - \overline{X}_i)^2} \right\}.$$

2. Show that the least squares estimates $a, b_1, b_2, \ldots, b_m$, of the parameters $\alpha, \beta_1, \beta_2, \ldots, \beta_m$ in the family of straight lines

$$E(Y_i) = \alpha + \beta_i X_i, \qquad i = 1, 2, \ldots, m,$$

are given by

$$a = \frac{\sum_{i=1}^m n_i (\overline{Y}_i - \overline{X}_i \{\sum_{u=1}^{n_i} X_{iu} Y_{iu} / \sum_{u=1}^{n_i} X_{iu}^2\})}{\sum_{i=1}^m n_i (1 - n_i \overline{X}_i^2 / \sum_{u=1}^{n_i} X_{iu}^2)},$$

$$b_i = \frac{\{\sum_{u=1}^{n_i} X_{iu} Y_{iu} - a \sum_{u=1}^{n_i} X_{iu}\}}{\sum_{u=1}^{n_i} X_{iu}^2},$$

where X_{iu}, Y_{iu} are as above. Also show that the residual sum of squares is

$$S^2 = \sum_{i=1}^{m}\sum_{u=1}^{n_i} Y_{iu}^2 - \sum_{i=1}^{m} b_i^2 \sum_{u=1}^{n_i} X_{iu}^2 + a^2 \sum_{i=1}^{m} n_i - 2a \sum_{i=1}^{m} n_i \overline{Y}_i$$

with $(\sum_{i=1}^{m} n_i - m - 1)$ degrees of freedom, that

$$\sigma_a^2 = \sigma^2 / \sum_{i=1}^{m} n_i (1 - n_i \overline{X}_i^2 / \sum_{u=1}^{n_i} X_{iu}^2),$$

and that

$$\sigma_{b_i}^2 = \left\{ \frac{1}{\sum_{u=1}^{n_i} X_{iu}^2} + \frac{(\sum_{u=1}^{n_i} X_{iu})^2/(\sum_{u=1}^{n_i} X_{iu}^2)^2}{\sum_{i=1}^{m} n_i (1 - n_i \overline{X}_i^2 / \sum_{u=1}^{n_i} X_{iu}^2)} \right\} \sigma^2.$$

Q. A test for the equality of the slopes β_i of m lines represented by the first-order models

$$Y_{iu} - \overline{Y}_i = \beta_i (X_{iu} - \overline{X}_i) + \epsilon_{iu}, \qquad i = 1, 2, \ldots, m,$$

can be conducted as follows. Suppose that

$$X_{i1}, X_{i2}, \ldots, X_{iu}, \ldots, X_{in_i} \quad \text{(fixed)}$$

$$Y_{i1}, Y_{i2}, \ldots, Y_{iu}, \ldots, Y_{in_i} \quad (\epsilon_{iu} \sim N(0, \sigma^2), \text{independent})$$

are available for estimation of the parameters of the ith line. The least squares estimate of β_i is

$$b_i = \left\{ \frac{\sum_{u=1}^{n_i} (X_{iu} - \overline{X}_i)(Y_{iu} - \overline{Y}_i)}{\sum_{u=1}^{n_i} (X_{iu} - \overline{X}_i)^2} \right\}$$

with sum of squares (1 df)

$$SS(b_i) = b_i^2 \left\{ \sum_{u=1}^{n_i} (X_{iu} - \overline{X}_i)^2 \right\}$$

and residual sum of squares $(n_i - 2 \text{ df})$

$$S_i = \sum_{u=1}^{n_i} (Y_{iu} - \overline{Y}_i)^2 - SS(b_i).$$

If we assume $\beta_i = \beta$, all i, then the least squares estimate of β is

$$b = \left\{ \frac{\sum_{i=1}^{m}\sum_{u=1}^{n_i} (X_{iu} - \overline{X}_i)(Y_{iu} - \overline{Y}_i)}{\sum_{i=1}^{m}\sum_{u=1}^{n_i} (X_{iu} - \overline{X}_i)^2} \right\}$$

with sum of squares (1 df),

$$SS(b) = b^2 \left\{ \sum_{i=1}^{m}\sum_{u=1}^{n_i} (X_{iu} - \overline{X}_i)^2 \right\}$$

and residual sum of squares $(\sum n_i - 2m \text{ df})$

$$S = \sum_{i=1}^{m}\sum_{u=1}^{n_i} (Y_{iu} - \overline{Y}_i)^2 - SS(b).$$

We can form an analysis of variance table as follows:

ANOVA

Source	df	SS	MS	F
b	1	SS(b)	M_1	$F_1 = \dfrac{M_1}{s^2}$
All $b_i\|b$	$m - 1$	$\sum\limits_{i=1}^{m}$ SS(b_i) − SS(b)	M_2	$F_2 = \dfrac{M_2}{s^2}$
Residual	$\sum\limits_{i=1}^{m} n_i - 2m$	by subtraction	s^2 (estimates σ^2 if first-order models are correct)	
Total	$\sum\limits_{i=1}^{m} n_i - m$	$\sum\limits_{i=1}^{m} \sum\limits_{u=1}^{n_i} (Y_{iu} - \bar{Y}_i)^2$		

The hypothesis $H_0 : \beta_i = \beta$ is tested by comparing F_2 with an appropriate percentage point of the $F\{(m - 1), (\sum_{i=1}^{m} n_i - 2m)\}$ distribution. If H_0 is not rejected, b is used as the common slope of the lines. (This is a special case of testing a linear hypothesis. A test for the equality of intercepts of two lines can also be constructed.) F_1 is used to test $H_0 : \beta = 0$.
   Apply the above procedure to the data below.

u	X_1	Y_1	X_2	Y_2	X_3	Y_3
1	3.5	24	3.2	22	3.0	32
2	4.1	32	3.9	33	4.0	36
3	4.4	37	4.9	39	5.0	47
4	5.0	40	6.1	44	6.0	49
5	5.5	43	7.0	53	6.5	55
6	6.1	51	8.1	57	7.0	59
7	6.6	62			7.3	64
8					7.4	64

R. The table shows data from a test to compare two types of outdoor boots coded A and B. Six subjects wore pairs of each type of the boot on each of four different occasions; thus there were eight tests per subject and 48 tests in all. For each test, the subject was placed in a room held at the ambient temperature shown and the drop in temperature in degrees Fahrenheit after 90 minutes at one (randomly selected) little toe was recorded. (The rest of the clothing worn was the same for each subject.) Analyze the data and answer the question: Which boot is better and is it significantly so?

Subject Number	Boot A				Boot B			
	Ambient Temperature (°F)				Ambient Temperature (°F)			
	20	0	−5	−22	20	0	−5	−22
1	4.5	10.3	8.4	12.6	8.3	9.0	9.9	8.6
2	2.1	7.6	7.3	13.6	11.6	10.6	11.2	17.0
3	7.9	11.9	11.9	14.5	11.3	12.0	15.1	16.0
4	7.9	15.2	10.8	16.5	5.9	18.2	15.3	9.0
5	5.0	5.9	14.1	12.3	8.6	11.3	15.6	16.1
6	5.3	10.2	12.8	10.2	6.0	10.1	13.3	13.0

S. (*Source*: "Productivity of field-grown soybeans exposed to simulated acidic rain," by L. S. Evans, K. F. Lewin, M. J. Patti, and E. A. Cunningham, *New Phytologist*, **93**, 1983, 377–388.) The data shown in the table result from some 1981 experiments performed to determine the effects of simulated acidic rain on soybean yields. The lower pH levels denote "more acidic" simulated rain. Some plants were shielded from ambient rainfall, and some were not, as indicated. The response values shown are y = seed mass per plant in grams.

Plants Were Shielded from Ambient Rainfall?	x = pH Level of Simulated Rainfall			
	2.7	3.3	4.1	5.6
Yes	10.1	10.9	11.7	13.1
No	10.55	10.62	11.11	11.42

Fit the straight line model $Y = \beta_0 + \beta_1 X + \epsilon$ to each set of data. Test H_0: "the slopes of the two lines are equal" versus H_1: "not so." What models would you adopt for the two sets of data and why?

T. (*Source*: "Effects of population density on sex expression in *Onoclea sensibilis* L. on agar and ashed soil," by G. Rubin, D. S. Robson, and D. J. Paolillo, Jr., *Annals of Botany*, **55**, 1985, 205–215.) The 1983 data in the accompanying table have been adapted from Figures 1 and 2 of the source reference. The symbol X denotes time in days and Y denotes either proportion female attained at that time when gametophytes of the fern *Onoclea sensibilis* L. are cultured on agar soil (denoted by f), or proportion male attained at that time when cultured on ashed soil (denoted by m). The data of the sex opposite to that indicated were sparse and are not given. The Y data have already been translated to probit values (a transformation not described here). In adapting the data we have (a) ignored all zero Y values in days prior to those indicated, (b) ignored all 1982 data, and (c) read off the data from the graphs to a crude accuracy of two figures. Please see the original source reference for a more complete presentation.

Define a suitable numerical dummy variable to separate the two groups (f and m) of data and investigate whether two parallel straight lines will provide a good fit in either of these two cases:

(*a*) Predictor variable X, response Y
(*b*) Predictor variable $U = \ln X$, response Y.

X	Y	f or m	X	Y	f or m
18	4.8	f	40	3.3	m
20	5.5	f	42	3.8	m
24	5.8	f	45	4.1	m
28	6.0	f	62	5.0	m
30	6.5	f	71	5.5	m
33	6.6	f	75	6.0	m
36	6.7	f			
48	7.0	f			
60	7.3	f			

U. Two types of tomatoes, cherry (six plants) and yellow oval (nine plants), were grown from seeds indoors in Madison, Wisconsin, in 1996 and then replanted at three separate planting times. The numbers of tomatoes produced were recorded as follows:

	Planting Time		
	Early	Middle	Late
Cherry	47, 36	19, 20, 27, 50	—
Yellow oval	39, 50	—	4, 8, 8, 10, 12, 14, 19

Set up one dummy variable to distinguish types of tomatoes and a pair of dummies to separate the planting times, and analyze the data. What do you conclude?

CHAPTER 15

Selecting the "Best" Regression Equation

15.0. INTRODUCTION

In this chapter we discuss some specific statistical procedures for selecting variables in regression. Suppose we wish to establish a linear regression equation for a particular response Y in terms of the basic predictor variables $X_1, X_2, \ldots, X_k$. Suppose further that $Z_1, Z_2, \ldots, Z_r$, all functions of one or more of the X's, represent the complete set of variables from which the equation is to be chosen and that this set includes any functions, such as squares, cross-products, logarithms, inverses, and powers, thought to be desirable and necessary. Two opposed criteria of selecting a resultant equation are usually involved:

1. To make the equation useful for predictive purposes we should like our model to include as many Z's as are necessary to keep bias errors small, so that reliable fitted values can be determined.

On the other hand:

2. (a) To keep the variance of the predictions reasonably small (recall that the average variance of the $\hat{Y}_i$ is $p\sigma^2/n$, where p is the number of parameters in the model and n is the number of observations), and (b) because of the costs involved in obtaining information on a large number of Z's and subsequently monitoring them, we should like the equation to include as few Z's as possible.

The practical compromise between these extremes is what is usually called *selecting the best regression equation*. There is no unique statistical procedure for doing this and many procedures have been suggested. If we knew the magnitude of σ^2 (the true random variance of the observations) for any single well-defined problem, our choice of a best regression equation would be much easier. Unfortunately, we are rarely in this position, so a great deal of personal judgment will be a necessary part of any of the methods discussed. We shall describe several procedures that have been proposed; all of these have their supporters. To add to the confusion they do not all necessarily lead to the same solution when applied to the same problem, although for many problems they will achieve the same answer. We shall discuss: (1) all possible regressions using three criteria: R^2, s^2, and the Mallows C_p; (2) best subset regression using R^2, R^2 (adjusted), and C_p; (3) stepwise regression; (4) backward elimination; and (5) some variations on previous methods. After each discussion we state a personal opinion. These opinions may stimulate the reader to agree or disagree.

Some Cautionary Remarks on the Use of Unplanned Data

When we do regression calculations on unplanned data (i.e., data arising from continuing operations and not from a designed experiment), some potentially dangerous possibilities can arise, as discussed by Box (1966). The error in the model may well not be random but may result from the joint effect of several variables not incorporated in the regression equation nor, perhaps, even measured. (Box calls these *latent* or *lurking* variables.) Due to the possibilities of bias in the estimates, discussed in Section 10.1, an observed false effect of a visible variable may, in fact, be caused by an unmeasured latent variable. Provided the system continues to run in the same way as when the data were recorded, this will not mislead. However, because the latent variable is not measured, its changes will not be seen or recorded, and such changes may well cause the predicted equation to become unreliable. Another defect in unplanned data is that, often, the most effective predictor variables are kept within quite a small range to keep the response(s) within specification limits. These small ranges will then frequently cause the corresponding regression coefficients to be found "nonsignificant," a conclusion that practical workers will interpret as ridiculous because they "know" the variable is effective. Both viewpoints are, of course, compatible; if an effective predictor variable is not varied much, it will show little or no effect. A third problem with unplanned data is that the operating policy (e.g., "if X_1 goes high, reduce X_2 to compensate") often causes large correlations between the predictors. This makes it impossible to see if changes in Y are associated with X_1, or X_2, or both. A carefully designed experiment can eliminate all the ambiguities described above. The effects of latent variables can be "randomized out," effective ranges of the predictor variables can be chosen, and correlations between predictors can be avoided. Where designed experiments are not feasible, happenstance data may still be analyzed via regression methods. However, the additional possibilities of jumping to erroneous conclusions must be kept in mind.

All of the cautions above imply that the use of *any* mechanical selection procedure is fraught with possible dangers for the unwary user. Selection procedures are valuable for quickly producing regression equations worth further consideration. However, common sense, basic knowledge of the data being analyzed, and considerations related to polynomial formation (Section 12.3) cannot ever be set aside.

Example Data

We use, for illustration, the data in a four-variable ($k = 4$) problem given by Hald (1952, p. 647). These are shown in Table 15.1. This particular problem has been chosen because it illustrates some typical difficulties that occur in regression analysis, and yet it has only four predictor variables and thirteen observations. There are only 15 possible regressions containing one or more of the X's and all of these 15 are shown in Appendix 15A, together with the original source of the data, some descriptive details, and a correlation matrix. Details from these printouts will be used in the various discussions that follow.

Comments

The data in the original reference have an additional predictor variable and one more observation. We have chosen to omit these so as not to become entangled here with intricacies not related to the selection procedures but related to the data. The omitted

T A B L E 15.1. The "Hald" Regression Data (Also See Appendix 15A)

X_1	X_2	X_3	X_4	Y	Row Sum of X's
7	26	6	60	78.5	99
1	29	15	52	74.3	97
11	56	8	20	104.3	95
11	31	8	47	87.6	97
7	52	6	33	95.9	98
11	55	9	22	109.2	97
3	71	17	6	102.7	97
1	31	22	44	72.5	98
2	54	18	22	93.1	96
21	47	4	26	115.9	98
1	40	23	34	83.8	98
11	66	9	12	113.3	98
10	68	8	12	109.4	98

predictor variable varied only slightly. The data are actually from a mixtures experiment on cement in which the predictors were intended to add to 100%. Contrary to some views we have heard expressed, this does not make the data ineligible for a selection procedure, particularly as we employ only first-degree terms. We shall also make use of these data in the discussion of Chapter 19, where the special requirements of linear models for mixtures must be understood, particularly for quadratic and higher-order models.

15.1. ALL POSSIBLE REGRESSIONS AND "BEST SUBSET" REGRESSION

The first procedure is a rather cumbersome one. It requires the fitting of every possible regression equation that involves Z_0 plus any number of the variables $Z_1, \ldots, Z_r$ (where we have added a dummy variable $Z_0 = 1$ to the set of Z's as usual). Since each Z_i can either be, or not be, in the equation (two possibilities) and this is true for every Z_i, $i = 1, 2, \ldots, r$ (r Z's), there are altogether 2^r equations. (The Z_0 term is always in the equation.) If $r = 10$, a not unusually excessive number, $2^r = 1024$ equations must be examined! Each regression equation is assessed according to some criterion. The three criteria most used are:

1. The value of R^2 achieved by the least squares fit.
2. The value of s^2, the residual mean square.
3. The C_p statistic.

(All these are related to one another, in fact.) The choice of which equation is best to use is then made by assessing the patterns observed, as we describe via our example. The predictor variables here are X_1, X_2, X_3, and X_4. In this particular problem, there are no transformations, so that $Z_i = X_i$, $i = 1, 2, 3, 4$. The response variable is Y. A β_0 term is *always* included. Thus there are $2^4 = 16$ possible regression equations, which involve X_0 and the X_i, $i = 1, 2, 3, 4$. One of these is the fit of $Y = \beta_0 + \epsilon$. The other 15 fits appear in Appendix 15A. We now look at the R^2, s^2, and C_p statistics and assess the various equations.

T A B L E 15.2. R^2 Values for All Possible Regressions, Hald Data

B	C	D	E
0.675 (4)	0.979 (1, 2)	0.98234 (1, 2, 4)	0.98237 (1, 2, 3, 4)
0.666 (2)	0.972 (1, 4)	0.98228 (1, 2, 3)	
0.534 (1)	0.935 (3, 4)	0.98128 (1, 3, 4)	
0.286 (3)	0.847 (2, 3)	0.97282 (2, 3, 4)	
	0.680 (2, 4)		
	0.548 (1, 3)		

Use of the R^2 Statistic

1. Divide the runs into five sets:

Set A consists of the run with only the mean value [model $E(Y) = \beta_0$]; this is not shown.

Set B consists of the four 1-variable runs [model $E(Y) = \beta_0 + \beta_i X_i$].

Set C consists of all the 2-variable runs [model $E(Y) = \beta_0 + \beta_i X_i + \beta_j X_j$].

Set D consists of all the 3-variable runs (and so on ...).

Set E consists of the run with 4 variables.

2. Order the runs within each set by the value of the square of the multiple correlation coefficient, R^2. The parenthetical numbers in Table 15.2 are the subscripts of the X's in each equation. Later, the separating commas will be dropped but we put them in for this first display.

3. Examine the leaders and see if there is any consistent pattern of variables in the leading equations in each set. We see that after two variables have been introduced, further gain in R^2 is minor. Examination of the correlation matrix for the data (Appendix 15A) reveals that the pairs (X_1 and X_3) and (X_2 and X_4) are highly correlated, since

$$r_{13} = -0.824 \quad \text{and} \quad r_{24} = -0.973.$$

Thus the addition of further variables when X_1 and X_2 or when X_1 and X_4 are already in the regression equation will remove very little of the unexplained variation in the response. This is clearly shown by comparing Set C to Set D. The gain in R^2 from Set D to Set E is extremely small. This is simply explained by the observation that the X's are four (of five) mixture ingredients and the sum of the X-values for any specific point is nearly a constant (actually between 95 and 99).

What equation should be selected for further attention? One of the equations in Set C is clearly indicated but which one? If $f(X_1, X_2)$ is chosen, there is some inconsistency because the best single-variable equation involves X_4. For this reason many workers would prefer to use $f(X_1, X_4)$. The examination of all possible regressions does not provide a clear-cut answer to the problem. Other information, such as knowledge of the characteristics of the product studied and the physical role of the X-variables, must as always be added to enable a decision to be made.

Use of the Residual Mean Square, s^2

If all regressions are done on a large problem, an assessment of the average magnitude, $s^2(p)$, say, of the residual mean square as the number of variables in regression increases sometimes indicates the best cutoff point for the number of variables in regression.

T A B L E 15.3. Residual Mean Squares for Hald Data Regressions

p	Residual Mean Squares	Average $s^2(p)$
2	115.06, 82.39, 176.31, 80.35	113.53
3	5.79, 122.71, 7.48, 41.54, 86.89, 17.57[a]	47.00
4	5.35, 5.33, 5.65, 8.20	6.13
5	5.98	5.98

[a] For example, 17.57 is the residual mean square obtained when fitting the model containing X_3 and X_4.

For the Hald data, the various residual mean squares for all sets of "p" variables, where p is the number of parameters in the model *including* β_0, are read off the computer displays of Appendix 15A and are shown in Table 15.3. The subscripts of the variables are ordered 1, 2, 3, 4, 12, 13, ..., 123, 124, ..., 1234.

When the number of potential variables in the model is large, $r > 10$, say, and when the number of data points is much larger than r, say, $5r$ to $10r$, the plot of $s^2(p)$ is usually quite informative. The fitting of regression equations that involve more predictor variables than are necessary to obtain a satisfactory fit to data is called overfitting. As more and more predictor variables are added to an already overfitted equation, the residual mean square will tend to stabilize and approach the true value of σ^2 as the number of variables increases, provided that all important variables have been included.

For small sets of data, such as in our example, we cannot expect this idea to work as effectively, of course, but it may provide a helpful first guideline. A plot of the average $s^2(p)$ against p is shown in Figure 15.1. It appears from this that an excellent estimate of σ^2 is about 6.00, and that four parameters (i.e., three predictor variables) should be included. However, looking at the s^2 values in more detail, we see that one of the runs with $p = 3$ had a residual mean square of 5.79, indicating that there exists a better run with three parameters (i.e., two predictor variables) than was indicated by the average residual mean square for $p = 3$, namely, 47.00. This is, in fact, the fitted equation with variables X_1 and X_2 in it. The next best $p = 3$ is the (X_1, X_4) combination with $s^2 = 7.48$. Thus this procedure has given us an "asymptotic" estimate of σ^2 with which we can choose a model or models whose residual variance estimate is approximately 6 and which contain the fewest predictor variables to achieve that.

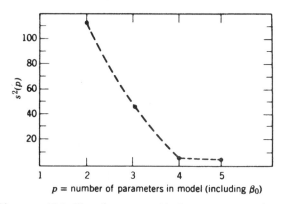

Figure 15.1. Plot of average residual mean square against p.

Use of the Mallows C_p Statistic

An alternative statistic is the C_p statistic, initially suggested by C. L. Mallows. This has the form

$$C_p = \text{RSS}_p/s^2 - (n - 2p), \tag{15.1.1}$$

where RSS_p is the residual sum of squares from a model containing p parameters, p is the number of parameters in the model *including* β_0, and s^2 is the residual mean square from the largest equation postulated containing all the Z's and is presumed to be a reliable unbiased estimate of the error variance σ^2.

Note that the notation C_p is not completely explicit, as there are several C_p statistics for each p, except for the C_p value when all variables are included, C_{r+1}. The latter value is, in fact, not a random variable at all, but is

$$C_{r+1} = \{(n - r - 1)s^2\}/s^2 - \{n - 2(r + 1)\} = r + 1. \tag{15.1.2}$$

Essentially we are comparing the various models with p parameters to the full model with $r + 1$ parameters in it. Thus C_{r+1} is always "perfect" in the C_p versus p plot described below and must be discounted.

As Kennard (1971) has pointed out, C_p is closely related to the adjusted R^2 statistic, R_a^2, and it is also related to the R^2 statistic. Now, if an equation with p parameters is adquate, that is, does not suffer from lack of fit, $E(\text{RSS}_p) = (n - p)\sigma^2$. Because we are also assuming that $E(s^2) = \sigma^2$, it is true, *approximately,* that the ratio RSS_p/s^2 has expected value $(n - p)\sigma^2/\sigma^2 = n - p$ so that, again approximately,

$$E(C_p) = p$$

for an adequate model. It follows that a plot of C_p versus p will show up the "adequate models" as points fairly close to the $C_p = p$ line. Equations with considerable lack of fit, that is, *biased equations,* will give rise to points above (often considerably above) the $C_p = p$ line. Because of random variation, points representing well-fitting equations can also fall below the $C_p = p$ line. The actual height C_p of each plotted point is also of importance because (it can be shown) it is an estimate of the overall total sum of squares of discrepancies (variance error plus bias error) of the fitted model from the true but unknown model. As terms are added to the model to reduce RSS_p, C_p usually increases. The "best" model is chosen after inspecting the C_p plot. We would look for a regression with a low C_p value about equal to p. When the choice is not clear-cut, it is a matter of personal judgment whether one prefers:

1. a biased equation that does not represent the actual data as well because it has larger RSS_p (so that $C_p > p$) but has a smaller estimate C_p of total discrepancy (variance error plus bias error) from the true but unknown model, or,

2. an equation with more parameters that fits the actual data better (i.e., $C_p \approx p$) but has a larger total discrepancy (variance error plus bias error) from the true but unknown model.

In other words, the smaller model has a smaller C_p value, but the C_p of the larger model (which has a larger value of p) is closer to its p.

Additional Reading. More detail on the considerations of judgment that apply in such cases will be found in Daniel and Wood (1980) and Gorman and Toman (1966). See also Mallows (1973). One quotation from the latter is worth repeating: "[C_p] cannot be expected to provide a single best equation when the data are intrinsically inadequate to support such a strong inference." Nor can *any* of the other selection

T A B L E 15.4. Values of C_p and p for the Hald Data Equations[a]

Subscripts of Variables in Equation				C_p Values in Same Order				p
				443.2				1
1,	2,	3,	4	202.5,	142.5,	315.2,	138.7	2
12,	13,	14		2.7,	198.1,	5.5		3
23,	24,	34		62.4,	138.2,	22.4		3
123,	124,	134,	234	3.0,	3.0	3.5,	7.3	4
	1234				5.0			5

[a] For example, for the equation with predictors X_2 and X_4 we look for 24 in the left column; the corresponding values of C_p and p are 138.2 and 3, respectively.

procedures. All selection procedures are essentially methods for the orderly displaying and reviewing of the data. Applied with common sense, they can produce useful results; applied thoughtlessly, and/or mechanically, they may be useless or even misleading. See, further, the discussion by Mallows (1995); the dangers in using C_p when there are many competing "good" C_p values are described. Also relevant is Gilmour (1996).

Example of Use of the C_p Statistic

For the Hald data (see Appendix 15A) we have $n = 13$ and $s^2 = 5.983$ from the model fitted to all four predictor variables. Thus, for example, for the model $Y = \beta_0 + \beta_1 X_1 + \epsilon$ (for which $p = 2$ notice) we find a value

$$C_p = 1265.687/5.983 - (13 - 4) = 202.5.$$

This and all the remaining C_p values are given in Table 15.4. (Recall that, for the equation with all predictors in, $C_p = p$, as must be true by definition, because in this case $RSS_p = (n - p)s^2$; here $p = r + 1 = 5$.)

A plot of the smaller C_p statistics is given in Figure 15.2. The larger ones come from models that are clearly so biased (in comparison with the ones remaining) that we can eliminate them from consideration immediately. Since we are interested only in C_p values close to p, we can use the same scale on both vertical and horizontal axes and omit the points that will not be of interest. On the basis of the C_p statistic we see that the fitted equation with X_1 and X_2 is to be preferred over all others. It not only provides the smallest C_p value overall but has the edge over the fitted equation with X_1 and X_4, which exhibits signs of bias. The conclusion that the X_1 and X_2 equation is preferable is consistent with what we have already decided by checking both the R^2 and $s^2(p)$ values for the various equations as described above. However, the conclusion comes somewhat more easily from the C_p plot in the present example.

Opinion. In general, the analysis of all regressions is quite unwarranted. While it means that the investigator has "looked at all possibilities" it also means he has examined a large number of regression equation that intelligent thought would often reject out of hand. The amount of computer time used is wasteful and the sheer physical effort of examining all the computer printouts is enormous when more than a few variables are being examined. Some sort of selection procedure that shortens this task is preferable.

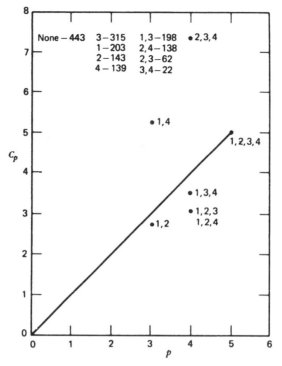

Figure 15.2. C_p plot for the Hald data.

Remark

We mentioned earlier (see Table 15.1, also) the fact that $X_1 + X_2 + X_3 + X_4$ is almost a constant. Thus any one X is almost linearly dependent on the other three. Thus if all four X's are put into the model, the $\mathbf{X'X}$ matrix has a small determinant value, 0.0010677. When an $\mathbf{X'X}$ determinant does have such a small value, it sometimes happens that the calculations involve primarily rounding errors and are meaningless. While this has not happened in the present case, the occurrence of a small determinant must always be regarded as a danger signal. Calculation of the Variance Inflation Factors (see Section 16.4) is one routine way to check for such problems.

"Best Subset" Regression

An alternative to performing all regressions is to use a program that provides a listing of the best K (you choose K) equations with one predictor variable in, two in, three in, ..., and so on. Remember that β_0 is in all equations. See, for example, Furnival and Wilson (1974). "Best" is interpreted via (1) maximum R^2 value, or (2) maximum adjusted R^2 value, or (3) the Mallows C_p statistic, your choice. The best K equations overall are indicated in some programs also. Not all $2^r - 1$ equations are examined, so equations that should appear in a list sometimes do not. Table 15.5 shows a MINI-TAB display of the "best three" ($K = 3$) selections for the Hald data. The crosses indicate which predictors are in the regression given in each line. It is clear that we would reach the same conclusions as earlier, namely, that equations with (X_1, X_2) and (X_1, X_4) look reasonable.

The MINITAB instructions for producing such a printout, with the predictor variables stored in C1–C4 and the response in C5 are simply:

T A B L E 15.5. Best Subsets Regression: MINITAB's breg Version, with K Selected as 3, for the Hald Data

Best Subsets Regression of C5

Vars	R-sq	Adj. R-sq	C-p	s	C1	C2	C3	C4
1	67.5	64.5	138.7	8.9639				X
1	66.6	63.6	142.5	9.0771		X		
1	53.4	49.2	202.5	10.727	X			
2	97.9	97.4	2.7	2.4063	X	X		
2	97.2	96.7	5.5	2.7343	X			X
2	93.5	92.2	22.4	4.1921			X	X
3	98.2	97.6	3.0	2.3087	X	X		X
3	98.2	97.6	3.0	2.3121	X	X	X	
3	98.1	97.5	3.5	2.3766	X		X	X
4	98.2	97.4	5.0	2.4460	X	X	X	X

```
breg  C5 C1-C4;
    best 3.
```

Opinion. There are some drawbacks to this type of procedure: (1) It tends to provide (in variations that supply the overall "best K" subset) equations with too many predictors included. (2) If K is chosen to be too small, the most sensible choice of fitted equation may not appear in the overall "best K" subset, though it will usually appear elsewhere in the printout. (3) No printed information is readily available concerning how the various subsets were obtained. However, provided these features are taken into account, this type of program can be useful as a quick alternative to doing all regressions.

15.2. STEPWISE REGRESSION

The stepwise regression procedure starts off by choosing an equation containing the single best X variable and then attempts to build up with subsequent additions of X's one at a time as long as these additions are worthwhile. The order of addition is determined by using the partial F-test values to select which variable should enter next. The highest partial F-value is compared to a (selected or default) F-to-enter value. After a variable has been added, the equation is examined to see if any variable should be deleted.

The basic procedure is as follows. First we select the Z most correlated with Y (suppose it is Z_1) and find the first-order, linear regression equation $\hat{Y} = f(Z_1)$. We check if this variable is significant. If it is not, we quit and adopt the model $Y = \overline{Y}$ as best; otherwise we search for the second predictor variable to enter regression.

We examine the partial F-values of all the predictor variables not in regression. The Z_j with the highest such value (suppose this is Z_2) is now selected and a second regression equation $\hat{Y} = f(Z_1, Z_2)$ is fitted. The overall regression is checked for significance, the improvement in the R^2 value is noted, and the partial F-values for *both variables now in the equation* (not just the one most recently entered[1]) are examined. The lower of these two partial F's is then compared with an appropriate F percentage point, F-to-remove, and the corresponding predictor variable is retained in the equation or rejected according to whether the test is significant or not significant.

This testing of the "least useful predictor currently in the equation" is carried out at every stage of the stepwise procedure. A predictor that may have been the best entry candidate at an earlier stage may, at a later stage, be superfluous because of the relationships between it and other variables now in the regression. To check on this, the partial F criterion for each variable in the regression at any stage of calculation is evaluated, and the lowest of these partial F-values (which may be associated with the most recent entrant or with a previous entrant) is then compared with a preselected percentage point of the appropriate F-distribution or a corresponding default F-value. This provides a judgment on the contribution of the least valuable variable in the regression at that stage, treated as though it had been the most recent variable entered, irrespective of its actual point of entry into the model. If the tested variable provides a nonsignificant contribution, it is removed from the model and the appropriate fitted regression equation is then computed for all the remaining variables still in the model.

The best of the variables not currently in the model (i.e., the one whose partial F-value, given the predictors already in the equation, is greatest) is then checked to see if it passes the partial F entry test. If it passes, it is entered, and we return to checking all the partial F's for variables in. If it fails, a further removal is attempted. Eventually (unless the α-levels for entry and removal are badly chosen to provide a cycling effect[2]), when no variables in the current equation can be remoded and the next best candidate variable cannot hold its place in the the equation, the process stops. As each variable is entered into the regression, its effect on R^2, the square of the multiple correlation coefficient, is usually recorded and printed.

We shall use the Hald data once again to illustrate the workings of the stepwise procedure. Both entry and exit tests will be made at the level $\alpha = 0.05$.

Stepwise Regression on the Hald Data

1. The first variable in is the predictor most correlated with the response. It is also the one with the largest partial F-value (same as the regression equation F-value) of all those obtained with only one variable in regression; see the first line in Table 15.6. The choice falls on X_4; $r_{4Y} = -0.821$ is the largest correlation with Y and $F_{4|-} = 22.80$ is the largest of the partial F's. (The notation $F_{4|-}$ means the partial F-value for variable X_4 when no other X's are in the equation; β_0 is always in.)

2. Test for X_4. The F-value of 22.80 exceeds $F(1, 11, 0.95) = 4.84$. Retain X_4.

3. Seek the next best X. From Table 15.6 or Appendix 15A we see it is X_1 because $F_{1|4} = 108.22$ exceeds $F_{2|4} = 0.17$, and $F_{3|4} = 40.29$. Enter X_1 and compute the equation $\hat{Y} = f(X_1, X_4)$ as

---

[1] A simpler, and less effective, procedure in which only the most recent entrant is tested is called the *forward selection procedure*. This was described in the first edition and remains an option in many stepwise computer routines. Forward selection ensures that variables entered are not subsequently removed, which may be desirable for specific applications.

[2] It is usually best to choose the same significance levels for both the entry and exit test. If a smaller α is chosen for the exit test than for the entry test, a cycling pattern may occur. Use of a larger α for the exit test is conservative and may cause variables whose contributions have weakened to be retained. Some workers find this a desirable characteristic; it is a matter of personal preference.

T A B L E 15.6. Partial F-Values for the Hald Data[a]

Subscripts of Variables in Regression	Partial F-Values of Variables Not in Regression			
	1	2	3	4
—	12.60	21.96	4.40	22.80
1	—	208.58	0.31	159.30
2	146.52	—	11.82	0.43
3	5.81	36.68	—	100.36
4	108.22	0.17	40.29	—
12	—	—	1.83	1.86
13	—	220.55	—	208.24
14	—	5.03	4.24	—
23	68.72	—	—	41.65
24	154.01	—	96.94	—
34	22.11	12.43	—	—
123	—	—	—	0.04
124	—	—	0.02	—
134	—	0.50	—	—
234	4.34	—	—	—

[a] For example, $F_{3|14} = 4.24$ is the partial F-value for b_3 given that $(b_1$ and $b_4)$ are already in the regression equation. Remember that an intercept β_0 is fitted in all of the equations but is not specifically mentioned in the table.

$$\hat{Y} = 103.10 + 1.440X_1 - 0.614X_4.$$

This has an $R^2 = 0.972$ and has an overall $F = 176.63$, clearly significant. This completes this entry step. (The notation $F_{2|4}$ indicates the partial F-value for X_2 in an equation with X_2, X_4 and intercept, and so on.)

4. Check for a possible exit. The two partial F's are $F_{1|4} = 108.22$ and $F_{4|1} = 159.30$. The recently entered X_1 is the weaker, but it is significant since $108.22 > F(1, 10, 0.95) = 4.96$. Retain both X_4 and X_1.

5. The stepwise method now selects, as the next variable to enter, the one most highly partially correlated with the response (given that variables X_4 and X_1 are already in regression). This is seen to be variable X_2. $F_{2|14} = 5.03$ exceeds $F_{3|14} = 4.24$. The recomputed equation is

$$\hat{Y} = 71.65 + 1.452X_1 + 0.416X_2 - 0.237X_4,$$

with $R^2 = 0.98234$ and a significant overall $F = 166.83$.

6. Check for a possible exit. The three partial F's are

$$F_{1|24} = 154.01, \qquad F_{2|14} = 5.03, \qquad F_{4|12} = 1.86.$$

Clearly, $F_{4|12} = 1.86 < F(1, 9, 0.95) = 5.12$, so we must remove X_4. [*Note:* Although $F_{2|14} = 5.03$ is also less than 5.12, we take no action on X_2, because we first recompute the $\hat{Y} = f(X_1, X_2)$ equation. We remove only one X at a time and then "think again."]

7. The fitted equation with X_1 and X_2 is

$$\hat{Y} = 52.58 + 1.468X_1 + 0.662X_2.$$

The partial F's are $F_{1|2} = 146.52$ and $F_{2|1} = 208.58$ from Table 15.6, so both variables are retained. $R^2 = 0.979$ and overall $F = 229.50$.

8. We now look for a new candidate variable and check $F_{3|12} = 1.83$ and $F_{4|12} = 1.86$. Although X_4 is slightly the better, it is useless to enter it, as this just takes us back to Step 6. So the stepwise procedure terminates with the result given in Step 7.

Opinion. We believe this to be one of the best of the variable selection procedures and recommend its use. It makes economical use of computer facilities, and it avoids working with more X's than are necessary while improving the equation at every stage. However, stepwise regression can easily be abused by the "amateur" statistician. As with all the procedures discussed, sensible judgment is still required in the initial selection of variables and in the critical examination of the model through examination of residuals. It is easy to rely too heavily on the automatic selection performed in the computer.

A discussion of the method is given by M. A. Efroymson in an article in Ralston and Wilf (1962).

Remarks

1. Some programs based on this precedure use a t-test on the square root of the partial F-value instead of an F-test, making use of the fact that $F(1, v) = t^2(v)$, where $F(1, v)$ is an F-variable with 1 and v degrees of freedom and $t(v)$ is a t-variable with v degrees of freedom.

2. The test values for the partial F statistics are often called "F to enter" and "F to remove." (In MINITAB, Fenter and Fremove.)

3. Some programs use a fixed F-test value (e.g., $F = 4$) rather than look up individual percentage points as the df change. Typically the default values can be reset by the user. For example, F to enter can be set either higher to restrict entry or lower to allow more variables to enter.

Forward Selection Result

Suppose we had conducted the forward selection procedure mentioned in footnote 1. What would have been different? At Step 4 only X_1 (the last X entered) would have been tested, but the net result would be unchanged. Step 5 would also be unchanged. At Step 6, however, only X_2 would have been tested (and, just, rejected) and we would not have seen that X_4 is now the weakest variable. (Had we tested at the $\alpha = 0.10$ level, however, X_2 would have been retained.) We do not recommend use of Forward Selection unless it is specifically desired never to remove variables that were retained at earlier stages.

MINITAB Version of Stepwise Regression

Table 15.7 shows the progress of the MINITAB stepwise regression method as applied to the Hald data. The instructions that produced this printout with the predictors in columns C1–C4 and the response in column C5 were as follows:

```
stepwise C5 C1-C4;
fenter = 4;
fremove = 4.
```

Note that the default values (preset and operative unless they are altered) *are*

T A B L E 15.7. Stepwise Method Applied to the Hald Data

STEPWISE REGRESSION OF   C5   ON 4 PREDICTORS, WITH N =   13

STEP	1	2	3	4
CONSTANT	117.57	103.10	71.65	52.58
C4	−0.738	−0.614	−0.237	
T-RATIO	−4.77	−12.62	−1.37	
C1		1.44	1.45	1.47
T-RATIO		10.40	12.41	12.10
C2			0.416	0.662
T-RATIO			2.24	14.44
S	8.96	2.73	2.31	2.41
R-SQ	67.45	97.25	98.23	97.87

fenter = 4 and fremove = 4, so that there was no need to specify them. We can study this printout and see the effect of lowering the fenter and fremove values. Suppose, for example, that we had set both to 1. These would correspond to t-values of $1^{1/2} = 1$. We see that X_4 would not have been removed from the equation under those circumstances and the procedure would have terminated at step 3. Note that many selection procedures have special instructions allowing one to retain specific variables if desired, even if they are not large contributors to the regression.

15.3. BACKWARD ELIMINATION

The backward elimination method is also an economical procedure. It *tries* to examine only the "best" regressions containing a certain number of variables. The basic steps in the procedure are these:

1. A regression equation containing all variables is computed.
2. The partial F-test value is calculated for every predictor variable[3] *treated as though it were the last variable to enter the regression equation.*
3. The lowest partial F-test value, say, F_L, is compared with a preselected or default significance level, say, F_0.
   a. If $F_L < F_0$, remove the variable Z_L, which gave rise to F_L, from consideration and recompute the regression equation in the remaining variables; reenter stage (2).
   b. If $F_L > F_0$, adopt the regression equation as calculated.

We illustrate this procedure using the same data (Hald, 1952) as in the previous section. Since no transformations are used here, $Z_i = X_i$ and we refer to the variables as X's in discussing the example. Printouts of all the regression equations are in Appendix 15A. Also see Table 15.6.

The regression containing all four X's is

---

[3] Of course, the partial F-value is associated with a test of $H_0: \beta = 0$ versus $H_1: \beta \neq 0$ for any particular regression coefficient but this loose phrasing, in which we talk of the F-statistic for a particular predictor variable, is convenient and colloquial, and we use it frequently here and elsewhere.

$$\hat{Y} = 62.41 + 1.551X_1 + 0.510X_2 + 0.102X_3 - 0.144X_4.$$

The four partial F-values are

$$F_{1|234} = 4.34, \qquad F_{2|134} = 0.50,$$

$$F_{3|124} = 0.02, \qquad F_{4|123} = 0.04,$$

where the X-variable subscript before the vertical line is the one examined, and those after the vertical line are the others in the regression equation. The extra sum of squares formula is used to obtain all these F's from details given in Appendix 15A. (Example calculation: $F_{1|234} = [(73.815 - 47.863)/1]/5.983 = 4.34$. Note that our calculation is based on residual sums of squares, one of three ways of calculation that could be used.) Recall that a complete table of all the partial F-values is shown as Table 15.6. The ones quoted above can also be obtained as the squares of the t-values in the regression of Y onto X_1, X_2, X_3, and X_4.

We now choose the smallest partial F-value and compare it to a critical value of F based on a predetermined α-risk (or to a default value). In this case, the critical F-value for, say $\alpha = 0.05$ is $F(1, 8, 0.95) = 5.32$. The smallest partial F is for variable X_3; namely, calculated $F = 0.018$. Since the calculated F is smaller than the critical value 5.32, we reject X_3, dropping it from further consideration.

Next, find the least squares equation, $\hat{Y} = f(X_1, X_2, X_4)$. This is

$$\hat{Y} = 71.65 + 1.452X_1 + 0.416X_2 - 0.237X_4$$

(see Appendix 15A). The overall F-value for the equation is $F = 166.83$, which is statistically significant and in fact exceeds $F(3, 9, 0.999) = 13.90$. Examining this equation for potential elimination, one sees that X_4 has the smallest partial F and is a candidate for removal. The procedure for this elimination is similar to the preceding elimination with one change: namely, the critical F-value is $F(1, 9, 0.95) = 5.12$. Because the partial F associated with X_4 is 1.86 (which is less than 5.12), we remove X_4.

We now find the least squares equation $\hat{Y} = f(X_1, X_2)$, namely,

$$\hat{Y} = 52.58 + 1.468X_1 + 0.662X_2.$$

This provides a statistically significant overall equation with an F-value of 229.50, which exceeds $F(2, 10, 0.99) = 14.91$. Both variables X_1 and X_2 are significant regardless of position; that is, each partial F-value exceeds $F(1, 10, 0.95) = 4.96$, and also both exceed higher percentage points. Thus the backward elimination selection procedure is terminated and yields the equation

$$\hat{Y} = 52.59 + 1.47X_1 + 0.66X_2.$$

Opinion. This is a satisfactory procedure, especially for statisticians who like to see all the variables in the equation once in order "not to miss anything." It is much more economical of computer and personnel time than the "all regressions" method. However, if the input data yield an $\mathbf{X'X}$ matrix that is ill conditioned—that is, nearly singular—then the overfitted equation may be nonsense due to rounding errors. With modern matrix inversion routines this is not usually a serious problem. One must also recognize that once a variable has been eliminated in this procedure *it is gone forever.* Thus all alternative models using the eliminated variable are not available for possible examination.

T A B L E 15.8. Backward Elimination Method Applied to the Hald Data

STEPWISE REGRESSION OF	C5	ON 4 PREDICTORS, WITH N = 13	
STEP	1	2	3
CONSTANT	62.41	71.65	52.58
C1	1.55	1.45	1.47
T-RATIO	2.08	12.41	12.10
C2	0.510	0.416	0.662
T-RATIO	0.70	2.24	14.44
C3	0.10		
T-RATIO	0.14		
C4	−0.14	−0.24	
T-RATIO	−0.20	−1.37	
S	2.45	2.31	2.41
R-SQ	98.24	98.23	97.87

Remarks

1. Some programs based on this procedure use a t-test on the square root of the partial F-value instead of an F-test as given above, making use of the fact that, if $F(1, \nu)$ is an F-variable with 1 and ν degrees of freedom, and if $t(\nu)$ is a t-variable with ν degrees of freedom, then $F(1, \nu) = t^2(\nu)$.

2. Some programs use the words "F to remove" in their computer printouts. This is exactly the same as the partial F.

3. Some programs use a fixed F-test value (e.g., $F = 4$) rather than look up individual percentage points as the df change. Typically the user can reset the default value to a higher value, to finally include fewer variables, or to a lower value, to leave more variables in the equation.

4. As long as the $\mathbf{X'X}$ matrix is not singular, nor badly conditioned, the residual error variance obtained from the full model is a good estimate of σ^2 in the "asymptotic" sense discussed in Section 15.1. The backward elimination procedure essentially attempts to remove all unneeded X-variables without substantially increasing the size of the "asymptotic" estimate of σ^2. Note that the C_p method takes, basically, the same approach, comparing estimates of σ^2 from various models with the estimate of σ^2 from the full model. Satisfactory models are those for which the two estimates are of about the same size.

Table 15.8 shows a printout, for the Hald data, of the MINITAB version of the backward elimination method. The program is actually implemented through the stepwise program described in the foregoing section, via instructions like the following:

```
stepwise C5 C1-C4;
enter C1-C4;
fenter = 10000;
fremove = 4.
```

The "enter" subcommand puts in all the X's to begin. The large value of fenter prevents rejected variables being restored to the equation, while the moderate value of fremove permits the removal of predictors. If fremove had been set to a number less than the square of the t-ratio for C4 at step 2, the procedure would have retained X_4 and terminated at step 2.

15.4. SIGNIFICANCE LEVELS FOR SELECTION PROCEDURES

Selecting Significance Levels in Stepwise Regression

In working the stepwise example for the Hald data in detail, we quoted the F percentage points for $\alpha = 0.05$. In Step 4, however, both F-values exceeded the 99.9% values, while in Step 5, the newly entering variable X_2 has an associated F-value that is significant at the $\alpha = 0.10$ level but not at the $\alpha = 0.05$ level. Once in, however, X_2 displaces X_4. Obviously it depends on how the program is written, and what α's are chosen for entering and removing variables whether or not a predictor even gets in. Thus choice of the α-level of the tests determines how the program will run.

1. If a small α (large F-test value) is selected, say, $\alpha \leq 0.05$, we reduce the chance of spurious variables entering. (Computer studies have shown that if we choose α as large as 0.15, columns of random numbers are often selected. There is good reason for this; see below.)

2. If a larger α (smaller F-test value) is selected, more predictor variables would typically be admitted. One can then look at the printout, however, and deduce what equation would have been offered if larger F-test values (smaller α's) had been inserted.

Clearly (2) is a more flexible approach. Most often, the entry and exit tests would be performed using the same α-value. It is also possible to use different levels for the entry and exit tests. If this is done, however, it is not wise to set the "exit α" smaller than the "entry α," or else one sometimes rejects predictors just admitted. Some workers like to set the "exit α" to be larger than the "entry α" to provide some "protection" for predictors already admitted to the equation. Such variations are a matter of personal taste, which, together with the α-values actually selected, have a great effect on the way a particular selection procedure behaves, and how many predictors are retained in the final equation. Some workers ignore the F-table altogether and simply compare the F-values with an arbitrary number such as 4. (This is used as the default value in MINITAB, for example.) To readers without strong personal opinions on this matter, we suggest setting $\alpha = 0.05$, or $\alpha = 0.10$ for both entry and exit tests, if the regression package being used allows this option (e.g., SAS). These levels can then be changed as experience dictates. (Alternatively, we can produce a larger printout by using smaller F-values, that is, larger α's.) As we discuss below the α-values are not accurate measures anyway, so that detailed agonizing over the precise choice of α is hardly worthwhile. Typically, the $\alpha = 0.05$ choice is conservative, that is, the actual value of α is much larger than 0.05 (some limited studies have been done and indicate this) and thus there will be a tendency to admit more predictors than the user might anticipate.

A Drawback to Understand But Not Be Overly Concerned About

Both the backward elimination and the stepwise procedures suffer from a difficulty that is perhaps not obvious at first sight. In the stepwise procedure, for example, the partial F-test made on the entry variable uses the largest partial F-value from all those partial F-values that might have been selected, arising from variables that are not in the regression at that stage. The correct "null" sampling and testing distribution for this is *not* the ordinary F-distribution as we implicitly assume and is very difficult to obtain, except in certain simple cases. Studies have shown, for example, that, in some cases where an entry F-test was made at the α-level, the appropriate probability was roughly $q\alpha$, where there were q entry candidates at that stage. What can be done about this? One possibility is to find out the correct probability levels for any given case, while another is to make use of a different test statistic instead of the partial F. These possibilities have been discussed in various research papers but the overall problem has not, at the time of writing, been resolved to the extent that would warrant amending the procedures. Until it is resolved, we suggest that the reader use the procedures as given, not be too concerned with the actual probability levels, and simply regard the procedures as making a series of internal comparisons that will produce what appears to be the most useful set of predictors.

As we have already mentioned, some programs simply ask the user to provide numbers F-to-enter and F-to-remove for making the tests (e.g., such as $F = 4$). A. B. Forsythe put it this way in several editions of the BMDP program book *Biomedical Computer Programs, P-Series,* W. J. Dixon, Chief Editor, University of California Press, Berkeley:

> Several users of the stepwise regression and discriminant function analysis programs have asked why we use the terms F-to-enter and F-to-remove throughout the writeup and output, rather than simply calling them F values. Some have suggested that we could ask the user for the level of significance (α) and have the program convert it to the appropriate F values. The computer programming to do this is simple enough but the statistics are difficult. Since the program is selecting the "best" variable, the usual F tables do not apply. The appropriate critical value is a function of the number of cases, the number of variables, and, unfortunately, the correlation structure of the predictor variables. This implies that the level of significance corresponding to an F-to-enter depends upon the particular set of data being used. For example, with several hundred cases and 50 potential predictors, an F-to-enter of 11 would roughly correspond to $\alpha = 5\%$ if all 50 predictors were uncorrelated. The usual use of the F table would erroneously suggest a value of 4.

(We add to this that use of a value 4 would not be "wrong" but would simply imply a higher α-value, as we have discussed. Often a *substantially* higher value is implied, however. For the example in the quotation with nominal $\alpha = 0.05$, and $m = 50$ uncorrelated predictors, the "actual" α-value, given by $1 - (1 - \alpha)^m$, would be 0.923. This formula can be used as a rough guide for the consequences in general. Note that the BMDP program book is revised periodically; a current version should be consulted for possible changes in procedure.)

(The paper by Grechanovsky and Pinsker (1995) gives a computationally intensive way to obtain accurate p-values in forward selection. Forward selection, however, is not a procedure we recommend, and even with more accuracy, the final equation for the Hald data is unchanged from what we have already given, that is, X_4 and X_1 are successively entered and the next candidate X_2 is rejected. Recall that, if the stepwise method is used, it is X_4 that is rejected, not X_2, at this stage.)

15.5. VARIATIONS AND SUMMARY

Variations on the Previous Methods

While the procedures discussed do not necessarily select the absolute best model, they usually select an acceptable one. However, alternative procedures have been suggested in attempts to improve the model selection. One proposal has been: Run the stepwise regression procedure with given levels for acceptance and rejection. When the selection procedure stops, determine the number of variables in the final selected model. Using this number of variables, say, q, do all possible sets of q variables from the r original variables and choose the best set.

Opinion. This procedure would reveal the situation pointed out in the discussion of the Hald data for the two-variable case: namely, that there are two close candidates for the model instead of one. When this happens, one could say that there is insufficient information in the data to make a clear-cut single choice. Other *a priori* considerations and the experimenter's judgment would be needed to select a final predictive model. This procedure also fails when a larger set of variables would be better but was never examined by the stepwise algorithm. In our experience, the added advantages of this procedure are minor.

Most selection programs permit the option of forcing variables into the equation, whether they would have been selected or not, ordinarily. This is often done to satisfy the fact that a variable is "known" to have an effect, even if such an effect is unlikely to be detectable (significant) in a "controlled" set of data. Forcing in variables is usually sensible and rarely does harm.

Summary

As we have seen, all the least squares selection procedures chose the fitted equation $\hat{Y} = 52.58 + 1.47X_1 + 0.66X_2$ as the "best" for the Hald data. The all-regressions method also provided $\hat{Y} = 103.10 + 1.44X_1 - 0.61X_4$ as a close second possibility, but subsequent procedures demonstrated it to be less desirable than the model involving X_1 and X_2. Although in many cases the same equation will arise from all least squares procedures, this is not guaranteed. Note that the selection of an equation with X_1 and X_2 in it means that we can explain the response values reasonably well by using only those two predictors. It does not imply that X_3 and X_4 are not needed in making the cement mixture.

In one sense the all-regressions procedure is best in that it enables us to "look at everything." The availability of best subsets routines eliminates the necessity of doing all regressions. However, for the Hald data, both the backward elimination and stepwise procedures also picked the same equation. Much would depend on the selection of rejection levels for the various F-tests, and also on the statistician's feeling about the level of increase desired in R^2 when all regressions are viewed. Inconsistent choices might lead to entirely different equations being reached by the various methods and this is not surprising.

Opinion. Our own preference for a practical regression method is the stepwise procedure. If exploration of equations "around" the stepwise choice is then desired, we prefer[4] the "best subsets" procedure, perhaps with the C_p statistic as a criterion for examination. To perform all regressions is not sensible, except when there are few predictors. If difficulties arise in the predictive interpretation of the fitted model or with the functional values of the estimated regression coefficients, then it is advisable to check the stability of the model in terms of the actual data to try to discover where the problem lies, rather than to blindly apply a rescue technique whose limitations may not be fully understood. Of course, with present day computer capabilities, one

---

[4]Some workers do the best subsets procedure first and follow it with stepwise or some alternative to stepwise.

can now do everything we have discussed with little effort. It is better to work with one selected technique and to master its particular idiosyncrasies; such an approach is probably more important than which technique *is* selected, in the long run. No technique will work well in all circumstances, no matter how good it may look on a particular example. No technique is always better than all the others. (If it were, this chapter would be shorter!) It must be continually kept in mind that, when the data are messy and not from a well-designed experiment, any fitted model is subject to constraints implied by both the patterns of the data and the practical limitations of the problem. The techniques discussed in this chapter can be useful tools. However, none of them can compensate for common sense and experience.

It is generally unwise to allow a selection procedure to choose from a set of polynomial terms; see Section 12.3 for a discussion. For an example that works out well, however, see Exercise P in "Exercises for Chapter 15."

Readers with a deeper interest in the topics of this chapter should consult the excellent and thought-provoking book by A. J. Miller (1990).

Another method of doing subset regression was proposed by Breiman (1995). In this paper, the least squares estimates are shrunk by factors whose sum is constrained. This produces a procedure "more stable" than the usual subset procedures and "less stable" than ridge regression (see Chapter 17), where stable means (here) that small changes in the data would not induce large changes in the equation selected. It is, of course, inevitable that permitting shrinkage and introducing restrictions would provide "stability." As long as these extra assumptions make practical sense, no harm is done. When the assumptions are wrong, however, a false sense of security is attained. See also Breiman (1996, e.g., p. 2374).

A more general constraint system, where the sum of the positive values of the β's to a power q could be used to restrict the parameter values, was suggested by Frank and Friedman (1993); see their p. 124. As $q \to 0$, the method tends to subset regression and $q = 1$ leads to the work of Tibshirani (1996). Again, a user must always consider whether such restrictions make practical sense. Ridge regression occurs when $q = 2$ (see Chapter 17). For other work look up "selection" in *Current Index to Statistics.*

15.6. SELECTION PROCEDURES APPLIED TO THE STEAM DATA

The steam data of Appendix 1A were subjected to selection procedure analysis by the MINITAB routines for stepwise regression, backward elimination, and best subsets regression using the following commands:

```
read C1-C10
[data from Appendix 1A]
end of data

stepwise C1 C2-C10;
fenter=1
fremove=1.

stepwise C1 C2-C10;
enter C2-C10;
fenter=10000;
fremove=1.

breg C1 C2-C10;
best 3.
```

Remarks

We have deliberately set three F-values to 1. This is to produce a larger printout than would typically arise with larger F-values.

The stepwise procedure passes through five steps and the final equation contains X_8, X_2, X_6, X_5 and X_{10}, entering in that order. If the default F-values of 4 were used, only X_8 and X_2 would enter because $F_{6|8,2} = (1.85)^2 = 3.42 < 4$, at stage 3.

The backward elimination procedure successively eliminates X_3, X_7, and X_5 and then stops at the point where the lowest F-value, for X_4, is 2.03. Note that, if we had set fremove $= 4$ rather than 1, the *same* equation, containing X_8, X_2, X_6, X_{10}, X_9, and X_4 would have been obtained! (The order of the X's is that of decreasing $|t|$ values; this facilitates comparison with the stepwise order above.) We see that the two procedures produce different equations but that X_8, X_2, X_6, and X_{10} occur in both sets.

The best subsets printout with a choice of $K = 3$ equations in each set provides some clarification. It shows that the best pair is X_8 and X_2. The best triple is X_8, X_6, and X_5, but the triple X_8, X_2, and X_6 is close behind in terms of R^2 value. Using four predictor variables contributes little to R^2, and the jump from two X's to three X's also adds little (about 2%). Both the equation with X_8 and X_2 and the equation with X_8, X_2, and X_6 seem like reasonable compromises here.

STEPWISE REGRESSION OF STEAM DATA C1 ON 9 PREDICTORS, C2-C10

STEP	1	2	3	4	5
CONSTANT	13.62299	9.47422	8.56626	0.09878	2.14331
C8	-0.0798	-0.0798	-0.0758	-0.0756	-0.0775
T-RATIO	-7.59	-10.59	-10.16	-10.50	-10.42
C2		0.76	0.49	0.30	0.43
T-RATIO		4.78	2.30	1.26	1.59
C6			0.108	0.142	0.150
T-RATIO			1.85	2.36	2.47
C5				0.29	0.22
T-RATIO				1.61	1.18
C10					-0.20
T-RATIO					-1.00
S	0.890	0.637	0.605	0.583	0.583
R-SQ	71.44	86.00	87.96	89.34	89.88

===

BACKWARD ELIMINATION ON STEAM DATA C1 USING 9 PREDICTORS, C2-C10

STEP	1	2	3	4
CONSTANT	1.8942	0.7743	-0.3667	4.3966
C2	0.71	0.48	0.54	0.66
T-RATIO	1.25	1.73	2.05	2.98
C3	-1.9			
T-RATIO	-0.46			
C4	1.13	1.28	1.27	1.30
T-RATIO	1.52	1.95	1.96	2.03
C5	0.12	0.15	0.16	
T-RATIO	0.58	0.79	0.87	

STEP	1	2	3	4	(continued)
C6	0.179	0.156	0.148	0.138	
T-RATIO	2.22	2.53	2.48	2.37	
C7	-0.018	-0.017			
T-RATIO	-0.74	-0.73			
C8	-0.077	-0.077	-0.068	-0.069	
T-RATIO	-4.67	-4.76	-6.39	-6.62	
C9	-0.086	-0.095	-0.094	-0.098	
T-RATIO	-1.65	-2.05	-2.06	-2.17	
C10	-0.35	-0.34	-0.34	-0.41	
T-RATIO	-1.64	-1.67	-1.69	-2.19	
S	0.570	0.555	0.548	0.544	
R-SQ	92.37	92.26	92.01	91.65	

```
======================================================================
```
Best Subsets Regression of Steam Data, K = 3

Vars	R-sq	Adj. R-sq	C-p	s	C2	C3	C4	C5	C6	C7	C8	C9	C10
1	71.4	70.2	35.1	0.89012							X		
1	41.0	38.5	94.9	1.2790						X			
1	28.7	25.6	119.1	1.4061					X				
2	86.0	84.7	8.5	0.63716	X						X		
2	84.9	83.5	10.7	0.66157						X	X		
2	84.7	83.3	11.1	0.66691		X					X		
3	88.5	86.8	5.6	0.59136					X	X	X		
3	88.0	86.2	6.7	0.60493	X					X	X		
3	86.5	84.6	9.5	0.64038	X						X		X
4	89.3	87.2	6.0	0.58318	X				X	X	X		
4	89.1	87.0	6.4	0.58877	X					X	X		X
4	89.0	86.8	6.6	0.59265					X	X	X	X	
5	90.0	87.3	6.7	0.58088	X	X			X		X		X
5	89.9	87.2	6.9	0.58304	X			X	X		X		X
5	89.8	87.1	7.1	0.58665			X	X	X		X	X	
6	91.7	88.9	5.4	0.54395	X		X		X		X	X	X
6	90.8	87.7	7.2	0.57264	X	X			X		X	X	X
6	90.7	87.6	7.4	0.57532	X		X	X	X		X	X	
7	92.0	88.7	6.7	0.54770	X		X	X	X		X	X	X
7	92.0	88.7	6.8	0.54917	X		X		X	X	X	X	X
7	91.9	88.5	7.0	0.55269	X	X	X		X		X	X	X
8	92.3	88.4	8.2	0.55548	X		X	X	X	X	X	X	X
8	92.2	88.3	8.3	0.55781	X	X	X		X	X	X	X	X
8	92.1	88.1	8.6	0.56168	X	X	X	X	X		X	X	X
9	92.4	87.8	10.0	0.56975	X	X	X	X	X	X	X	X	X

T A B L E 15A.1. The Hald Data

X_1	X_2	X_3	X_4	Y	Row Sum of X's
7	26	6	60	78.5	99
1	29	15	52	74.3	97
11	56	8	20	104.3	95
11	31	8	47	87.6	97
7	52	6	33	95.9	98
11	55	9	22	109.2	97
3	71	17	6	102.7	97
1	31	22	44	72.5	98
2	54	18	22	93.1	96
21	47	4	26	115.9	98
1	40	23	34	83.8	98
11	66	9	12	113.3	98
10	68	8	12	109.4	98

APPENDIX 15A. HALD DATA, CORRELATION MATRIX, AND ALL 15 POSSIBLE REGRESSIONS

(*Source:* The data were first given in "Effect of composition of Portland cement on heat evolved during hardening," by H. Woods, H. H. Steinour, and H. R. Starke, *Industrial and Engineering Chemistry,* **24**, 1932, 1207–1214, Table I.) The variables shown in Table 15A.1 are:

X_1 = amount of tricalcium aluminate, $3\,CaO \cdot Al_2O_3$.

X_2 = amount of tricalcium silicate, $3\,CaO \cdot SiO_2$.

X_3 = amount of tetracalcium alumino ferrite, $4\,CaO \cdot Al_2O_3 \cdot Fe_2O_3$.

X_4 = amount of dicalcium silicate, $2\,CaO \cdot SiO_2$.

(Response)Y = heat evolved in calories per gram of cement.

X_1, X_2, X_3, and X_4 are measured as percent of the weight of the clinkers from which the cement was made.

Table 15A.2 shows the MINITAB version of the correlation matrix for the Hald data.

The following pages (MINITAB derived) show all 15 regressions containing an

T A B L E 15A.2. Correlation Matrix for Hald Data (MINITAB Version)

```
read cl-c5
(enter Hald data now)
correlate cl-c5 ml
print ml
end
stop


                        Matrix M1
                        1.00000    0.22858   -0.82413   -0.24545    0.73072
                        0.22858    1.00000   -0.13924   -0.97295    0.81625
                       -0.82413   -0.13924    1.00000    0.02954   -0.53467
                       -0.24545   -0.97295    0.02954    1.00000   -0.82131
                        0.73072    0.81625   -0.53467   -0.82131    1.00000
```

intercept and one or more of the Hald predictors. In these, Y = column C50 and X_1–X_4 correspond to C1–C4.

```
MTB > regress c50 1 c1
```

The regression equation is
C50 = 81.5 + 1.87 C1

Predictor	Coef	Stdev	t-ratio	p
Constant	81.479	4.927	16.54	0.000
C1	1.8687	0.5264	3.55	0.005

s = 10.73   R-sq = 53.4%   R-sq(adj) = 49.2%

Analysis of Variance

SOURCE	DF	SS	MS	F	p
Regression	1	1450.1	1450.1	12.60	0.005
Error	11	1265.7	115.1		
Total	12	2715.8			

Unusual Observations

Obs.	C1	C50	Fit	Stdev.Fit	Residual	St.Resid
10	21.0	115.90	120.72	7.72	-4.82	-0.65 X

X denotes an obs. whose X value gives it large influence.

```
MTB > regress c50 1 c2
```

The regression equation is
C50 = 57.4 + 0.789 C2

Predictor	Coef	Stdev	t-ratio	p
Constant	57.424	8.491	6.76	0.000
C2	0.7891	0.1684	4.69	0.000

s = 9.077   R-sq = 66.6%   R-sq(adj) = 63.6%

Analysis of Variance

SOURCE	DF	SS	MS	F	p
Regression	1	1809.4	1809.4	21.96	0.000
Error	11	906.3	82.4		
Total	12	2715.8			

Unusual Observations

Obs.	C2	C50	Fit	Stdev.Fit	Residual	St.Resid
10	47.0	115.90	94.51	2.53	21.39	2.45R

R denotes an obs. with a large st resid.

The regression equation is
C50 = 110 - 1.26 C3

Predictor	Coef	Stdev	t-ratio	p
Constant	110.203	7.948	13.87	0.000
C3	-1.2558	0.5984	-2.10	0.060

s = 13.28   R-sq = 28.6%   R-sq(adj) = 22.1%

Analysis of Variance

SOURCE	DF	SS	MS	F	p
Regression	1	776.4	776.4	4.40	0.060
Error	11	1939.4	176.3		
Total	12	2715.8			

MTB > regress c50 1 c4

The regression equation is
C50 = 118 − 0.738 C4

Predictor	Coef	Stdev	t-ratio	p
Constant	117.568	5.262	22.34	0.000
C4	-0.7382	0.1546	-4.77	0.000

s = 8.964   R-sq = 67.5%   R-sq(adj) = 64.5%

Analysis of Variance

SOURCE	DF	SS	MS	F	p
Regression	1	1831.9	1831.9	22.80	0.000
Error	11	883.9	80.4		
Total	12	2715.8			

Unusual Observations

Obs.	C4	C50	Fit	Stdev.Fit	Residual	St.Resid
10	26.0	115.90	98.38	2.56	17.52	2.04R

R denotes an obs. with a large st. resid.

MTB > regress c50 2 c1,c2

The regression equation is
C50 = 52.6 + 1.47 C1 + 0.662 C2

Predictor	Coef	Stdev	t-ratio	p
Constant	52.577	2.286	23.00	0.000
C1	1.4683	0.1213	12.10	0.000
C2	0.66225	0.04585	14.44	0.000

s = 2.406   R-sq = 97.9%   R-sq(adj) = 97.4%

Analysis of Variance

SOURCE	DF	SS	MS	F	p
Regression	2	2657.9	1328.9	229.50	0.000
Error	10	57.9	5.8		
Total	12	2715.8			

SOURCE	DF	SEQ SS
C1	1	1450.1
C2	1	1207.8

MTB > regress c50 2 c1,c3

The regression equation is
C50 = 72.3 + 2.31 C1 + 0.494 C3

```
Predictor         Coef        Stdev      t-ratio          p
Constant         72.35        17.05         4.24      0.002
C1              2.3125       0.9598         2.41      0.037
C3              0.4945       0.8814         0.56      0.587


s = 11.08      R-sq = 54.8%      R-sq(adj) = 45.8%

Analysis of Variance

SOURCE           DF          SS           MS          F          p
Regression        2       1488.7        744.3       6.07      0.019
Error            10       1227.1        122.7
Total            12       2715.8


SOURCE           DF       SEQ SS
C1                1       1450.1
C3                1         38.6



MTB > regress c50 2 c1,c4

The regression equation is
C50 = 103 + 1.44 C1 - 0.614 C4

Predictor         Coef        Stdev      t-ratio          p
Constant       103.097        2.124        48.54      0.000
C1              1.4400       0.1384        10.40      0.000
C4            -0.61395      0.04864       -12.62      0.000


s = 2.734      R-sq = 97.2%      R-sq(adj) = 96.7%

Analysis of Variance

SOURCE           DF          SS           MS          F          p
Regression        2       2641.0       1320.5     176.63      0.000
Error            10         74.8          7.5
Total            12       2715.8

SOURCE           DF       SEQ SS
C1                1       1450.1
C4                1       1190.9

Unusual Observations
Obs.     C1        C50       Fit  Stdev.Fit   Residual   St.Resid
   8    1.0     72.500    77.523      1.241     -5.023     -2.06R

R denotes an obs. with a large st. resid.



MTB > regress c50 2 c2,c3

The regression equation is
C50 = 72.1 + 0.731 C2 - 1.01 C3

Predictor         Coef        Stdev      t-ratio          p
Constant        72.075        7.383         9.76      0.000
C2              0.7313       0.1207         6.06      0.000
C3             -1.0084       0.2934        -3.44      0.006
```

```
s = 6.445      R-sq = 84.7%      R-sq(adj) = 81.6%
```

Analysis of Variance

SOURCE	DF	SS	MS	F	p
Regression	2	2300.3	1150.2	27.69	0.000
Error	10	415.4	41.5		
Total	12	2715.8			

SOURCE	DF	SEQ SS
C2	1	1809.4
C3	1	490.9

Unusual Observations

Obs.	C2	C50	Fit	Stdev.Fit	Residual	St.Resid
10	47.0	115.90	102.41	2.92	13.49	2.35R

R denotes an obs. with a large st. resid.

MTB > regress c50 2 c2,c4

The regression equation is
C50 = 94.2 + 0.311 C2 − 0.457 C4

Predictor	Coef	Stdev	t-ratio	p
Constant	94.16	56.63	1.66	0.127
C2	0.3109	0.7486	0.42	0.687
C4	−0.4569	0.6960	−0.66	0.526

```
s = 9.321      R-sq = 68.0%      R-sq(adj) = 61.6%
```

Analysis of Variance

SOURCE	DF	SS	MS	F	p
Regression	2	1846.88	923.44	10.63	0.003
Error	10	868.88	86.89		
Total	12	2715.76			

SOURCE	DF	SEQ SS
C2	1	1809.43
C4	1	37.46

Unusual Observations

Obs.	C2	C50	Fit	Stdev.Fit	Residual	St.Resid
10	47.0	115.90	96.89	4.46	19.01	2.32R

R denotes an obs. with a large st. resid.

MTB > regress c50 2 c3,c4

The regression equation is
C50 = 131 − 1.20 C3 − 0.725 C4

Predictor	Coef	Stdev	t-ratio	p
Constant	131.282	3.275	40.09	0.000
C3	−1.1999	0.1890	−6.35	0.000
C4	−0.72460	0.07233	−10.02	0.000

```
s = 4.192      R-sq = 93.5%      R-sq(adj) = 92.2%
```

Analysis of Variance

SOURCE	DF	SS	MS	F	p
Regression	2	2540.0	1270.0	72.27	0.000
Error	10	175.7	17.6		
Total	12	2715.8			

SOURCE	DF	SEQ SS
C3	1	776.4
C4	1	1763.7

Unusual Observations

Obs.	C3	C50	Fit	Stdev.Fit	Residual	St.Resid
10	4.0	115.90	107.64	1.89	8.26	2.21R

R denotes an obs. with a large st. resid.

MTB > regress c50 3 c1,c2,c3

The regression equation is
C50 = 48.2 + 1.70 C1 + 0.657 C2 + 0.250 C3

Predictor	Coef	Stdev	t-ratio	p
Constant	48.194	3.913	12.32	0.000
C1	1.6959	0.2046	8.29	0.000
C2	0.65691	0.04423	14.85	0.000
C3	0.2500	0.1847	1.35	0.209

s = 2.312   R-sq = 98.2%   R-sq(adj) = 97.6%

Analysis of Variance

SOURCE	DF	SS	MS	F	p
Regression	3	2667.65	889.22	166.34	0.000
Error	9	48.11	5.35		
Total	12	2715.76			

SOURCE	DF	SEQ SS
C1	1	1450.08
C2	1	1207.78
C3	1	9.79

MTB > regress c50 3 c1,c2,c4

The regression equation is
C50 = 71.6 + 1.45 C1 + 0.416 C2 − 0.237 C4

Predictor	Coef	Stdev	t-ratio	p
Constant	71.65	14.14	5.07	0.000
C1	1.4519	0.1170	12.41	0.000
C2	0.4161	0.1856	2.24	0.052
C4	-0.2365	0.1733	-1.37	0.205

s = 2.309   R-sq = 98.2%   R-sq(adj) = 97.6%

Analysis of Variance

SOURCE	DF	SS	MS	F	p
Regression	3	2667.79	889.26	166.83	0.000
Error	9	47.97	5.33		
Total	12	2715.76			

SOURCE	DF	SEQ SS
C1	1	1450.08
C2	1	1207.78
C4	1	9.93

MTB > regress c50 3 c1,c3,c4

The regression equation is
C50 = 112 + 1.05 C1 − 0.410 C3 − 0.643 C4

Predictor	Coef	Stdev	t-ratio	p
Constant	111.684	4.562	24.48	0.000
C1	1.0519	0.2237	4.70	0.000
C3	−0.4100	0.1992	−2.06	0.070
C4	−0.64280	0.04454	−14.43	0.000

s = 2.377   R-sq = 98.1%   R-sq(adj) = 97.5%

Analysis of Variance

SOURCE	DF	SS	MS	F	p
Regression	3	2664.93	888.31	157.27	0.000
Error	9	50.84	5.65		
Total	12	2715.76			

SOURCE	DF	SEQ SS
C1	1	1450.08
C3	1	38.61
C4	1	1176.24

MTB > regress c50 3 c2,c3,c4

The regression equation is
C50 = 204 − 0.923 C2 − 1.45 C3 − 1.56 C4

Predictor	Coef	Stdev	t-ratio	p
Constant	203.64	20.65	9.86	0.000
C2	−0.9234	0.2619	−3.53	0.006
C3	−1.4480	0.1471	−9.85	0.000
C4	−1.5570	0.2413	−6.45	0.000

s = 2.864   R-sq = 97.3%   R-sq(adj) = 96.4%

Analysis of Variance

SOURCE	DF	SS	MS	F	p
Regression	3	2641.95	880.65	107.38	0.000
Error	9	73.81	8.20		
Total	12	2715.76			

SOURCE	DF	SEQ SS
C2	1	1809.43
C3	1	490.89
C4	1	341.63

MTB > regress c50 4 c1,c2,c3,c4

The regression equation is
C50 = 62.4 + 1.55 C1 + 0.510 C2 + 0.102 C3 − 0.144 C4

Predictor	Coef	Stdev	t-ratio	p	VIF
Constant	62.41	70.07	0.89	0.399	
C1	1.5511	0.7448	2.08	0.071	38.5
C2	0.5102	0.7238	0.70	0.501	254.4
C3	0.1019	0.7547	0.14	0.896	46.9
C4	−0.1441	0.7091	−0.20	0.844	282.5

s = 2.446   R-sq = 98.2%   R-sq(adj) = 97.4%

Analysis of Variance

SOURCE	DF	SS	MS	F	p
Regression	4	2667.90	666.97	111.48	0.000
Error	8	47.86	5.98		
Total	12	2715.76			

SOURCE	DF	SEQ SS
C1	1	1450.08
C2	1	1207.78
C3	1	9.79
C4	1	0.25

EXERCISES FOR CHAPTER 15

A. A production plant cost-control engineer is responsible for cost reduction. One of the costly items in his plant is the amount of water used by the production facilities each month. He decided to investigate water usage by collecting 17 observations on his plant's water usage and other variables. He had heard about multiple regression, but since he was quite skeptical he added a column of random numbers to his original observations. The complete set of data is shown in Table A.
1. Fit a suitable model to these data for the prediction of water usage.
2. Discuss the pros and cons of your selected equation.
3. Comment on the role of the random vector X_5.
4. Check the residuals.

T A B L E A. Water Usage Data and Correlation Matrix

Data Code
X_1 = average monthly temperature (°F)
X_2 = average of production (M pounds)
X_3 = number of plant operating days in the month
X_4 = number of persons on the monthly plant payroll
X_5 = two-digit random number
$X_6 = Y$ is the monthly water usage (gallons)

Original Data

	X_1	X_2	X_3	X_4	X_5	$X_6 = Y$
1	58.8	7107	21	129	52	3067
2	65.2	6373	22	141	68	2828
3	70.9	6796	22	153	29	2891
4	77.4	9208	20	166	23	2994
5	79.3	14792	25	193	40	3082

Original Data

	X_1	X_2	X_3	X_4	X_5	$X_6 = Y$
6	81.0	14564	23	189	14	3898
7	71.9	11964	20	175	96	3502
8	63.9	13526	23	186	94	3060
9	54.5	12656	20	190	54	3211
10	39.5	14119	20	187	37	3286
11	44.5	16691	22	195	42	3542
12	43.6	14571	19	206	22	3125
13	56.0	13619	22	198	28	3022
14	64.7	14575	22	192	7	2922
15	73.0	14556	21	191	42	3950
16	78.9	18573	21	200	33	4488
17	79.4	15618	22	200	92	3295

Correlation Matrix (Multiplied by 100)

	1	2	3	4	5	Y
1	100	-2	44	-8	11	29
2	-2	100	11	92	-11	63
3	44	11	100	3	4	-9
4	-8	92	3	100	-16	41
5	11	-11	4	-16	100	-6
Y	29	63	-9	41	-6	100

B. The demand for a consumer product is affected by many factors. In one study, measurements on the relative urbanization, educational level, and relative income of nine geographic areas were made in an attempt to determine their effect on the product usage. The data collected were as follows:

Area Number	Relative Urbanization X_1	Educational Level X_2	Relative Income X_3	Product Usage Y
1	42.2	11.2	31.9	167.1
2	48.6	10.6	13.2	174.4
3	42.6	10.6	28.7	160.8
4	39.0	10.4	26.1	162.0
5	34.7	9.3	30.1	140.8
6	44.5	10.8	8.5	174.6
7	39.1	10.7	24.3	163.7
8	40.1	10.0	18.6	174.5
9	45.9	12.0	20.4	185.7
Means	41.86	10.62	22.42	167.07
Standard deviations	s_1 4.1765	s_2 0.7463	s_3 7.9279	s_4 12.6452

The correlation matrix is

	X_1	X_2	X_3	Y
X_1	1	0.684	-0.616	0.802
X_2	0.684	1	-0.172	0.770
X_3	-0.616	-0.172	1	-0.629
Y	0.802	0.770	-0.629	1

Requirements

1. Use the stepwise procedure to determine a fitted first-order model using $F = 2.00$ for both entering and rejecting variables.

2. Write out the analysis of variance table and comment on the adequacy of the final fitted equation after examining residuals.

C. A parcel packing crew consists of five workers numbered 1 through 5 plus a foreman who works at all times. In the data below,

$$X_j = 1 \text{ if worker } j \text{ is on duty, and } 0 \text{ otherwise,}$$

$$Y = \text{number of parcels dispatched that day.}$$

By employing a regression selection procedure of your choice, find an equation of form $Y = b_0 + \Sigma \, b_j X_j$, where the summation sign may *not* be over all j, which, in your opinion, suitably explains the data. Also, obtain the usual analysis of variance table and perform other useful analyses and comment as extensively as you can on your selected equation and on any other relationships you observe. In particular, having seen your results, what advice would you give?

Run	X_1	X_2	X_3	X_4	X_5	Y
1	1	1	1	0	1	246
2	1	0	1	0	1	252
3	1	1	1	0	1	253
4	0	1	1	1	0	164
5	1	1	0	0	1	203
6	0	1	1	1	0	173
7	1	1	0	0	1	210
8	1	0	1	0	1	247
9	0	1	0	1	0	120
10	0	1	1	1	0	171
11	0	1	1	1	0	167
12	0	0	1	1	0	172
13	1	1	1	0	1	247
14	1	1	1	0	1	252
15	1	0	1	0	1	248
16	0	1	1	1	0	169
17	0	1	0	0	0	104
18	0	1	1	1	0	166
19	0	1	1	1	0	168
20	0	1	1	0	0	148
Sum	9	16	16	9	9	3880
SS	9	16	16	9	9	795364

Repeat runs: (1, 3, 13, 14), (2, 8, 15), (4, 6, 10, 11, 16, 18, 19), (5, 7)

D. (*Sources*: "Selection of variables for fitting equations to data," by J. W. Gorman and R. J. Toman, *Technometrics*, **8**, 1966, 27–51. *Fitting Equations to Data,* 2nd edition, by C. Daniel and F. S. Wood, Wiley, New York, 1980, pp. 95–103, 109–117.) Gorman and Toman discussed an experiment in which the "rate of rutting" was measured on 31 experimental asphalt pavements. Five independent or predictor variables were used to specify the conditions under which each asphalt was prepared, while a sixth "dummy" variable was used to express the difference between the two separate "blocks" of runs into which the experiment was divided. The equation used to fit the data was

$$Y = \beta_0 + \beta_1 X_1 + \beta_2 X_2 + \beta_3 X_3 + \beta_4 X_4 + \beta_5 X_5 + \beta_6 X_6 + \epsilon \tag{1}$$

where

Y = log (change of rut depth in inches per million wheel passes),

X_1 = log (viscosity of asphalt),

X_2 = percent asphalt in surface course,

X_3 = percent asphalt in base course,

X_4 = dummy variable to separate two sets of runs,

X_5 = percent fines in surface course,

X_6 = percent voids in surface course.

You may assume that Eq. (1) is "complete" in the sense that it includes all the relevant terms. Your assignment is to select a suitable subset of these terms as the "best" regression equation in the circumstances.

T A B L E D. Residual Sums of Squares (RSS) for All Possible Fitted Models, to Three Decimal Places

Variables[a]	RSS	Variables[a]	RSS	Variables[a]	RSS	Variables[a]	RSS
—	11.058	5	9.922	6	9.196	56	7.680
1	0.607	15	0.597	16	0.576	156	0.574
2	10.795	25	9.479	26	9.192	256	7.679
12	0.499	125	0.477	126	0.367	1256	0.364
3	10.663	35	9.891	36	8.848	356	7.678
13	0.600	135	0.582	136	0.567	1356	0.561
23	10.168	235	9.362	236	8.838	2356	7.675
123	0.498	1235	0.475	1236	0.365	12356	0.364
4	1.522	45	1.397	46	1.507	456	1.352
14	0.582	145	0.569	146	0.558	1456	0.553
24	1.218	245	1.030	246	1.192	2456	1.024
124	0.450	1245	0.413	1246	0.323	12456	0.313
34	1.453	345	1.383	346	1.437	3456	1.342
134	0.581	1345	0.561	1346	0.555	13456	0.545
234	1.041	2345	0.958	2346	0.995	23456	0.939
1234	0.441	12345	0.412	12346	0.311	123456	0.307

[a]Variables included in the equation; a β_0 term is included in all equations.

You will find the residual sums of squares for all the possible regressions in the accompanying table. This information will permit you to use any of the following procedures:
1. Backward elimination.
2. Forward selection.
3. Stepwise regression.
4. Examination of C_p.
5. Variations of 1–4.

You are welcome to use your own ingenuity instead of, or together with, any of these "standard" procedures.

Hint: The residual for the regression containing *no* predictors but only β_0 will give you the corrected total sum of squares.

E. (*Source:* "Sex differentials in teachers' pay," by P. Turnbull and G. Williams, *Journal of the Royal Statistical Society,* **A-137**, 1974, 245–258.) Note carefully: The data we provide were obtained by (first) a systematic selection of 100 observations (every 30th up to 3000 from an original set of 3414 observations) and (then) a weeding down to 90 observations by a removal of 10 handpicked observations. Two predictor variables were also deleted at this stage. Thus, while the data given below are in some sense representative, they do not necessarily reflect properly the behavior to be observed in the original, larger set. For this, the given reference should be consulted.

The data in the attached table consist of 90 observations on a number of characteristics of a selected group of British teachers in 1971.

Y = salary in pounds sterling.

X_1 = service in months, minus 12

X_2 = dummy variable for sex, 1 for a man, 0 for a women.

X_3 = dummy variable for sex, 1 for a man or single woman, 0 for a married women.

[Note that the combination $(X_2, X_3) = (1, 0)$ is never used. The remaining three combinations distinguish between a man, (1, 1); a single woman, (0, 1); and a married woman, (0, 0).]

X_4 = class of degree for graduates, coded 0 to 6.

X_5 = type of school in which employed, coded 0 or 1.

X_6 = 1 for trained graduate, 0 for untrained graduate or trained nongraduate teacher.

X_7 = 1 for a break in service of more than two years, 0 otherwise.

1. Using a selection procedure of your choice, fit the "best" model of the form

$$\log_{10} Y = \beta_0 + \Sigma\beta_j X_j + \epsilon$$

to the data, where the summation may not include all the possible X's.
2. Check the residuals. What other possible term for the model suggests itself?
3. Add this new term to the model and fit the "best" model now, using the same selection procedure.
4. What model would you, in fact, use if you wanted to explain a usefully large amount of the variation in the salary data? Why?

(We are grateful to Dr. P. Turnbull who kindly provided the original set of data.)

Y	X_1	X_2	X_3	X_4	X_5	X_6	X_7
998	7	0	0	0	0	0	0
1015	14	1	1	0	0	0	0
1028	18	1	1	0	1	0	0
1250	19	1	1	0	0	0	0
1028	19	0	1	0	1	0	0
1028	19	0	0	0	0	0	0
1018	27	0	0	0	0	0	1
1072	30	0	0	0	0	0	0
1290	30	1	1	0	0	0	0
1204	30	0	1	0	0	0	0
1352	31	0	1	2	0	1	0
1204	31	0	0	0	1	0	0
1104	38	0	0	0	0	0	0
1118	41	1	1	0	0	0	0
1127	42	0	0	0	0	0	0
1259	42	1	1	0	1	0	0
1127	42	1	1	0	0	0	0
1127	42	0	0	0	1	0	0
1095	47	0	0	0	0	0	1
1113	52	0	0	0	0	0	1
1462	52	0	1	2	0	1	0
1182	54	1	1	0	0	0	0
1404	54	0	0	0	1	0	0
1182	54	0	0	0	0	0	0

Y	X_1	X_2	X_3	X_4	X_5	X_6	X_7
1594	55	1	1	2	1	1	0
1459	66	0	0	0	1	0	0
1237	67	1	1	0	1	0	0
1237	67	0	1	0	1	0	0
1496	75	0	1	0	0	0	0
1424	78	1	1	0	1	0	0
1424	79	0	1	0	0	0	0
1347	91	1	1	0	1	0	0
1343	92	0	0	0	0	0	1
1310	94	0	0	0	1	0	0
1814	103	0	0	2	1	1	0
1534	103	0	0	0	0	0	0
1430	103	1	1	0	0	0	0
1439	111	1	1	0	1	0	0
1946	114	1	1	3	1	1	0
2216	114	1	1	4	1	1	0
1834	114	1	1	4	1	1	1
1416	117	0	0	0	0	0	1
2052	139	1	1	0	1	0	0
2087	140	0	0	2	1	1	1
2264	154	0	0	2	1	1	1
2201	158	1	1	4	0	1	1
2992	159	1	1	5	1	1	1
1695	162	0	1	0	0	0	0
1792	167	1	1	0	1	0	0
1690	173	0	0	0	0	0	1
1827	174	0	0	0	0	0	1
2604	175	1	1	2	1	1	0
1720	199	0	1	0	0	0	0
1720	209	0	0	0	0	0	0
2159	209	0	1	4	1	0	0
1852	210	0	1	0	0	0	0
2104	213	1	1	0	1	0	0
1852	220	0	0	0	0	0	1
1852	222	0	0	0	0	0	0
2210	222	1	1	0	0	0	0
2266	223	0	1	0	0	0	0
2027	223	1	1	0	0	0	0
1852	227	0	0	0	1	0	0
1852	232	0	0	0	0	0	1
1995	235	0	0	0	0	0	1
2616	245	1	1	3	1	1	0
2324	253	1	1	0	1	0	0
1852	257	0	1	0	0	0	1
2054	260	0	0	0	0	0	0
2617	284	1	1	3	1	1	0
1948	287	1	1	0	0	0	0
1720	290	0	1	0	0	0	1
2604	308	1	1	2	1	1	0

Y	X_1	X_2	X_3	X_4	X_5	X_6	X_7
1852	309	1	1	0	1	0	1
1942	319	0	0	0	1	0	0
2027	325	1	1	0	0	0	0
1942	326	1	1	0	1	0	0
1720	329	1	1	0	1	0	0
2048	337	0	0	0	0	0	0
2334	346	1	1	2	1	1	1
1720	355	0	0	0	0	0	1
1942	357	1	1	0	0	0	0
2117	380	1	1	0	0	0	1
2742	387	1	1	2	1	1	1
2740	403	1	1	2	1	1	1
1942	406	1	1	0	1	0	0
2266	437	0	1	0	0	0	0
2436	453	0	1	0	0	0	0
2067	458	0	1	0	0	0	0
2000	464	1	1	2	1	1	0

F. Twenty (20) mature wood samples of slash pine cross sections were prepared, 30 μm in thickness. These sections were stained jet-black in Chlorozol E. The following determinations were made.

Data on Anatomical Factors and Wood Specific Gravity of Slash Pine

Number of Fibers/mm² in Springwood	Number of Fibers/mm² in Summerwood	Spring-wood (%)	Light Absorption Springwood (%)	Light Absorption Summerwood (%)	Wood Specific Gravity
573	1059	46.5	53.8	84.1	0.534
651	1356	52.7	54.5	88.7	0.535
606	1273	49.4	52.1	92.0	0.570
630	1151	48.9	50.3	87.9	0.528
547	1135	53.1	51.9	91.5	0.548
557	1236	54.9	55.2	91.4	0.555
489	1231	56.2	45.5	82.4	0.481
685	1564	56.6	44.3	91.3	0.516
536	1182	59.2	46.4	85.4	0.475
685	1564	63.1	56.4	91.4	0.486
664	1588	50.6	48.1	86.7	0.554
703	1335	51.9	48.4	81.2	0.519
653	1395	62.5	51.9	89.2	0.492
586	1114	50.5	56.5	88.9	0.517
534	1143	52.1	57.0	88.9	0.502
523	1320	50.5	61.2	91.9	0.508
580	1249	54.6	60.8	95.4	0.520
448	1028	52.2	53.4	91.8	0.506
476	1057	42.9	53.2	92.9	0.595
528	1057	42.4	56.6	90.0	0.568

Source: "Anatomical factors influencing wood specific gravity of slash pines and the implications for the development of a high-quality pulpwood," by J. P. Van Buijtenen, *Tappi,* **47**(7), 1964, 401–404.

Requirements. Develop a prediction equation for wood specific gravity. Examine the residuals for your model and state conclusions.

G. Fit to the steam data of Appendix 1A a model for predicting monthly steam usage and which is such that:

1. R^2 exceeds 0.8.
2. All b coefficients are statistically significant (take $\alpha = 0.05$).
3. No evidence exists of patterns in the residuals obtained from the fitted model.
4. The residual standard deviation expressed as a percentage of the mean response is less than 7%.

H. In the CAED Report 17, Iowa State University, 1963, the following data are shown for the state of Iowa, 1930–1962.

Year		X_1	X_2 Pre season Precipitation (in.)	X_3 May Temperature (°F)	X_4 June Rain (in.)	X_5 June Temperature (°F)	X_6 July Rain (in.)	X_7 July Temperature (°F)	X_8 August Rain (in.)	X_9 August Temperature (°F)	Y Corn Yield (bu/acre)
1930	1	17.75	60.2	5.83	69.0	1.49	77.9	2.42	74.4	34.0
	2	14.76	57.5	3.83	75.0	2.72	77.2	3.30	72.6	32.9
	3	27.99	62.3	5.17	72.0	3.12	75.8	7.10	72.2	43.0
	4	16.76	60.5	1.64	77.8	3.45	76.1	3.01	70.5	40.0
	5	11.36	69.5	3.49	77.2	3.85	79.7	2.84	73.4	23.0
	6	22.71	55.0	7.00	65.9	3.35	79.4	2.42	73.6	38.4
	7	17.91	66.2	2.85	70.1	0.51	83.4	3.48	79.2	20.0
	8	23.31	61.8	3.80	69.0	2.63	75.9	3.99	77.8	44.6
	9	18.53	59.5	4.67	69.2	4.24	76.5	3.82	75.7	46.3
	10	18.56	66.4	5.32	71.4	3.15	76.2	4.72	70.7	52.2
	11	12.45	58.4	3.56	71.3	4.57	76.7	6.44	70.7	52.3
	12	16.05	66.0	6.20	70.0	2.24	75.1	1.94	75.1	51.0
	13	27.10	59.3	5.93	69.7	4.89	74.3	3.17	72.2	59.9
	14	19.05	57.5	6.16	71.6	4.56	75.4	5.07	74.0	54.7
	15	20.79	64.6	5.88	71.7	3.73	72.6	5.88	71.8	52.0
	16	21.88	55.1	4.70	64.1	2.96	72.1	3.43	72.5	43.5
	17	20.02	56.5	6.41	69.8	2.45	73.8	3.56	68.9	56.7
	18	23.17	55.6	10.39	66.3	1.72	72.8	1.49	80.6	30.5
	19	19.15	59.2	3.42	68.6	4.14	75.0	2.54	73.9	60.5
	20	18.28	63.5	5.51	72.4	3.47	76.2	2.34	73.0	46.1
	21	18.45	59.8	5.70	68.4	4.65	69.7	2.39	67.7	48.2
	22	22.00	62.2	6.11	65.2	4.45	72.1	6.21	70.5	43.1
	23	19.05	59.6	5.40	74.2	3.84	74.7	4.78	70.0	62.2
	24	15.67	60.0	5.31	73.2	3.28	74.6	2.33	73.2	52.9
	25	15.92	55.6	6.36	72.9	1.79	77.4	7.10	72.1	53.9
	26	16.75	63.6	3.07	67.2	3.29	79.8	1.79	77.2	48.4
	27	12.34	62.4	2.56	74.7	4.51	72.7	4.42	73.0	52.8
	28	15.82	59.0	4.84	68.9	3.54	77.9	3.76	72.9	62.1
	29	15.24	62.5	3.80	66.4	7.55	70.5	2.55	73.0	66.0
	30	21.72	62.8	4.11	71.5	2.29	72.3	4.92	76.3	64.2
	31	25.08	59.7	4.43	67.4	2.76	72.6	5.36	73.2	63.2
	32	17.79	57.4	3.36	69.4	5.51	72.6	3.04	72.4	75.4
1962	33	26.61	66.6	3.12	69.1	6.27	71.6	4.31	72.5	76.0
Averages:		17	19.09	60.8	4.85	70.3	3.55	75.2	3.82	73.2	50.0

Requirements. Construct a predictive model for corn yield (bu/acre). Comment on the relative importance of the predictor variables concerned and suggest what further investigations could be made.

I. The density of a finished product is an important performance characteristic. It can be controlled to a large extent by four main manufacturing variables:

X_1 = the amount of water in the product mix,
X_2 = the amount of reworked material in the product mix,
X_3 = the temperature of the mix,
X_4 = the air temperature in the drying chamber.

In addition, the raw material received from the supplier is important to the process, and a measure of its quality is X_5, the temperature rise. The following data were collected:

X_1	X_2	X_3	X_4	X_5	Y
0	800	135	578	13.195	104
0	800	135	578	13.195	102
0	800	135	578	13.195	100
0	800	135	578	13.195	96
0	800	135	578	13.195	93
0	800	135	578	13.195	103
0	800	150	585	13.180	118
0	800	150	585	13.180	113
0	800	150	585	13.180	107
0	800	150	585	13.180	114
0	800	150	585	13.180	110
0	800	150	585	13.180	114
0	1000	135	590	13.440	97
0	1000	135	590	13.440	87
0	1000	135	590	13.440	92
0	1000	135	590	13.440	85
0	1000	135	590	13.440	94
0	1000	135	590	13.440	102
0	1000	150	590	13.600	104
0	1000	150	590	13.600	102
0	1000	150	590	13.600	101
0	1000	150	590	13.600	104
0	1000	150	590	13.600	98
0	1000	150	590	13.600	101
75	800	135	550	12.745	103
75	800	135	550	12.745	111
75	800	135	550	12.745	111
75	800	135	550	12.745	107
75	800	135	550	12.745	112
75	800	135	550	12.745	106
75	800	150	595	13.885	111
75	800	150	595	13.885	107
75	800	150	595	13.885	104
75	800	150	595	13.885	103
75	800	150	595	13.885	104
75	800	150	595	13.885	103
75	1000	135	530	11.705	116
75	1000	135	530	11.705	108
75	1000	135	530	11.705	104
75	1000	135	530	11.705	116
75	1000	135	530	11.705	116
75	1000	135	530	11.705	112
75	1000	150	590	13.835	111
75	1000	150	590	13.835	110
75	1000	150	590	13.835	115
75	1000	150	590	13.835	114
75	1000	150	590	13.835	114
75	1000	150	590	13.835	114

Requirements

1. Examine the data carefully and state preliminary conclusions.

2. Estimate the coefficients in the model

$$Y = \beta_0 + \beta_1 X_1 + \beta_2 X_2 + \beta_3 X_3 + \beta_4 X_4 + \beta_5 X_5 + \epsilon.$$

3. Is the above model adequate?

4. Would you recommend an alternative model?

5. Draw some general conclusions about this experiment.

J. An experiment was conducted to determine the effect of six factors on rate. From the data shown below, develop a prediction equation for rate, and use it to propose a set of operating values that may provide an increased rate.

X_1	X_2	X_3	X_4	X_5	X_6	Y (Rate)
149	66	−15	150	105	383	267
143	66	−5	115	105	383	269
149	73	−5	150	105	383	230
143	73	−15	115	105	383	233
149	73	−15	115	78	383	222
143	66	−15	150	78	383	267
143	73	−5	150	78	383	231
149	66	−5	115	78	383	260
149	73	−15	150	78	196	238
149	66	−5	150	78	196	262
143	73	−5	115	78	196	252
143	66	−15	115	78	196	263
143	66	−5	150	105	196	263
149	73	−5	115	105	196	236
149	66	−15	115	105	196	268
143	73	−15	150	105	196	242

K. The data below consist of 16 indexed observations on rubber consumptions from 1948 to 1963. Use these data to develop suitable fitted equations for predicting Y_1 and Y_2 separately in terms of the predictor variables X_1, X_2, X_3, and X_4.

Observation Number	Total Rubber Consumption Y_1	Tire Rubber Consumption Y_2	Car Production X_1	Gross national Product X_2	Disposable Personal Income X_3	Motor Fuel Consumption X_4
1	0.909	0.871	1.287	0.984	0.987	1.046
2	1.252	1.220	1.281	1.078	1.064	1.081
3	0.947	0.975	0.787	1.061	1.007	1.051
4	1.022	1.021	0.796	1.013	1.012	1.046
5	1.044	1.002	1.392	1.028	1.029	1.036
6	0.905	0.890	0.893	0.969	0.993	1.020
7	1.219	1.213	1.400	1.057	1.047	1.057
8	0.923	0.918	0.721	1.001	1.024	1.034
9	1.001	1.014	1.032	0.996	1.003	1.014
10	0.916	0.914	0.685	0.972	0.993	1.013
11	1.173	1.170	1.291	1.046	1.027	1.037
12	0.938	0.952	1.170	1.004	1.001	1.007
13	0.965	0.946	0.817	1.002	1.014	1.008
14	1.106	1.096	1.231	1.049	1.032	1.024
15	1.011	0.999	1.086	1.023	1.020	1.030
16	1.080	1.093	1.001	1.035	1.053	1.029

T A B L E L. Student Questionnaire Averages for 12 Instructors

	X_1	X_2	X_3	X_4	X_5	X_6	X_7	Y
Instructor No.	Course No.	Well Org.	Log. Devel.	Questions Helpful	Out of Class	Text book	Fair Exams	Overall Grade
1	201	4.46	4.42	4.23	4.10	4.56	4.37	4.11
2	224	4.11	3.82	3.29	3.60	3.99	3.82	3.38
3	301	3.58	3.31	3.24	3.76	4.39	3.75	3.17
4	301	4.42	4.37	4.34	4.40	3.63	4.27	4.39
5	301	4.62	4.47	4.53	4.67	4.63	4.57	4.69
6	309	3.18	3.82	3.92	3.62	3.50	4.14	3.25
7	311	2.47	2.79	3.58	3.50	2.84	3.84	2.84
8	311	4.29	3.92	4.05	3.76	2.76	4.11	3.95
9	312	4.41	4.36	4.27	4.75	4.59	4.41	4.18
10	312	4.59	4.34	4.24	4.39	2.64	4.38	4.44
11	333	4.55	4.45	4.43	4.57	4.45	4.40	4.47
3	351	3.71	3.41	3.39	4.18	4.06	4.06	3.17
4	411	4.28	4.45	4.10	4.07	3.76	4.43	4.15
9	424	4.24	4.38	4.35	4.48	4.15	4.50	4.33
12	424	4.67	4.64	4.52	4.39	3.48	4.21	4.61

L. (*Source*: Steven Lemire. We are grateful to Steve for making this study available to us.) Table L, reproduced by Associated Students of Madison at the University of Wisconsin, shows average scores on a 1–5 scale (5 is the most favorable response) for 12 teachers in a particular University of Wisconsin Department in a single semester, based on results from student questionnaires in certain low-numbered courses. (Note that three teachers taught two courses each.) The response Y is the average overall grade on the student questionnaires and is often quoted alone as the crucial number for the teacher's performance. The predictor X_1 is the course number, and X_2 to X_7 are scores relating to a specific question as follows:

X_2: Were the lectures well organized?

X_3: Was the subject matter developed logically?

X_4: Were responses to questions helpful?

X_5: Was out-of-class contact helpful?

X_6: Was the textbook useful?

X_7: Were examinations graded fairly?

For Y, the question was: Overall the teacher was _____? with 1 = poor, 3 = good, 5 = excellent. Only integer responses 1, 2, 3, 4, or 5 could be made to all questions.

1. Using a selection procedure of your choice, or otherwise, find a useful equation that relates Y to $X_1, X_2, \ldots, X_7$, or to some subset of these X's.
2. Do any X's appear to be superfluous in the questionnaire? Explain how you reached that conclusion.
3. Your final equation may or may not imply a causal relationship. Suppose it did, however. What advice would you give to the teachers, to enable them to improve their scores in the future?

M. Refer to the table of Exercise D and then answer the following questions.
1. If variables 4 and 5 were already in the model, which variable would be the best one to add? Does it make a "significant" (at $\alpha = 0.05$) contribution?
2. If variables 1, 2, 3, 4, and 5 were already in the model, which would be a candidate for removal. At the $\alpha = 0.05$ level, would you remove it?
3. What is R^2 for the model with variables 1, 2, 3, and 4 in it?

4. What is the C_p statistic value for the model with variables 3, 4, 5, and 6 in it?

5. What is s^2 for the model with variables 1, 2, 4, and 6 in it?

N. A set of data given on p. 454 of K. A. Brownlee's *Statistical Theory and Methodology in Science and Engineering* (2nd edition), 1965, Wiley, New York, is reproduced below. The figures are from 21 days of operation of a plant oxidizing NH_3 to HNO_3 and the variables are:

X_1 = rate of operation,

X_2 = temperature of the cooling water in the coils of the absorbing tower for the nitric oxides,

X_3 = concentration of HNO_3 in the absorbing liquid (coded by minus 50, times 10),

Y = the percent of the ingoing NH_3 that is lost by escaping in the unabsorbed nitric oxides ($\times 10$). This is an inverse measure of the yield of HNO_3 for the plant.

Data from Operation of a Plant for the Oxidation of Ammonia to Nitric Acid

Run Number	Air Flow X_1	Cooling Water Inlet Temperature X_2	Acid Concentration X_3	Stack Loss Y
1	80	27	89	42
2	80	27	88	37
3	75	25	90	37
4	62	24	87	28
5	62	22	87	18
6	62	23	87	18
7	62	24	93	19
8	62	24	93	20
9	58	23	87	15
10	58	18	80	14
11	58	18	89	14
12	58	17	88	13
13	58	18	82	11
14	58	19	93	12
15	50	18	89	8
16	50	18	86	7
17	50	19	72	8
18	50	19	79	8
19	50	20	80	9
20	56	20	82	15
21	70	20	91	15

Investigation of plant operations indicates that the following sets of runs can be considered as replicates: (1, 2), (4, 5, 6), (7, 8), (11, 12), and (18, 19). While the runs in each set are not *exact* replicates, the points *are* sufficiently close to each other in the X-space for them to be used as such.

Requirements

1. Use a selection procedure to pick a planar model for these data and check for lack of fit.

2. Examine the residuals and comment.

O. This is a practical exercise on selecting the "best" regression equation. Coffee extract sold in the stores is obtained from coffee beans via a certain industrial process. An important response variable is

$$Y = \text{coffee density},$$

and it is anticipated that it may depend on six predictor variables,

T A B L E O. Residual Sums of Squares (RSS) for All Possible Fitted Models, to Three Decimal Places

Variables[a]	RSS	Variables[a]	RSS	Variables[a]	RSS
—	41.685	123	27.359	1234	27.199
1	40.085	124	31.242	1235	14.084
2	31.302	125	15.987	1236	27.305
3	32.663	126	30.080	1245	15.777
4	41.659	134	31.063	1246	29.747
5	33.817	135	24.746	1256	15.979
6	34.457	136	31.562	1345	23.176
12	31.276	145	31.066	1346	30.482
13	31.855	146	32.833	1356	24.680
14	39.425	156	28.017	1456	25.537
15	32.482	234	27.213	2345	13.885
16	34.437	235	14.688	2346	27.206
23	27.366	236	27.305	2356	12.695
24	31.245	245	16.081	2456	16.078
25	16.748	246	29.760	3456	23.886
26	30.251	256	16.684	12345	13.727
34	32.503	345	24.805	12346	27.189
35	25.395	346	31.211	12356	12.542
36	31.862	356	25.035	12456	15.773
45	33.517	456	26.050	13456	22.884
46	33.178			23456	12.587
56	28.042			123456	12.508

[a] In the equation; a β_0 term is included in all equations.

X_1 = static pressure,
X_2 = air inlet temperature,
X_3 = air outlet temperature,

X_4 = extract pump pressure,
X_5 = extract pump concentration,
X_6 = extract temperature,

through the model

$$Y = \beta_0 + \beta_1 X_1 + \beta_2 X_2 + \beta_3 X_3 + \beta_4 X_4 + \beta_5 X_5 + \beta_6 X_6 + \epsilon. \tag{1}$$

A set of 26 observations on $(X_1, X_2, \ldots, X_6, Y)$ is available and all possible regressions have been run; see Table O.

You may assume that model (1) is "complete" in the sense that it includes all the relevant terms. Your assignment is to select a suitable subset of these terms as the "best" regression equation in the circumstances.

You may try up to 23 different equations. For each subset of variables that you specify, you will be given (only) the residual sum of squares for that regression. This information will permit you to use any of the following selection procedures: backward elimination, forward selection, stepwise regression, examination of C_p, or variations of the above. You should now decide on the method of your choice and proceed. Remember that, to test for entry or exit of a single variable at any stage, you can use this formula for the test statistic F:

$$\frac{\{(\text{Residual SS with that variable out}) - (\text{Residual SS with that variable in})\}/1}{s^2 \text{ from larger model with that variable in}}.$$

The appropriate degrees of freedom for the test are (1, df of s^2).

(*Hint*: The residual for the regression containing *no* predictors but only β_0 will give you the corrected total sum of squares.)

TABLE P. Kilgo Data

x_1 Pressure (bar)	x_2 Temperature (°C)	x_3 Moisture (% by wt.)	x_4 Flow Rate (liters/min)	x_5 Average Size (mm)	Y Solubility
−1	−1	−1	−1	−1	29.2
1	−1	−1	−1	1	23.0
−1	1	−1	−1	1	37.0
1	1	−1	−1	−1	139.7
−1	−1	1	−1	1	23.3
1	−1	1	−1	−1	38.3
−1	1	1	−1	−1	42.6
1	1	1	−1	1	141.4
−1	−1	−1	1	1	22.4
1	−1	−1	1	−1	37.2
−1	1	−1	1	−1	31.3
1	1	−1	1	1	48.6
−1	−1	1	1	−1	22.9
1	−1	1	1	1	36.2
−1	1	1	1	1	33.6
1	1	1	1	−1	172.6

Source: The table is adapted from pp. 48 and 49 of the source reference by courtesy of Marcel Dekker, Inc.

P. (*Source*: "An application of fractional factorial experimental design." by M. B. Kilgo, *Quality Engineering,* **1,** 1988–89, 45–54.) Results from a 2_V^{5-1} fractional factorial design are shown in Table P in so-called standard order. The levels of the predictor variables $x_1, x_2, \ldots x_5$ are all coded to ± 1 and the response Y is the "solubility" measured in an experiment involving the extraction of oil from peanuts. These data are probably best analyzed by the standard factorial analysis, but consider the following selection procedure approach. Consider the predictor variables $x_1, x_2, \ldots x_5, x_1x_2, x_1x_3, \ldots x_4x_5$ and fit a useful linear model via the selection procedure of your choice. When you have decided on a suitable model, check the residuals and then recheck the data themselves. Notice anything? If yes, suggest a remedy, and then fit your chosen model again. What are your conclusions? (*Hint*: One of the observations may be faulty.)

Q. (*Source*: Dr. Jim Douglas, University of New South Wales, Sydney, Australia.) Because of a number of road alterations in Sydney, Dr. Douglas was faced with the problem of choosing between alternative routes for driving into the University in the most attractive way. The criteria were not altogether well defined, but certainly involved short travel time, short total distance, small exposure to fast multi-lane traffic, and pleasant surroundings. Dr. Douglas chose $Y =$ travel time in minutes as the response variable and his predictors were specially coded in the following manner: $X_1 = 1, 2$ for into and out of UNSW; $X_2 = 0, 1$ for no school holiday; $X_3 =$ time of departure, coded to reflect the way accident reports relate to the time of day; $X_4 =$ day of week, coded 0 for Sunday, 1.7 for Monday and Tuesday, 2.3 for Wednesday, 3.6 for Thursday, 7 for Friday and 3.9 for Saturday; these numbers were proportional to the average daily numbers of accidents; $X_5 =$ coded weather; $X_6 =$ route; $X_7 =$ number of persons in car. Nineteen observations, shown on page 565, were obtained on these variables. The runs were not fully randomized, said Dr. Douglas; "If it's a beautiful day, how can I not drive past Botany Bay!" A perfectly reasonable decision! Analyze the data using any of the selection procedures or analyses you have learned, and state your conclusions.

CHAPTER 16

Ill-Conditioning in Regression Data

16.1. INTRODUCTION

Suppose we wish to fit the model $Y = X\beta + \epsilon$. The solution $b = (X'X)^{-1}X'Y$ would usually be sought. However, if $X'X$ is singular, we cannot perform the inversion and the normal equations do not have a unique solution. (An infinity of solutions exists instead.) When this happens, it stems from the fact that there is at least one linear combination of the columns of the X matrix that is zero. Or to put in another way, at least one column of X is linearly dependent on (i.e., is a linear combination of) the other columns. We would say that collinearity (or *multicollinearity*) exists among the columns of X.

Confusingly, the words collinearity and multicollinearity are also often used in regression when there is a "near dependency" in the columns of X. (What "near" means is also a problem.) This is why the situation when $\det(X'X) = 0$ is sometimes called "exact collinearity" or "exact multicollinearity." The adjective "exact" is needed only because of the weakening in the use of the nouns. In general, we assume that the word "multicollinearity" means a "near dependence" in the X columns. One can also say that the data are *ill-conditioned* in such a case. (Alternative phrases are *poorly conditioned* or *badly conditioned*.)

Ill-conditioning is undesirable in regression analysis. It usually leads to unreliable estimates of the regression coefficients, which then have large variances and covariances. Whether the data are well-conditioned or ill-conditioned depends on the way the data relate *to the specific model under consideration*. Data are often collected with a particular model in mind, in fact.

A Simple Example

Consider the data:

X	4	4	7	7	7.1	7.1
Y	19	20	37	39	36	38

From the plot in Figure 16.1 we see that, although the data are at three "distinct" locations, $X = 4, 7$, and 7.1, the locations 7 and 7.1 are relatively very close, compared to the gap between $X = 4$ and 7. With three X-sites, we usually would be able to fit a quadratic model $Y = \beta_0 + \beta_1 X + \beta_{11} X^2 + \epsilon$. If we try to do this here, we obtain normal equations $X'Xb = X'Y$ of the form

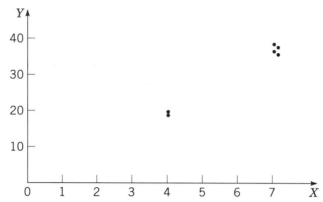

Figure 16.1. A simple example of data ill-conditioned for fitting a quadratic model $Y = \beta_0 + \beta_1 X + \beta_{11} X^2 + \epsilon$.

$$\begin{bmatrix} 6 & 36.2 & 230.82 \\ 36.2 & 230.82 & 1529.82 \\ 230.82 & 1529.82 & 10396.3362 \end{bmatrix} \begin{bmatrix} b_0 \\ b_1 \\ b_{11} \end{bmatrix} = \begin{bmatrix} 189 \\ 1213.4 \\ 8078.34 \end{bmatrix}.$$

The **X'X** matrix is close to singular. (One well-known regression program simply rejects entry of the X^2 term, in fact.) The fitted equation is

$$\hat{Y} = -151 + 63.5X - 5.2X^2$$
$$(113) \quad (44) \quad (4.0)$$

with very large standard errors shown in the parentheses. Obviously the regression is meaningless, this arising from the strong near-dependence in the columns of **X**. A look at Figure 16.1 makes it clear that a lot of different quadratic fits will pick up most of the variation here.

The data are well explained ($R^2 = 0.984$) by the least squares line $\hat{Y} = -4.03 + 5.89X$, with standard errors for the coefficients of 2.3 and 0.38, respectively. So the data are *not* inadequate for a straight line fit, but they are completely inadequate for a quadratic fit. Putting this another way, we can say that the data are well-conditioned for a straight line model fit, but very ill-conditioned for a quadratic model fit. Ill-conditioning in data is thus not just an attribute of the data alone. The level of conditioning of data depends on the model to which the data are supposed to relate. (One of the basic aspects of designing good experiments is, naturally enough, assuring that the data are adequate for actually estimating the model being considered for them. When the columns of **X** are orthogonal, the best possible conditioning is attained, and **X'X** is a diagonal matrix.)

Demonstrating Dependence in X Via Regression

To show the dependence of the three columns in our example, we can regress any one column of **X** on all the others. If we use X^2 as the "response," for example, we get a fitted equation $\hat{X}^2 = -28.217(1) + 11.053(X)$, which accounts for over 99.99% of the variation about the mean of the X^2 values. This equation thus holds almost exactly and exhibits the near linear dependence of the three columns 1, X, and X^2.

We next discuss the concepts of *centering* and *scaling* data and then discuss how multicollinearity might be detected in a set of data.

16.2. CENTERING REGRESSION DATA

Suppose we have n observations on a response Y and r predictors $X_1, X_2, \ldots, X_r$. For the moment, let us assume that the X's are separately measured, unconnected variables (e.g., one X is not the square or logarithm of another, or a cross-product of two others). Suppose the model of interest is (for $u = 1, 2, \ldots, n$)

$$Y_u = \beta_0 + \beta_1 X_{1u} + \beta_2 X_{2u} + \cdots + \beta_r X_{ru} + \epsilon. \qquad (16.2.1)$$

The **X** matrix for this fit consists of the columns:

$X_0 = 1$	X_1	X_2	$\cdots$	X_r
1	X_{11}	X_{21}	$\cdots$	X_{r1}
1	X_{12}	X_{22}	$\cdots$	X_{r2}
$\cdots$				
1	X_{1n}	X_{2n}	$\cdots$	X_{rn}

$$(16.2.2)$$

with sums $\quad n \quad\quad \Sigma X_{1u} \;\; \Sigma X_{2u} \quad\quad\quad \Sigma X_{ru}$ (summed over $u = 1, 2, \ldots, n$)
and averages $\quad 1 \quad\quad \overline{X}_1 \quad\;\; \overline{X}_2 \quad \cdots \quad \overline{X}_r$

The **X'X** matrix and **X'Y** vector for this model are (all sums are over $u = 1, 2, \ldots, n$)

$$\begin{bmatrix} n & \Sigma X_{1u} & \Sigma X_{2u} & \cdots & \Sigma X_{ru} \\ \Sigma X_{1u} & \Sigma X_{1u}^2 & \Sigma X_{1u} X_{2u} & \cdots & \Sigma X_{1u} X_{ru} \\ \vdots & \vdots & \vdots & & \vdots \\ \Sigma X_{ru} & \Sigma X_{ru} X_{1u} & \Sigma X_{ru} X_{2u} & \cdots & \Sigma X_{ru}^2 \end{bmatrix} \quad \begin{bmatrix} \Sigma Y_u \\ \Sigma X_{1u} Y_u \\ \vdots \\ \Sigma X_{ru} Y_u \end{bmatrix} \qquad (16.2.3)$$

The **X'X** matrix in (16.2.3) will be singular if and only if the columns of (16.2.2) are linearly dependent, that is, if and only if there exist constants $c_0, c_1, \ldots, c_r$, say, not all zero, such that

$$c_0 + c_1 X_{1u} + c_2 X_{2u} + \cdots + c_r X_{ru} = 0 \qquad (16.2.4)$$

for every $u = 1, 2, \ldots, n$. Or we can use vectors of (16.2.2) to write the same equation as

$$c_0 \mathbf{1} + c_1 \mathbf{X}_1 + c_2 \mathbf{X}_2 + \cdots + c_r \mathbf{X}_r = 0. \qquad (16.2.5)$$

Centering

We can rewrite the model (16.2.1) as

$$Y_u = (\beta_0 + \beta_1 \overline{X}_1 + \beta_2 \overline{X}_2 + \cdots + \beta_r \overline{X}_r)$$
$$+ \beta_1(X_{1u} - \overline{X}_1) + \cdots + \beta_r(X_{ru} - \overline{X}_r) + \epsilon, \qquad (16.2.6)$$
$$= \beta_0' + \beta_1(X_{1u} - \overline{X}_1) + \cdots + \beta_r(X_{ru} - \overline{X}_r) + \epsilon, \qquad (16.2.7)$$

where $n \overline{X}_i = \Sigma_u X_{iu}$ and where

$$\beta_0' = \beta_0 + \beta_1 \overline{X}_1 + \cdots + \beta_r \overline{X}_r. \qquad (16.2.8)$$

The $\mathbf{X}$ matrix for the model (16.2.7) consists of columns

$$
\begin{array}{cccc}
1 & (X_{11} - \overline{X}_1) & \cdots & (X_{r1} - \overline{X}_r) \\
1 & (X_{12} - \overline{X}_1) & \cdots & (X_{r2} - \overline{X}_r) \\
\vdots & \vdots & & \vdots \\
1 & (X_{1n} - \overline{X}_1) & \cdots & (X_{rn} - \overline{X}_r)
\end{array}
\tag{16.2.9}
$$

Note that the first column is orthogonal to all the others now, and that the $\mathbf{X}'\mathbf{X}$ matrix and $\mathbf{X}'\mathbf{Y}$ vector for this fit are

$$
\begin{bmatrix}
n & 0 & 0 & \cdots & 0 \\
0 & S_{11} & S_{12} & \cdots & S_{1r} \\
0 & S_{21} & S_{22} & \cdots & S_{2r} \\
\vdots & \vdots & \vdots & & \vdots \\
0 & S_{r1} & S_{r2} & \cdots & S_{rr}
\end{bmatrix}
\quad \text{and} \quad
\begin{bmatrix}
\Sigma Y_u \\
\Sigma(X_{1u} - \overline{X}_1)Y_u \\
\Sigma(X_{2u} - \overline{X}_2)Y_u \\
\vdots \\
\Sigma(X_{ru} - \overline{X}_r)Y_u
\end{bmatrix}.
\tag{16.2.10}
$$

Here $S_{ij} = \Sigma(X_{iu} - \overline{X}_i)(X_{ju} - \overline{X}_j)$ and all summations are over $u = 1, 2, \ldots, n$ unless otherwise stated. The first normal equation thus immediately gives $b_0' = \overline{Y}$. Moreover, the remaining normal equations have the solution

$$
\begin{bmatrix}
b_1 \\
b_2 \\
\vdots \\
b_r
\end{bmatrix}
=
\begin{bmatrix}
S_{11} & S_{12} & \cdots & S_{1r} \\
S_{21} & S_{22} & \cdots & S_{2r} \\
\vdots & \vdots & & \vdots \\
S_{r1} & S_{r2} & \cdots & S_{rr}
\end{bmatrix}^{-1}
\begin{bmatrix}
\Sigma(X_{1u} - \overline{X}_1)(Y_u - \overline{Y}) \\
\Sigma(X_{2u} - \overline{X}_2)(Y_u - \overline{Y}) \\
\vdots \\
\Sigma(X_{ru} - \overline{X}_r)(Y_u - \overline{Y})
\end{bmatrix}.
\tag{16.2.11}
$$

Where do the extra $\overline{Y}$'s come from as we go from (16.2.10) to (16.2.11)? They come from the fact that

$$
\Sigma(X_{iu} - \overline{X}_i)Y_u = \Sigma(X_{iu} - \overline{X}_i)(Y_u - \overline{Y})
\tag{16.2.12}
$$

because the term we have subtracted is $\overline{Y}\Sigma(X_{iu} - \overline{X}_i) = 0$. All of this implies that, if we fit the model in the form

$$
Y_u - \overline{Y} = \beta_1(X_{1u} - \overline{X}_1) + \cdots + \beta_r(X_{ru} - \overline{X}_r) + \epsilon'
\tag{16.2.13}
$$

to give estimates $b_1, b_2, \ldots, b_r$, and if [see (16.2.8) and the fact that $b_0' = \overline{Y}$] we recover b_0 from

$$
b_0 = \overline{Y} - b_1\overline{X}_1 - b_2\overline{X}_2 - \cdots - b_r\overline{X}_r,
\tag{16.2.14}
$$

the estimates will be *exactly* the same as we would have obtained from fitting (16.2.1).

The data $(Y_u - \overline{Y}), (X_{1u} - \overline{X}_1), \ldots, (X_{ru} - \overline{X}_r)$ are said to be *centered around their average values*, or often just *centered*. (One can also "center" about values other than the mean, but we shall not use the term in that way.)

If the model contains terms other than simple, separate X's, for example, if it contains X_1^2 or X_1X_2 or $\ln X_1$, the same arguments as above apply except that the centering takes place around the column averages $\overline{X_1^2}$ (the average of the X_1^2 values) or $\overline{X_1X_2}$ (the average of the X_1X_2 values) or $\overline{\ln X_1}$ (the average of the $\ln X_1$ values).

Singularity and Centering

Suppose the centered X's are linearly dependent, that is, there exist constants $c_1, c_2, \ldots, c_r$ not all zero such that

$$c_1(X_{1u} - \overline{X}_1) + \cdots + c_r(X_{ru} - \overline{X}_r) = 0. \qquad (16.2.15)$$

This implies that

$$-(c_1\overline{X}_1 + \cdots + c_r\overline{X}_r) + c_1 X_{1u} + \cdots + c_r X_{ru} = 0, \qquad (16.2.16)$$

which, if we denote the first term by c_0, is (16.2.4). Thus if the centered X's are dependent, the uncentered X's are also dependent. The reverse of this is not necessarily true, however. For example, suppose c_0 and c_1 are nonzero and

$$c_0 + c_1 X_{1u} = 0 \qquad (16.2.17)$$

is the only relationship among the **X**-columns. This implies that all $X_{1u} = \overline{X}_1$ so that $X_{iu} - \overline{X} = 0$ for all u. Examining columns of centered $(X_{iu} - \overline{X}_i)$'s for linear relationships will thus be useless in detecting (16.2.17); recall that no other relationships exist by our assumption. More generally, two vectors $\mathbf{X}_1$ and $\mathbf{X}_2$, both close to the **1** vector, and thus to each other, could be such that the "residual" vectors $\mathbf{X}_1 - \overline{X}_1\mathbf{1}$ and $\mathbf{X}_2 - \overline{X}_2\mathbf{1}$ were orthogonal. Again, examination of the centered vectors would reveal no dependency relationship. For (a lot) more on this, see Chapter 6 (especially Section 6.3) of Belsley (1991).

The basic message from this section is that assessment of collinearity should be made on *uncentered* X-data, including the **1** column, not centered data.

16.3. CENTERING AND SCALING REGRESSION DATA

Suppose we not only center the variables as in (16.2.9) but divide each column there, except the first, by the sum of squares of its members, namely, by $S_{11}^{1/2}, S_{22}^{1/2}, \ldots, S_{rr}^{1/2}$, respectively. Let us also do the same to the centered Y's (without worrying about the effect of this on the error structure of the new responses). If we write

$$x_{iu} = \frac{X_{iu} - \overline{X}_i}{S_{ii}^{1/2}}, \quad i = 1, 2, \ldots, r, \quad \text{and} \quad y_u = \frac{Y_u - \overline{Y}}{S_{YY}^{1/2}}, \qquad (16.3.1)$$

where $S_{YY} = \Sigma(Y_u - \overline{Y})^2$, and substitute these in (16.2.13), we shall obtain

$$y S_{YY}^{1/2} = \beta_1 S_{11}^{1/2} x_1 + \cdots + \beta_r S_{rr}^{1/2} x_r + \epsilon'. \qquad (16.3.2)$$

Dividing through by $S_{YY}^{1/2}$, again without worrying about the effect of this on the error structure, gives

$$y = \alpha_1 x_1 + \alpha_2 x_2 + \cdots + \alpha_r x_r + \epsilon'', \qquad (16.3.3)$$

where $\alpha_1 = \beta_1(S_{11}/S_{YY})^{1/2}, \ldots, \alpha_r = \beta_r(S_{rr}/S_{YY})^{1/2}$ are now coefficients to be estimated from the manipulated data $y, x_1, \ldots, x_r$, by least squares. The form of the normal equations for such a fit are

$$
\begin{bmatrix}
1 & c_{12} & c_{13} & \cdots & c_{1r} \\
c_{21} & 1 & c_{23} & \cdots & c_{2r} \\
\vdots & \vdots & \vdots & & \vdots \\
c_{r1} & c_{r2} & c_{r3} & \cdots & 1
\end{bmatrix}
\begin{bmatrix}
a_1 \\ a_2 \\ \vdots \\ a_r
\end{bmatrix}
=
\begin{bmatrix}
c_{1Y} \\ c_{2Y} \\ \vdots \\ c_{rY}
\end{bmatrix},
\tag{16.3.4}
$$

where

$$
c_{ij} = \frac{S_{ij}}{(S_{ii}S_{jj})^{1/2}}, \qquad c_{jY} = \frac{S_{jY}}{(S_{jj}S_{YY})^{1/2}}
\tag{16.3.5}
$$

are correlations between (X_i, X_j) and (X_j, Y), respectively, with $S_{jY} = \Sigma_u (X_{ju} - \overline{X}_j)(Y_u - \overline{Y})$, and where $a_1, a_2, \ldots, a_r$ are estimates of $\alpha_1, \alpha_2, \ldots, \alpha_r$. Equations (16.3.4) are said to be *the normal equations in correlation form*. The $r \times r$ square matrix on the left of (16.3.4) is the correlation matrix of the predictors and will be referred to as **C** below.

It can be shown that, if we take the solutions $a_1, a_2, \ldots, a_r$ from (16.3.4), and calculate

$$
b_1 = a_1 \left(\frac{S_{YY}}{S_{11}} \right)^{1/2},
$$

$$
\vdots
$$

$$
b_r = a_r \left(\frac{S_{YY}}{S_{rr}} \right)^{1/2},
\tag{16.3.6}
$$

$$
b_0 = \overline{Y} - b_1 \overline{X}_1 - \cdots - b_r \overline{X}_r,
$$

we recover the appropriate solution to the original least squares problem. Why would we choose this route, however? Here are two reasons:

1. Workers in some fields argue that the coefficients $a_1, a_2, \ldots a_r$ are more meaningful to them in interpreting the regression. While our own experiences with physical science data do not confirm this, we would not argue against it. Readers can form their own opinions, depending on their field of interest. For some connected references, see Bring (1994).
2. Centering and scaling of the predictor variables is routine in a basic form of ridge regression discussed in Chapter 17. In that application, the response variable Y is usually not centered.

Centering and Scaling and Singularity

The inverse of the correlation matrix **C** on the left of (16.3.4) involves the determinant of this matrix. This determinant value lies between 0, when an exact dependence between the centered columns of the **X** matrix exists, as in (16.2.15) and (16.2.16), and 1, when all the correlations are zero. The size of the determinant is thus some guide to the extent of overall collinearity in the regression problem, although we are unsure how to calibrate it.

The sizes of individual correlations provide no dependable guide to pairwise dependence of **X** columns in general, unfortunately, because of the effect of centering. As noted at the end of Section 6.2, two vectors that are nearly linearly dependent can be little correlated. Conversely, two highly correlated vectors can be orthogonal, as the following extreme example, due to Belsley, shows: $\mathbf{X}_1 = (1, 1, 1, 3^{1/2})'$, $\mathbf{X}_2 = (-1, -1, -1, 3^{1/2})'$ are orthogonal vectors but perfectly correlated! So a high correlation in $\mathbf{C}$ is perhaps best regarded as an invitation to check in more detail. More generally, pairwise correlations typically provide little or no clue to more complicated dependencies between several columns of $\mathbf{X}$.

16.4. MEASURING MULTICOLLINEARITY

Several criteria have been used to check on multicollinearity. We refer the reader to Belsley (1991, particularly Sections 1.3, 1.4, 2.3, and 5.2), for an exceptionally detailed discussion, and concentrate here on the highlights adequate for an appreciation of what to do.

Suggestion 1. Check if some regression coefficients have the wrong sign, based on prior knowledge.

Suggestion 2. Check if predictors anticipated to be important based on prior knowledge have regression coefficients with small t-statistics.

Suggestion 3. Check if deletion of a row or column of the $\mathbf{X}$ matrix produces surprisingly large changes in the fitted model.

Suggestion 4. Check the correlations between all pairs of predictor variables to see if any are surprisingly high. These correlations can be obtained by viewing the correlation matrix $\mathbf{C}$ shown on the left of (16.3.4).

Suggestion 5. Examine the *variance inflation factors*, usually abbreviated to VIF. The name is due to D. W. Marquardt. When the model

$$Y = \beta_0 + \beta_1 X_1 + \beta_2 X_2 + \cdots + \beta_r X_r + \epsilon \tag{16.4.1}$$

is fitted by least squares, the variances of the estimates $b_1, b_2, \ldots, b_r$ are

$$V(b_i) = \text{VIF}_i(\sigma^2/S_{ii}), \qquad i = 1, 2, \ldots, r \tag{16.4.2}$$

where $S_{ii} = \sum_{u=1}^{n} (X_{iu} - \overline{X}_i)^2$ is the usual corrected sum of squares of the $\mathbf{X}_i = (X_{i1}, X_{i2}, \ldots X_{in})'$ column. If an $\mathbf{X}_i$ column is orthogonal to all other columns of the $\mathbf{X}$ matrix, $\text{VIF}_i = 1$. Thus VIF_i is a measure of how much σ^2/S_{ii} is inflated by the relationship of other columns of $\mathbf{X}$ to the $\mathbf{X}_i$ column. [As always, σ^2 would be replaced by s^2 to get an estimated $\hat{V}(b_i)$ and the square root of the result would be the appropriate standard error $\text{se}(b_i)$.]

The VIFs are the diagonal elements of the inverse $\mathbf{C}^{-1}$ of the correlation matrix $\mathbf{C}$. The VIFs can be defined specifically in the following way. Suppose that R_i^2 is the multiple correlation coefficient obtained when the ith predictor variable column $\mathbf{X}_i$, $i = 1, 2, \ldots, r$, is regressed against all the remaining predictors $\mathbf{X}_j$ with $j \neq i$. (*Note:* The intercept column $\mathbf{X}_0 = \mathbf{1}$ is also in the regression but does not feature in R_i^2. Using centered X's also gives the same R_i^2 values, of course.) Then it can be shown that

$$\text{VIF}_i = (1 - R_i^2)^{-1}. \tag{16.4.3}$$

Suggestion 6. Belsley's suggestion, described in Section 16.5.

Recommendations on Suggestions 1–6

1–3. These may or may not arise as a result of multicollinearity. Thus they are unreliable indicators of its presence or absence.

4. Each high pairwise correlation may indicate a (near) linear relationship between the pair of predictors involved or may not, as described at the end of Section 16.3. Low pairwise correlations can occur when multicollinearity involves several predictors, however, so low pairwise correlations are *not* a contrary indicator. Examination of correlations, while not decisive, is probably worthwhile, because it sometimes gives clues to the behavior of selection procedures.

5. When VIF_i is large, $1 - R_i^2$ is small and R_i^2 is close to 1. A large VIF_i thus indicates that there is a (near) dependence of the ith column of $\mathbf{X}$ on the other columns of $\mathbf{X}$, but excluding the $\mathbf{1}$ column, which, by the nature of the calculation made, is excluded. Thus even if all the usual VIFs were "not large" a dependency involving the $\mathbf{1}$ vector could exist, unseen by the VIFs. This could be overcome by calculating an uncentered R^2; see Belsley (1991, pp. 28–29). Obviously, how large a VIF value has to be to be "large enough" comes back to the question of when an R_i^2 is large enough and perhaps should be thought of in that manner. In some writings, specific numerical guidelines for VIF values are seen, but they are essentially arbitrary. Each person must decide for himself or herself.

6. See Section 16.5.

What Are the Relationships?

Even if suggestions 4 and/or 5 give indications of relationships, they do not in general tell how many different relationships might exist. It would then be necessary to discover this. The number of relationships can be found by finding the eigenvalues of the $\mathbf{X}'\mathbf{X}$ matrix. Zero eigenvalues indicate that an exact relationship exists between the columns of $\mathbf{X}$, and small eigenvalues (again the question "how small?" arises!) indicate approximate relationships.

It is also possible to use regression formulas to systematically transform the $\mathbf{X}$ matrix to a form in which all columns are orthogonal. A column transformed to all zeros would indicate that the column it replaces is exactly dependent on previous (former) columns, but not *which* previous columns, unless the regression equation were examined. A column transformed to "all small near-zero values" (again, how small?) would indicate an approximate dependency. See Appendix 16A for some additional details. To invoke a (near) dependency, it would be necessary to make some judgment on how small the length of a transformed vector should be.

Belsley (1991) has discussed the difficulties of all this extensively. His conclusions are that the method of the next section has merit as a way to discover and study multicollinearity. We think he is right.

16.5. BELSLEY'S SUGGESTION FOR DETECTING MULTICOLLINEARITY

1. Determine the n by p "$\mathbf{X}$ matrix" for the regression.
2. Column-equilibrate "$\mathbf{X}$." This means divide each column of "$\mathbf{X}$" by the square root of the sum of squares of its elements. In the column-equilibrated $\mathbf{X}$, then, each column will have sum of squares unity. (In what follows, we simply use the notation $\mathbf{X}$ for the column-equilibrated form, rather than introduce another letter.)
3. Find the *singular value decomposition* of $\mathbf{X}$, that is, find matrices $\mathbf{U}(n$ by $p)$, $\mathbf{D}(p$ by p and diagonal), and $\mathbf{V}(p$ by $p)$ such that

$$\mathbf{X} = \mathbf{UDV}' \tag{16.5.1}$$

and where

$$\mathbf{U'U} = \mathbf{V'V} = \mathbf{VV'} = \mathbf{I}_p. \qquad (16.5.2)$$

The matrix $\mathbf{D}$ = diagonal $(\mu_1, \mu_2, \ldots, \mu_p)$. The μ_i are called the *singular values* of $\mathbf{X}$ and are non-negative, although some may be zero. The matrix $\mathbf{D}^2 = \mathbf{DD}$ = diagonal $(\mu_1^2, \mu_2^2, \ldots, \mu_p^2)$ = diagonal $(\lambda_1, \lambda_2, \ldots, \lambda_p)$ = Λ, say, is the matrix of the *eigenvalues* of $\mathbf{X'X}$. Note the following algebra:

$$\mathbf{X'X} = \mathbf{V}\,\mathbf{DU'UDV'} = \mathbf{VD}^2\mathbf{V'} \qquad (16.5.3)$$

so that

$$\mathbf{X'XV} = \mathbf{VD}^2 = \mathbf{V}\Lambda. \qquad (16.5.4)$$

Thus the matrix $\mathbf{V}$ consists of the p eigenvectors of $\mathbf{X'X}$. These could also be obtained in the more usual manner by solving the equation in λ of degree p

$$|\mathbf{X'X} - \lambda\mathbf{I}| = 0 \qquad (16.5.5)$$

for the roots $\lambda_1, \lambda_2, \ldots, \lambda_p$ and then solving

$$\mathbf{X'Xv}_i = \lambda_i\mathbf{v}_i \qquad (16.5.6)$$

for a normalized (sum of squares of elements = 1) vector $\mathbf{v}_i$, $i = 1, 2, \ldots, p$. Then $\mathbf{V} = (\mathbf{v}_1, \mathbf{v}_2, \ldots, \mathbf{v}_p)$. The ordering of the columns of $\mathbf{V}$ is the same as the order selected for the corresponding $\lambda_1, \lambda_2, \ldots, \lambda_p$. Computer program outputs usually list the λ's in order of decreasing absolute magnitude. If $\mathbf{X'X}$ is invertible,

$$(\mathbf{X'X})^{-1} = (\mathbf{V'})^{-1}\mathbf{D}^{-2}\mathbf{V}^{-1} = \mathbf{VD}^{-2}\mathbf{V'}. \qquad (16.5.7)$$

4. Obtain the p *condition indices*. These are defined as

$$\eta_j = \mu_{\max}/\mu_j, \qquad j = 1, 2, \ldots, p. \qquad (16.5.8)$$

Thus a (relatively) near-zero μ_j will be associated with a large condition index. A zero μ_j would imply the existence of an *exact* linear relationship among the columns of $\mathbf{X}$. (Zero $\lambda_j = \mu_j^2$ values sometimes appear as negative rounding errors, even though the eigenvalues λ_j cannot be negative for $\mathbf{X'X}$.) (*Note*: The ratio $\mu_{\max}/\mu_{\min}$ is called the *condition number*.)

5. Decompose the variance structure of the parameter estimates. When $\mathbf{X'X}$ can be inverted, we can write

$$\sigma^{-2}\mathbf{V(b)} = (\mathbf{X'X})^{-1} = \mathbf{VD}^{-2}\mathbf{V'} \qquad (16.5.9)$$

$$= \mathbf{V}\Lambda^{-1}\mathbf{V'} \qquad (16.5.10)$$

T A B L E 16.1. Proportional Decompositions of $V(b_i)$

Condition Index	Proportions of			
	$V(b_1)$	$V(b_2)$	$\cdots$	$V(b_p)$
η_1	q_{11}	q_{21}	$\cdots$	q_{p1}
η_2	q_{12}	q_{22}	$\cdots$	q_{p2}
$\vdots$	$\vdots$	$\vdots$		$\vdots$
η_p	q_{1p}	q_{2p}	$\cdots$	q_{pp}
Column sums	1	1		1

from (16.5.7). Writing $\mathbf{V} = (v_{ij})$, we see that

$$\sigma^{-2}V(b_i) = \frac{v_{i1}^2}{\mu_1^2} + \frac{v_{i2}^2}{\mu_2^2} + \cdots + \frac{v_{ip}^2}{\mu_p^2}$$

$$= (q_{i1} + q_{i2} + \cdots + q_{ip})\sum_{j=1}^{p}\frac{v_{ij}^2}{\mu_j^2}, \tag{16.5.11}$$

where the q's are simply the proportions (adding to 1) of the total amount of the expression represented by $\sigma^{-2}V(b_i)$. This enables the display of Table 16.1 to be formed.

6. Examine the table (of form 16.1) to detect possible dependencies. Large condition indices η_j indicate dependencies; large proportions q_{ij} within the corresponding rows indicate the **X** columns that are candidates for the dependencies. (Once again we enter the world of "How large is large?"). Low condition indices rows are examined also. High q_{ij} in these rows indicate the *non*-involvement in dependencies of the corresponding columns of **X**.

Computational Note: Use of (16.5.1) on **X**, as recommended by Belsley, gives more accurate answers, particularly in cases where exact or very nearly exact dependencies exist. Except for such cases, the more conveniently available (e.g., in MINITAB) eigenvalue programs can be used directly to get Λ and **V** from **X'X**. When $(\mathbf{X'X})^{-1}$ can be computed, it is convenient to evaluate [see (16.5.9) and (16.5.10)]

$$(\mathbf{X'X})^{-1} - \mathbf{V\Lambda}^{-1}\mathbf{V'} \tag{16.5.12}$$

as a check; it should be **0**. If it is considerably off-zero, or if $(\mathbf{X'X})^{-1}$ cannot be obtained, this is a signal that exact (or very close to exact) dependencies exist.

Example 1. The original "**X** matrix" is

$$\begin{bmatrix} -74 & 80 & 18 & -56 & -112 \\ 14 & -69 & 21 & 52 & 104 \\ 66 & -72 & -5 & 764 & 1528 \\ -12 & 66 & -30 & 4096 & 8192 \\ 3 & 8 & -7 & -13276 & -26552 \\ 4 & -12 & 4 & 8421 & 16842 \end{bmatrix}. \tag{16.5.13}$$

This matrix is called "the modified Bauer matrix" by Belsley who took four columns given by F. L. Bauer (in his article on pp. 119–133 of *Handbook for Automatic*

T A B L E 16.2. Proportional Decompositions of $V(b_i)$ for the "Modified Bauer Data"

Condition[a] Index	Proportions of				
	$V(b_1)$	$V(b_2)$	$V(b_3)$	$V(b_4)$	$V(b_5)$
1	.000	.000	.000	.000	.000
1	.005	.005	.000	.000	.000
1	.001	.001	.047	.000	.000
16	.994	.994	.953	.000	.000
5799	.000	.000	.000	1.000	1.000
Column sums	1.000	1.000	1.000	1.000	1.000

[a] Rounded to the nearest integer.

Computation, Volume II: Linear Algebra, edited by J. H. Wilkinson and C. Reisch, published by Springer-Verlag, 1971) and added a fifth column equal to double the fourth. So "the $\mathbf{X'X}$ is obviously singular. Dividing each entry by the appropriate root sum of squares gives 10^{-3} times the following column-equilibrated figures:

$$10^3 \mathbf{X} = \begin{bmatrix} -733 & 553 & 430 & -3 & -3 \\ 139 & -477 & 501 & 3 & 3 \\ 654 & -498 & -119 & 47 & 47 \\ -119 & 456 & -716 & 252 & 252 \\ 30 & 55 & -167 & -816 & -816 \\ 40 & -83 & 95 & 518 & 518 \end{bmatrix}. \tag{16.5.14}$$

The figures are rounded for display. The equivalence of the last two columns is obvious. One could be dropped immediately, once this is noticed, but we continue with both in. The resulting display of the type shown in Table 16.1 is as in Table 16.2.

Comparison with Belsley's (1991, Exhibit 5.8, p. 148) table shows a large difference in the numerical value of the fifth condition index, undoubtedly because his computations are more accurate than ours, but no difference in any other figure nor in the message transmitted. The condition index 5799 (Belsley gets 8×10^{19}; in practical terms, both numbers are infinity!) indicates a relationship between X_4 and X_5. The next largest index 16 indicates a possible relationship between X_1, X_2, and X_3. We now pick one variable from each set, X_5 from the first, X_1 from the second, say, and regress these as "responses" against the other three variables X_2, X_3, and X_4 to show up the nature of the column dependencies or near dependencies. We obtain the exact relationship

$$X_5 = 0.0X_2 + 0.0X_3 + 2X_5$$

and the approximate relationship

$$X_1 = -0.701X_2 - 1.269X_3 - 0.0X_4.$$

Of the total (not corrected for the mean) SS variation in X_5, 100% is accounted for by the first equation; and for X_1, 98.20% is accounted for by the second equation.

Example 2. The Hald Data (see Appendix 15A). The original "**X** matrix" consists of five columns of $[1, X_1, X_2, X_3, X_4]$. After dividing each entry by the corresponding column root sum of squares we get 10^{-3} times the following figures:

$$10^3\mathbf{X} = \begin{bmatrix} 277 & 207 & 143 & 125 & 489 \\ 277 & 30 & 160 & 313 & 424 \\ 277 & 326 & 308 & 167 & 163 \\ 277 & 326 & 171 & 167 & 383 \\ 277 & 207 & 286 & 125 & 269 \\ 277 & 326 & 303 & 188 & 179 \\ 277 & 89 & 391 & 355 & 49 \\ 277 & 30 & 171 & 459 & 359 \\ 277 & 59 & 297 & 376 & 179 \\ 277 & 622 & 259 & 84 & 212 \\ 277 & 30 & 220 & 480 & 277 \\ 277 & 326 & 363 & 188 & 98 \\ 277 & 296 & 374 & 167 & 98 \end{bmatrix}. \tag{16.5.15}$$

The variances of the b's decompose into the proportions shown in Table 16.3. As always, each column sums to 1, apart from rounding error in the $V(b_3)$ column.

Obviously only one linear relationship is indicated and it links all the columns. Via regression we get the almost exact relationship $X_0 = 0.0103\ X_1 + 0.0103\ X_2 + 0.0105\ X_3 + 0.0101\ X_4$, accounting for 99.99% of the uncorrected SS of 13. Clearly this is showing up the original near-mixture relationship that leads to the fact that (nearly) $X_1 + X_2 + X_3 + X_4 = 1 \equiv 0.277X_0$ for the equilibrated columns of (16.5.15).

Example 3. The Steam Data (see Appendix 1A). We proceed directly to Table 16.4. Again each column sums to 1 (i.e., 1000 times 10^{-3}) apart from rounding error. We have dropped the decimal points by using a factor of 10^{-3} and replaced 000 by a dash to simplify the look at the table and to show an alternative display format. We recall that $X_1 = Y$ is the response, so that the factors are numbered 0 for the b_0 term, and 2–10 for the nine (other) X's. Working up from the bottom of the table, line by line, we need to check possible relations between

T A B L E 16.3. Proportional Decompositions of $V(b_i)$ for the Hald Data of Appendix 15A

Condition[a] Index	Proportions of				
	$V(b_0)$	$V(b_1)$	$V(b_2)$	$V(b_3)$	$V(b_4)$
1	0.000	0.000	0.000	0.000	0.000
3	0.000	0.010	0.000	0.003	0.000
4	0.000	0.001	0.000	0.002	0.002
10	0.000	0.057	0.003	0.046	0.001
250	1.000	0.932	0.997	0.950	0.997

[a] Rounded to the nearest integer.

T A B L E 16.4. Proportional Decompositions of $V(b_i)$ for the Steam Data of Appendix 1A

Condition[a] Index	10^{-3} Times Proportions of $V(b_i)$ for $i =$									
	0	2	3	4	5	6	7	8	9	10
1	—	—	—	—	—	—	1	—	—	—
4	—	—	—	—	—	—	113	4	—	—
7	—	—	—	—	—	—	135	—	7	3
13	—	2	4	—	—	4	114	118	—	40
25	—	7	16	—	—	18	7	10	—	721
31	5	4	5	3	10	66	345	404	5	6
43	1	50	27	2	1	574	230	129	10	8
96	23	612	595	13	2	160	20	106	14	18
148	—	13	134	789	134	138	1	74	809	139
253	970	311	220	193	853	40	34	156	155	66

[a] Rounded to nearest integer.

$$X_0 \quad \text{and} \quad X_5 \quad \text{(Condition index 253)}$$
$$X_4 \quad \text{and} \quad X_9 \quad (148) \tag{16.5.16}$$
$$X_2 \quad \text{and} \quad X_3 \quad (96)$$

The condition indices just quoted (253, 148, and 96) are all of the same order of magnitude, as will be described more fully after this example. This indicates possible *coexisting near dependencies*. For example, variables X_0 and X_5, which are linked by the strongest indicated near dependency (condition index 253), might also be involved (but masking, i.e., concealing, each other) in the other near dependencies; and similarly for the other pairs of variables. Thus, while the most simple pairwise involvements of variables (X_0, X_5), (X_4, X_9), and (X_2, X_3) *might* be all that is indicated by Table 16.3 (and we note, in passing, that $r_{49} = 0.990$ and $r_{23} = 0.944$), it is incumbent on us to investigate more complicated possibilities. We select one variable from each set (here, each pair), say, X_5, X_9, and X_2, and then regress these against all of the remaining seven X's. This leads to the fitted equations

$$X_5 = 33.011X_0 + 2.717X_3 - 0.158X_4 - 0.033X_6 - 0.025X_7 - 0.025X_8$$
$$- 0.280X_{10} \quad (99.95\%),$$

$$X_9 = -51.278X_0 + 14.962X_3 + 14.202X_4 - 0.379X_6 + 0.017X_7 + 0.134X_8$$
$$- 1.290X_{10} \quad (99.78\%),$$

and

$$X_2 = 1.947X_0 + 6.480X_3 - 0.002X_4 - 0.040X_6 - 0.008X_7 - 0.008X_8$$
$$+ 0.073X_{10} \quad (99.81\%).$$

The percentages of the uncorrected SS accounted for are shown in parentheses after the equations. An examination of the usual t-statistics for the regression coefficients confirms the major entanglement within the set (X_0, X_5) and suggests that the second set should be expanded from (X_4, X_9) to (X_0, X_4, X_9), and that the third set should be expanded from (X_2, X_3) to (X_0, X_2, X_3).

Comments

In the end, the choice of which X's would be retained for regression purposes would usually be made on the basis of practicalities such as ease of measurement, value in terms of assessing subsequent regression equations, and a desire to keep an intercept term $\beta_0 X_0$ always in the final fitted equation. We noted in Section 15.6 that a useful equation for the steam data could be obtained by using the set of predictors (X_0, X_2, X_8) or (X_0, X_2, X_6, X_8). Both equations include X_0 because it was forced to be in all the selection procedures, and neither equation includes X_5 (entangled with X_0), nor the entangled with X_0 pair (X_4, X_9) nor X_3 (entangled with X_0 and X_2). So the steam data results in Chapter 15 and in this chapter present complementary and consistent pictures.

How Large Is a "Large" Condition Index?

The assessment of when a condition index is "large" is difficult. Let us talk about the sequences in our examples:

    Example 1: 1, 1, 1, 16, 5799.

    Example 2: 1, 3, 4, 10, 250.

    Example 3: 1, 4, 7, 13, 25, 31, 43, 96, 148, 253.

Belsley notes (1991, p. 139) that condition indices increase in order of magnitude along a scale of 1, 3, 10, 30, 100, 300, 1000, and so on. Moreover, his experiments show that the number 30 provides a reasonable threshold for indicating the presence of collinearity; that is, condition indices of 30 or more can be considered large. With this in mind, Belsley suggests (pp. 139–141) assessing a table of condition indices by beginning with the largest ones. If the largest condition index is 5 or 10, "collinearity is not really a major problem," while if it is in the range of 30–100, "then there are collinearity problems." If the largest condition index is in the range of 1000–3000, there are severe collinearity problems, and "even ones like 30 are not necessarily of major concern." Belsley also suggests that gaps "in the 10/30 progression" are of interest. Interpreting all this in our examples makes us focus on 5799 and 16 in Example 1; on 250 and 10 in Example 2; and in Example 3 (a more difficult case) to back up along 253, 148, 96, until we see that the "big" proportions in the body of the table are dying out. Thus we lose interest in the index 43. In spite of the fact that 43 might otherwise appear big, it is simply dwarfed by the larger numbers below it. Going back up the sequence too far would not be a mistake but would involve one in looking at linear combinations of X's that are not close to zero and then discarding them as possibilities.

    For a more detailed discussion the reader should consult *Conditioning Diagnostics, Collinearity and Weak Data in Regression*, by D. A. Belsley, Wiley, 1991. The methods described by Belsley, and discussed in this section, can be implemented by using a package available in the SAS computing system.

APPENDIX 16A. TRANSFORMING X MATRICES TO OBTAIN ORTHOGONAL COLUMNS

The **X** matrix in a regression problem must be such that none of the columns can be expressed as a linear combination of the other columns. This effectively implies also that there must be at least as many rows not dependent on other rows, as there are parameters to estimate, or else a dependence will appear in the columns also. As an example, suppose observations Y are recorded at only three levels of X, namely, $X = a, b,$ and c, but that the model $Y = \beta_0 + \beta_1 X + \beta_2 X^2 + \beta_3 X^3 + \epsilon$ is postulated. The **X** matrix takes the form

$$\begin{bmatrix} 1 & a & a^2 & a^3 \\ 1 & b & b^2 & b^3 \\ 1 & c & c^2 & c^3 \end{bmatrix}$$

and the columns are dependent since (column 4) $-$ $(a + b + c)$ (column 3) $+$ $(ab + bc + ca)$ (column 2) $-$ abc(column 1) $= 0$. To spot such a dependence in a general regression problem is often very difficult. When it exists the $\mathbf{X'X}$ matrix will always be singular and thus cannot be inverted. When the columns of the $\mathbf{X}$ matrix are almost dependent, the $\mathbf{X'X}$ matrix will be almost singular and difficulties in inversion, including large round-off errors, are likely.

One procedure that can be programmed and used as a routine check on $\mathbf{X}$ matrices consists of successively transforming the columns so that each new column is orthogonal to all previously transformed columns. If a column dependence exists we shall eventually obtain a new column that consists entirely of zeros. If the columns are nearly dependent, a new column will contain all very small numbers perhaps with some zeros. The column transformation takes the following form:

$$\begin{aligned} \mathbf{Z}_{iT} &= (\mathbf{I} - \mathbf{Z}(\mathbf{Z'Z})^{-1}\mathbf{Z}')\mathbf{Z}_i \\ &= \mathbf{Z}_i - \mathbf{Z}(\mathbf{Z'Z})^{-1}\mathbf{Z'Z}_i \end{aligned} \tag{16A.1}$$

where $\mathbf{Z} =$ the matrix of column vectors already transformed,

  $\mathbf{Z}_i =$ the next column vector of $\mathbf{X}$ to be transformed, and

  $\mathbf{Z}_{iT} =$ the transformed vector orthogonal to vectors already in $\mathbf{Z}$.

Note that $\mathbf{Z}_{iT}$ is actually the residual vector of $\mathbf{Z}_i$ after $\mathbf{Z}_i$ has been regressed on the columns of $\mathbf{Z}$.

The column transformation continues until a dependence or near dependence occurs, which causes $\mathbf{Z'Z}$ to be singular or so ill-conditioned that $(\mathbf{Z'Z})^{-1}$ cannot be obtained. Removal of the current last column of $\mathbf{Z}$ would allow the process to continue.

Example

We illustrate using a special case that will lead us to the orthogonal polynomials (see Chapter 22) for $n = 5$. Suppose values of Y are recorded at $X = 1, 2, 3, 4$, and 5 and the model

$$Y = \beta_0 + \beta_1 X + \beta_2 X^2 + \beta_3 X^3 + \epsilon \tag{16A.2}$$

is postulated. The original $\mathbf{X}$ matrix is

$$\begin{array}{cccc} 1 & X & X^2 & X^3 \end{array}$$
$$\mathbf{X} = \begin{bmatrix} 1 & 1 & 1 & 1 \\ 1 & 2 & 4 & 8 \\ 1 & 3 & 9 & 27 \\ 1 & 4 & 16 & 64 \\ 1 & 5 & 25 & 125 \end{bmatrix}. \tag{16A.3}$$

We set

$$\mathbf{Z}_{1T} = \begin{bmatrix} 1 \\ 1 \\ 1 \\ 1 \\ 1 \end{bmatrix} = \mathbf{Z} \tag{16A.4}$$

at this stage. (One column vector must be chosen to begin the process.) Choose

$$\mathbf{Z}_2 = \begin{bmatrix} 1 \\ 2 \\ 3 \\ 4 \\ 5 \end{bmatrix}. \tag{16A.5}$$

Then by Eq. (16A.1),

$$\mathbf{Z}_{2T} = \begin{bmatrix} 1 \\ 2 \\ 3 \\ 4 \\ 5 \end{bmatrix} - \begin{bmatrix} 1 \\ 1 \\ 1 \\ 1 \\ 1 \end{bmatrix} (5)^{-1}(15)$$

$$= \begin{bmatrix} 1 \\ 2 \\ 3 \\ 4 \\ 5 \end{bmatrix} - \begin{bmatrix} 3 \\ 3 \\ 3 \\ 3 \\ 3 \end{bmatrix} = \begin{bmatrix} -2 \\ -1 \\ 0 \\ 1 \\ 2 \end{bmatrix}. \tag{16A.6}$$

At this stage

$$\mathbf{Z} = \begin{bmatrix} 1 & -2 \\ 1 & -1 \\ 1 & 0 \\ 1 & 1 \\ 1 & 2 \end{bmatrix} = [\mathbf{Z}_{1T}, \mathbf{Z}_{2T}] \tag{16A.7}$$

and the third column of $\mathbf{X}$ is used as $\mathbf{Z}_3$. We find that

$$(\mathbf{Z}'\mathbf{Z})^{-1} = \begin{bmatrix} 5 & 0 \\ 0 & 10 \end{bmatrix}^{-1} = \begin{bmatrix} \frac{1}{5} & 0 \\ 0 & \frac{1}{10} \end{bmatrix},$$

$$\mathbf{Z}'\mathbf{Z}_i = \begin{bmatrix} 55 \\ 60 \end{bmatrix}, \quad \text{so} \quad (\mathbf{Z}'\mathbf{Z})^{-1}\mathbf{Z}'\mathbf{Z}_i = \begin{bmatrix} 11 \\ 6 \end{bmatrix}. \tag{16A.8}$$

Hence

$$
\mathbf{Z}_{3T} = \begin{bmatrix} 1 \\ 4 \\ 9 \\ 16 \\ 25 \end{bmatrix} - \begin{bmatrix} 11 - 12 \\ 11 - 6 \\ 11 + 0 \\ 11 + 6 \\ 11 + 12 \end{bmatrix} = \begin{bmatrix} 2 \\ -1 \\ -2 \\ -1 \\ 2 \end{bmatrix}.
\tag{16A.9}
$$

We leave the evaluation of $\mathbf{Z}_{4T}$ as an exercise. The final, orthogonal matrix is

$$
\begin{bmatrix} 1 & -2 & 2 & -1.2 \\ 1 & -1 & -1 & 2.4 \\ 1 & 0 & -2 & 0 \\ 1 & 1 & -1 & -2.4 \\ 1 & 2 & 2 & 1.2 \end{bmatrix}.
\tag{16A.10}
$$

Note that the first three columns are ψ_0, ψ_1, and ψ_2—the orthogonal polynomials of zero, first, and second order for $n = 5$. The fourth column is 1.2 times ψ_3, the orthogonal polynomial of third order for $n = 5$. See Chapter 22.

The above procedure is usually known as a Gram–Schmidt orthogonalization of the columns. Another way of tackling this problem is to find the eigenvalues (or characteristic values, or latent roots—all terms mean the same thing) of $\mathbf{X}'\mathbf{X}$. See Section 16.5. Zero eigenvalues arise if linear dependencies exist among the $\mathbf{Z}$'s and small eigenvalues (small in relation to the range 0 to 1) are indicative of possible near dependencies. See, for example, p. 78 of Snee (1973). For related algorithms see Clayton (1971) (Fortran) and Farebrother (1974) (Algol 60).

EXERCISES FOR CHAPTER 16

A. Can we use the data below to get a unique fit to the model $Y = \beta_0 + \beta_1 X_1 + \beta_2 X_2 + \beta_3 X_3 + \epsilon$? If not, what model(s) can be fitted?

X_1	X_2	X_3	Y
1	-2	4	81
2	-7	11	88
4	3	5	94
7	1	13	95
8	-1	17	123

B. Can we use the data below to get a unique fit to the model $Y = \beta_0 + \beta_1 X_1 + \beta_2 X_2 + \beta_3 X_3 + \epsilon$? If not, what model(s) can be fitted?

X_1	X_2	X_3	Y
-4	1	3	7.4
3	2	-5	14.7
1	3	-4	13.9
4	4	-8	18.2
-3	5	-2	12.1
-1	6	-5	14.8

C. Can we use the data below to fit, uniquely, the model $Y = \beta_0 + \beta_1 X_1 + \beta_2 X_2 + \beta_{11} X_1^2 + \beta_{22} X_2^2 + \beta_{12} X_1 X_2 + \epsilon$? If not, what model(s) can be fitted?

X_1	X_2	Y
-1	-1	38
1	-1	45
-1	1	41
1	1	40
0	0	47
0	0	42
0	0	48

D. Consider the data below for the model $Y = \beta_0 + \beta_1 X_1 + \beta_2 X_2 + \epsilon$. Center and scale the X_1, X_2, and Y columns, and write down the normal equations for the centered and scaled model. Solve these. What is the value of the determinant of the correlation matrix?

	X_1	X_2	Y
	-2	-4	12
	-1	-1	9
	0	0	9
	1	1	14
	2	4	16
Sum	0	0	60
SS	10	34	758

CHAPTER 17

Ridge Regression

17.1. INTRODUCTION

The *ridge trace* procedure, first suggested by A. E. Hoerl in 1962, is discussed at length in two seminal papers by Hoerl and Kennard (1970a,b). The procedure is intended to overcome "ill-conditioned" situations where near dependencies between various columns in $\mathbf{X}$ cause the $\mathbf{X'X}$ matrix to be close to singular, giving rise to unstable parameter estimates, typically with large standard errors. Because ill-conditioning is a function of model and data, it would usually be wise to find out how the ill-conditioning occurs, as discussed in Chapter 16. Ridge regression is a way of proceeding that adds specific additional information to the problem to remove the ill-conditioning. Not all users realize this and ridge regression is sometimes used when it is entirely inappropriate. This is not the fault of the originators, who applied the technique to industrial problems in sensible ways.

17.2. BASIC FORM OF RIDGE REGRESSION

The ridge regression procedure, in its simplest form, is as follows. Let $\mathbf{F}$ represent the appropriate centered and scaled "$\mathbf{X}$ matrix" when the regression problem under study is in "correlation form" (as discussed in Chapter 16). Thus, if the original model is

$$Y = \beta_0 + \beta_1 Z_1 + \beta_2 Z_2 + \cdots + \beta_r Z_r + \epsilon, \tag{17.2.1}$$

the new centered and scaled predictor variables are

$$f_{ju} = \frac{Z_{ju} - \overline{Z}_j}{S_{jj}^{1/2}}, \tag{17.2.2}$$

where $\overline{Z}_j$ is the average of the Z_{ju}, $u = 1, 2, \ldots, n$, and $S_{jj} = \Sigma_u (Z_{ju} - \overline{Z}_j)^2$. Thus

$$\mathbf{F} = \begin{bmatrix} f_{11} & f_{21} & \cdots & f_{r1} \\ \vdots & \vdots & & \vdots \\ f_{1u} & f_{2u} & \cdots & f_{ru} \\ \vdots & \vdots & & \vdots \\ f_{1n} & f_{2n} & \cdots & f_{rn} \end{bmatrix} \tag{17.2.3}$$

and $\mathbf{F'F}$ is the correlation matrix of the Z's. The size of $\mathbf{F'F}$ is r by r; the original model had $(r + 1)$ parameters in it, including β_0. Note that $\mathbf{F}'\mathbf{1} = \mathbf{0}$, because of the centering.

In Chapter 16 we also put $\mathbf{Y}$ in centered and scaled form. Usually $\mathbf{Y}$ is only centered in ridge regression, but it could be scaled, too, if desired. If $\mathbf{Y}$ is not scaled, the least squares model (17.2.1) can be fitted as

$$Y_u - \overline{Y} = \beta_{1F}f_{1u} + \beta_{2F}f_{2u} + \cdots + \beta_{rF}f_{ru} + \epsilon_u \tag{17.2.4}$$

with a vector of least squares estimates we shall refer to as

$$\mathbf{b}_F(0) = \{b_{1F}(0), b_{2F}(0), \ldots, b_{rF}(0)\}', \tag{17.2.5}$$

say. We can think of $\overline{Y}$ as $b_{0F}(0)$. The reason for the zeros in parentheses will become apparent in a moment. The ridge regression estimates of the r elements of $\boldsymbol{\beta}_F = (\beta_{1F}, \beta_{2F}, \ldots, \beta_{rF})'$ are the r elements of $\mathbf{b}_F(\theta) = \{b_{1F}(\theta), b_{2F}(\theta), \ldots, b_{rF}(\theta)\}'$ given by the equation

$$\mathbf{b}_F(\theta) = (\mathbf{F}'\mathbf{F} + \theta\mathbf{I}_r)^{-1}\mathbf{F}'\mathbf{Y}, \tag{17.2.6}$$

where θ is a positive number. [In applications, the interesting values of θ usually lie in the range $(0, 1)$. An exception is described by Brown and Payne (1975).] Note that, when $\theta = 0$, we obtain the least squares estimates. The reader may wonder why $\mathbf{Y}$, rather than $\mathbf{Y} - \overline{Y}\mathbf{1}$ (the centered vector), appears on the right of (17.2.6). The reason is that $\overline{Y}\mathbf{1}$ makes no difference to the calculations at all; the centering of the predictor variables causes $\mathbf{F}'\mathbf{1} = \mathbf{0}$, so that $\mathbf{F}'(\mathbf{Y} - \overline{Y}\mathbf{1}) = \mathbf{F}'\mathbf{Y}$. To recover estimates of the original parameters in (17.2.1) we refer to (16.3.6) but remember that, because $\mathbf{Y}$ is centered but not scaled, the factor $S_{YY}^{1/2}$ is not needed. (It will be needed only if the Y's are centered *and* scaled.) The conversions are

$$b_j(\theta) = b_{jF}(\theta)/S_{jj}^{1/2}, \quad j = 1, 2, \ldots, r, \tag{17.2.7}$$

and

$$b_0(\theta) = \overline{Y} - \sum_{j=1}^{r} b_j(\theta)\overline{Z}_j \tag{17.2.8}$$

to provide the $(r + 1) \times 1$ vector $\mathbf{b}(\theta) = \{b_0(\theta), b_1(\theta), \ldots, b_r(\theta)\}'$. When $\theta = 0$, the $b_j(0)$, $j = 0, 1, 2, \ldots, r$, are the usual least squares estimates for the parameters in (17.2.1), as is clear from setting $\theta = 0$ in Eqs. (17.2.6) to (17.2.8).

We can now plot the $b_{jF}(\theta)$ or the $b_j(\theta)$ against θ for $j = 1, 2, \ldots, r$ and examine the resulting figure; the intercept is usually not plotted. Such a plot is called a *ridge trace*. Typically this plot is drawn in "correlation" units [i.e., the $b_{jF}(\theta)$], to enable a direct comparison to be made between the relative effects of the various coefficients, and to remove the effects of the various Z scalings that affect the sizes of the original estimated coefficients. [In our example below, however, we have converted the plot to "original" units–that is, the $b_j(\theta)$—so that, when $\theta = 0$, we obtain the least squares estimates that would be obtained if the original Z's were *not* scaled.] As θ is increased to values beyond 1, the estimates typically become smaller in absolute value, and they tend to zero as θ tends to infinity. A value of θ, say, θ^*, is then selected. Hoerl and Kennard (1970a, p. 65) say:

These kinds of things can be used to guide one to a choice:

1. At a certain value of θ the system will stabilize and have the general characteristics of an orthogonal system.

2. Coefficients will not have unreasonable absolute values with respect to the factors for which they represent rates of change.

3. Coefficients with apparently incorrect signs at $\theta = 0$ will have changed to have the proper signs.

4. The residual sum of squares will not have been inflated to an unreasonable value. It will not be large relative to the minimum residual sum of squares or large relative to what would be a reasonable variance for the process generating the data.

The first point must be interpreted in the context of small θ, $0 \le \theta \le 1$, of course, because as $\theta \to \infty$, all the $b_{bF} \to 0$. Points 2 and 3 require some prior knowledge for proper application. Point 4 raises the question of what is an unreasonable value. Fortunately, as we shall see in our example, an automatic (but sensible) way of choosing θ is available for those whose prior knowledge is insufficient.

Once θ has been selected (equal to θ^*) the values $b_j(\theta^*)$ are used in the prediction equation. The resulting equation is made up of estimates that are not least squares and are biased but that are more stable in the sense described and that (it is hoped; see Appendix 17B) provide a smaller overall mean square error, since the reduction they achieve in variance error will more than compensate for the bias introduced.

(Note that the biased estimates, which can be selected if desired through the Mallows C_p procedure, are biased by coefficients that have not been taken into account in the model fitted. The biased estimates in the Hoerl and Kennard procedure are biased by the presence of θ in the estimation equation and this bias will occur even if the postulated equation contains all the "right" predictor variables. In other words, the two sorts of biases are of different characters.)

17.3. RIDGE REGRESSION OF THE HALD DATA

We now apply this method to the Hald data, to illustrate the mechanics of the procedure. Because of the mixture relationship between the four predictor variables, these data could possibly give rise to unstable estimates, as we have already discussed.

Figure 17.1 shows a ridge trace for the Hald data for $0 \le \theta \le 1$ while Figure 17.2 shows an enlargement of the same figure for $0 \le \theta \le 0.03$. What value θ^* should be selected?

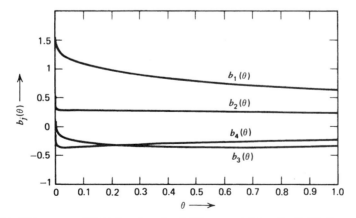

Figure 17.1. Ridge trace of Hald data, $0 \le \theta \le 1$. We are grateful to Dr. R. W. Kennard who kindly provided the results from which this figure was drawn.

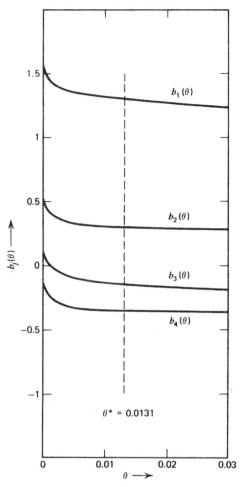

F i g u r e 17.2. Ridge trace of Hald data, $0 \le \theta \le 0.03$. We are grateful to Dr. R. W. Kennard who kindly provided the results from which this figure was drawn.

Automatic Choice of θ^*

One suggested automatic way of choosing θ^* is given by Hoerl, Kennard, and Baldwin (1975). They argue that a reasonable choice might be

$$\theta^* = rs^2/\{\mathbf{b}_F(0)\}'\{\mathbf{b}_F(0)\}, \tag{17.3.1}$$

where

1. r is the number of parameters in the model *not counting* β_0.
2. s^2 is the residual mean square in the analysis of variance table obtained from the standard least squares fit of (17.2.1), or that of (17.2.4); the results are the same.
3.
$$\begin{aligned}
\{\mathbf{b}_F(0)\}' &= \{b_{1F}(0), b_{2F}(0), \ldots, b_{rF}(0)\} \\
&= \{\sqrt{S_{11}}b_1(0), \sqrt{S_{22}}b_2(0), \ldots, \sqrt{S_{rr}}b_r(0)\}.
\end{aligned} \tag{17.3.2}$$
(These two may differ slightly in practice due to rounding error.)

[Note that θ^* in Eq. (17.3.1) is s^2 (an estimate of σ^2) divided by the average value of the squares of the least squares estimates $b_{jF}(0)$. The latter can be regarded as an estimate of σ_β^2, the variance of the true but unknown β_F values. It follows that, from a Bayesian viewpoint (see Section 17.4), the value of θ^* in (17.3.1) is a sensible choice.]

For these data, we have $r = 4$, $s^2 = 5.983$ (see Appendix 15A), and $\mathbf{b}_F(0) = (31.633, 27.516, 2.241, -8.338)'$, so that $\theta^* = 4(5.983)/1832.4 = 0.0131$. A vertical line is drawn at this θ-value in Figure 17.2, and the corresponding values of the coefficients can be read off the graph, or calculated more accurately. The following equation results, in fact:

$$\hat{Y} = 83.414 + 1.300X_1 + 0.300X_2 - 0.142X_3 - 0.349X_4. \qquad (17.3.3)$$

This equation can be used in its present form.

Possible Use of Ridge Regression as a Selection Procedure; Other θ^*

Alternatively, consideration can be given to the possible deletion of one or more predictor variables. An obvious first choice is X_3. The value $b_3(0.0131) = -0.142$ is not only the smallest coefficient but, in addition, the values of X_3 in the data are such that the *maximum* effect on the response from X_3 is only $-0.142(23) = -3.266$. Probably the most sensible first steps to take would be to eliminate X_3 and then to refit and perform a new ridge trace for an equation containing X_1, X_2, and X_4. We do not pursue this matter here but refer the reader to the applications paper by Hoerl and Kennard (1970b) for comments on the considerations involved. It should be noted that, in applications, ridge regression has *not* usually been used as a selection-type procedure. We mention this use only as a possibility.

There are other ways of choosing θ^*. One, for example, is to use an iterative procedure. The basic idea is this: In the choice of θ^* above, a denominator of $\{\mathbf{b}_F(0)\}'\{\mathbf{b}_F(0)\}$ was used. For that reason let us call that value θ_0^*. Consider the iterative formula

$$\theta_{j+1}^* = rs^2[\mathbf{b}_F(\theta_j^*)\}'\{\mathbf{b}_F(\theta_j^*)\}]. \qquad (17.3.4)$$

Use of θ_0^* on the right-hand side of this formula will enable us to obtain a revised value θ_1^*, which can then be substituted into the right-hand side of the formula to update to θ_2^*, and so on. Updating continues until

$$(\theta_{j+1}^* - \theta_j^*)/\theta_j^* \le \delta, \qquad (17.3.5)$$

where δ is appropriately chosen. For more on this procedure, see Hoerl and Kennard (1976). The authors note that $\theta_{j+1}^* \ge \theta_j^*$ always so that $\delta \ge 0$ and suggest

$$\delta = 20\{\text{trace}(\mathbf{F}'\mathbf{F})^{-1}/r\}^{-1.30} \qquad (17.3.6)$$

as a suitable practical value, for reasons they describe on (their) pp. 79–80.

There is no mechanically best way of choosing θ^*. As we have already noted, Eq. (17.3.1) has some intuitive appeal because it can be regarded as a rough estimate of the ratio σ^2/σ_β^2 mentioned in the Bayesian formulation that follows. The iterative Eq. (17.3.4) can be regarded as giving another such estimate.

17.4. IN WHAT CIRCUMSTANCES IS RIDGE REGRESSION ABSOLUTELY THE CORRECT WAY TO PROCEED?

Opinions about the value of the ridge regression method vary over a large range. Before setting out our own opinions, we describe two (rather restrictive) situations in which we can wholeheartedly recommend ridge regression as the best treatment of the problem described.

1. A Bayesian formulation of a regression problem with specific prior knowledge of a certain type on the parameters. This is discussed by Goldstein and Smith (1974, p. 291), and also mentioned in the original Hoerl and Kennard (1970a) paper. Essentially, ridge regression can be regarded as an estimate of $\boldsymbol{\beta}_F$ from the data, subject to the prior knowledge (or belief) that smaller values (in modulus, i.e., in size ignoring sign) of the β_{jF} are more likely than larger values, and that larger and larger values of the β_{jF} are more and more unlikely. More accurately, this prior knowledge would be expressed by imagining a multivariate normal distribution of prior belief on the values of the β_{jF}, a distribution whose spread would depend on a parameter σ_β^2. The parameter θ of the ridge regression would, in fact, be the ratio σ^2/σ_β^2 (where σ^2 is the usual variance of a single observation). Thus choice of a value of θ in ridge regression is equivalent to an expression of how big we believe the β_{jF} might be. Use of a very small θ (<0.01, say) would imply that we believed quite large values of the β_{jF} were not unreasonable. (The least squares estimates correspond to $\theta = 0$, or $\sigma_\beta^2 = \infty$, which means that, *a priori*, we do not feel we wish to restrict the β_{jF} at all.) Choice of a large θ (>1, say) would imply that we believed the β_{jF} were more likely to be quite small than otherwise. If we think of ridge regression in this way, then, we see that the "stabilization" of the ridge trace as θ increases is really *not* something implied by the data but is created by an *a priori* restriction of the possibilities for the parameter values.

2. A formulation of a regression problem as one of least squares subject to a specific type of restriction on the parameters. Suppose we perform least squares subject to the spherical restriction

$$\boldsymbol{\beta}_F'\boldsymbol{\beta}_F \le c^2, \tag{17.4.1}$$

where c^2 is a specified value. If we use the method of Lagrange multipliers (see Appendix 9A) to do this, we can form the objective function

$$f \equiv (\mathbf{Y} - \overline{Y}\mathbf{1} - \mathbf{F}\boldsymbol{\beta}_F)'(\mathbf{Y} - \overline{Y}\mathbf{1} - \mathbf{F}\boldsymbol{\beta}_F) + \theta(\boldsymbol{\beta}_F'\boldsymbol{\beta}_F - c^2), \tag{17.4.2}$$

using the equality in the restriction, whereupon setting $\partial f/\partial \boldsymbol{\beta}_F = 0$ provides the equations

$$(\mathbf{F}'\mathbf{F} + \theta\mathbf{I})\boldsymbol{\beta}_F = \mathbf{F}'(\mathbf{Y} - \overline{Y}\mathbf{1}) = \mathbf{F}'\mathbf{Y}, \tag{17.4.3}$$

which leads to the ridge solution Eq. (17.2.6). This must be solved together with the assumed restriction $\boldsymbol{\beta}_F'\boldsymbol{\beta}_F = c^2$. If we solve for $\boldsymbol{\beta}_F$ in Eq. (17.4.3) and substitute into the restriction, we get

$$\mathbf{Y}'\mathbf{F}(\mathbf{F}'\mathbf{F} + \theta\mathbf{I}_r)^{-2}\mathbf{F}'\mathbf{Y} = c^2 \tag{17.4.4}$$

an equation that determines θ in terms of the given c value. Thus ridge regression can be viewed as least squares subjected to a spherical restriction on the parameters, the appropriate value of θ being determined by the radius c of the restriction. Figure 17.3 shows a geometric representation of this for the two-parameter (in addition to β_0) case. The usual (unrestricted) least squares solution occurs as the "bottom of the bowl" defined by the elliptical sum of squares contours. (An explanation of these is given in Chapter 24. We ask the reader who has not read those details to bear with us and gather a "general impression" of what is happening without worrying about the details here.) The spherical (here circular) restriction is shown and the restricted solution is on the innermost elliptical contour that just touches the spherical restriction. We obtain the whole sequence of ridge regression solutions as we "close in" the circle,

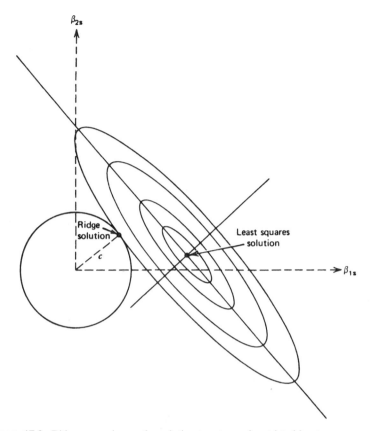

Figure 17.3. Ridge regression as the solution to a type of restricted least squares problem.

that is, reduce the radius from the length where the circle would just pass through the least squares solution (which would correspond to $\theta = 0$) to the "point circle" where $c = 0$ (which would correspond to $\theta = \infty$). Larger radius values would return us to the least squares solution, of course. The "ridge" of ridge regression is the path traced by the solution point as the circle radius is contracted or expanded from one extreme to the other of its effective range.

An alternative viewpoint is that ridge regression is the minimax solution to least squares when $\beta'\beta \leq \sigma^2/\theta$; see Bunke (1975).

Comments

We have said that situations (1) and (2) have the ridge regression formula as their solution. Provided we feel we can take one of these two approaches then, ridge regression is *exactly* the appropriate solution. There are two troubling aspects of this, however.

1. Can we really specify our prior knowledge in either of the ways given above? Are we *that* sure? For example, if we want to specify an ellipsoidal restriction

$$\beta_F'\mathbf{U}\beta_F \leq c^2 \qquad\qquad (17.4.5)$$

instead of the spherical one, Eq. (17.4.1), we shall be led to a ridge-type solution of different form,

$$(\mathbf{F'F} + \theta\mathbf{U})^{-1}\mathbf{F'Y}, \tag{17.4.6}$$

instead. Thus the precise form of our restriction can be critical to the solution we obtain.

2. When we use the standard form [or any other form as in Eq. (17.4.6)] of ridge regression, we are—*whether we like it or not*—essentially making some sort of assumption of the type (1) or (2) given above, without really specifying it. The ridge regression solution is not a miraculous panacea. It is a least squares solution restricted by the addition of some external information about the parameters. The blind use of ridge regression without this realization can be dangerous and misleading. If the external information is sensible and is not contradicted by the data, then ridge regression is also sensible. A simple test of compatibility is to check whether the ridge estimates of the original parameters lie within, or on, the 95% (or other selected percentage) ellipsoidal confidence contour for those parameters. The contour is given by Eq. (5.3.7). When the appropriate ridge estimate vector is inserted in the positions occupied by $\boldsymbol{\beta}$, the left-hand side of (5.3.7) will be smaller than the right-hand side if the ridge estimate point falls inside the contour; equality means it falls on the contour, and left > right in (5.3.7) means it falls outside the contour.

An additional worrying feature of ridge regression is the following. The characteristic effect of the ridge regression procedures is to change (from the least squares values) the nonsignificant estimated regression coefficients (whose values are statistically doubtful, anyway) to a far greater extent than the significant estimated coefficients. It is questionable that much real improvement in estimation can be achieved by such a procedure. (Most variable selection procedures, in the same context, would assign a value zero to a nonsignificant coefficient rather than adjusting it to a different but also nonsignificant value and such action would be no less reasonable and would have the virtue of concentrating attention on the apparently important predictors.)

Canonical Form of Ridge Regression

Much of the published technical work on ridge regression is done on the "canonical form," obtained by applying a transformation to convert the "$\mathbf{F'F}$ matrix" into a diagonal matrix. This simplifies the algebra and provides a convenient way of making ridge trace calculations. The canonical form of ridge regression is discussed in Appendix 17C.

17.5. THE PHONEY DATA VIEWPOINT

A third interpretation of ridge regression is possible. We can introduce additional information into a regression problem by adding "prior data" of the form $\mathbf{F}_0$, $\mathbf{Y}_0$. If we give a weight of 1 to each real data point and a weight θ to each prior data point, the weighted least squares estimator of $\boldsymbol{\beta}_F$ is

$$\mathbf{b}_{FPD} = (\mathbf{F'F} + \theta\mathbf{F}_0'\mathbf{F}_0)^{-1}(\mathbf{F'Y} + \theta\mathbf{F}_0'\mathbf{Y}_0). \tag{17.5.1}$$

Suppose our prior data are "phoney" and are chosen so that $\mathbf{F}_0'\mathbf{F}_0 = \mathbf{I}$ and $\mathbf{Y}_0 = \mathbf{0}$; then we obtain the ridge estimator Eq. (17.2.6). Thus another viewpoint of ridge regression is as the tempering of the data by some zero observations on an orthogonal design of unit radius, appropriately weighted.

An alternative viewpoint is to choose $\mathbf{F}_0'\mathbf{F} = \theta\mathbf{I}$ for selected θ, $\mathbf{Y}_0 = \mathbf{0}$ and perform an unweighted regression analysis. Thus readers who do not have a ridge regression package can obtain the same results by using a standard regression package and adding one dummy case for each of the "r" variables in the regression model with $\theta^{1/2}$ as the value for the corresponding variable and zeros for the rest of the variables in the model. In other words, $\mathbf{F}_0 = \theta^{1/2}\mathbf{I}_r$ and $\mathbf{Y}_0 = \mathbf{0}$.

In either event, we are adjoining some zero observations onto the data.

This method must be used with caution. The correct ridge regression coefficients are produced by it, but most of the other traditional computer output is no longer meaningful for the original data, providing a dangerous trap to the unwary. (On the other hand, it can be argued that this output correctly represents the effects of the assumption made.)

17.6. CONCLUDING REMARKS

Ridge Regression Simulations—A Caution

A number of papers containing simulations of regression situations claim to show that ridge regression estimates are better than least squares estimates when judged via mean square error. Such claims must be viewed with caution. Careful study typically reveals that the simulation has been done with effective restrictions on the true parameter values—precisely the situations where ridge regression is the appropriate technique theoretically. The extended inference that ridge regression is "always" better than least squares is, typically, completely unjustified.

Summary

From this discussion, we can see that use of ridge regression is perfectly sensible in circumstances in which it is believed that large β-values are unrealistic from a practical point of view. However, it must be realized that the choice of θ is essentially equivalent to an expression of how big one believes those β's to be. In circumstances where one cannot accept the idea of restrictions on the β's, ridge regression would be completely inappropriate.

Note that, in many sets of data, where the sizes of the least squares estimates are acceptable as they are, the ridge trace procedure would result in a choice of $\theta = 0$. A value of $\theta \neq 0$ would be used only when the least squares results were not regarded as satisfactory.

Opinion

Ridge regression is useful and completely appropriate in circumstances where it is believed that the values of the regression parameters are unlikely to be "large" (as interpreted through σ_β^2 or c^2 above). In viewing the ridge traces, a subjective judgment is needed that either (1) effectively specifies one's Bayesian prior beliefs as to the likely sizes of the parameters, or (2) effectively places a spherical restriction on the parameter space. The procedure is very easy to apply and a standard least squares regression program could easily be adapted by a skilled programmer. Overall, however, we would advise against the indiscriminate use of ridge regression unless its limitations are fully appreciated. (The reader should be aware that many writers disagree with our somewhat pessimistic assessment of ridge regression.)

17.7. REFERENCES

There is a great deal of published work on ridge regression and we give only some selected references. In the body of the text we have already mentioned Hoerl and Kennard (1970a,b, 1976); Hoerl, Kennard, and Baldwin (1975); Goldstein and Smith (1974); and Bunke (1975). For applications see Anderson and Scott (1974), Füle (1995), Marquardt (1970), and Marquardt and Snee (1975), which we especially recommend. For further work, see Coniffe and Stone (1973), Draper and Herzberg (1987), Draper and Van Nostrand (1979), Egerton and Laycock (1981), Gibbons (1981), Hoerl and Kennard (1975, 1981), Lawless and Wang (1976), McDonald (1980), Oman (1981), Park (1981), Smith (1980), and Swindel (1981). For ridge regression in other circumstances, see Chaturvedi (1993), Kozumi and Ohtani (1994), le Cessie and van Houwelingen (1992), Nyquist (1988), and Segerstedt (1992). For variations of ridge regression and links with other things, see Copas (1983); Crouse, Jin, and Hanumara (1995), Obenchain (1978); Panopoulos (1989); Saleh and Kibria (1993); Srivastava and Giles (1991); and Walker and Birch (1988).

APPENDIX 17A. RIDGE ESTIMATES IN TERMS OF LEAST SQUARES ESTIMATES

How do the ridge regression estimates relate to the least squares estimates? We recall from Eq. (17.2.6) that

$$\mathbf{b}_F(\theta) = (\mathbf{F}'\mathbf{F} + \theta\mathbf{I}_r)^{-1}\mathbf{F}'\mathbf{Y}. \tag{17A.1}$$

Using the fact that $(\mathbf{AB})^{-1} = \mathbf{B}^{-1}\mathbf{A}^{-1}$, we can remove a factor $(\mathbf{F}'\mathbf{F})^{-1}$ from the right of the inverted matrix in (17A.1) to obtain

$$\begin{aligned}\mathbf{b}_F(\theta) &= \{\mathbf{I} + \theta(\mathbf{F}'\mathbf{F})^{-1}\}^{-1}\mathbf{b}_F(0) \\ &= \mathbf{T}\mathbf{b}_F(0) \\ &= \mathbf{T}\mathbf{b}_F,\end{aligned} \tag{17A.2}$$

say. Because

$$\mathbf{b}_F = \mathbf{b}_F(0) = \{b_{1F}(0), b_{2F}(0), \dots, b_{rF}(0)\}' \tag{17A.3}$$

is the vector of least squares estimates after centering and scaling, except for the intercept term, we see that the ridge estimators are all linear combinations of the corresponding least squares estimators with coefficients determined by the matrix

$$\mathbf{T} = \{\mathbf{I} + \theta(\mathbf{F}'\mathbf{F})^{-1}\}^{-1}. \tag{17A.4}$$

APPENDIX 17B. MEAN SQUARE ERROR ARGUMENT

Ridge regression is sometimes "justified" as a practical technique by the claim that it produces lower mean square error. The basic result is as follows (e.g., Hoerl and Kennard, 1970a, p. 62). The mean square error of the ridge estimator can be written

$$\begin{aligned}\text{MSE}(\theta) &= E\{\mathbf{b}_F(\theta) - \boldsymbol{\beta}_F\}'\{\mathbf{b}_F(\theta) - \boldsymbol{\beta}_F\} = E\{\mathbf{T}\mathbf{b}_F - \boldsymbol{\beta}_F\}'\{\mathbf{T}\mathbf{b}_F - \boldsymbol{\beta}_F\} \\ &= E\{(\mathbf{b}_F - \boldsymbol{\beta}_F)'\mathbf{T}'\mathbf{T}(\mathbf{b}_F - \boldsymbol{\beta}_F) \\ &\quad + \boldsymbol{\beta}_F'(\mathbf{T} - \mathbf{I})'(\mathbf{T} - \mathbf{I})\boldsymbol{\beta}_F.\end{aligned} \tag{17B.1}$$

To get this, we have substituted for $\mathbf{b}_F(\theta) = \mathbf{T}\mathbf{b}_F$ from Eq. (17A.2), where $\mathbf{T} = \{\mathbf{I} + \theta(\mathbf{F}'\mathbf{F})^{-1}\}^{-1}$, isolated a quadratic term in $(\mathbf{b}_F - \boldsymbol{\beta}_F)$, taken the expectation $E(\mathbf{b}_F) = \boldsymbol{\beta}_F$ in the remaining terms, performed some cancellations, and regrouped. Application of the matrix results of Appendix 5A then leads to

$$\text{MSE}(\theta) = \sigma^2 \, \text{trace} \, \{\mathbf{T}(\mathbf{F}'\mathbf{F})^{-1}\mathbf{T}'\} + \boldsymbol{\beta}_F'(\mathbf{T} - \mathbf{I})'(\mathbf{T} - \mathbf{I})\boldsymbol{\beta}_F. \qquad (17\text{B}.2)$$

The first term, being the sum of the diagonal terms of $V(\mathbf{T}\mathbf{b}_F) = \sigma^2\mathbf{T}(\mathbf{F}'\mathbf{F})^{-1}\mathbf{T}'$, is the sum of the variances of the elements of the ridge estimator $\mathbf{T}\mathbf{b}$. The second term is a ridge "squared bias" term. [Note that, if $\theta = 0$, then $\mathbf{T} = \mathbf{I}$ and the first term is the sum of the variances of the least squares estimators of the coefficients, while the second term vanishes. The value so attained would be MSE(0).] An existence theorem then states that there always exists a $\theta^* > 0$ such that $\text{MSE}(\theta^*) < \text{MSE}(0)$. The catch in this result is that its value depends on σ^2 and $\boldsymbol{\beta}_F$, neither of which is known. Thus although θ^* exists, there is no way of knowing whether or not we have attained a θ-value that provides lower MSE than MSE(0) in a specific practical problem.

APPENDIX 17C. CANONICAL FORM OF RIDGE REGRESSION

The canonical form of ridge regression requires a rotation of the $\boldsymbol{\beta}_F$-axes to new axes that are parallel to the principal axes of the sum of squares contours (See Figure 17C.1.) Let $\lambda_1, \lambda_2, \ldots, \lambda_r$ be the eigenvalues of $\mathbf{F}'\mathbf{F}$ and let

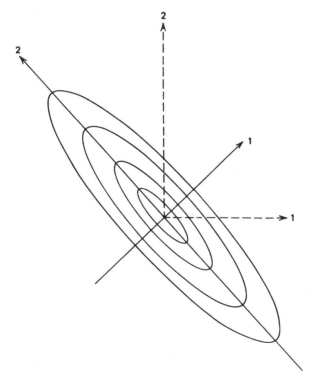

Figure 17C.1. The sum of squares surface contours are shown for $r = 2$: note that the center of the axial system has been moved to the point defined in the $\boldsymbol{\beta}_F$-space by the least squares estimates. The directions of the original axes are shown dashed, while the new rotated axes are solid.

$$G = (g_1, g_2, \ldots, g_r) \tag{17C.1}$$

be the $r \times r$ matrix whose $r \times 1$ columns g_1, g_2, ..., g_r are the eigenvectors of $F'F$. This means that the λ_i and g_i have the property that

$$F'Fg_i = \lambda_i g_i, \qquad i = 1, 2, \ldots, r. \tag{17C.2}$$

To obtain these we solve the polynomial equation

$$\det (F'F - \lambda I) = 0 \tag{17C.3}$$

for its r roots λ_1, λ_2, ..., λ_r. (In practical cases these roots are usually distinct. When they are not, which happens in some experimental design situations and occasionally in other cases, it means that certain sections of the sum of squares surface are circular or spherical, and no rotation is needed for axes already lying in that section.) We then substitute the λ_i in turn into the dependent Eqs. (17C.2) and choose the corresponding solutions g_i that are normalized so that $g_i' g_i = 1$. The following properties can be shown to hold:

$$G'G = GG' = I, \tag{17C.4}$$

$$G'F'FG = \Lambda, \tag{17C.5}$$

where $\Lambda = \text{diagonal } (\lambda_1, \lambda_2, \ldots, \lambda_r)$ is a diagonal matrix consisting of λ_i in the ith main diagonal position and zeros everywhere off the diagonal. Define Q and γ as follows:

$$Q = FG \quad \text{and} \quad \gamma = G'\beta_F, \tag{17C.6}$$

so that

$$F = QG' \quad \text{and} \quad \beta_F = G\gamma. \tag{17C.7}$$

Then

$$F'F + \theta I = G(\Lambda + \theta I)G' \tag{17C.8}$$

and, because $G^{-1} = G'$, due to Eq. (17C.4),

$$(F'F + \theta I)^{-1} = G(\Lambda + \theta I)^{-1}G'. \tag{17C.9}$$

The ridge estimator is thus

$$b_F(\theta) = G(\Lambda + \theta I)^{-1}G'F'Y, \tag{17C.10}$$

or, if we premultiply this equation by G',

$$\hat{\gamma}(\theta) = (\Lambda + \theta I)^{-1}Q'Y. \tag{17C.11}$$

If we write

$$c_i = (Q'Y)_i \tag{17C.12}$$

for the ith component of the vector $Q'Y$ and invert the diagonal matrix $\Lambda + \theta I$, we obtain, for the components of $\hat{\gamma}(\theta)$,

$$\hat{\gamma}_i(\theta) = c_i/(\lambda_i + \theta), \qquad i = 1, 2, \ldots, r. \tag{17C.13}$$

This is the *canonical form* of ridge regression. The ridge trace is usually plotted in terms of the β_F or β's, however. We can convert back to these via

$$b_F(\theta) = G\hat{\gamma}(\theta), \tag{17C.14}$$

$$b(\theta) = \{\text{Diagonal}(S_{jj}^{-1/2})\}G\hat{\gamma}(\theta). \tag{17C.15}$$

The canonical form is often seen in papers discussing ridge regression and provides a convenient computational form for making the ridge trace calculations.

Residual Sum of Squares

This can be written

$$\text{RSS}_\theta = \text{RSS}_0 + \theta^2 \sum_{i=1}^{r} \hat{\gamma}_i^2(\theta)/\lambda_i, \tag{17C.16}$$

where

$$\text{RSS}_0 = \left\{\sum Y_i^2 - n\overline{Y}^2\right\} - \sum_{i=1}^{r} c_i^2/\lambda_i \tag{17C.17}$$

is the minimum value attained by the least squares estimator $\hat{\gamma}(0)$, which is given by $\theta = 0$. A possible way to select a value θ^* for θ is so that the difference $\text{RSS}_\theta - \text{RRS}_0$ is equal to some selected number or expression. One possibility is to set it equal to $s^2(r + 1)$. The rationale for this is that $\sigma^2(r + 1)$ is the expected value $E(\text{RRS}_\gamma - \text{RRS}_0)$ when the true value γ is used instead of $\hat{\gamma}(\theta)$. This means that θ is selected in such a way that the squared distance represented by $\text{RRS}_\theta - \text{RSS}_0$ is the size we "expect" it to be. [Thus we attempt to keep the estimate on roughly the same confidence contour as that on which the true value falls on average. The direction from $\hat{\gamma}(0)$ may, however, be different.] This provides

$$\theta^* = \left[s^2(r + 1)\left/\left\{\sum_{i=1}^{r} \gamma_i^2(\theta^*)/\lambda_i\right\}\right]^{1/2}, \tag{17C.18}$$

which could be solved for θ^*, perhaps iteratively in the way described around Eq. (17.3.4), provided convergence was achieved as in Eq. (17.3.5). Substitution of $\theta^* = 0$ on the right-hand side of Eq. (17C.18) provides the value from the first iteration, which might be good enough in some circumstances.

Mean Square Error

Equation (17B.2) becomes, in this notation

$$\text{MSE}(\theta) = \sum_{i=1}^{r} (\lambda_i\sigma^2 + \theta^2\gamma_i^2)/(\lambda_i + \theta)^2. \tag{17C.19}$$

Some Alternative Formulas

If we define

$$\hat{\gamma}(0) = \Lambda^{-1}Q'Y \tag{17C.20}$$

for the least squares estimate vector, and

$$\Delta = (\Lambda + \theta I)^{-1}\Lambda \tag{17C.21}$$

for the diagonal matrix with elements $\delta_i = \lambda_i/(\lambda_i + \theta)$, we see that Eqs. (17.11), (17C.13), (17C.14), and (17C.19) become, respectively,

$$\hat{\gamma}(\theta) = \boldsymbol{\Delta}\hat{\gamma}(0), \tag{17C.22}$$

$$\hat{\gamma}_i(\theta) = \delta_i c_i / \lambda_i, \tag{17C.23}$$

$$\mathbf{b}_F(\theta) = \mathbf{G}\boldsymbol{\Delta}\hat{\gamma}(0), \tag{17C.24}$$

$$\text{MSE}(\theta) = \text{trace}\{\sigma^2\boldsymbol{\Delta}^2\boldsymbol{\Lambda}^{-1} + (\mathbf{I} - \boldsymbol{\Delta})\gamma\gamma'(\mathbf{I} - \boldsymbol{\Delta})\}. \tag{17C.25}$$

These formulas exhibit the conversion from the least squares estimators to the ridge estimators.

EXERCISES FOR CHAPTER 17

A. Is ridge regression a least squares procedure?

B. Find the ridge regression solution for the data below for a general value of θ and for the straight line model $Y = \beta_0 + \beta_1 X + \epsilon$. Show that when θ is chosen as 0.4, the ridge solution fit is $\hat{Y} = 40 + 6.25Z = 40 + 1.5625X$, where Z is a centered and scaled form of X.

X	Y
-2	35
-1	40
-1	36
-1	38
0	40
1	43
2	45
2	43

C. Suggested readings for more on ridge regression (in addition to the basic Hoerl and Kennard papers mentioned in Section 17.1) are: D. W. Marquardt (1970), "Generalized inverses, ridge regression, biased linear estimation and nonlinear estimation," *Technometrics,* **12,** 591–612; D. W. Marquardt and R. D. Snee (1975), "Ridge regression in practice," *The American Statistician,* **29,** 3–19; and "Why regression coefficients have the wrong sign," by G. M. Mullett, *Journal of Quality Technology,* **8,** 1976, 121–126.

CHAPTER 18

Generalized Linear Models (GLIM)

18.1. INTRODUCTION

In Section 9.2, we discussed *generalized least squares* (of which *weighted least squares* is a special case). That section concerned the application of least squares methods in situations where $\mathbf{Y} = \mathbf{X}\boldsymbol{\beta} + \boldsymbol{\epsilon}$, where the errors were normal, and $E(\boldsymbol{\epsilon}) = \mathbf{0}$, but the variance–covariance matrix of the errors was $\mathbf{V}(\mathbf{Y}) = \mathbf{V}\sigma^2$, not merely $\mathbf{I}\sigma^2$. The topic of this chapter is different from the Section 9.2 topic, although there are two points of confusion. One is the use of the word "generalized" in both titles. The other is that the maximum likelihood solution for estimating parameters in generalized linear models (this chapter!) involves the use of generalized (weighted) least squares in an iterative manner.

Generalized linear models (GLIM) analysis comes into play when the error distribution is not normal and/or when a vector of nonlinear *functions* of the responses, $\boldsymbol{\eta}(\mathbf{Y}) = \{\eta(Y_1), \eta(Y_2), \ldots, \eta(Y_n)\}'$, say, and not $\mathbf{Y}$ itself, has expection $\mathbf{X}\boldsymbol{\beta}$. GLIM analysis provides a larger framework for estimation, which *includes* the case of normally distributed errors. When the errors *are* normal and $\eta(Y_i) = Y_i$, the GLIM analysis applies perfectly well but simplifies to the least squares case with which we have become familiar. Because of limitations in the general GLIM procedure, direct use of least squares is most appropriate for regression data with normally distributed errors, employing techniques already described.

In this chapter we offer a brief self-contained introduction to GLIM. This is intended to set the least squares methods already discussed into a more general context and to prepare the reader for specialized GLIM textbooks. Actual implementation of GLIM on data requires a complicated calculation for which a number of standard programs are available. Becoming familiar with one of these is advisable for further study.

Acronym

We are going to use the acronym GLIM in this chapter. There is a slight confusion, when we do, with the GLIM computer package, which is only *one* of several possible packages that perform the appropriate calculations. (See Hilbe, 1994.) (However, both GLIM acronyms apply to the same thing, and this is not true of another possible acronym, GLM.)

18.2. THE EXPONENTIAL FAMILY OF DISTRIBUTIONS

We start by talking about a very general family of error distributions, which includes the normal distribution as a special case. A random variable u that belongs to the *exponential family* with a single parameter θ has a probability density function

$$f(u, \theta) = s(u)t(\theta)e^{a(u)b(\theta)}, \tag{18.2.1}$$

where the letters s, t, a, and b are all known functions. It is convenient to rewrite this as

$$f(u, \theta) = \exp\{a(u)b(\theta) + c(\theta) + d(u)\}, \tag{18.2.2}$$

where $d(u) = \ln s(u)$ and $c(\theta) = \ln t(\theta)$. If there were other parameters not of direct interest, they could occur within the functions a, b, c, and d.

Some Definitions

When $a(u) = u$ in (18.2.2), the distribution is said to be in *canonical form* and $b(\theta)$ is then called the *natural parameter* of the distribution. Parameters other than the parameter of interest θ can be involved in the functions a, b, c, and d. These are *nuisance* parameters and are regarded as known. For example, for the normal variable u in (18.2.4), given below, the natural parameter is μ/σ^2, and σ^2 is a nuisance parameter if μ is of interest. For the binomial u in (18.2.5), given below, the natural parameter is $\ln[p/(1 - p)]$, the natural logarithm of the odds ratio $p/(1 - p)$. Although the canonical parameterization has convenient computational and theoretical properties, it is not necessarily the best choice in applications.

[Nuisance parameters are usually scale parameters and, asymptotically in most cases, exactly in the normal case, do not need to be estimated along with the parameters of the linear model to be fitted in (18.3.2). However, when there is concern about the possibility of dependence in the observations, scale parameters may be estimated even when not required. This enables confidence intervals to be widened, for example, and so made more robust against the possible *overdispersion* of the observations due to dependence. We do not further discuss this complication.]

Some Members of the Exponential Family

1. The normal distribution $N(\mu, \sigma^2)$.

$$f(u, \mu) = (2\pi\sigma^2)^{-1/2}\exp\{-(u - \mu)^2/(2\sigma^2)\}, \qquad -\infty \le u \le \infty, \tag{18.2.3}$$

where μ is the parameter of interest and σ^2 is regarded as a nuisance parameter (or as known). By the contortional rewrite

$$f(u, \mu) = \exp\{u(\mu/\sigma^2) + [-\mu^2/(2\sigma^2) - \tfrac{1}{2}\ln(2\pi\sigma^2)] - u^2/(2\sigma^2)\}, \tag{18.2.4}$$

we recognize that with $a(u) = u$, $b(\theta) = \mu/\sigma^2$, and so on, the form (18.2.2) appears.

2. The binomial distribution $B(n, p)$.

$$f(u, p) = \binom{n}{u} p^u(1 - p)^{n-u}, \qquad u = 0, 1, 2, \ldots, n,$$

$$= \exp\left\{u \ln[p/(1 - p)] + n \ln(1 - p) + \ln\binom{n}{u}\right\}. \tag{18.2.5}$$

3. The Poisson distribution.

$$f(u, \lambda) = \frac{\lambda^u e^{-\lambda}}{u!}, \qquad u = 0, 1, 2, \dots$$

$$= \exp\{u \ln \lambda - \lambda - \ln(y!)\}. \tag{18.2.6}$$

The natural parameter is $\ln \lambda$.

4. The gamma distribution with parameter θ of interest and ϕ as nuisance parameter.

$$f(u, \theta) = \{u^{\phi-1}\theta^\phi \exp(-u\theta)\}/\Gamma(\phi), \qquad 0 \le u \le 1$$

$$= \exp\{-u\theta + [\phi \ln u - \ln \Gamma(\phi)] + (\phi - 1)\ln u\}. \tag{18.2.7}$$

5. The Pareto distribution.

$$f(u, \theta) = \theta u^{-(1+\theta)}, \qquad u \ge 0$$

$$= \exp\{-(1 + \theta)\ln u + \ln \theta\}. \tag{18.2.8}$$

This is not of canonical form.

6. The exponential distribution.

$$f(u, \theta) = \theta \exp(-u\theta), \qquad u \ge 0$$

$$= \exp\{-u\theta + \ln \theta\}. \tag{18.2.9}$$

The natural parameter is θ. This distribution is a special case of (18.2.7) when $\phi = 1$.

7. The negative binomial distribution. The variable u is the number of failures observed to attain a given number (r) of successes in binomial trials with probability θ of success. In the GLIM context, the distribution more often arises via an overdispersed Poisson variable (Feller, 1957).

$$f(u, \theta) = \binom{u + r - 1}{r - 1} \theta^r (1 - \theta)^u, \qquad u = 1, 2, \dots$$

$$= \exp\left\{ u \ln(1 - \theta) + r \ln \theta + \ln \binom{u + r - 1}{r - 1} \right\}. \tag{18.2.10}$$

The natural parameter is $\ln(1 - \theta)$.

Expected Value and Variance of $a(u)$

The following results can be obtained:

$$E[a(u)] = -c'(\theta)/b'(\theta),$$

$$V[a(u)] = [b''(\theta)c'(\theta) - c''(\theta)b'(\theta)]/[b'(\theta)]^3, \tag{18.2.11}$$

where the prime indicates differentiation with respect to θ. For the binomial, for example, $a(u) = u$, $\theta = p$, and

$$b(\theta) = \ln[p/(1 - p)], \qquad b'(\theta) = 1/[p(1 - p)], \qquad b''(\theta) = (2p - 1)/[p(1 - p)]^2,$$

$$c(\theta) = n \ln(1 - p), \qquad c'(\theta) = -n/(1 - p), \qquad c''(\theta) = -n/(1 - p)^2. \tag{18.2.12}$$

and so

$$E(u) = np, \qquad V(u) = np(1 - p). \tag{18.2.13}$$

For the normal distribution the formulas (18.2.11) lead to μ and σ^2, of course.

Joint Probability Density of a Sample

If we have a sample $Y_1, Y_2, \ldots, Y_n$ of independent observations from (18.2.2), the joint probability density function is

$$\exp\left\{ b(\theta) \sum_{i=1}^{n} a(Y_i) + nc(\theta) + \sum_{i=1}^{n} d(Y_i) \right\}. \tag{18.2.14}$$

The quantity $\sum_{i=1}^{n} a(Y_i)$ is a *sufficient statistic* for $b(\theta)$. This implies that it carries all the information the sample provides about $b(\theta)$; knowing the individual Y_i, for example, would not tell us any more about $b(\theta)$.

18.3. FITTING GENERALIZED LINEAR MODELS (GLIM)

The fitting of generalized linear models via iteratively reweighted least squares, which has proved a useful practical technique over the years, stems from work of Nelder and Wedderburn (1972). Suppose we have a set of independent observations $Y_1, Y_2, \ldots, Y_n$ each from the same exponential type distribution (e.g., all binomials) of canonical form [i.e., $a(Y) = Y$]. Then the joint probability density function is

$$f(Y_i, \boldsymbol{\theta}, \boldsymbol{\phi}) = \exp\{\Sigma Y_i b(\theta_i) + \Sigma c(\theta_i) + \Sigma d(Y_i)\}, \tag{18.3.1}$$

where all summations are from $i = 1, \ldots, n$ unless otherwise specified, where $\boldsymbol{\phi}$ is a vector of nuisance parameters that occur within $b(\theta)$, $c(\theta)$, and $d(\theta)$ although not explicitly shown therein, and where $\boldsymbol{\theta} = (\theta_1, \theta_2, \ldots, \theta_n)'$ is a vector of parameters of interest. Note that the θ_i *could* be all different. In general, however, we typically consider a smaller set of parameters $\boldsymbol{\beta} = (\beta_1, \beta_2, \ldots, \beta_p)'$, which relate some function $g(\mu_i)$ of μ_i [which is $E(Y_i)$] to a linear combination of β's via

$$g(\mu_i) = \mathbf{x}_i' \boldsymbol{\beta}, \tag{18.3.2}$$

where $\mathbf{x}_i = (x_{i1}, x_{i2}, \ldots, x_{ip})'$ is a p by 1 vector of predictor variables observed along with Y_i. The function $g(\cdot)$ is monotone, differentiable, and called the *link function*.

Example: Binomial Distributions, Indices n_i, Parameters p_i

Suppose we have data $Y_i, \mathbf{x}_i'$ from a binomial distribution with index n_i and parameter p_i. The single observation Y_i is of the form r_i/n_i, where r_i is the number of successes in n_i trials, each having probability p_i of success, and

$$\mathbf{x}_i = (x_{i1}, x_{i2}, \ldots, x_{ip})' \tag{18.3.3}$$

is a set of observations of p predictor variables associated with Y_i. As we have seen, the binomial distribution is a member of the exponential family. The probability distribution function of n such sets of data for $i = 1, 2, \ldots, n$ is then

$$\exp\left\{ \sum Y_i \ln[p_i/(1 - p_i)] + \sum n_i \ln(1 - p_i) + \sum \ln\binom{n_i}{Y_i} \right\}, \tag{18.3.4}$$

with summations over $i = 1, 2, \ldots, n$. We would typically hope that the variation in the p_i values could be explained in terms of the $\mathbf{x}_i$ values; that is, we would hope that we could find some suitable link function $g(\cdot)$ such that the model

$$g(p_i) = \mathbf{x}_i' \boldsymbol{\beta} \qquad (18.3.5)$$

held, where $\boldsymbol{\beta} = (\beta_1, \beta_2, \ldots, \beta_p)'$ was a set of parameters, which was the same for each i value. A link function that is often regarded as a sensible one for binomial data is the natural parameter, that is, the *log odds* or *logit* given by

$$g(p_i) = \ln\{p_i/(1 - p_i)\}, \qquad (18.3.6)$$

and this would imply that we wish to fit a model in which

$$\ln\{p_i/(1 - p_i)\} = \mathbf{x}_i' \boldsymbol{\beta} = \beta_1 x_{i1} + \beta_2 x_{i2} + \cdots + \beta_p x_{ip}. \qquad (18.3.7)$$

Note that this is equivalent to saying that p_i is representable as

$$p_i = \exp(\mathbf{x}_i' \boldsymbol{\beta})/\{1 + \exp(\mathbf{x}_i' \boldsymbol{\beta})\}. \qquad (18.3.8)$$

[When $\mathbf{x}_i' \boldsymbol{\beta} = \beta_1 + x_{i2}\beta_2$, (18.3.8) is sometimes called the logistic function.]

Estimation Via Maximum Likelihood

To estimate the elements of the parameter vector $\boldsymbol{\beta}$ in (18.3.2), which becomes (18.3.7) for the binomial example, we use the method of maximum likelihood. This requires regarding each of (18.3.1), (18.3.4), and similar appropriate formulas for other distributions, as a function of the parameters in $\boldsymbol{\beta}$, given a specific set of Y_i; we then call such a quantity the *likelihood function* and maximize it with respect to the elements of $\boldsymbol{\beta}$. More specifically, we usually maximize the *log likelihood function* obtained by taking the natural logarithm of formulas like (18.3.4); this leads to the same estimate $\mathbf{b}$ of $\boldsymbol{\beta}$.

The actual maximization procedure, while mathematically interesting, is tedious to work through. [See details in, for example, Dobson (1990).] It results in some normal equations of the general form

$$\mathbf{X}'\mathbf{WXb} = \mathbf{X}'\mathbf{Wz}, \qquad (18.3.9)$$

which, we can quickly appreciate, see Eq. (9.2.9), are the normal equations for a generalized least squares model fit. There is an additional catch here, however, in that $\mathbf{W}$ and $\mathbf{z}$ involve $\mathbf{b}$ also. Thus the solution of (18.3.9) usually requires an iterative calculation. An initial value for $\mathbf{b}$, say, $\mathbf{b}_0$, is inserted into $\mathbf{W}$ and $\mathbf{z}$, and (18.3.9) is solved for $\mathbf{b}$ to give solution $\mathbf{b}_1$, say. Then $\mathbf{b}_1$ replaces $\mathbf{b}_0$ and a new solution $\mathbf{b}_2$ is calculated, and so on, until convergence is attained. Standard programs exist for doing these *iteratively reweighted least squares* calculations. They typically permit specification of the distribution (e.g., binomial, Poisson), the link function [e.g., $\ln\{p/(1 - p)\}$ for the binomial, $\ln \lambda$ or λ for the Poisson with mean and variance λ], and the $\mathbf{x}_i$ values, whereupon the solution $\hat{\boldsymbol{\beta}}$ is given. Perhaps the GLIM program (produced by Numerical Algorithms Group, Oxford, U.K.) is the best known program in some parts of the world, but versions are also available in BMDP, GENSTAT, SAS, SPSS, SPLUS, and so on.

Deviance

Comparisons between nested models in ordinary least squares are usually made via extra sums of squares F-tests, which essentially compare residual sums of squares. In the case of generalized linear models, the residual sum of squares role is played by the deviance

$$-2\{l(\hat{\boldsymbol{\beta}}) - l(\hat{\boldsymbol{\theta}})\}, \tag{18.3.10}$$

where $l(\cdot)$ is the log likelihood function, $l(\hat{\boldsymbol{\beta}})$ is its value for the maximum likelihood value $\hat{\boldsymbol{\beta}}$, and $l(\hat{\boldsymbol{\theta}})$ is its value when $\hat{\theta}_i = Y_i$ is used, that is, when each θ_i in (18.3.1) is estimated by its corresponding Y_i value. Thus we can rewrite (18.3.10) as

$$\text{Deviance} = -2\{l(\hat{\boldsymbol{\beta}}) - l(\mathbf{Y})\}. \tag{18.3.11}$$

The deviance is asymptotically distributed as a χ^2 variable with $(n - p)$ degrees of freedom when the model is correct, and so large deviance values are taken to be indicative of a bad fit for the $\mathbf{x}'\boldsymbol{\beta}$ model. We compare the deviance value with a selected percentage point of the appropriate χ^2 distribution whose df correspond to the number of parameters in $\boldsymbol{\beta}$. Differences between deviances can also be used in a way parallel to the use of extra sums of squares.

It should be noted that the χ^2 approximation can be inaccurate for small samples, for example, if the Y_i are Poisson variables with means close to 1, or binomial. See McCullagh and Nelder (1989, p. 121).

18.4. PERFORMING THE CALCULATIONS: AN EXAMPLE

Table 18.1 shows a set of data on reported occurrences of a communicable disease in two areas of the country at ten 2 month intervals, 2, 4, ..., 20. There are 20 data points, so that $i = 1, 2, \ldots, 20$ below.

We assume that the occurrences Y_i are Poisson variables with $E(Y_i) = \mu_i$ and that

$$g(\mu_i) = \ln \mu_i = \beta_0 + \beta_1 X_{1i} + \beta_2 X_{2i} \tag{18.4.1}$$

is the model under consideration. We take $X_{1i} = \ln(2i)$ corresponding to ln(month indicator) and $X_{2i} = 0$ for the observations from area A and $X_{2i} = 1$ for observations in area B. Other choices of X_{1i} and X_{2i} are possible. Also, another link function could be used; the one chosen, $g(\mu_i) = \ln \mu_i$, is the "natural" link, corresponding to the "natural parameter."

T A B L E 18.1. Reported Occurrences of a Communicable Disease in Two Areas of the Country

	Occurrences in		
	Area A	Area B	Month, e^{X_1}
	8	14	2
	8	19	4
	10	16	6
	11	21	8
	14	23	10
	17	27	12
	13	28	14
	15	29	16
	17	33	18
	15	31	20
Dummy X_2	0	1	

To fit (18.4.1) in GLIM, one needes these instructions:

```
glim
  ?   $units 20$              (the ? prompt appears automatically. The instruction
                               indicated 20 observations are coming)
  ?   $data y x w $           (Names the variables as y, x, and w.)
      $read                   (No $ closure here.)
      $REA?  8 0.693 0        ($REA? appears automatically on each line. Data are
                               entered line by line)
      $REA?  8 1.386 0
      $REA? 10 1.792 0
      $REA? 11 2.079 0
      $REA? 14 2.303 0
      $REA? 17 2.485 0
      $REA? 13 2.639 0
      $REA? 15 2.773 0
      $REA? 17 2.890 0
      $REA? 15 2.996 0
      $REA? 14 0.693 1
      $REA? 19 1.386 1
      $REA? 16 1.792 1
      $REA? 21 2.079 1
      $REA? 23 2.303 1
      $REA? 27 2.485 1
      $REA? 28 2.639 1
      $REA? 29 2.773 1
      $REA? 33 2.890 1
      $REA? 31 2.996 1        (Last data group. The ? prompt now reappears.)
  ?   $yvar y $               (Declares y as the response.)
  ?   $error poisson $        (Declares the error distribution.)
  ?   $link L $               (Can be omitted for the natural link, here $g(\mu_i) =$
                               $\ln \mu_i$)[1]
  ?   $fit x + w $            (Fit additive terms in x and w.)
scaled deviance = 3.1258 at cycle 3 (Printed out.)
d.f. = 17
  ?   $display e $            (Gives printout as follows.)
        estimate s.e.        parameter
  1     1.684    0.2202        1 (Estimate of $\beta_0$)
  2     0.3784   0.08524       X (Estimate of $\beta_1$)
  3     0.6328   0.1094        W (Estimate of $\beta_2$)
scale parameter taken as 1.000
                               (By retyping the $link, $fit, and $display e
                               instructions, other fits can be made using the
                               same data set.)
```

T A B L E 18.2. Some Reasonable Choices of Link Functions

Symbol to be Entered	Name of Link	Form of $g(u)$	Used for
C	Complementary log-log	$\ln\{-\ln(1-u)\}$	Binomial
E	Exponential	e^u	
G	Logit	$\ln\{u/(1-u)\}$	Binomial
I	Identity	u	Normal
L	Logarithm	$\ln u$	Poisson, Gamma
P	Probit	$\Phi^{-1}\{(u-\mu)/\sigma\}$	Binomial
R	Reciprocal	u^{-1}	Gamma, Poisson
S	Square root	$u^{1/2}$	Poisson

[1]Available link choices are given in Table 18.2. It is not always clear in practice what link should be employed and data are often analyzed by comparing several alternative choices.

The fitted model is

$$\widehat{\ln Y} = 1.684 + 0.3784X_1 + 0.6328X_2 \qquad (18.4.2)$$

Compared with $\chi^2(3, 0.95) = 7.81$, the scaled deviance of 3.13 is nonsignificant, indicating a reasonable fit.

18.5. FURTHER READING

There are several excellent books in this area. Dobson (1990) provides a good basic introduction. McCullagh and Nelder (1989) is the technical bible. Aitken et al. (1989) and Healy (1988) provide good practical examples. Francis, Green, and Payne (1993) is about the GLIM program. The paper by Hilbe (1994) reviews seven software packages. These are the places to start, the individual selection depending on your objective. Other useful books are Collett (1991); Cox and Snell (1989); Fahrmeir and Tutz (1994); Freund and Littell (1991); Green and Silverman (1994); Hosmer and Lemeshow (1989); and Littell, Freund, and Spector (1991). For comments on interpretation of fitted models, see the note by Chappell (1994).

EXERCISE FOR CHAPTER 18

A. Suppose we have n observations of variables $X_1, X_2, \ldots, X_k, Y$, where the X's are predictors and Y is a response variable. If the Y's are binomial ratios $Y_i = r_i/m$, say, where m is constant, what types of analyses are feasible?

CHAPTER 19

Mixture Ingredients as Predictor Variables

19.1 MIXTURE EXPERIMENTS: EXPERIMENTAL SPACES

Response surface experiments often involve mixtures of ingredients. For example, fuels can consist of a mixture of petroleum and various additives; fish patties may contain several types of fish; a fruit juice drink may consist of a mixture of orange, pineapple, and grapefruit juices; or a regional wine may be blended from several grape varieties. This adds a restriction to the problem, often of type

$$x_1 + x_2 + \cdots + x_q = 1, \tag{19.1.1}$$

where the predictor variables are proportions, $0 \leq x_i \leq 1$, $i = 1, 2, \ldots, q$. Note that a more general restriction such as $c_1 t_1 + c_2 t_2 + \cdots + c_q t_q = 1$, where the c_i are specified constants, can be converted to (19.1.1) by recoding $c_i t_i = x_i$ for $i = 1, 2, \ldots, q$. The restriction displayed above means that the complete x-space is not available for collecting data.

Two Ingredients

Consider the simplest case, $q = 2$, for which $x_1 + x_2 = 1$. This defines (see Figure 19.1) a straight line in the (x_1, x_2) space. Because $0 \leq x_1, x_2 \leq 1$, however, only that part of the line between and including the points $(0, 1)$ and $(1, 0)$ defines the appropriate mixture space. All points on this segment take the form $(x_1, 1 - x_1)$ or $(1 - x_2, x_2)$. If x_1 is the proportion of gin in a mixed drink and x_2 is the proportion of tonic water, all possible gin and tonic mixtures will lie on the line segment described. (One might question whether drinks with no gin, or drinks with only gin, can properly be called "mixtures," but it is conventional to do so, in this context.)

Three Ingredients

When $x_1 + x_2 + x_3 = 1$, the mixture space is defined by a portion of a plane containing the three extreme mixture points $(x_1, x_2, x_3) = (1, 0, 0)$, $(0, 1, 0)$, and $(0, 0, 1)$ and such that $0 \leq x_i \leq 1$. It is thus a two-dimensional equilateral triangle with the extreme points as the vertices, as depicted in Figure 19.2. Note that each side of the triangle is a two-ingredient subspace, with the third ingredient set equal to zero. Compare the (x_2, x_3) axes part of Figure 19.2 with Figure 19.1, for example. The Dietzgen Corporation

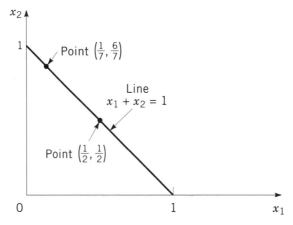

Figure 19.1. A one-dimensional experimental mixture space $x_1 + x_2 = 1$ embedded in the two-dimensional (x_1, x_2) space. Only the region for which $0 \le x_i \le 1$ is valid, as shown.

manufactures a triangular coordinate (TC) graph paper (No. 340-TC) that has three sets of lines—one set parallel to each of the three sides of the triangle. The depiction of Figure 19.3a illustrates one of these three sets of lines, and Figure 19.3b illustrates how the intersection of any two lines from different sets defines a point in the mixture space. Naturally $x_3 = 1 - x_1 - x_2$, so we do not actually need the third set of lines (which *does* appear on Dietzgen 340-TC paper, however) to fully define the point.

Four Ingredients

The original four-dimensional space cannot be drawn but the mixture space is now a regular (equal-sided) tetrahedron, shown in Figure 19.4a. Any point (x_1, x_2, x_3, x_4) in or on the boundaries of the tetrahedron is such that $x_1 + x_2 + x_3 + x_4 = 1$. Each triangular face is defined by $x_i = 0$ for $i = 1, 2, 3,$ and 4. It is easy to construct a tetrahedron in three dimensions for yourself, as indicated in Figure

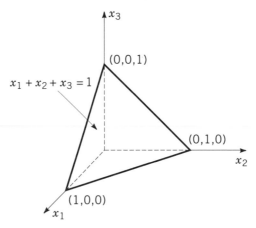

Figure 19.2. A two-dimensional experimental mixture space $x_1 + x_2 + x_3 = 1$ embedded in the three-dimensional (x_1, x_2, x_3) space. Only the region for which $0 \le x_i \le 1$ is valid, as shown.

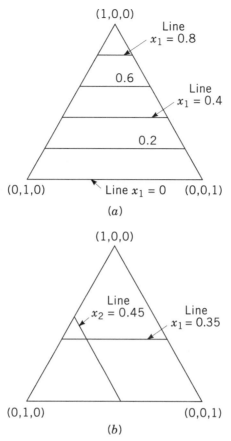

Figure 19.3. (a) Parallel coordinate lines $x_1 = 0$, 0.2, 0.4, 0.6, and 0.8 in the mixture space $x_1 + x_2 + x_3 = 1$. (b) The point (0.35, 0.45, 0.20) defined by the intersection of the two coordinate lines $x_1 = 0.35$ and $x_2 = 0.45$. Note that $x_3 = 1 - x_1 - x_2 = 0.20$.

19.4b. First, draw and then cut out an equilateral triangle indicated by the 1's. Next, join up, in pairs, the midpoints 2, 3, and 4 of the sides to produce four equal-sized triangles as shown. Use the triangle 234 as the base and fold along the lines 23, 24, and 34. Bring the three points marked 1, 1, and 1 together into one point 1. Tape the two sides 12 together where they meet and do the same for 13 and 14. The point 1 is defined by (1, 0, 0, 0), point 2 by (0, 1, 0, 0), and so on. The 234 face is defined by $x_1 = 0$ and similarly for the other faces. One can also imagine the tetrahedron built up in triangular slices $x_1 = c$, where $c = 0$ on the base 234 and where c increases to 1 at the vertex marked 1. This slicing can be imagined relative to any of the x_i dimensions, of course. For example, Figure 19.4c shows slices at about $x_1 = 0.15$ and $x_1 = 0.85$.

Five or More Ingredients

For five ingredients, the best we can do pictorially is to imagine a series of tetrahedra of diminishing size. The largest would be defined by $x_5 = 0$, and successive ones would be $x_5 = c$, where c increases from 0 to 1. When $x_5 = 1$ we have the point (0, 0, 0, 0, 1), which is the limiting case of these tetrahedra.

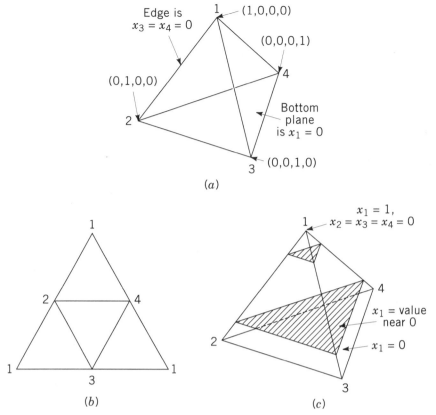

Figure 19.4. (*a*) The mixture space for four ingredients is an equal-sided tetrahedron (pyramid). (*b*) Construction of an equal-sided tetrahedron model. (*c*) Slices of four-ingredient mixture space for which x_1 is nearer 0 or nearer 1.

19.2. MODELS FOR MIXTURE EXPERIMENTS

We discuss models in detail for $q = 3$ ingredients; the pattern of models for general q will then become easy to understand.

Three Ingredients, First-Order Model

It is quickly obvious that the usual polynomial response surface models no longer can be used. Consider, for example, the planar model function

$$\beta_0 + \beta_1 x_1 + \beta_2 x_2 + \beta_3 x_3. \tag{19.2.1}$$

The **X** matrix consists of four columns $[1, \mathbf{x}_1, \mathbf{x}_2, \mathbf{x}_3]$ For each line of this matrix, however, the mixture restriction requires that $1 = x_1 + x_2 + x_3$ so that a linear relationship holds between the four columns of **X**. Although we have a three-dimensional x-space, the space of the experimental domain is only two-dimensional; see Figure 19.2. Thus, in the usual least squares regression fit formulation $\mathbf{b} = (\mathbf{X}'\mathbf{X})^{-1}\mathbf{X}'\mathbf{Y}$, where **X** is the matrix of observed x-data, **Y** is the vector of responses, and **b** is the vector of estimates of the parameter vector $\boldsymbol{\beta}$ in the model $\mathbf{Y} = \mathbf{X}\boldsymbol{\beta} + \boldsymbol{\epsilon}$, the matrix **X'X** is singular and its inverse does not exist. This difficulty can be overcome in either of two (essentially the same but mechanically different) ways:

1. Use the restriction (19.1.1) to transform the model symmetrically into *canonical form*.
2. Transform the q original x-coordinates to $q - 1$ new z-coordinates. (This is regarded as somewhat tedious to do, especially for $k \geq 4$.) See Appendix 19A.

Canonical Form

Method 1 was pioneered by Scheffé (1958, 1963) and is now illustrated for $q = 3$. By rewriting

$$\beta_0 = \beta_0(x_1 + x_2 + x_3), \tag{19.2.2}$$

using the fact that $x_1 + x_2 + x_3 = 1$, we can rewrite (19.2.1) as

$$(\beta_0 + \beta_1)x_1 + (\beta_0 + \beta_2)x_2 + (\beta_0 + \beta_3)x_3 \tag{19.2.3}$$

or

$$\alpha_1 x_1 + \alpha_2 x_2 + \alpha_3 x_3, \tag{19.2.4}$$

where $\alpha_i = \beta_0 + \beta_i$. This canonical form is simply a no-intercept plane. One way of looking at this is that, since (19.2.1) is overparameterized because of the mixture restriction (19.1.1), the least squares problem has an infinity of solutions. One of these solutions is the one for which we choose $\beta_0 = 0$ in (19.2.1), which gives us a model of form (19.2.4). This type of model is appealing because of its symmetry in the x's. Less appealing, but just as valid, would be any model obtained by deletion of any one term in (19.2.1). There are also other valid choices. All valid choices give the same fitted values.

q Ingredients, First-Order Model

The extension to q ingredients leads to the canonical form model function (we revert now to using β's)

$$\eta = \beta_1 x_1 + \beta_2 x_2 + \cdots + \beta_q x_q, \tag{19.2.5}$$

and of course we then fit $y = \eta + \epsilon$ by least squares.

Example: The Hald Data

The 13 observations of Table 15.1 (or see Appendix 15A) are used to fit the model

$$Y = \beta_1 X_1 + \beta_2 X_2 + \beta_3 X_3 + \beta_4 X_4 + \epsilon \tag{19.2.6}$$

by least squares. The X's here are percentages, not proportions, we recall. Moreover, the data are not "perfect" mixture data in the sense that the X's do not add to the same total *exactly*. The fitted equation is

$$\hat{Y} = 2.1930 X_1 + 1.1533 X_2 + 0.7585 X_3 + 0.4863 X_4 \tag{19.2.7}$$

and the corresponding (no intercept, remember) analysis of variance table is:

Source	df	SS	MS	F
b_1, b_2, b_3, b_4	4	121,035	30,259	5176
Residual	9	53	6	
Total	13	121,088		

Obviously such a model explains nearly all the variation in the data. We do not calculate an R^2 value because there is no intercept in the model and so the R^2 statistic is not defined. We could, of course, drop any one $\beta_i X_i$ term and replace it by β_0. If *the data were "perfect" mixture data*, that is, if the X's added *exactly* to the same total, any of the resulting four fitted models and the model (19.2.7) would all provide the same $\hat{Y}$ predictions. As we have already observed in Appendix 15A, there is some slight variation in the R^2 values for the four three-X models with intercepts and there will be variation in the fitted values also.

Test for Overall Regression

Note that in fitting the model function (19.2.5), it makes little sense to test that $\beta_1 = \beta_2 = \cdots = \beta_q = 0$, leading to the reduced model $Y = \epsilon$. The appropriate null hypothesis is $H_0 : \beta_1 = \beta_2 = \cdots = \beta_q \, (= \beta_0$, say) whereupon the reduced model would be simply $Y = \beta_0 + \epsilon$ due to (19.1.1). We would have

$$\text{SS}(H_0) = \text{SS}(b_1, b_2, \ldots, b_q) - n\overline{Y}^2 \tag{19.2.8}$$

and would perform the usual test comparing $\text{SS}(H_0)/q$ with the residual mean square from the full model.

This will not work perfectly for the Hald data, but approximately we can say that $\text{SS}(H_0) = 121,035 - 118,372 = 2663$ with 3 df, leading to an approximate F-value of $(2663/3)/6 = 148$, indicating the need for more than an overall mean value.

As we have seen in Chapter 15, a three-parameter model containing an intercept and X_1 and X_2 terms appears to be satisfactory for representing the data. The no-intercept model (19.2.7) provides an alternative possibility with four terms instead of three, and this alternative would provide the same fitted values as all the three-X equations if the data were "perfect" mixture data in the sense described above.

Three Ingredients, Second-Order Model

The usual second-order model

$$g(\mathbf{x}, \boldsymbol{\beta}) = \beta_0 + \beta_1 x_1 + \beta_2 x_2 + \beta_3 x_3 + \beta_{11} x_1^2 + \beta_{22} x_2^2 + \beta_{33} x_3^2 + \beta_{12} x_1 x_2 + \beta_{13} x_1 x_3 + \beta_{23} x_2 x_3 \tag{19.2.9}$$

produces a singular $\mathbf{X'X}$ matrix. As before, we replace β_0 with $\beta_0 x_1 + \beta_0 x_2 + \beta_0 x_3$. In addition, because $x_1 = 1 - x_2 - x_3$, we have $x_1^2 = x_1 - x_1 x_2 - x_1 x_3$, and similarly for x_2^2 and x_3^2. Substituting these four relationships in (19.2.9), gathering like terms, and renaming coefficients gives

$$h(\mathbf{x}, \boldsymbol{\alpha}) = \alpha_1 x_1 + \alpha_2 x_2 + \alpha_3 x_3 + \alpha_{12} x_1 x_2 + \alpha_{13} x_1 x_3 + \alpha_{23} x_2 x_3. \tag{19.2.10}$$

This is called the canonical form of the three-ingredient second-order model.

q Ingredients, Second-Order Model

The general second-order canonical form model is clearly (reverting to β's)

$$\beta_1 x_1 + \beta_2 x_2 + \cdots + \beta_q x_q + \beta_{12} x_1 x_2 + \beta_{13} x_1 x_3 + \cdots + \beta_{q-1,q} x_{q-1} x_q. \quad (19.2.11)$$

A null hypothesis of no quadratic effects returns us to the reduced model (19.2.5) with first-order terms only.

Example: The Hald Data

Adding second-order terms does nothing much for us with these data since the residual SS from the first-order model is already tiny (53). Several near dependencies are created, not surprisingly. The addition of only the $x_2 x_3$ term reduces the residual SS to zero (to the nearest integer). This fit looks perhaps better than the reality. When we check where the data points are in the mixture space, we see the following. Apart from one point where $X_1 = 21$, all the data have $X_1 < 11$ so that they are squashed into a thin set of planes between $X_1 = 0$ and $X_1 = 11$. Moreover, eight of the X_3 values are small, less than or equal to 9. The remaining five points have $X_1 = 1, 2$, or 3 (near 0) and fan out somewhat away from the $X_3 = 0$ wall of the tetrahedron. In short, most of the points cover a rather restricted region of the space and provide information adequate for fitting only a few parameters. So a fit with only three or four parameters is not unreasonable. Moreover, extrapolation into the rest of the mixture space is likely to be quite dangerous. An approximate plot of the data is shown in Figure 19.5.

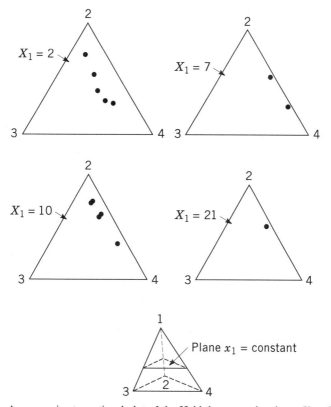

Figure 19.5. An approximate sectional plot of the Hald data onto the planes $X_1 = 2, 7, 10$, and 21.

Because we have plotted the planes $X_1 = 2, 7, 10$, and 21 only, some points are slightly above or below the planes shown. Also, we recall that the X-values for each point do not add perfectly to 100% anyway. Nevertheless, the diagram does clearly indicate the restricted nature of the data.

Third-Order Canonical Model Form

Third-order models of two main types are sometimes used. They also possess canonical forms. For $q = 4$, for example, we can fit the *general cubic* model function

$$
\begin{aligned}
h(\mathbf{x}, \boldsymbol{\beta}) = {} & \beta_1 x_1 + \beta_2 x_2 + \beta_3 x_3 + \beta_4 x_4 + \beta_{12} x_1 x_2 + \beta_{13} x_1 x_3 \\
& + \beta_{14} x_1 x_4 + \beta_{23} x_2 x_3 + \beta_{24} x_2 x_4 + \beta_{34} x_3 x_4 \\
& + \alpha_{12} x_1 x_2 (x_1 - x_2) + \alpha_{13} x_1 x_3 (x_1 - x_3) + \alpha_{14} x_1 x_4 (x_1 - x_4) \quad (19.2.12) \\
& + \alpha_{23} x_2 x_3 (x_2 - x_3) + \alpha_{24} x_2 x_4 (x_2 - x_4) + \alpha_{34} x_3 x_4 (x_3 - x_4) \\
& + \beta_{123} x_1 x_2 x_3 + \beta_{124} x_1 x_2 x_4 + \beta_{134} x_1 x_3 x_4 + \beta_{234} x_2 x_3 x_4 .
\end{aligned}
$$

For the *special cubic*, the six α_{ij} terms are deleted from (19.2.12). Note that when $q = 3$ [so that all terms containing x_4 are deleted in (19.2.12)], the general cubic contains four terms more than the quadratic while the special cubic contains only one more, namely, $\beta_{123} x_1 x_2 x_3$.

Scheffé Designs

While the usual response surface designs can be employed for mixture experiments, special designs linked to the canonical forms were suggested by Scheffé. These designs typically consist of combinations of symmetrical point sets. Some examples are given below. For more detail, and for a comprehensive treatment of the mixtures area in general, see Cornell (1990).

Design 1. The {3, 2} lattice. (Three factors and six points, which use the levels $\frac{0}{2}$, $\frac{1}{2}$, and $\frac{2}{2}$.) (1, 0, 0), (0, 1, 0), (0, 0, 1), (0.5, 0.5, 0), (0.5, 0, 0.5), and (0, 0.5, 0.5). Sometimes, the centroid $(\frac{1}{3}, \frac{1}{3}, \frac{1}{3})$ is added, also.

Design 2. The {3, 3} lattice. (Three factors and ten points, which use the levels $\frac{0}{3}$, $\frac{1}{3}$, $\frac{2}{3}$, and $\frac{3}{3}$.) (1, 0, 0), (0, 1, 0), (0, 0, 1), $(\frac{2}{3}, \frac{1}{3}, 0)$, $(\frac{1}{3}, \frac{2}{3}, 0)$, $(\frac{2}{3}, 0, \frac{1}{3})$, $(\frac{1}{3}, 0, \frac{2}{3})$, $(0, \frac{2}{3}, \frac{1}{3})$, $(0, \frac{1}{3}, \frac{2}{3})$, and $(\frac{1}{3}, \frac{1}{3}, \frac{1}{3})$. The design *includes* the centroid.

Design 3. The {4, 2} lattice. Ten points: (1, 0, 0, 0), (0, 1, 0, 0), (0, 0, 1, 0), (0, 0, 0, 1), (0.5, 0.5, 0, 0)(0.5, 0, 0.5, 0), (0.5, 0, 0, 0.5)(0, 0.5, 0.5, 0), (0, 0.5, 0, 0.5), (0, 0, 0.5, 0.5). Sometimes the centroid (0.25, 0.25, 0.25, 0.25) is added also.

Design 4. The {4, 3} lattice. Twenty points: (1, 0, 0, 0), (0, 1, 0, 0), (0, 0, 1, 0), (0, 0, 0, 1), $(\frac{2}{3}, \frac{1}{3}, 0, 0)$, $(\frac{1}{3}, \frac{2}{3}, 0, 0)$, $(\frac{2}{3}, 0, \frac{1}{3}, 0)$, $(\frac{1}{3}, 0, \frac{2}{3}, 0)$, $(\frac{2}{3}, 0, 0, \frac{1}{3})$, $(\frac{1}{3}, 0, 0, \frac{2}{3})$, $(0, \frac{2}{3}, \frac{1}{3}, 0)$, $(0, \frac{1}{3}, \frac{2}{3}, 0)$, $(0, \frac{2}{3}, 0, \frac{1}{3})$, $(0, \frac{1}{3}, 0, \frac{2}{3})$, $(0, 0, \frac{2}{3}, \frac{1}{3})$, $(0, 0, \frac{1}{3}, \frac{2}{3})$, $(\frac{1}{3}, \frac{1}{3}, \frac{1}{3}, 0)$, $(\frac{1}{3}, \frac{1}{3}, 0, \frac{1}{3})$, $(\frac{1}{3}, 0, \frac{1}{3}, \frac{1}{3})$, $(0, \frac{1}{3}, \frac{1}{3}, \frac{1}{3})$. Sometimes the centroid (0.25, 0.25, 0.25, 0.25) is added also.

19.3. MIXTURE EXPERIMENTS IN RESTRICTED REGIONS

In many mixture experiments, exploration of the entire mixture domain (e.g., for $q = 3$, this is the triangle in Figure 19.2) is not feasible or perhaps undesirable. A common type of restriction is that each ingredient has a lower bound, such as $x_1 \geq$

a_1, $x_2 \geq a_2$, $x_3 \geq a_3$, as in Figure 19.6a. Obviously this produces a restricted space of the same shape as the original mixture space, so that any design considerations can be translated to the smaller space exactly. To do this one simply transforms into new *pseudo-coordinates*

$$x_1' = \frac{x_1 - a_1}{1 - A}, \qquad x_2' = \frac{x_2 - a_2}{1 - A}, \qquad x_3' = \frac{x_3 - a_3}{1 - A}, \qquad (19.3.1)$$

where $A = a_1 + a_2 + a_3 < 1$. The inner triangle of Figure 19.6 is then defined by the sides $x_1' = 0$ (corresponding to $x_1 = a_1$), $x_2' = 0$ ($x_2 = a_2$), and $x_3' = 0$ ($x_3 = a_3$). Moreover, $x_1' + x_2' + x_3' = 1$. The extension to more ingredients is obvious.

When upper *and* lower bounds are specified, for example, $b_1 \geq x_1 \geq a_1$ and so on, the domain shape is different from the original one, as shown in Figure 19.6b. One simple way of choosing some experimental points is to select all or some of the extreme vertices of the domain in an optimum way via some selected criterion. The extreme vertices in our example are the black dots in Figure 19.6b. The criterion used might be any of those used in experimental design situations that do not involve mixture ingredients (e.g., D-optimality). Other points, centroids of the boundaries, for example, can be added to the set of points under consideration for the design. In dimensions higher than three, the boundaries will be of several kinds. In addition to one-dimensional edges, as in the $q = 3$ case, there will also be two-dimensional polyhedra

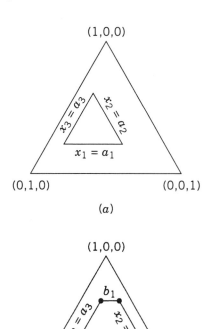

(a)

(b)

Figure 19.6. Restrictions in the $q = 3$ mixture space. (This space is pulled from Figure 19.2 and shown in two dimensions.) (a) Each ingredient must exceed a lower bound. (b) Each ingredient has an upper and a lower bound.

T A B L E 19.1. Propellent Data

Mixture Point (i)	x_1	x_2	x_3	Y_i
1	0.40	0.40	0.20	2.35
2	0.20	0.60	0.20	2.45
3	0.20	0.40	0.40	2.65
4	0.30	0.50	0.20	2.40
5	0.30	0.40	0.30	2.75
6	0.20	0.50	0.30	2.95
7	0.267	0.467	0.267	3.00
8	0.333	0.433	0.233	2.69
9	0.233	0.533	0.233	2.77
10	0.233	0.433	0.333	2.98

faces, and so on, depending on the number of ingredients. For more details, consult Cornell (1990).

19.4. EXAMPLE 1

[*Source:* Kurotori, 1966. The original paper had a different emphasis to that given in our adaptation. It presented a basic set of data, to which the model was fitted, plus some checkpoints to test lack of fit. It also illustrated the use of *pseudo-components;* a restricted region of same shape as the mixture region was recoded as in (19.3.1) so that the new (pseudo-component) coordinates mimicked the original mixture space. Here, we use all the data together to fit a second-order mixture model in canonical form in the original mixture proportion coordinates and find a best region of operation in the restricted space directly.]

In making a certain type of propellant, three mixture ingredients have their proportions restricted as follows:

$$\text{Binder } (x_1) \geq 0.2,$$
$$\text{Oxidizer } (x_2) \geq 0.4, \tag{19.4.1}$$
$$\text{Fuel } (x_3) \geq 0.2.$$

We thus have the situation of Figure 19.6a with

$$a_1 = 0.2, \quad a_2 = 0.4, \quad a_3 = 0.2. \tag{19.4.2}$$

The data are given in Table 19.1. The original responses, values of modulus of elasticity, have been divided by 1000. It is desired to find, within the permissible subspace, mixtures that will provide response values of at least 3 and that will also involve the minimum possible amount of x_1. The basic design chosen was an "augmented {3, 2} simplex lattice" (see Design 1 at the end of Section 19.2). When translated into the restricted space, it consists of the six points numbered 1 to 6 in Table 19.1, augmented by the centroid point numbered 7. The points numbered 8, 9, and 10 were additional checkpoints, and these will also be included in our regression fit. The responses observed are shown in the Y column of Table 19.1.

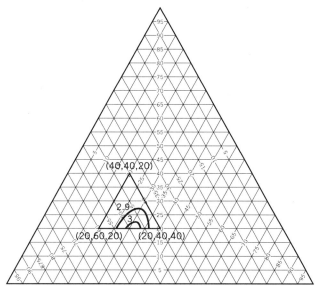

Figure 19.7. Estimated response contours within the restricted region.

The model we select for this problem is the second-order canonical polynomial

$$Y = \beta_1 x_1 + \beta_2 x_2 + \beta_3 x_3 + \beta_{12} x_1 x_2 + \beta_{13} x_1 x_3 + \beta_{23} x_2 x_3 + \epsilon. \tag{19.4.3}$$

The fitted least squares equation, with standard errors of coefficients in parentheses, is

$$\hat{Y} = -2.756 x_1 - 3.352 x_2 - 17.288 x_3 + 9.38 x_1 x_2 + 34.76 x_1 x_3 + 49.49 x_2 x_3.$$
$$\quad (4.1) \qquad (2.0) \qquad (4.1) \qquad (10.7) \qquad (10.7) \qquad (10.7)$$

It is an excellent fit; $s = 0.1$. Figure 19.7 shows the restricted region and the $\hat{Y} = 2.9$ and 3.0 contours. Desirable readings are at the lower boundary around the point $x_1 = 0.20$, $x_2 = 0.49$, and $x_3 = 0.31$, approximately. (A confirmatory run later showed that the response level predicted *could* actually be attained at this location.)

19.5. EXAMPLE 2

(*Source:* Draper et al., 1993.) The data in Table 19.2 come from an investigation of four bread flours that were combined in proportions denoted by x_1, x_2, x_3, and x_4 and were baked into loaves. The response values are specific volumes (mL/100 g), and higher specific volumes are more desirable. Four blocks of experiments were performed. Blocking is a normal procedure at Spillers Milling Limited (then a member of the Dalgety Group of companies) where the experiment was performed, to reduce the possible disturbances from within-day time effects. It was anticipated, as a result of previous studies, that these baking-session effects would alter the mean level of the response but would not interact with the x's. It was further anticipated that a second-order model function of form

$$h(\mathbf{x}, \boldsymbol{\beta}) = \beta_1 x_1 + \beta_2 x_2 + \beta_3 x_3 + \beta_4 x_4$$
$$+ \beta_{12} x_1 x_2 + \beta_{13} x_1 x_3 + \beta_{14} x_1 x_4 + \beta_{23} x_2 x_3 + \beta_{24} x_2 x_4 + \beta_{34} x_3 x_4 \tag{19.5.1}$$

would be satisfactory. (All these assumptions were to be checked, of course.) The model to be fitted was thus chosen as

TABLE 19.2. Orthogonally Blocked Design and Response Values ($b = 0.25$, $d = 0.75$)

	Blocks 1 and 3			Responses from Block			Blocks 2 and 4			Responses from Block	
x_1	x_2	x_3	x_4	1	3	x_1	x_2	x_3	x_4	2	4
0	b	0	d	403	381	0	d	0	b	423	404
b	0	d	0	425	422	b	0	d	0	417	425
0	d	0	b	442	412	0	b	0	d	398	391
d	0	b	0	433	413	d	0	b	0	407	426
0	d	b	0	445	398	0	0	b	d	388	362
b	0	0	d	435	412	b	d	0	0	435	427
0	0	d	b	385	371	0	b	d	0	379	390
d	b	0	0	425	428	d	0	0	b	406	411
0.25	0.25	0.25	0.25	433	393	0.25	0.25	0.25	0.25	439	409

$$Y = z_1 h(\mathbf{x}, \boldsymbol{\beta}) + \gamma z_2 + \delta z_3 + \omega z_4 + \epsilon, \tag{19.5.2}$$

where the z's are dummy variables to separate blocks, allocated as in Table 19.3. (The design used is, in fact, orthogonally blocked and nearly D-optimal although these points do not specifically concern us here.)

An initial fit to the model (19.5.2) showed nonsignificant t-values for the coefficients of $x_2 x_3$, $x_2 x_4$, and $x_3 x_4$. An *extra sum of squares* test, for all three as a unit, confirmed that it was reasonable to omit all three from the model. The resulting fitted model was

$$\hat{Y} = 397.6x_1 + 444.5x_2 + 389.4x_3 + 395.8x_4$$
$$\quad (11.1) \quad\;\; (6.8) \quad\;\; (7.5) \quad\;\; (6.8)$$

$$+ 107.8x_1 x_2 + 217.9x_1 x_3 + 169.7x_1 x_4$$
$$\quad\;\; (41.7) \quad\quad\; (41.6) \quad\quad\; (41.7)$$

$$- 14.9z_2 - 21.8z_3 - 20.1z_4.$$
$$\quad (5.2) \quad\;\; (5.2) \quad\;\; (5.2)$$

$$\tag{19.5.3}$$

Standard errors are shown in parentheses below their respective coefficients. An examination of residuals did not reveal any alarming characteristics. Figure 19.8 provides the $x_1 = 0$, 0.25, and 0.75 cross sections of the $\hat{Y}$ contours when $z_2 = z_3 = z_4 = 0$ in the four-dimensional mixture space. We see that these cross sections show planes, but that the actual contours curve as the value of x_1 is changed. The maximum estimated response within the mixture space is 453.1 at the edge point (0.283, 0.717, 0, 0), which is close to the corner point (0.25, 0.75, 0, 0) shown in Figure 19.8b. Note that the responses actually observed at this corner point (453, 427) were lower than the 453.0 predicted there when $z_2 = z_3 = z_4 = 0$ because they were depressed by the block

TABLE 19.3. Dummy Variables for Four Blocks

Baking Session	z_1	z_2	z_3	z_4
1	1	0	0	0
2	1	1	0	0
3	1	0	1	0
4	1	0	0	1

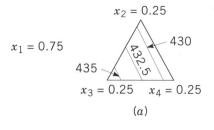

(a)

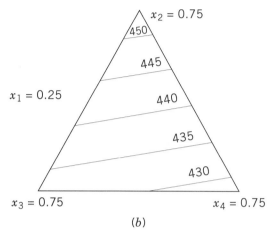

(b)

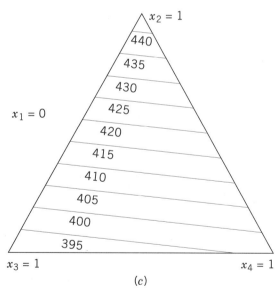

(c)

Figure 19.8. Contours of the fitted surface (19.5.3) when $z_2 = z_3 = z_4 = 0$. Sections are shown at (a) $x_1 = 0.75$, (b) $x_1 = 0.25$, and (c) $x_1 = 0$.

effects for blocks 2 and 4, estimated as -14.9 and -20.1, respectively. Allowing for blocks in the analysis enables us to make proper response comparisons.

The design used in this example consists mostly of blends of only two ingredients. Some experiments use single ingredients and these are not really mixtures at all. Whether such runs are adequate or need to be supplemented by three- and four-component blends is a question that needs to be carefully considered before any experiment is conducted.

References

Cornell (1990), Draper et al. (1993), Kurotori (1966), Scheffé (1958, 1963).

APPENDIX 19A. TRANSFORMING q MIXTURE VARIABLES TO $q - 1$ WORKING VARIABLES

We first illustrate the transformation from basic variables x to working variables z for the case $q = 3$. In this case we write

$$z_1 = -\frac{\sqrt{3}}{2}x_1 + \frac{\sqrt{3}}{2}x_2,$$

$$z_2 = -\tfrac{1}{2}x_1 - \tfrac{1}{2}x_2 + x_3,$$

$$z_3 = x_1 + x_2 + x_3 = 1,$$

or $z = Tx$, where

$$z = \begin{bmatrix} z_1 \\ z_2 \\ z_3 \end{bmatrix}, \qquad T = \begin{bmatrix} -\dfrac{\sqrt{3}}{2} & \dfrac{\sqrt{3}}{2} & 0 \\ -\tfrac{1}{2} & -\tfrac{1}{2} & 1 \\ 1 & 1 & 1 \end{bmatrix}, \qquad x = \begin{bmatrix} x_1 \\ x_2 \\ x_3 \end{bmatrix}.$$

The *inverse* transformation, which takes z to x, is $x = T^{-1}z$, where

$$T^{-1} = \frac{1}{3}\begin{bmatrix} -\sqrt{3} & -1 & 1 \\ \sqrt{3} & -1 & 1 \\ 0 & 2 & 1 \end{bmatrix}$$

so that the z to x reverse transformation for $q = 3$ takes the form

$$x_1 = -\frac{\sqrt{3}}{3}z_1 - \tfrac{1}{3}z_2 + \tfrac{1}{3}z_3,$$

$$x_2 = \frac{\sqrt{3}}{3}z_1 - \tfrac{1}{3}z_2 + \tfrac{1}{3}z_3,$$

$$x_3 = \tfrac{2}{3}z_2 + \tfrac{1}{3}z_3.$$

Figure 19A.1 illustrates how the corner points and the center point of our triangular region when $q = 3$ are described both in (x_1, x_2, x_3) coordinates and in (z_1, z_2) coordinates. The (z_1, z_2) axes are also shown. Note that, since $z_3 = 1$ always, we do not need to give it. The z_3 axis would be drawn from the $z_1 = z_2 = 0$ point out perpendicularly from the plane of the paper. [The z-origin, given by $z_1 = z_2 = z_3 = 0$, lies *below* the plane of the paper by a distance of one unit. Since $z_3 = 1$ always on the mixture space, we can work from the $(z_1, z_2) = (0, 0)$ origin in the $z_3 = 1$ plane in practice.]

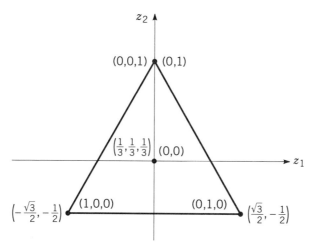

Figure 19A.1. Some points of the simplex region for $q = 3$ in both (x_1, x_2, x_3) and (z_1, z_2) notation.

General q

For general q we can still write $\mathbf{z} = \mathbf{Tx}$ and $\mathbf{x} = \mathbf{T}^{-1}\mathbf{z}$ but now

$$
\mathbf{z} = \begin{bmatrix} z_1 \\ z_2 \\ \vdots \\ z_q \end{bmatrix}, \qquad
\mathbf{x} = \begin{bmatrix} x_1 \\ x_2 \\ \vdots \\ x_q \end{bmatrix},
$$

$$
\mathbf{T} = \begin{bmatrix}
-a & a & 0 & 0 & 0 & \cdots & 0 \\
-b & -b & 2b & 0 & 0 & \cdots & 0 \\
-c & -c & -c & 3c & 0 & \cdots & 0 \\
-d & -d & -d & -d & 4d & \cdots & 0 \\
\vdots & \vdots & \vdots & \vdots & \vdots & & \vdots \\
-t & -t & -t & -t & -t & \cdots & (q-1)t \\
1 & 1 & 1 & 1 & 1 & \cdots & 1
\end{bmatrix}.
$$

The values of $a, b, c, \ldots, t$ can be chosen as follows. The vertices of the simplex region are given by $(x_1, x_2, x_3, x_4, \ldots, x_q) = (1, 0, 0, 0, \ldots, 0), (0, 1, 0, 0, \ldots, 0), (0, 0, 1, 0, \ldots, 0), \ldots, (0, 0, 0, 0, \ldots, 1)$. These translate into z-points, which have coordinates that are the columns of $\mathbf{T}$. For example,

$$
\mathbf{x}' = (1, 0, 0, 0, \ldots, 0, 0) \rightarrow \mathbf{z}' = (-a, -b, -c, -d, \ldots, -t, 1),
$$

$$
\begin{array}{cccc}
\vdots & \vdots & \vdots & \vdots
\end{array}
$$

$$
\mathbf{x}' = (0, 0, 0, 0, \ldots, 0, 1) \rightarrow \mathbf{z}' = (0, 0, \ldots, (q-1)t, 1).
$$

Suppose we choose the scale factors $a, b, c, \ldots, t$ so that, in the z-space, all points are one unit away from the point $(0, 0, \ldots, 0, 1)$, which will be our origin in the space of the factors $(z_1, z_2, \ldots, z_{q-1})$ since $x_q = 1$ always. Then the sum of squares of the first $(q-1)$ elements in every column of $\mathbf{T}$ must be equal to 1. Thus

$$a^2 + b^2 + c^2 + d^2 + e^2 + \cdots + t^2 = 1$$

$$4b^2 + c^2 + d^2 + e^2 + \cdots + t^2 = 1$$

$$9c^2 + d^2 + e^2 + \cdots + t^2 = 1$$

$$16d^2 + e^2 + \cdots + t^2 = 1$$

$$\cdots$$

$$(q-1)^2 t^2 = 1.$$

From these equations it follows that

$$a^2 = 3b^2, 2b^2 = 4c^2, 3c^2 = 5d^2, 4d^2 = 6e^2, \ldots, (q-2)s^2 = qt^2, \qquad t^2 = 1/(q-1)^2,$$

which implies that

$$2a^2 = 6b^2 = 12c^2 = 20d^2 = \cdots = (q-1)(q-2)s^2 = q(q-1)t^2 = q/(q-1).$$

This is precisely what we would obtain if we set the sum of squares of the elements of each of the first $(q-1)$ *rows* of **T** equal to $q/(q-1)$. This last, then, is the easiest way to choose the scale factors in **T** and it sets all the transformed simplex points the same distance, 1, from the new origin in the $(q-1)$-dimensional space of $z_1, z_2, \ldots, z_{q-1}$.

Example: $q = 3$. Here $2a^2 = 6b^2 = \frac{3}{2}$, so that $a = \sqrt{3}/2$, $b = \frac{1}{2}$, as given earlier.

Example: $q = 4$. Here $2a^2 = 6b^2 = 12c^2 = \frac{4}{3}$, so that $a = \sqrt{\frac{2}{3}}$, $b = \sqrt{2}/3$, $c = \frac{1}{3}$.

Example: $q = 5$. Here $2a^2 = 6b^2 = 12c^2 = 20d^2 = \frac{5}{4}$, so that $a = \sqrt{\frac{5}{8}}$, $b = \sqrt{\frac{5}{24}}$, $c = \sqrt{\frac{5}{48}}$, $d = \sqrt{\frac{5}{80}}$.

The general inverse transformation from z back to x is given by $\mathbf{x} = \mathbf{T}^{-1}\mathbf{z}$, where

$$\mathbf{T}^{-1} = \frac{q-1}{q}
\begin{bmatrix}
-a & -b & -c & -d & \cdots & -t & \frac{1}{q-1} \\
a & -b & -c & -d & \cdots & -t & \frac{1}{q-1} \\
0 & 2b & -c & -d & \cdots & -t & \frac{1}{q-1} \\
0 & 0 & 3c & -d & \cdots & -t & \frac{1}{q-1} \\
0 & 0 & 0 & 4d & \cdots & -t & \frac{1}{q-1} \\
\vdots & \vdots & \vdots & & \vdots & & \vdots \\
0 & 0 & 0 & 0 & \cdots & (q-1)t & \frac{1}{q-1}
\end{bmatrix}.$$

Comment. The presentation can be made somewhat simpler by taking the z-origin to be the center of the simplex, in other words choosing the point $(z_1, z_2, \ldots, z_q) = (0, 0, \ldots, 0)$ to be the center of the simplex instead of the point $(z_1, z_2, \ldots, z_{q-1}, z_q) = (0, 0, \ldots, 0, 1)$, which we *have* chosen. Then **T** can be chosen to be an orthogonal matrix so that $\mathbf{T}^{-1} = \mathbf{T}'$, the transpose of **T**. The above, of course, comes close to this, except for scale factors and this is why $\mathbf{T}^{-1}$ is so easy to write down, above. The variation of $\mathbf{T}$ given simply keeps the situation a little closer to the practical mixture problem in holding $z_q = 1$, the mixture total.

EXERCISES FOR CHAPTER 19

A. The fit of a second-order model to some mixture data resulted in the fitted equation $\hat{Y} = 5x_1 + 6x_2 + 7x_3 + 3x_1x_2 + x_1x_3 + 2x_2x_3$. Substitute for the z-coordinates in Appendix 19A, and find the resulting equation in terms of z_1 and z_2. (Set $z_3 = 1$.)

B. Provide the appropriate transformation matrix for transforming four mixture variables to three "working" variables. Hence convert the points $(0, 1, 0, 0)$, $(\frac{1}{2}, \frac{1}{2}, 0, 0)$, $(\frac{1}{3}, \frac{1}{3}, \frac{1}{3}, 0)$ and $(\frac{1}{4}, \frac{1}{4}, \frac{1}{4}, \frac{1}{4})$ to the new coordinates.

C. Four mixture ingredients are restricted to the space $0.10 \le x_1 \le 0.80$, $0.25 \le x_2 \le 0.45$, $0.20 \le x_3 \le 0.40$, $0.15 \le x_4 \le 0.55$. Find the extreme vertices of the restricted space, the face centroids, and the overall centroid.

D. Fit the quadratic model $Y = \beta_1x_1 + \beta_2x_2 + \beta_3x_3 + \beta_{12}x_1x_2 + \beta_{13}x_1x_3 + \beta_{23}x_2x_3 + \epsilon$ to the data below. Predict the response at the centroid $(\frac{1}{3}, \frac{1}{3}, \frac{1}{3})$. Plot the contours on triangular graph paper.

x_1	x_2	x_3	Y
1	0	0	40.9
0	1	0	25.5
0	0	1	28.6
0.5	0.5	0	31.1
0.5	0	0.5	24.9
0	0.5	0.5	29.1
0.2	0.6	0.2	27.0
0.3	0.5	0.2	28.4

E. Plot on triangular graph paper (e.g., Dietzgen 340-TC) these mixtures of three ingredients (x_1, x_2, x_3): $(1, 0, 0)$, $(0, 1, 0)$, $(0, 0, 1)$, $(\frac{1}{2}, \frac{1}{2}, 0)$, $(\frac{1}{2}, 0, \frac{1}{2})$, $(0, \frac{1}{2}, \frac{1}{2})$, $(\frac{1}{3}, \frac{1}{3}, \frac{1}{3})$.

F. Scheffé's *simplex lattice* design type uses $(m + 1)$ equally spaced levels of each mixture ingredient. (For example, if $m = 2$, the levels are $0, \frac{1}{2}, 1$.) Suppose we have q ingredients and we write down all $(m + 1)$ combinations of ingredients. Are they all suitable design points? Try this for $m = 2$, $q = 3$ and see why the answer is no. How many points *are* in the design in general?

G. Scheffé's *simplex centroid* type design consists of all vertices of the mixture space, plus all centroids (averages) of two or more of these points. Show that for $q = 3$ ingredients we get the points given in Exercise E.

H. How many regression coefficients are there in the second-order mixture model for $q = 3$ ingredients? What does that mean if (a) Scheffé's simplex lattice design is used or (b) Scheffé's simplex centroid design is used.

I. A number of examples of the analysis of mixture data are exhibited and discussed in "Mixture experiment approaches: examples, discussion, and recommendations," by G. F. Piepel and J. A. Cornell, *Journal of Quality Technology*, **26**, 1995, No. 3, 177–196. Also see "A catalog of mixture experiment examples," by G. F. Piepel and J. A. Cornell, Report BN-SA-3298 (Revision 4), available from G. F. Piepel, Battelle, Pacific Northwest Laboratories, Richland, WA.

C H A P T E R   20

The Geometry of Least Squares

Comment: Philosophies on teaching the geometry of least squares vary. An early introduction is certainly possible, as demonstrated by the successful presentation by Box, Hunter, and Hunter (1978, e.g., pp. 179, 197–201) for specific types of designs. For general regression situations, more general explanations are required. Our view is that, while regression can be taught perfectly well without the geometry, understanding of the geometry provides much better understanding of, for example, the difficulties associated with singular or nearly singular regressions, and of the difficulties associated with interpretations of the R^2 statistic. For an advanced understanding of least squares, knowledge of the geometry is essential.

20.1. THE BASIC GEOMETRY

We want to fit, by least squares, the model

$$\mathbf{Y} = \mathbf{X}\boldsymbol{\beta} + \boldsymbol{\epsilon}, \tag{20.1.1}$$

where $\mathbf{Y}$ and $\boldsymbol{\epsilon}$ are both $n \times 1$, $\mathbf{X}$ is $n \times p$, and $\boldsymbol{\beta}$ is $p \times 1$. Consider a Euclidean (i.e., "ordinary") space of n dimensions, call it S. The n numbers $Y_1, Y_2, \ldots, Y_n$ within $\mathbf{Y}$ define a point Y in this space. They also define a vector (a line with length and direction) usually represented by the line joining the origin O, $(0, 0, 0, \ldots, 0)$, to the point Y. (In fact, any parallel line of the same length can also be regarded as the vector $\mathbf{Y}$, but most of the time we think about the vector from O.) The columns of $\mathbf{X}$ also define vectors in the n-dimensional space. Let us assume for the present that all p of the $\mathbf{X}$ columns are linearly independent, that is, none of them can be represented as a linear combination of any of the others. This implies that $\mathbf{X}'\mathbf{X}$ is nonsingular. Then the p columns of $\mathbf{X}$ define a subspace (we call it the estimation space) of S of p ($<n$) dimensions. Consider

$$\mathbf{X}\boldsymbol{\beta} = [\mathbf{x}_0, \mathbf{x}_1, \mathbf{x}_2, \mathbf{x}_3, \ldots, \mathbf{x}_{p-1}] \begin{bmatrix} \beta_0 \\ \beta_1 \\ \vdots \\ \beta_{p-1} \end{bmatrix} = \beta_0 \mathbf{x}_0 + \beta_1 \mathbf{x}_1 + \cdots + \beta_{p-1} \mathbf{x}_{p-1}, \tag{20.1.2}$$

where each $\mathbf{x}_i$ is an $n \times 1$ vector and the β_i are scalars. (Usually $\mathbf{x}_0 = \mathbf{1}$.) This defines a vector formed as a linear combination of the $\mathbf{x}_i$, and so $\mathbf{X}\boldsymbol{\beta}$ is a vector in the estimation space. Precisely where it is depends on the values chosen for the β_i. We can now draw

427

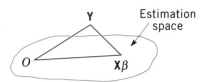

Figure 20.1. A general point $\mathbf{X}\boldsymbol{\beta}$ in the estimation space.

Figure 20.1. We see that the points marked O, $\mathbf{Y}$, $\mathbf{X}\boldsymbol{\beta}$ form a triangle in n-space. In general, the angles will be determined by the values of the β_i, given $\mathbf{Y}$ and $\mathbf{X}$. The three sides of the triangle are the vectors $\mathbf{Y}$, $\mathbf{X}\boldsymbol{\beta}$, and $\boldsymbol{\epsilon} = \mathbf{Y} - \mathbf{X}\boldsymbol{\beta}$. So the model (20.1.1) simply says that $\mathbf{Y}$ can be split up into two vectors, one of which, $\mathbf{X}\boldsymbol{\beta}$, is completely in the estimation space and one of which, $(\mathbf{Y} - \mathbf{X}\boldsymbol{\beta})$, is partially not, in general. When we estimate $\boldsymbol{\beta}$ by least squares, the solution $\mathbf{b} = (\mathbf{X}'\mathbf{X})^{-1}\mathbf{X}'\mathbf{Y}$ is the one that minimizes the sum of squares function

$$S(\boldsymbol{\beta}) = (\mathbf{Y} - \mathbf{X}\boldsymbol{\beta})'(\mathbf{Y} - \mathbf{X}\boldsymbol{\beta}). \tag{20.1.3}$$

At this point, we need to know the fact that, if $\mathbf{z}$ is any vector, $\mathbf{z}'\mathbf{z}$ is the squared length of that vector. So the sum of squares function $S(\boldsymbol{\beta})$ is just the squared length of the vector joining $\mathbf{Y}$ and $\mathbf{X}\boldsymbol{\beta}$ in Figure 20.1. When is this a minimum? It is when $\boldsymbol{\beta}$ is set equal to $\mathbf{b} = (\mathbf{X}'\mathbf{X})^{-1}\mathbf{X}'\mathbf{Y}$. It follows that the vector $\mathbf{X}\mathbf{b} = \hat{\mathbf{Y}}$ joins the origin O to the foot of the perpendicular from $\mathbf{Y}$ to the estimation space. See Figure 20.2. Compare it with Figure 20.1. When $\boldsymbol{\beta}$ takes the special value $\mathbf{b}$, the least squares value, the triangle of Figure 20.1 becomes a right-angled one as in Figure 20.2. If that is true, the vectors $\hat{\mathbf{Y}} = \mathbf{X}\mathbf{b}$ and $\mathbf{e} = \mathbf{Y} - \mathbf{X}\mathbf{b}$ should be orthogonal. Two vectors $\mathbf{p}$ and $\mathbf{q}$ are orthogonal if $\mathbf{p}'\mathbf{q} = 0 = \mathbf{q}'\mathbf{p}$. Here,

$$0 = (\mathbf{X}\mathbf{b})'\mathbf{e} = \mathbf{b}'\mathbf{X}'\mathbf{e}$$
$$= \mathbf{b}'\mathbf{X}'(\mathbf{Y} - \mathbf{X}\mathbf{b}) \tag{20.1.4}$$
$$= \mathbf{b}'(\mathbf{X}'\mathbf{Y} - \mathbf{X}'\mathbf{X}\mathbf{b}).$$

Thus the orthogonality requirement in (20.1.4) implies either that $\mathbf{b} = \mathbf{0}$, in which case the vector $\mathbf{Y}$ is orthogonal to the estimation space and $\hat{\mathbf{Y}} = \mathbf{0}$, or that the normal equations $\mathbf{X}'\mathbf{X}\mathbf{b} = \mathbf{X}'\mathbf{y}$ hold. The space in which $\mathbf{e} = \mathbf{Y} - \mathbf{X}\mathbf{b}$ lies is called the error space, and it has $(n - p)$ dimensions. The estimation space of p dimensions and the error space together constitute S. So the least squares fitting of the regression model splits the space S up into two orthogonal spaces; every vector in the estimation space is orthogonal to every vector in the error space.

Exercise 1. Let $\mathbf{Y} = (3.1, 2.3, 5.4)'$, $\mathbf{x}_0 = (1, 1, 1)'$ and $\mathbf{x}_1 = (2, 1, 3)'$, so that we are

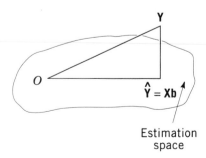

Figure 20.2. The right-angled triangle of vectors $\mathbf{Y}$, $\mathbf{X}\mathbf{b}$, and $\mathbf{e} = \mathbf{Y} - \mathbf{X}\mathbf{b}$.

going to fit the straight line $Y = \beta_0 + \beta_1 x_1 + \epsilon$ via least squares. First, do the regression fit without worrying about the geometry. Show that $\hat{Y} = 0.50x_0 + 1.55x_1$. Then show that $\hat{\mathbf{Y}} = \mathbf{Xb} = (3.60, 2.05, 5.15)'$ and $\mathbf{e} = (-0.50, 0.25, 0.25)'$; confirm that these two vectors are orthogonal and furthermore that $\mathbf{e}$ is orthogonal to any vector of the form $\mathbf{X\beta} = \beta_0 \mathbf{x}_0 + \beta_1 \mathbf{x}_1$, for all β, because $\mathbf{e}$ is orthogonal to both (in general, all) the individual columns of $\mathbf{X}$. Write the coordinates of the various points on a diagram like Figure 20.2.

20.2. PYTHAGORAS AND ANALYSIS OF VARIANCE

Every analysis of variance table arising from a regression is an application of Pythagoras's Theorem about the "square of the hypotenuse of a right-angled triangle equals the sum of squares of the other two sides." Usually, repeated applications of Pythagoras's result are needed. Look again at Figure 20.2. It implies that

$$\mathbf{Y'Y} = \hat{\mathbf{Y}}'\hat{\mathbf{Y}} + (\mathbf{Y} - \mathbf{Xb})'(\mathbf{Y} - \mathbf{Xb}) \tag{20.2.1}$$

or

Total sum of squares = Sum of squares due to regression + Residual sum of squares.

The corresponding degrees of freedom equation

$$n = p + (n - p) \tag{20.2.2}$$

corresponds to the dimensional split of S into two orthogonal spaces of dimensions p and $(n - p)$, respectively. The orthogonality of $\hat{\mathbf{Y}}$ and $\mathbf{e}$ is essential for a clear split-up of the total sum of squares. Where a split-up is required of a particular sum of squares that does not have a natural orthogonal split-up, orthogonality must be introduced to achieve it.

Exercise 2. Show that for the data in Exercise 1, the two equations (20.2.1) and (20.2.2) correspond to

$$44.06 = 43.685 + 0.375,$$
$$3 = 2 + 1.$$

Further Split-up of a Regression Sum of Squares

If the $\mathbf{X}_i$ vectors that define the estimation space happen to be all orthogonal to one another, the regression sum of squares can immediately be split down into orthogonal pieces. Suppose, for example, that $\mathbf{X} = (\mathbf{1}, \mathbf{x})$ and that $\mathbf{1'x} = 0 = \mathbf{x'1}$. The model $\mathbf{Y} = \mathbf{X\beta} + \epsilon$ is of a straight line, Y versus x. Figure 20.3 shows the estimation space defined by the two *orthogonal* vectors $\mathbf{1}$ and $\mathbf{x}$.

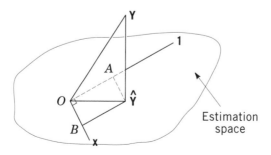

Figure 20.3. Orthogonal vectors in the model function.

Because these vectors are orthogonal, perpendiculars from $\hat{\mathbf{Y}}$ to these lines form a (right-angled) rectangle, so that

$$O\hat{Y}^2 = OA^2 + OB^2 \qquad\qquad (20.2.3)$$

(or $OA^2 + A\hat{Y}^2$, etc.). The division is unique because of the orthogonality of the base vectors $\mathbf{1}$, $\mathbf{x}$; this result extends in the obvious way to any number of orthogonal base vectors, in which case the rectangle of Figure 20.3 becomes a rectangular block in the p-dimensional estimation space.

Exercise 3. For the least squares regression problem where

$$\mathbf{Y}' = (6, 9, 17, 18, 26) \quad \text{and} \quad \mathbf{X}' = \begin{bmatrix} 1 & 1 & 1 & 1 & 1 \\ -2 & -1 & 0 & 1 & 2 \end{bmatrix}$$

show that the normal equations $\mathbf{X'Xb} = \mathbf{X'Y}$ split up into two separate pieces and that A and B in Figure 20.3 are the two "individual $\hat{Y}$'s" that come from looking at each piece separately. Hence evaluate Eqs. (20.2.1) and (20.2.2) for these data ($1406 = 1395.3 + 10.7$, and $4 = 2 + 2$).

An Orthogonal Breakup of the Normal Equations

Let us redraw Figure 20.3 with some additional detail; see Figure 20.4, in which lines OY and $O\hat{Y}$ are omitted but YA and YB have been drawn. In other respects the diagrams are intended to be identical. The points marked O, A, $\hat{Y}$, and B form a rectangle, and thus $O\hat{Y}^2 = OA^2 + OB^2$, as mentioned. Also, YA is perpendicular to OA, and YB is perpendicular to OB. Thus OA is "the $\hat{Y}$ for the regression of $\mathbf{Y}$ on $\mathbf{1}$ alone" and OB is "the $\hat{Y}$ for the regression of $\mathbf{Y}$ on $\mathbf{x}$ alone." This happens only because (here) $\mathbf{1}$ and $\mathbf{x}$ are orthogonal. The result also extends to any number of $\mathbf{X}$ vectors provided they are mutually orthogonal.

Exercise 4. For the data of Exercise 3, show that $OA^2 = 1155.2$, $OB^2 = 240.1$, and their sum is $O\hat{Y}^2 = 1395.3$. Also show that OA is the vector $(15.2, 15.2, 15.2, 15.2, 15.2)'$, OB is $(-9.8, -4.9, 0, 4.9, 9.8)'$, and $O\hat{Y}$ is the sum of these orthogonal vectors.

Orthogonalizing the Vectors of X in General

It is frequently desired to break down a regression sum of squares into components even when the vectors of $\mathbf{X}$ are not mutually orthogonal. Such a breakdown can

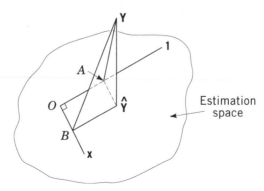

Figure 20.4. Orthogonal breakup of the normal equations when model vectors are orthogonal.

provide a set of sequential sums of squares, each being an extra sum of squares to the entries above it in the sequence. Geometrically, we need to perform a succession of orthogonalization procedures to do this. Figure 20.5 shows the idea for two vectors $\mathbf{1} = \mathbf{x}_0$ and $\mathbf{x}$, for which $\mathbf{1}'\mathbf{x} \neq 0$. We first pick one vector as a starter (we shall pick $\mathbf{1}$, but either would work) and then construct $\mathbf{x}_{\cdot 0}$, the portion of $\mathbf{x}$ (the second vector) that is orthogonal to $\mathbf{1}$. How? We use the standard result of regression that residuals are orthogonal to fitted values. We fit $\mathbf{x}$ (thinking of it as a "$\mathbf{Y}$") on to $\mathbf{1}$ and take residuals, giving $\hat{\mathbf{x}} = b\mathbf{1}$, where $b = $ "$(\mathbf{X}'\mathbf{X})^{-1}\mathbf{X}'\mathbf{Y}$" $= (\mathbf{1}'\mathbf{1})^{-1}\mathbf{1}'\mathbf{x} = \bar{x}$. Thus $\mathbf{x}_{\cdot 0} = \mathbf{x} - \bar{x}\mathbf{1}$. It is easy to check that $\mathbf{x}'_{\cdot 0}\mathbf{1} = 0$ so that these two vectors are orthogonal. Now the original regression equation was of the form

$$\hat{\mathbf{Y}} = b_0\mathbf{1} + b_1\mathbf{x}, \tag{20.2.4}$$

where

$$\begin{pmatrix} b_0 \\ b_1 \end{pmatrix} = (\mathbf{X}'\mathbf{X})^{-1}\mathbf{X}'\mathbf{Y}.$$

The orthogonalized regression equation is

$$\begin{aligned} \hat{\mathbf{Y}} &= \bar{Y}\mathbf{1} + b_1(\mathbf{x} - \bar{x}\mathbf{1}) \\ &= \bar{Y}\mathbf{1} + b_1\mathbf{x}_{\cdot 0}. \end{aligned} \tag{20.2.5}$$

We have used the same symbol b_1 in both equations (20.2.4) and (20.2.5) because the values will be identical. What we have in (20.2.4) and (20.2.5) are two different descriptions of the *same* vector $O\hat{Y}$ in Figure 20.5. This vector, $\hat{\mathbf{Y}}$, can be represented either as the sum of the nonorthogonal vectors OA^* and OB^*, or as the sum of the orthogonal vectors OA and OB. Either description is valid but only the second permits us to split up the sum of squares $O\hat{Y}^2$ as $OA^2 + OB^2$ via the Pythagoras result. Note that $\mathbf{Y}$ and $\hat{\mathbf{Y}}$ are unaffected by what we have done. All we have done is alter the description of $\hat{\mathbf{Y}}$ in terms of a linear combination of vectors in the space. Exercises 5 and 6 explore the fact that we can choose either vector first.

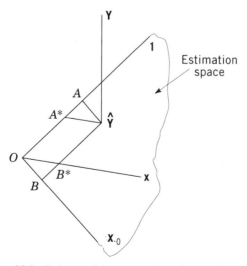

Figure 20.5. Orthogonalizing a second predictor with respect to **1**.

Exercise 5. For the data of Exercise 1, namely, $\mathbf{Y} = (3.1, 2.3, 5.4)'$, $\mathbf{x}_0 = (1, 1, 1)'$, $\mathbf{x} = (2, 1, 3)'$, and $\hat{\mathbf{Y}} = 0.501 + 1.55\mathbf{x}$, show that $\mathbf{x}_{\cdot 0} = (0, -1, 1)'$, and that $\hat{\mathbf{Y}} = 3.51 + 1.55\mathbf{x}_{\cdot 0}$ produces the same fitted values and residuals. Draw a (fairly) accurate diagram of the form of Figure 20.5, working out the lengths of OA^*, OA, OB, OB^*, $O\hat{Y}$, and $Y\hat{Y}$ beforehand to get it (more or less) right. Note that YA is the orthogonal projection of $\mathbf{Y}$ onto $\mathbf{1}$ and YB is the orthogonal projection of $\mathbf{Y}$ on to $\mathbf{x}_{\cdot 0}$. Show that

$$O\hat{Y}^2 = OA^2 + OB^2 \neq OA^{*2} + OB^{*2} \tag{20.2.6}$$

or

$$43.685 = 38.88 + 4.805 \neq 0.125 + 33.635.$$

The quantity $OA^2 = SS(b_0) = n\bar{Y}^2 = 38.88$, while $OB^2 = SS(b_1|b_0) = 4.805$. Note that, above, we picked $\mathbf{x}_0 = \mathbf{1}$ first and found $\mathbf{x}_{\cdot 0}$.

Exercise 6. Pick, for the data of Exercise 5, $\mathbf{x}$ first and determine $\mathbf{x}_1$ such that $\mathbf{x}_1'\mathbf{x} = 0$. Let A and B be replaced by perpendiculars from $\hat{\mathbf{Y}}$ to $\mathbf{x}_1$ and $\mathbf{x}$ at C and D, say. Show that $43.685 = O\hat{Y}^2 = OC^2 + OD^2 = OA^2 + OB^2$ but that the split-up is different in the two cases, that is, $OC \neq OA$ and $OD \neq OB$. Specifically, $0.107 = OC^2 \neq OA^2 = 38.88$ and $43.578 = OD^2 \neq OB^2 = 4.805$.

In what we have done above, we see, geometrically, a fact we already know algebraically. Unless all the vectors in $\mathbf{X}$ are mutually orthogonal, the sequential sums of squares will depend on the sequence selected for the entry of the predictor variables. After the entry of the variable selected first, the others are (essentially) successively orthogonalized to all the ones entered before them. We reemphasize a point of great importance in this. $\hat{\mathbf{Y}}$ is unique and fixed no matter what the entry sequence may be. We merely give $\hat{\mathbf{Y}}$ different descriptions according to the vectors we select in the orthogonalized sequence. That is all.

20.3. ANALYSIS OF VARIANCE AND *F*-TEST FOR OVERALL REGRESSION

We reconsider the case of Figure 20.5 where the model is $\mathbf{Y} = \beta_0\mathbf{1} + \beta_1\mathbf{x} + \boldsymbol{\epsilon}$ and $\mathbf{1}'\mathbf{x} \neq 0$. The standard analysis of variance table takes the form of Table 20.1.

Note that the analysis of variance table is simply a Pythagoras split-up of (first)

$$O\hat{Y}^2 = OA^2 + OB^2$$

followed by

$$(OA^2 + OB^2) + Y\hat{Y}^2 = OY^2,$$

and the *F*-test for $H_0: \beta_1 = 0$ versus $\beta_1 \neq 0$ is simply a comparison of the "length-squared per df of OB" versus the "length-squared per df of $Y\hat{Y}$." Then rejection of

T A B L E 20.1. Analysis of Variance Table for a Straight Line Fit

Source	SS	df	MS	F
b_0	$OA^2 = n\bar{Y}^2$	1	—	—
$b_1\|b_0$	$OB^2 = S_{XY}^2/S_{XX}$	1	S_{XY}^2/S_{XX}	$F = S_{XY}^2/(s^2 S_{XX})$ $= \{OB^2/1\}/\{Y\hat{Y}^2/(n-2)\}$
Residual	$Y\hat{Y}^2 =$ By subtraction	$n-2$	s^2	
Total	$OY^2 = \mathbf{Y}'\mathbf{Y}$	n		

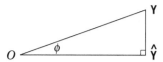

Figure 20.6. Basic geometry of the F-test for no regression at all, not even β_0.

H_0 is caused by larger values of OB^2 rather than smaller, so we are essentially asking: "Is B close to O compared with the size of s?" (do not reject H_0) or "Is B *not* close to O compared with the size of s"? (reject H_0). [The fact that the F-ratio follows the $F(1, n-2)$ distribution when H_0 is true can be established algebraically, but it also has a geometric interpretation, given below.]

An F-test for $H_0: \beta_0 = \beta_1 = 0$ versus H_1: not so, would similarly involve

$$F = \{(OA^2 + OB^2)/2\}/\{Y\hat{Y}^2/(n-2)\}.$$

Note that (see Figure 20.6) this involves a comparison of the "per-df lengths" of $O\hat{Y}^2$ with $Y\hat{Y}^2$. Significant regression will be one in which $O\hat{Y}$ is "large" compared with $Y\hat{Y}$. This implies that the angle ϕ on Figure 20.6 is "small." If we think of OY (the data) as being a fixed axis and $O\hat{Y}$ as one possible position of the fitted vector, which could lie anywhere at the same angle ϕ to OY (in positive or negative direction), we have a geometrical interpretation of the F-test (see Figure 20.7). The tail probability of the F-statistic is the proportional area of the two circular "caps," defined as $\hat{Y}$ rotates around Y, on a sphere of any radius.

What if the test were on nonzero values of β_0 and β_1? Suppose we wished to test $H_0: \beta_0 = \beta_{00}$ and $\beta_1 = \beta_{10}$? Then the point O^* defined by $\mathbf{Y}_0 = \beta_{00}\mathbf{1} + \beta_{10}\mathbf{x}$ would replace O in what we have said above but the other elements of the geometry would essentially be preserved. (See Figure 20.8.) We leave this as an exercise, noting only that in nearly all such problems $\beta_{10} = 0$ anyway, even when $\beta_{00} \neq 0$.

20.4. THE SINGULAR X'X CASE: AN EXAMPLE

What happens to the geometry in a regression where the problem is singular, that is, $\det(\mathbf{X}'\mathbf{X}) = 0$? The answer is "very little" and the understanding of this answer will reduce one's fear of meeting such problems in practice. Look at Figure 20.9. Although discussed in the framework of three dimensions, the issues discussed are general ones. Suppose that the estimation space is the plane shown, but the $\mathbf{X} = (\mathbf{1}, \mathbf{x}_1, \mathbf{x}_2)$ matrix consists of three vectors. It is obvious that the $\mathbf{X}'\mathbf{X}$ matrix is singular because the three vectors in $\mathbf{X}$ must be dependent. If we suppose that any two of the vectors define the plane (i.e., the three vectors are distinct ones) then the third vector must be a linear combination of the other two. Now it is sometimes mistakenly thought

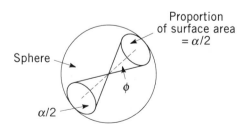

Figure 20.7. A geometrical interpretation of the F-test probability.

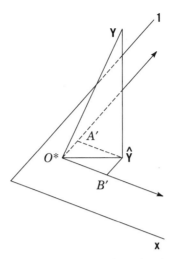

Figure 20.8. Testing that $(\beta_0, \beta_1) = (\beta_{00}, \beta_{10})$.

that the least squares problem has no solution in these circumstances, because $(\mathbf{X'X})^{-1}$ does not exist. We can see from Figure 20.9 that not only is there a least squares solution but, as always, it is unique. That is, we can drop a perpendicular onto the estimation space at $\hat{Y}, \hat{Y}$ is unique, and we have a unique vector $\mathbf{Y} - \hat{\mathbf{Y}}$ orthogonal to the estimation space, and thus orthogonal to all the columns of $\mathbf{X}$. What is not unique is the description of $\hat{\mathbf{Y}}$ in terms of $\mathbf{1}$, $\mathbf{x}_1$, and $\mathbf{x}_2$. Because we have (one, here) too many base vectors, there are an infinite number of ways in which $\hat{\mathbf{Y}}$ can be described. The normal equations exist and can be solved, but the solution for the parameter estimates is not unique. We illustrate this with the smallest possible example, two parameters and two vectors $\mathbf{1}$, $\mathbf{x}$ that are multiples of each other.

Example

Suppose we have n data values at $X = X^*$. (See Figure 20.10 where $n = 5$, although any n can be used.) Consider fitting the line $Y = \beta_0 + \beta_1 X + \epsilon$ by least squares. Clearly there are an infinite number of solutions. *Any* line through the point $(X, Y) = (X^*, \overline{Y})$ will provide a least squares fit. Geometrically the problem is that the two vectors that define the predictor space, $\mathbf{1}$ and $\mathbf{x} = X^*\mathbf{1}$, are multiples of one

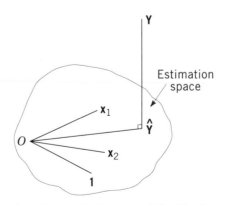

Figure 20.9. A two-dimensional estimation space defined by three vectors, one too many.

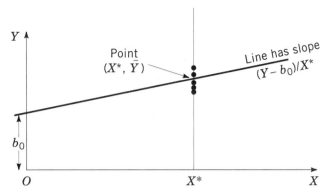

Figure 20.10. Singular regression problem with multiple descriptions for a unique $\hat{Y}$.

another. (See Figure 20.11.) Thus the foot of the perpendicular from **Y** to the predictor space can be described by an infinite number of linear combinations of the vectors **1** and $X*\mathbf{1}$. Note that $\hat{\mathbf{Y}} = \bar{Y}\mathbf{1}$ is unique, however. We now write down the two normal equations $\mathbf{X'Xb} = \mathbf{X'Y}$ as

$$nb_0 + nX*b_1 = n\bar{Y},$$
$$nX*b_0 + nX*^2b_1 = nX*\bar{Y},$$
(20.4.1)

and see immediately they are not independent but are one equation whose solution is $(b_0, b_1 = (\bar{Y} - b_0)/X*)$ for any b_0. The fitted "line" is thus

$$\hat{Y} = b_0 + b_1X = b_0 + X(\bar{Y} - b_0)/X*,$$
(20.4.2)

representing an infinity of lines with different slopes and intercepts, all passing through the point $(X, Y) = (X*, \bar{Y})$. Whatever value is assigned to b_0, we have a least squares solution. At $X = X*$, $\hat{\mathbf{Y}} = \bar{Y}\mathbf{1}$, which is unique whatever value b_0 takes.

20.5 ORTHOGONALIZING IN THE GENERAL REGRESSION CASE

We consider the fit of a general linear model $\mathbf{Y} = \mathbf{X}\boldsymbol{\beta} + \boldsymbol{\epsilon}$ and write $\mathbf{X} = (\mathbf{Z}_1, \mathbf{Z}_2)$, representing any division of **X** into two parts that would, in general, not be orthogonal. We divide $\boldsymbol{\beta'} = (\boldsymbol{\theta}_1', \boldsymbol{\theta}_2')$ in a corresponding fashion. Thus we write

$$\mathbf{Y} = \mathbf{Z}_1\boldsymbol{\theta}_1 + \mathbf{Z}_2\boldsymbol{\theta}_2 + \boldsymbol{\epsilon},$$
(20.5.1)

where **Y** is n by 1, $\mathbf{Z}_1$ is n by p_1, $\boldsymbol{\theta}_1$ is p_1 by 1, $\mathbf{Z}_2$ is n by p_2, and $\boldsymbol{\theta}_2$ is p_2 by 1. If we fit to just the $\mathbf{Z}_1\boldsymbol{\theta}_1$ part of the model, we get

$$\hat{\mathbf{Y}} = \mathbf{Z}_1(\mathbf{Z}_1'\mathbf{Z}_1)^{-1}\mathbf{Z}_1'\mathbf{Y} = \mathbf{P}_1\mathbf{Y},$$
(20.5.2)

say, where $\mathbf{P}_1$ is the *projection matrix,* that is, the matrix that projects the vector **Y**

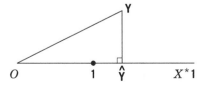

Figure 20.11. The estimation space is overdefined by the vectors **1** and $X*\mathbf{1}$.

down into the estimation space defined by the columns of $\mathbf{Z}_1$. The residual vector is $\mathbf{e} = \mathbf{Y} - \hat{\mathbf{Y}} = (\mathbf{I} - \mathbf{P}_1)\mathbf{Y}$. Note that in previous discussions we called $\mathbf{P}_1$ the *hat* matrix because it converts the Y's into the $\hat{Y}$'s. We now rename it to emphasize its geometrical properties.

It is obvious that the fitted $\hat{\mathbf{Y}}$ and the residual $\mathbf{Y} - \hat{\mathbf{Y}}$ are orthogonal because

$$\hat{\mathbf{Y}}'\mathbf{e} = \mathbf{Y}'\mathbf{P}_1'(\mathbf{I} - \mathbf{P}_1)\mathbf{Y}$$

$$= \mathbf{Y}'(\mathbf{P}_1' - \mathbf{P}_1'\mathbf{P}_1)\mathbf{Y} \qquad (20.5.3)$$

$$= 0,$$

because $\mathbf{P}_1$ is symmetric (so $\mathbf{P}_1' = \mathbf{P}_1$) and idempotent (so $\mathbf{P}_1^2 = \mathbf{P}_1$). This calculation is a repeat of (20.1.4) with different notation. The matrix $\mathbf{Z}_2$, when orthogonalized to $\mathbf{Z}_1$, becomes, by analogy to $\mathbf{Y} - \hat{\mathbf{Y}}$,

$$\mathbf{Z}_{2\cdot 1} = \mathbf{Z}_2 - \hat{\mathbf{Z}}_2$$

$$= (\mathbf{I} - \mathbf{P}_1)\mathbf{Z}_2$$

$$= \mathbf{Z}_2 - \mathbf{Z}_1(\mathbf{Z}_1'\mathbf{Z}_1)^{-1}\mathbf{Z}_1'\mathbf{Z}_2 \qquad (20.5.4)$$

$$= \mathbf{Z}_2 - \mathbf{Z}_1\mathbf{A}.$$

$\mathbf{A}$ is usually called the *alias* or *bias* matrix.

Exercise 7. Prove that $\mathbf{Z}_{2\cdot 1}$ and $\mathbf{Z}_1$ are orthogonal matrices.

We can now write the full model in the orthogonalized form

$$\mathbf{Y} = \mathbf{Z}_1\boldsymbol{\theta}_1 + \mathbf{Z}_2\boldsymbol{\theta}_2 + \boldsymbol{\epsilon}$$

$$= \mathbf{Z}_1(\boldsymbol{\theta}_1 + \mathbf{A}\boldsymbol{\theta}_2) + (\mathbf{Z}_2 - \mathbf{Z}_1\mathbf{A})\boldsymbol{\theta}_2 + \boldsymbol{\epsilon} \qquad (20.5.5)$$

$$= \mathbf{Z}_1\boldsymbol{\theta} + \mathbf{Z}_{2\cdot 1}\boldsymbol{\theta}_2 + \boldsymbol{\epsilon}.$$

We see immediately that $\hat{\boldsymbol{\theta}} = (\mathbf{Z}_1'\mathbf{Z}_1)^{-1}\mathbf{Z}_1'\mathbf{Y}$ estimates *not* just $\boldsymbol{\theta}_1$ but $\boldsymbol{\theta}_1 + \mathbf{A}\boldsymbol{\theta}_2$. Thus $\mathbf{A}\boldsymbol{\theta}_2$ provides the *biases* in the estimates of $\boldsymbol{\theta}_1$ if only the model $\mathbf{Y} = \boldsymbol{\theta}_1\mathbf{Z}_1 + \boldsymbol{\epsilon}$ is fitted.

Exercise 8. Show this by evaluating $E(\hat{\mathbf{Y}}) = E(\mathbf{Z}_1\hat{\boldsymbol{\theta}})$ with $E(\mathbf{Y}) = \boldsymbol{\theta}_1\mathbf{Z}_2$.

For the full regression we have the analysis of variance table of Table 20.2, where $\boldsymbol{\eta} = \mathbf{Z}_1\boldsymbol{\theta}_1 + \mathbf{Z}_2\boldsymbol{\theta}_2$.

Note the following points:

1. The $\hat{\boldsymbol{\theta}}$ obtained from fitting $\mathbf{Y} = \mathbf{Z}_1\boldsymbol{\theta}_1 + \mathbf{Z}_{2\cdot 1}\boldsymbol{\theta}_2 + \boldsymbol{\epsilon}$ is identical to the $\hat{\boldsymbol{\theta}}_1$ obtained by fitting $\mathbf{Y} = \mathbf{Z}_1\boldsymbol{\theta}_1 + \boldsymbol{\epsilon}$.
2. The $\hat{\boldsymbol{\theta}}_2$ obtained from fitting $\mathbf{Y} = \mathbf{Z}_1\boldsymbol{\theta}_1 + \mathbf{Z}_2\boldsymbol{\theta}_2 + \boldsymbol{\epsilon}$ is identical to the $\hat{\boldsymbol{\theta}}_2$ obtained by fitting just $\mathbf{Y} = \mathbf{Z}_{2\cdot 1}\boldsymbol{\theta}_2 + \boldsymbol{\epsilon}$.
3. The values $\boldsymbol{\theta}_1$ and $\boldsymbol{\theta}_2$ in the analysis of variance table can be thought of as "test

T A B L E 20.2. Analysis of Variance Table for the Orthogonalized General Regression

Source	df	SS
Response for $\mathbf{Z}_1$ only	p_1	$(\boldsymbol{\theta}_1 - \hat{\boldsymbol{\theta}}_1)'\mathbf{Z}_1'\mathbf{Z}_1(\boldsymbol{\theta}_1 - \hat{\boldsymbol{\theta}}_1)$
Extra for $\mathbf{Z}_2$	p_2	$(\boldsymbol{\theta}_2 - \hat{\boldsymbol{\theta}}_2)'\mathbf{Z}_{2\cdot 1}'\mathbf{Z}_{2\cdot 1}(\boldsymbol{\theta}_2 - \hat{\boldsymbol{\theta}}_2)$
Residual	$n - p_1 - p_2$	By subtraction
Total	n	$(\mathbf{Y} - \boldsymbol{\eta})'(\mathbf{Y} - \boldsymbol{\eta})$

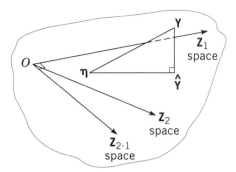

Figure 20.12. $\hat{\mathbf{Y}}$ can be described either as a linear combination of the spaces spanned (defined) by $\mathbf{Z}_1$ and $\mathbf{Z}_2$, which are not orthogonal, or of those spanned by $\mathbf{Z}_1$ and $\mathbf{Z}_{2\cdot1}$, which are orthogonal.

values" and the geometry is modified by a move from the origin to $\boldsymbol{\eta} = \mathbf{Z}_1\boldsymbol{\theta}_1 + \mathbf{Z}_2\boldsymbol{\theta}_2$ (see Figure 20.12).

4. The need for $\mathbf{Z}_2\boldsymbol{\theta}_2$ in the model will be indicated by a large "extra for $\mathbf{Z}_2$" sum of squares.

5. The "extra for $\mathbf{Z}_2$" sum of squares can also be obtained by fitting $\mathbf{Y} = \mathbf{Z}_1\boldsymbol{\theta}_1 + \mathbf{Z}_2\boldsymbol{\theta}_2 + \boldsymbol{\epsilon}$, then fitting $\mathbf{Y} = \mathbf{Z}_1\boldsymbol{\theta}_1 + \boldsymbol{\epsilon}$, and finding the difference between:

   a. The two regression sums of squares.

   b. The two regression sums of squares but with both corrected by $n\overline{Y}^2$.

   c. The two residual sums of squares in reverse order (to give a positive sign result).

More on the geometry of this is given in Chapter 21.

20.6. RANGE SPACE AND NULL SPACE OF A MATRIX M

The range space of a matrix $\mathbf{M}$, written $R(\mathbf{M})$, is the space of all vectors defined by the columns of $\mathbf{M}$. The dimension of the space is the column rank of $\mathbf{M}$, $cr(\mathbf{M})$, that is, the number of linearly independent columns of $\mathbf{M}$. (Thus the number of columns in $\mathbf{M}$ may exceed the dimension of the space they define.) The null space of $\mathbf{M}$, written $N(\mathbf{M})$, consists of the range space of all vectors $\mathbf{v}$ such that $\mathbf{M}\mathbf{v} = 0$, that is, all vectors $\mathbf{v}$ orthogonal to the *rows* of $\mathbf{M}$. If we want to define the null space of all vectors orthogonal to the *columns* of $\mathbf{M}$ we must write $N(\mathbf{M}')$.

Projection Matrices

Let E_n be n-dimensional Euclidean space (i.e., "ordinary" n-dimensional space). Let Ω be a p-dimensional subspace of E_n. Let $\Omega^\perp$ be the rest of E_n, that is, the subspace of E_n orthogonal to Ω. Let $\mathbf{P}_\Omega$ be an n by n projection matrix that projects a general n-dimensional vector $\mathbf{Y}$ entirely into the space Ω. (We shall write simply $\mathbf{P}$ in statements involving only the Ω-space.) Then a number of statements can be proved, as follows.

1. Every vector $\mathbf{Y}$ can be expressed uniquely in the form $\hat{\mathbf{Y}} + \mathbf{e}$, where $\hat{\mathbf{Y}}$ is wholly in Ω (we write $\hat{\mathbf{Y}} \in \Omega$) and $\mathbf{e} \in \Omega^\perp$.

2. If $\hat{\mathbf{Y}} = \mathbf{P}\mathbf{Y}$, then $\mathbf{P}$ is unique.

3. $\mathbf{P}$ can be written as $\mathbf{P} = \mathbf{T}\mathbf{T}'$, where the p columns of the n by p matrix $\mathbf{T}$ form an orthonormal basis (*not* unique) for the Ω-space. $\mathbf{T}$ is *not* unique and many choices are possible, even though $\mathbf{P}$ is unique.

[*Note:* A *basis* of Ω is a set of vectors that span the space of Ω, that is, permit every vector of Ω to be expressed as a linear combination of the base vectors. An *orthogonal* basis is one in which all basis vectors are orthogonal to one another. An *orthonormal* basis is an orthogonal basis for which the basis vectors have length (and so squared length) one, that is, $\mathbf{v}'\mathbf{v} = 1$ for all basis vectors $\mathbf{v}$.]

4. $\mathbf{P}$ is symmetric ($\mathbf{P}' = \mathbf{P}$) and idempotent ($\mathbf{P}^2 = \mathbf{P}$).

5. The vectors of $\mathbf{P}$ span the space Ω, that is, $R(\mathbf{P}) = \Omega$.

6. $\mathbf{I} - \mathbf{P}$ is the projection matrix for $\Omega^{\perp}$, the orthogonal part of E_n *not* in Ω. Thus $R(\mathbf{I} - \mathbf{P}) = \Omega^{\perp}$.

7. Any symmetric idempotent n by n matrix $\mathbf{P}$ represents an orthogonal projection matrix onto the space spanned by the columns of $\mathbf{P}$, that is, onto $R(\mathbf{P})$.

Statements 1–7 are generally true for any Ω even though we have given them in a notation that fits into the concept of regression. Statements 8 and 9 now make the connection with regression. We think of fitting the model $\mathbf{Y} = \mathbf{X}\boldsymbol{\beta} + \boldsymbol{\epsilon}$ by least squares, and $\Omega = R(\mathbf{X})$ will now be defined by the p columns of $\mathbf{X}$ and so will constitute the estimation space for our regression problem.

8. Suppose Ω is a space spanned by the columns of the n by p matrix $\mathbf{X}$. Suppose $(\mathbf{X}'\mathbf{X})^{-}$ is any generalized inverse of $\mathbf{X}'\mathbf{X}$ (see Appendix 20A). Then $\mathbf{P} = \mathbf{X}(\mathbf{X}'\mathbf{X})^{-}\mathbf{X}'$ is the unique projection matrix for Ω. Note carefully that $\mathbf{X}$ is not unique, nor is $(\mathbf{X}'\mathbf{X})^{-}$, but $\mathbf{P} = \mathbf{X}(\mathbf{X}'\mathbf{X})^{-}\mathbf{X}'$ is unique and so is $\hat{\mathbf{Y}} = \mathbf{P}\mathbf{Y}$.

9. If $\mathrm{cr}(\mathbf{X}) = p$ so that the columns of $\mathbf{X}$ are linearly independent, and $(\mathbf{X}'\mathbf{X})^{-1}$ exists, then $\mathbf{P} = \mathbf{X}(\mathbf{X}'\mathbf{X})^{-1}\mathbf{X}'$ and $R(\mathbf{X}) = R(\mathbf{P}) = \Omega$. This is the situation that will typically hold.

For proofs of these statements, see Seber (1977, pp. 394–395).

The practical consequences of these statements are as follows.

Given an estimation space Ω (and so, necessarily, an error space $\Omega^{\perp}$), any vector $\mathbf{Y}$ can uniquely be regressed into Ω, via a unique projection matrix $\mathbf{P} = \mathbf{X}(\mathbf{X}'\mathbf{X})^{-}\mathbf{X}'$. The facts that, even when $\mathbf{X}$ and $(\mathbf{X}'\mathbf{X})^{-}$ are not unique, $\mathbf{P}$ is unique and so is $\hat{\mathbf{Y}} = \mathbf{P}\mathbf{Y}$ for a given Ω are completely obvious when thought of geometrically. For a given Ω and $\mathbf{Y}$, there is a unique projection of $\mathbf{Y}$ onto Ω that defines $\hat{\mathbf{Y}}$ uniquely in Ω. It doesn't matter *how* Ω is described [i.e., what $\mathbf{X}$ is chosen, or what choice of $(\mathbf{X}'\mathbf{X})^{-}$ is made when $\mathbf{X}'\mathbf{X}$ is singular] or *how* $\hat{\mathbf{Y}}$ is described as a function $\mathbf{X}\mathbf{b} = \mathbf{X}(\mathbf{X}'\mathbf{X})^{-}\mathbf{X}'\mathbf{Y}$ of the b's. The basic triangle of Figure 20.13 joining $\mathbf{0}$, $\mathbf{Y}$, and $\hat{\mathbf{Y}}$ remains fixed forever, given Ω and $\mathbf{Y}$.

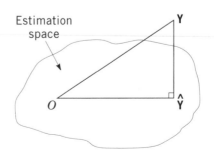

Figure 20.13. The triangle is unique, given Ω and $\mathbf{Y}$.

20.7. THE ALGEBRA AND GEOMETRY OF PURE ERROR

We consider a general linear regression situation with

$$Y = \beta_0 + \beta_1 X_1 + \beta_2 X_2 + \cdots + \beta_{p-1} X_{p-1} + \epsilon,$$

$$\text{or} \quad Y = \mathbf{X}\boldsymbol{\beta} + \epsilon. \tag{20.7.1}$$

Our model has p parameters. Some of the X's could be transformations of other X's. Thus, for example, the model could be a polynomial of second order. We assume $\mathbf{X}'\mathbf{X}$ is nonsingular; if it were not, we would redefine $\mathbf{X}$ to ensure that it was. Suppose we have m data sites with $n_1, n_2, n_3, \ldots, n_m$ repeats at these sites, with $n_1 + n_2 + \cdots + n_m = n$. Naturally, $p \leq m$ or we cannot estimate the parameters; preferably $p < m$, or we cannot test for lack of fit. Without loss of generality, some n_j could equal 1, but not all of them, or we would have no pure error with which to test lack of fit. We now define an n by m matrix $\mathbf{X}_e$ of form

$$\mathbf{X}_e = \begin{bmatrix} 1 \\ 1 \\ \cdots \\ 1 \\ \hline 1 \\ 1 \\ \cdots \\ \hline 1 \\ \cdots \\ \hline 1 \\ 1 \\ \cdots \\ 1 \end{bmatrix}, \tag{20.7.2}$$

which is such that the jth column contains only n_j ones and $n - n_j$ zeros, the ones positioned in the row locations $n_1 + n_2 + \cdots + n_{j-1} + 1$ to $n_1 + n_2 + \cdots + n_j$. For a site j with no repeats there will be only a single one in the corresponding column. Consider the model

$$Y = \mathbf{X}_e \boldsymbol{\mu} + \epsilon \tag{20.7.3}$$

where $\boldsymbol{\mu} = (\mu_1, \mu_2, \ldots, \mu_m)'$. If $p \leq m$, model (20.7.3) is inclusive of model (20.7.1) because we can reexpress the m μ's in terms of p β's, so $R(\mathbf{X})$ is included within $R(\mathbf{X}_e)$. Geometrically the m column vectors of $\mathbf{X}_e$ are linearly combined to the p column vectors of $\mathbf{X}$. This implies that, if we define $\mathbf{P} = \mathbf{X}(\mathbf{X}'\mathbf{X})^{-1}\mathbf{X}'$ and $\mathbf{P}_e = \mathbf{X}_e(\mathbf{X}'_e\mathbf{X}_e)^{-1}\mathbf{X}'_e$ as the respective projection matrices, the space $R(\mathbf{P}) = R(\mathbf{X})$ lies within the space $R(\mathbf{P}_e) = R(\mathbf{X}_e)$.

Now $\mathbf{X}'_e\mathbf{X}_e = \text{diagonal} \, (n_1, n_2, \ldots, n_m)$, a matrix with terms n_i along the upper-left to lower-right main diagonal and zeros elsewhere, so that $(\mathbf{X}'_e\mathbf{X}_e)^{-1}$ = diagonal $(n_1^{-1}, n_2^{-1}, \ldots, n_m^{-1})$. It follows (an exercise for the reader here!) that $\mathbf{P}_e$ consists of a matrix with m main-diagonal blocks of sizes $n_1, n_2, \ldots, n_m$. The jth of these has the form

$$\mathbf{B}_j = \begin{bmatrix} n_j^{-1} & n_j^{-1} & \cdots & n_j^{-1} \\ n_j^{-1} & n_j^{-1} & \cdots & n_j^{-1} \\ \vdots & \vdots & & \vdots \\ n_j^{-1} & n_j^{-1} & \cdots & n_j^{-1} \end{bmatrix} = n_j^{-1}\mathbf{11}' \tag{20.7.4}$$

where the $\mathbf{1}$ is an n_j by 1 vector.

We now consider the breakup of the residual sum of squares into lack of fit and pure error. We can write the residual vector as

$$\mathbf{Y} - \hat{\mathbf{Y}} = \tilde{\mathbf{Y}} - \hat{\mathbf{Y}} + \mathbf{Y} - \tilde{\mathbf{Y}}, \tag{20.7.5}$$

where $\tilde{\mathbf{Y}} = \mathbf{P}_e\mathbf{Y}$. Recalling that $\hat{\mathbf{Y}} = \mathbf{P}\mathbf{Y}$, we thus have

$$(\mathbf{I} - \mathbf{P})\mathbf{Y} = (\mathbf{P}_e - \mathbf{P})\mathbf{Y} + (\mathbf{I} - \mathbf{P}_e)\mathbf{Y}, \tag{20.7.6}$$

that is,

Residual vector = Lack of fit vector + Pure error vector.

The pure error vector consists of deviations of the individual observations from their own pure error group averages. If an $n_j = 1$, a zero appears in the appropriate position, of course. Note that

$$(\mathbf{P}_e - \mathbf{P})'(\mathbf{I} - \mathbf{P}_e) = \mathbf{P}_e - \mathbf{P}_e^2 - \mathbf{X}(\mathbf{X}'\mathbf{X})^{-1}\mathbf{X}'(\mathbf{I} - \mathbf{P}_e) = \mathbf{0}. \tag{20.7.7}$$

The first pair of terms cancel each other because $\mathbf{P}_e$ is a projection matrix and so idempotent. In the third term $\mathbf{X}'(\mathbf{I} - \mathbf{P}_e) = \mathbf{0}$ due to the special forms of $\mathbf{X}$ and $\mathbf{I} - \mathbf{P}_e$ when repeats occur. Specifically, the product $\mathbf{X}'(\mathbf{I} - \mathbf{P}_e)$ consists of a series of products, the jth of which takes the form

$$\begin{bmatrix} 1 & 1 & 1 & \cdots & 1 \\ a & a & a & \cdots & a \\ \vdots & \vdots & \vdots & & \vdots \\ z & z & z & & z \end{bmatrix} \begin{bmatrix} 1 - n_j^{-1} & -n_j^{-1} & \cdots & -n_j^{-1} \\ -n_j^{-1} & 1 - n_j^{-1} & \cdots & -n_j \\ \vdots & \vdots & & \vdots \\ -n_j^{-1} & -n_j^{-1} & \cdots & 1 - n_j^{-1} \end{bmatrix}, \tag{20.7.8}$$

where $(a, \ldots, z)$ are the values of $(X_1, \ldots, X_k)$ in the jth group of repeats; each letter $a, \ldots, z$ occurs n_j times. We could also write this product more compactly as

$$\begin{bmatrix} 1 \\ a \\ \vdots \\ z \end{bmatrix} \mathbf{1}'(\mathbf{I} - n_j^{-1}\mathbf{11}') = \mathbf{0} \tag{20.7.9}$$

because $\mathbf{1}'\mathbf{1} = n_j$ for the jth group.

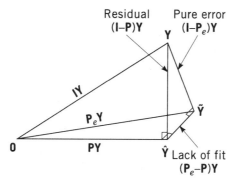

F i g u r e 20.14. The orthogonal breakup of the residual vector into the lack of fit and pure error vectors. All three of these vectors are orthogonal to the estimation space $R(\mathbf{X})$.

The Geometry of Pure Error

From the argument above we see that the geometry of pure error is an orthogonal decomposition of the residual vector $\mathbf{Y} - \hat{\mathbf{Y}} = (\mathbf{I} - \mathbf{P})\mathbf{Y}$ into orthogonal pieces $(\mathbf{P}_e - \mathbf{P})\mathbf{Y}$ for lack of fit and $(\mathbf{I} - \mathbf{P}_e)\mathbf{Y}$, for pure error, as in (20.7.6). If we define

$$n_e = (n_1 - 1) + (n_2 - 1) + \cdots + (n_m - 1) = n - m \qquad (20.7.10)$$

as the pure error degrees of freedom, the corresponding dimensional breakup is

$$n - p = (n - p - n_e) + n_e \qquad (20.7.11)$$

for the lack of fit and pure error spaces. As always, the corresponding sums of squares are the squared lengths of the vectors, namely,

$$\mathbf{Y}'(\mathbf{I} - \mathbf{P})\mathbf{Y} = \mathbf{Y}'(\mathbf{P}_e - \mathbf{P})\mathbf{Y} + \mathbf{Y}'(\mathbf{I} - \mathbf{P}_e)\mathbf{Y}, \qquad (20.7.12)$$

where we have reduced the matrix powers of the quadratic forms by invoking the idempotency of $\mathbf{I} - \mathbf{P}$, $\mathbf{P}_e - \mathbf{P}$, and $\mathbf{I} - \mathbf{P}_e$. Figure 20.14 illustrates the geometry, and Figure 20.15 shows the dimensions (degrees of freedom) of the various spaces.

APPENDIX 20A. GENERALIZED INVERSES M⁻

(Sometimes these are also called pseudo-inverses, as well as other terms.) Suppose, first, that $\mathbf{M}$ is a nonsingular square matrix. Then $\mathbf{M}^- = \mathbf{M}^{-1}$. If $\mathbf{M}$ is singular, $\mathbf{M}^-$ is any matrix for which

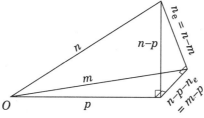

F i g u r e 20.15. The corresponding degrees of freedom of the residual vector breakup are shown as the dimensions of the subspaces in which the various vectors lie. For all subspaces to exist, we must have $n > m > p$, that is, there must be fewer sites than observations but more sites than parameters.

$$\mathbf{M}\mathbf{M}^{-}\mathbf{M} = \mathbf{M}. \tag{20A.1}$$

This concept of a generalized inverse is an interesting extension of the idea of the inverse matrix to singular matrices. (Although it *is* interesting, it can usually be circumvented in practical regression situations!) A generalized inverse satisfying $\mathbf{M}\mathbf{M}^{-}\mathbf{M} = \mathbf{M}$ always exists and is not unique.

(*Note:* It is also possible to define $\mathbf{M}^{-}$ in the same way when $\mathbf{M}$ is not square. For regression purposes, we do not need this extension.)

Moore–Penrose Inverse

A unique definition (the so-called Moore–Penrose inverse) can be obtained by insisting on $\mathbf{M}^{-}$ satisfying three more conditions, namely,

$$\mathbf{M}^{-}\mathbf{M}\mathbf{M}^{-} = \mathbf{M}^{-},$$

$$(\mathbf{M}\mathbf{M}^{-})' = \mathbf{M}\mathbf{M}^{-}, \tag{20A.2}$$

$$(\mathbf{M}^{-}\mathbf{M})' = \mathbf{M}^{-}\mathbf{M}.$$

Some writers write $\mathbf{M}^{+}$ for the Moore–Penrose inverse.

Getting a Generalized Inverse

We assume $\mathbf{M}$ is square and, for our regression applications, is of the symmetric form $\mathbf{X}'\mathbf{X}$. Let $\mathbf{M}$ be p by p and have rank (row or column rank) $r < p$. Probably the easiest method for getting an $\mathbf{M}^{-}$ is the following.

A Method for Getting M⁻

Examine $\mathbf{M}$ to find an r by r submatrix that has full rank, that is, is nonsingular. If it occupies the upper-left corner, then

$$\mathbf{M} = \begin{bmatrix} \mathbf{M}_{11} & \mathbf{M}_{12} \\ \mathbf{M}_{21} & \mathbf{M}_{22} \end{bmatrix}, \tag{20A.3}$$

where $\mathbf{M}_{11}$ is nonsingular, so that $\mathbf{M}_{11}^{-1}$ exists. Then

$$\mathbf{M}^{-} = \begin{bmatrix} \mathbf{M}_{11}^{-1} & \mathbf{0} \\ \mathbf{0} & \mathbf{0} \end{bmatrix} \tag{20A.4}$$

will be a generalized inverse for $\mathbf{M}$.

Proof: Form the product $\mathbf{M}\mathbf{M}^{-}\mathbf{M}$ to get

$$\begin{bmatrix} \mathbf{M}_{11} & \mathbf{M}_{12} \\ \mathbf{M}_{21} & \mathbf{M}_{21}\mathbf{M}_{11}^{-1}\mathbf{M}_{12} \end{bmatrix}, \tag{20A.5}$$

which has three submatrices correct. Because of the implicit assumption that the $(p - r)$ later columns of $\mathbf{M}$ depend on the first r, there exists an r by $(p - r)$ matrix $\mathbf{Q}$, say, such that

$$\mathbf{M}_{11}\mathbf{Q} = \mathbf{M}_{12} \quad \text{and} \quad \mathbf{M}_{21}\mathbf{Q} = \mathbf{M}_{22}. \tag{20A.6}$$

Solving for $\mathbf{Q}$ in the first of these gives $\mathbf{Q} = \mathbf{M}_{11}^{-1}\mathbf{M}_{12}$ whence $\mathbf{M}_{21}\mathbf{M}_{11}^{-1}\mathbf{M}_{12} = \mathbf{M}_{21}\mathbf{Q} = \mathbf{M}_{22}$. The method works in exactly the same way *wherever* the elements of the nonsingular ($\mathbf{M}_{11}$) matrix are located. It is inverted where it stands and zeros occupy all other places.

Our regression application is the following. If $\mathbf{X}'\mathbf{X}$ is singular and $(\mathbf{X}'\mathbf{X})^-$ is any generalized inverse, the normal equations $\mathbf{X}'\mathbf{Xb} = \mathbf{X}'\mathbf{Y}$ are satisfied by $\mathbf{b} = (\mathbf{X}'\mathbf{X})^-\mathbf{X}'\mathbf{Y}$. See Seber (1977, pp. 76 and 391). Note that although different choices of $(\mathbf{X}'\mathbf{X})^-$ produce different $\mathbf{b}$ estimates, $\hat{\mathbf{Y}} = \mathbf{Xb}$ is invariant due to the geometry.

Example

Consider the fitting of a straight line $Y = \beta_0 + \beta_1 X + \epsilon$ to n data points all at the same location $X = X^*$. The least squares solution is any line through the point $(\bar{X}, \bar{Y}) = (X^*, \bar{Y})$, because a unique solution is obtained only when there are two or more X-sites, and here we have only one. The general solution is thus of the form

$$\hat{Y} = b_0 + (\bar{Y} - b_0)(X/X^*) \tag{20A.7}$$

for any choice of b_0. [Note that when $X = X^*$, $\hat{Y} = \bar{Y}$ so that $\hat{\mathbf{Y}} = (\bar{Y}, \bar{Y}, \ldots, \bar{Y})'$ and is unique, as we know it must be from the geometry.] For this problem, the normal equations are

$$\begin{bmatrix} n & nX^* \\ nX^* & nX^{*2} \end{bmatrix} \begin{bmatrix} b_0 \\ b_1 \end{bmatrix} = \begin{bmatrix} \Sigma Y_i \\ X^*\Sigma Y_i \end{bmatrix} \tag{20A.8}$$

and do not have a unique solution. We now look at what is achieved by specific choices of $(\mathbf{X}'\mathbf{X})^-$, when evaluating $\mathbf{b} = (\mathbf{X}'\mathbf{X})^-\mathbf{X}'\mathbf{Y}$.

Choice 1. Let

$$(\mathbf{X}'\mathbf{X})^- = \begin{bmatrix} n^{-1} & 0 \\ 0 & 0 \end{bmatrix}. \tag{20A.9}$$

Then $b_0 = \bar{Y}$ and $b_1 = 0$. We obtain a horizontal straight line through $(X^*, \bar{Y})$.

Choice 2. Let

$$(\mathbf{X}'\mathbf{X})^- = \begin{bmatrix} 0 & 0 \\ 0 & (nX^{*2})^{-1} \end{bmatrix}. \tag{20A.10}$$

Then $b_0 = 0$, $b_1 = \bar{Y}/X^*$. We have a straight line joining the origin to the point $(X^*, \bar{Y})$.

Choice 3. Let

$$(\mathbf{X}'\mathbf{X})^- = \begin{bmatrix} 0 & (nX^*)^{-1} \\ 0 & 0 \end{bmatrix}. \tag{20A.11}$$

Then $b_0 = \bar{Y}$ and $b_1 = 0$, which is the same solution as Choice 1.

Choice 4. Let

$$(\mathbf{X}'\mathbf{X})^- = \begin{bmatrix} 0 & 0 \\ (nX^*)^{-1} & 0 \end{bmatrix}. \tag{20A.12}$$

Then $b_0 = 0$, $b_1 = \bar{Y}/X^*$, the same solution as Choice 2.

We note two features from this (somewhat limited) example:

1. Any specific choice of $(\mathbf{X'X})^-$ merely leads to one of the infinity of solutions provided by (20A.7). (This point is true in the general case also.)

2. Only the two most obvious solutions arise, those assuming that: $b_0 = \overline{Y}$ or $b_0 = 0$. To get other solutions [which all still satisfy (20A.7)], other choices of $(\mathbf{X'X})^-$ are needed. It is, however, pointless to follow this up, as other choices essentially ask us to make other assumptions about the b's. Obviously we could apply any assumption we chose *directly* to the general solution (20A.7) and not use a generalized inverse at all.

What Should One Do?

Our overall recommendation is that using a generalized inverse for a practical regression problem is usually a waste of time. Four alternative choices are:

1. Keep the original data but modify the model to make the new $\mathbf{X'X}$ non-singular.
2. Keep the original model but get more data to make the new $\mathbf{X'X}$ nonsingular.
3. Keep both data and model and decide to implement sensible linear restrictions on the parameters to make the new $\mathbf{X'X}$ nonsingular. [Computer programs typically set equal to zero all parameters associated with the later (in sequence, as given to the computer) dependent columns of the original $\mathbf{X}$ matrix.]
4. Add nonlinear restrictions and solve the least squares problem subject to them. Ridge regression is an example of this.

      Choice 3 will often be the most practical.

EXERCISES FOR CHAPTER 20

A. Fit the model $Y = \beta_0 + \beta_1 X + \epsilon$ by least squares to the four data points $(X, Y) = (1, 1.4)$, $(2, 2.2)$, $(3, 2.3)$, $(4, 3.1)$.
  1. Write down the SS function $S(\beta_0, \beta_1)$.
  2. Find the least squares estimate $\mathbf{b}$.
  3. Find the vector $\hat{\mathbf{Y}}$ of fitted values and $\mathbf{e} = \mathbf{Y} - \hat{\mathbf{Y}}$.
  4. Find the projection matrix $\mathbf{P} = \mathbf{X}(\mathbf{X'X})^{-1}\mathbf{X'}$.
  5. For the sample space (E_4 space), make a plot, as best you can, showing what you have done, and identifying all relevant details.
  6. Write down an analysis of variance table "appropriate for checking $H_0 : \beta_0 = \beta_1 = 0$ versus $H_1 :$ "not so," for this regression situation. Specify the test statistic and get its observed value. Identify what this means in your figure also.
  7. Repeat 6 for $H_0 : \beta_0 = 1$, $\beta_1 = 0.5$.
  8. The two vectors of $\mathbf{X}$ and the four vectors of the least squares projection matrix $\mathbf{P}$ span the same space. What specific linear combinations of the two vectors of $\mathbf{X}$ will give $\mathbf{P}$?
  9. What specific linear combinations of the four vectors of $\mathbf{P}$ will give $\mathbf{X}$?
  10. The vectors of $\mathbf{X}$ span the estimation space. Write $\mathbf{X} = (\mathbf{X}_0, \mathbf{X}_1)$, where $\mathbf{X}_0 = (1, 1, 1, 1)'$. Find $\mathbf{X}_{1.0}$, a vector orthogonal to $\mathbf{X}_0$, such that $(\mathbf{X}_0, \mathbf{X}_{1.0})$ also spans the estimation space.
  11. Hence (see 10) or otherwise, find a suitable sum of squares and F-value for testing $H_0 : \beta_1 = 0$, irrespective of the value of β_0.

B. Consider (very carefully) the least squares regression problem with model $\mathbf{Y} = \mathbf{X}\boldsymbol{\beta} + \boldsymbol{\epsilon}$ where

$$
\begin{array}{ccc} \mathbf{1} & \mathbf{X}_1 & \mathbf{X}_2 \end{array}
$$

$$
\mathbf{X} = \begin{bmatrix} 1 & -3 & 1 \\ 1 & -1 & 7 \\ 1 & 1 & 13 \\ 1 & 3 & 19 \end{bmatrix}, \quad \boldsymbol{\beta} = \begin{bmatrix} \beta_0 \\ \beta_1 \\ \beta_2 \end{bmatrix}, \quad \mathbf{Y} = \begin{bmatrix} 4 \\ 7 \\ 2 \\ 3 \end{bmatrix}.
$$

1. Write down the normal equations.
2. Find a general solution to the normal equations.
3. Determine $\hat{\mathbf{Y}}$.
4. Make a few appropriate comments.

C. To be estimable, a linear combination $\mathbf{c}'\boldsymbol{\beta}$ (say) of the elements of $\boldsymbol{\beta}$ must have a $\mathbf{c}'$ vector that is a linear combination of the rows of $\mathbf{X}$. If $\mathbf{c}'$ cannot be expressed that way, $\mathbf{c}'\boldsymbol{\beta}$ is not estimable. If the regression is of full rank, all $\mathbf{c}'\boldsymbol{\beta}$ are estimable because $\boldsymbol{\beta}$ itself can be uniquely estimated. If the regression is not full rank, then some $\mathbf{c}'\boldsymbol{\beta}$ are estimable and some are not. With this in mind, consider the following:

Four observations of Y are taken at each of $X = -1, 0, 1$ with the intention of fitting the cubic model $Y = \beta_0 + \beta_1 X + \beta_2 X^2 + \beta_3 X^3 + \epsilon$ via least squares, under the usual error assumptions.

1. What is the projection matrix $\mathbf{P} = \mathbf{X}(\mathbf{X}'\mathbf{X})^{-}\mathbf{X}'$.
2. Is $\mathbf{P}$ unique? Is $\mathbf{PY}$ unique?
3. Which of the following are estimable?

$$
\beta_0 + \beta_2, \quad \beta_1, \quad \beta_0 - \beta_1, \quad \beta_1 + \beta_3, \quad \beta_0 + \beta_1 + \beta_2 + \beta_3 + \beta_4.
$$

D. Consider the least squares fit $Y = \beta_1 X_1 + \beta_2 X_2 + \epsilon$ (no intercept) to the data $(X_1, X_2, Y) = (1, 2, 19)$, $(2, 1, 13)$, and $(0, 0, 16)$. Using axes (U_1, U_2, U_3), say, for three-dimensional space, do the following:

1. Draw a diagram showing the basic least squares fit. On this diagram, name the various spaces and explain how they are defined, and label any points with actual coordinates.
2. Show what the numbers in the ANOVA table mean in your figure.
3. The regression is not a significant one. What feature(s) of the data make(s) it so?
4. Is there anything "special" about the estimation space?
5. Evaluate $\mathbf{X}_{2 \cdot 1}$ and draw a new diagram explaining what the ANOVA numbers mean in this diagram. $\mathbf{X}_{2 \cdot 1}$ is the part of $\mathbf{X}_2$ orthogonal to $\mathbf{X}_1$.
6. Suppose that, instead of the number 16, we substitute 2. Geometrically, what does that do?
7. Provide a new ANOVA table and a new F-value for the situation when the number 16 is replaced by 2.

E. 1. Find the (unique) projection matrix $\mathbf{P}$ onto a space Ω spanned by the vectors of $\mathbf{A}$, where

$$
\mathbf{A} = \begin{bmatrix} 1 & -3 & 1 \\ 1 & -1 & -1 \\ 1 & 1 & -1 \\ 1 & 3 & 1 \end{bmatrix}.
$$

2. Find a basis for the space orthogonal to Ω, that is, find vectors that span that orthogonal space.
3. Find the projections into Ω of all the vectors you gave in (2), and specify the space spanned by the $\mathbf{P}$ you gave in (1).

F. Which of the matrices below are generalized inverses of the matrix $\mathbf{11}'$, namely,

$$\begin{bmatrix} 1 & 1 \\ 1 & 1 \end{bmatrix}?$$

1. $\begin{bmatrix} 1 & 0 \\ 0 & 1 \end{bmatrix}.$

2. $\begin{bmatrix} 0 & 1 \\ 0 & 0 \end{bmatrix}.$

3. $\begin{bmatrix} 0.25 & 0.25 \\ 0.25 & 0.25 \end{bmatrix}.$

4. $\begin{bmatrix} 0.1 & 0.2 \\ 0.3 & 0.4 \end{bmatrix}.$

G. A straight line $Y = \beta_0 + \beta_1 X + \epsilon$ is to be fitted to the data below. Show a "tree diagram" of the allocation of the 10 degrees of freedom and find 10 orthogonal vectors that span the whole space, saying which correspond to your degrees of freedom split-up, and so divide the whole space up into three distinct orthogonal spaces.

Y	1	X
Y_1	1	−2
Y_2	1	−2
Y_3	1	−1
Y_4	1	−1
Y_5	1	0
Y_6	1	0
Y_7	1	1
Y_8	1	1
Y_9	1	2
Y_{10}	1	2

H. If $\mathbf{P}$ is a projection matrix for a regression with a β_0 in the model, its row and column sums should all be 1. Explain geometrically why this must *obviously* be true?

CHAPTER 21

More Geometry of Least Squares

The basic geometry of least squares appears in the foregoing chapter. Here we take things a little further by considering what happens geometrically when we test linear hypotheses of the form H_0: $\mathbf{A}\boldsymbol{\beta} = \mathbf{c}$ in the model $\mathbf{Y} = \mathbf{X}\boldsymbol{\beta} + \boldsymbol{\epsilon}$. (The alternative hypothesis is always that H_0 is false.) We suppose that $\mathbf{A}$ is a q by p matrix ($q < p$) of full rank so that the rows of $\mathbf{A}$ are linearly independent; $\boldsymbol{\beta}$ is p by 1 and $\mathbf{c}$ is a q by 1 vector of constants; $\mathbf{X}$ is n by p, and assumed to be of full rank p.

21.1. THE GEOMETRY OF A NULL HYPOTHESIS: A SIMPLE EXAMPLE

We first consider the simple example of fitting a straight line $Y = \beta_0 + \beta_1 X + \epsilon$ using a set of n data points represented by two n by 1 vectors $\mathbf{Y}$ and $\mathbf{X}_1$. In matrix terms $\mathbf{Y} = \mathbf{X}\boldsymbol{\beta} + \boldsymbol{\epsilon}$, we thus have the model function

$$\mathbf{X}\boldsymbol{\beta} = (\mathbf{1}, \mathbf{X}_1) \begin{bmatrix} \beta_0 \\ \beta_1 \end{bmatrix} = \beta_0 \mathbf{1} + \beta_1 \mathbf{X}_1. \tag{21.1.1}$$

The estimation space Ω is a plane spanned by $\mathbf{1}$ and $\mathbf{X}_1$. Suppose $\mathbf{A} = (2, -1)$ and $\mathbf{c} = 4$ so that $\mathbf{A}\boldsymbol{\beta} = \mathbf{c}$ implies $2\beta_0 - \beta_1 = 4$. Obviously $p = 2$ and $q = 1$. We substitute in (21.1.1) for β_1 to obtain for the model under $\mathbf{A}\boldsymbol{\beta} = \mathbf{c}$,

$$\mathbf{X}\boldsymbol{\beta} = \beta_0(\mathbf{1} + 2\mathbf{X}_1) - 4\mathbf{X}_1. \tag{21.1.2}$$

The estimation space ω for this model function is a straight line swept out by combining the constant vector $-4\mathbf{X}_1$ with the variable length vector $\beta_0(\mathbf{1} + 2\mathbf{X}_1)$, as shown in Figure 21.1. The three black dots show points for which $\beta_0 = 0, 1$, and 1.5 on ω. Clearly ω is part of Ω. The space $\Omega - \omega$ is spanned by any set of vectors in Ω that are all orthogonal to ω. For our example, $p = 2$ and $q = 1$, so there is only one such vector. If we write, in (21.1.2),

$$\mathbf{u} = \beta_0 \mathbf{1} + (2\beta_0 - 4)\mathbf{X}_1$$

for the vector that spans ω, an obviously orthogonal vector is

$$(\mathbf{u}'\mathbf{X}_1)\mathbf{1} - (\mathbf{u}'\mathbf{1})\mathbf{X}_1, \tag{21.1.3}$$

which spans $\Omega - \omega$.

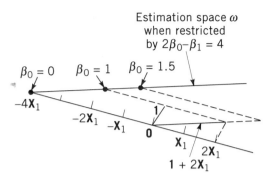

F i g u r e 21.1. The estimation space Ω is the plane spanned by vectors $\mathbf{1}$ and $\mathbf{X}_1$. When restricted by $2\beta_0 - \beta_1 = 4$, the reduced estimation space ω is a straight line parallel to $\mathbf{1} + 2\mathbf{X}_1$ but displaced a distance equal to the length of $-4\mathbf{X}_1$.

21.2. GENERAL CASE $H_0 : A\beta = c$: THE PROJECTION ALGEBRA

The constant $\mathbf{c}$ is essentially an "origin choice" on ω. For purposes of defining the spaces ω and $\Omega - \omega$, we can temporarily get rid of it. Suppose β^* is *any* numerical choice that satisfies $A\beta^* = c$. We can rewrite the model as

$$\mathbf{Y} - \mathbf{X}\beta^* = \mathbf{X}(\beta - \beta^*) + \epsilon$$
$$= \mathbf{X}\theta + \epsilon \tag{21.2.1}$$

and for the new parameter vector θ,

$$A\theta = A\beta - A\beta^* = A\beta - c = 0. \tag{21.2.2}$$

We now rewrite $A\theta = 0$ as

$$A(\mathbf{X}'\mathbf{X})^{-1}\mathbf{X}'(\mathbf{X}\theta) = 0. \tag{21.2.3}$$

This makes it obvious that all $\mathbf{X}\theta$ points in the $(p - q)$-dimensional space ω are orthogonal to the *columns* of the n by q matrix $\mathbf{U} = \mathbf{X}(\mathbf{X}'\mathbf{X})^{-1}A'$.

This implies that the q-dimensional $\Omega - \omega$ space is defined by the columns of $\mathbf{U}$, and so a unique projection matrix for $\Omega - \omega$ is given by

$$\mathbf{P}_1 = \mathbf{U}(\mathbf{U}'\mathbf{U})^{-1}\mathbf{U}' = \mathbf{X}(\mathbf{X}'\mathbf{X})^{-1}A'[A(\mathbf{X}'\mathbf{X})^{-1}A']^{-1}A(\mathbf{X}'\mathbf{X})^{-1}\mathbf{X}'. \tag{21.2.4}$$

Because

$$\mathbf{P} = \mathbf{P}_\Omega = \mathbf{X}(\mathbf{X}'\mathbf{X})^{-1}\mathbf{X}' \tag{21.2.5}$$

is the unique projection matrix for Ω, the projection matrix for ω is

$$\mathbf{P}_\omega = \mathbf{P} - \mathbf{P}_1. \tag{21.2.6}$$

We now project $\mathbf{Y} - \mathbf{X}\beta^*$ via (21.2.6) to give

$$\mathbf{P}_\omega\mathbf{Y} - \mathbf{P}_\omega\mathbf{X}\beta^* = \mathbf{P}\mathbf{Y} - \mathbf{P}\mathbf{X}\beta^* - \mathbf{P}_1(\mathbf{Y} - \mathbf{X}\beta^*). \tag{21.2.7}$$

We note that:

(i) $\mathbf{P}_\omega\mathbf{Y} = \mathbf{X}\mathbf{b}_H$, where $\mathbf{b}_H$ is the least squares estimate of β in the restricted space ω.
(ii) $\mathbf{P}_\omega\mathbf{X}\beta^* = \mathbf{X}\beta^* = \mathbf{c}$, because the projection into ω of a vector already in ω (namely, $\mathbf{X}\beta^*$) leaves it untouched.

(iii) PY = Xb, where $\mathbf{b} = (\mathbf{X}'\mathbf{X})^{-1}\mathbf{X}'\mathbf{Y}$ is the usual (unrestricted) least squares estimator.

(iv) PXβ* = Xβ* = c; the argument is similar to (ii).

(v) $\mathbf{P}_1(\mathbf{Y} - \mathbf{X}\boldsymbol{\beta}^*) = \mathbf{X}(\mathbf{X}'\mathbf{X})^{-1}\mathbf{A}'[\mathbf{A}(\mathbf{X}'\mathbf{X})^{-1}\mathbf{A}']^{-1}(\mathbf{Ab} - \mathbf{c})$. (21.2.8)

Putting the pieces back into (21.2.7), canceling two **c**'s, and multiplying through by $(\mathbf{X}'\mathbf{X})^{-1}\mathbf{X}'$ to "cancel" **X** throughout, gives the restricted (by $\mathbf{A}\boldsymbol{\beta} = \mathbf{c}$) least squares estimate vector

$$\mathbf{b}_H = \mathbf{b} - (\mathbf{X}'\mathbf{X})^{-1}\mathbf{A}'[\mathbf{A}(\mathbf{X}'\mathbf{X})^{-1}\mathbf{A}']^{-1}(\mathbf{Ab} - \mathbf{c}). \quad (21.2.9)$$

The form of this is **b** adjusted by an amount that depends on **X**, **A**, and how far off **Ab** is from **c**.

Properties

All three of the projection matrices are symmetric and idempotent. Note also that

$$\text{(a)} \quad \mathbf{PP}_\omega = \mathbf{P}_\omega = \mathbf{P}_\omega\mathbf{P}$$

$$\text{(b)} \quad \mathbf{PP}_1 = \mathbf{P}_1 = \mathbf{P}_1\mathbf{P} \quad (21.2.10)$$

$$\text{(c)} \quad \mathbf{P}_\omega\mathbf{P}_1 = \mathbf{0} = \mathbf{P}_1\mathbf{P}_\omega$$

Geometrically, (a) means that a vector **Y** projected first into ω and then into Ω stays in ω, or that, if projected first into Ω and then into ω, finishes up in ω; (b) is a similar result. Part (c), which can be proved by writing $\mathbf{P}_1\mathbf{P}_\omega = \mathbf{P}_1(\mathbf{P} - \mathbf{P}_1) = \mathbf{P}_1\mathbf{P} - \mathbf{P}_1^2 = \mathbf{P}_1 - \mathbf{P}_1 = \mathbf{0}$, means that the split of Ω into the two subspaces, ω created by $\mathbf{A}\boldsymbol{\beta} = \mathbf{c}$, and $\Omega - \omega$, is an orthogonal split.

21.3 GEOMETRIC ILLUSTRATIONS

Figure 21.2 shows the case $n \geq 3$, $p = 2$, $q = 1$. The base plane of the figure is Ω defined by the two vectors in **X** (which are not specifically shown, but define the plane). The space ω is a straight line (shown) and the space $\Omega - \omega$ is a perpendicular straight line (not shown). The vertical dimension of the figure represents the other $(n - 2)$ dimensions. The points $\hat{\mathbf{Y}} = \mathbf{Xb}$ and $\hat{\mathbf{Y}}_H = \mathbf{Xb}_H$ are the unrestricted and restricted least squares points on Ω and ω, respectively. Note that we also show a general point $\mathbf{X}\boldsymbol{\beta}_H$ on ω. The sum of squares due to the hypothesis, $\text{SS}(H_0)$, is the squared distance between $\hat{\mathbf{Y}}$ and $\hat{\mathbf{Y}}_H$. Via Pythagoras's theorem,

$$\text{SS}(H_0) = (\mathbf{Y} - \hat{\mathbf{Y}}_H)'(\mathbf{Y} - \hat{\mathbf{Y}}_H) - (\mathbf{Y} - \hat{\mathbf{Y}})'(\mathbf{Y} - \hat{\mathbf{Y}}), \quad (21.3.1)$$

that is, the difference between the two residual sums of squares. Also, if $\mathbf{c} = \mathbf{0}$ (so that ω includes the origin **O**, and *not* as in Figure 21.1) we can write, alternatively,

$$\text{SS}(H_0) = \hat{\mathbf{Y}}'\hat{\mathbf{Y}} - \hat{\mathbf{Y}}_H'\hat{\mathbf{Y}}_H. \quad (21.3.2)$$

[*Note:* If in the example of Section 21.1, **c** were zero, than ω would consist of the line $(1 + 2\mathbf{X}_1)$ and would contain the origin.]

Figure 21.3 shows the case $n > 3$, $p = 3$, $q = 1$. The point $\hat{\mathbf{Y}} = \mathbf{Xb}$ lies in the three-dimensional Ω space, and $\hat{\mathbf{Y}}_H = \mathbf{Xb}_H$ is in the base plane ω. The lines from these two points back to **Y** (not seen) are orthogonal to their associated respective spaces,

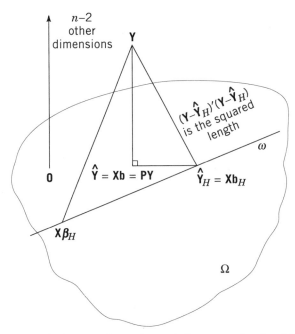

Figure 21.2. Case $n \geq 3$, $p = 2$, $q = 1$. Geometry related to $\mathbf{A}\boldsymbol{\beta} = \mathbf{c}$.

although this cannot be visualized directly in the figure. Again, the line joining $\hat{\mathbf{Y}}$ and $\hat{\mathbf{Y}}_H$ (which is in $\Omega - \omega$) is orthogonal to ω. The two representations of SS(H_0) apply as before.

21.4. THE *F*-TEST FOR H_0, GEOMETRICALLY

The *F*-test (see Section 9.1) is carried out on the ratio

$$F = \{SS(H_0)/q\}/s^2, \tag{21.4.1}$$

where s^2 is the residual from the full model. The appropriate degrees of freedom are

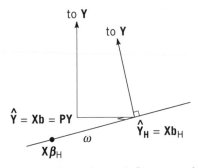

Figure 21.3. Case $n > 3$, $p = 3$, $q = 1$. Geometry related to $\mathbf{A}\boldsymbol{\beta} = \mathbf{c}$.

$\{q, n - p\}$. Figure 21.4 is a simplified version of Figure 21.3, and the letters *A*, *B*, *C*, *D*, *E*, and *F* in Figure 21.4*a* denote lengths. The dimensions in which these vectors lie, in the general case, are shown in Figure 21.4*b*. Thus

$$F = \{C^2/q\}/\{B^2/(n - p)\} \tag{21.4.2}$$

is a per degree of freedom comparison of the squared lengths C^2 and B^2. The hypothesis $\mathbf{A}\boldsymbol{\beta} = \mathbf{c}$ would *not* be rejected if *F* were small, and would be rejected if the ratio were large.

All the routine *F*-tests in regression can be set up via the Ω, ω framework. For example:

1. In the pure error/lack of fit test of Section 20.7, the roles of $\hat{\mathbf{Y}}$ and $\hat{\mathbf{Y}}_H$ in Figure 21.4*a* are played by $\tilde{\mathbf{Y}}$ and $\hat{\mathbf{Y}}$. See Figures 20.14, 20.15, and 21.4*c*.

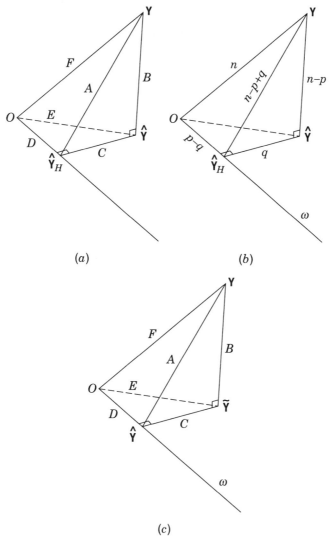

Figure 21.4. Geometry of the *F*-test for $H_0: \mathbf{A}\boldsymbol{\beta} = \mathbf{c}$: (*a*) lengths of vectors whose squares provide the sums of squares; (*b*) the dimensions (degrees of freedom of the spaces) in which those vectors lie, in the general case; and (*c*) an amended version of (*a*) appropriate for the pure error test of Section 20.7.

2. In the test for overall regression, $H_0: \beta_1 = \beta_2 = \cdots = \beta_{p-1} = 0$, $\hat{\mathbf{Y}}_H$ is given by $\overline{Y}\mathbf{1}$, and $q = p - 1$ in Figure 21.4b. See Section 21.5 and Figure 21.5.

21.5. THE GEOMETRY OF R^2

Figure 21.5 has the same general appearance as Figure 21.4. It is in fact a special case where the initial model is

$$Y = \beta_0 + \beta_1 X_1 + \cdots \beta_{p-1}X_{p-1} + \epsilon \tag{21.5.1}$$

and the hypothesis to be tested is that of "no regression," interpreted as $H_0: \beta_1 = \beta_2 = \cdots = \beta_{p-1} = 0$. Thus our restriction $\mathbf{A}\boldsymbol{\beta} = \mathbf{c}$ becomes

$$[\mathbf{0}, \mathbf{I}_{p-1}]\boldsymbol{\beta} = \mathbf{0}. \tag{21.5.2}$$

The reduced model is just $Y = \beta_0 + \epsilon$, or $\mathbf{Y} = \mathbf{1}\beta_0 + \epsilon$, so that ω is defined by the n by 1 vector $\mathbf{1}$. Thus $\hat{\mathbf{Y}}_H = \overline{Y}\mathbf{1}$. The R^2 statistic is defined as

$$R^2 = \frac{\Sigma(\hat{Y}_i - \overline{Y})^2}{\Sigma(Y_i - \overline{Y})^2} = \frac{(\hat{\mathbf{Y}} - \overline{Y}\mathbf{1})'(\hat{\mathbf{Y}} - \overline{Y}\mathbf{1})}{(\mathbf{Y} - \overline{Y}\mathbf{1})'(\mathbf{Y} - \overline{Y}\mathbf{1})}$$

$$= \frac{G^2}{K^2}$$

in Figure 21.5. Special cases are $R^2 = 1$, which results when $B = 0$ (zero residual vector), and $R^2 = 0$, occurring when $G = 0$, that is, when $\hat{\mathbf{Y}} = \overline{Y}\mathbf{1}$ and there is no regression in excess of $\hat{Y}_i = \overline{Y}$.

21.6. CHANGE IN R^2 FOR MODELS NESTED VIA $\mathbf{A}\boldsymbol{\beta} = 0$, NOT INVOLVING β_0

Figure 21.6 shows $\hat{\mathbf{Y}}$, and $\hat{\mathbf{Y}}_H$ developed through imposing the full rank hypothesis $\mathbf{A}\boldsymbol{\beta} = \mathbf{0}$, where $\mathbf{A}$ is q by p and does not involve β_0. We have

$$R^2 = G^2/K^2 \quad \text{and} \quad R_H^2 = H^2/K^2. \tag{21.6.1}$$

So

$$R^2 - R_H^2 = (G^2 - H^2)/K^2 = C^2/K^2, \tag{21.6.2}$$

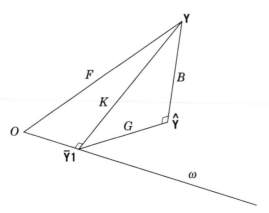

Figure 21.5. The geometry of $R^2 = G^2/K^2$.

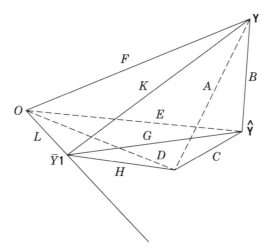

Figure 21.6. The geometry of changes in R^2 involving models nested via $\mathbf{A}\boldsymbol{\beta} = \mathbf{0}$, not involving β_0.

because the lines with lengths H and C lie in orthogonal spaces spanned by P_ω and $P_\Omega - P_\omega$. Now

$$C^2 = A^2 - B^2 \tag{21.6.3}$$

is the sum of squares due to the hypothesis $\mathbf{A}\boldsymbol{\beta} = \mathbf{0}$ of this section, and is tested via the F-statistic

$$F = \frac{C^2/q}{B^2/(n-p)} = \frac{n-p}{q} \cdot \frac{C^2}{B^2} \tag{21.6.4}$$

and

$$B^2 = K^2 - G^2 = K^2(1 - R^2). \tag{21.6.5}$$

Thus, from (21.6.2), (21.6.4), and (21.6.5),

$$F = \frac{n-p}{q} \cdot \frac{R^2 - R_H^2}{1 - R^2}. \tag{21.6.6}$$

This shows how the F-statistic for testing such an $\mathbf{A}\boldsymbol{\beta} = \mathbf{0}$ is related to the difference in the R^2 statistics. A special case is when the points $\bar{Y}\mathbf{1}$ and $\hat{\mathbf{Y}}_H$ coincide as in Section 21.5, where $\mathbf{A} = (\mathbf{0}, \mathbf{I}_{p-1})$. We then have $R_H^2 = 0$ so that now

$$F = \frac{n-p}{q} \cdot \frac{R^2}{1 - R^2} \tag{21.6.7}$$

linking the F for testing $H_0: \beta_1 = \beta_2 = \cdots = \beta_{p-1} = 0$ with R^2 for the full model. If we rewrite this as

$$R^2 = \frac{qF/(n-p)}{\{qF/(n-p)\} + 1},$$

we revert to Eq. (5.3.3) with $\nu_1 = q = p - 1$ and $\nu_2 = n - p$.

21.7. MULTIPLE REGRESSION WITH TWO PREDICTOR VARIABLES AS A SEQUENCE OF STRAIGHT LINE REGRESSIONS

The stepwise selection procedure discussed in Chapter 15 involves the addition of one variable at a time to an existing equation. In this section we discuss, algebraically and geometrically, how a composite equation can be built up through a series of simple straight line regressions. Although this is not the best practical way of obtaining the final equation, it is instructive to consider how it is done. We illustrate using the steam data with the two variables X_8 and X_6. The equation obtained from the joint regression is given in Section 6.2 as

$$\hat{Y} = 9.1266 - 0.0724X_8 + 0.2029X_6.$$

Another way of obtaining this solution is as follows:

1. Regress Y on X_8. This straight line regression was performed in Chapter 1, and the resulting equation was

$$\hat{Y} = 13.6230 - 0.0798X_8.$$

This fitted equation predicts 71.44% of the variation about the mean. Adding a new variable, say, X_6 (the number of operating days), to the prediction equation might improve the prediction significantly.

In order to accomplish this, we desire to relate the number of operating days to the amount of unexplained variation in the data after the atmospheric temperature effect has been removed. However, if the atmospheric temperature variations are in any way related to the variability shown in the number of operating days, we must correct for this first. Thus we need to determine the relationship between the unexplained variation in the amount of steam used after the effect of atmospheric temperature has been removed, and the remaining variation in the number of operating days after the effect of atmospheric temperature has been removed from it.

2. Regress X_6 on X_8; calculate residuals $X_{6i} - \hat{X}_{6i}$, $i = 1, 2, \ldots, n$. A plot of X_6 against X_8 is shown in Figure 21.7. The fitted equation is

$$\hat{X}_6 = 22.1685 - 0.0367X_8.$$

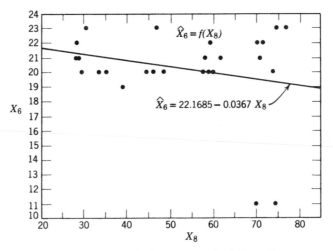

Figure 21.7. The least squares fit of X_6 on X_8.

TABLE 21.1. Residuals: $X_{6i} - \hat{X}_{6i}$

Observation Number i	X_{6i}	$\hat{X}_{6i}$	$X_{6i} - \hat{X}_{6i}$	Observation Number i	X_{6i}	$\hat{X}_{6i}$	$X_{6i} - \hat{X}_{6i}$
1	20	20.87	−0.87	14	19	20.73	−1.73
2	20	21.08	−1.08	15	23	20.45	2.55
3	23	21.04	1.96	16	20	20.39	−0.39
4	20	20.01	−0.01	17	22	19.99	2.01
5	21	19.92	1.08	18	22	19.60	2.40
6	22	19.55	2.45	19	11	19.60	−8.60
7	11	19.44	−8.44	20	23	19.44	3.56
8	23	19.36	3.64	21	20	19.53	0.47
9	21	19.58	1.42	22	21	20.04	0.96
10	20	20.06	−0.06	23	20	20.53	−0.53
11	20	20.47	−0.47	24	20	20.94	−0.94
12	21	21.11	−0.11	25	22	21.12	0.88
13	21	21.14	−0.14				

Fitted values and residuals are shown in Table 21.1. We note that there are two residuals −8.44 and −8.60 that have absolute values considerably greater than the other residuals. They arise from months in which the number of operating days was unusually small, 11 in each case. We can, of course, take the attitude that these are "outliers" and that months with so few operating days should not even be considered in the analysis. However, if we wish to obtain a satisfactory prediction equation that will be valid for *all* months, irrespective of the number of operating days, then it is important to take account of these particular results and develop an equation that makes use of the information they contain. As can be seen from the data and from Figure 21.7 and Table 21.2, if these particular months were ignored, the apparent effect of the number of operating days on the response would be small. This would *not* be because the variable did not affect the response but because the variation actually observed in the variable was so slight that the variable could not exert any appreciable effect on the response. If a variable appears to have a significant effect on the response in one analysis but not in a second, it may well be that it varied over

TABLE 21.2. Deviations of $\hat{Y}_i = f(X_8)$ and $\hat{X}_{6i} = f(X_8)$ from Y_i and X_{6i}, Respectively

Observation Number i	$Y_i - \hat{Y}_i$	$X_{6i} - \hat{X}_{6i}$	Observation Number i	$Y_i - \hat{Y}_i$	$X_{6i} - \hat{X}_{6i}$
1	0.17	−0.87	14	−0.93	−1.73
2	−0.12	−1.08	15	1.05	2.55
3	1.34	1.96	16	−0.17	−0.39
4	−0.53	−0.01	17	1.20	2.01
5	0.55	1.08	18	0.08	2.40
6	0.80	2.45	19	−1.20	−8.60
7	−1.32	−8.44	20	1.20	3.56
8	1.00	3.64	21	−0.19	0.47
9	−0.16	1.42	22	−0.51	0.96
10	0.11	−0.06	23	−1.20	−0.53
11	−1.68	−0.47	24	−0.60	−0.94
12	0.87	−0.11	25	−0.26	0.88
13	0.50	−0.14			

a wider range in the first set of data than in the second. This, incidentally, is one of the drawbacks of using plant data "as it comes." Quite often the normal operating range of a variable is so slight that no effect on response is revealed, even when the variable does, over larger ranges of operation, have an appreciable effect. Thus designed experiments, which assign levels wider than normal operating ranges, often reveal effects that had not been noticed previously.

  3. We now regress $Y - \hat{Y}$ against $X_6 - \hat{X}_6$ by fitting the model

$$(Y_i - \hat{Y}_i) = \beta(X_{6i} - \hat{X}_{6i}) + \epsilon_i.$$

No "β_0" term is required in this first-order model since we are using two sets of residuals whose sums are zero, and thus the line must pass through the origin. (If we did put a β_0 term in, we should find $b_0 = 0$, in any case.) For convenience the two sets of residuals used as data are extracted from Tables 1.2 and 21.1 and are given in Table 21.2. A plot of these residuals is shown in Figure 21.8. The fitted equation takes the form

$$(\widehat{Y - \hat{Y}}) = 0.2015(X_6 - \hat{X}_6).$$

Within the parentheses we can substitute for $\hat{Y}$ and $\hat{X}_6$ as functions of X_8, and the large caret on the left-hand side can then be attached to Y to represent the overall fitted value $\hat{Y} = \hat{Y}(X_6, X_8)$ as follows:

$$[\hat{Y} - (13.6230 - 0.0798X_8)] = 0.2015[X_6 - (22.1685 - 0.0367X_8)]$$

or

$$\hat{Y} = 9.1560 - 0.0724X_8 + 0.2015X_6.$$

The previous result was

$$\hat{Y} = 9.1266 - 0.0724X_8 + 0.2029X_6.$$

In theory these two results are identical; practically, as we can see, discrepancies have occurred due to rounding errors. Ignoring rounding errors for the moment, we shall

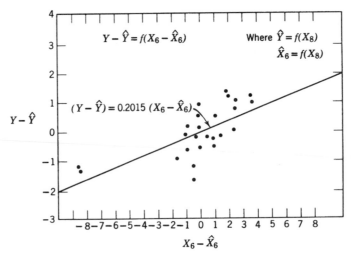

Figure 21.8. A plot of the residuals in Table 21.2.

now show, geometrically, through a simple example, why the two methods should provide us with identical results.

Geometrical Interpretation

Consider an example in which we have $n = 3$ observations of the response Y, namely, Y_1, Y_2, and Y_3 taken at the three sets of conditions: (X_1, Z_1), (X_2, Z_2), (X_3, Z_3). We can plot in three dimensions on axes labeled 1, 2, and 3, with origin at O, the points $Y \equiv (Y_1, Y_2, Y_3)$, $X \equiv (X_1, X_2, X_3)$, and $Z \equiv (Z_1, Z_2, Z_3)$. The geometrical interpretation of regression is as follows. To regress Y on X we drop a perpendicular YP onto OX. The coordinates of the point P are the fitted values $\hat{Y}_1$, $\hat{Y}_2$, $\hat{Y}_3$. The length OP^2 is the sum of squares due to the regression, OY^2 is the total sum of squares, and YP^2 is the residual sum of squares. By Pythagoras, $OP^2 + YP^2 = OY^2$, which provides the analysis of variance breakup of the sums of squares (see Figure 21.9).

If we complete the parallelogram, which has OY as diagonal and OP and PY as sides, we obtain the parallelogram $OP'YP$ as shown. Then the coordinates of P' are the values of the residuals from the regression of variable Y on variable X. In vector terms we could write

$$\overrightarrow{OP} + \overrightarrow{OP'} = \overrightarrow{OY},$$

or, in "statistical" vector notation,

$$\hat{\mathbf{Y}} + (\mathbf{Y} - \hat{\mathbf{Y}}) = \mathbf{Y}.$$

This result is true in general for n dimensions. (The only reason we take $n = 3$ is so we can provide a diagram.)

Suppose we wish to regress variable Y on variables X and Z simultaneously. The lines OX and OZ define a plane in three dimensions. We drop a perpendicular YT onto this plane. Then the coordinates of the point T are the fitted values $\hat{Y}_1$, $\hat{Y}_2$, $\hat{Y}_3$ for *this* regression. OT^2 is the regression sum of squares, YT^2 is the residual sum of squares, and OY^2 is the total sum of squares. Again, by Pythagoras, $OY^2 = OT^2 + YT^2$, which, again, gives the sum of squares breakup we see in the analysis of variance table. Completion of the parallelogram $OT'YT$ with diagonal OY and sides OT and TY provides OT', the vector of residuals of this regression, and the coordinates of T' give the residuals $\{(Y_1 - \hat{Y}_1), (Y_2 - \hat{Y}_2), (Y_3 - \hat{Y}_3)\}$ of the regression of Y on X and Z simultaneously. Again, in vector notation,

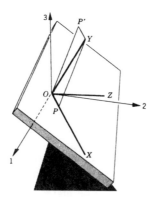

Figure 21.9. Geometrical interpretation of the regression of Y on X.

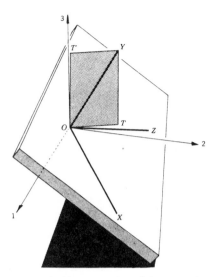

Figure 21.10. Geometrical interpretation of the regression of Y on X and Z.

$$\overrightarrow{OT} + \overrightarrow{OT'} = \overrightarrow{OY}$$

or, in "statistical" vector notation,

$$\hat{\mathbf{Y}} + (\mathbf{Y} - \hat{\mathbf{Y}}) = \mathbf{Y}$$

for this regression (see Figure 21.10).

As we saw in the numerical example above, the same final residuals should arise (ignoring rounding) if we do the regressions (1) Y on X, and (2) Z on X, and then regress the residuals of (1) on the residuals of (2). That this is true can be seen geometrically as follows. Figure 21.11 shows three parallelograms in three-dimensional space.

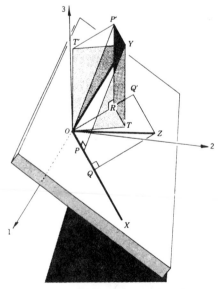

Figure 21.11. The regression of Y on X and Z can also be viewed as a two-step procedure as described in the text.

1. $OP'YP$ from the regression of Y on X.
2. $OQ'ZQ$ from the regression of Z on X.
3. $OT'YT$ from the regression of Y on X and Z simultaneously.

Now the regression of the residuals of (1) onto the residuals of (2) is achieved by dropping the perpendicular from P' onto OQ'. Suppose the point of impact is R. Then a line through O parallel to RP' and of length RP' will be the residual vector of the two-step regression of Y on X and Z. However, the points O, Q', Z, P, Q, X, and T all lie in the plane π defined by OZ and OX. Thus so does the point R. Since $OP'YP$ is a parallelogram, and $P'R$ and YT are perpendicular to plane π, $P'R = YT$ in length. Since $TY = OT'$, it follows that $OT' = RP'$. But OT', RP', and TY are all parallel and perpendicular to plane π. Hence $OT'P'R$ is a parallelogram from which it follows that $\overrightarrow{OT'}$ is the vector of residuals from the two-step regression. Since it originally resulted from the regression of Y on Z and X together, the two methods must be equivalent. Thus we can see that the planar regression of Y on X and Z together can be regarded as the totality of successive straight line regressions of:

1. Y on X,
2. Z on X, and
3. Residuals of (1) on the residuals of (2).

The same result is obtained if the roles of Z and X are interchanged. All linear regressions can be broken down into a series of simple regressions in this way.

EXERCISES FOR CHAPTER 21

A. Suppose

$$\mathbf{X} = (\mathbf{X}_1, \mathbf{X}_2) = \begin{array}{ccc} 1 & X_1 & X_2 \\ \begin{bmatrix} 1 & -3 & 1 \\ 1 & -1 & -1 \\ 1 & 1 & -1 \\ 1 & 3 & 1 \end{bmatrix} \end{array}$$

Let ω be the space defined by the $(1, X_1)$ columns and Ω be the space defined by the $(1, X_1, X_2)$ columns.
1. Evaluate $\mathbf{P}_\Omega$, $\mathbf{P}_\omega$ and their difference. Give the dimensions of the spaces spanned by their columns.
2. Show, through your example data, that the general result

$$\mathbf{P}_\Omega - \mathbf{P}_\omega = \mathbf{P}_{\omega^\perp \cap \Omega},$$

where $\omega^\perp$ is the complement of ω with respect to the full four-dimensional space E_4, is true.
3. Give a basis for (i.e., a set of vectors that span) $R(\mathbf{P}_{\omega^\perp \cap \Omega})$.
4. What are the eigenvalues of $\mathbf{P}_\Omega - \mathbf{P}_\omega$? (Write them down without detailed calculations if you wish, but explain how you did this.) What theorem does your answer confirm?
B. Use Eq. (21.2.9) to fit the model $Y = \beta_0 + \beta_1 X + \beta_2 X^2 + \epsilon$, subject to $\beta_0 - 2\beta_1 = 4$, to the seven data points:

X	Y
−1	15, 18
0	18, 21, 22
1	28, 30

C. In *The American Statistician*, Volume 45, No. 4, November 1991, pp. 300–301, A. K. Shah notes, in a numerical example based on Exercise 15H with $Y = X_6$ and $X = X_5$, that if he regresses Y versus X, he gets $R^2 = 0.0782$, while if he regresses $Y - X$ versus X, he gets $R^2 = 0.6995$. He also notes that he gets the same residual SS in both cases and that the fitted relationship between Y and X is exactly the same, namely, $\hat{Y} = 4.6884 - 0.2360X$.

Look at this geometrically and draw diagrams showing how this sort of thing can happen. Show also that we can have a "reverse" case, where R^2 will decrease, not increase.

D. Show (via the method of Lagrange's undetermined multipliers) that Eq. (21.2.9) can also be obtained by minimizing the sum of squares function $(\mathbf{Y} - \mathbf{X}\boldsymbol{\beta})'(\mathbf{Y} - \mathbf{X}\boldsymbol{\beta})$ subject to $\mathbf{A}\boldsymbol{\beta} = \mathbf{c}$.

E. See (21.2.10). Confirm that the projection matrices $\mathbf{P}$, $\mathbf{P}_1 = \mathbf{P}_{\Omega-\omega}$, and $\mathbf{P}_\omega$ are all symmetric and idempotent. Also confirm that $\mathbf{P}_\omega$ and $\mathbf{P}_1$ are orthogonal.

F. Show, for Section 21.5, when $(\mathbf{0}, \mathbf{I}_{p-1})\boldsymbol{\beta} = \mathbf{0}$ determines ω, that ω is spanned by $\mathbf{1}$. Do this via the (unnecessarily complicated for this example) method of Eqs. (21.2.4) to (21.2.6), and using the formula for the partitioned inverse of $\mathbf{X}'\mathbf{X}$, when $\mathbf{X} = [\mathbf{1}, \mathbf{X}_1]$, where here $\mathbf{X}_1$ is n by $(p - 1)$. Refer to Appendix 5A.

G. (*Sources:* The data below are quoted from p. 165 of *Linear Models* by S. R. Searle, published in 1971 by John Wiley & Sons. Searle's source is a larger set of data given by W. T. Federer in *Experimental Design*, published in 1955 by MacMillan, p. 92. The "Searle data" also feature in these papers: "A matrix identity and its applications to equivalent hypotheses in linear models," by N. N. Chan and K.-H. Li, *Communications in Statistics, Theory and Methods*, **24**, 1995, 2769–2777; and "Nontestable hypotheses in linear models," by S. R. Searle, W. H. Swallow, and C. E. McCulloch, *SIAM Journal of Algebraic and Discrete Methods*, **5**, 1984, 486–496.) Consider the model $\mathbf{Y} = \mathbf{X}\boldsymbol{\beta} + \boldsymbol{\epsilon}$, where

$$\mathbf{X} = \begin{bmatrix} 1 & 1 & 0 & 0 \\ 1 & 1 & 0 & 0 \\ 1 & 1 & 0 & 0 \\ 1 & 0 & 1 & 0 \\ 1 & 0 & 1 & 0 \\ 1 & 0 & 0 & 1 \end{bmatrix}, \quad \boldsymbol{\beta} = \begin{bmatrix} \beta_0 \\ \beta_1 \\ \beta_2 \\ \beta_3 \end{bmatrix}, \quad \mathbf{Y} = \begin{bmatrix} 101 \\ 105 \\ 94 \\ 84 \\ 88 \\ 32 \end{bmatrix}.$$

It is desired to test the hypothesis $H_0 : \beta_1 = 7$, $\beta_2 = 4$, versus the alternative "not so." Is this possible? If not, what *is* it possible to test?

H. Suppose we have five observations of Y at five coded X-values, $-2, -1, 0, 1, 2$. Consider the problem of fitting the quadratic equation $Y = \beta_0 + \beta_1 X + \beta_2 X^2 + \epsilon$ and testing the null hypothesis $H_0 : \beta_2 = 0$ versus $H_1 : \beta_2 \neq 0$. The null hypothesis divides the estimation space Ω (whose projection matrix is $\mathbf{P}$) into the subspaces: ω (defined by H_0 and with projection matrix $\mathbf{P}_\omega$) and $\Omega - \omega$ (with projection matrix $\mathbf{P}_1$).

1. Find $\mathbf{P}$.

2. Find $\mathbf{P}_\omega$.

3. Find $\mathbf{P}_1$.

4. Confirm that equations (21.2.10) are true for this example.

5. Evaluate $\mathbf{P}_1\mathbf{x}$, where $\mathbf{x} = (-2, -1, 0, 1, 2)'$, and then explain why your answer is obvious.

CHAPTER 22

Orthogonal Polynomials and Summary Data

22.1. INTRODUCTION

In this chapter we bring together two topics with specialized usages in regression situations.

1. Orthogonal polynomials were much used in precomputer days to fit polynomial models of any order to a single predictor X, mostly to equally spaced predictor values. This was achieved by replacing the columns of the $\mathbf{X}$ matrix that would arise from the predictors $(1, X, X^2, X^3, \ldots)$ by alternative columns that were orthogonal and generated by special polynomials of (zero, first, second, third, ...) order in X. The maximum possible order is, of course, $n - 1$ if n observations are available. Such polynomials could be worked out once and for all. The best-known published tables are by Fisher and Yates (1964) and Pearson and Hartley (1958). Once the orthogonal polynomial fit has been made, the original polynomial form can then be reconstructed, if desired.

Even in the computer age, orthogonal polynomials could be useful, particularly with computer programs that have troublesome round-off errors. These polynomials are also useful more generally, however. They can be used in any situation where orthogonal columns with integer values are needed, for example, to span an estimation space or an error space, to split a sum of squares up into orthogonal components, or to obtain a set of orthogonal dummy variable columns.

Orthogonal polynomials are discussed in Section 22.2.

2. Summary data are data in which the Y's from individual repeat runs are not available because the data have been reduced to sample averages and sample variance estimates in writing a paper or report. It is possible to carry out a large portion of the regression calculations nevertheless. This is discussed in Section 22.3.

22.2. ORTHOGONAL POLYNOMIALS

Orthogonal polynomials are used to fit a polynomial model of any order in one variable. The idea is as follows. Suppose we have n observations (X_i, Y_i), $i = 1, 2, \ldots, n$, where X is a predictor variable and Y is a response variable, and we wish to fit the model

$$Y = \beta_0 + \beta_1 X + \beta_2 X^2 + \cdots + \beta_p X^p + \epsilon. \tag{22.2.1}$$

In general, the columns of the resulting $\mathbf{X}$ matrix will not be orthogonal. If later we wish to add another term $\beta_{p+1}X^{p+1}$, changes will usually occur in the estimates of all the other coefficients. However, if we can construct polynomials of the form

$$\psi_0(X_i) = 1 \qquad\qquad\qquad \text{zero-order polynomial}$$

$$\psi_1(X_i) = P_1X_i + Q_1 \qquad\qquad \text{first-order}$$

$$\psi_2(X_i) = P_2X_i^2 + Q_2X_i + R_2 \qquad \text{second-order}$$

$$\vdots$$

$$\psi_r(X_i) = P_rX_i^r + Q_rX_i^{r-1} + \cdots + T_r \quad \text{rth order}$$

$$\vdots$$

with the property that they are *orthogonal polynomials*, that is,

$$\sum_{i=1}^{n} \psi_j(X_i)\psi_l(X_i) = 0 \qquad (j \neq l), \tag{22.2.2}$$

for all $j, l < n - 1$, we can rewrite the model as

$$Y = \alpha_0\psi_0(X) + \alpha_1\psi_1(X) + \cdots + \alpha_p\psi_p(X) + \epsilon. \tag{22.2.3}$$

In this case we shall have

$$\mathbf{X} = \begin{bmatrix} 1 & \psi_1(X_1) & \psi_2(X_1) & \cdots & \psi_p(X_1) \\ 1 & \psi_1(X_2) & \psi_2(X_2) & \cdots & \psi_p(X_2) \\ \vdots & \vdots & \vdots & & \vdots \\ 1 & \psi_1(X_n) & \psi_2(X_n) & \cdots & \psi_p(X_n) \end{bmatrix} \tag{22.2.4}$$

so that

$$\mathbf{X'X} = \begin{bmatrix} A_{00} & & & & \mathbf{0} \\ & A_{11} & & & \\ & & A_{22} & & \\ & & & \ddots & \\ \mathbf{0} & & & & A_{pp} \end{bmatrix}, \tag{22.2.5}$$

where $A_{jj} = \sum_{i=1}^{n}\{\psi_j(X_i)\}^2$ since all off-diagonal terms vanish, by Eq. (22.2.2). Since the inverse matrix $(\mathbf{X'X})^{-1}$ is also diagonal and is obtained by inverting each element separately, the least squares procedure provides, as an estimate of α_j, the quantity

$$a_j = \frac{\sum_{i=1}^{n} Y_i\psi_j(X_i)}{\sum_{i=1}^{n} [\psi_j(X_i)]^2}, \qquad j = 0, 1, 2, \ldots, p,$$

$$= \frac{A_{jY}}{A_{jj}} \tag{22.2.6}$$

with an obvious notation. Since $V(\mathbf{b}) = (\mathbf{X'X})^{-1}\sigma^2$ for the general regression model, it follows that the variance of a_j is

$$V(a_j) = \frac{\sigma^2}{A_{jj}}, \tag{22.2.7}$$

where σ^2 is usually estimated from the analysis of variance table. To obtain the entries in the table we evaluate the sum of squares due to a_j from

$$SS(a_j) = \frac{A_{jY}^2}{A_{jj}} \tag{22.2.8}$$

and so obtain an analysis of variance table as in Table 22.1. If the model is correct, s^2 estimates σ^2. Usually we pool, with the residual, sums of squares whose mean squares are not significantly larger than s^2 to obtain an estimate of σ^2 based on more degrees of freedom. Note that if it were desired to add another term $\alpha_{p+1}\psi_{p+1}(X)$ to Eq. (22.2.3) no recomputation would be necessary for the coefficients already obtained, due to the orthogonality of the polynomials. Thus higher and higher order polynomials can be fitted with ease and the process terminated when a suitably fitting equation is found.

The $\psi_j(X_i)$ can be constructed and the procedure above can be carried out for any values of the X_i. However, when the X_i are *not* equally spaced, polynomials must be specially constructed. See, for example, Wishart and Metakides (1953) and Robson (1959). If the X_i are equally spaced, published tables can be employed. The actual numerical values of $\psi_j(X_j)$ and A_{jj} as well as the general functional forms of $\psi_j(X)$, for $j = 1, 2, \ldots, 6$ and for $n \leq 52$, are given in Pearson and Hartley (1958). This means that provided $p \leq 6$ we have the elements of the **X** and **X'X** matrices. Similar values for $j \leq 5$ and $n \leq 75$ appear in Fisher and Yates (1964). When $p > 6$, more extensive tables such as De Lury (1960) are needed. See also Wilkie (1965). A short table of orthogonal polynomials for equally spaced data for values of n as high as 12 is given later in this section.

Although orthogonal polynomials are often recommended only when pocket calculators are used, Bright and Dawkins (1965) concluded that even when a computer is available, and especially when the polynomial is of high order, orthogonal polynomials are worthwhile. Using them provides greater computing accuracy and reduced computing times and can avoid rejection of columns of powers of X that are highly correlated with other columns of powers of X.

We illustrate the application of orthogonal polynomials by an example.

Example

The Archer Daniels Midland Company reported net income per share for the years 1986–1993 as follows:

T A B L E 22.1. Analysis of Variance Table

Source	df	SS	MS
a_0 (mean)	1	$SS(a_0)$	—
a_1	1	$SS(a_1)$	$SS(a_1)$
a_2	1	$SS(a_2)$	$SS(a_2)$
$\vdots$	$\vdots$	$\vdots$	$\vdots$
a_p	1	$SS(a_p)$	$SS(a_p)$
Residual	$n - p - 1$	By subtraction	s^2
Total	n	$\sum_{i=1}^{n} Y_i^2$	

Year (Z_i)	1986	1987	1988	1989	1990	1991	1992	1993
Income per share ($\$Y_i$)	0.70	0.77	1.02	1.24	1.41	1.35	1.47	1.66

Fit a polynomial of suitable order that will provide a satisfactory approximating function for these data.

Solution. We ignore the year values Z_i for the moment except to note that they are equally spaced. From a table of orthogonal polynomials we find values $\psi_j(X_i)$ corresponding to $n = 8$ observations as shown in Table 22.2.

Note that $A_{00} = n = 8$. We consider the model

$$Y = \sum_{j=0}^{6} \alpha_j \psi_j(X).$$

Using Eq. (22.2.6) we obtain

$$a_0 = 1.2025, \qquad a_1 = 0.067738, \qquad a_2 = -0.009524,$$

$$a_3 = 0.001515, \qquad a_4 = 0.006721, \qquad a_5 = -0.000559,$$

$$a_6 = -0.002879.$$

The fitted model (with $R^2 = 0.997$) is thus

$$\hat{Y} = \sum_{j=0}^{6} a_j \psi_j(X),$$

where the a_j are as shown and the $\psi_j(X)$ are found in tables. (We shall do this in a moment, after reducing the equation.) The analysis of variance table is shown in Table 22.3. All the sums of squares for $a_1, a_2, \ldots, a_6$ are orthogonal and their values do not depend on the order of entry of the polynomial terms.

Note. When the order of the fitted polynomial is $p = n - 1$, the maximum order, the model fits the data exactly and no residual occurs. Here $p = n - 2$ and so the residual is SS(a_7) in fact.

If an external estimate of σ^2 were available we could compare all the mean squares with it. Alternatively, we note that the fourth-, fifth-, and sixth-order sums of squares are very small and the corresponding terms can be dropped immediately. It is mechanically tempting to declare the fourth-order term significant by performing the F-test with (1, 3) degrees of freedom $F = \{0.028/1\}\{(0.001 + 0.002 + 0.003)/3\} = 14$, which exceeds $F(1, 3, 0.95) = 10.13$. Fitting a fourth-order polynomial to eight data points

T A B L E 22.2. The Calculation Table for the Example

Y	ψ_0	ψ_1	ψ_2	ψ_3	ψ_4	ψ_5	ψ_6
0.70	1	−7	7	−7	7	−7	1
0.77	1	−5	1	5	−13	23	−5
1.02	1	−3	−3	7	−3	−17	9
1.24	1	−1	−5	3	9	−15	−5
1.41	1	1	−5	−3	9	15	−5
1.35	1	3	−3	−7	−3	17	9
1.47	1	5	1	−5	−13	−23	−5
1.66	1	7	7	7	7	7	1
A_{jj}	8	168	168	264	616	2184	264

T A B L E 22.3. ANOVA Table for the Example

Source	df	SS	MS
a_1	1	0.771	(same as SS)
a_2	1	0.015	
a_3	1	0.001	
a_4	1	0.028	
a_5	1	0.001	
a_6	1	0.002	
Residual	1	0.003	
Total, corrected	7		
		0.820	

is not very sensible in general, however, and once we get past this point, it seems sensible to reduce further. The straight line model $Y = 1.2025 + 0.0067738\psi_1(X)$ explains $R^2 = 0.940$ of the variation about the mean and the residual mean square from this model is $s^2 = 0.00818$. A plot of the data and fitted line shows this model to be a reasonable fit.

In order to obtain the fitted equation in terms of the original variables we have first to substitute for the ψ_j and relate these to the Z's. From a table of orthogonal polynomials with $n = 8$ (e.g., the information provided later in this section),

$$\psi_0(X) = 1, \qquad \psi_1(X) = 2X, \qquad \psi_2(X) = X^2 - \frac{21}{4},$$

$$\psi_3(X) = \frac{2}{3}\left[X^3 - \frac{37}{4}X\right], \qquad \text{and so on.}$$

Z and X correspond as follows:

$\psi_1(X_i) = 2X_i$:	-7	-5	-3	-1	1	3	5	7
X_i:	$-\frac{7}{2}$	$-\frac{5}{2}$	$-\frac{3}{2}$	$-\frac{1}{2}$	$\frac{1}{2}$	$\frac{3}{2}$	$\frac{5}{2}$	$\frac{7}{2}$
Z_i:	1986	1987	1988	1989	1990	1991	1992	1993

Clearly the required coding is

$$X = Z - 1989\tfrac{1}{2}.$$

The fitted polynomial is thus

$$\hat{Y} = 1.2025 + 0.0067738(2)(Z - 1989\tfrac{1}{2})$$

or

$$\hat{Y} = 1.2025 + 0.0135476(Z - 1989\tfrac{1}{2}).$$

This can be rearranged as a straight line fit in Z but the above form is as convenient to substitute into to obtain fitted values and residuals, in this case. It is recommended that the reader plot the data and fitted values and also examine the residuals, to complete the analysis.

Orthogonal Polynomials for $n = 3, \ldots, 12$

The formulas for the orthogonal polynomials $\psi_j(X)$ up to order six for equally spaced X's and for any value of n are as follows:

$$\psi_0(X) = 1,$$

$$\psi_1(X) = \lambda_1 X,$$

$$\psi_2(X) = \lambda_2\{X^2 - \tfrac{1}{12}(n^2 - 1)\},$$

$$\psi_3(X) = \lambda_3\{X^3 - \tfrac{1}{20}(3n^2 - 7)X\},$$

$$\psi_4(X) = \lambda_4\{X^4 - \tfrac{1}{14}(3n^2 - 13)X^2 + \tfrac{3}{560}(n^2 - 1)(n^2 - 9)\},$$

$$\psi_5(X) = \lambda_5\{X^5 - \tfrac{5}{18}(n^2 - 7)X^3 + \tfrac{1}{1008}(15n^4 - 230n^2 + 407)X\},$$

$$\psi_6(X) = \lambda_6\{X^6 - \tfrac{5}{44}(3n^2 - 31)X^4 + \tfrac{1}{176}(5n^4 - 110n^2 + 329)X^2$$
$$- \tfrac{5}{14784}(n^2 - 1)(n^2 - 9)(n^2 - 25)\}.$$

(If $n \le 6$, we go only as far as $j = n - 1$.) The appropriate λ-values are given at the bottom of each column and are chosen to ensure that the values in the tables are all integers. Above the λ's are the sums of squares of the column entries, that is, the A_{jj} of Eq. (22.2.5). Note that although X does not appear as such in the tables, it is always true that $\psi_1 = \lambda_1 X$, where λ_1 is 1 for odd n and 2 for even n, so we can read off X as ψ_1/λ_1, when needed. All ψ_j columns are symmetric for even j, antisymmetric for odd j.

For a Fortran subroutine that generates orthogonal polynomials, see Cooper (1971).]

Table of Orthogonal Polynomial Coefficients

$n = 3$		$n = 4$			$n = 5$				$n = 6$				
ψ_1	ψ_2	ψ_1	ψ_2	ψ_3	ψ_1	ψ_2	ψ_3	ψ_4	ψ_1	ψ_2	ψ_3	ψ_4	ψ_5
-1	1	-3	1	-1	-2	2	-1	1	-5	5	-5	1	-1
0	-2	-1	-1	3	-1	-1	2	-4	-3	-1	7	-3	5
1	1	1	-1	-3	0	-2	0	6	-1	-4	4	2	-10
		3	1	1	1	-1	-2	-4	1	-4	-4	2	10
					2	2	1	1	3	-1	-7	-3	-5
									5	5	5	1	1
2	6	20	4	20	10	14	10	70	70	84	180	28	252
1	3	2	1	$\tfrac{10}{3}$	1	1	$\tfrac{5}{6}$	$\tfrac{35}{12}$	2	$\tfrac{3}{2}$	$\tfrac{5}{3}$	$\tfrac{7}{12}$	$\tfrac{21}{10}$

$n = 7$						$n = 8$					
ψ_1	ψ_2	ψ_3	ψ_4	ψ_5	ψ_6	ψ_1	ψ_2	ψ_3	ψ_4	ψ_5	ψ_6
-3	5	-1	3	-1	1	-7	7	-7	7	-7	1
-2	0	1	-7	4	-6	-5	1	5	-13	23	-5
-1	-3	1	1	-5	15	-3	-3	7	-3	-17	9
0	-4	0	6	0	-20	-1	-5	3	9	-15	-5
1	-3	-1	1	5	15	1	-5	-3	9	15	-5
2	0	-1	-7	-4	-6	3	-3	-7	-3	17	9
3	5	1	3	1	1	5	1	-5	-13	-23	-5
						7	7	7	7	7	1
28	84	6	154	84	924	168	168	264	616	2184	264
1	1	$\tfrac{1}{6}$	$\tfrac{7}{12}$	$\tfrac{7}{20}$	$\tfrac{77}{60}$	2	1	$\tfrac{2}{3}$	$\tfrac{7}{12}$	$\tfrac{7}{10}$	$\tfrac{11}{60}$

Table of Orthogonal Polynomial Coefficients (*Continued*)

		$n = 9$						$n = 10$			
ψ_1	ψ_2	ψ_3	ψ_4	ψ_5	ψ_6	ψ_1	ψ_2	ψ_3	ψ_4	ψ_5	ψ_6
-4	28	-14	14	-4	4	-9	6	-42	18	-6	3
-3	7	7	-21	11	-17	-7	2	14	-22	14	-11
-2	-8	13	-11	-4	22	-5	-1	35	-17	-1	10
-1	-17	9	9	-9	1	-3	-3	31	3	-11	6
0	-20	0	18	0	-20	-1	-4	12	18	-6	-8
1	-17	-9	9	9	1	1	-4	-12	18	6	-8
2	-8	-13	-11	4	22	3	-3	-31	3	11	6
3	7	-7	-21	-11	-17	5	-1	-35	-17	1	10
4	28	14	14	4	4	7	2	-14	-22	-14	-11
						9	6	42	18	6	3
60	2772	990	2002	468	1980	330	132	8580	2860	780	660
1	3	$\frac{5}{6}$	$\frac{7}{12}$	$\frac{3}{20}$	$\frac{11}{60}$	2	$\frac{1}{2}$	$\frac{5}{3}$	$\frac{5}{12}$	$\frac{1}{10}$	$\frac{11}{240}$

		$n = 11$						$n = 12$			
ψ_1	ψ_2	ψ_3	ψ_4	ψ_5	ψ_6	ψ_1	ψ_2	ψ_3	ψ_4	ψ_5	ψ_6
-5	15	-30	6	-3	15	-11	55	-33	33	-33	11
-4	6	6	-6	6	-48	-9	25	3	-27	57	-31
-3	-1	22	-6	1	29	-7	1	21	-33	21	11
-2	-6	23	-1	-4	36	-5	-17	25	-13	-29	25
-1	-9	14	4	-4	-12	-3	-29	19	12	-44	4
0	-10	0	6	0	-40	-1	-35	7	28	-20	-20
1	-9	-14	4	4	-12	1	-35	-7	28	20	-20
2	-6	-23	-1	4	36	3	-29	-19	12	44	4
3	-1	-22	-6	-1	29	5	-17	-25	-13	29	25
4	6	-6	-6	-6	-48	7	1	-21	-33	-21	11
5	15	30	6	3	15	9	25	-3	-27	-57	-31
						11	55	33	33	33	11
110	858	4290	286	156	11220	572	12012	5148	8008	15912	4488
1	1	$\frac{5}{6}$	$\frac{1}{12}$	$\frac{1}{40}$	$\frac{11}{120}$	2	3	$\frac{2}{3}$	$\frac{7}{24}$	$\frac{3}{20}$	$\frac{11}{360}$

22.3. REGRESSION ANALYSIS OF SUMMARY DATA

Suppose we have k sets of repeat observations $\{Y_{iu}, u = 1, 2, \ldots, n_i\}, i = 1, 2, \ldots, k$, but, because of condensation of the data, the individual repeats are not shown and only the k averages $\overline{Y}_i$ and the k sample variance estimates (of σ^2)

$$s_i^2 = \sum_{u=1}^{n_i} (Y_{iu} - \overline{Y}_i)^2/(n_i - 1)$$

are given. How do we proceed? *For purposes of obtaining the regression coefficients, we can simply work as if $Y_{iu} = \overline{Y}_i$,* that is, as if every set of repeat observations consisted of an equal number of observations all equal to the average value. This is clear if we think of what happens to the ith set of repeats in the matrix product $\mathbf{b} = (\mathbf{X'X})^{-1}\mathbf{X'Y}$. For the $\mathbf{X'Y}$ portion we have

$$\mathbf{X'Y} = \begin{bmatrix} \cdots & \cdots & 1 & 1 & \cdots & 1 & \cdots & \cdots \\ \cdots & \cdots & a & a & \cdots & a & \cdots & \cdots \\ \cdots & \cdots & b & b & \cdots & b & \cdots & \cdots \\ & & \vdots & & & & & \\ \cdots & \cdots & z & z & \cdots & z & \cdots & \cdots \end{bmatrix} \begin{bmatrix} \cdots \\ \\ \cdots \\ \\ Y_{i1} \\ Y_{i2} \\ \vdots \\ Y_{in_i} \\ \vdots \\ \cdots \end{bmatrix}.$$

All the a's, b's, and so on, are equal because of the repeat runs. Thus all the contributions of the ith set of repeat observations to the $\mathbf{X'Y}$ product take forms like,

$$aY_{i1} + aY_{i2} + \cdots + aY_{in_i} = a \sum_{u=1}^{n_i} Y_{iu}$$

$$= an_i \overline{Y}_i$$

$$= a\overline{Y}_i + a\overline{Y}_i + \cdots + a\overline{Y}_i$$

so that replacement of all Y_{iu} by $\overline{Y}_i$ will not alter the calculation for the vector $\mathbf{b}$.

Use of the $\overline{Y}_i$ instead of the individual Y_{iu} will, however, lead to the wrong total corrected sum of squares. For example, if $\overline{Y}$ is the overall mean of the observations, the contribution to the (wrong) corrected sum of squares from the ith set of observations will be

$$\sum_{u=1}^{n_i} (\overline{Y}_i - \overline{Y})^2 = n_i(\overline{Y}_i - \overline{Y})^2,$$

whereas we actually need

$$\sum_{u=1}^{n_i} (Y_{iu} - \overline{Y})^2.$$

However, we can easily show that

$$\sum_{u=1}^{n_i} (Y_{iu} - \overline{Y})^2 = \sum_{u=1}^{n_i} \{(Y_{iu} - \overline{Y}_i) + (\overline{Y}_i - \overline{Y})\}^2$$

$$= \sum_{u=1}^{n_i} (Y_{iu} - \overline{Y}_i)^2 + n_i(\overline{Y}_i - \overline{Y})^2,$$

the cross-product term canceling on applying the summation. Thus to obtain the *correct* contribution to the corrected sum of squares for the ith set of repeats, we need to add to $n_i(\overline{Y}_i - \overline{Y})^2$ the quantity

$$\sum_{u=1}^{n_i} (Y_{iu} - \overline{Y}_i)^2 = (n_i - 1)s_i^2.$$

(In some data, only the s_i are given, and it must be remembered that these must be squared before multiplying by $n_i - 1$.)

The quantities $(n_i - 1)s_i^2$, $i = 1, 2, \ldots, k$, also have a further role. They provide the k contributions with $(n_i - 1)$ degrees of freedom, $i = 1, 2, \ldots, k$, respectively, that must be added together to give the pure error sum of squares.

In summary, then, we can still perform the basic regression calculations if only the means and sample variance estimates are provided for each set of repeat runs. We do not need to know the individual observations themselves.

Note. The above comments will help you work Exercise G.

EXERCISES FOR CHAPTER 22

A. Here is a suggestion for a project. Consult, for example, company annual reports, or look up the same information in publications available at many libraries. Obtain the annual earnings per share $\$Y_i$ for several consecutive years for one or more companies, your choice. For each company, fit a polynomial model $Y_i = f(\text{Year}) + \epsilon$ of suitable order that will explain the data satisfactorily, using the method of orthogonal polynomials.

Investors like to see the companies whose shares they own increase earnings per share at a satisfactory *rate* measured in terms of percentage (or proportional) increase per year *over the previous year's results*. For this reason, it makes sense, provided earnings are positive, to examine not only earnings per share, but also the logarithms of these earnings. If earnings increased at a constant rate, for example, log(earnings) would exhibit a straight line trend. Take logarithms to the base e, or to the base 10 (the results differ only by a constant factor), of the data you selected and use the method of orthogonal polynomials to fit a polynomial of suitable order to the results. Compare, for each individual company you examine, the results of the fits to the unlogged and logged data. Check the residuals to see if the error assumptions appear to be violated. Comment on your results.

The following example data are genuine, but are given anonymously:

Company A: 1.93, 0.20, 2.55, 4.31, 8.51, 9.20, 5.10, 3.46, 0.99, 2.33, 3.37.

Company B: 2.17, −0.43, −0.72, −1.81, 5.48, 4.46, 4.37, −6.09, −9.29, 2.44, 2.20.

B. Fit a cubic equation using orthogonal polynomials to the Y-values 13, 4, 3, 4, 10, 22, which are equally spaced in the respective X-values given by $2X = -5, -3, -1, 1, 3, 5$, that is, $X = -2.5, -1.5, \ldots$. Is the cubic term needed? If not, what is the best quadratic, both in orthogonal polynomial form and in terms of X?

If the model $Y = \beta_0 + \beta_1 X + \beta_2 X^2 + \beta_3 X^3 + \epsilon$ had been fitted directly, how would the extra sums of squares $SS(b_1|b_0) = 58.51$, $SS(b_2|b_0, b_1) = 210.58$, $SS(b_3|b_0, b_1, b_2) = 0.006$ relate to the sums of squares for the first-, second-, and third-order orthogonal polynomials?

C. A newly born baby was weighed weekly, the figure adopted in each case being the average of the weights on three successive days. Twenty such weights are shown below, recorded in ounces. Fit to the data, using orthogonal polynomials, a polynomial model of a degree justified by the accuracy of the figures; that is, test as you go along for the significance of the linear, quadratic, and so forth, terms.

Number of week	1	2	3	4	5	6	7	8	9	10
Weight	141	144	148	150	158	161	166	170	175	181

Number of week	11	12	13	14	15	16	17	18	19	20
Weight	189	194	196	206	218	229	234	242	247	257

Source: Cambridge Diploma, 1950. Published with permission of Cambridge University Press.

D. A finished product is known to lose weight after it is produced. The following data demonstrate this drop in weight.

Time After Production, t	Weight Difference (in $\frac{1}{16}$ oz), Y
0	0.21
0.5	−1.46
1.0	−3.04
1.5	−3.21
2.0	−5.04
2.5	−5.37
3.0	−6.03
3.5	−7.21
4.0	−7.46
4.5	−7.96

Requirements

1. Using orthogonal polynomials, develop a second-order fitted equation that represents the loss in weight as a function of time after production.
2. Analyze the residuals from this model and draw conclusions about its adequacy.

E. Nine equally spaced levels of a dyestuff were applied to apparently identical pieces of cloth. The color ratings awarded, in order of increasing dyestuff levels, were

$$Y = 11, \quad 12, \quad 10, \quad 12, \quad 11, \quad 14, \quad 16, \quad 22, \quad 28.$$

Find a suitable polynomial relationship between Y and the level of dyestuff using orthogonal polynomials.

F. A mother keeps records of the monthly number of ice-cream portions served to her family over a 12-month period, with the results below. She wonders if the pattern can be explained by a polynomial regression model, and asks you to carry out such an analysis using the orthogonal polynomials of sixth and smaller orders. Do this, provide all the standard analyses, as indicated below, and set out the conclusions to which your solution leads.

Month	Portions Served
January	2
February	72
March	106
April	116
May	103
June	82
July	54
August	30
September	14
October	23
November	64
December	129

Details Required

Model fitting

ANOVA table

Decision on model order to use and telescoping of ANOVA, and so on

R^2, F-tests, and so on.

Estimated variance–covariance matrix

Evaluation of residuals *from chosen model*

Estimated variance–covariance matrix of residuals, diagonal elements only, and plots of residuals

Tests of residuals, including Durbin–Watson, and time sequence runs

Figures, as appropriate

Conclusions

G. The data below are selected from "'Health' and 'Ailments' of M.S.U. Alumni: A Study in Retrospect Over a Five Year Period (1961–62 to 1965–66)" by Mrs. J. V. Bhanot and Shri R. C. Patel, published by the Department of Statistics at the Maharaja Sayajirao University of Baroda, in October 1968. The data show average weights in pounds and standard deviations of various categories of students for 1965–1966. Only a portion of the data is reproduced here, and it is not necessarily fully representative of the whole. Because individual data points are not available, treat each set of repeat runs as the same number of all equal runs for regression fitting purposes, but use the corresponding s.d. (standard deviation) figure, squared and multiplied by the corresponding degrees of freedom (namely, number − 1) to obtain a suitable contribution to the pure error sum of squares. Be careful, because you will have to make suitable adjustments to recover proper sums of squares entries elsewhere.

Fit a model of the form $Y = \beta_0 + \beta_1(\text{age}) + \epsilon$ to the data, where Y = weight of student, adding suitable dummy variables to distinguish the different categories of persons. Provide a detailed analysis. What are your conclusions?

If you feel the model fitted is unsuitable, what model would you suggest instead, and why? Fit any alternative model you have suggested, perform the appropriate analysis, and state your conclusions.

Age in Years	Number of Students	Average Weight, Pounds	Est. s.d.	Category[a]
16	19	103.82	12.70	A
17	19	105.39	12.50	A
18	16	107.50	15.30	A
19	8	103.12	7.68	A
20	6	105.83	16.75	A
21	6	107.50	5.77	A
22	1	102.50	—	A
23	1	117.50	—	A
16	12	99.17	9.20	B
17	29	107.67	19.55	B
18	28	103.57	13.05	B
19	32	112.19	17.60	B
20	22	110.00	12.04	B
21	8	113.13	14.87	B
22	4	112.00	11.18	B
23	7	113.21	7.28	B
16	18	100.83	16.75	C
17	33	99.32	14.50	C

Age in Years	Number of Students	Average Weight, Pounds	Est. s.d.	Category[a]
18	24	100.83	19.67	C
19	18	96.39	13.29	C
20	6	101.67	18.12	C
21	4	100.00	15.21	C
22	0	—	—	C
23	1	112.50	—	C
Total	322			

[a] A = Hindu eggitarian male students; B = non-Hindu mixed (diet) male students; C = Hindu mixed (diet) female students.

C H A P T E R 23

Multiple Regression Applied to Analysis of Variance Problems

23.1. INTRODUCTION

Many carefully designed experiments fall into the category of *analysis of variance problems*. Although some of these are, in fact, regression problems, they usually are not discussed in that way. Special computer programs exist for analyzing the data that are collected. The methods of analysis depend on special linear models that are traditionally adopted for particular designs.

Analysis of variance models can involve *fixed effects*, *random effects*, or a combination of both types of effects (when the model is usually referred to as a *mixed model*). We shall be concerned here only with the connections between regression and fixed effects analysis of variance models. For the other types of analysis of variance problems see, for example, Hocking (1996, Chapters 15–17).

Fixed Effects, Variable Effects

Suppose we have I groups of data defined, for example, by I different levels of a nutritional diet. Suppose the ith data group, $i = 1, 2, \ldots, I$, consisted of observations $Y_{i1}, Y_{i2}, \ldots, Y_{iJ_i}$ with average value $\overline{Y}_i$, and that this $\overline{Y}_i$ estimated the theoretical response θ_i (say). Then we could write

$$Y_{ij} = \theta_i + \epsilon_{ij}, \qquad \epsilon_{ij} \sim N(0, \sigma^2), \text{iid} \qquad (23.1.1)$$

to indicate that the diet levels had different effects $\theta_1, \theta_2, \ldots, \theta_I$, and that individual Y_{ij} varied about θ_i with errors that were independently and identically distributed (iid) normally, with mean zero and variance σ^2. Here, we would typically think of the θ_i as being *fixed effects*, each associated specifically with one of the diet levels.

In other applications, the θ_i of (23.1.1) could be *random effects*, that is, random variables. Suppose, for example, that the different i-values represented simply different batches of raw material. Then the specific θ_i would not be of interest in themselves, because another set of batches in a subsequent experiment would provide different θ_i. What *might* be of interest is the size of the variation in the θ_i. We would now most likely tentatively assume that $\theta_i \sim N(\mu, \sigma_\theta^2)$ and then estimate μ and σ_θ^2.

We link our standard regression methods to the fixed effects model in what follows. Analysis of variance problems that involve fixed effects *could* be handled by a general regression program. We do *not* recommend actually doing this in general, particularly where all the data have become available, and where there was no hitch in carrying out the design. It is useful to know how to apply regression methods directly, however, for a number of reasons:

1. Using regression methods makes us think more carefully about the model being assumed. Because analysis of variance models typically have more parameters in them than can be estimated (they are *overparameterized*), the singular regressions that arise must be circumvented if regression methods are applied. Dummy variables can be selected in a variety of ways and some selections may be better than others, computationally, although all valid selections of dummies will lead to the same predictive values for the response.

2. Analysis of variance problems provide fitted values and residuals. Often, insufficient emphasis is placed on examining the residuals in analysis of variance problems.

3. If the data are incomplete, regression methods can still be applied, even if the neater analysis of variance methods cannot be. So regression provides a backup analysis for partially botched experimental designs.

4. Additional predictor variables (usually called *covariates*) can be added to the analysis of variance structure via regression. For example, in animal experiments where "final weight" is the response, the variable "initial weight" is often included as a covariate. This would be a continuous covariate addition. Or a class variable with several categories can be adjoined by using a suitable set of dummy variables to represent the discrete levels of the covariate. Orthogonal polynomials (see Chapter 22) might be useful for equally spaced classes and might lead to a low-order polynomial fit for the effects of the class variable.

In subsequent sections, we first discuss the one-way classification, and later the two-way classification, with equal numbers of observations in each cell, using a practical example with data for each case to give the reader some feel for the considerations involved and the advantages and disadvantages of the regression approach. Theoretical presentations for more general one- and equal-celled two-way classifications are also provided, together with another two-way, equal-celled example. Essentially, the procedures needed for the regression treatments are all choices of suitable dummy variable systems. As always, dummy variables can be chosen in an infinite number of ways.

23.2. THE ONE-WAY CLASSIFICATION: STANDARD ANALYSIS AND AN EXAMPLE

Suppose we have data in I groups, $i = 1, 2, \ldots, I$, with J_i observations in each group as given below:

Group 1 $Y_{11}, Y_{12}, \ldots, Y_{1J_1},$ mean $\overline{Y}_1$

Group 2 $Y_{21}, Y_{22}, \ldots, Y_{2J_2},$ mean $\overline{Y}_2$

$\vdots$

Group I $Y_{I1}, Y_{I2}, \ldots, Y_{IJ_I},$ mean $\overline{Y}_I$

T A B L E 23.1. Standard ANOVA Table for a One-Way Classification

Source	df	SS	MS
Between groups	$I - 1$	$\sum_{i=1}^{I} J_i(\overline{Y}_i - \overline{Y})^2$	s_B^2
Within groups	$\sum_{i=1}^{I} (J_i - 1) = n - I$	$\sum_{i=1}^{I} \sum_{j=1}^{J_i} (Y_{ij} - \overline{Y}_i)^2$	s_W^2
Mean	1	$n\overline{Y}^2$	
Total	$\sum_{i=1}^{I} J_i = n$	$\sum_{i=1}^{I} \sum_{j=1}^{J_i} Y_{ij}^2$	

The usual fixed-effects analysis of variance model for such a situation is

$$Y_{ij} = \mu + t_i + \epsilon_{ij}, \qquad i = 1, 2, \ldots, I$$
$$j = 1, 2, \ldots, J_i \tag{23.2.1}$$

where $t_1, t_2, \ldots, t_I$ are parameters such that

$$J_1 t_1 + J_2 t_2 + \cdots + J_I t_I = 0, \tag{23.2.2}$$

and where $\epsilon_{ij} \sim N(0, \sigma^2)$, iid. Some sort of restriction on the parameters is necessary since Eq. (23.2.1) contains more parameters than are really needed. It is usual to regard μ as the overall mean level and t_i as the difference between the ith group mean and the overall mean level. Thus the total of all differences between groups and the overall level is zero, and that is what Eq. (23.2.2) expresses. The usual analysis of variance table (where $\overline{Y} = \sum_{i=1}^{I} J_i \overline{Y}_i / \sum_{i=1}^{I} J_i$) is given in Table 23.1.

It is usual to test the hypothesis that there are no differences between means of groups; that is, $H_0: t_1 = t_2 = \cdots = t_I = 0$ by comparing the ratio $F = s_B^2/s_W^2$ to a suitable percentage point of the $F(I - 1, \sum_{i=1}^{I}(J_i - 1))$ distribution.

An Example

Caffeine, orally ingested, is said to be a stimulant, the amount of stimulation and its basic variability differing according to the dose ingested. In order to get some idea of the effect of caffeine on a physical task, the following simple experiment was performed.

1. *The Experiment.* Three treatment levels, 0, 100, and 200, in milligrams (mg) of caffeine were used. Thirty healthy male college students of the same age and with essentially the same physical ability were selected and trained in finger tapping. After the training was completed, ten men were randomly assigned to each treatment. Neither the men nor the physiologist knew which treatment the men received; only the statistician knew this. Two hours after the treatment was administered, each man was required to do finger tapping. The number of finger taps per minute was recorded, as shown in Table 23.2.

2. *The Analysis of Variance (ANOVA) Model for This Experiment.* Let:

Y_{ij} = the number of finger taps per minute of the "jth" man on the "ith" treatment.

μ = true value for the average number of finger taps in a population of males of which the selected 30 form a random sample.

T A B L E 23.2. Number of Finger Taps Per Minute Achieved by Thirty Male Students Receiving Designated Doses of Caffeine

		Treatment (Row)		
		Totals $T_i = \sum_j Y_{ij}$	Means $\bar{Y}_i$	Effects $\hat{t}_i = \bar{Y}_i - \bar{Y}$
Treatments, i	Observations			
$i = 1$, 0 mg caffeine (placebo)	242, 245, 244, 248, 247, 248, 242, 244, 246, 242	2448	244.8	−1.7
$i = 2$, 100 mg caffeine	248, 246, 245, 247, 248, 250, 247, 246, 243, 244	2464	246.4	−0.1
$i = 3$, 200 mg caffeine	246, 248, 250, 252, 248, 250, 246, 248, 245, 250	2483	248.3	1.8
		$\sum_i \sum_j Y_{ij} = 7395$	$246.5 = \bar{Y}$	

t_i = the ith treatment effect, that is, the additive effect of the "ith" treatment over and above (or below) μ. To preserve μ as the true mean of the overall sample we must assume that $t_1 + t_2 + t_3 = 0$. [All J_i of Eq. (23.2.2) are equal.]

ϵ_{ij} = the random effect; this is a measure of the failure of student j tested with treatment i to have done exactly $\mu + t_i$ number of taps, due to random error.

With these definitions, the ANOVA model is

$$Y_{ij} = \mu + t_i + \epsilon_{ij}, \qquad (23.2.3)$$

and we make the usual normality assumptions on the errors.

3. *The Standard ANOVA Calculations.* The usual ANOVA calculations lead to Table 23.3. The split indicated by the asterisks is explained below.

4. *Test for Equality of Treatments.* To test H_0: $t_1 = t_2 = t_3$ versus H_1: not so, we compare $F = 6.18$ with $F(2, 27, 0.95) = 3.35$ and so reject H_0. This indicates differences between the treatments.

T A B L E 23.3. Analysis of Variance Table for the Finger Taps Example

Source of Variation	df	SS	MS	F
Between treatments	2	61.40	30.70	6.18
Linear*	1	61.25	61.25	12.32
Quadratic*	1	0.15	0.15	0.03
Within treatments	27	134.10	$s^2 = 4.97$	
Total corrected	29	195.50		
Mean	1	1,822,867.50		
Total	30	1,823,063.00		

* This split of the "between treatments" SS is explained in Section 23.5. The designations "among treatments," "between treatments," and "treatments" all mean the same thing.

5. *Linear and Quadratic Contrasts.* The "between treatments" sum of squares with 2 degrees of freedom can be split up in various ways. One standard way, which takes advantage of the equal spacing of the treatments, is to construct orthogonal linear and quadratic contrasts. Because of the orthogonality, the sums of squares of the two individual contrasts will add to the sum of squares between treatments. The contrasts use coefficients attached to the treatment totals $T_i = \Sigma_j Y_{ij}$ as follows:

	Caffeine (mg) in the Treatment		
Contrast	0	100	200
Linear (L)	-1	0	1
Quadratic (Q)	1	-2	1

These are, in fact, the linear and quadratic orthogonal polynomials shown in Section 22.2. Thus the contrasts are

$$\mathbf{c}'_L\mathbf{Y} = L = -T_1 \qquad + T_3 = 35.$$

$$\mathbf{c}'_Q\mathbf{Y} = Q = T_1 - 2T_2 + T_3 = 3.$$

If $i = 1, 2, \ldots, I$ and $j = 1, 2, \ldots, J$ (for our example, $I = 3$, $J = 10$), the sums of squares attributable to L and Q are, by the general formula for individual contrasts $\mathbf{c}'\mathbf{Y}$, $\mathrm{SS}(\mathbf{c}'\mathbf{Y}) = (\mathbf{c}'\mathbf{Y})^2/\mathbf{c}'\mathbf{c}$,

$$\mathrm{SS}(L) = \frac{L^2}{J\{(-1)^2 + 0^2 + 1^2\}} = \frac{(35)^2}{10(2)} = 61.25,$$

$$\mathrm{SS}(Q) = \frac{Q^2}{J\{1^2 + (-2)^2 + 1^2\}} = \frac{(3)^2}{10(6)} = 0.15.$$

These are already entered in Table 23.3. In the denominators of these SS, J is the number of observations contributing to the totals T_i used to obtain L and Q, and the other term is the sum of squares of the coefficients that multiply the T_i. Because the contrasts $L = \mathbf{c}'_L\mathbf{Y}$ and $Q = \mathbf{c}'_Q\mathbf{Y}$ are orthogonal ($\mathbf{c}'_L\mathbf{c}_Q = 0$), the two SS add to 61.40.

Tests for linear and quadratic effects reveal a significant linear effect [because $12.32 > F(1, 27, 0.95) = 4.21$] and a nonsignificant quadratic effect. Thus we conclude that, within the range 0–200 mg of caffeine used, the true number of finger taps increases (because $\overline{Y}_1 < \overline{Y}_2 < \overline{Y}_3$) linearly with the amount of caffeine ingested.

This concludes the standard ANOVA approach to this example and we now discuss the regression approach to it.

23.3. REGRESSION TREATMENT OF THE ONE-WAY CLASSIFICATION EXAMPLE

The model given in Eq. (23.2.1) involves four parameters μ, t_1, t_2, t_3. Only two of these, μ and one t_i, are involved with any one Y_{ij}; also, $t_1 + t_2 + t_3 = 0$. A natural first step in a regression approach is to write down the model

$$Y = \mu X_0 + t_1 X_1 + t_2 X_2 + t_3 X_3 + \epsilon \tag{23.3.1}$$

and consider what choice of variables X_i would reproduce Eq. (23.2.1). A little thought shows that use of the dummies $X_0 = 1$ and

$$X_i = \begin{cases} 1, & \text{if the } i\text{th treatment is applied to provide } Y_{ij}, \\ 0, & \text{otherwise}, \end{cases} \qquad (23.3.2)$$

for $i = 1, 2, 3$, will do the trick. This gives the setup:

$$
\begin{array}{ccc}
\mathbf{Y} & \mathbf{X} & \boldsymbol{\beta} \\
(30 \times 1) & (30 \times 4) & (4 \times 1)
\end{array}
$$

		$\mathbf{X_0}$	$\mathbf{X_1}$	$\mathbf{X_2}$	$\mathbf{X_3}$		
242		1	1	0	0		
245		1	1	0	0		
⋮		⋮	⋮	⋮	⋮		
246		1	1	0	0		
242		1	1	0	0		
248		1	0	1	0		μ
246		1	0	1	0		t_1
⋮		⋮	⋮	⋮	⋮		t_2
243		1	0	1	0		t_3
244		1	0	1	0		
246		1	0	0	1		
248		1	0	0	1		
⋮		⋮	⋮	⋮	⋮		
245		1	0	0	1		
250		1	0	0	1		

$$(23.3.3)$$

A drawback is immediately apparent. Because of the column dependence

$$\mathbf{X_0} = \mathbf{X_1} + \mathbf{X_2} + \mathbf{X_3} \qquad (23.3.4)$$

in $\mathbf{X}$, the $\mathbf{X'X}$ will necessarily be singular and so the normal equations will not have a unique solution. We have so far not taken into account the ANOVA model restriction $t_1 + t_2 + t_3 = 0$. If this is applied, the solution to the normal equations *is* unique. However, this provides an added complication (which we discuss further in Section 23.4) and means that we have not obtained a standard nonsingular regression format for our problem. How do we do that? There are many possibilities—an infinite number, technically—as there always are whenever dummy variables are used. One of these ways, which we now describe, enables us to reproduce the linear and quadratic contrasts split-up featured in Table 23.3. Define dummy variables Z_1 and Z_2 to replace $X_1, X_2,$ and X_3 in the following correspondence:

Z_1	Z_2	for	X_1	X_2	X_3
−1	1		1	0	0
0	−2		0	1	0
1	1		0	0	1

$$(23.3.5)$$

The model can then be written in regression format as

$$Y = \beta_0 X_0 + \beta_1 Z_1 + \beta_2 Z_2 + \varepsilon \qquad (23.3.6)$$

with regression setup:

$$
\underset{(30 \times 1)}{\mathbf{Y}} =
\begin{bmatrix}
242 \\
245 \\
\vdots \\
242 \\
\hline
248 \\
246 \\
\vdots \\
244 \\
\hline
246 \\
248 \\
\vdots \\
250
\end{bmatrix}
\quad
\underset{(30 \times 3)}{\mathbf{X}} =
\begin{array}{ccc}
\mathbf{X}_0 & \mathbf{Z}_1 & \mathbf{Z}_2 \\
\begin{bmatrix}
1 & -1 & 1 \\
1 & -1 & 1 \\
\vdots & \vdots & \vdots \\
1 & -1 & 1 \\
1 & 0 & -2 \\
1 & 0 & -2 \\
\vdots & \vdots & \vdots \\
1 & 0 & -2 \\
1 & 1 & 1 \\
1 & 1 & 1 \\
\vdots & \vdots & \vdots \\
1 & 1 & 1
\end{bmatrix}
\end{array}
\quad
\underset{(3 \times 1)}{\boldsymbol{\beta}} =
\begin{bmatrix}
\beta_0 \\
\beta_1 \\
\beta_2
\end{bmatrix}
\qquad (23.3.7)
$$

Note that the vectors $\mathbf{X}_0$, $\mathbf{Z}_1$, and $\mathbf{Z}_2$ are mutually orthogonal. This makes the least squares solution extremely simple, as follows:

$$
\begin{bmatrix}
b_0 \\
b_1 \\
b_2
\end{bmatrix}
= \mathbf{b} = (\mathbf{X}'\mathbf{X})^{-1}\mathbf{X}'\mathbf{Y} =
\begin{bmatrix}
\frac{1}{30} & 0 & 0 \\
0 & \frac{1}{20} & 0 \\
0 & 0 & \frac{1}{60}
\end{bmatrix}
\begin{bmatrix}
7395 \\
35 \\
3
\end{bmatrix}
=
\begin{bmatrix}
246.50 \\
1.75 \\
0.05
\end{bmatrix},
\qquad (23.3.8)
$$

$$SS(b_0, b_1, b_2) = \mathbf{b}'\mathbf{X}'\mathbf{Y} = 1{,}822{,}867.50 + 61.25 + 0.15$$

$$(= 1{,}822{,}928.90) \qquad (23.3.9)$$

$$= SS(b_0) + SS(b_1) + SS(b_2).$$

The clean split of the SS is due to the orthogonality previously mentioned. The details of Table 23.3 are thus reproduced exactly. The fitted regression model is

$$\hat{Y} = 246.50 + 1.75 Z_1 + 0.05 Z_2. \qquad (23.3.10)$$

If we now drop the nonsignificant "$0.05 Z_2$" term and note that X_1 can be regarded as a coding

$$Z_1 = (C - 100)/100, \qquad (23.3.11)$$

where C = amount of caffeine ingested in milligrams, of the caffeine level, we see that the fitted equation becomes

$$\hat{Y} = 244.75 + 0.0175 C. \qquad (23.3.12)$$

Thus, within the ranges of C observed, the number of finger taps attained per minute by trained college students can be predicted in terms of C via Eq. (23.3.12).

(Note that, when a variable such as Z_2 is dropped, we would usually have to refit the regression equation. It would not be correct in general just to drop the term out. Here, however, because $\mathbf{Z}_2$ is orthogonal to the other columns of $\mathbf{X}$, the result is the same either way.)

A Caution

Careful choice of the levels of the dummies Z_1 and Z_2 to produce a diagonal $\mathbf{X'X}$ matrix and the equality of the three sample sizes made the above example an extremely neat one. In general, we cannot always achieve such tidy results from a regression approach. Let us look again at Eq. (23.3.3). Suppose we had opted, instead, to just drop one of the parameters, say, t_3, from $\boldsymbol{\beta}$, and thus the corresponding column $\mathbf{X}_3$ from $\mathbf{X}$. The corresponding least squares solution would then have been (using $\hat{\mu}$ to denote the estimate of μ, and so on)

$$
\begin{bmatrix} \hat{\mu} \\ \hat{t}_1 \\ \hat{t}_2 \end{bmatrix} = \mathbf{b} = (\mathbf{X'X})^{-1}\mathbf{X'Y} = (10)^{-1} \begin{bmatrix} 3 & 1 & 1 \\ 1 & 1 & 0 \\ 1 & 0 & 1 \end{bmatrix}^{-1} \begin{bmatrix} 7395 \\ 2448 \\ 2464 \end{bmatrix}
$$

$$
= \tfrac{1}{10} \begin{bmatrix} 1 & -1 & -1 \\ -1 & 2 & 1 \\ -1 & 1 & 2 \end{bmatrix} \begin{bmatrix} 7395 \\ 2448 \\ 2464 \end{bmatrix} = \begin{bmatrix} 248.3 \\ -3.5 \\ -1.9 \end{bmatrix}, \tag{23.3.13}
$$

$$
\mathrm{SS}(\hat{\mu}, \hat{t}_1, \hat{t}_2) = \mathbf{b'X'Y} = (248.3, -3.5, -1.9) \begin{bmatrix} 7395 \\ 2448 \\ 2464 \end{bmatrix}
$$

$$
= 1{,}822{,}928.90. \tag{23.3.14}
$$

This value is exactly the same as the total of the components in Eq. (23.3.9) as, of course, it should be, but the informative split-up into three pieces achieved there does not happen here due to the lack of orthogonality, that is, because $\mathbf{X'X}$ is not diagonal here. We can, of course, obtain the extra sum of squares for $\hat{t}_1$, $\hat{t}_2$ by subtraction of $n\bar{Y}^2$ to give

$$
\mathrm{SS}(\hat{t}_1, \hat{t}_2 \mid \hat{\mu}) = 1{,}822{,}928.90 - 1{,}822{,}867.50 = 61.40 \tag{23.3.15}
$$

in the usual way, but we would have to perform additional "extra sums of squares" calculations to proceed further. The moral is that, although many regression setups of an ANOVA problem are possible, some are more informative than others, and care in choosing a good setup will be repaid.

Proceeding further with the current (nonorthogonal) setup, we obtain the fitted model

$$
\hat{Y} = \hat{Y}(X_1, X_2) = 248.3 - 3.5X_1 - 1.9X_2, \tag{23.3.16}
$$

which produces predicted values as follows:

$$\hat{Y}(1, 0) = 244.8, \quad \text{when 0 mg of caffeine is ingested,}$$

$$\hat{Y}(0, 1) = 246.4, \quad \text{when 100 mg of caffeine is ingested,}$$

$$\hat{Y}(0, 0) = 248.3, \quad \text{when 200 mg of caffeine is ingested.}$$

The reader might initially be surprised to see negative coefficients in Eq. (23.3.16), because the number of finger taps *increases* with the dose level of caffeine. The results are correct, however. The dummy variable setup chosen here has caused $\hat{Y}(0, 0)$ to be the "base level" prediction and $\hat{Y}(1, 0)$ and $\hat{Y}(0, 1)$ require *negative* adjustments to achieve the proper values, that is all. As we have already said, care is needed in these applications.

In Sections 23.4 and 23.5, we discuss the one-way classification more generally. After that we turn to the two-way classification.

Relationship to the Underlying Geometry

The usual analysis of variance model, (23.2.1) in our example, is overparameterized and so the estimation space is described by more vectors, shown in (23.3.3), than are needed. An infinite number of descriptions of (the unique) $\hat{\mathbf{Y}}$ is thus possible unless extra restrictions are added. The restriction $t_1 + t_2 + t_3$ of our example simply selects one of that infinite number of descriptions. From the regression point of view it is simpler to define the estimation space by a minimal set of vectors, as in (23.3.7), or by dropping an $\mathbf{X}_i$ column, for our example. (Many alternative choices of the minimal set of vectors are, of course, feasible.) This allows us to proceed without any restrictions and leads to a unique description of $\hat{\mathbf{Y}}$. See Section 20.4 for additional discussion.

23.4. REGRESSION TREATMENT OF THE ONE-WAY CLASSIFICATION USING THE ORIGINAL MODEL

We now look at the general one-way classification without reduction of parameters and show that, while the original normal equations cannot be solved uniquely, addition of the usual restriction $J_1 t_1 + J_2 t_2 + \cdots + J_I t_I = 0$ provides a unique solution for the parameter estimates. Essentially, the restriction can replace any one of the original normal equations. Instead of Eq. (23.3.1), write

$$E(Y) = \mu X_0 + t_1 X_1 + t_2 X_2 + \cdots + t_I X_I. \tag{23.4.1}$$

We want this to express the fact that if we consider an observation Y_{ij} from the *i*th group it must have expectation $\mu + t_i$. We define

$$\mathbf{Y}' = (Y_{11}, Y_{12}, \ldots, Y_{1J_1}; Y_{21}, Y_{22}, \ldots, Y_{2J_2}; \ldots; Y_{I1}, Y_{I2}, \ldots, Y_{IJ_I})$$

and

$$
\begin{array}{cccccc}
\mathbf{X}_0 & \mathbf{X}_1 & \mathbf{X}_2 & \mathbf{X}_3 & \cdots & \mathbf{X}_I
\end{array}
$$

$$
\mathbf{X} = \left[
\begin{array}{cccccc}
1 & 1 & 0 & 0 & \cdots & 0 \\
1 & 1 & 0 & 0 & \cdots & 0 \\
\vdots & \vdots & \vdots & \vdots & \cdots & \vdots \\
1 & 1 & 0 & 0 & \cdots & 0 \\
\hline
1 & 0 & 1 & 0 & \cdots & 0 \\
1 & 0 & 1 & 0 & \cdots & 0 \\
\vdots & \vdots & \vdots & \vdots & \cdots & \vdots \\
1 & 0 & 1 & 0 & \cdots & 0 \\
\hline
\vdots & \vdots & \vdots & \vdots & \cdots & \vdots \\
\hline
1 & 0 & 0 & 0 & \cdots & 1 \\
1 & 0 & 0 & 0 & \cdots & 1 \\
\vdots & \vdots & \vdots & \vdots & \cdots & \vdots \\
1 & 0 & 0 & 0 & \cdots & 1
\end{array}
\right]
$$

where the dashes divide the matrix sets of rows, there being $\mathbf{J}_1, \mathbf{J}_2, \ldots, \mathbf{J}_I$ rows in successive sets. The headings show to which X the columns relate. Furthermore, define

$$
\boldsymbol{\beta}' = (\mu; t_1, t_2, \ldots, t_I),
$$

then

$$
E(\mathbf{Y}) = \mathbf{X}\boldsymbol{\beta}
$$

expresses the model in matrix notation. Now

$$
\mathbf{X}'\mathbf{X} = \begin{bmatrix}
n & J_1 & J_2 & \cdots & J_I \\
J_1 & J_1 & 0 & \cdots & 0 \\
J_2 & 0 & J_2 & \cdots & 0 \\
\vdots & & & \cdots & \\
J_I & 0 & 0 & \cdots & J_I
\end{bmatrix}, \qquad
\mathbf{X}'\mathbf{Y} = \begin{bmatrix}
n\bar{Y} \\
J_1\bar{Y}_1 \\
J_2\bar{Y}_2 \\
\vdots \\
J_I\bar{Y}_I
\end{bmatrix}
\qquad (23.4.2)
$$

If we write b_0, b_i for the least squares estimates of μ and t_i, we can write the normal equations $(\mathbf{X}'\mathbf{X})\mathbf{b} = \mathbf{X}'\mathbf{Y}$ as

$$
\begin{aligned}
nb_0 + J_1b_1 + J_2b_2 + \cdots + J_Ib_I &= n\bar{Y}, \\
J_1b_0 + J_1b_1 \qquad\qquad\qquad\quad &= J_1\bar{Y}_1, \\
J_2b_0 \qquad + J_2b_2 \qquad\qquad\quad &= J_2\bar{Y}_2, \\
\vdots \qquad\qquad\qquad \cdots \qquad\qquad & \\
J_Ib_0 \qquad\qquad\qquad + J_Ib_I &= J_I\bar{Y}_I.
\end{aligned}
\qquad (23.4.3)
$$

Here the $(\mathbf{X'X})^{-1}$ matrix does not exist since $\mathbf{X'X}$ is singular. This is due to the fact that Eqs. (23.4.3) are *not* independent because the first equation is the sum of the other I equations. There are in fact only I equations in the $(I + 1)$ unknowns b_0, $b_1, \ldots, b_I$, because the original model (23.3.1) contained more parameters than were actually necessary. This "singularity" of the $\mathbf{X'X}$ matrix is also clear from examination of the $\mathbf{X}$ matrix, where the X_0 column is equal to the sum of the X_1, X_2, $\ldots$, X_I columns, a dependence that becomes transmitted into the normal equations of (23.4.3) as we have noted. How can we proceed then? A condition we have not so far taken into account is Eq. (23.2.2), which, if true of the parameters, must also hold for the estimates of the parameters. Hence

$$J_1 b_1 + J_2 b_2 + \cdots + J_I b_I = 0. \tag{23.4.4}$$

This provides the additional independent equation we require. We now take *any I* equations from (23.4.3) together with Eq. (23.4.4) and use these as the normal equations. It is most convenient to drop the first equation of (23.4.3) since it has the most terms. This leaves, as the equations to solve,

$$
\begin{aligned}
J_1 b_1 + J_2 b_2 + \cdots + J_I b_I &= 0, \\
J_1 b_0 + J_1 b_1 &= J_1 \overline{Y}_1, \\
J_2 b_0 \qquad + J_2 b_2 &= J_2 \overline{Y}_2, \\
\vdots \qquad\qquad \cdots \\
J_I b_0 \qquad\qquad + J_I b_I &= J_I \overline{Y}_I.
\end{aligned}
\tag{23.4.5}
$$

To maintain symmetry we have not divided through the second to $(I + 1)$st equations by their common factors. In matrix form we can write Eq. (23.4.5) as

$$
\begin{bmatrix}
0 & J_1 & J_2 & \cdots & J_I \\
J_1 & J_1 & 0 & \cdots & 0 \\
J_2 & 0 & J_2 & \cdots & \\
\vdots & & & \cdots & \\
J_I & 0 & 0 & \cdots & J_I
\end{bmatrix}
\begin{bmatrix}
b_0 \\
b_1 \\
b_2 \\
\vdots \\
b_I
\end{bmatrix}
=
\begin{bmatrix}
0 \\
J_1 \overline{Y}_1 \\
J_2 \overline{Y}_2 \\
\vdots \\
J_I \overline{Y}_I
\end{bmatrix}.
\tag{23.4.6}
$$

Since we cannot express these in the form $(\mathbf{X'X})\mathbf{b} = \mathbf{X'Y}$ it will usually be impractical to use this procedure when the work is done by a computer routine, which requires this form.

From Eq. (23.4.5),

$$b_i = \overline{Y}_i - b_0, \qquad i = 1, 2, \ldots, I.$$

Substituting in the first equation

$$0 = \sum_{i=1}^{I} J_i b_i = \sum_{i=1}^{I} J_i(\overline{Y}_i - b_0)$$

$$= \sum_{i=1}^{I} J_i \overline{Y}_i - b_0 \sum_{i=1}^{I} J_i$$

$$= n\overline{Y} - n b_0.$$

Thus

$$b_0 = \overline{Y}$$

and

$$b_i = \overline{Y}_i - \overline{Y}.$$

The sum of squares due to a vector of estimates $\mathbf{b}$ determined from equations $\mathbf{X'Xb} = \mathbf{X'Y}$ is defined by $\mathbf{b'X'Y}$ even if $\mathbf{X'X}$ is singular and cannot be inverted and extra conditions of the form $\mathbf{Qb} = 0$ are needed to give a unique solution. No matter what form $\mathbf{Q}$ takes [here it is a vector $(0, J_1, J_2, \ldots, J_I)$], $\mathbf{b'X'Y}$ is invariant. This is so since if $\mathbf{b}_1$, $\mathbf{b}_2$ are two solutions arising from different "extra conditions"

$$\mathbf{b}_1'(\mathbf{X'Y}) = \mathbf{b}_1'(\mathbf{X'Xb}_2)$$

$$= (\mathbf{X'Xb}_1)'\mathbf{b}_2$$

[regrouping and using the matrix theory fact that $(\mathbf{AB})' = \mathbf{B'A'}$],

$$= (\mathbf{X'Y})'\mathbf{b}_2$$

$$= \mathbf{b}_2'\mathbf{X'Y}.$$

Of course, this is obvious geometrically. We are evaluating the squared distance from the origin O to the point $\hat{\mathbf{Y}}$ and this remains the same however we choose to describe it.

Thus the sum of squares due to regression is

$$\mathbf{b'X'Y} = n\overline{Y}^2 + \sum_{i=1}^{I} J_i \overline{Y}_i (\overline{Y}_i - \overline{Y})$$

$$= n\overline{Y}^2 + \sum_{i=1}^{I} J_i (\overline{Y}_i - \overline{Y})^2$$

with I degrees of freedom, since the additional term,

$$\sum_{i=1}^{I} (-\overline{Y}) J_i (\overline{Y}_i - \overline{Y}),$$

added to the right-hand size, is zero by definition of the means. If the model had only a term μ in it we should have

$$\mathrm{SS}(b_0) = n\overline{Y}^2$$

with one degree of freedom. Thus

$$\mathrm{SS}(b_1, b_2, \ldots, b_I \mid b_0) = \mathbf{b'X'Y} - n\overline{Y}^2$$

$$= \sum_{i=1}^{I} J_i (\overline{Y}_i - \overline{Y})^2,$$

with $(I - 1)$ degrees of freedom.

These provide the sums of squares due to the "mean" and "between groups," of the analysis of variance table of Table 23.1. The "within groups" sum of squares is found by the difference $\mathbf{Y'Y} - \mathbf{b'X'Y}$ as usual and is the same as the form given in Table 23.1 if evaluated. The test for $H_0 : t_1 = t_2 = \cdots = t_I = 0$ is made exactly as in the analysis of variance case.

We have shown here that the one-way analysis of variance problem can be done formally by regression using the original model. However, to perform the calculations on a computer, it is probably best to remove the singularity in the problem by choosing the model more carefully prior to computation.

Note. The work above shows how we must proceed in general in a regression problem, if there are more parameters to estimate than there are independent normal equations. If no natural restrictions are available, as in the analysis of variance case, we must make restrictions in an arbitrary fashion. While the choice of restrictions will influence the actual values of the regression coefficients, it will not affect the sum of squares due to regression. Usually we would select restrictions that would make the normal equations easier to solve.

Example. Suppose the normal equations were

$$22b_1 + 10b_2 + 12b_3 + 5b_4 + 8b_5 + 9b_6 = 34.37,$$

$$10b_1 + 10b_2 \quad\quad + 3b_4 + 4b_5 + 3b_6 = 21.21,$$

$$12b_1 \quad\quad + 12b_3 + 2b_4 + 4b_5 + 6b_6 = 13.16,$$

$$5b_1 + 3b_2 + 2b_3 + 5b_4 \quad\quad\quad = 10.28,$$

$$8b_1 + 4b_2 + 4b_3 \quad\quad + 8b_5 \quad\quad = 14.23,$$

$$9b_1 + 3b_2 + 6b_3 \quad\quad\quad 9b_6 = 9.86.$$

[These equations appear in Plackett (1960, p. 44).] They arise from a two-way classification with unequal numbers of observations in the cells. Such data might also arise from an intended "equal cells" analysis if several observations were missing. We discuss the "equal cells" analysis in the next section.)

Only four of the six equations are independent since the second and third equations add to give the first as do the fourth, fifth, and sixth. Thus two additional equations are required to give six equations in six unknowns. We must add two independent restrictions on $b_1, b_2, \ldots, b_6$, which must not be linear combinations of the existing equations.

Since there are actually only four independent normal equations we can drop two dependent ones, say, the first and sixth. The remaining four can be written in matrix form as follows:

$$\begin{matrix} 1 & 2 & 3 & 4 & 5 & 6 \end{matrix}$$
$$\begin{bmatrix} 10 & 10 & 0 & 3 & 4 & 3 \\ 12 & 0 & 12 & 2 & 4 & 6 \\ 5 & 3 & 2 & 5 & 0 & 0 \\ 8 & 4 & 4 & 0 & 8 & 0 \end{bmatrix} \begin{bmatrix} b_1 \\ b_2 \\ b_3 \\ b_4 \\ b_5 \\ b_6 \end{bmatrix} = \begin{bmatrix} 21.21 \\ 13.16 \\ 10.28 \\ 14.23 \end{bmatrix}.$$

Since the original matrix was symmetric, the row (or equation) dependence noticed earlier is also reflected in the fact that the first column is the sum of the second and third, and also the sum of the fourth, fifth, and sixth. When adding the two restrictions on the b's we must be careful on two counts. The restrictions, when placed below the four selected equations, will contribute two additional rows to the matrix and two zeros to the right-hand side vector (we would usually take the restrictions to be of the form $\Sigma c_i b_i = 0$). The final matrix must be such that there is no dependence between either rows *or* columns, if a unique solution is required. [There are more elegant matrix ways of expressing this in general; see, for example, Plackett (1960), but we do not provide them in our more elementary development.] For example, we *cannot* take restrictions

$$7b_1 + 6b_2 + b_3 + b_4 + b_5 + 5b_6 = 0,$$
$$11b_1 + 9b_2 + 2b_3 + 4b_4 + 4b_5 + 3b_6 = 0,$$

since the original column dependency is preserved. Even if only one column dependency is preserved, for example, by restrictions

$$3b_4 + 4b_5 + 3b_6 = 0,$$
$$9b_1 + 5b_2 + 4b_3 \qquad\qquad = 0,$$

which allow column one to be the sum of columns two and three, the restrictions are useless. However,

$$3b_4 + 4b_5 + 3b_6 = 0,$$
$$b_2 + b_3 \qquad\qquad = 0$$

would be usable, since no dependence will occur, and we shall have six equations in six unknowns as required. (Different restrictions were used by Plackett, whose book should be consulted for the subsequent solution.)

The idea of adding arbitrary restrictions may seem somewhat peculiar at first. One must remember that it is required whenever more parameters are used than are actually necessary to express the model. At some stage this "looseness" must be removed and that is the purpose of the added restrictions.

Again this becomes obvious geometrically. When an infinite number of least squares solutions exist because of overdescription of the estimation space, we must add enough extra restrictions ("information") to particularize to one solution, *if* we wish to describe only one specific solution.

23.5. REGRESSION TREATMENT OF THE ONE-WAY CLASSIFICATION: INDEPENDENT NORMAL EQUATIONS

In this section we tackle the one-way classification yet again with a reparameterization, reducing the parameter dimensions from $I + 1$ to I. An alternative way to look at this is that we choose to drop the column of 1's associated with the parameter μ. However it is described, the underlying regression framework remains the same, with vectors $\mathbf{Y}$ and $\hat{\mathbf{Y}}$ permanently fixed but $\hat{\mathbf{Y}}$ variously defined.

The analysis of variance model is

$$E(Y_{ij}) = \mu + t_i. \tag{23.5.1}$$

Let us write

$$\beta_i = \mu + t_i, \qquad i = 1, 2, \ldots, I.$$

Then, in regression terms, we can write

$$E(\mathbf{Y}) = \mathbf{X}\boldsymbol{\beta}, \tag{23.5.2}$$

where, comparing with the definitions in Section 23.4, $\mathbf{Y}$ is as before, $\mathbf{X}$ is the matrix formed by dropping the X_0 column of the previous $\mathbf{X}$ matrix, and $\boldsymbol{\beta}' = (\beta_1, \beta_2, \ldots,$

β_I). Let $\mathbf{b}' = (b_1, b_2, \ldots, b_I)$. Then

$$
\mathbf{X}'\mathbf{X} =
\begin{bmatrix}
J_1 & & & & \\
& J_2 & & & \mathbf{0} \\
& & \cdot & & \\
& \mathbf{0} & & \cdot & \\
& & & & J_I
\end{bmatrix}, \quad
\mathbf{X}'\mathbf{Y} =
\begin{bmatrix}
J_1 \bar{Y}_1 \\
J_2 \bar{Y}_2 \\
\vdots \\
J_I \bar{Y}_I
\end{bmatrix}.
\tag{23.5.3}
$$

Since $\mathbf{X}'\mathbf{X}$ is a diagonal matrix with J_i in the ith diagonal position and zero elsewhere, its inverse is a diagonal matrix with $1/J_i$ in the ith diagonal position. From this it is easy to see that

$$
b_i = \bar{Y}_i.
\tag{23.5.4}
$$

The sum of squares due to $\mathbf{b}$ is

$$
\mathbf{b}'\mathbf{X}'\mathbf{Y} = \sum_{i=1}^{I} \bar{Y}_i(J_i\bar{Y}_i) = \sum_{i=1}^{I} J_i \bar{Y}_i^2
\tag{23.5.5}
$$

and the residual sum of squares is

$$
\mathbf{Y}'\mathbf{Y} - \mathbf{b}'\mathbf{X}'\mathbf{Y} = \sum_{i=1}^{I} \sum_{j=1}^{J_i} Y_{ij}^2 - \sum_{i=1}^{I} J_i \bar{Y}_i^2
$$

$$
= \sum_{i=1}^{I} \left\{ \sum_{j=1}^{J_i} Y_{ij}^2 - J_i \bar{Y}_i^2 \right\}
\tag{23.5.6}
$$

$$
= \sum_{i=1}^{I} \sum_{j=1}^{J_i} (Y_{ij} - \bar{Y}_i)^2,
$$

with $(n - I)$ degrees of freedom.

The hypothesis $H_0 : t_1 = t_2 = \cdots = t_I = 0$ is expressed in our present notation as

$$
H_0 : \beta_1 = \beta_2 = \cdots = \beta_I = \mu.
$$

If H_0 were true, the model would be

$$
E(Y_{ij}) = \mu
$$

or

$$
E(\mathbf{Y}) = \mathbf{j}\mu,
\tag{23.5.7}
$$

where $\mathbf{j}$ is a vector of ones of the same length as $\mathbf{Y}$. The (single) normal equation would be

$$
n\mu = \mathbf{j}'\mathbf{j}\mu = \mathbf{j}'\mathbf{Y} = \sum_{i=1}^{I} \sum_{j=1}^{J_i} Y_{ij} = n\bar{Y}.
$$

Thus the estimate of μ would be

$$
b_0 = \bar{Y}
\tag{23.5.8}
$$

and

$$
\mathrm{SS}(b_0) = n\bar{Y}^2,
\tag{23.5.9}
$$

providing a residual sum of squares of

$$\sum_{i=1}^{I} \sum_{j=1}^{J_i} Y_{ij}^2 - n\overline{Y}^2 = \sum_{i=1}^{I} \sum_{j=1}^{J_i} (Y_{ij} - \overline{Y})^2 \tag{23.5.10}$$

as can be shown, with $(n - 1)$ degrees of freedom.

The sum of squares due to H_0 is the difference between Eqs. (23.5.10) and (23.5.6), namely,

$$SS(H_0) = \sum_{i=1}^{I} \sum_{j=1}^{J_i} \{(Y_{ij} - \overline{Y})^2 - (Y_{ij} - \overline{Y}_i)^2\}$$

$$= \sum_{i=1}^{I} J_i(\overline{Y}_i - \overline{Y})^2$$

after reduction, with $(n - 1) - (n - I) = (I - 1)$ degrees of freedom. The test statistic for H_0 is thus

$$F = \frac{SS(H_0)}{I - 1} \Big/ \frac{\sum_{i=1}^{I} \sum_{j=1}^{J_i} (Y_{ij} - \overline{Y}_i)^2}{n - I},$$

which is exactly what we obtain from the analysis of variance. Thus if we express the model for the one-way classification as $E(Y_{ij}) = \beta_i$ and test the hypothesis $H_0 : \beta_1 = \beta_2 = \cdots = \beta_I = \mu$, we can reproduce the analysis of variance through regression analysis using standard programs. We can obtain estimates of the parameters t_i from $b_i - b_0$.

23.6. THE TWO-WAY CLASSIFICATION WITH EQUAL NUMBERS OF OBSERVATIONS IN THE CELLS: AN EXAMPLE

The principles given in Sections 23.2 and 23.3 become even more important for more complicated experimental designs. We will consider an example of a two-way classification model with main effects and an interaction term, discussed by Smith (1969).

A manufacturer was having trouble in his catalyst plant with production rates. After extensive discussion with the research unit, it was decided to investigate the effects of twelve different combinations of four reagents and three catalysts. One of the problems that the plant had been encountering was an inability to reproduce production rates under what seemed to be identical conditions. In order to obtain an estimate of this inherent variability, it was decided to do each of the experimental runs twice. Thus the experiment consisted of 24 experimental runs, and these were done in a random order. The data are given in Table 23.4; they have been coded and rounded.

T A B L E 23.4. Twenty-four Production Rates (Coded and Rounded) for Twelve Combinations of Reagent and Catalyst

Reagent	Catalyst		
	1	2	3
A	4,6	11,7	5,9
B	6,4	13,15	9,7
C	13,15	15,9	13,13
D	12,12	12,14	7,9

T A B L E 23.5. Analysis of Variance for Model of Eq. (23.6.1)

Source	df	SS	MS	F
Among reagents	3	120	40	10.0[a]
Among catalysts	2	48	24	6.0[b]
Reagents × catalysts	6	84	14	3.5[b]
Pure error	12	48	4	
Total, corrected	23	200		

[a] Significant at the $\alpha = 0.05$ level.
[b] Significant at the $\alpha = 0.01$ level.

The ANOVA model for this experiment was taken as

$$Y_{ijk} = \mu + R_i + C_j + (RC)_{ij} + \epsilon_{ijk}, \tag{23.6.1}$$

where

μ = overall mean level (1 parameter).
R_i = effect of the ith reagent (4 parameters).
C_j = effect of the jth catalyst (3 parameters).
$(RC)_{ij}$ = interaction effect of reagent i and catalyst j (12 parameters).
ϵ_{ijk} = random error within the (i, j)th cell for the k observations within the cell, assumed to be distributed $N(0, \sigma^2)$. Errors are assumed to be pairwise uncorrelated.

Also,

$$i = 1, 2, \ldots, I \, (I = 4, \text{here}),$$

$$j = 1, 2, \ldots, J \, (J = 3, \text{here}),$$

$$k = 1, 2, \ldots, K \, (K = 2, \text{here}).$$

Note that this model involves 20 parameters and is overparameterized. The following restrictions are therefore assumed:

$$\sum_i R_i = \sum_j C_j = 0; \tag{23.6.2}$$

$$\sum_i (RC)_{ij} = 0, \quad j = 1, 2, \ldots, J; \tag{23.6.3}$$

$$\sum_j (RC)_{ij} = 0, \quad i = 1, 2, \ldots, I. \tag{23.6.4}$$

The number of independent parameters is thus reduced to $1 + 3 + 2 + 6 = 12$ [or $1 + (I - 1) + (J - 1) + (I - 1)(J - 1) = IJ$, in general], the number of cells. The usual analysis provides the ANOVA table of Table 23.5.
We now discuss how to analyze these data using a regression approach.

23.7. REGRESSION TREATMENT OF THE TWO-WAY CLASSIFICATION EXAMPLE

As we have already mentioned, twelve independent model parameters are needed for this example. Assignment of more than twelve to a regression model will thus produce a singular $\mathbf{X'X}$ matrix. Of these twelve, we need one for the mean, three for reagents, two for catalysts, and six for interaction. As always, many choices are possible for the set of dummy variables used to represent the regression model. Our choice will produce a diagonal $\mathbf{X}'\mathbf{X}$ (many choices will not, as in the one-way classification case). We shall write the regression model as

$$Y = \beta_0 X_0 \qquad \text{(constant term; } X_0 = 1)$$
$$+ \beta_1 X_1 + \beta_2 X_2 + \beta_3 X_3 \qquad \text{(for reagents)}$$
$$+ \beta_4 X_4 + \beta_5 X_5 \qquad \text{(for catalysts)}$$
$$+ \beta_6 X_6 + \beta_7 X_7 + \beta_8 X_8 + \beta_9 X_9 + \beta_{10} X_{10} + \beta_{11} X_{11} \qquad \text{(for } R \times C \text{ interaction)}$$
$$+ \epsilon, \qquad \text{(error)} \qquad (23.7.1)$$

and define the dummy variables as follows:

	Corresponding Values of Dummies		
Reagent Used	X_1	X_2	X_3
A	-1	0	-1
B	1	0	-1
C	0	-1	1
D	0	1	1

These are not arbitrary choices; they allow special comparisons to be made between reagents

$$A \text{ and } B, \quad \text{via } X_1,$$
$$C \text{ and } D, \quad \text{via } X_2, \qquad (23.7.2)$$
$$(A + B) \text{ and } (C + D), \quad \text{via } X_3,$$

and these independent comparisons or contrasts "carry" 3 degrees of freedom. Any other type of comparison can be generated from these columns. For example, to compare A and D we could use a column (dependent on X_1, X_2, and X_3 columns, of course) generated by $X_1 + X_2 + X_3$ whose entries will be $(-2, 0, 0, 2)'$. Also, because of the symmetry of the design as well, the $\mathbf{X}$ matrix columns generated by X_1, X_2, and X_3 will be orthogonal to one another. Provided they are also orthogonal to other $\mathbf{X}$ columns, as indeed they will be, individual additive sums of squares can be ascribed to each of the three contrasts generated by the X_1, X_2, and X_3 columns.

For the catalyst dummy variables, we assign as follows:

	Corresponding Values of Dummies	
Catalyst	X_4	X_5
1	-1	1
2	0	-2
3	1	1

The two orthogonal columns allow comparisons between catalysts

$$1 \text{ and } 3, \quad \text{via } X_4,$$
$$(1 + 3) \text{ and } 2, \quad \text{via } X_5. \qquad (23.7.3)$$

Because of the perfect balance of this two-way design and the fact that there are equal numbers of observations in each cell, we can construct the interaction dummies by multiplying up corresponding elements of the other dummies, taking the six combinations of (three reagent dummies) times (two catalyst dummies). Thus

$$X_6 = X_1 X_4, \qquad X_7 = X_1 X_5,$$
$$X_8 = X_2 X_4, \qquad X_9 = X_2 X_5, \qquad (23.7.4)$$
$$X_{10} = X_3 X_4, \qquad X_{11} = X_3 X_5.$$

We can now write down the $\mathbf{Y}$ vector and $\mathbf{X}$ matrix as given below.

$\mathbf{Y}$	$\mathbf{X_0}$	$\mathbf{X_1}$	$\mathbf{X_2}$	$\mathbf{X_3}$	$\mathbf{X_4}$	$\mathbf{X_5}$	$\mathbf{X_6}$	$\mathbf{X_7}$	$\mathbf{X_8}$	$\mathbf{X_9}$	$\mathbf{X_{10}}$	$\mathbf{X_{11}}$
4	1	-1	0	-1	-1	1	1	-1	0	0	1	-1
6	1	-1	0	-1	-1	1	1	-1	0	0	1	-1
11	1	-1	0	-1	0	-2	0	2	0	0	0	2
7	1	-1	0	-1	0	-2	0	2	0	0	0	2
5	1	-1	0	-1	1	1	-1	-1	0	0	-1	-1
9	1	-1	0	-1	1	1	-1	-1	0	0	-1	-1
6	1	1	0	-1	-1	1	-1	1	0	0	1	-1
4	1	1	0	-1	-1	1	-1	1	0	0	1	-1
13	1	1	0	-1	0	-2	0	-2	0	0	0	2
15	1	1	0	-1	0	-2	0	-2	0	0	0	2
9	1	1	0	-1	1	1	1	1	0	0	-1	-1
7	1	1	0	-1	1	1	1	1	0	0	-1	-1
13	1	0	-1	1	-1	1	0	0	1	-1	-1	1
15	1	0	-1	1	-1	1	0	0	1	-1	-1	1
15	1	0	-1	1	0	-2	0	0	0	2	0	-2
9	1	0	-1	1	0	-2	0	0	0	2	0	-2
13	1	0	-1	1	1	1	0	0	-1	-1	1	1
13	1	0	-1	1	1	1	0	0	-1	-1	1	1
12	1	0	1	1	-1	1	0	0	-1	1	-1	1
12	1	0	1	1	-1	1	0	0	-1	1	-1	1
12	1	0	1	1	0	-2	0	0	0	-2	0	-2
14	1	0	1	1	0	-2	0	0	0	-2	0	-2
7	1	0	1	1	1	1	0	0	1	1	1	1
9	1	0	1	1	1	1	0	0	1	1	1	1

$$\mathbf{Y} = \qquad\qquad \mathbf{X} = \qquad\qquad\qquad\qquad\qquad\qquad\qquad (23.7.5)$$

Examination shows that the columns of $\mathbf{X}$ are all orthogonal to one another so that $\mathbf{X'X}$ is a 12×12 diagonal matrix with entries

$$\mathbf{X'X} = \text{diag}\{24, 12, 12, 24, 16, 48, 8, 24, 8, 24, 16, 48\} \qquad (23.7.6)$$

and $(\mathbf{X'X})^{-1}$ is also diagonal with entries $\{\frac{1}{24}, \frac{1}{12}, \ldots, \frac{1}{16}, \frac{1}{48}\}$. Now

$$\mathbf{X'Y} = \{240, 12, -12, 48, 0, -48, 2, -18, -6, -18, -20, 36\}'. \qquad (23.7.7)$$

T A B L E 23.6. Anova Table for the Two-Way Classification Example Using a Regression Approach

Source of Variation	df	SS	MS	F
Among reagents	3	120	40	10^a
A vs B	1	12	12	3
C vs D	1	12	12	3
$(A + B)$ vs $(C + D)$	1	96	96	24^a
Among catalysts	2	48	24	6^b
1 vs 3	1	0^c	0	0
$(1 + 3)$ vs 2	1	48	48	12^a
Reagents $\times$ catalysts	6	84^d	14	3.5^b
Pure error	12	48	$s^2 = 4$	
Total, corrected	23	300		
b_0	1	2400		
Total	24	2700		

[a] Significant at the $\alpha = 0.01$ level.
[b] Significant at the $\alpha = 0.05$ level.
[c] A sum of squares that is exactly zero rarely occurs with real data and is often a sign that the data are constructed. Here it happens because we coded and rounded the original numbers to simplify the arithmetic.
[d] This sum of squares can be partitioned as shown in Eq. (23.7.9) into single df sums of squares. The significant contributors are then b_{10} and b_{11}, indicating the existence of "real" interactions between (X_3 and X_4) and (X_3 and X_5).

It follows that $\mathbf{b} = (\mathbf{X'X})^{-1}\mathbf{X'Y}$ has value

$$\mathbf{b} = \{10, 1, -1, 2, 0, -1, \tfrac{1}{4}, -\tfrac{3}{4}, -\tfrac{3}{4}, -\tfrac{3}{4}, -\tfrac{5}{4}, \tfrac{3}{4}\}', \tag{23.7.8}$$

and the sum of squares for regression, $\mathbf{b'X'Y}$, consists of the following twelve independent contributions, grouped for convenience, each ascribable as the sum of squares due to one b-coefficient:

$$\mathbf{b'X'Y} = 2400 + \{12 + 12 + 96\} + \{0 + 48\}$$

$$+ \{\tfrac{1}{2} + 13\tfrac{1}{2} + 4\tfrac{1}{2} + 13\tfrac{1}{2} + 25 + 27\} \tag{23.7.9}$$

$$= 2400 + 120 + 48 + 84$$

$$= 2652.$$

These results allow us to write the ANOVA table for this example as shown in Table 23.6.

From Eq. (23.7.8) we obtain the fitted regression model as

$$\hat{Y} = 10 + X_1 - X_2 + 2X_3 + 0X_4 - X_5 + \tfrac{1}{4}X_6 - \tfrac{3}{4}X_7$$

$$- \tfrac{3}{4}X_8 - \tfrac{3}{4}X_9 - \tfrac{5}{4}X_{10} + \tfrac{3}{4}X_{11}, \tag{23.7.10}$$

and this can be used to predict the production rates in the various cells and to obtain residuals. For example, for reagent C and catalyst 2, we see that

$$X_1 = 0, \quad X_4 = 0, \quad X_6 = 0, \quad X_7 = 0,$$

$$X_2 = -1, \quad X_5 = -2, \quad X_8 = 0, \quad X_9 = 2,$$

$$X_3 = 1, \qquad\qquad X_{10} = 0, \quad X_{11} = -2,$$

so that

$$\hat{Y} = 10 + 1 + 2 + 2 - \tfrac{3}{4}(2) + \tfrac{3}{4}(-2) = 12.$$

In that cell, the actual observations are 15 and 9, so the corresponding residuals are 3 and -3, which add to zero. The residuals add to zero in every cell, in fact, the model being essentially fitted to the means of the repeats. This happens in all regression situations in which the same number of repeat runs occur at every location.

Variances of estimated parameters and of fitted values can be obtained via the usual regression formulas. We leave further analysis of this example to the reader.

In Section 23.8, we present the standard general algebra of the two-way equal-celled analysis of variance, and in Section 23.9 we show two alternative regression approaches. One makes use of the extra sum of squares principle; the other is the "unsymmetrical dropping of parameters" approach to achieve a nonsingular $\mathbf{X'X}$ matrix. The first of these methods is illustrated by an example in Section 23.10.

23.8. THE TWO-WAY CLASSIFICATION WITH EQUAL NUMBERS OF OBSERVATIONS IN THE CELLS

Suppose we have a two-way classification with I rows and J columns, with K observations Y_{ijk}, $k = 1, 2, \ldots, K$ in the cells so defined. The usual fixed-effects analysis of variance model is

$$E(Y_{ijk}) = \mu + \alpha_i + \beta_j + \gamma_{ij}, \quad i = 1, 2, \ldots, I,$$
$$j = 1, 2, \ldots, J, \quad (23.8.1)$$
$$k = 1, 2, \ldots, K,$$

subject to the restrictions

$$\sum_{i=1}^{I} \alpha_i = \sum_{j=1}^{J} \beta_j = \sum_{i=1}^{I} \gamma_{ij} \, (\text{all } j) = \sum_{j=1}^{J} \gamma_{ij} \, (\text{all } i) = 0. \quad (23.8.2)$$

These restrictions enable μ to be regarded as the overall mean level while the α_i, β_i, and γ_{ij} are differences between row level and overall mean, column level and overall mean, and cell level and the joint row plus column level, respectively. The usual analysis of variance table is as follows:

ANOVA

Source	df	SS	MS
Rows	$I - 1$	$JK \sum_{i=1}^{I} (\overline{Y}_{i..} - \overline{Y})^2$	s_r^2
Columns	$J - 1$	$IK \sum_{j=1}^{J} (\overline{Y}_{.j.} - \overline{Y})^2$	s_c^2
Cells (interaction)	$(I - 1)(J - 1)$	$K \sum_{i=1}^{I} \sum_{j=1}^{J} (\overline{Y}_{ij.} - \overline{Y}_{i..} - \overline{Y}_{.j.} + \overline{Y})^2$	s_{rc}^2
Residual	$IJ(K - 1)$	By subtraction	s^2
Mean	1	$IJK\overline{Y}^2$	
Total	IJK	$\sum_{i=1}^{I} \sum_{j=1}^{J} \sum_{k=1}^{K} Y_{ijk}^2$	

Where $\overline{Y}_{i..}$ = the mean of all observations in row i
$\overline{Y}_{.j.}$ = the mean of all observations in column j
$\overline{Y}_{ij.}$ = the mean of all observations in the cell (i, j)
$\overline{Y}$ = the mean of all observations

The usual tests made are

H_0: all $\alpha_i = 0$ $F = s_r^2/s^2$ is compared with $F[(I-1), IJ(K-1)]$.

H_0: all $\beta_j = 0$ $F = s_c^2/s^2$ is compared with $F[(J-1), IJ(K-1)]$.

H_0: all $\gamma_{ij} = 0$ $F = s_{rc}^2/s^2$ is compared with $F[(I-1)(J-1), IJ(K-1)]$.

23.9. REGRESSION TREATMENT OF THE TWO-WAY CLASSIFICATION WITH EQUAL NUMBERS OF OBSERVATIONS IN THE CELLS

We could, if desired, handle this problem in a way similar to that given in Section 23.4, by writing down the *dependent* normal equations in the parameters μ, α_i, β_j, γ_{ij}, adding the equations given by the restrictions of Eq. (23.8.2), and solving a selected set of $(I+1)(J+1)$ independent equations. We shall instead deal with the situation in another way, so that nonsingular $\mathbf{X'X}$ matrices will always occur in the calculations.

There are $1 + I + J + IJ = (I+1)(J+1)$ parameters in all, but these are dependent through the $1 + 1 + J + I - 1 = I + J + 1$ restrictions defined by Eq. (23.8.2). We have to take one off the (at first sight) apparent number of restrictions to allow for the fact that, if we know all the *row* sums of the γ_{ij} are zero, so is the total sum of all γ_{ij}. This means that it is necessary to specify only that $J-1$ of the column sums of the γ_{ij} be zero, in order to achieve zero for the final column sum. Thus we actually need only IJ parameters to describe the model and we can define these as

$$\delta_{ij} = \mu + \alpha_i + \beta_j + \gamma_{ij}, \qquad i = 1, 2, \ldots, I,$$
$$j = 1, 2, \ldots, J.$$

Consider the following models:

1. $E(Y_{ijk}) = \delta_{ij}$.
2. $E(Y_{ijk}) = \delta_j$ independent of i.
3. $E(Y_{ijk}) = \delta_{i\cdot}$ independent of j.
4. $E(Y_{ijk}) = \delta$ independent of i and j.

We can represent all these in matrix form. Let

$$\mathbf{Y} = (Y_{111}, Y_{112}, \ldots, Y_{11K}; Y_{121}, Y_{122}, \ldots, Y_{12K}; \ldots; Y_{IJ1}, Y_{IJ2}, \ldots, Y_{IJK})',$$

where we order the cells in the sequence

$$(11), (12), \ldots, (1J); (21), (22), \ldots, (2J); \ldots; (I1), (I2), \ldots, (IJ)$$

and order the observations within the cells in the numerical order of the third subscript. Then, with the matrices indicated below, we can write all the models (1), (2), (3), and (4) in the form $E(\mathbf{Y}) = \mathbf{X}\boldsymbol{\beta}$.

Model (1)

$$\mathbf{X} = \begin{array}{c} \begin{array}{cccccccccccc} \delta_{11} & \delta_{12} & \cdots & \delta_{1J} & \delta_{21} & \delta_{22} & \cdots & \delta_{2J} & \cdots & \delta_{I1} & \delta_{I2} & \cdots & \delta_{IJ} \end{array} \\ \left[\begin{array}{cccccccccccc} 1 & 0 & \cdots & 0 & 0 & 0 & \cdots & 0 & \cdots & 0 & 0 & \cdots & 0 \\ 1 & 0 & \cdots & 0 & 0 & 0 & \cdots & 0 & \cdots & 0 & 0 & \cdots & 0 \\ \vdots & \vdots & & \vdots & \vdots & \vdots & & \vdots & & \vdots & \vdots & & \vdots \\ 1 & 0 & \cdots & 0 & 0 & 0 & \cdots & 0 & \cdots & 0 & 0 & \cdots & 0 \\ \hline 0 & 1 & \cdots & 0 & 0 & 0 & \cdots & 0 & \cdots & 0 & 0 & \cdots & 0 \\ 0 & 1 & \cdots & 0 & 0 & 0 & \cdots & 0 & \cdots & 0 & 0 & \cdots & 0 \\ \vdots & \vdots & & \vdots & \vdots & \vdots & & \vdots & & \vdots & \vdots & & \vdots \\ 0 & 1 & \cdots & 0 & 0 & 0 & \cdots & 0 & \cdots & 0 & 0 & \cdots & 0 \\ \hline \vdots & \vdots & & \vdots & \vdots & \vdots & & \vdots & & \vdots & \vdots & & \vdots \\ \hline 0 & 0 & \cdots & 0 & 0 & 0 & \cdots & 0 & \cdots & 0 & 0 & \cdots & 1 \\ 0 & 0 & \cdots & 0 & 0 & 0 & \cdots & 0 & \cdots & 0 & 0 & \cdots & 1 \\ \vdots & \vdots & & \vdots & \vdots & \vdots & & \vdots & & \vdots & \vdots & & \vdots \\ 0 & 0 & \cdots & 0 & 0 & 0 & \cdots & 0 & \cdots & 0 & 0 & \cdots & 1 \end{array}\right] \end{array},$$

where each row segment is of depth K.

$$\boldsymbol{\beta}' = (\delta_{11}, \delta_{12}, \ldots, \delta_{1J}; \delta_{21}, \delta_{22}, \ldots, \delta_{2J}; \ldots; \delta_{I1}, \delta_{I2}, \ldots, \delta_{IJ}).$$

Model (2)

$$\mathbf{X} = \begin{array}{c} \begin{array}{ccccc} \delta_{\cdot 1} & \delta_{\cdot 2} & \delta_{\cdot 3} & \cdots & \delta_{\cdot J} \end{array} \\ \left[\begin{array}{ccccc} \mathbf{j} & \mathbf{0} & \mathbf{0} & \cdots & \mathbf{0} \\ \mathbf{0} & \mathbf{j} & \mathbf{0} & \cdots & \mathbf{0} \\ \mathbf{0} & \mathbf{0} & \mathbf{j} & \cdots & \mathbf{0} \\ \vdots & \vdots & \vdots & \cdots & \vdots \\ \mathbf{0} & \mathbf{0} & \mathbf{0} & \cdots & \mathbf{j} \\ \hline \multicolumn{5}{c}{(I-1) \text{ more blocks}} \\ \multicolumn{5}{c}{\text{exactly as above}} \\ \\ \end{array}\right] \end{array},$$

where $\mathbf{j}$ denotes a $K \times 1$ vector of unities.

$$\boldsymbol{\beta}' = (\delta_{\cdot 1}, \delta_{\cdot 2}, \ldots, \delta_{\cdot J}).$$

Model (3)

$$X = \begin{matrix} \delta_{1\cdot} & \delta_{2\cdot} & \delta_{3\cdot} & \cdots & \delta_{I\cdot} \\ \begin{bmatrix} j & 0 & 0 & \cdots & 0 \\ 0 & j & 0 & \cdots & 0 \\ 0 & 0 & j & \cdots & 0 \\ \vdots & \vdots & \vdots & \cdots & \vdots \\ 0 & 0 & 0 & \cdots & j \end{bmatrix} \end{matrix},$$

where **j** here denotes a $JK \times 1$ vector of unities.

$$\boldsymbol{\beta'} = (\delta_{1\cdot}, \delta_{2\cdot}, \ldots, \delta_{I\cdot}).$$

Model (4)

$$X = j,$$

where **j** here denotes a $IJK \times 1$ vector of unities.

$$\boldsymbol{\beta} = \delta, \text{ a scalar.}$$

We can construct the standard analysis of variance table through regression methods as follows. Denote by S_1, S_2, S_3, and S_4 the regression sums of squares, which arise from the four regression models given above and let $S = \Sigma_{i=1}^{I} \Sigma_{j=1}^{J} \Sigma_{k=1}^{K} \Sigma Y_{ijk}^2$. Then using the "extra sum of squares" principle given in Section 6.1, we can construct the table below.

ANOVA

Source	df	MS
Rows	$I - 1$	$S_3 - S_4$
Columns	$J - 1$	$S_2 - S_4$
Interaction	$(I - 1)(J - 1)$	$S_1 - S_2 - S_3 + S_4$
Residual	$IJ(K - 1)$	$S - S_1$
Mean	1	S_4
Total	IJK	S

(The interaction sum of squares is actually obtained from

$$S_1 - (S_2 - S_4) - (S_3 - S_4) - S_4,$$

which reduces to the form given in the table.)

The equivalence between these sums of squares and the ones given in the analysis of variance procedure can easily be demonstrated mathematically but we shall not do this.

We are usually interested in obtaining estimates m, a_i, b_j, c_{ij} of the original parameters $\mu, \alpha_i, \beta_j, \gamma_{ij}$ in the analysis of variance model. These estimates can be obtained from the estimated $d_{ij}, d_{\cdot j}, d_{i\cdot}, d$ of the regression coefficients $\delta_{ij}, \delta_{\cdot j}, \delta_{i\cdot}, \delta$ in the four models. They are

$$m = d,$$

$$a_i = d_{i\cdot} - d,$$

$$b_j = d_{\cdot j} - d,$$

$$c_{ij} = d_{ij} - d_{i\cdot} - d_{\cdot j} + d.$$

An Alternative Method

Our suggested method of dealing with the two-way analysis of variance classification has involved four symmetric regression analyses and the use of the extra sum of squares principle. To deal with this situation in a single regression analysis we must write an unsymmetric model that omits some of the dependent parameters of the standard analysis of variance model. We illustrate this with an example. Consider the two-way classification below, which has two observations in each cell:

	Column $j = 1$	$j = 2$
Row $i = 1$	Y_1, Y_2	Y_3, Y_4
$i = 2$	Y_5, Y_6	Y_7, Y_8
$i = 3$	Y_9, Y_{10}	Y_{11}, Y_{12}

The standard analysis of variance model is

$$E(Y_{ij}) = \mu + \alpha_i + \beta_j + \gamma_{ij},$$

where

$$\alpha_1 + \alpha_2 + \alpha_3 = 0 \qquad \gamma_{11} + \gamma_{12} = 0$$
$$\beta_1 + \beta_2 = 0 \qquad \gamma_{21} + \gamma_{22} = 0$$
$$\gamma_{31} + \gamma_{32} = 0$$
$$\gamma_{11} + \gamma_{21} + \gamma_{31} = 0$$
$$\gamma_{12} + \gamma_{22} + \gamma_{32} = 0$$

Thus if (for example) μ, α_1, α_2, β_1, γ_{11}, and γ_{21} are known or estimated, all other parameters or their estimates, respectively, can be found from the restrictions. We can thus write the regression model

$$E(Y_{ij}) = \mu + \alpha_1 X_1 + \alpha_2 X_2 + \beta_1 X_3 + \gamma_{11} X_4 + \gamma_{21} X_5$$

or

$$E(\mathbf{Y}) = \mathbf{X}\boldsymbol{\beta},$$

where

$$
\mathbf{Y} =
\begin{bmatrix}
Y_1 \\
Y_2 \\
Y_3 \\
Y_4 \\
Y_5 \\
Y_6 \\
Y_7 \\
Y_8 \\
Y_9 \\
Y_{10} \\
Y_{11} \\
Y_{12}
\end{bmatrix},
\qquad
\mathbf{X} =
\begin{array}{cccccc}
\mu & \alpha_1 & \alpha_2 & \beta_1 & \gamma_{11} & \gamma_{21} \\
\end{array}
\begin{bmatrix}
1 & 1 & 0 & 1 & 1 & 0 \\
1 & 1 & 0 & 1 & 1 & 0 \\
1 & 1 & 0 & -1 & -1 & 0 \\
1 & 1 & 0 & -1 & -1 & 0 \\
1 & 0 & 1 & 1 & 0 & 1 \\
1 & 0 & 1 & 1 & 0 & 1 \\
1 & 0 & 1 & -1 & 0 & -1 \\
1 & 0 & 1 & -1 & 0 & -1 \\
1 & -1 & -1 & 1 & -1 & -1 \\
1 & -1 & -1 & 1 & -1 & -1 \\
1 & -1 & -1 & -1 & 1 & 1 \\
1 & -1 & -1 & -1 & 1 & 1
\end{bmatrix},
$$

$$\boldsymbol{\beta}' = (\mu, \alpha_1, \alpha_2, \beta_1, \gamma_{11}, \gamma_{21}).$$

Note: The elements of the γ_{ij} column are obtained as the product of corresponding elements of the α_i and β_j columns.

Any independent subset of the parameters can be used for such a model and there are many possible alternative forms. From this point, the usual regression methods are used to estimate β. Because of the orthogonality in $\mathbf{X}$, we can obtain separate, orthogonal sums of squares for the estimates of (1) μ, (2) α_1 and α_2, (3) β_1, (4) γ_{11} and γ_{21}. These will be the usual sums of squares for (1) mean, (2) rows, (3) columns, (4) interaction, in the standard analysis of variance setup.

23.10. EXAMPLE: THE TWO-WAY CLASSIFICATION

The data in the two-way classification below appear in Brownlee (1965, p. 475). Descriptive details are omitted.

	Column 1	Column 2	Column 3
Row 1	17, 21, 49, 54	64, 48, 34, 63	62, 72, 61, 91
Row 2	33, 37, 40, 16	41, 64, 34, 64	56, 62, 57, 72

Following the procedure given in Section 23.9 we can calculate the quantities below through regression methods.

(1) $S_1 = 65,863$ $\begin{pmatrix} d_{11} & d_{12} & d_{13} \\ d_{21} & d_{22} & d_{23} \end{pmatrix} = \begin{pmatrix} 35.25 & 52.25 & 71.50 \\ 31.50 & 50.75 & 61.75 \end{pmatrix}.$

(2) $S_2 = 65,640.25$ $(d_{\cdot 1}, d_{\cdot 2}, d_{\cdot 3}) = (33.375, 51.5, 66.625).$

(3) $S_3 = 61,356$ $\begin{pmatrix} d_{1\cdot} \\ d_{2\cdot} \end{pmatrix} = \begin{pmatrix} 53 \\ 48 \end{pmatrix}.$

(4) $S_4 = 61,206$ $d = 50.5.$

Using the formulas given in the previous section we obtain an analysis of variance table as follows:

ANOVA

Source	df	SS
Rows	1	150.00
Columns	2	4,434.25
Interaction	2	72.75
Residual	18	3,495.00
Mean	1	61,206.00
Total (uncorrected)	24	69,358.00

The same table was obtained in Brownlee (1965) through the usual analysis of variance calculations. The estimates of the usual analysis of variance parameters are given by

$$m = 50.5,$$

$$a_1 = 53 - 50.5 = 2.5, \qquad a_2 = 48 - 50.5 = -2.5.$$

Note: $a_1 + a_2 = 0$, as it should.

$$b_1 = 33.375 - 50.5 = -17.125, \qquad b_2 = 51.5 - 50.5 = 1.0,$$

$$b_3 = 66.625 - 50.5 = 16.125.$$

Note: $b_1 + b_2 + b_3 = 0$ as it should.

$$c_{11} = 35.25 - 53 - 33.375 + 50.5 = -0.625,$$

$$c_{12} = 52.25 - 53 - 51.5 + 50.5 = -1.750,$$

$$c_{13} = 71.50 - 53 - 66.625 + 50.5 = 2.375,$$

$$c_{21} = 31.50 - 48 - 33.375 + 50.5 = 0.625,$$

$$c_{22} = 50.75 - 48 - 51.5 + 50.5 = 1.750,$$

$$c_{23} = 61.75 - 48 - 66.625 + 50.5 = -2.375.$$

Note: $\Sigma_{i=1}^2 c_{ij} = \Sigma_{j=1}^3 c_{ij} = 0$, as they should.

The residuals from this analysis of variance model would be

$$e_{ijk} = Y_{ijk} - m - a_i - b_j - c_{ij}$$

$$= Y_{ijk} - d - (d_{i.} - d) - (d_{.j} - d) - (d_{ij} - d_{i.} - d_{.j} + d)$$

$$= Y_{ijk} - d_{ij},$$

which are the residuals from the regression analysis using Model (1). These would be examined in the usual ways. Also, plots of residuals for each row and column can be examined.

23.11. RECAPITULATION AND COMMENTS

We have seen, in the specific cases discussed, that analysis of variance can, if necessary, be conducted by standard regression techniques. If the model is examined carefully and reparameterized properly, the analysis of variance table for other models could also be obtained in a similar way. The proper selection of the dummy variables is crucial to the proper presentation of the detailed results and can ease the computations considerably. However, many reparameterizations are valid, as is generally true in dummy variable situations; some reparameterizations are simply better than others for certain purposes. The more complicated the design is, the more complicated will be the regression form conversion, and the larger the **X** matrix. The effort required can be considerable and, when the design is standard and all the data are available (i.e., there are no missing observations), is usually not worthwhile. It is usually better to employ an appropriate analysis of variance computation method, or computer routine, where it exists. Nevertheless, it is useful to appreciate the connection between the two methods of analysis for several reasons:

1. It focuses attention on the fact that a model is necessary in analysis of variance problems.

2. It points up the fact that the residuals in analysis of variance models play the same role as residuals in regression models and *must* be examined for the information they contain on the possible inadequacy of the model under consideration. (There seems to be a tacit assumption in most variance analysis that the model is correct.)

3. When observations are missing from analysis of variance data, they can often be "estimated" by standard formulas. If this is inconvenient, or too many observations are missing, the data can usually be analyzed by a regression routine setting up models as illustrated above but deleting rows of the **X** matrices for which no observations are available. (The word estimated is placed in quotes because no real estimation takes place. The "estimates" are simply numbers inserted for calculation purposes, which lead to the same estimates of parameters that would otherwise have been obtained from the incomplete data via regression analysis.)

4. Adding additional predictor variables (covariates) to an analysis of variance structure is simple via regression.

EXERCISES FOR CHAPTER 23

A. Below, denoted by a and b, are two-way classification experiments. On each of these experiments, perform these analyses:
 1. Analyze these data using any of the methods described in Chapter 23.
 2. Evaluate the fitted values and residuals and examine the residuals in all reasonable ways. State any defects you find.
 3. Confirm that the alternative regression methods given in Chapter 23 lead to the same results.
   a. An experiment was conducted to determine the effect of steam pressure and blowing time on the percentage of foreign matter left in filter earth. The data are as follows:

Steam Pressure (pounds)	Percentage of Foreign Matter		
	Blowing Time (hours)		
	1	2	3
10	45.2, 46.0	40.0, 39.0	35.9, 34.1
20	41.8, 20.6	27.8, 19.0	22.5, 17.7
30	23.5, 33.1	44.6, 52.2	42.7, 48.6

   b. An experiment was conducted on the effect of premixing speed and finish mixer speed on the center heights of cakes. Three different levels of speed were chosen for each of the two variables. The data collected were as follows:

(Premix speed -5)	(Finish mix speed -3.5) $\times$ 2	(Center height -2) $\times$ 100
X_1	X_2	Y
-1	-1	4, -3
-1	0	3, 2
-1	1	$-1, -5$
0	-1	3, 10
0	0	2, 2
0	1	0, 0
1	-1	$-1, -10$
1	0	1, 2
1	1	7, 9

B. A chemical experiment was performed to investigate the effect of extrusion temperature X_1 and cooling temperature X_2 on the compressibility of a finished product. Knowledge of the process suggested that a model of the form

$$Y = \beta_0 + \beta_1 X_1 + \beta_2 X_2 + \beta_{12} X_1 X_2 + \epsilon$$

would satisfactorily explain the variation observed. Two levels of extrusion temperature and two levels of cooling temperature were chosen and all four of the combinations were performed. Each of the four experiments was carried out four times and the data yielded the following information:

ANOVA

Source of Variation	df	SS	MS
Total	16	921.0000	
Due to regression		881.2500	
b_0	1	798.0625	
b_1	1	18.0625	
b_2			
b_{12}		5.0625	
Residual			

1. a. Complete the above ANOVA table.
   b. Using $\alpha = 0.05$, examine the following questions:
   (1) Is the overall regression equation given b_0 statistically significant?
   (2) Are all the b-coefficients significant?
2. Given the following additional information,

$$\Sigma Y_i = 113, \quad \Sigma x_{1i} y_i = 31, \quad \text{and} \quad \Sigma x_{1i} x_{2i} y_i = -9,$$

where $x_{ji} = X_{ji} - \bar{X}_j$, $j = 1, 2$, and $y_i = Y_i - \bar{Y}$:
   a. Determine b_0, b_1, b_2, and b_{12}, and write out the prediction equation.
   b. The predicted value of $\hat{Y}$ at $X_1 = 70°$ and $X_2 = 150°$ is 54. The variance of this predicted value is 0.6875. What is the variance of a single predicted observation at the point $X_1 = 70$, $X_2 = 150$?
   c. Place 95% confidence limits on the true mean value of Y at the point $X_1 = 70$, $X_2 = 150$.
3. What conclusions can be drawn from your analysis?
C. Analyze the data given in Section 23.10 by the alternative method of Section 23.9.
D. (*Source: Introduction to Statistical Inference* by E. S. Keeping, Van Nostrand, Princeton, NJ, 1962, p. 216.) Apply the regression method of either Section 23.4 or Section 23.5 to the one-way classification data below. Actually carry out the calculation steps of your selected method, rather than just going to the analysis of variance table. The data were taken to test the effect of adding a small percentage of coal dust to the sand used for making concrete. Several batches were mixed under practically identical conditions except for the variation in the percentage of coal. From each batch, four cylinders were made and tested for breaking strength in pounds per square inch (lb/in.²). One cylinder in the third sample was defective, so there were only three items in this sample.

Sample Number:	1	2	3	4	5
Percentage coal:	0	0.05	0.1	0.5	1.0
Breaking strengths:	1690	1550	1625	1725	1530
	1580	1445	1450	1550	1545
	1745	1645	1510	1430	1565
	1685	1545		1445	1520

(*Hint:* Work with $Y_{ij} - 1430$, it is easier; ask yourself what difference it will make to the analysis.)

E. An important step in paper manufacturing is the removal of water from the paper. In a particular process, four factors were thought to affect the amount of water removed, and it was decided to run a 2^4 factorial design to examine them. The factors and the two levels (in appropriate units) chosen for each, were as follows:

Factor	Designation	Low Level	High Level
Vacuum on pressure roll 2 (hg)	A	0	18
Vacuum on pressure roll 1 (hg)	B	0	19
Weight of the paper (lb)	C	10.0	13.7
Process line speed (fpm)	D	1700	2000

The data are given in the table below. The notation *bcd*, for example, designates the run with *A* at its low level and *B*, *C*, and *D* at their high levels; the "1" indicates all factors at their low levels.

Run Number	Factor Combination	Percentage of Water Removed, Y
1	bcd	39.7
2	abcd	41.1
3	cd	40.6
4	acd	40.4
5	ad	41.0
6	bd	37.6
7	d	38.7
8	abd	39.0
9	bc	38.9
10	ac	40.0
11	abc	41.0
12	c	42.9
13	a	40.2
14	b	35.4
15	ab	39.4
16	1	39.0

Requirements

1. Analyze these data using regression techniques.
2. Show that the factorial design approach and the analysis of variance approach to these data provide exactly the same results and/or conclusions.

F. The data below came from a one way analysis of variance problem with (overparameterized) model

$$Y_{ij} = \mu + t_i + \epsilon_{ij}, \quad i,j = 1, 2, 3.$$

Group	Y_{ij}'s	Row Sum	Row SS
1	3, 4, 5	12	50
2	4, 6, 8	18	116
3	6, 8, 10	24	200
Column sums		54	366

1. Write down the **X** matrix implied by this model, and explain why using this **X** may not be a good idea in a regression context.
2. Suggest any one way of resolving the difficulty, perform the analysis, and test the null hypothesis that all groups have the same effects.

G. (*Source*: "Obtaining a sterilized soil for the growth of *Onoclea* gametophytes," by G. Rubin and D. J. Paolillo, *The New Phytologist* **97**, 1984, 621–628. *Onoclea sensibilis* L. is a fern.) On p. 623 of the source reference, we read: "The sterilized soil within three tubes was divided into five levels, and the soil from each level was placed into individual Petri plates. Spore germination was assessed for each plate. No clear pattern emerged for the variation within and between tubes."

1. The data are given in Table G1. Use regression methods with dummy variables (any workable set will do) to confirm the results quoted below Table G1.
2. If, instead, the data had been as in Table G2, what would the conclusion have been? (*Note*: Table G2 has no basis in fact but is simply a partial rearrangement of the numbers in Table G1, made in order to provide a second exercise.)

T A B L E G1. Germination of *Onoclea* Spores on Gamma-Irradiated Soil

Level Within Tube	Germination, % in Tube Number			Level Means
	1	2	3	
1	78.4	84.5	62.1	75.0
2	10.6	77.5	94.4	60.8
3	62.7	4.4	24.8	30.6
4	92.5	17.5	0.9	37.0
5	11.6	0.9	20.4	11.0
Tube means	51.2	37.0	40.5	42.9

$F_{tube}^{2,8} = 0.21$, $F_{level}^{4,8} = 1.48$. Neither is significant at the 5% level.

T A B L E G2. Fake (Rearranged) Data

Level Within Tube	Germination, % in Tube Number			Level Means
	1	2	3	
1	78.4	84.5	62.1	75.0
2	92.5	77.5	94.4	88.1
3	62.7	20.4	24.8	36.0
4	10.6	17.5	11.6	13.2
5	0.9	0.9	4.4	2.1
Tube means	49.0	40.2	39.5	42.9

H. (*Source*: "Hybridization and mating behavior in *Aedes aegypti* (diptera: culicidae)," by D. F. Moore, *Journal of Medical Entomology*, **16**, 1979, No. 3, 223–226.) A study on two forms of the yellow fever mosquito, the light-colored domestic (D) and the dark-colored sylvan (S) in Kenya, examined various matings. Table H shows the cumulative mean hatchabilities of DD, DS, SD, and SS matings where the first letter refers to female and the second to male mosquitos. In the original paper, equality of population means of the four matings was tested using (nonparametric) Kruskal–Wallis tests because "the hatchabilities are not known to be normally distributed" (p. 225). (In particular, there is correlation column-wise.) Nevertheless:

1. Perform a regression analysis with model $Y = \beta_0 + \beta_1 X_1 + \beta_2 X_2 + \beta_3 X_3 + \beta_{23} X_2 X_3 + \epsilon$ using these variables

$$Y = \text{hatchability,}$$
$$X_1 = \text{number of immersions in water}$$
$$X_2 = -1 \text{ for female, 1 for male, domestic (D),}$$
$$X_3 = 1 \text{ for female, } -1 \text{ for male, sylvan (S).}$$

Are there differences between the four groups, do you think? Also try these regression analyses as well:

2. Replace Y by $n_i^{1/2} \sin^{-1}(Y^{1/2})$.
3. Replace X_1 by $\ln X_1$
4. Make both replacements (2) and (3).

T A B L E H. Cumulative Mean Hatchabilities of Eggs Resulting from Matings Among Forms of *Aedes aegypti*

Number of Immersions in Water	Matings			
	DD	DS	SD	SS
1	0.665	0.469	0.284	0.127
2	0.736	0.504	0.354	0.183
3	0.782	0.559	0.404	0.195
8	0.899	0.770	0.577	0.297
n_i	2024	541	1100	222

n_i = total number of eggs per column.

CHAPTER 24

An Introduction to Nonlinear Estimation

24.1. LEAST SQUARES FOR NONLINEAR MODELS

Introduction

This chapter introduces nonlinear estimation, which involves the fitting of nonlinear models by least squares. The normal equations are not linear in this application and are, in general, difficult to solve. Direct minimization of the sum of squares function is usually performed. This typically requires heavy iterative calculations and the use of a special program. Although the details and outputs of such programs vary, the basic material in this chapter applies whichever program the reader selects. The exercises (and their solutions) provide a variety of examples of nonlinear data and the models fitted to them.

Nonlinear Models

In previous chapters we have mostly fitted, by least squares, models that were *linear in the parameters* and were of the type

$$Y = \beta_0 + \beta_1 Z_1 + \beta_2 Z_2 + \cdots + \beta_{p-1} Z_{p-1} + \epsilon, \tag{24.1.1}$$

where the Z_i can represent any functions of the basic predictor variables $X_1, X_2, \ldots,$ X_k. While Eq. (24.1.1) can represent a wide variety of relationships, there are many situations in which a model of this form is not appropriate; for example, when definite information is available about the form of the relationship between the response and the predictor variables. Such information might involve direct knowledge of the actual form of the true model or might be represented by a set of differential equations that the model must satisfy. Sometimes the information leads to several alternative models (in which case methods for discriminating between them will be of interest). When we are led to a model of nonlinear form, we would usually prefer to fit such a model whenever possible, rather than to fit an alternative, perhaps less realistic, linear model.

Any model that is *not* of the form given in Eq. (24.1.1) will be called a *nonlinear model,* that is, nonlinear in the *parameters.* Two examples of such models are

$$Y = \exp(\theta_1 + \theta_2 t + \epsilon), \tag{24.1.2}$$

$$Y = \frac{\theta_1}{\theta_1 - \theta_2} [e^{-\theta_2 t} - e^{-\theta_1 t}] + \epsilon. \tag{24.1.3}$$

In these examples the parameters to be estimated are denoted by θ's rather than β's as used previously, t is the single predictor variable, and ϵ is a random error term with $E(\epsilon) = 0$, $V(\epsilon) = \sigma^2$. (We could also write these models without ϵ and replacing Y by η. Then, the models would show how *true* values of the response, η, depend on t. Here, we wish to be specific about how the error enters the model to permit the discussion that follows.)

The models in Eqs. (24.1.2) and (24.1.3) are both nonlinear in the sense that they involve θ_1 and θ_2 in a nonlinear way but they are of essentially different characters. Equation (24.1.2) can be transformed, by taking logarithms to the base e, into the form

$$\ln Y = \theta_1 + \theta_2 t + \epsilon, \tag{24.1.4}$$

which is the form of Eq. (24.1.1) and is *linear* in the parameters. We can thus say that the model given in Eq. (24.1.2) is *intrinsically linear* since it can be transformed into linear form. (Some writers use the phrase *nonintrinsically nonlinear,* but we shall not.)

However, it is impossible to convert Eq. (24.1.3) into a form linear in the parameters. Such a model is said to be *intrinsically nonlinear.* While, at times, it may be useful to transform a model of this type so that it can be more easily fitted, it will remain a nonlinear model, whatever the transformation applied. Unless specifically noted, all models mentioned in this chapter will be intrinsically nonlinear.

Note: In models in which the error is additive, an intrinsically linear model is one that can be made linear by a transformation of parameters; for example, $Y = e^\theta X + \epsilon$ is of this type since, if we transform by $\beta = e^\theta$, the model becomes $Y = \beta X + \epsilon$. Other authors use the words *intrinsically linear* in this sense only.

Least Squares in the Nonlinear Case

The standard notation for nonlinear least squares situations is different from that for linear least squares cases. This may seem confusing to the reader at first, but the notation is well established in the literature. The differences are shown in Table 24.1.

Suppose the postulated model is of the form

$$Y = f(\xi_1, \xi_2, \ldots, \xi_k; \theta_1, \theta_2, \ldots, \theta_p) + \epsilon. \tag{24.1.5}$$

If we write

$$\boldsymbol{\xi} = (\xi_1, \xi_2, \ldots, \xi_k)',$$
$$\boldsymbol{\theta} = (\theta_1, \theta_2, \ldots, \theta_p)',$$

T A B L E 24.1. Standard Notations for Linear and Nonlinear Least Squares

	Linear	Nonlinear
Response	Y	Y
Subscripts of observations	$i = 1, 2, \ldots, n$	$u = 1, 2, \ldots, n$
Predictor variables	$X_1, X_2, \ldots, X_k$	$\xi_1, \xi_2, \ldots, \xi_k$
		(Sometimes $t = $ time, or $T = $ temperature, etc., sometimes even $X_1, X_2, \ldots, X_k$)
Parameters	$\beta_0, \beta_1, \ldots, \beta_p$	$\theta_0, \theta_1, \ldots, \theta_p$
		(Sometimes, $\alpha, \beta, \ldots, \phi, \ldots,$ etc.)

we can shorten Eq. (24.1.5) to

$$Y = f(\boldsymbol{\xi}, \boldsymbol{\theta}) + \epsilon$$

or

$$E(Y) = f(\boldsymbol{\xi}, \boldsymbol{\theta}) \tag{24.1.6}$$

if we assume that $E(\epsilon) = 0$. We shall also assume that errors are uncorrelated, that $V(\epsilon) = \sigma^2$, and, usually, that $\epsilon \sim N(0, \sigma^2)$ so that errors are independent.

When there are n observations of the form

$$Y_u, \xi_{1u}, \xi_{2u}, \ldots, \xi_{ku},$$

for $u = 1, 2, \ldots, n$, available, we can write the model in the alternative form

$$Y_u = f(\xi_{1u}, \xi_{2u}, \ldots, \xi_{ku}; \theta_1, \theta_2, \ldots, \theta_p) + \epsilon_u, \tag{24.1.7}$$

where ϵ_u is the uth error, $u = 1, 2, \ldots, n$. This can be abbreviated to

$$Y_u = f(\boldsymbol{\xi}_u, \boldsymbol{\theta}) + \epsilon_u, \tag{24.1.8}$$

where $\boldsymbol{\xi}_u = (\xi_{1u}, \xi_{2u}, \ldots, \xi_{ku})'$. The assumption of normality and independence of the errors can now be written as $\boldsymbol{\epsilon} \sim N(\mathbf{0}, \mathbf{I}\sigma^2)$, where $\boldsymbol{\epsilon} = (\epsilon_1, \epsilon_2, \ldots, \epsilon_n)'$, and as usual $\mathbf{0}$ is a vector of zeros and $\mathbf{I}$ is a unit matrix, both of appropriate sizes. We define the *error sum of squares* for the nonlinear model and the given data as

$$S(\boldsymbol{\theta}) = \sum_{u=1}^{n} \{Y_u - f(\boldsymbol{\xi}_u, \boldsymbol{\theta})\}^2. \tag{24.1.9}$$

Note that since Y_u and $\boldsymbol{\xi}_u$ are fixed observations, the sum of squares is a function of $\boldsymbol{\theta}$. We shall denote by $\hat{\boldsymbol{\theta}}$, a *least squares esimate* of $\boldsymbol{\theta}$, that is a value of $\boldsymbol{\theta}$ which minimizes $S(\boldsymbol{\theta})$. [It can be shown that, if $\boldsymbol{\epsilon} \sim N(\mathbf{0}, \mathbf{I}\sigma^2)$, the least squares estimate of $\boldsymbol{\theta}$ is also the maximum likelihood estimate of $\boldsymbol{\theta}$. This is because the likelihood function for this problem can be written

$$\ell(\boldsymbol{\theta}, \sigma^2) = (2\pi\sigma^2)^{-n/2} e^{-S(\boldsymbol{\theta})/2\sigma^2}$$

so that if σ^2 is known, maximizing $\ell(\boldsymbol{\theta}, \sigma^2)$ with respect to $\boldsymbol{\theta}$ is equivalent to minimizing $S(\boldsymbol{\theta})$ with respect to $\boldsymbol{\theta}$.]

To find the least squares estimate $\hat{\boldsymbol{\theta}}$ we need to differentiate Eq. (24.1.9) with respect to $\boldsymbol{\theta}$. This provides the p *normal equations*, which must be solved for $\hat{\boldsymbol{\theta}}$. The normal equations take the form

$$\sum_{u=1}^{n} \{Y_u - f(\boldsymbol{\xi}_u, \hat{\boldsymbol{\theta}})\} \left[\frac{\partial f(\boldsymbol{\xi}_u, \boldsymbol{\theta})}{\partial \theta_i} \right]_{\boldsymbol{\theta} = \hat{\boldsymbol{\theta}}} = 0 \tag{24.1.10}$$

for $i = 1, 2, \ldots, p$, where the quantity denoted by brackets is the derivative of $f(\boldsymbol{\xi}_u, \boldsymbol{\theta})$ with respect to θ_i with all θ's replaced by the corresponding $\hat{\theta}$'s, which have the same subscript. We recall that when the function $f(\boldsymbol{\xi}_u, \boldsymbol{\theta})$ was linear this quantity was a function of the $\boldsymbol{\xi}_u$ only and did not involve the $\hat{\theta}$'s at all. For example, if

$$f(\boldsymbol{\xi}_u, \boldsymbol{\theta}) = \theta_1 \xi_{1u} + \theta_2 \xi_{2u} + \cdots + \theta_p \xi_{pu},$$

then

$$\frac{\partial f}{\partial \theta_i} = \xi_{iu}, \qquad i = 1, 2, \ldots, p,$$

and is independent of $\boldsymbol{\theta}$. This leaves the normal equations in the form of linear equations in $\theta_1, \theta_2, \ldots, \theta_p$ as we saw in previous chapters. When the model is nonlinear in the θ's, so will be the normal equations. We now illustrate this with a simple example involving the estimation of a single parameter θ in a nonlinear model.

Example. Suppose we wish to find the normal equation for obtaining the least squares estimate $\hat{\theta}$ of θ for the model $Y = f(\theta, t) + \epsilon$, where $f(\theta, t) = e^{-\theta t}$, and where n pairs of observations $(Y_1, t_1), (Y_2, t_2), \ldots, (Y_n, t_n)$ are available. We find that

$$\frac{\partial f}{\partial \theta} = -t e^{-\theta t}.$$

Applying Eq. (24.1.10) leads to the single normal equation

$$\sum_{u=1}^{n} [Y_u - e^{-\hat{\theta} t_u}][-t_u e^{-\hat{\theta} t_u}] = 0$$

or

$$\sum_{u=1}^{n} Y_u t_u e^{-\hat{\theta} t_u} - \sum_{u=1}^{n} t_u e^{-2\hat{\theta} t_u} = 0. \qquad (24.1.11)$$

We see that even with one parameter and a comparatively simple nonlinear model, finding $\hat{\theta}$ by solving the (only) normal equation is not easy. When more parameters are involved and the model is more complicated, the solution of the normal equations can be extremely difficult to obtain, and iterative methods must be employed in nearly all cases. To compound the difficulties it may happen that multiple solutions exist, corresponding to multiple stationary values of the function $S(\hat{\boldsymbol{\theta}})$. We now discuss methods that have been used to estimate the parameters in nonlinear systems.

24.2. ESTIMATING THE PARAMETERS OF A NONLINEAR SYSTEM

In some nonlinear problems it is most convenient to write down the normal Eqs. (24.1.10) and develop an iterative technique for solving them. Whether this works satisfactorily or not depends on the form of the equations and the iterative method used. In addition to this approach there are several currently employed methods available for obtaining the parameter estimates by a routine computer calculation. We shall mention three of these: (1) linearization, (2) steepest descent, and (3) Marquardt's compromise.

The *linearization* (or Taylor series) method uses the results of linear least squares in a succession of stages. Suppose the postulated model is of the form of Eq. (24.1.8). Let $\theta_{10}, \theta_{20}, \ldots, \theta_{p0}$ be initial values for the parameters $\theta_1, \theta_2, \ldots, \theta_p$. These initial values may be intelligent guesses or preliminary estimates based on whatever information is available. (For example, they may be values suggested by the information gained in fitting a similar equation in a different laboratory or suggested as "about right" by the experimenter based on his/her experience and knowledge.) These initial values will, hopefully, be improved upon in the successive iterations to be described below. If we carry out a Taylor series expansion of $f(\xi_u, \boldsymbol{\theta})$ about the point $\boldsymbol{\theta}_0$, where $\boldsymbol{\theta}_0 =$

$(\theta_{10}, \theta_{20}, \ldots, \theta_{p0})'$, and curtail the expansion at the first derivatives, we can say that, approximately, when $\boldsymbol{\theta}$ is close to $\boldsymbol{\theta}_0$,

$$f(\boldsymbol{\xi}_u, \boldsymbol{\theta}) = f(\boldsymbol{\xi}_u, \boldsymbol{\theta}_0) + \sum_{i=1}^{p} \left[\frac{\partial f(\boldsymbol{\xi}_u, \boldsymbol{\theta})}{\partial \theta_i} \right]_{\theta = \theta_0} (\theta_i - \theta_{i0}). \tag{24.2.1}$$

If we set

$$f_u^0 = f(\boldsymbol{\xi}_u, \boldsymbol{\theta}_0),$$

$$\beta_i^0 = \theta_i - \theta_{i0},$$

$$Z_{iu}^0 = \left[\frac{\partial f(\boldsymbol{\xi}_u, \boldsymbol{\theta})}{\partial \theta_i} \right]_{\theta = \theta_0}, \tag{24.2.2}$$

we can see that Eq. (24.1.8) is of the form, approximately,

$$Y_u - f_u^0 = \sum_{i=1}^{p} \beta_i^0 Z_{iu}^0 + \epsilon_u; \tag{24.2.3}$$

in other words it is of the linear form shown in Eq. (24.1.1), to the selected order of approximation. We can now estimate the parameters β_i^0, $i = 1, 2, \ldots, p$, by applying linear least squares theory. If we write

$$\mathbf{Z}_0 = \begin{bmatrix} Z_{11}^0 & Z_{21}^0 & \cdots & Z_{p1}^0 \\ Z_{12}^0 & Z_{22}^0 & \cdots & Z_{p2}^0 \\ \vdots & \vdots & & \vdots \\ Z_{1u}^0 & Z_{2u}^0 & \cdots & Z_{pu}^0 \\ \vdots & \vdots & & \vdots \\ Z_{1n}^0 & Z_{2n}^0 & \cdots & Z_{pn}^0 \end{bmatrix} = \{Z_{iu}^0\}, \qquad n \times p, \tag{24.2.4}$$

$$\mathbf{b}_0 = \begin{bmatrix} b_1^0 \\ b_2^0 \\ \vdots \\ b_p^0 \end{bmatrix} \quad \text{and} \quad \mathbf{y}_0 = \begin{bmatrix} Y_1 - f_1^0 \\ Y_2 - f_2^0 \\ \vdots \\ Y_u - f_u^0 \\ \vdots \\ Y_n - f_n^0 \end{bmatrix} = \mathbf{Y} - \mathbf{f}^0, \tag{24.2.5}$$

say, with an obvious notation, then the estimate of $\boldsymbol{\beta}_0 = (\beta_1^0, \beta_2^0, \ldots, \beta_p^0)'$ is given by

$$\mathbf{b}_0 = (\mathbf{Z}_0' \mathbf{Z}_0)^{-1} \mathbf{Z}_0' (\mathbf{Y} - \mathbf{f}^0). \tag{24.2.6}$$

The vector $\mathbf{b}_0$ will therefore minimize the sum of squares

$$SS(\boldsymbol{\theta}) \equiv \sum_{u=1}^{n} \left\{ Y_u - f(\boldsymbol{\xi}_u, \boldsymbol{\theta}_0) - \sum_{i=1}^{p} \beta_i^0 Z_{iu}^0 \right\}^2 \tag{24.2.7}$$

with respect to the β_i^0, $i = 1, 2, \ldots, p$, where $\beta_i^0 = \theta_i - \theta_{i0}$. Let us write $b_i^0 = \theta_{i1} - \theta_{i0}$. Then the θ_{i1}, $i = 1, 2, \ldots, p$, can be thought of as the revised best estimates of $\boldsymbol{\theta}$.

Note carefully the difference between the sum of squares $S(\boldsymbol{\theta})$ in Eq. (24.1.9), where the appropriate *nonlinear* model is used, and the sum of squares $SS(\boldsymbol{\theta})$ in Eq. (24.2.7), where the *approximating linear expansion* of the model is employed.

We can now place the values θ_{i1}, the revised estimates, in the same roles as were played above by the values θ_{i0} and go through exactly the same procedure described above by Eqs. (24.2.1) through (24.2.7), but replacing all the zero subscripts by ones. This will lead to another set of revised estimates θ_{i2}, and so on. In vector form, extending the previous notation in an obvious way, we can write

$$\boldsymbol{\theta}_{j+1} = \boldsymbol{\theta}_j + \mathbf{b}_j \tag{24.2.8}$$

$$\boldsymbol{\theta}_{j+1} = \boldsymbol{\theta}_j + (\mathbf{Z}_j' \mathbf{Z}_j)^{-1} \mathbf{Z}_j' (\mathbf{Y} - \mathbf{f}^j),$$

where

$$\mathbf{Z}_j = \{Z_{iu}^j\},$$
$$\mathbf{f}^j = (f_1^j, f_2^j, \ldots, f_n^j)', \tag{24.2.9}$$
$$\boldsymbol{\theta}_j = (\theta_{1j}, \theta_{2j}, \ldots, \theta_{pj})'.$$

This iterative process is continued until the solution converges, that is, until in successive iterations j, $(j + 1)$,

$$|\{\theta_{i(j+1)} - \theta_{ij}\}/\theta_{ij}| < \delta, \qquad i = 1, 2, \ldots, p,$$

where δ is some prespecified amount (e.g., 0.000001). At each stage of the iterative procedure, $S(\boldsymbol{\theta}_j)$ can be evaluated to see if a reduction in its value has actually been achieved.

The linearization procedure has possible drawbacks for some problems in that:

1. It may converge very slowly; that is, a very large number of iterations may be required before the solution stabilizes even though the sum of squares $S(\boldsymbol{\theta}_j)$ may decrease consistently as j increases. This sort of behavior is not common but can occur.
2. It may oscillate widely, continually reversing direction, and often increasing, as well as decreasing the sum of squares. Nevertheless, the solution may stabilize eventually.
3. It may not converge at all, and even diverge, so that the sum of squares increases iteration after iteration without bound.

To combat these deficiencies, a program written by G. W. Booth and T. I. Peterson (1958), Non-linear estimation, IBM, SHARE Program Pa. No. 687 (WLNLI), under the direction of G. E. P. Box amended the correction vector $\mathbf{b}_j$ in Eq. (24.2.8) by halving it if

$$S(\boldsymbol{\theta}_{j+1}) > S(\boldsymbol{\theta}_j)$$

or doubling it if

$$S(\boldsymbol{\theta}_{j+1}) < S(\boldsymbol{\theta}_j).$$

This halving and/or doubling process is continued until three points between $\boldsymbol{\theta}_j$ and $\boldsymbol{\theta}_{j+1}$ are found, which include between them a local minimum of $S(\boldsymbol{\theta})$. A quadratic interpolation is used to locate the minimum, and the iterative cycle begins again.

Although in theory this method always converges (see Hartley, 1961), in practice difficulties may occur. The linearization method is, in general, a useful one and will successfully solve many nonlinear problems. Where it does not, consideration should be given to reparameterization of the model (see Section 24.4) or to the use of Marquardt's compromise.

A Remark on Derivatives. Many computer programs that use a method needing the values of the derivatives of a function at certain points do not use the functional values of the derivatives at all. Instead they compute ratios such as

$$\{f(\xi_u, \theta_{10}, \theta_{20}, \ldots, \theta_{i0} + h_i, \ldots, \theta_{p0}) - f(\xi_u, \theta_{10}, \theta_{20}, \ldots, \theta_{p0})\}/h_i,$$

$$i = 1, 2, \ldots, p,$$

where h_i is a selected small increment. The ratio given above is an approximation to the expression

$$\left[\frac{\partial f(\xi_u, \theta)}{\partial \theta_i} \right]_{\theta = \theta_0}$$

since, if h_i tends to zero, the limit of the ratio is this expression by definition.

A Geometrical Interpretation of Linearization

The sum of squares function $S(\theta)$ is a function of the parameter elements of θ only; the data provide the numerical coefficients in $S(\theta)$ and these are fixed for any specific nonlinear estimation problem. In the *parameter space,* that is, the p-dimensional geometrical space of $\theta_1, \theta_2, \ldots, \theta_p$, the function $S(\theta)$ can be represented by the contours of a surface. If the model were linear in the θ's, the surface contours would be ellipsoidal and would have a single local (and so a single global) minimum height, $S(\hat{\theta})$, at the location defined by the least squares estimator $\hat{\theta}$. If the model is nonlinear, the contours are not ellipsoidal but tend to be irregular and often "banana-shaped," perhaps with several local minima and perhaps with more than one global minimum, that is, the minimum height may be attained at more than one θ-location. Figure 24.1 provides examples for $p = 2$. Figure 24.1a shows the "elliptical bowl" $S(\theta)$ contours for a linear model, while Figure 24.1b shows the "irregular bowl" $S(\theta)$ contours for a nonlinear model.

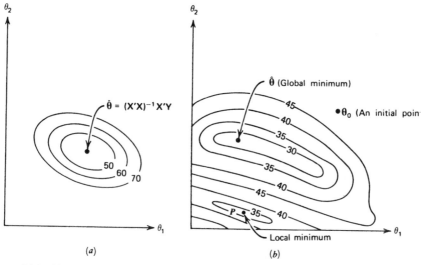

(a) (b)

Figure 24.1. (a) Elliptical $S(\theta)$ contours for a linear model $Y = \theta_1 X_1 + \theta_2 X_2 + \epsilon$; this bowl has a unique minimum point $\hat{\theta}$. (b) Irregular $S(\theta)$ contours for a nonlinear model, with two local minima. The desired solution is at $\hat{\theta}$ but iteration into point P is possible. Iterative solutions should be obtained starting from several well spread initial points θ_0 as a check.

The precise shape and orientation of the $S(\boldsymbol{\theta})$ contours depend on the model and the data. When the contours surrounding the least squares estimator $\hat{\boldsymbol{\theta}}$ are greatly elongated, and many possible $\boldsymbol{\theta}$-values are "nearly as good" as $\hat{\boldsymbol{\theta}}$ in the sense that their $S(\boldsymbol{\theta})$ "bowl height" values are close to $S(\hat{\boldsymbol{\theta}})$, the problem is said to be ill-conditioned and $\hat{\boldsymbol{\theta}}$ may be difficult to obtain computationally. Ill-conditioning could indicate a model that is overparameterized, that is, one that has more parameters than are needed, or inadequate data that will not allow us to estimate the parameters postulated. Because these are two sides of the same coin, the choice of whether one or the other is the culprit depends on one's prior knowledge about the practical problem and one's point of view. For example, consider $f(t;\ \theta_1,\ \theta_2)$ in Eq. (24.1.3), which represents a curve that begins (at $t = 0$) and ends (at $t = \infty$) at height zero, and rises to a peak somewhere in between. The slope at $t = 0$ is θ_1, and the peak is at

$$t_{\text{peak}} = \ln(\theta_1/\theta_2)/(\theta_1 - \theta_2).$$

It follows that, if our data cover just the early part of the curve (see Figure 24.2a) we shall be able to estimate θ_1 well, but not θ_2. For the latter we must obtain information on where the curve peaks as in Figure 24.2b. A one-parameter model $Y = \theta t + \epsilon$ would be adequate for the data of Figure 24.2a; these data are inadequate for estimating the two-parameter model, Eq. (24.1.3).

The linearization method converts the problem of finding the minimum height of $S(\boldsymbol{\theta})$ for a nonlinear model starting from an initial point $\boldsymbol{\theta}_0$, into a series of linear model problems. The initial linearization Eq. (24.2.1) of $f(\boldsymbol{\xi}, \boldsymbol{\theta})$ about $\boldsymbol{\theta}_0$ replaces the irregular $S(\boldsymbol{\theta})$ bowl by an elliptical bowl $SS(\boldsymbol{\theta})$ that "looks the right shape," that is, has the same first derivatives of the corresponding model function right at $\boldsymbol{\theta}_0$. As we shall see in the example of Section 24.3, it may approximate the actual $S(\boldsymbol{\theta})$ contours badly or well, depending on the actual circumstances, namely, the model assumed, the data available, and the relative positions of $\boldsymbol{\theta}_0$ and $\hat{\boldsymbol{\theta}}$ in the $\boldsymbol{\theta}$-space. In any event we solve the "linearized at $\boldsymbol{\theta}_0$" problem by moving to the bottom point of the linearized bowl at $\boldsymbol{\theta}_0$ (a relatively easy linear-least-squares calculation) to reach $\boldsymbol{\theta}_1$ as shown in Figure 24.3. Then we repeat the whole linearization process at $\boldsymbol{\theta}_1$. Our hope is that the successive iterations will converge to $\hat{\boldsymbol{\theta}}$, as indicated in Figure 24.4, rather than diverge. Typically, linearization does well when the starting point of the iteration is close to $\hat{\boldsymbol{\theta}}$, because the actual contours are then usually well approximated by the linearized ones.

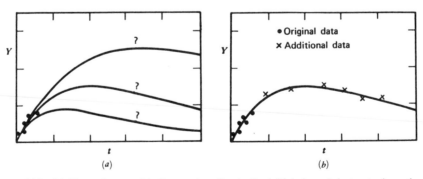

Figure 24.2. (a) These data would allow us to estimate the initial slope θ_1 but not where the curve peaked, which involves θ_2 as well. The $S(\boldsymbol{\theta})$ surface would be ill-conditioned if Eq. (24.1.3) were used. (b) Additional data shown would enable us to estimate both θ_1 and θ_2 well in Eq. (24.1.3) and the $S(\boldsymbol{\theta})$ surface would now be comparatively well-conditioned.

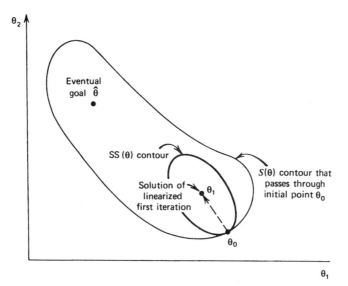

Figure 24.3. A first iteration in the linearization process. The ellipse of the SS(θ) contour "fits closely" to the $S(\theta)$ contour at θ_0, in the sense described in the text. The linearized problem provides the solution θ_1, and the procedure is then repeated from there.

More detail on the geometry of linear and nonlinear least squares is given in Sections 24.5 and 24.6.

Steepest Descent

The steepest descent method involves concentration on the sum of squares function, $S(\theta)$ as defined by Eq. (24.1.9), and use of an iterative process to find the minimum

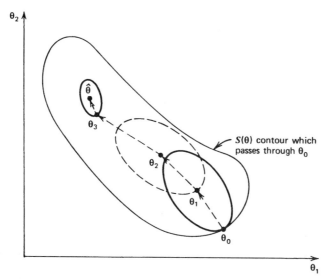

Figure 24.4. Successive iterations of a linearization procedure shown converging to the point $\hat{\theta}$, where minimum $S(\theta)$ is attained. Convergence is guaranteed by theory but, in practice, the process may diverge.

of this function. The basic idea is to move, from an initial point $\boldsymbol{\theta}_0$, along the vector with components

$$-\frac{\partial S(\boldsymbol{\theta})}{\partial \theta_1}, -\frac{\partial S(\boldsymbol{\theta})}{\partial \theta_2}, \ldots, -\frac{\partial S(\boldsymbol{\theta})}{\partial \theta_p},$$

whose values change continuously as the path is followed. One way of achieving this in practice, without evaluating functional derivatives, is to estimate the vector slope components at various places on the surface $S(\boldsymbol{\theta})$ by fitting planar approximating functions. This is a technique of great value in experimental work for finding stationary values of response surfaces. A full description of the method is given in Box and Draper (1987) and we shall discuss it only briefly here.

The procedure is as follows. Starting in one particular region of the $\boldsymbol{\theta}$-space (or the *parameter space* as we shall call it) several *runs* are made, by selecting n (say) combinations of levels of $\theta_1, \theta_2, \ldots, \theta_p$ and evaluating $S(\boldsymbol{\theta})$ at those combinations of levels. The runs are usually chosen in the pattern of a two-level factorial design. Using the evaluated $S(\boldsymbol{\theta})$ values as observations of a dependent variable and the combinations of levels of $\theta_1, \theta_2, \ldots, \theta_p$ as the observations of corresponding predictor variables, we fit the model

$$\text{``Observed } S(\boldsymbol{\theta})\text{''} = \beta_0 + \sum_{i=1}^{p} \beta_i (\theta_i - \bar{\theta}_i)/s_i + \epsilon$$

by standard least squares. Here $\bar{\theta}_i$ denotes the mean of the levels θ_{iu}, $u = 1, 2, \ldots, n$, of θ_i used in the runs, and s_i is a scaling factor chosen so that $\sum_{u=1}^{n} (\theta_{iu} - \bar{\theta}_i)^2/s_i^2 = $ constant. This implies that we believe the true surface defined by $S(\boldsymbol{\theta})$ can be approximately represented by a plane in the region of the parameter space in which we made our runs. The estimated coefficients

$$b_1, b_2, \ldots, b_p$$

indicate the direction of steepest ascent so the negatives of these, namely,

$$-b_1, -b_2, \ldots, -b_p$$

indicate the direction of steepest descent. This means that as long as the linear approximation is realistic the maximum decrease in $S(\boldsymbol{\theta})$ will be obtained by moving along the line which contains points such that

$$(\theta_i - \bar{\theta}_i)/s_i \propto -b_i.$$

Denoting the proportionality factor by λ, the path of steepest descent contains points $(\theta_1, \theta_2, \ldots, \theta_p)$ such that

$$\frac{(\theta_i - \bar{\theta}_i)}{s_i} = -\lambda b_i,$$

where $\lambda > 0$, or

$$\theta_i = \bar{\theta}_i - \lambda b_i s_i.$$

By giving λ selected values the path of steepest descent can be followed. A number of values of λ are selected and the path of steepest descent is followed as long as $S(\boldsymbol{\theta})$ decreases. When it does not, another experimental design is set down and the process is continued until it converges to the value $\hat{\boldsymbol{\theta}}$ that minimizes $S(\boldsymbol{\theta})$.

While, theoretically, the steepest descent method will converge, it may do so in practice with agonizing slowness after some rapid initial progress. Slow convergence is particularly likely when the $S(\theta)$ contours are attenuated and banana-shaped (as they often are in practice), and it happens when the path of steepest descent zigzags slowly up a narrow ridge, each iteration bringing only a slight reduction in $S(\theta)$. (This is less of a problem in laboratory-type investigations where human intervention can be permitted at each stage of calculation since then the experimental design can be revised, the scales of the independent variables can be changed, and so on.) This difficulty has led to modifications of the basic steepest descent procedure when used for nonlinear fitting. For some references see Spang (1962). (One possible modification is to use a second-order approximating function rather than a first-order or planar approximation. While this provides better graduation of the true surface, it also re-quires additional computation in the iterative procedures.)

A further disadvantage of the steepest descent method is that it is not scale invariant. The indicated direction of movement changes if the scales s_i of the variables are changed, unless all are changed by the same factor. The steepest descent method is, on the whole, slightly less favored than the linearization method but will work satisfac-torily for many nonlinear problems, especially if modifications are made to the ba-sic technique.

On the whole, steepest descent works well when the current position in the θ-space is far from the desired $\hat{\theta}$, which is usually so in the early iterations. As $\hat{\theta}$ is approached, the "zigzagging" behavior of steepest descent previously described is likely, and linear-ization tends to work better. The Marquardt procedure, based on work by Levenberg (1944), takes account of these facts.

Marquardt's Compromise

A method developed by Marquardt (1963) enlarged considerably the number of practi-cal problems that can be tackled by nonlinear estimation. Marquardt's method repre-sents a compromise between the linearization (or Taylor series) method and the steepest descent method and appears to combine the best features of both while avoiding their most serious limitations. It is good in that it almost always converges and does not "slow down" as the steepest descent method often does. However, as we again emphasize, the other methods will work perfectly well on many practical problems that do not violate the limitations of the methods. (In general, we must keep in mind that, given a particular method, a problem can usually be constructed to defeat it. Alternatively, given a particular problem and a suggested method, ad hoc modifications can often provide quicker convergence than an alternative method. The Marquardt method is one that appears to work well in many circumstances and thus is a sensible practical choice. For the reasons stated above, no method can be called "best" for all nonlinear problems.)

The idea of Marquardt's method can be explained briefly as follows. Suppose we start from a certain point in the parameter space, θ. If the method of steepest descent is applied, a certain vector direction, δ_g, where g stands for gradient, is obtained for movement away from the initial point. Because of attenuation in the $S(\theta)$ contours this may be the best *local* direction in which to move to attain smaller values of $S(\theta)$ but may not be the best *overall* direction. However, the best direction must be within 90° of δ_g or else $S(\theta)$ will get larger locally. The linearization (or Taylor series) method leads to another correction vector δ given by a formula like Eq. (24.2.6). Marquardt found that for a number of practical problems he studied, the angle, ϕ say, between

$\boldsymbol{\delta}_g$ and $\boldsymbol{\delta}$ fell in the range $80° < \phi < 90°$. In other words, the two directions were almost at right angles! The Marquardt algorithm provides a method for interpolating between the vectors $\boldsymbol{\delta}_g$ and $\boldsymbol{\delta}$ and for obtaining a suitable step size as well.

We shall not go into the detail of the method here. The basic algorithm is given in the quoted reference. See also Bates and Watts (1988) and Seber and Wild (1989).

Confidence Contours

Some idea of the nonlinearity in the model under study can be obtained, after the estimation of $\boldsymbol{\theta}$, by evaluating the ellipsoidal confidence region obtained on the assumption that the linearized form of the model is valid around $\hat{\boldsymbol{\theta}}$, the final estimate of $\boldsymbol{\theta}$. This is given by the formula

$$(\boldsymbol{\theta} - \hat{\boldsymbol{\theta}})'\hat{\mathbf{Z}}'\mathbf{Z}(\boldsymbol{\theta} - \hat{\boldsymbol{\theta}}) \le ps^2F(p, n - p, 1 - \alpha),$$

where $\hat{\mathbf{Z}}$ denotes a matrix of the form shown in Eq. (24.2.4) but with $\hat{\boldsymbol{\theta}}$ substituted into the elements in place of $\boldsymbol{\theta}_0$ everywhere, and where

$$s^2 = S(\hat{\boldsymbol{\theta}})/(n - p).$$

Note that when the difference between successive values $\boldsymbol{\theta}_{j+1}$ and $\boldsymbol{\theta}_j$ is sufficiently small so that the linearization procedure terminates with $\boldsymbol{\theta}_{j+1} = \hat{\boldsymbol{\theta}}$ ($= \boldsymbol{\theta}_j$ for practical purposes), then $S(\hat{\boldsymbol{\theta}})$ is a minimum value of $S(\boldsymbol{\theta})$ in Eq. (24.1.9) to the accuracy imposed by the termination procedure selected. This can be seen by examining Eq. (24.2.7) with $\hat{\boldsymbol{\theta}}$, β_i^{j+1}, and Z_{iu}^{j+1} replacing $\boldsymbol{\theta}_0$, β_i^0, and Z_{iu}^0, respectively, and remembering that, to the order of accuracy imposed by the termination procedure, $\mathbf{b}_{j+1} = \mathbf{0}$. The ellipsoid above will *not* be a true confidence region when the model is nonlinear. We can, however, determine the end points on the major axes of this ellipsoid by canonical reduction (see, e.g., Box and Draper, 1987). The *actual* values of $S(\boldsymbol{\theta})$ can be evaluated at these points and compared with each other. Under linear theory the values would all be the same.

An exact confidence contour is defined by taking $S(\boldsymbol{\theta}) = $ constant, but since we do not know the correct distribution properties in the general nonlinear case, we are unable to obtain a specified probability level. However, we can, for example, choose the contour such that

$$S(\boldsymbol{\theta}) = S(\hat{\boldsymbol{\theta}})\left\{1 + \frac{p}{n - p}F(p, n - p, 1 - \alpha)\right\},$$

which, if the model is linear, provides an *exact*, ellipsoidal $100(1 - \alpha)\%$ boundary, and label it as an approximate $100(1 - \alpha)\%$ confidence contour in the nonlinear case. Note that the contour so determined *will be a proper correct confidence contour in this case* (and will not be elliptical in general), *and it is only the probability level that is approximate.* When only two parameters are involved the confidence contour can be drawn. For more parameters, sectional drawings can be constructed if desired.

In general, when a linearized form of a nonlinear model is used, all the usual formulas and analyses of linear regression theory can be applied. Any results obtained, however, are valid only to the extent that the linearized form provides a good approximation to the true model.

Grids and Plots

Two obvious ways of examining the sum of squares surface $S(\boldsymbol{\theta})$ are often overlooked; they can be particularly useful when an iterative procedure, beginning from chosen values, does not satisfactorily converge.

The first of these is to select a grid of points, that is, a factorial design, in the space of the parameters $(\theta_1, \theta_2, \ldots, \theta_p)$ and to evaluate (usually on a computer) the sum of squares function at every point of the grid. These values will provide some idea of the form of the sum of squares surface and may reveal, for example, that multiple minima are possible. In any case, the grid point at which the smallest sum of squares is found can be used as the starting point of an iterative parameter estimation procedure, or a reduced grid can be examined in the best neighborhood, to obtain a better starting point. The simplest type of grid available is that in which *two* levels of every parameter are selected. In this case, the grid points are those of a 2^p factorial design, and it is possible to use standard methods to evaluate the factorial effects and interactions and so provide information on the effects of changes in the parameters on the sum of squares function $S(\boldsymbol{\theta})$.

The second possibility is to draw sum of squares contours in any particular region of the parameter space in which difficulty in convergence occurs or in which additional information would be helpful. This is usually straightforward when only one or two parameters are involved. When there are more than two parameters, two-dimensional slices of the contours can be obtained for selected values of all but two of the parameters, and a composite picture can be built up.

The Importance of Good Starting Values

All iterative procedures require initial values $\theta_{10}, \theta_{20}, \ldots, \theta_{p0}$, of the parameters θ_1, $\theta_2, \ldots, \theta_p$, to be selected. All available prior information should be used to make these starting values as reliable as they possibly can be. Good starting values will often allow an iterative technique to converge to a solution much faster than would otherwise be possible. Also, if multiple minima exist or if there are several local minima in addition to an absolute minimum, poor starting values may result in convergence to an unwanted stationary point of the sum of squares surface. This unwanted point may have parameter values that are physically impossible or that do not provide the true minimum value of $S(\boldsymbol{\theta})$. As suggested above, a preliminary evaluation of $S(\boldsymbol{\theta})$ at a number of grid points in the parameter space is often useful.

Getting Initial Estimates θ_0

There is no standard "crank the handle" mechanism for getting initial estimates θ_0 for every nonlinear estimation problem. Methods that have worked in various problems are the following:

1. If there are p parameters, substitute for p sets of observations (Y_u, ξ_u) into the postulated model *ignoring the error.* Solve the resulting p equations for the parameters (if possible). Widely separated ξ_u often work best.

2. In approach (1), or sometimes as an alternative to it, consider the behavior of the response function as the ξ_i go to zero or infinity, and substitute in for observations that most nearly represent those conditions in the scale and context of the problem. Solve (if possible) the resulting equations.

3. Check the form of the model to see if (were it not for the additive error) the model could be transformed approximately. For example, if the model is $Y = \theta_1 e^{-\theta_2 t} + \epsilon$, a plot of $\ln Y$ versus t will usually give good initial estimates of $\ln \theta_1$ and $-\theta_2$ obtained from the intercept and slope of the $\ln Y$ versus t plot. (This is based on

the idea that if the model had been $Y = \theta_1 e^{-\theta_2 t}\epsilon$ instead, the transformation $\ln Y = \ln \theta_1 - \theta_2 t + \ln \epsilon$ would work.)

4. A more complicated example of (3) is given by the model $\ln W = -\theta_1^{-1} \ln(\theta_2 + \theta_3 X^{\theta_4}) + \epsilon$, a model useful in growth studies. (for source details, see Mead, 1970.) A plot of W^{-1} versus X would be a straight line with intercept θ_2 and slope θ_3, if $\theta_1 = \theta_4 = 1$. If this plot is curved, then trial and error can be used to search for values of θ_1 and θ_4 that make the plot of $W^{-\theta_1}$ versus X^{θ_4} into a "good" straight line. Once this has been achieved to whatever extent possible, the corresponding θ-values are the initial estimates. Note that, in models like this, poor initial estimates may cause the argument $\theta_2 + \theta_3 X^{\theta_4}$ to go negative, creating computing problems; good initial estimates will often prevent the occurrence of such problems.

5. If all else fails, grids and plots can be used—see above.

Note: When small initial values are chosen initially, or are expected in subsequent iterations, for some parameters, care must be taken to see that the intervals h_i used to evaluate the corresponding numerical partial derivatives are taken suitably small. Some routines fail if this is not done appropriately.

24.3. AN EXAMPLE

The example that follows is taken from an investigation performed at Procter & Gamble and reported by Smith and Dubey (1964). We shall use this example to illustrate how a solution can be obtained for a nonlinear estimation problem by solving the normal equations directly, or alternatively by the linearization method. We shall not provide an example of the use of steepest descent (but see, e.g., Box and Coutie, 1956), nor an example of Marquardt's compromise procedure. The investigation involved a product A, which must have a fraction of 0.50 of Available Chlorine at the time of manufacture. The fraction of Available Chlorine in the product decreases with time; this is known. In the 8 weeks before the product reaches the consumer a decline to a level 0.49 occurs but since many uncontrolled factors then arise (such as warehousing environments, handling facilities), theoretical calculations are not reliable for making extended predictions of the Available Chlorine fraction present at later times. To assist management in decisions—such as (1) When should warehouse material be scrapped? and (2) When should store stocks be replaced?—cartons of the product were analyzed over a period to provide the data of Table 24.2. (Note that the product is made only every other week and code-dated only by the week of the year. The predicted values shown in the table are obtained from the fitted equation to be found in what follows.) It was postulated that a nonlinear model of the form

$$Y = \alpha + (0.49 - \alpha)e^{-\beta(X-8)} + \epsilon \qquad (24.3.1)$$

would suitably account for the variation observed in the data, for $X \geq 8$. This model provides a true level, without error, of $\eta = 0.49$ when $X = 8$, and it exhibits the proper sort of decay. An additional point of information, agreed upon by knowledgeable chemists, was that an equilibrium, asymptotic level of Available Chlorine somewhere above 0.30 should be expected. The problem is to estimate the parameters α and β of the nonlinear model given in Eq. (24.3.1) using the data given in the table. The residual sum of squares for this model can be written as

$$S(\alpha, \beta) = \sum_{u=1}^{n} [Y_u - \alpha - (0.49 - \alpha)e^{-\beta(X_u-8)}]^2, \qquad (24.3.2)$$

T A B L E 24.2. Percent of Available Chlorine in a Unit of Product

Length of Time Since Produced (weeks) X	Available Chlorine Y	Average Available Chlorine $\overline{Y}$	Predicted Y, Using the Model $\hat{Y}$
8	0.49, 0.49	0.490	0.490
10	0.48, 0.47, 0.48, 0.47	0.475	0.472
12	0.46, 0.46, 0.45, 0.43	0.450	0.457
14	0.45, 0.43, 0.43	0.437	0.445
16	0.44, 0.43, 0.43	0.433	0.435
18	0.46, 0.45	0.455	0.427
20	0.42, 0.42, 0.43	0.423	0.420
22	0.41, 0.41, 0.40	0.407	0.415
24	0.42, 0.40, 0.40	0.407	0.410
26	0.41, 0.40, 0.41	0.407	0.407
28	0.41, 0.40	0.405	0.404
30	0.40, 0.40, 0.38	0.393	0.401
32	0.41, 0.40	0.405	0.399
34	0.40	0.400	0.397
36	0.41, 0.38	0.395	0.396
38	0.40, 0.40	0.400	0.395
40	0.39	0.390	0.394
42	0.39	0.390	0.393

where (X_u, Y_u), $u = 1, 2, 3, \ldots, 44$, are the corresponding pairs of observations from the table (e.g., $X_1 = 8$, $Y_1 = 0.49, \ldots, X_{44} = 42$, $Y_{44} = 0.39$).

A Solution Through the Normal Equations

Differentiating Eq. (24.3.2) first with respect to α, and then with respect to β, and setting the results equal to zero provides two normal equations. After removal of a factor of 2 from the first equation and a factor of $2(0.49 - \alpha)$ from the second equation and some rearrangement, the equations reduce to

$$\alpha = \frac{\Sigma Y_u - \Sigma Y_u e^{-\beta t_u} - 0.49 \Sigma e^{-\beta t_u} + 0.49 \Sigma e^{-2\beta t_u}}{n - 2\Sigma e^{-\beta t_u} + \Sigma e^{-2\beta t_u}} \qquad (24.3.3)$$

and

$$\alpha = \frac{0.49 \Sigma t_u e^{-2\beta t_u} - \Sigma Y_u t_u e^{-\beta t_u}}{\Sigma t_u e^{-2\beta t_u} - \Sigma t_u e^{-\beta t_u}}, \qquad (24.3.4)$$

where all summations are from $u = 1$ to $u = 44$, and $t_u = X_u - 8$. We see that these normal equations have a particular simplification in that the parameter α can be eliminated by subtracting one equation from the other. If this is done, a single nonlinear equation of the form $f(\beta) = 0$ in β results. This can be solved by applying the Newton–Raphson technique, first guessing a value for β, call it β_0, and then "correcting" it by h_0 obtained as follows. If the root of $f(\beta) = 0$ is at $(\beta_0 + h_0)$, then

$$0 = f(\beta_0 + h_0) = f(\beta_0) + h_0 \left[\frac{df(\beta)}{d\beta} \right]_{\beta=\beta_0} \qquad (24.3.5)$$

approximately, or

$$h_0 = -f(\beta_0) \bigg/ \left[\frac{df(\beta)}{d\beta}\right]_{\beta=\beta_0} \tag{24.3.6}$$

approximately. We can now use $\beta_1 = \beta_0 + h_0$ instead of β_0 and repeat the correction procedure to find h_1 and so $\beta_2 = \beta_1 + h_1$. This process can be continued until it converges to a value $\hat{\beta}$, which is then the least squares estimate of β. The value of $\hat{\alpha}$, the least squares estimate of α, can be obtained from Eqs. (24.3.3) and (24.3.4) by substituting $\hat{\beta}$ in the right-hand sides. As a check, the same value should be obtained from both equations.

We can guess β_0 initially by noting, for example, that when $X_{44} = 42$, $Y_{44} = 0.39$. If Y_{44} contained no error then we should have

$$0.39 = \alpha + (0.49 - \alpha)e^{-34\beta_0},$$

whereas if we assume that Y tends to 0.30 as X tends to infinity (on the basis of the prior information given by the chemists) then we can make an initial guess of $\alpha_0 = 0.30$ for α, from which it follows that

$$0.39 = 0.30 + (0.49 - 0.30)e^{-34\beta_0}$$

or

$$e^{-34\beta_0} = \frac{0.09}{0.19}$$

so that

$$\beta_0 = \frac{-[\ln(0.09/0.19)]}{34} = 0.02$$

approximately. (Note that we must have $\beta > 0$ or else a decay cannot be represented by the function.)

If we denote Eqs. (24.3.3) and (24.3.4) by

$$\alpha = f_1(\beta), \tag{24.3.7}$$

$$\alpha = f_2(\beta),$$

respectively, then

$$f(\beta) \equiv f_1(\beta) - f_2(\beta) = 0 \tag{24.3.8}$$

and

$$\frac{\partial f(\beta)}{\partial \beta} \equiv \frac{\partial f_1(\beta)}{\partial \beta} - \frac{\partial f_2(\beta)}{\partial \beta}. \tag{24.3.9}$$

Rather than write down the rather lengthy expressions that result from the differentiation of $f_1(\beta)$ and $f_2(\beta)$ we shall adopt a simpler method of finding $\hat{\alpha}$, $\hat{\beta}$.

An alternative procedure for estimating α and β in this case is to plot the functions (24.3.7) over a reasonable range of β and note where the two curves intersect. This provides both $\hat{\alpha}$ and $\hat{\beta}$ immediately since the point of intersection is at $(\hat{\beta}, \hat{\alpha})$. Some

T A B L E 24.3. Points on the
Curves $f_i(\beta)$

β	$f_1(\beta)$	$f_2(\beta)$
0.06	0.3656	0.3627
0.07	0.3743	0.3720
0.08	0.3808	0.3791
0.09	0.3857	0.3847
0.10	0.3896	0.3894
0.11	0.3927	0.3935
0.12	0.3953	0.3970
0.13	0.3975	0.4002
0.14	0.3993	0.4031
0.15	0.4009	0.4057
0.16	0.4023	0.4082

values of $f_1(\beta)$ and $f_2(\beta)$ for a range of values of β are shown in Table 24.3 and the
resulting plot is shown in Figure 24.5. The estimates can be read off squared paper
to sufficient accuracy as $\hat{\alpha} = 0.30$, $\hat{\beta} = 0.10$. We can see from the figure that the two
curves plotted are quite close together for a comparatively large range of β and for
a somewhat smaller range of α. This indicates that β is somewhat less well determined
than α. For example, $|f_1(\beta) - f_2(\beta)| < 0.0025$ is achieved by a range of values of β
between about 0.07 and 0.12 and a range of values of α between about 0.37 and 0.40.
Many pairs of values of (β, α) such as, for example, $(0.09, 0.385)$, $(0.11, 0.393)$ bring
the two curves of the figure almost together, and thus do not appear unreasonable
estimates for (β, α) in the light of the data even though they do not actually minimize
$S(\alpha, \beta)$. We shall see that these comments are substantiated by the confidence regions
for the true (β, α), which can be constructed for this problem (see Figure 24.8). The
fitted equation now takes the form

$$\hat{Y} = 0.39 + 0.10e^{-0.10(X-8)}. \tag{24.3.10}$$

The observed values of X can be inserted in this equation to give the fitted values

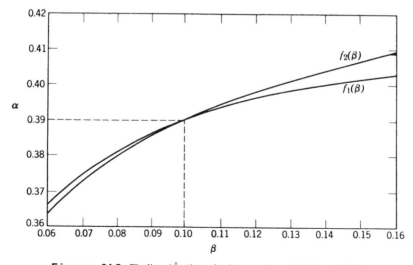

Figure 24.5. Finding $(\hat{\beta}, \hat{\alpha})$ as the intersection of $f_1(\beta)$ and $f_2(\beta)$.

shown in Table 24.2. The fitted curve and the observations are shown in Figure 24.6. The usual analysis of residuals could now be carried out. (The large residuals at $X =$ 18 strike the eye immediately. According to the authors no assignable cause could be found for these.)

A Solution Through the Linearization Technique

To linearize the model into the form of Eq. (24.2.1) we need to evaluate the first derivatives of

$$f(\xi_u, \boldsymbol{\theta}) = f(\mathbf{X}_u; \alpha, \beta)$$

$$= \alpha + (0.49 - \alpha)e^{-\beta(X_u-8)}, \tag{24.3.11}$$

namely,

$$\frac{\partial f}{\partial \alpha} = 1 - e^{-\beta(X_u-8)},$$

$$\frac{\partial f}{\partial \beta} = -(0.49 - \alpha)(X_u - 8)e^{-\beta(X_u-8)}. \tag{24.3.12}$$

Thus if $\alpha = \alpha_j$, $\beta = \beta_j$ are the values inserted at the jth stage, as described in Section 24.2, we have, in the notation implied in that section, a model of form (at the jth stage)

$$Y_u - f_u^i = [1 - e^{-\beta_j(X_u-8)}](\alpha - \alpha_j)$$
$$+ [-(0.49 - \alpha_j)(X_u - 8)e^{-\beta_j(X_u-8)}](\beta - \beta_j) + \epsilon,$$

or in matrix form

$$\mathbf{Y} - \mathbf{f}^j = \mathbf{Z}_j \begin{bmatrix} \alpha - \alpha_j \\ \beta - \beta_j \end{bmatrix} + \boldsymbol{\epsilon},$$

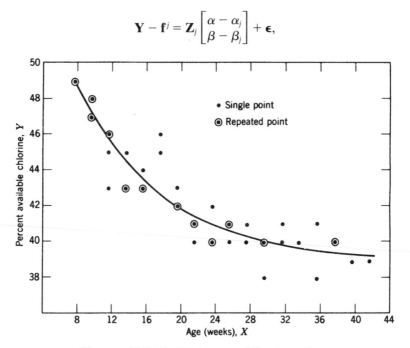

Figure 24.6. The fitted curve and the observations.

where

$$f^i_u = \alpha_j + (0.49 - \alpha_j)e^{-\beta_j(X_u-8)} \tag{24.3.13}$$

$$\mathbf{Z}_j = \begin{bmatrix} 1 - e^{-\beta_j(X_1-8)} & -(0.49 - \alpha_j)(X_1 - 8)e^{-\beta_j(X_1-8)} \\ \vdots & \vdots \\ 1 - e^{-\beta_j(X_u-8)} & -(0.49 - \alpha_j)(X_u - 8)e^{-\beta_j(X_u-8)} \\ \vdots & \vdots \\ 1 - e^{-\beta_j(X_n-8)} & -(0.49 - \alpha_j)(X_n - 8)e^{-\beta_j(X_n-8)} \end{bmatrix}, \tag{24.3.14}$$

and the vector of quantities to be estimated is

$$\begin{bmatrix} \alpha - \alpha_j \\ \beta - \beta_j \end{bmatrix} \tag{24.3.15}$$

with estimate given by

$$\begin{bmatrix} \alpha_{j+1} - \alpha_j \\ \beta_{j+1} - \beta_j \end{bmatrix} = (\mathbf{Z}'_j \mathbf{Z}_j)^{-1} \mathbf{Z}'_j \begin{bmatrix} Y_1 - f^j_1 \\ Y_2 - f^j_2 \\ \vdots \\ Y_n - f^j_n \end{bmatrix}. \tag{24.3.16}$$

If we begin the iterations with initial guesses $\alpha_0 = 0.30$ and $\beta_0 = 0.02$ as before, and apply Eq. (24.3.16) iteratively, we obtain estimates as follows:

Iteration	α_j	β_j	$S(\alpha_j, \beta_j)$
0	0.30	0.02	0.0263
1	0.8416	0.1007	4.4881
2	0.3901	0.1004	0.0050
3	0.3901	0.1016	0.0050
4	0.3901	0.1016	0.0050

Note: These figures were rounded from the end results of computer calculations, which carried more significant figures. Numerical differences might occur if a parallel calculation were made on a pocket calculator.

This process converges to the same least squares estimates as before, namely, $\hat{\alpha} = 0.39$ and $\hat{\beta} = 0.10$ to give the fitted model, Eq. (24.3.10). Note that this happens in spite of the rather alarming fact that, after the first stage, $S(\alpha_1, \beta_1) = 4.4881$, which is about 170 times the initial $S(\alpha_0, \beta_0) = 0.0263$. The reduction in the next iteration is dramatic and practically final, the subsequent reduction in $S(\alpha, \beta)$ being in the sixth place of decimals, which is not shown. In some nonlinear problems no correction, dramatic or otherwise, occurs and the process diverges, providing larger and larger values for $S(\theta)$. (For a possible reason for this sort of behavior, see Section 24.6.)

Figure 24.7a shows the nonlinear sum of squares contours $S(\alpha, \beta)$ in the region $0 \le \beta \le 0.20$, $0.28 \le \alpha \le 0.90$ together with the progress from the initial point (β_0, α_0) to (β_2, α_2). The reason for this path is shown in Figures 24.7b and 24.7c. In Figure 24.7b the SS(θ) contours are relatively ill-conditioned and lead to a minimum SS(α, β) value at a point (β_1, α_1) well away from the actual minimum of $S(\alpha, \beta)$. In the next iteration the linearized sum of squares contours are relatively well-conditioned (Figure 24.7c) and also their center happens to be close to the minimum point of the

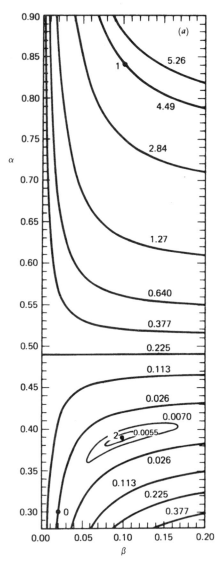

Figure 24.7. (a) Nonlinear sum of squares contours $S(\alpha, \beta)$ for the example. The points marked 0, 1, 2 are the (β_j, α_j) for the initial point and the first and second iterations.

actual sum of squares contours, as we have seen. Note that, in both Figures 24.7b and 24.7c, the direction of steepest descent, which would be perpendicular to the $S(\alpha, \beta)$ contours at the starting point, would take us off in a different direction from that determined by linearization.

Further Analysis

The usual tests appropriate in the linear model case are, in general, *not* appropriate when the model is nonlinear. As a practical procedure we can compare the unexplained variation with an estimate of $V(Y_u) = \sigma^2$ but cannot use the F-statistic to obtain conclusions at any stated level. The unexplained variation is $S(\hat{\alpha}, \hat{\beta}) = 0.0050$. In the absence of exact results for the nonlinear case, we can regard this

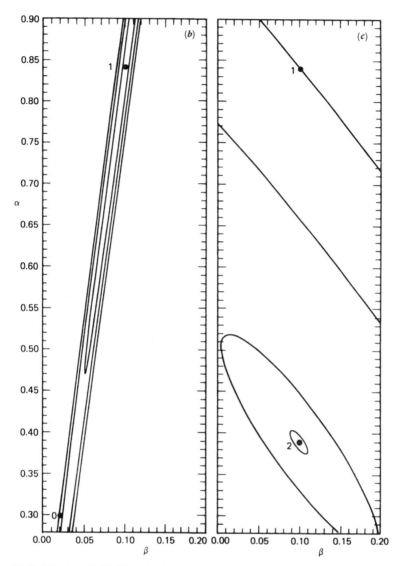

Figure 24.7. (*Continued*) (*b*) Linearized sum of squares contours SS(α, β) at the point (β, α) = (β_0, α_0) = (0.02, 0.30). The center of the elliptical system is at (β_1, α_1) = (0.1007, 0.8416) and the next iteration begins from there. (*c*) Linearized sum of squares contours SS(α, β) at the point (β, α) = (β_1, α_1) = (0.1007, 0.8416). The center of the elliptical system is at (β_2, α_2) = (0.1004, 0.3901) and the next iteration begins from there.

sum of squares as being based on approximately $44 - 2 = 42$ degrees of freedom (since two parameters have been estimated). In the nonlinear case this does not, in general, lead to an unbiased estimate of σ^2 as in the linear case, even when the model is correct.

A pure error estimate of σ^2 (see Section 2.1) can be obtained from the repeat observations. This provides a sum of squares $S_{pe} = 0.0024$ with 26 degrees of freedom.

An appropriate idea of possible lack of fit can be obtained by evaluating

$$S(\hat{\alpha}, \hat{\beta}) - S_{pe} = 0.0026 \quad \text{with} \quad 42 - 26 = 16 \text{ degrees of freedom}$$

and comparing the mean squares

$$\frac{(S(\hat{\alpha}, \hat{\beta}) - S_{pe})}{16} = 0.00016$$

$$\frac{S_{pe}}{26} = 0.00009.$$

An F-test is *not* applicable here but we can use the value of $F(16, 26, 0.95) = 2.08$ as a measure of comparison. We see that $16/9 = 1.8$, which would make us tentatively feel that the model does not fit badly.

Confidence Regions

We can calculate approximate $100(1 - q)\%$ confidence contours (described in Section 24.2) by finding points (α, β) that satisfy

$$S(\alpha, \beta) = S(\hat{\alpha}, \hat{\beta}) \left[1 + \frac{p}{n - p} F(p, n - p, 1 - q) \right]$$

$$= 0.0050[1 + F(2, 42, 1 - q)/21]$$

$$= S_q.$$

From Eq. (24.3.2) we can write this as

$$\sum_{u=1}^{n} \{(Y_u - 0.49e^{-\beta(X_u - 8)}) + \alpha(e^{-\beta(X_u - 8)} - 1)\}^2 = S_q$$

or

$$A\alpha^2 + 2B\alpha + C - S_q = 0,$$

where

$$A = \sum_{u=1}^{n} (e^{-\beta(X_u - 8)} - 1)^2,$$

$$B = \sum_{u=1}^{n} (Y_u - 0.49e^{-\beta(X_u - 8)})(e^{-\beta(X_u - 8)} - 1),$$

$$C = \sum_{u=1}^{n} (Y_u - 0.49e^{-\beta(X_u - 8)})^2$$

are all functions of β alone. We can thus select a value for q and then evaluate

$$\alpha = \frac{\{-B \pm [B^2 - A(C - S_q)]^{1/2}\}}{A}$$

for a range of β to obtain points on the boundaries of confidence regions with approximately $100(1 - q)\%$ confidence coefficients. The 75%, 95%, 99%, and 99.5% regions, obtained from $q = 0.25, 0.05, 0.01,$ and 0.005, respectively, are shown in Figure 24.8. The dot denotes the point $(\hat{\beta}, \hat{\alpha})$.

Points (β, α) that lie within the contour marked (say) 95% are considered, by the data, as not unreasonable for the true values of (β, α), at an approximate 95% level of confidence. The orientation and shape of the contours indicate that $\hat{\beta}$ is less well determined than $\hat{\alpha}$. (For discussion on this point in the linear case, see Section 24.5; see also Section 24.4.)

Some Typical Nonlinear Program Output Features

The printed output from nonlinear least squares programs varies from program to program. However, some features are common to many outputs and we now describe and illustrate these briefly, using the output from a Marquardt-type analysis of the example given in this section.

1. The *singular values* are the square roots of the eigenvalues of $\mathbf{Z}_j'\mathbf{Z}_j$, where $\mathbf{Z}_j$ is derived from Eq. (24.2.4) but employs the current $\boldsymbol{\theta}_j$ value and not $\boldsymbol{\theta}_0$. Widely disparate singular values indicate a tendency toward ill-conditioning. The corresponding singular vectors provide the orientation of the axes of the "linearization approximation bowl" with respect to the θ-axes.

Example 1. At $\hat{\theta}$ we have the following:

Singular values:	4.964	0.773
α:	-0.946	0.326
β:	0.326	0.946

The ratio of the largest and smallest singular values (there *are* only two singular values here, of course, but in general there would be p) is about 6.4, implying a fairly well-conditioned problem. (In an ill-conditioned problem, this ratio can be in the thousands. A ratio of 1 would imply circular linearizing approximating contours.) To achieve the same change in $SS(\boldsymbol{\theta})$, one would have to move about 6.4 units in the β direction compared with one unit in the α direction. Reading down the columns, we see that the first singular direction

$$\phi_1 = -0.946\alpha + 0.326\beta$$

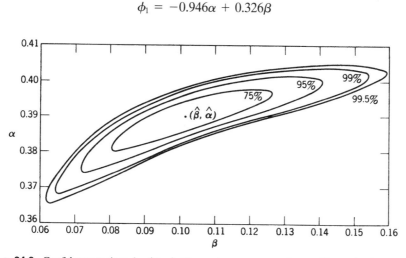

Figure 24.8. Confidence regions for (β, α). The regions are exact, the confidence levels approximate.

is mostly in the $-\alpha$ direction (the sign is not important here and is determined by the direction of the labeling only), while

$$\phi_2 = 0.326\alpha + 0.946\beta$$

is mostly in the β direction, as Figure 24.8 confirms should be the case. Note that the sum of squares of the coefficients (e.g., -0.946 and 0.326) is 1, apart from rounding error, and this sort of result is true in general for all the columns.

2. The *normalizing elements* are the square roots of the diagonal entries of the $\mathbf{Q} = ((q_{il})) = (\mathbf{Z}_j'\mathbf{Z}_j)^{-1}$ matrix. The linearized approximate *correlation matrix* $\mathbf{C} = ((c_{il}))$ of the $\hat{\theta}$'s is obtained by dividing each row and column of $(\mathbf{Z}_j'\mathbf{Z}_j)^{-1}$ by the two corresponding normalizing elements, which correspond to that row and column, namely, $c_{il} = q_{il}/(q_{ii}q_{ll})^{1/2}$.

Example 2.

Parameter:	α	β
Normalizing elements:	0.462	1.226
Parameter: α	1.000	
β	0.888	1.000

The correlation between α and β is positive, indicating that variations in (α, β) from $(\hat{\alpha}, \hat{\beta})$ in such a way that *both* increase, or *both* decrease, will give roughly similar $SS(\theta)$ values. We can see from the $S(\alpha, \beta)$ plot (Figure 24.8) how this can happen. [Remember that the $SS(\theta)$ ellipses around $\hat{\theta}$ will mimic the $S(\theta)$ contours but not represent them perfectly.] Although 0.888 is a high correlation, it is not high enough for us to feel that the model is overparameterized. (For more on this point, see the "overparameterization" subsection, below.)

3. Confidence limits for the true values of the θ's can be evaluated on the basis of the linearized approximation, evaluated at $\hat{\theta}$. The limits are, typically, $\hat{\theta}_i \pm 2\,se(\hat{\theta}_i)$, where $se(\hat{\theta}_i) = \{\text{appropriate diagonal element of } (\hat{\mathbf{Z}}'\hat{\mathbf{Z}})^{-1}s^2\}^{1/2}$, where $\hat{\mathbf{Z}}$ is the $\mathbf{Z}$ of Eq. (24.2.4) with $\hat{\theta}$ replacing θ_0 and $s^2 = S(\hat{\theta})/(n-p)$. These are, roughly, 95% limits in general.

Example 3.

	Lower Limit	Final Value	Upper Limit
α	0.380	0.390	0.400
β	0.075	0.102	0.128

The final estimate of α is $\hat{\alpha} = 0.390$ and an approximate 95% confidence band for α is $(0.380, 0.400)$. The final estimate of β is $\hat{\beta} = 0.102$ with an approximate 95% confidence band for β of $(0.075, 0.128)$. Note that the message implied by the width of the bands is confirmed by Figure 24.8. (The reader may wish to reread Section 5.4. Not only is the caution there valid but also the confidence statements made here apply only to the extent that the linearized model is a reliable approximation to the nonlinear model.) Both bands exclude zero, indicating nonzero values for α and β.

Curvature Measures

How close to reality is a linearized form of the nonlinear model in providing confidence statements on the parameters? The answer varies from problem to problem in a complicated way. Useful measures are described in Chapter 7 of Bates and Watts

(1988). These measures depend on second-order derivatives of the model function. A distinction can be made between *intrinsic nonlinearity* (which is dependent on the nature of the problem itself) and parameter effects nonlinearity (which may be reducible by a better choice of the parameterization—*if* a good choice can be found). A study of 67 specific data–model combinations with real sets of data, many of which are from "Exercises for Chapter 24" (H, L, M, N, and P), brought Bates and Watts to the following conclusions (their p. 256):

1. "Intrinsic curvature is smaller, and in almost all cases very much smaller, than the parameter effects curvature."
2. "Parameter effects curvatures . . . are often bad."
3. "Linear approximation inference regions . . . can be very misleading in many practical situations."

We do not explore these complications here and recommend the reference cited to readers with deeper interest in these topics. Many readers will find the usual linearized approximations perfectly adequate for making decisions on a day-to-day basis. Those who are involved in larger projects, or in a deep exploration of a specific few sets of data, may wish to delve further. In any event, mastery of the topics we have already dealt with is a prerequisite for moving on.

Overparameterization

Situations in which there are more parameters in the model than are needed to represent the data generally show up in the pattern of the (linearized) correlation matrix of the estimated coefficients. This matrix is given by taking $(\hat{\mathbf{Z}}'\hat{\mathbf{Z}})^{-1}$ and dividing the element in the ith row and jth column by

$$\{[i\text{th diagonal entry of } (\hat{\mathbf{Z}}'\hat{\mathbf{Z}})^{-1}] \times [j\text{th diagonal entry of } (\hat{\mathbf{Z}}'\hat{\mathbf{Z}})^{-1}]\}^{1/2},$$

which reduces the diagonal entries to 1's and the off-diagonal entries to correlations. When some of these correlations are large, it indicates that one or more parameters may not be useful or, more accurately, that a reparameterized model involving fewer parameters might do almost as well. Note that this does *not* necessarily mean that the original model is inappropriate for the physical situation under study. It may simply be an indication that the data in hand are not adequate to the task of estimating all the original parameters. (For example, in a linear model, when a parameter β does not appear to be different from zero, it does not always imply that the corresponding X is ineffective; it may be that, *in the particular set of data under study, X does not change enough for its effect to be discernible*. Similar, but more complicated, parallels arise in nonlinear work.)

In overparameterized situations, the $S(\theta)$ contours are often greatly elongated in the current θ-metric, and, in such cases, examination of an $S(\theta)$ plot, or sections of it according to the number of dimensions involved, can be very helpful.

24.4. A NOTE ON REPARAMETERIZATION OF THE MODEL

When the sum of squares surface, defined by Eq. (24.1.9), is attenuated and contains long ridges, slow convergence of any iterative estimation procedure is likely. As a simple example of attenuation in the *linear* case, consider the model $Y_u = \theta_0 + \theta_1 X_u + \epsilon_u$ and suppose we have three observations Y_1, Y_2, and Y_3 at $X = 9$, 10, and 11. Then

$$S(\boldsymbol{\theta}) = (Y_1 - \theta_0 - 9\theta_1)^2 + (Y_2 - \theta_0 - 10\theta_1)^2 + (Y_3 - \theta_0 - 11\theta_1)^2$$

$$= \sum_{u=1}^{3} Y_u^2 - 2\theta_0 \sum_{u=1}^{3} Y_u - 2\theta_1(9Y_1 + 10Y_2 + 11Y_3)$$

$$+ 3\theta_0^2 + 302\theta_1^2 + 60\theta_0\theta_1.$$

In coordinates (θ_0, θ_1), the contours of $S(\boldsymbol{\theta}) = $ constant are long thin ellipses. Such a sum of squares surface can be called *poorly conditioned* or ill-conditioned. However, if we rewrite the model as

$$Y_u = (\theta_0 + \theta_1\overline{X}) + \theta_1(X_u - \overline{X}) + \epsilon_u$$

$$= \phi_0 + \phi_1 x_u + \epsilon_u,$$

where $\phi_0 = \theta_0 + \theta_1\overline{X}$, $\phi_1 = \theta_1$, $x_u = X_u - \overline{X}$ ($= -1, 0, 1$ for $u = 1, 2, 3$), we obtain a sum of squares in terms of $\boldsymbol{\phi} = (\phi_1, \phi_2)'$ given by

$$S(\boldsymbol{\phi}) = \sum (Y_u - \phi_0 - \phi_1 x_u)^2$$

$$= (Y_1 - \phi_0 + \phi_1)^2 + (Y_2 - \phi_0)^2 + (Y_3 - \phi_0 - \phi_1)^2$$

$$= \sum_{u=1}^{3} Y_u^2 - 2\phi_0 \sum_{u=1}^{3} Y_u + 2\phi_1(Y_1 - Y_3) + 3\phi_0^2 + 2\phi_1^2.$$

In coordinates (ϕ_0, ϕ_1) these contours are "well-rounded" ellipses—the surface is said to be *well-conditioned*.

A similar sort of ill-conditioning can occur in nonlinear models of the form

$$Y_u = \theta_0 e^{\theta_1 X_u} + \epsilon_u$$

if the mean of the X_u, $\overline{X}$, is not close to zero. When expressions of this type occur it is sometimes better to consider the model in the alternative form

$$Y_u = (\theta_0 e^{\theta_1 \overline{X}})(e^{\theta_1(X_u - \overline{X})}) + \epsilon_u$$

$$= \phi_0 e^{\phi_1 x_u} + \epsilon_u,$$

where $\phi_0 = \theta_0 e^{\theta_1 \overline{X}}$, $\phi_1 = \theta_1$, and $x_u = X_u - \overline{X}$.

Note: In our example in Section 24.3 we did not do this. There it would have complicated the model since α occurred in two places and more than a simple replacement of one parameter by another is involved.

Suitable reparameterizations that will improve the conditioning of a sum of squares surface in a general case are not always apparent. Simple transformations that permit a "centering" of some variables, as in the examples above, may often be beneficial, however, and are often, at worst, harmless. For additional comments on reparameterization see the reference list at the end of the chapter.

24.5. THE GEOMETRY OF LINEAR LEAST SQUARES

There is a certain amount of overlap between this section (which has been preserved from the second edition) and the material in Chapters 20 and 21. However, the notation here is intended to conform more to what we need for Section 24.6, where the geometry of nonlinear least squares is discussed. Sections 24.5 and 24.6 are self-contained and do not require Chapters 20 and 21 as prerequisite.

To understand why iterative methods applied to nonlinear problems are not always successful, it is helpful to consider the geometrical interpretation of *linear* least squares first of all. In the linear case, in the notation of this chapter we can write the model as

$$Y = f(\xi, \boldsymbol{\theta}) + \epsilon$$
$$= \theta_1 X_1 + \theta_2 X_2 + \cdots + \theta_p X_p + \epsilon$$

where the X_i are functions of ξ. If we have observations Y_u containing errors ϵ_u when the X_i take the values $X_{1u}, X_{2u}, \ldots, X_{pu}$, for $u = 1, 2, \ldots, n$, then we can write the model in the alternative form:

$$\mathbf{Y} = \mathbf{X}\boldsymbol{\theta} + \boldsymbol{\epsilon},$$

where

$$\mathbf{Y} = \begin{bmatrix} Y_1 \\ Y_2 \\ \vdots \\ Y_n \end{bmatrix}, \quad \mathbf{X} = \begin{bmatrix} X_{11} & X_{21} & \cdots & X_{p1} \\ X_{12} & X_{22} & \cdots & X_{p2} \\ \vdots & \vdots & & \vdots \\ X_{1n} & X_{2n} & \cdots & X_{pn} \end{bmatrix}, \quad \boldsymbol{\theta} = \begin{bmatrix} \theta_1 \\ \theta_2 \\ \vdots \\ \theta_p \end{bmatrix}, \quad \boldsymbol{\epsilon} = \begin{bmatrix} \epsilon_1 \\ \epsilon_2 \\ \vdots \\ \epsilon_n \end{bmatrix}.$$

(Note that we can obtain a "β_0 term" in the model in this form by taking $X_{1u} = 1$ for $u = 1, 2, \ldots, n$.) The sum of squares surface in Eq. (24.1.9) can be written as

$$S(\boldsymbol{\theta}) = \sum_{u=1}^{n} \left[Y_u - \sum_{i=1}^{p} \theta_i X_{iu} \right]^2$$
$$= (\mathbf{Y} - \mathbf{X}\boldsymbol{\theta})'(\mathbf{Y} - \mathbf{X}\boldsymbol{\theta})$$
$$= \mathbf{Y}'\mathbf{Y} - 2\boldsymbol{\theta}'\mathbf{X}'\mathbf{Y} + \boldsymbol{\theta}'\mathbf{X}'\mathbf{X}\boldsymbol{\theta}.$$

If we differentiate this expression with respect to $\boldsymbol{\theta}$, set the result equal to $\mathbf{0}$, and write $\hat{\boldsymbol{\theta}}$ for $\boldsymbol{\theta}$, we obtain the normal *equations*

$$\mathbf{X}'\mathbf{X}\hat{\boldsymbol{\theta}} = \mathbf{X}'\mathbf{Y},$$

with solution, if $\mathbf{X}'\mathbf{X}$ is nonsingular, given by

$$\hat{\boldsymbol{\theta}} = (\mathbf{X}'\mathbf{X})^{-1}\mathbf{X}'\mathbf{Y}.$$

We recall that the regression sum of squares is $\hat{\boldsymbol{\theta}}'\mathbf{X}'\mathbf{Y}$ and the residual sum of squares is $\mathbf{Y}'\mathbf{Y} - \hat{\boldsymbol{\theta}}'\mathbf{X}'\mathbf{Y}$. Now

$$S(\hat{\boldsymbol{\theta}}) = \mathbf{Y}'\mathbf{Y} - 2\hat{\boldsymbol{\theta}}'\mathbf{X}'\mathbf{Y} + \hat{\boldsymbol{\theta}}'\mathbf{X}'\mathbf{X}\hat{\boldsymbol{\theta}}$$
$$= \mathbf{Y}'\mathbf{Y} - \hat{\boldsymbol{\theta}}'\mathbf{X}'\mathbf{Y} - \hat{\boldsymbol{\theta}}'(\mathbf{X}'\mathbf{Y} - \mathbf{X}'\mathbf{X}\hat{\boldsymbol{\theta}})$$
$$= \mathbf{Y}'\mathbf{Y} - \hat{\boldsymbol{\theta}}'\mathbf{X}'\mathbf{Y}$$

since $\hat{\boldsymbol{\theta}}$ satisfies the normal equations. Thus $S(\hat{\boldsymbol{\theta}})$, the smallest value of $S(\boldsymbol{\theta})$, is equal to the residual sum of squares in the analysis of variance table. We can also write

$$S(\boldsymbol{\theta}) - S(\hat{\boldsymbol{\theta}}) = \boldsymbol{\theta}'\mathbf{X}'\mathbf{X}\boldsymbol{\theta} - 2\boldsymbol{\theta}'\mathbf{X}'\mathbf{Y} + \hat{\boldsymbol{\theta}}'\mathbf{X}'\mathbf{X}\hat{\boldsymbol{\theta}}$$
$$= (\boldsymbol{\theta} - \hat{\boldsymbol{\theta}})'\mathbf{X}'\mathbf{X}(\boldsymbol{\theta} - \hat{\boldsymbol{\theta}}).$$

If the errors ϵ_u are independent and each follows the distribution $N(0, \sigma^2)$; that is, if

$\boldsymbol{\epsilon} \sim N(\mathbf{0}, \mathbf{I}\sigma^2)$, then it can be shown that, if the model is correct, the following results are true:

1. $\hat{\boldsymbol{\theta}} \sim N[\boldsymbol{\theta}, (\mathbf{X}'\mathbf{X})^{-1}\sigma^2]$.
2. $S(\hat{\boldsymbol{\theta}}) \sim \sigma^2 \chi^2_{n-p}$.
3. $S(\boldsymbol{\theta}) - S(\hat{\boldsymbol{\theta}}) \sim \sigma^2 \chi^2_p$.
4. $S(\boldsymbol{\theta}) - S(\hat{\boldsymbol{\theta}})$ and $S(\hat{\boldsymbol{\theta}})$ are distributed independently so that the ratio

$$\frac{[S(\boldsymbol{\theta}) - S(\hat{\boldsymbol{\theta}})]/p}{S(\hat{\boldsymbol{\theta}})/(n-p)} \sim F(p, n-p).$$

The contours defined by $S(\boldsymbol{\theta}) = $ constant can be examined in two different but related ways. We can examine them in the *sample space* (in which the mechanism of linear least squares can best be understood) or in the *parameter space* [in which we concentrate on the contours of $S(\boldsymbol{\theta})$ alone]. We shall now discuss these two representations.

The Sample Space

The sample space is an n-dimensional space. The vector of observations $\mathbf{Y} = (Y_1, Y_2, \ldots, Y_n)'$ defines a vector $\overrightarrow{OY}$ from the origin O to the point Y with coordinates $(Y_1, Y_2, \ldots, Y_n)$. The $\mathbf{X}$ matrix has p column vectors, each containing n elements. The elements of the jth column define the coordinates $(X_{j1}, X_{j2}, \ldots, X_{jn})$ of a point X_j in the sample space and the jth column vector of $\mathbf{X}$ defines the vector $\overrightarrow{OX_j}$ in the sample space. The p vectors $\overrightarrow{OX_1}, \overrightarrow{OX_2}, \ldots, \overrightarrow{OX_p}$ define a subspace of p dimensions, called the *estimation space,* which is contained within the sample space. Any point of this subspace can be represented by the end point of a vector, which is a linear combination of the vectors defining the space—that is, which is a linear combination of the columns of $\mathbf{X}$, such as, for example, $\mathbf{X}\boldsymbol{\theta}$ where $\boldsymbol{\theta} = (\theta_1, \theta_2, \ldots, \theta_p)'$ is a $p \times 1$ vector. Suppose the vector $\mathbf{X}\boldsymbol{\theta}$ defines the point T. Then the squared distance YT^2 is given by

$$(\mathbf{Y} - \mathbf{X}\boldsymbol{\theta})'(\mathbf{Y} - \mathbf{X}\boldsymbol{\theta}) = S(\boldsymbol{\theta})$$

as defined earlier. Thus the sum of squares $S(\boldsymbol{\theta})$ represents, in the sample space, the squared distance of Y from a general point T of the estimation space. Minimization of $S(\boldsymbol{\theta})$ with respect to $\boldsymbol{\theta}$ implies finding that value of $\boldsymbol{\theta}$, say, $\hat{\boldsymbol{\theta}}$, which provides a point P (defined by the vector $\hat{\mathbf{Y}} = \mathbf{X}\hat{\boldsymbol{\theta}}$) of the estimation space closest to the point Y. Geometrically, then, P must be the foot of the perpendicular from Y to the estimation space, that is, the foot of a line passing through Y and orthogonal to all the vectors defined by the columns of the $\mathbf{X}$ matrix. In terms of vectors from the origin, we can write

$$\mathbf{Y} = \hat{\mathbf{Y}} + (\mathbf{Y} - \hat{\mathbf{Y}})$$
$$= \hat{\mathbf{Y}} + \mathbf{e},$$

where $\mathbf{e}$ is the vector of *residuals.* The vector $\mathbf{Y}$ is thus divided into two orthogonal components: (1) $\hat{\mathbf{Y}}$, which lies entirely in the estimation space, and (2) $\mathbf{Y} - \hat{\mathbf{Y}} = \mathbf{e}$, the vector of residuals, which lies in what is called the *error space.* The error space is defined as the $(n - p)$-dimensional subspace that remains of the full n-dimensional space, after the p-dimensional estimation space has been defined. The estimation and error spaces are thus orthogonal. We can confirm algebraically that $\hat{\mathbf{Y}}$ and $\mathbf{e}$ are orthogonal as follows:

$$\hat{\mathbf{Y}}'\mathbf{e} = (\mathbf{X}\hat{\boldsymbol{\theta}})'(\mathbf{Y} - \mathbf{X}\hat{\boldsymbol{\theta}})$$
$$= \hat{\boldsymbol{\theta}}'\mathbf{X}'(\mathbf{Y} - \mathbf{X}\hat{\boldsymbol{\theta}})$$
$$= \hat{\boldsymbol{\theta}}'(\mathbf{X}'\mathbf{Y} - \mathbf{X}'\mathbf{X}\hat{\boldsymbol{\theta}})$$
$$= \mathbf{0}$$

since $\hat{\boldsymbol{\theta}}$ satisfies the normal equations, thus causing the parentheses to vanish. The vector $\mathbf{e}$ is a vector $\overrightarrow{OR}$, say, from the origin O, with length $OR = YP$, and with OR parallel to PY.

If T is a general point of the estimation space and YP is orthogonal to the space, then

$$YT^2 = YP^2 + PT^2$$

or

$$S(\boldsymbol{\theta}) = S(\hat{\boldsymbol{\theta}}) + PT^2.$$

Thus the contours for which $S(\boldsymbol{\theta}) = $ constant must be such that

$$PT^2 = S(\boldsymbol{\theta}) - S(\hat{\boldsymbol{\theta}}) = \text{a constant.}$$

In the sample space, then, the contours defined by $S(\boldsymbol{\theta}) = $ constant consist of all points T such that $PT^2 = $ constant, that is, points in the estimation space and of the form $\mathbf{X}\boldsymbol{\theta}$ that lie on a p-dimensional sphere centered at the point P defined by $\mathbf{X}\hat{\boldsymbol{\theta}}$. The radius of this sphere is $[S(\boldsymbol{\theta}) - S(\hat{\boldsymbol{\theta}})]^{1/2}$. By using the fact, given earlier, that

$$\frac{[S(\boldsymbol{\theta}) - S(\hat{\boldsymbol{\theta}})]/p}{S(\hat{\boldsymbol{\theta}})/(n-p)} \sim F(p, n-p),$$

we can define the boundary of a $100(1-\alpha)\%$ confidence region for the point $\mathbf{X}\boldsymbol{\theta}$, which arises from the true (but unknown) value of $\boldsymbol{\theta}$, by

$$\frac{[S(\boldsymbol{\theta}) - S(\hat{\boldsymbol{\theta}})]/p}{S(\hat{\boldsymbol{\theta}})/(n-p)} = F(p, n-p, 1-\alpha),$$

that is, by

$$S(\boldsymbol{\theta}) = S(\hat{\boldsymbol{\theta}})\left[1 + \frac{p}{n-p}F(p, n-p, 1-\alpha)\right],$$

which is of the sensible form $S(\hat{\boldsymbol{\theta}})(1+q^2)$ indicating values of $S(\boldsymbol{\theta})$ somewhat greater than the minimum value $S(\hat{\boldsymbol{\theta}})$. The confidence region will thus consist of the inside of a sphere in the estimation space centered at P and with radius

$$[S(\boldsymbol{\theta}) - S(\hat{\boldsymbol{\theta}})]^{1/2} = \left[S(\hat{\boldsymbol{\theta}})\frac{p}{n-p}F(p, n-p, 1-\alpha)\right]^{1/2}.$$

The Sample Space When $n = 3$, $p = 2$

In order to illustrate the foregoing remarks by a diagram, we shall suppose that $n = 3$. When $n > 3$ the complete situation cannot be drawn but the mental extension to higher dimensions is not difficult.

Figure 24.9 shows the sample space when $n = 3$, the coordinate axes being labeled 1, 2, and 3 to correspond to the three components (Y_1, Y_2, Y_3) of the vector $\mathbf{Y}'$. We shall suppose that there are $p = 2$ parameters θ_1 and θ_2 so that $\mathbf{X}$ is a 3×2 matrix of form

$$\mathbf{X} = \begin{bmatrix} X_{11} & X_{21} \\ X_{12} & X_{22} \\ X_{13} & X_{23} \end{bmatrix}.$$

The columns of $\mathbf{X}$ define two points P_1 and P_2 with coordinates (X_{11}, X_{12}, X_{13}) and (X_{21}, X_{22}, X_{23}), respectively, and the vectors $\overrightarrow{OP}_1$ and $\overrightarrow{OP}_2$ define a plane that represents the two-dimensional estimation space in which the vector $\hat{\mathbf{Y}} = \mathbf{X}\hat{\boldsymbol{\theta}}$ must lie. The point Y lies above this plane and the perpendicular YP from Y to the plane OP_1P_2 hits the plane at P. Thus YP is the shortest distance from Y to any point in the estimation space, P is defined by $\hat{\mathbf{Y}} = \mathbf{X}\hat{\boldsymbol{\theta}}$, and $S(\hat{\boldsymbol{\theta}}) = YP^2$. In addition, since $OY^2 = \mathbf{Y}'\mathbf{Y}$, the standard analysis of variance breakup

$$\mathbf{Y}'\mathbf{Y} = \hat{\boldsymbol{\theta}}'\mathbf{X}'\mathbf{Y} + (\mathbf{Y}'\mathbf{Y} - \hat{\boldsymbol{\theta}}'\mathbf{X}'\mathbf{Y})$$

or

$$\mathbf{Y}'\mathbf{Y} = \hat{\boldsymbol{\theta}}'\mathbf{X}'\mathbf{Y} + S(\hat{\boldsymbol{\theta}})$$

is equivalent to the Pythagoras result:

$$OY^2 = OP^2 + YP^2.$$

If we draw a line OR through O, equal in length [so that $OR^2 = S(\hat{\boldsymbol{\theta}})$] and parallel to PY, then $\overrightarrow{OR}$ represents the vector of residuals $\mathbf{e} = \mathbf{Y} - \hat{\mathbf{Y}}$. The vector $\overrightarrow{OP}$ is $\hat{\mathbf{Y}}$

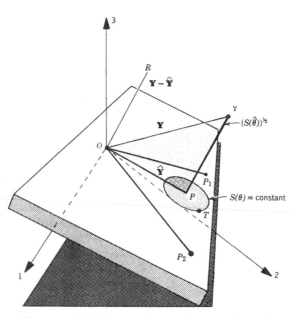

Figure 24.9. The sample space when $n = 3$, $p = 2$.

so we have the vector equation

$$\vec{OY} = \vec{OP} + \vec{OR}$$

or

$$\mathbf{Y} = \hat{\mathbf{Y}} + (\mathbf{Y} - \hat{\mathbf{Y}}).$$

We recall that, in general, contours of constant $S(\boldsymbol{\theta})$ are represented by p-dimensional spheres in the estimation space. Here, then, the contours must be circles on the plane OP_1P_2. This is easy to see, for if T is a general point $\mathbf{X}\boldsymbol{\theta}$ on the plane, $S(\boldsymbol{\theta}) = $ constant means $YT^2 = $ constant, so that $PT^2 = YT^2 - YP^2 = $ constant. We thus obtain circles about P. One such circle is shown on the figure. The circle that provides a $100(1 - \alpha)\%$ confidence interval for the true point $\mathbf{X}\boldsymbol{\theta}$ has radius given by

$$[2S(\hat{\boldsymbol{\theta}})F(2, 1, 1 - \alpha)]^{1/2},$$

obtained by putting $n = 3$, $p = 2$ in the general formula.

The Sample Space Geometry When the Model Is Wrong

Suppose $\mathbf{Y} = \mathbf{X}\boldsymbol{\theta} + \boldsymbol{\epsilon}$ is the postulated linear model containing p parameters but that the true linear model is

$$\mathbf{Y} = \mathbf{X}\boldsymbol{\theta} + \mathbf{X}_2\boldsymbol{\theta}_2 + \boldsymbol{\epsilon}$$

and contains additional terms $\mathbf{X}_2\boldsymbol{\theta}_2$ not considered. Then since the estimation space consists only of points of the form $\mathbf{X}\boldsymbol{\theta}$, the true point $\eta = \mathbf{X}\boldsymbol{\theta} + \mathbf{X}_2\boldsymbol{\theta}_2$ cannot lie in the estimation space. In this case the perpendicular YP from Y onto the estimation space (whose foot P is given by $\hat{\mathbf{Y}} = \mathbf{X}\boldsymbol{\theta}$) will be longer than it would have been if the correct model and estimation space had been used. To illustrate this point we shall give, in Figure 24.10, a diagram for the case $n = 3$, $p = 1$, where the true model contains two parameters θ and θ_2. The true model takes the form

$$\begin{bmatrix} Y_1 \\ Y_2 \\ Y_3 \end{bmatrix} = \begin{bmatrix} X_{11} \\ X_{12} \\ X_{13} \end{bmatrix}\theta + \begin{bmatrix} X_{21} \\ X_{22} \\ X_{23} \end{bmatrix}\theta_2 + \begin{bmatrix} \epsilon_1 \\ \epsilon_2 \\ \epsilon_3 \end{bmatrix}$$

$$= \mathbf{X}\theta \quad + \mathbf{X}_2\theta_2 \quad + \boldsymbol{\epsilon}$$

and the postulated model is given when $\theta_2 = 0$. The single column of the $\mathbf{X}$ matrix defines the point P_1 and the line OP_1, which is the estimation space for the postulated model. The line YP is the perpendicular from Y onto OP_1 and P is the point $\hat{\mathbf{Y}} = \mathbf{X}\hat{\theta}$. Therefore the shortest squared distance $S(\hat{\theta})$, of all the squared distances $S(\theta)$ from the point Y to points $\mathbf{X}\theta$ on the line OP_1, is represented by the square of the length of YP. The true value of θ defines an unknown point $\mathbf{X}\theta$ on the line OP_1. A confidence interval for the true value $\mathbf{X}\theta$ can be constructed on OP_1 and around the point P.

Now the second vector $\mathbf{X}_2$ in the true model defines a line OP_2 and the lines OP_1 and OP_2 define a plane in which the true point $\mathbf{X}\theta + \mathbf{X}_2\theta_2$ lies. Suppose YP^* is the perpendicular from Y to the *correct* estimation space given by the plane OP_1P_2. Then P^* represents the point that would have given the correct fitted value $\hat{\mathbf{Y}}^*$, say, *if* the correct model had been used. This is always of length less than or equal to YP since a perpendicular to a space (a plane here, OP_1P_2) cannot be longer than the

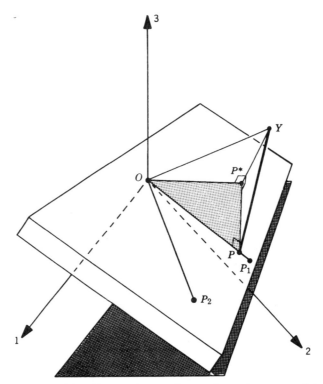

Figure 24.10. The sample space for $n = 3$, $p = 1$; wrong model.

perpendicular to an included space (here the line OP_1). If the model is incorrect, therefore, $S(\hat{\boldsymbol{\theta}}) = YP^2$ will be too long, if anything. [Note that it *could* happen that P and P^* coincide, so that the same minimum value $S(\hat{\boldsymbol{\theta}})$ would occur, whichever model was used. This would be very unusual, of course.]

When the postulated model is correct, in the general case, $S(\hat{\boldsymbol{\theta}})$ has expected or mean value $(n - p)\sigma^2$. If a pure error or prior estimate of σ^2 is available we know how big, roughly, the quantity YP^2 should be. However, if the postulated model is inadequate, YP^2 will probably be too long. The standard lack of fit test is thus examining the question: "Is the squared length YP^2 greater than we should expect on the basis of the good information we have about the size of the random error?" How much greater is *too great* is determined through the distribution properties involved, as formalized earlier.

Geometrical Interpretation of Pure Error

A geometrical interpretation of pure error is shown in Figure 24.11. In the sample space, O is the origin, Y is the end point of the vector $\mathbf{Y}$ of observations, and P is the foot of the perpendicular from Y onto the estimation space defined by the columns of the $\mathbf{X}$ matrix. Thus OP is the vector $\hat{\mathbf{Y}} = \mathbf{X}\hat{\boldsymbol{\theta}}$. The point $\tilde{Y}$ is the end point of the vector $\tilde{\mathbf{Y}}$ whose ith element, $i = 1, 2, \ldots, n$, is

$$\tilde{Y}_i = (\text{average response in the group of repeats to which } Y_i \text{ belongs})$$

$$= \overline{Y}_{i0},$$

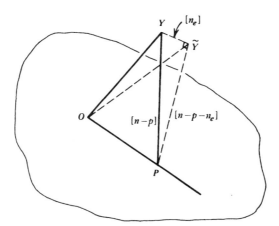

Figure 24.11. Geometrical representation of pure error. (The symbol $[n_e]$ means that the vector $\mathbf{Y} - \tilde{\mathbf{Y}}$, which is parallel to the line $Y\tilde{Y}$, lies in a subspace of the sample space of dimension n_e, etc.)

say. For nonrepeated points, $\overline{Y}_{i0} = Y_i$. It is easy to show that vectors $\mathbf{Y} - \tilde{\mathbf{Y}}$ and $\hat{\mathbf{Y}} - \tilde{\mathbf{Y}}$ are orthogonal so that the (respectively parallel) lines $Y\tilde{Y}$ and $\tilde{Y}P$ are perpendicular to each other, whereupon, by Pythagoras's theorem $Y\tilde{Y}^2 + \tilde{Y}P^2 = YP^2$. We can identify

YP^2 = squared length of residual vector $\mathbf{Y} - \mathbf{X}\boldsymbol{\theta}$ = residual sum of squares for the fitted model,

$Y\tilde{Y}^2$ = squared length of vector $\mathbf{Y} - \tilde{\mathbf{Y}}$ = pure error sum of squares,

$\tilde{Y}P^2$ = squared length of vector $\hat{\mathbf{Y}} - \tilde{\mathbf{Y}}$ = lack of fit sum of squares.

The F-test for lack of fit thus numerically compares $\tilde{Y}P^2/(n - p - n_e)$ with $Y\tilde{Y}^2/n_e$, where n_e is the pure error degrees of freedom and $(n - p - n_e)$ is the lack of fit degrees of freedom, each squared vector length being divided by the dimension of the subspace in which the corresponding vector lies. The F-distribution properties follow from the usual normality assumptions on the errors.

The Parameter Space

The parameter space is a p-dimensional space in which a set of values $(\theta_1, \theta_2, \ldots, \theta_p)$ of the parameters defines a point. The minimum value of $S(\boldsymbol{\theta})$ is attained at the point $\hat{\boldsymbol{\theta}} = (\hat{\theta}_1, \hat{\theta}_2, \ldots, \hat{\theta}_p)$. We recall that

$$S(\boldsymbol{\theta}) - S(\hat{\boldsymbol{\theta}}) = (\boldsymbol{\theta} - \hat{\boldsymbol{\theta}})'\mathbf{X}'\mathbf{X}(\boldsymbol{\theta} - \hat{\boldsymbol{\theta}}).$$

All values of $\boldsymbol{\theta}$ that satisfy $S(\boldsymbol{\theta})$ = constant = K are given by

$$(\boldsymbol{\theta} - \hat{\boldsymbol{\theta}})'\mathbf{X}'\mathbf{X}(\boldsymbol{\theta} - \hat{\boldsymbol{\theta}}) = K - S(\hat{\boldsymbol{\theta}})$$

and it can be shown that this is the equation of a closed ellipsoidal contour surrounding the point $\hat{\boldsymbol{\theta}}$. When $K_1 > K_2$ the contour $S(\boldsymbol{\theta}) = K_1$ completely encloses the contour $S(\boldsymbol{\theta}) = K_2$ and $\hat{\boldsymbol{\theta}}$ lies in the center of these nested p-dimensional "eggs." A $100(1 - \alpha)\%$ confidence region for the true (but unknown) value of $\boldsymbol{\theta}$ is enclosed by the contour, which is such that

$$\frac{[S(\boldsymbol{\theta}) - S(\hat{\boldsymbol{\theta}})]/p}{S(\hat{\boldsymbol{\theta}})/(n - p)} = F(p, n - p, 1 - \alpha)$$

if the errors are normally distributed, that is, $\epsilon \sim N(0, I\sigma^2)$. This can be rearranged as

$$S(\boldsymbol{\theta}) = S(\hat{\boldsymbol{\theta}}) \left\{ 1 + \frac{p}{n-p} F(p, n-p, 1-\alpha) \right\}$$

in which the expression on the right-hand side is the constant value that defines the contour.

The Parameter Space When $p = 2$

We again use a simple case to illustrate the situation. Figure 24.12 shows some possible contours of the form $S(\boldsymbol{\theta}) = $ constant for three values of the constant when $p = 2$. The outer contour is labeled as a $100(1-\alpha)\%$ confidence contour, defined as above. In the two-dimensional space of (θ_1, θ_2), the contours are concentric ellipses about the point $(\hat{\theta}_1, \hat{\theta}_2)$. Note that contours of this type are obtained no matter what the value of n (the number of observations) may be, since the dimension of the parameter space depends on p alone.

In general, the orientation and the shape of the ellipses are both of importance. If the axes of the ellipses are parallel to the θ_1 and θ_2 axes, then the value $\hat{\theta}_1$ that makes $S(\theta_1, \theta_2)$ a minimum has no dependence on θ_2; that is, if we fix θ_2 at any value the same value of $\theta_1 = \hat{\theta}_1$ minimizes $S(\theta_1, \theta_2 \mid \theta_2$ fixed). This means that specific information about θ_2, which fixed its value, would not alter the least squares estimate $\hat{\theta}_1$. This situation occurs when the expression for $S(\theta_1, \theta_2)$ can be written without a cross-product term in $\theta_1 \theta_2$. The model when $p = 2$ can be written as

$$Y_u = \theta_1 X_{1u} + \theta_2 X_{2u} + \epsilon_u, \qquad u = 1, 2, \ldots, n.$$

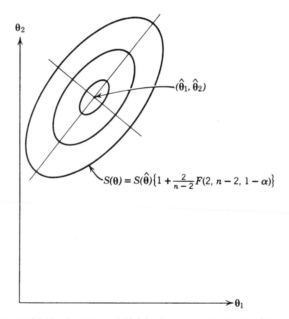

Figure 24.12. Contours of $S(\boldsymbol{\theta})$ in the parameter space when $p = 2$.

Thus

$$S(\boldsymbol{\theta}) = S(\theta_1, \theta_2) = \sum_{u=1}^{n} (Y_u - \theta_1 X_{1u} - \theta_2 X_{2u})^2$$

$$= \sum Y_u^2 - \theta_1 2 \sum X_{1u} Y_u - \theta_2 2 \sum X_{2u} Y_u$$

$$+ \theta_1^2 \sum X_{1u}^2 + \theta_2^2 \sum X_{2u}^2 + \theta_1 \theta_2 2 \sum X_{1u} X_{2u},$$

where all summations are over $u = 1, 2, \ldots, n$. It is clear from this that the minimizing value of θ_1, namely, $\hat{\theta}_1$, which satisfies $\partial S(\boldsymbol{\theta})/\partial \theta_1 = 0$, will not depend on θ_2 (and vice versa) if the coefficient of $\theta_1 \theta_2$ vanishes; that is, if $\sum X_{1u} X_{2u} = 0$, when the columns of the **X** matrix are orthogonal.

When the X_1 and X_2 columns of the **X** matrix are not orthogonal, a $\theta_1 \theta_2$ term occurs in $S(\theta_1, \theta_2)$, and the ellipses are obliquely oriented with respect to the θ_1 and θ_2 axes.

The shape of the $S(\theta_1, \theta_2)$ contours shows the relative precisions with which the estimates $\hat{\theta}_1$ and $\hat{\theta}_2$ are determined. Figure 24.13 illustrates some of the possibilities. The single contour shown is intended to represent the 95% confidence region boundary, and the point $\hat{\boldsymbol{\theta}}$ with coordinates $(\hat{\theta}_1, \hat{\theta}_2)$ is the least squares estimate of $\boldsymbol{\theta}$, in each case.

24.6. THE GEOMETRY OF NONLINEAR LEAST SQUARES

The Sample Space

When the model is nonlinear rather than linear there is no **X** matrix in the linear model sense. While there is still an estimation space it is not one defined by a set of vectors and may be very complex. The estimation space (also called the *solution locus*) consists of all points with coordinates expressible as

$$\{f(\boldsymbol{\xi}_1, \boldsymbol{\theta}), f(\boldsymbol{\xi}_2, \boldsymbol{\theta}), \ldots, f(\boldsymbol{\xi}_n, \boldsymbol{\theta})\}.$$

Since the sum of squares function $S(\boldsymbol{\theta})$ still represents the square of the distance from the point $(Y_1, Y_2, \ldots, Y_n)$ to a point of the estimation space, minimization of $S(\boldsymbol{\theta})$ still corresponds geometrically to finding a point P of the estimation space nearest to Y. The sample space for a very simple nonlinear example involving only $n = 2$ observations Y_1 and Y_2 taken at $\boldsymbol{\xi} = \boldsymbol{\xi}_1$ and $\boldsymbol{\xi} = \boldsymbol{\xi}_2$, respectively, and a single parameter θ, is shown in Figure 24.14. The estimation space consists of the curved line that contains points

$$\{f(\boldsymbol{\xi}_1, \theta), f(\boldsymbol{\xi}_2, \theta)\}$$

as θ varies, where $\boldsymbol{\xi}_1, \boldsymbol{\xi}_2$ are fixed. Y has coordinates (Y_1, Y_2), and P is the point of the estimation space nearest to Y.

Figure 24.15 shows the sample space for an example involving $n = 3$ observations Y_1, Y_2, and Y_3 taken at $\boldsymbol{\xi} = \boldsymbol{\xi}_1, \boldsymbol{\xi}_2$, and $\boldsymbol{\xi}_3$, respectively, and two parameters θ_1 and θ_2. The curved lines indicate the coordinate systems of the parameters on the estimation space or solution locus, which consists of all points of the form

$$\{f(\boldsymbol{\xi}_1, \theta_1, \theta_2), f(\boldsymbol{\xi}_2, \theta_1, \theta_2), f(\boldsymbol{\xi}_3, \theta_1, \theta_2)\}$$

as θ_1 and θ_2 vary, where $\boldsymbol{\xi}_1, \boldsymbol{\xi}_2$, and $\boldsymbol{\xi}_3$ are fixed. Y has coordinates (Y_1, Y_2, Y_3) and P is the point of the estimation space nearest to Y. When we apply the linearization technique to nonlinear problems we are selecting a point of the estimation space $\boldsymbol{\theta}_0$, say, as new origin, defining a linearized estimation space in the form of the tangent

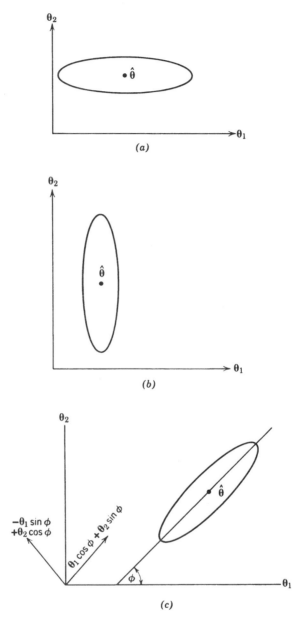

F i g u r e 24.13. The interpretation of some possible 95% confidence regions for parameters (θ_1, θ_2). (*a*) $\hat{\theta}_1$ not well-determined; $\hat{\theta}_2$ well-determined; no dependence between $\hat{\theta}_1$ and $\hat{\theta}_2$. (*b*) $\hat{\theta}_1$ well-determined; $\hat{\theta}_2$ not well-determined; no dependence between $\hat{\theta}_1$ and $\hat{\theta}_2$. (*c*) $\hat{\theta}_1 \cos \phi + \hat{\theta}_2 \sin \phi$ not well-determined; $-\hat{\theta}_1 \sin \phi + \hat{\theta}_2 \cos \phi$ well-determined; dependence between $\hat{\theta}_1$ and $\hat{\theta}_2$.

space at $\boldsymbol{\theta}_0$ and solving the linearized least squares problem so defined. The solution to this (which will be given in units of rate of change of $\boldsymbol{\theta}$ that are appropriate at $\boldsymbol{\theta}_0$ only) are applied to the nonlinear problem, where they may not be correct, and another iteration is attempted. For a nonlinear problem involving only two observations and one parameter the effect will be as shown in Figure 24.16. Figure 24.16 shows the estimation space or solution locus with units of θ shown upon it. We assume here that $\theta_0 = 0$ and that the point marked $\theta = 1$ is the point of the estimation space attained

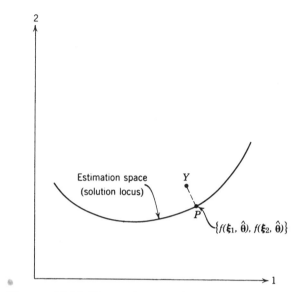

Figure 24.14. The sample space when $n = 2$, $f(\xi, \theta)$ nonlinear.

where $\theta = 1$, and so on. Note that the markings for θ are *not* equally spaced due to the nonlinearity and the nonuniformity of the coordinate system. The line tangent to the estimation space curve at $\theta = \theta_0 = 0$ is shown, graduated with units $\theta = 0, 1, 2, \ldots$, which are obtained from the rate of change found at θ_0. These units are equally spaced. We now find the least squares estimate of θ based on the linear assumption.

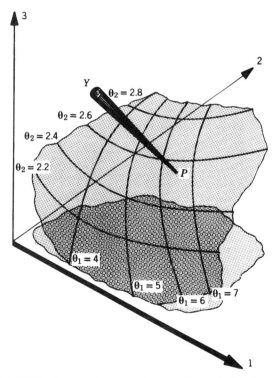

Figure 24.15. The sample space when $n = 3$, $p = 2$, $f(\xi, \theta)$ nonlinear.

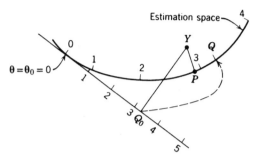

Figure 24.16. Geometrical interpretation of linearization method ($n = 2, p = 1$).

Geometrically, this means finding the point Q_0 so that YQ_0 is perpendicular to the tangent line. We see that, in the linearized units, a value of θ of about 3.2 (at Q_0) is indicated. In the next iteration of the linearization procedure we thus use the tangent line at the point where $\theta = 3.2$ on the *estimation space curve,* that is, at the point Q.

It is easy to see from this one reason why the linearization procedure sometimes fails. If the rate of change of $f(\xi, \theta)$ is small at θ_0, but increases rapidly, the units on the tangent line may be quite unrealistic. For example, in Figure 24.17, the rate of change at $\theta_0 = 0$ is small and so the linearized units of θ are small. The actual units increase sharply, however. Thus, if we begin a further iteration using the indicated value of θ of about 26 at Q_0, our starting point on the estimation space will be farther from the best point P than was our original guess $\theta = \theta_0 = 0$. The situation may or may not be corrected in successive iterations. (Although we have used $\theta_0 = 0$ and units 1, 2, ... , for simplicity, similar remarks apply in general whatever the value of the initial guess θ_0 and whatever the system of units near θ_0 may be.)

When there are more observations than two and more than one parameter the same ideas hold but the situation is more complicated and is difficult or impossible to draw.

When the model is linear, contours of constant $S(\boldsymbol{\theta})$ in the sample space consist of spheres. In nonlinear problems this is no longer true and quite irregular contours may

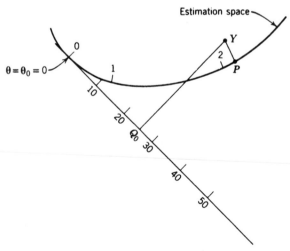

Figure 24.17. The effect on the linearization method of gross inequities in the systems of units ($n = 2, p = 1$).

arise consisting of all points of the estimation space equidistant, a selected distance, from the point $Y:(Y_1, Y_2, \ldots, Y_n)$.

The Parameter Space

In the linear model case, contours of constant $S(\boldsymbol{\theta})$ in the parameter space, or $\boldsymbol{\theta}$-space, consist of concentric ellipses. When the model is nonlinear the contours are sometimes banana-shaped, often elongated. Sometimes the contours stretch to infinity and do not even close, or they may have multiple loops surrounding a number of stationary values. When several stationary values exist they may have various levels or provide alternative minima for $S(\boldsymbol{\theta})$. Consider, for example, the model

$$Y = \frac{(\theta_1 e^{-\theta_2 t} - \theta_2 e^{-\theta_1 t})}{(\theta_2 - \theta_1)} + \epsilon.$$

Interchange of θ_1 and θ_2 leaves the model unaltered. Thus if the minimum $S(\boldsymbol{\theta})$ is attained at $(\theta_1, \theta_2) = (\hat{\theta}_1, \hat{\theta}_2)$, the same minimum value is given at $(\theta_1, \theta_2) = (\hat{\theta}_2, \hat{\theta}_1)$, so that a double solution exists. Multiple solutions are not easy to spot in this way in general. An example of banana-shaped contours is given in Section 24.3.

Confidence Contours in the Nonlinear Case

When the model is nonlinear a number of results that are true for the linear case no longer apply. When the error ϵ of the nonlinear model (24.1.5) is assumed to be normally distributed, $\hat{\boldsymbol{\theta}}$ is no longer normally distributed, $s^2 = S(\hat{\boldsymbol{\theta}})/(n - p)$ is no longer an unbiased estimate of σ^2, and there is no variance–covariance matrix of form $(\mathbf{X}'\mathbf{X})^{-1}\sigma^2$ in general.

Although confidence regions can still be *defined* by the expression

$$S(\boldsymbol{\theta}) = S(\hat{\boldsymbol{\theta}}) \left\{ 1 + \frac{p}{n - p} F(p, n - p, 1 - \alpha) \right\},$$

which provides a $100(1 - \alpha)\%$ confidence region in the linear model, normal error situation, the confidence coefficient will not be $1 - \alpha$, in the nonlinear case. We do not know in general what the confidence will be but we can call such regions *approximate* $100(1 - \alpha)\%$ *confidence regions* for $\boldsymbol{\theta}$. The banana-shaped regions for the example in Section 24.3 were obtained in this manner. While suitable comparisons of mean squares can still be made visually, the usual F-tests for regression and lack of fit are not valid, in general, in the nonlinear case.

Measuring Nonlinearity

Several suggestions have been made to meet the need for a measure of "the amount of nonlinearity" in nonlinear problems. Such a measure would help us decide when linearized results provide acceptable approximations, for example. For a brief discussion see the subsection "Curvature Measures" near the end of Section 24.3.

24.7. NONLINEAR GROWTH MODELS

This section is concerned with some examples of nonlinear models that have been used to describe growth behavior, as it varies in time. Growth models are applied in many fields. In biology, botany, forestry, zoology, and ecology, growth occurs in organisms, plants, trees and bushes, animals, and human beings. In chemistry and chemical engineering, growth occurs as a result of chemical reactions. In economics and political science, growth of organizations, supplies of food and material, and nations occurs.

Types of Models

The type of model needed in a specific area and in a specific problem depends on the type of growth that occurs. In general, growth models are *mechanistic* rather than *empirical* ones. A mechanistic model usually arises as a result of making assumptions about the type of growth, writing down differential or difference equations that represent these assumptions, and then solving these equations to obtain a growth model. (An empirical model, on the other hand, is a model chosen to empirically approximate an unknown mechanistic model. Typically, the empirical model is a polynomial of some suitable order.)

An Example of a Mechanistic Growth Model

Consider a growth situation in which it is believed that the *rate* of growth at a particular time t is directly proportional to the amount of growth yet to be achieved. If we denote the limiting (i.e., the maximum possible) growth size by α, and if ω is the size at time t, then

$$\frac{d\omega}{dt} = k(\alpha - \omega), \tag{24.7.1}$$

where k is the *rate constant* of the growth pattern. Integrating Eq. (24.7.1) gives

$$\omega = \alpha(1 - \beta e^{-kt}), \tag{24.7.2}$$

usually known as the *monomolecular* growth function. This function rises steadily from a point $\alpha(1 - \beta)$ (at $t = 0$) to the limiting value of α. It has no point of inflection [i.e., there is no change in sign of the second derivative $(d^2\omega/dt^2)$ for any t], and it climbs steadily at a decreasing rate as in Eq. (24.7.1). It has been used in the past to represent the later portions of a life history (see, e.g., Gregory, 1928). Suppose we now wish to get some idea of what the curve described by Eq. (24.7.2) looks like. Because α is simply a scale factor, we can set $\alpha = 1$; also, because k and t only occur together as a unit, kt, we can take $k = 1$, for this purpose. Then, varying β, we get curves like those in Figure 24.18. Choice of a different value of α would change the vertical scale of the curve; choice of a different value of k would stretch out or contract the curve horizontally. Each curve starts at the value, when $t = 0$, of $\omega = \alpha(1 - \beta)$, which is simply $1 - \beta$ when $\alpha = 1$ as in Figure 24.18.

  The curves drawn in Figure 24.18 are, of course, theoretical ones, based on the theoretical function given in Eq. (24.7.2). If we assume w_i, $i = 1, 2, \ldots, n$, to be observations of ω at times $t_1, t_2, \ldots, t_n$, we can postulate the model

$$w_i = \alpha(1 - \beta e^{-kt_i}) + \epsilon_i \tag{24.7.3}$$

where ϵ_i is a random error such that $E(\epsilon_i) = 0$, $V(\epsilon_i) = \sigma_i^2$, say. Suppose we assume that the errors ϵ_i all have the same variance and are uncorrelated. Then it is reasonable to use the least squares method to fit Eq. (24.7.3) to the data. Unless otherwise stated, we shall *always* assume our errors to be additive ones in the work that follows and

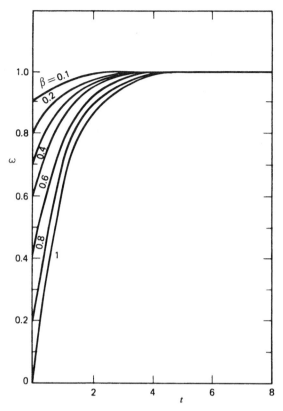

Figure 24.18. Theoretical curves of form $\omega = 1 - \beta e^{-t}$ for various β, as indicated.

will not repeat the above discussion. For example, "fitting the model given in Eq. (24.7.2)" will always mean use of Eq. (24.7.3), and so on for other models mentioned, unless otherwise specified.

Querying the Least Squares Assumptions

Growth data do not always satisfy the "usual least squares assumptions." For example, if the observations w_i are all taken on the same plant, animal, or organism, there is little reason to expect that they will be uncorrelated. Ideally, the experimenter should use different, independent items for each observation, but this is not always sensible or possible due to limitations on the experimental material or equipment. Also the $\sigma_i^2 = V(\epsilon_i)$ may well not all be the same and may depend, for example, on the growth size or some function of it. When specific information on such points is available, one can take advantage of it—for example, perhaps by using weighted least squares. Otherwise, the standard procedure is to fit the model in as common sense a manner as possible, and then to examine the residuals from the fit to see if they exhibit characteristics that give clues of invalid assumptions. These clues would then be used to iterate to an improved fitting procedure and/or an improved model.

The Logistic Model

Suppose the growth rate is such that (for $k > 0$)

$$\frac{d\omega}{dt} = \frac{k\omega(\alpha - \omega)}{\alpha},$$

(24.7.4)

that is, it is proportional to the product of the present size and the future amount of growth, α being some limiting growth value. If we compare with Eq. (24.7.1) we see that it is now the growth rate *relative to present size,* $(d\omega/dt)/\omega$, that declines linearly with increasing ω. Integrating Eq. (24.7.4) we find

$$\omega = \alpha/\{1 + \beta e^{-kt}\}, \tag{24.7.5}$$

known as the *logistic* or *autocatalytic* growth function. This curve has an S-shape. Note that, for $t = 0$, $\omega = \alpha/(1 + \beta)$ so that this is the starting growth value; also, for $t = \infty$, $\omega = \alpha$, the limiting growth value. It follows that $\beta > 0$. Also it is clear that $k > 0$. From Eq. (24.7.4) it is obvious that the slope of the curve is always positive, and the second derivative,

$$\frac{d^2\omega}{dt^2} = \frac{k}{\alpha}(\alpha - 2\omega), \tag{24.7.6}$$

is positive for $\omega < \frac{1}{2}\alpha$, vanishes at the point of inflection where $\omega = \omega_I = \frac{1}{2}\alpha$, and is negative for $\omega > \frac{1}{2}\alpha$. At the point of inflection, substituting in Eq. (24.7.5), we find $t_I = (\ln \beta)/k$. Writing $t = t_I + u$ we can write Eq. (24.7.5) as $\omega = \alpha/(1 + e^{-ku})$ so that

$$\omega - \omega_I = \frac{\alpha}{2}\left\{\frac{1 - e^{-ku}}{1 + e^{-ku}}\right\} = g(u), \tag{24.7.7}$$

say, from which it is clear that the curve shape is symmetric about its point of inflection, because $g(-u) = -g(u)$. Of course, the curve extends only to $t = 0$ (i.e., $u = -t_I$) to the left, but to $t = \infty$ (i.e., $u = \infty$) to the right. Note that, if $0 < \beta < 1$, the curve begins *above* the point of inflection, while if β is very large and positive, the point of inflection is attained at a high value of t and may not appear in the t-range of the data. Of course, $\beta > 0$ because we must have $\alpha > \alpha/(1 + \beta)$.

To get some idea of how this family of curves looks, we can set $\alpha = 1$ without loss of generality and plot Eq. (24.7.5) for various β and k values. Some illustrative curves are shown in Figures 24.19 and 24.20. Changing β alters the starting point on the vertical scale at $t = 0$. Changing k alters the steepness of the curve. Of course, because

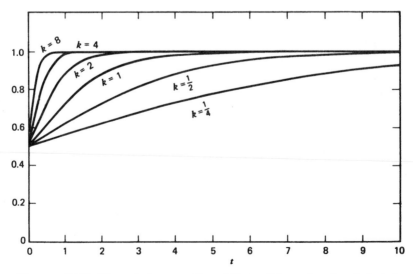

Figure 24.19. Theoretical curves of form $1/\{1 + e^{-kt}\}$ for various k, as indicated.

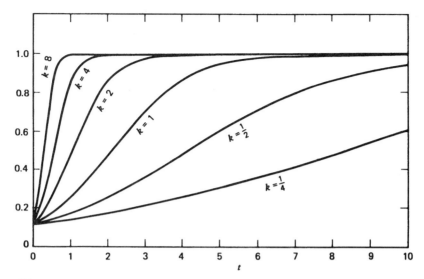

Figure 24.20. Theoretical curves of form $1/\{1 + 8e^{-kt}\}$ for various k, as indicated.

kt occurs as a product, a change of k can essentially be compensated by a recalibration of the t-axis; for example, $kt = (\frac{1}{2}k)(2t) = KT$, where $K = \frac{1}{2}k$ and $T = 2t$.

Another Form of the Logistic Model

Consider the model

$$\eta = \delta - \ln(1 + \beta e^{-kt}), \tag{24.7.8}$$

obtained by taking natural logarithms in Eq. (24.7.5), and setting $\eta = \ln \omega$ and $\delta = \ln \alpha$. Essentially, the curve shapes are similar to those of Figures 24.19 and 24.20 except that the vertical scale would be altered by the log transformation. Fitting Eqs. (24.7.5) and (24.7.8) by least squares involves completely different assumptions about the deviations from these models. According to Nelder (1961), the assumption that observations of η, $y_i = \ln w_i$, where the w_i are growth observations, have constant variance is usually a sensible one for the case of growing plants. In practice, both Eqs. (24.7.5) and (24.7.8) could be fitted and the residuals checked in both cases to see which model (if either) was suitable for the experimenter's purposes.

How Do We Get the Initial Parameter Estimates?

As we already know, nonlinear estimation procedures require initial parameter estimates and the better these initial estimates are, the faster will be the convergence to the fitted values. In fact, experience with growth models shows that, if the initial estimates are poor, convergence to the wrong final values can easily occur.

There is no general method for obtaining initial estimates. One uses whatever information is available. For example, for the logistic model (24.7.8), we can argue in this manner:

Step 1. When $t = \infty$, $\eta = \delta$. So take $\delta_0 = y_{max}$.

Step 2. For any two other observations, the ith and jth, say, set

$$y_i = \delta_0 - \ln(1 + \beta_0 e^{-k_0 t_i}),$$

$$y_j = \delta_0 - \ln(1 + \beta_0 e^{-k_0 t_j}),$$

acting as though (24.7.8) were true without error for these observations. Then, developing, we find that

$$\exp(\delta_0 - y_i) - 1 = \beta_0 \exp(-k_0 t_i),$$

$$\exp(\delta_0 - y_j) - 1 = \beta_0 \exp(-k_0 t_j),$$

whereupon by division, taking natural logarithms, and rearrangement, we obtain

$$k_0 = \frac{1}{t_j - t_i} \ln \left\{ \frac{\exp(\delta_0 - y_i) - 1}{\exp(\delta_0 - y_j) - 1} \right\}.$$

In general i and j should be more widely spaced rather than otherwise, to lead to stable estimates.

Step 3. From the ith equation above we can evaluate

$$\beta_0 = \exp(k_0 t_i)\{\exp(\delta_0 - y_i) - 1\}.$$

(Either the ith or jth equation can be used; it makes no difference.)

Step 4. Substitution of $\delta_0 = y_{\max}$ in the two foregoing equations provides us with values for k_0 and β_0.

An alternative at the first step would be to take δ_0 somewhat bigger than $y_{\max}$, say, 110% of it, or whatever experience suggests. This would usually provide slightly better initial estimates. In some problems we can set $t = 0$ and set η at $t = 0$ to $y_{\min}$, or perhaps 90% of $y_{\min}$, and so on. In general, whatever method leads to simple equations to solve is employed to get the initial estimates.

Initial estimates for fitting the model (24.7.5) can similarly be obtained in the order α_0, k_0, β_0 from the equations

$$\alpha_0 = y_{\max},$$

$$\beta_0 = \{(\alpha_0 - w_i)/\alpha_0\}\exp(k_0 t_i),$$

$$k_0 = \frac{1}{t_i - t_j} \ln \left\{ \frac{\alpha_0 - w_j}{\alpha_0 - w_i} \right\}.$$

The Gompertz Model

If the growth rate

$$\frac{d\omega}{dt} = k\omega \log(\alpha/\omega), \tag{24.7.9}$$

we obtain, by integration, the Gompertz model

$$\omega = \alpha \exp\{-\beta e^{-kt}\}. \tag{24.7.10}$$

Although this curve is an S-shaped one like the logistic, it is not symmetrical about its point of inflection, the latter being where $d^2\omega/dt^2 = 0$, namely, where $\omega_I = \alpha/e =$

0.368α, which implies that $t_I = (\log \beta)/k$. Note that Eqs. (24.7.9) and (24.7.10) imply the relationships

$$\frac{d\omega/dt}{\omega} = k(\log \alpha - \log \omega), \qquad (24.7.11)$$

$$\frac{d\omega/dt}{\omega} = k\beta e^{-kt}. \qquad (24.7.12)$$

The latter can also be written as

$$\log\left\{\frac{d\omega/dt}{\omega}\right\} = \log(k\beta) - kt. \qquad (24.7.13)$$

Thus Eq. (24.7.11) implies a linear relationship between the relative growth rate and $\log \omega$; Eq. (24.7.13) implies a linear relationship between the relative growth rate and time. According to Richards (1959), this curve has been used more for population studies and animal growth than for botanical applications; it was employed, for example, by Medawar (1940) in a study of the growth of a chicken's heart. It has also been applied, however, to the growth of *Pelargonium* leaves by Amer and Williams (1957).

As $t \to \infty$, $\omega \to \alpha$, the limiting growth. When $t = 0$, $\omega = \alpha e^{-\beta}$, the initial growth value. If, without loss of generality, we set $\alpha = 1$, we obtain, for $k = 1$ and selected β, the shapes shown in Figure 24.21. For each fixed value of β, variation in k produces sets of curves emanating from the same initial point, similar to the behavior shown in Figures 24.19 and 24.20 for the logistic model.

Von Bertalanffy's Model

This four-parameter model has the form

$$\omega = \{\alpha^{1-m} - \theta e^{-kt}\}^{1/(1-m)}, \qquad (24.7.14)$$

where α, θ, k, and m are parameters to be estimated. In its original derivation by von Bertalanffy (1941, 1957) limits were imposed on m but Richards (1959) has pointed out its value over other ranges of m. Of special interest are these facts:

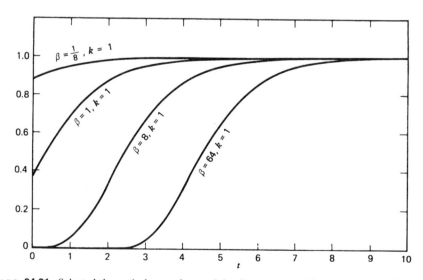

Figure 24.21. Selected theoretical curve forms of the Gompertz model $\omega = \alpha \exp(-\beta e^{-kt})$ for $\alpha = 1$.

1. When $m = 0$, we obtain the monomolecular function with a definition of $\theta = \alpha\beta$.
2. When $m = 2$, we obtain the logistic function with a definition of $\theta = \beta/\alpha$.
3. When $m \to 1$, the curve tends to the Gompertz form as can be shown by examining the limiting behavior of the growth rate, exact substitution leading to a breakdown. Thus estimated values of m close to 1 would indicate the usefulness of the Gompertz curve in a particular situation.
4. When $m > 1$, θ is negative; when $m < 1$, θ is positive.

Problems can arise when fitting this model if the current parameter values cause a negative quantity to be raised to a noninteger power and a constraint may have to be introduced to avoid this possibility in some cases. This can make the von Bertalanffy model somewhat awkward to fit compared with previous models.

24.8. NONLINEAR MODELS: OTHER WORK

Once we have methods of estimating the parameters of a nonlinear model (as earlier in this chapter) we can turn our attention to other problems. We discuss some important topics briefly and give references for further reading.

Design of Experiments in the Nonlinear Case

Reference. Box and Lucas (1959).

If we use the linearization method to estimate the parameters of a nonlinear model, we are led to the iterative formula

$$\mathbf{b}_j = (\mathbf{Z}_j' \mathbf{Z}_j)^{-1} \mathbf{Z}_j (\mathbf{Y} - \mathbf{f}^j)$$

to get $\theta_{j+1} = \theta_j + \mathbf{b}_j$. It can be shown that the approximate confidence region (see p. 516) at this stage for θ has volume proportional to $|(\mathbf{Z}_j' \mathbf{Z}_j)^{-1}|$. Thus, if we regard the best design, that is, the best set of runs to take, as being the design that minimizes the volume of the confidence region, we need to

$$\text{Maximize } |\mathbf{Z}_j' \mathbf{Z}_j|.$$

How do we use this idea in practice?
If no runs have been performed, we choose the set of n runs (assume n is given) that

$$\text{Maximizes } |\mathbf{Z}_0' \mathbf{Z}_0|.$$

(Or, given several designs to choose from, we choose the one with the biggest $|\mathbf{Z}_0' \mathbf{Z}_0|$.)
If n runs have already been performed, and we wish to choose an $(n + 1)$st run, we can write down $|\mathbf{Z}_j' \mathbf{Z}_j|$ as a function of the $(n + 1)$st run and maximize $|\mathbf{Z}_j' \mathbf{Z}_j|$ with respect to the $(n + 1)$st run.
In general, this maximization has to be done numerically on a computer. Analytical solution is possible only in simple cases.

Example. Given $\theta_0' = (\theta_{10}, \theta_{20}) = (0.7, 0.2)$ and the model

$$Y = \frac{\theta_1}{\theta_1 - \theta_2} \{ {}^{-\theta_2\xi} - e^{-\theta_1\xi} \} + \epsilon,$$

select two runs ξ_1 and ξ_2 that maximize $|\mathbf{Z}_0'\mathbf{Z}_0|$. Differentiating $f(\theta_1, \theta_2, \xi)$ with respect to θ_1 and θ_2 and setting $\boldsymbol{\theta} = \boldsymbol{\theta}_0$ we obtain

$$Z_{1u}^0 = (0.8 + 1.4\xi_u)e^{-0.7\xi_u} - 0.8e^{-0.2\xi_u},$$

$$Z_{2u}^0 = -2.8e^{-0.7\xi_u} + (2.8 - 1.4\xi_u)e^{-0.2\xi_u}.$$

Thus

$$\mathbf{Z}_0 = \begin{bmatrix} Z_{11}^0 & Z_{21}^0 \\ Z_{12}^0 & Z_{22}^0 \end{bmatrix}.$$

Now here, $n = p = 2$. So we can write $|\mathbf{Z}_0'\mathbf{Z}_0| = |\mathbf{Z}_0'||\mathbf{Z}_0| = |\mathbf{Z}_0|^2$. So all we have to do in this case is to maximize

$$|\mathbf{Z}_0| = Z_{11}^0 Z_{22}^0 - Z_{12}^0 Z_{21}^0.$$

It can be shown numerically that this is a maximum when $\xi_1 = 1.23$, $\xi_2 = 6.86$, so that is our design. We would now find the Y-values at these two ξ's and go on to estimate θ_1 and θ_2 beginning with the initial estimates given, namely, $(0.7, 0.2)$. For examples and discussion of the $n \rightarrow (n + 1)$ case see Box and Hunter (1965).

At stage n the experimenter would supply the computer with (1) the model, (2) the data, and (3) the current parameter estimates.

The computer would then produce:

1. The new least squares estimates.
2. The best conditions for the next experiment.
3. Information on the stability of the best conditions.
4. Any other items of special interest requested, for example, the covariance structure of the estimates.

A Useful Model-Building Technique

Reference. Box and Hunter (1962).

Suppose we wish to fit the model

$$Y = f(\theta_1, \theta_2, \ldots, \theta_p; W_1, W_2, \ldots, W_l) + \epsilon$$
$$= f(\boldsymbol{\theta}, \mathbf{W}) + \epsilon.$$

Let $X_1, X_2, \ldots, X_k$ be a set of predictor variables *not* in the model above. Write

$$X_{1j}, X_{2j}, \ldots, X_{kj} \text{ for the } j\text{th setting of these, } j = 1, 2, \ldots, n.$$

Suppose that, at each of these n settings, we have several runs with different sets of W's that we can estimate the parameters $\theta_1, \theta_2, \ldots, \theta_p$ at each of the n settings of the X's

This gives a table as follows:

					$\hat{\theta}_i$ column					
X_{11}	X_{21}	X_{31}	$\cdots$	X_{k1}	$\hat{\theta}_{11}$	$\hat{\theta}_{21}$	$\cdots$	$\hat{\theta}_{i1}$	$\cdots$	$\hat{\theta}_{p1}$
X_{12}	X_{22}	X_{32}	$\cdots$	X_{k2}	$\hat{\theta}_{12}$	$\hat{\theta}_{22}$	$\cdots$	$\hat{\theta}_{i2}$	$\cdots$	$\hat{\theta}_{p2}$
$\vdots$	$\vdots$	$\vdots$		$\vdots$	$\vdots$	$\vdots$		$\vdots$		$\vdots$
X_{1n}	X_{2n}	X_{3n}	$\cdots$	X_{kn}	$\hat{\theta}_{1n}$	$\hat{\theta}_{2n}$	$\cdots$	$\hat{\theta}_{in}$	$\cdots$	$\hat{\theta}_{pn}$

We would expect each $\hat{\theta}_{ij}$ *column* to be "stable" if the X's had nothing to do with the response function. It follows that a way of checking whether the response function depends on the X's is to do a regression of the $\hat{\theta}_i$ column on the set of X's and see if any regression coefficients are significant. That is, we fit the model

$$Y = \beta_0 + \beta_1 X_1 + \beta_2 X_2 + \cdots + \beta_k X_k + \epsilon$$

or

$$\mathbf{Y} = \mathbf{X}\boldsymbol{\beta} + \epsilon,$$

where $\mathbf{Y} = \hat{\boldsymbol{\theta}}_i$, and $\mathbf{X}$ = the entire block of X's shown above. If the estimated coefficient of X_q is significantly different from zero, we conclude that the parameter θ_i depends on the variable X_q and so X_q should be in the original model. (So should any other X's whose coefficients are significant.) In other words, the original model $Y = f(\boldsymbol{\theta}, \mathbf{W}) + \epsilon$ is inadequate and needs reconsideration (see Figure 24.22).

For examples of the use of this technique when the X-values form a 2^{k-p} fractional factorial ($p \neq 0$) or a full factorial ($p = 0$) design see Hunter and Mezaki (1964) and Box and Hunter (1962).

Multiple Responses

References. Box and Draper (1965); Erjavec, Box, Hunter and MacGregor (1973); Bates and Watts (1988, Chapter 4).

In some situations, several response variables can be observed simultaneously and the models for these responses contain some or all of the same parameters. A good example of this is an "A goes to B goes to C" reaction with response functions

$$\eta_1 = \exp(-\phi_1 t),$$

$$\eta_2 = \{\exp(-\phi_1 t) - \exp(-\phi_2 t)\}\phi_1/(\phi_2 - \phi_1),$$

$$\eta_3 = 1 + \{-\phi_2 \exp(-\phi_1 t) + \phi_1 \exp(-\phi_2 t)\}/(\phi_2 - \phi_1),$$

all dependent on the predictor variable time (t) and one (η_1) or both of the parameters

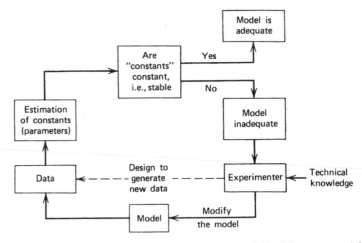

F i g u r e 24.22. A diagrammatic representation of the adaptive model-building process. Adapted, with permission, from "A useful method of model building," by G. E. P. Box and W. G. Hunter, *Technometrics,* **4**, 1962, p. 302.

ϕ_1 and ϕ_2. Note that $\eta_1 + \eta_2 + \eta_3 = 1$ here; however, the corresponding observations (y_{1u}, y_{2u}, y_{3u}), $u = 1, 2, \ldots, n$, for (η_1, η_2, η_3) may be determined individually, in which case, $y_{1u} + y_{2u} + y_{3u}$ is not necessarily 1, due to random error. In such a case, the appropriate way of estimating the parameters is to minimize, with respect to those parameters, the determinant $\|v_{ij}\|$, where

$$v_{ij} = \sum_{u=1}^{n} (y_{iu} - \eta_{iu})(y_{ju} - \eta_{ju}),$$

are the sums of squares and the sums of products of the deviations of the observed y_{ij} from their respective models.

Problems can arise with this criterion when one or more of the responses is determined arithmetically from others. In our example, for instance, this would occur if y_{1u} and y_{2u} were actually measured but y_{3u} were determined as $1 - y_{1u} - y_{2u}$. Ways of detecting this, and of allowing for it, are discussed in the second reference mentioned above.

24.9. REFERENCES

Bates and Watts (1988); Bunke and Bunke (1989); Gallant (1987); Ratkowsky (1983, 1990); Ross (1990); Seber and Wild (1989).

EXERCISES FOR CHAPTER 24

A. Estimate the parameter θ in the nonlinear model

$$Y = e^{-\theta t} + \epsilon$$

from the following observations:

t	Y
1	0.80
4	0.45
16	0.04

Construct an approximate 95% confidence interval for θ.

B. Estimate the parameter θ in the nonlinear model

$$Y = e^{-\theta t} + \epsilon$$

from the following observations:

t	Y
0.5	0.96, 0.91
1	0.86, 0.79
2	0.63, 0.62
4	0.48, 0.42
8	0.17, 0.21
16	0.03, 0.05

Construct an approximate 95% confidence interval for θ.

C. Estimate the parameters α, β in the nonlinear model

$$Y = \alpha + (0.49 - \alpha)e^{-\beta(X-8)} + \epsilon$$

from the following observations:

X	Y
10	0.48
20	0.42
30	0.40
40	0.39

Construct an approximate 95% confidence region for (α, β).

D. The relationship between the yield of a crop, Y, and the amount of fertilizer, X, applied to that crop has been formulated as $Y = \alpha - \beta\rho^X + \epsilon$, where $0 < \rho < 1$. Given:

X	Y
0	44.4
1	54.6
2	63.8
3	65.7
4	68.9

obtain estimates of α, β, and ρ. Also construct an approximate 95% confidence region for (α, β, ρ).

E. The relationship between pressure and temperature in saturated steam can be written as

$$Y = \alpha(10)^{\beta t/(\gamma+t)} + \epsilon,$$

where

Y = pressure,
t = temperature,
α, β, γ = unknown constants.

The following data were collected:

$t(°C)$	Y(Pressure)
0	4.14
10	8.52
20	16.31
30	32.18
40	64.62
50	98.76
60	151.13
70	224.74
80	341.35
85	423.36
90	522.78
95	674.32
100	782.04
105	920.01

Obtain estimates of α, β, and γ. Also construct an approximate 95% confidence region for (α, β, γ).

F. Consider the model

$$Y = \theta + \alpha X_1 X_3 + \beta X_2 X_3 + \alpha\gamma X_1 + \beta\gamma X_2 + \epsilon.$$

Is this model nonlinear? How can estimates of the parameters be obtained by using only linear regression methods, if data $(Y_u, X_{1u}, X_{2u}, X_{3u})$ are available?

G. (*Source: Exploring the Atmosphere's First Mile*, Vol. 1, H. H. Lettau and B. Davidson, eds., Pergamon Press, New York, 1957, pp. 332–336.) Under adiabatic conditions, the wind speed Y is given by the nonlinear model

$$Y = \theta_1 \log(\theta_2 X + \theta_3) + \epsilon,$$

where

$$\theta_1 = \text{friction velocity (cm-sec}^{-1}),$$

$$\theta_2 = 1 + (\text{zero point displacement})/(\text{roughness length}),$$

$$\theta_3 = (\text{roughness length})^{-1} (\text{cm}^{-1}),$$

$$X = \text{nominal height of anemometer (cm)}.$$

Estimate θ_1, θ_2, and θ_3 from the following data. Also construct an approximate 95% confidence region for $(\theta_1, \theta_2, \theta_3)$.

X	Y
40	490.2
80	585.3
160	673.7
320	759.2
640	837.5

H. (*Source:* "Planning experiments for fundamental process characterization," by W. G. Hunter and A. C. Atkinson, Technical Report No. 59, Statistics Dept., University of Wisconsin, Madison, Wisconsin, December 1965. For a shortened version of this paper see "Statistical designs for pilot-plant and laboratory experiments—Part II," *Chemical Engineering*, June 6, 1966, 159–164. The data originally appeared in "Kinetics of the thermal isomerization of bicyclo [2.1.1] hexane," by R. Srinivasan and A. A. Levi, *Journal of the American Chemical Society*, **85**, November 5, 1963, 3363–3365. Data on p. 3364 with "Reactant pressure, mm" values of less than unity have been omitted for the exercise.)

A certain chemical reaction can be described by the nonlinear model

$$Y = \exp\left\{-\theta_1 X_1 \exp\left[-\theta_2\left(\frac{1}{X_2} - \frac{1}{620}\right)\right]\right\} + \epsilon,$$

where θ_1 and θ_2 are parameters to be estimated, Y is the fraction of original material remaining, X_1 is the reaction time in minutes, and X_2 is the temperature in degrees Kelvin. Using the data below, which arose from an unplanned experiment, estimate θ_1 and θ_2 and construct a 95% confidence region for the point (θ_1, θ_2). You may use the preliminary estimates $\boldsymbol{\theta}_0' = (\theta_{10}, \theta_{20}) = (0.01155, 5000)$ if you wish.

Run	X_1	X_2	Y
1	120.0	600	0.900
2	60.0	600	0.949
3	60.0	612	0.886
4	120.0	612	0.785
5	120.0	612	0.791
6	60.0	612	0.890
7	60.0	620	0.787

Run	X_1	X_2	Y
8	30.0	620	0.877
9	15.0	620	0.938
10	60.0	620	0.782
11	45.1	620	0.827
12	90.0	620	0.696
13	150.0	620	0.582
14	60.0	620	0.795
15	60.0	620	0.800
16	60.0	620	0.790
17	30.0	620	0.883
18	90.0	620	0.712
19	150.0	620	0.576
20	60.0	620	0.802
21	60.0	620	0.802
22	60.0	620	0.804
23	60.0	620	0.794
24	60.0	620	0.804
25	60.0	620	0.799
26	30.0	631	0.764
27	45.1	631	0.688
28	40.0	631	0.717
29	30.0	631	0.802
30	45.0	631	0.695
31	15.0	639	0.808
32	30.0	639	0.655
33	90.0	639	0.309
34	25.0	639	0.689
35	60.1	639	0.437
36	60.0	639	0.425
37	30.0	639	0.638
38	30.0	639	0.659

I. (*Source*: See the Hunter and Atkinson reference in Exercise H.) Using the data below, which arose from a simulated planned experiment, estimate θ_1 and θ_2 in the model given in Exercise H and construct a 95% confidence region for (θ_1, θ_2). [It is interesting to note that the eight runs below, which were sequentially planned, produce a slightly smaller confidence region for (θ_1, θ_2) than the 38 unplanned experiments in Exercise H. This is a striking demonstration of the advantage of conducting planned experimentation whenever possible. Note particularly that the range of X_1 is now much greater than before. This, by itself, would lead to an improved confidence region if 38 unplanned runs were used. The planning of the experiment in addition makes it possible to use even fewer runs while maintaining about the same precision.]

Run	X_1	X_2	Y
1	109	600	0.912
2	65	640	0.382
3	1180	600	0.397
4	66	640	0.376
5	1270	600	0.342
6	69	640	0.358
7	1230	600	0.348
8	68	640	0.376

J. (*Source*: "Sequential design of experiments for nonlinear models," by G. E. P. Box and W. G. Hunter, *Proceedings of the IBM Scientific Computing Symposium on Statistics, October 21–23, 1963*, published in 1965, pp. 113–137). A certain chemical reaction can be described by the nonlinear model

$$Y = \theta_1\theta_3 X_1/(1 + \theta_1 X_1 + \theta_2 X_2) + \epsilon,$$

where Y is the reaction rate, X_1 and X_2 are partial pressures of reactant and product, respectively, θ_1 and θ_2 are adsorption equilibrium constants for reactant and product, respectively, and θ_3 is the effective reaction rate constant. Use the data below to estimate θ_1, θ_2, and θ_3 and construct a 95% confidence region for $(\theta_1, \theta_2, \theta_3)$. You may use the preliminary estimates $\boldsymbol{\theta}'_0 = (\theta_{10}, \theta_{20}, \theta_{30}) = (2.9, 12.2, 0.69)$ if you wish. These data are reprinted by permission of International Business Machines Corporation.

Run	X_1	X_2	Y
1	1.0	1.0	0.126
2	2.0	1.0	0.219
3	1.0	2.0	0.076
4	2.0	2.0	0.126
5	0.1	0.0	0.186
6	3.0	0.0	0.606
7	0.2	0.0	0.268
8	3.0	0.0	0.614
9	0.3	0.0	0.318
10	3.0	0.8	0.298
11	3.0	0.0	0.509
12	0.2	0.0	0.247
13	3.0	0.8	0.319

K. (*Source*: *Statistical Theory with Engineering Applications*, by A. Hald, Wiley, New York, 1960 p. 564.) Fit, to the constructed representative data below, the nonlinear model

$$Y = \theta_1 X^{\theta_2} + \epsilon$$

and provide a 95% confidence region for the parameters (θ_1, θ_2). For a larger body of data, see the reference above.

Speed of Automobile X	Stopping Distance Y
4	5
10	20
17	45
22	66
25	85

L. (*Source*: "Biochemical oxygen demand data interpretation using the sum of squares surface," by Donald Marske, M.S. thesis in Civil Engineering, University of Wisconsin, Madison, Wisconsin, 1967.) Fit, to each set of data below, the nonlinear model

$$Y = \theta_1(1 - e^{-\theta_2 t}) + \epsilon$$

and provide an approximate 95% confidence region for (θ_1, θ_2).

Set 1:

t	Y
1	82
2	112
3	153
4	163
5	176
6	192
7	200

Set 2:

t	Y
1	0.47
2	0.74
3	1.17
4	1.42
5	1.60
7	1.84
9	2.19
11	2.17

Set 3:

t	Y
1	168
2	336
3	468
5	660
6	708
7	696

Set 4:

t	Y
1	9
2	9
3	16
4	20
5	21
7	22

Set 5:

t	Y
1	4.3
2	8.2
3	9.5
4	10.4
5	12.1
7	13.1

Set 6:

t	Y
1	6.8
2	12.7
3	14.8
4	15.4
5	17.0
7	19.9

Set 7:

t	Y
1	109
2	149
3	149
5	191
7	213
10	224

Set 8:

t	Y
1	8.3
2	10.3
3	19
4	16
5	15.6
7	19.8

Set 9:

t	Y
1	4710
2	7080
3	8460
4	9580

M. To the ice crystal data of Exercise 13E fit the nonlinear model $M = \alpha T^\beta + \epsilon$. Examine the residuals and state your conclusions.

N. The data below arose from five orange trees grown at Riverside, California, during the period 1969–1973. The response w in the body of the table is the trunk circumference in millimeters, and the predictor variable t is the time in days, with an arbitrary origin taken on December 31, 1968. Fit the models given in Eqs. (24.7.2), (24.7.5), (24.7.8), and (24.7.10) to these data. Based on a visual inspection of the fitted models, which seems to be most useful?

	Response w for Tree No.				
t	1	2	3	4	5
118	30	33	30	32	30
484	58	69	51	62	49
664	87	111	75	112	81
1004	115	156	108	167	125
1231	120	172	115	179	142
1372	142	203	139	209	174
1582	145	203	140	214	177

O. Estimate the parameters α, β in the nonlinear model $Y = \alpha + X^\beta + \epsilon$, using the data $(X, Y) = (0, -1.1)$, $(1, 0)$, $(2, 2.9)$, $(3, 8.1)$. [This can be done either by (1) nonlinear least squares or (2) fixing β, setting $\hat{\alpha}(\beta) = \{$average of the $(Y_u - X_u^\beta)\}$, the appropriate linear least squares estimate of α for specified β, plotting $S(\alpha, \beta) = \Sigma_{u=1}^n \{Y_u - \hat{\alpha}(\beta) - X_u^\beta\}^2$ versus β and estimating $\hat{\beta}$ as the value of β that minimizes $S(\alpha, \beta)$; the appropriate $\hat{\alpha}$ is then $\hat{\alpha}(\hat{\beta})$ the value of $\hat{\alpha}$ that corresponds to $\hat{\beta}$.]

P. Three tensile properties of ductile cast iron are

$$x = \text{percentage elongation,}$$

$$y = \text{tensile strength, kg/in.}^2$$

and

$$z = \text{yield strength, kg/in.}^2$$

It has been suggested[1] that models of the form

$$y = \alpha + \beta/x^\gamma + \epsilon,$$

$$z = \delta + \theta/x^\phi + \epsilon$$

hold for these properties and that the model parameters might be suitable measures of the quality of the iron.

The figures in the table below represent minimum values of the qualities indicated, as given in a particular industry specification ASTM A53 6-70 to which ductile iron is manufactured. Fit the models above to these data and so obtain estimates of the parameters.

[*Hint:* The two models are both linear in the transformed predictor variables $x^{-\gamma}$ and $x^{-\phi}$. For the first model, assume for the moment that γ is fixed. Then we can solve the least squares normal equations for $\hat{\alpha}$ and $\hat{\beta}$, both of which depend on γ and so write an expression, depending only on the unknown γ, for the sum of squares function

$$S(\gamma) = \sum_{u=1}^n (y_u - \hat{\alpha} - \hat{\beta}/x_u^\gamma)^2.$$

By letting γ vary and plotting, or printing out a table of $S(\gamma)$, we find the $\hat{\gamma}$ that minimizes $S(\gamma)$. This $\hat{\gamma}$ and the corresponding $\hat{\alpha}$ and $\hat{\beta}$ are then the appropriate least squares estimators to use. The same idea is employed to get $\hat{\delta}$, $\hat{\theta}$, and $\hat{\phi}$.]

Percentage Elongation	Tensile Strength, kg/in.^{2}	Yield Strength, kg/in.^{2}
x	y	z
2	120	90
3	100	70
6	80	55
12	65	45
18	60	40

Q. Fit the model $Y = \alpha X_1^\beta X_2^\gamma + \epsilon$ to the data of Exercise 13F.

R. (*Source:* "The utilization of dietary energy by steers during periods of restricted food intake and subsequent realimentation. Part 1," by H. P. Ledger and A. R. Sayers, *Journal of Agricultural Science, Cambridge,* **88**, 1977, 11–26. "Part 2," by H. P. Ledger is on pp. 27–33 of the same journal issue. Adapted with the permission of Cambridge University Press.)

---

[1] By C. R. Loper, Jr., and R. M. Kotschi of the University of Wisconsin to whom we are grateful for this exercise.

T A B L E R. Group Mean Daily Dry Matter Intake as Percentage of Live Weight

Reference Designation of Data Set		3	6	9	12	15	18	21	24
B 185 kg	%	1.835	1.255	1.037	1.045	0.900	0.901	0.836	0.896
	sd	0.412	0.315	0.692	0.712	0.149	0.197	0.156	0.234
B 275 kg	%	1.702	1.313	1.037	0.952	0.863	0.879	0.897	0.800
	sd	0.292	0.262	0.245	0.187	0.161	0.145	0.160	0.199
B × H 275 kg	%	1.545	1.134	0.941	0.840	0.791	0.822	0.855	0.773
	sd	0.206	0.205	0.215	0.211	0.293	0.095	0.072	0.124
$\frac{3}{4}$ B 450 kg	%	0.999	0.803	0.790	0.797	0.734	0.687	0.687	0.716
	sd	0.287	0.191	0.168	0.093	0.127	0.145	0.162	0.224
$\frac{3}{4}$ H 450 kg	%	0.818	0.753	0.737	0.713	0.664	0.660	0.706	0.670
	sd	0.249	0.185	0.133	0.140	0.196	0.122	0.122	0.114

The table shows values of the response Y = group mean daily dry matter intake as percentage of live weight of steers, measured at eight equally spaced values of X = weeks on maintenance. Also shown are the standard deviations (sd) of the Y's. There are five groups of such data, as designated in the left column. (Thus, for example, in the fourth group, the sixth response observation is $Y = 0.687$, with sd $= 0.145$, and $X = 18$.) To each group of data fit, by weighted least squares, the model

$$Y = \beta_0 + \beta_1 \theta^X + \epsilon,$$

and provide the usual analyses.

[*Hints*: If a weighted nonlinear least squares program is not available to you, the following approaches are possible:

1. Use weighted linear least squares over a selected range of values of θ (such as $0 \le \theta \le 1$) and estimate θ as the value that gives the smallest residual sum of squares. For that value of θ, estimate β_0 and β_1.

2. Write the model as

$$Y/sd = (1/sd)\beta_0 + (\theta^X/sd)\beta_1 + \text{error}$$

and use ordinary (unweighted) nonlinear least squares to estimate β_0, β_1, and θ.

3. Write the model as in (2) and use ordinary (unweighted) linear least squares over a selected range of values of θ (such as $0 \le \theta \le 1$) and estimate θ as the value that gives the smallest residual sum of squares. For that estimate of θ, estimate β_0 and β_1,

Note that $V(\mathbf{Y})$ is a diagonal matrix in this problem.]

S. (*Source*: Carol C. House, U.S. Department of Agriculture, Washington, DC, 20250.) The data in the table consist of eight samples taken in a corn growth study, in which were measured:

$$Y = \text{mean dry kernel weight of four plants,}$$

$$t = \text{mean time since silking of four plants.}$$

1. Fit, to each individual sample, the nonlinear model

$$\ln Y = \delta - \ln(1 + \beta e^{-kt}) + \epsilon,$$

where δ, β, and k are parameters to be estimated from the data.

2. Plot each data sample and show the corresponding fitted curve on the same diagram. Comment on what you see, both in the diagrams and in the computer output.

3. Plot a dot diagram of all eight $\hat{\delta}$'s; do the same for the $\hat{\beta}$'s; and the $\hat{k}$'s. Comment on what you see in these plots. What would you *expect* to see, if there were no differences from sample to sample?

4. Fit the model in (1) to *all* the data at once and add the resulting parameter estimates to your dot diagrams in (3). What do you conclude?

5. The data values in the table are all means of four readings. If the individual readings that provided these means were available, would you use them in the model-fitting procedure instead of the means? Why or why not? Give the advantages and disadvantages.

Sample	Y	t	Sample	Y	t
8	11.44	13.625	24	37.84	18.625
	29.51	19.750		67.34	29.125
	69.05	28.625		157.10	49.250
	98.79	41.750	32	7.84	13.375
	138.01	49.625		15.12	16.875
	162.82	69.625		73.97	28.250
14	35.88	19.750		110.58	43.125
	106.89	30.250		114.36	47.500
	168.58	35.625		185.87	58.000
	136.84	43.375		115.18	60.625
	164.50	49.500	52	10.60	8.750
16	3.52	6.500		15.13	14.875
	10.56	14.875		38.74	22.500
	41.55	24.625		120.19	37.625
	94.55	35.875		126.32	41.750
	122.52	49.875		171.75	51.625
	130.19	56.875		156.67	63.875
22	14.26	17.125	54	11.92	13.625
	50.51	25.625		66.82	24.750
	60.83	29.625		28.29	24.875
	104.78	39.625		106.92	39.000
	96.46	46.375		129.83	53.875
	97.02	54.250		143.26	60.875
	172.41	62.125			

T. (*Source*: "Relative curvature measures of nonlinearity," by D. M. Bates and D. G. Watts, *Journal of the Royal Statistical Society*, **B-42**, 1980, 1–16, discussion 16–25.) Fit the nonlinear model

$$Y = \theta_1 X / (\theta_2 + X) + \epsilon$$

to the (simulated) data in the table by least squares, and provide a follow-up analysis of your results, including a plot of the data and fitted equation. Suppose you were asked to plan one more run. At what X-value would you take an observation Y? If you could plan *two* more runs instead, what would the best X-values be?

X	Y	X	Y
2.000	0.0615	0.286	0.0129
2.000	0.0527	0.286	0.0183
0.667	0.0334	0.222	0.0083
0.667	0.0258	0.222	0.0169
0.400	0.0138	0.200	0.0129
0.400	0.0258	0.200	0.0087

U. Refer to the model in Exercise T. A more general model of that type is

$$Y = (\theta_4 + \theta_1 X^{\theta_3}) / (\theta_2 + X^{\theta_3}) + \epsilon.$$

Note that, when $\theta_4 = 0$, this model function passes through the origin; in such form it is called the Hill model. When $\theta_4 = 0$ and $\theta_3 = 1$, the so-called Michaelis–Menten form (shown in Exercise T) is reached. (See the paper "General model for nutritional responses of higher organisms," by P. H. Morgan, L. P. Mercer, and N. W. Flodin, *Proceedings of the National Academy of Sciences, USA,* **72**, 1975, November, 4327–4331. For additional examples of the use of this model, see "New methods for comparing the biological efficiency of alternate nutrient sources," by L. P. Mercer, N. W. Flodin, and P. H. Morgan, *Journal of Nutrition,* **108**, 1978, August, 1244–1249.)

First fit the more general model to the data in Exercise T and then find "extra sum of squares" contributions for (1) θ_4 given θ_1, θ_2, θ_3; (2) θ_3 given θ_1, θ_2, assuming $\theta_4 = 0$; (3) θ_3, θ_4 given θ_1, θ_2. (Get these by difference of the two $S(\hat{\theta})$ values for two models appropriately chosen, for each "extra SS.") Compare these "extra SS" values with the pure error mean square value s_e^2. What do you conclude? What model would you use to represent the data and why?

V. Fit the model (24.1.3) to each of the two sets of data below, and perform the standard analyses. What is the correlation between $\hat{\theta}_1$ and $\hat{\theta}_2$ in each case? How would you characterize these two data sets?

Data Set 1		Data Set 2	
t	Y	t	Y
0.2	0.142	1	0.445
0.4	0.240	2	0.585
0.6	0.329	3	0.601
0.8	0.381	4	0.532
1.0	0.455	5	0.470

W. You wish to fit the nonlinear model $Y = \theta_1\{1 - \exp(-\theta_2 t)\} + \epsilon$ to these data:

t	1	3	4	9
Y	0.47	1.17	1.42	2.19

The experimenter says: "I want you to use $\theta_{20} = 0.25$ as an initial estimate for θ_2. Please get me an initial estimate of θ_1?" What value θ_{10}, to one decimal place, would you suggest?

X. (*Source:* "Plant density and crop yield," by R. Mead, *Applied Statistics,* **19**, 1970, 64–81.) Fit the nonlinear model

$$\ln W_i = -\theta_1^{-1}\ln(\theta_2 + \theta_3 X_i^{\theta_4}) + \epsilon_i$$

to each set of data in the table. The variables are

$$W = \text{mean yield per plant in grams,}$$

$$X = \text{planting density of an onion variety in plants per square foot.}$$

Provide all the usual analyses and check the residuals. Can you suggest a simpler model that might fit pretty well?

Variety No. 1		Variety No. 2		Variety No. 3	
W	X	W	X	W	X
105.6	3.07	131.6	2.14	116.8	2.48
89.4	3.31	109.1	2.65	91.6	3.53
71.0	5.97	93.7	3.80	72.7	4.45
60.3	6.99	72.2	5.24	52.8	6.23
47.6	8.67	53.1	7.83	48.8	8.23
37.7	13.39	49.7	8.72	39.1	9.59
30.3	17.86	37.8	10.11	30.3	16.87
24.2	21.57	33.3	16.08	24.2	18.69
20.8	28.77	24.5	21.22	20.0	25.74
18.5	31.08	18.3	25.71	16.3	30.33

Y. (*Source*: "ROSFIT: An enzyme kinetics nonlinear regression curve fitting package for a microcomputer," by W. R. Greco, R. L. Priore, M. Sharma, and W. Korytnyk, *Computers and Biomedical Research*, **15**, 1982, 39–45. Copyright © 1982 Academic Press, Inc. Thanks go to William R. Greco, who supplied the data.) The data in the table come from an enzyme kinetics study described in the source article. Fit the "competitive inhibition" nonlinear model

$$Y = \theta_1 X_1 / \{\theta_2(1 + X_2/\theta_3) + X_1\} + \epsilon,$$

and provide the usual analyses. In particular, use the pure error to make an approximate test for lack of fit. (*Note*: Alternative models that were assessed as being poorer fits to the data are described in the source article.)

X_1	X_2	n_i	$Y_u \times 1000$
0.25	0	4	303, 310, 323, 310
0.5	0	4	451, 465, 479, 454
1.25	0	4	752, 694, 756, 723
2.5	0	3	950, 929, 964
3.75	0	4	1020, 1013, 1054, 1040
5.	0	3	1072, 1059, 1094
0.25	0.2	2	48, 48
0.5	0.2	2	88, 89
1.25	0.2	2	188, 177
2.5	0.2	2	318, 318
3.75	0.2	2	447, 447
5.	0.2	2	553, 545
0.25	0.5	2	21, 22
0.5	0.5	2	38, 36
1.25	0.5	2	81, 85
2.5	0.5	2	162, 150
3.75	0.5	2	225, 210
5.	0.5	2	294, 269
		$n = 46$	

Z. (*Source*: "On the use of mist nets for population studies of birds," by R. H. MacArthur and A. T. MacArthur, *Proceedings of the National Academy of Sciences, Washington, DC*, **71**, 1974, 3230–3233.) Mist nets are very fine thread nets used in bird-trapping studies. Birds so trapped are tagged and released after data are recorded on them. A study usually continues for a week or two at a chosen site. Suppose that:

N = number of birds in the local population,
p = the probability of a bird being trapped during the study,
t_u = time in days at which nets are checked, $u = 1, 2, \ldots, n$,
Y_u = number of birds caught for the first time at time t_u,
Q = number of drifting birds in the locality; these do not belong to the local population but may be trapped also.

In one such study, the following data were obtained:

t_u	0.5	1.5	2.5	3.5	4.5	5.5	6.5	7.5	8.5	9.5
Y_u	13	5	9	8	2	5	1	0	1	1

Fit by least squares to these data the models

(1) $Y_u = Q + pN \exp(-pt_u) + \epsilon_u,$
(2) As (1) but with $Q = 0,$

and thus obtain estimates of the parameters Q, p, and N. The usual nonlinear estimation analyses should also be provided. Comment on what you find. (For more on this topic, see also "The analysis of trapping records for birds trapped in mist nets," by B. F. J. Manly, *Biometrics*, **33**, 1977, 404–410; and "On a nonlinear regression problem in ornithology," by A. P. Gore and K. S. Madhava Rao, Technical Report No. 38, Department of Statistics, University of Poona, Pune—411 007, India.)

Table for Exercise 15Q, page 368

X_1	X_2	X_3	X_4	X_5	X_6	X_7	Y
1	0	0.33	7	2	1	1	36
1	0	0.75	2.3	1	2	1	33
1	0	0.45	1.7	1	3	1	33
1	1	0.52	1.7	1	3	1	33
1	1	0.52	3.6	1	2	1	35
1	1	1.25	1.7	1	2	1	36
2	0	2.67	7	1	1	2	40
2	0	0.83	2.3	1	1	2	38
2	0	2.00	7	1	1	1	43
2	0	2.25	1.7	1	1	1	36
2	0	1.50	1.7	2	1	1	37
2	1	0.38	3.6	1	1	1	41
2	1	0.25	1.7	1	1	2	40
1	0	0.47	1.7	1	4	1	36
1	0	0.88	7	1	1	1	39
2	0	0.40	1.7	1	4	1	38
2	0	1.73	7	1	4	1	43
2	0	0.58	1.7	1	1	2	38
2	0	0.75	3.6	1	1	2	38

Answer to 15Q.

Q. A model containing X_1 and X_4 explains $R^2 = 0.778$ of the variation, and the addition of other variables increases this but not by much. A two-way table of the observations arranged with respect to these two variables makes it clear that the trip home takes longer on all days and that Fridays should be avoided if possible.

CHAPTER 25

Robust Regression

Why Use Robust Regression?

When we perform least squares regression using n observations, for a p-parameter model $\mathbf{Y} = \mathbf{X}\boldsymbol{\beta} + \boldsymbol{\epsilon}$, we make certain idealized assumptions about the vector of errors $\boldsymbol{\epsilon}$, namely, that it is distributed $N(\mathbf{0}, \mathbf{I}\sigma^2)$. In practice, departures from these assumptions occur. If the departures are serious, we hope to spot them in the behavior of the residuals and so be led to make suitable adjustments to the model and/or the variables' metrics. For example, we may transform the response variable or one or more of the predictors, or adjust the model by adding higher-order terms. For many sets of data, the departures, if they exist at all, are not serious enough for corrective actions, and we proceed with the analysis in the usual way.

If our analysis seems to point to the errors having a non-normal distribution, we might consider a robust regression method, particularly in cases where the error distribution is heavier-tailed then the normal, that is, has more probability in the tails than the normal. Such heavier-tailed distributions are likely to generate more "large" errors than the normal. A least squares analysis weights each observation equally in getting parameter estimates. Robust methods enable the observations to be weighted unequally. Essentially, observations that produce large residuals are down-weighted by a robust estimation method. A number of methods are available. In general, robust regression methods require much more computing than least squares, and also require some assumptions to be made about the down-weighting procedure to be employed.

25.1. LEAST ABSOLUTE DEVIATIONS REGRESSION (L_1 REGRESSION)

When we can make a specific (non-normal) assumption about the errors, we can apply maximum likelihood methods directly. As already discussed in Section 5.1, the assumption of a double exponential error distribution

$$(2\sigma)^{-1} \exp\{-|\epsilon|/\sigma\} \qquad (-\infty \le \epsilon \le \infty)$$

leads to minimizing

$$\sum_{i=1}^{n} |\epsilon_i|,$$

the sum of absolute errors. This is also called L_1 regression (or L_1-norm regression), the subscript 1 referring to the power used in the function to be minimized. This method gives less weight to larger errors than the least squares method, where the error distribution is

$$(2\pi\sigma^2)^{-1/2} \exp\{-\epsilon^2/(2\sigma^2)\}$$

and the function minimized is the sum of squares of errors, namely,

$$\sum_{i=1}^{n} \epsilon_i^2.$$

Thus L_1 regression is more robust against large errors than least squares; least squares is sometimes called L_2 regression (or L_2-norm regression). There exist also L_p regression methods that minimize

$$\sum_{i=1}^{n} |\epsilon_i|^p.$$

It has been suggested (by Forsythe, 1972, on the basis of a 400-replicates Monte Carlo study) that a value of $p = 1.5$ could be a good general choice. It led to estimates that were no worse than 95% as efficient as least squares when the errors were actually normal. Efficiency was defined as the ratio (mean square error for least squares)/ (mean square error for power p), and Forsythe's Monte Carlo experiments used $p = 1.25, 1.50$, and 1.75 on sets of straight line regression data with errors generated from a normal distribution contaminated with another off-center normal to produce a skewed distribution of errors. See Forsythe (1972, p. 160) for details.

25.2. *M*-ESTIMATORS

M-Estimators are "maximum likelihood type" estimators. Suppose the errors are independently distributed and all follow the same distribution, $f(\epsilon)$. Then the maximum likelihood estimator (MLE) of $\boldsymbol{\beta}$ is given by $\hat{\boldsymbol{\beta}}$, which maximizes the quantity

$$\prod_{i=1}^{n} f(Y_i - \mathbf{x}_i'\boldsymbol{\beta}), \tag{25.2.1}$$

where $\mathbf{x}_i'$ is the ith row of $\mathbf{X}$, $i = 1, 2, \ldots, n$, in the model $\mathbf{Y} = \mathbf{X}\boldsymbol{\beta} + \boldsymbol{\epsilon}$. Equivalently, the MLE of $\boldsymbol{\beta}$ maximizes

$$\sum_{i=1}^{n} \ln f(Y_i - \mathbf{x}_i'\boldsymbol{\beta}). \tag{25.2.2}$$

As we have seen, this leads to minimizing the sum of squares function

$$\sum_{i=1}^{n} (Y_i - \mathbf{x}_i'\boldsymbol{\beta})^2 \tag{25.2.3}$$

in the normal case. In the double exponential case we minimize

$$\sum_{i=1}^{n} |Y_i - \mathbf{x}_i'\boldsymbol{\beta}|. \tag{25.2.4}$$

We can extend this idea as follows. Suppose $\rho(u)$ is a defined function of u (examples soon) and suppose s is an estimate of scale (not necessarily the usual least squares estimate). We define a robust estimator as one that minimizes

$$\sum_{i=1}^{n} \rho\left(\frac{e_i}{s}\right) = \sum_{i=1}^{n} \rho\left\{\frac{Y_i - \mathbf{x}_i'\boldsymbol{\beta}}{s}\right\}. \tag{25.2.5}$$

We see that if $\rho(u) = u^2$, the criterion minimized is the same as that of (25.2.3), while if $\rho(u) = |u|$ we get (25.2.4). So in these specific cases, the form of ρ and the underlying distribution are specifically related. In fact, the *knowledge* of an appropriate distribution for the errors could tell us what $\rho(u)$ to use.

Here, we encounter the first practical difficulty in using robust regression. In general, if we do not know what distribution to assume for the errors, we are not led naturally to a $\rho(u)$. What we have essentially in the literature are a number of suggestions for $\rho(u)$ that have worked out well in specific studies. It is not clear which should be used for a given set of data. All we can do is look at the weighting characteristics produced by $\rho(u)$ and choose accordingly, all else being equal.

Table 25.1 (a)–(f) shows some of the suggestions given in the literature for $\rho(u)$. Choices of specific values for the constants a, b, and c need to be made; values shown are suggested in the literature. [Row (g), attached here for convenience, shows an appropriate weight function $w(u)$, for the situation where it is assumed that the underlying error distribution is a t-distribution with f degrees of freedom; see Box and Draper, 1987, pp. 79–90.]

In Eq. (25.2.5), the role of u is played by the scaled residual $(Y_i - \mathbf{x}_i'\boldsymbol{\beta})/s$. We shall soon see, from Figure 25.1, that the least squares method gives the highest weight of 1 in (25.2.5) to larger scaled residuals, and the purpose of the various $\rho(u)$ functions is to (comparatively) down-weight larger scaled residuals in various ways, compared with least squares. We discuss this more in a moment. First we examine how the estimation procedure is carried out.

The M-Estimation Procedure

We need to minimize Eq. (25.2.5) with respect to the parameters $\beta_j, j = 0, 1, 2, \ldots,$ k, say. Differentiation of Eq. (25.2.5) with respect to each parameter leads to $p = k + 1$ equations of form

$$\sum_{i=1}^{n} x_{ij} \psi\left\{\frac{Y_i - \mathbf{x}_i'\boldsymbol{\beta}}{s}\right\} = 0, \qquad j = 0, 1, 2, \ldots, k, \tag{25.2.6}$$

where $\psi(u)$ is the partial derivative $\partial\rho/\partial u$ and x_{ij} is the jth entry of $\mathbf{x}_i' = (1, x_{i1}, x_{i2}, \ldots, x_{ik})$. These equations do not have an explicit solution in general, and iterative numerical solution is necessary. We can follow Beaton and Tukey (1974, p. 151) and define weights

$$w_{i\beta} = \frac{\psi\{(Y_i - \mathbf{x}_i'\boldsymbol{\beta})/s\}}{(Y_i - \mathbf{x}_i'\boldsymbol{\beta})/s}, \qquad i = 1, 2, \ldots, n, \tag{25.2.7}$$

defining them to be 1 if it happens that $Y_i - \mathbf{x}_i'\boldsymbol{\beta} = 0$ exactly. Then (25.2.6) becomes

$$\sum_{i=1}^{n} x_{ij} w_{i\beta}(Y_i - \mathbf{x}_i'\boldsymbol{\beta}) = 0, \qquad j = 0, 1, 2, \ldots, k, \tag{25.2.8}$$

T A B L E 25.1. Some Functions that Have Been Suggested for M-Estimation

Criterion	$\rho(u)$	Corresponding Ranges of u	$\psi(u) = \dfrac{\partial \rho(u)}{\partial u}$	$w(u) = \psi(u)/u$	Suggested Parameter or Breakpoint Values (Tuning Constants) and Reference
(a) Least squares (normal errors)	$\frac{1}{2}u^2$	$-\infty \le u \le \infty$	u	1	Not applicable
(b) Huber's, with breakpoint $a > 0$	$\begin{cases} \frac{1}{2}u^2 \\ a\lvert u\rvert - \frac{1}{2}a^2 \end{cases}$	$-a \le u \le a$ $u \le -a$ and $u \ge a$	u $a\,\text{sign}(u)$	1 $a/\lvert u\rvert$	$a = 2$ (Huber, 1964)
(c) Ramsay's, with parameter $a > 0$	$\{1 - e^{-a\lvert u\rvert}(1 + a\lvert u\rvert)\}/a^2$	$-\infty \le u \le \infty$	$u e^{-a\lvert u\rvert}$ (maximum at a^{-1})	$e^{-a\lvert u\rvert}$	$a = 0.3$ (Ramsay, 1977)
(d) Andrew's wave, with breakpoint $a\pi > 0$	$\begin{cases} a\{1 - \cos(u/a)\} \\ 2a \end{cases}$	$-a\pi \le u \le a\pi$ $u \le -a\pi$ and $u \ge a\pi$	$\sin(u/a)$ 0	$\{\sin(u/a)\}/(u/a)$ 0	$a = 1.339$ (Andrews et al., 1972)
(e) Tukey's biweight, with breakpoint $a > 0$	$\begin{cases} \frac{1}{2}u^2 - \dfrac{u^4}{4a^2} \\[4pt] \frac{1}{4}a^2 \end{cases}$	$-a \le u \le a$ $u \le -a$ and $u \ge a$	$u(1 - u^2/a^2)$ 0	$1 - u^2/a^2$ 0	$5 \le a \le 6$ (Beaton and Tukey, 1974)
(f) Hampel's, with breakpoints a, b, $c > 0$	$\begin{cases} \frac{1}{2}u^2 \\ a\lvert u\rvert - \frac{1}{2}a^2 \\ a\dfrac{(c\lvert u\rvert - \frac{1}{2}u^2)}{c - b} - \dfrac{7a^2}{6} \\ a(b + c - a) \end{cases}$	$-a \le u \le a$ $-b \le u \le -a$ and $a \le u \le b$ $-c \le u \le -b$ and $b \le u \le c$ $u \le -c$ and $u \ge c$	u $a\,\text{sign}(u)$ $\{c\,\text{sign}(u) - u\}a/(c - b)$ 0	1 $a/\lvert u\rvert$ $\{c/\lvert u\rvert - 1\}a/(c - b)$ 0	$a = 1.7$, $b = 3.4$, $c = 8.5$ (Andrews et al., 1972, p. 14)
(g) Assume errors follow t_f distribution and apply maximum likelihood		$-\infty \le u \le \infty$		$\dfrac{f+1}{f+u^2}$	Not applicable. As $f \to \infty$, the t-distribution $\to$ normal distribution

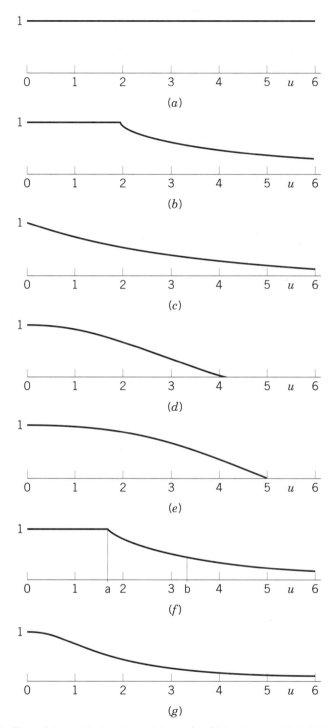

Figure 25.1. Plots of the weight functions $w(u) = u^{-1} \, \partial\rho(u)/\partial u$ for the $\rho(u)$ in Table 25.1. The plots arise from these choices of "tuning constants" from Table 25.1: (b) $a = 2$; (c) $a = 0.3$; (d) $a = 1.339$; (e) $a = 5$; (f) $a = 1.7, b = 3.4, c = 8.5$; (g) $f = 2$. Other choices provide different weight plots of essentially the same nature.

or

$$\sum_{i=1}^{n} x_{ij} w_{i\beta} \mathbf{x}_i' \boldsymbol{\beta} = \sum_{i=1}^{n} x_{ij} w_{i\beta} Y_i, \qquad j = 0, 1, 2, \ldots, k, \qquad (25.2.9)$$

which we can write in matrix form as

$$\mathbf{X}'\mathbf{W}_{\beta}\mathbf{X}\boldsymbol{\beta} = \mathbf{X}'\mathbf{W}_{\beta}\mathbf{Y}, \qquad (25.2.10)$$

where $\mathbf{W}_{\beta}$ = diagonal $(w_{1\beta} \ w_{2\beta}, \ldots, w_{n\beta})$. These equations are obviously of the form of generalized least squares normal equations. (More accurately, they are *weighted* least squares type equations because $\mathbf{W}_{\beta}$ is a diagonal matrix of weights.) The difficulty in solving them is that $\mathbf{W}_{\beta}$ depends on β. This is circumvented in a traditional way. We somehow get an initial estimate $\hat{\boldsymbol{\beta}}_0$ of $\boldsymbol{\beta}$ (perhaps by using least squares), calculate the value $\mathbf{W}_0$, say, of $\mathbf{W}_{\beta}$ for that $\hat{\boldsymbol{\beta}}_0$, and solve (25.2.10) to get solution $\hat{\boldsymbol{\beta}}_1$. Using $\hat{\boldsymbol{\beta}}_1$ in $\mathbf{W}_{\beta}$ gets us $\mathbf{W}_1$. Then $\mathbf{W}_1$ is used to get $\hat{\boldsymbol{\beta}}_2$, and so on. Typically, convergence occurs quite quickly. (The procedure can also be halted after a selected number of steps, if desired.) Only a standard weighted least squares program is needed, as in MINITAB, for example. We can write the iterative solution as

$$\hat{\boldsymbol{\beta}}_{q+1} = (\mathbf{X}'\mathbf{W}_q\mathbf{X})^{-1}\mathbf{X}'\mathbf{W}_q\mathbf{Y}, \qquad q = 0, 1, \ldots, \qquad (25.2.11)$$

and the procedure may be stopped when all the estimates change by less than some selected preset amount, say, 0.1% or 0.01%, or after a selected number of steps.

We see from (25.2.7) that the weights depend on the values of $[\partial\rho(u)/\partial u]/u$, where $u = (Y_i - \mathbf{x}_i'\boldsymbol{\beta})/s$. Thus we can look at the function $w(u) = [\partial\rho(u)/\partial u]/u$ for all the various choices of $\rho(u)$ in Table 25.1. Figure 25.1 shows plots of the weight functions $w(u) = [\partial\rho(u)/\partial u]/u$ for $u \geq 0$. (The $u \leq 0$ portion is symmetric with $u \geq 0$.) Obviously, choice of a particular $\rho(u)$ is basically a choice of what weight function the user wishes to assign to scaled residuals of various sizes. [One could even simply make up one's own weight function without defining a $\rho(u)$.]

The particular plots (b)–(g) shown in Figure 25.1 are based on specific assumptions that could be varied. For example, (f) assumes cutoff points $a = 1.7$, $b = 3.4$, and $c = 8.5$, at which point the weight becomes zero. If other a, b, c values were used, the diminishing-to-the-right three-part nature of (f) would remain, but the weight function would be somewhat compressed or expanded. Similarly, plot (b) is drawn with the change point at $u = 2$; this point could move left or right, and so on. The weight functions (d) and (e) are very similar in nature and could be adjusted to end at the same cutoff point. Moreover (e) is easier to deal with than (d), which involves trigonometric functions. Choice of the weight function to use, and of the constants (often called the *tuning constants*), is a second practical difficulty of robust regression. See, for example, Kelly (1996).

We next look at a couple of examples using robust regression. We recall that the weights are defined in terms of $u = (Y_i - \mathbf{x}_i'\boldsymbol{\beta})/s$ in (25.2.7), and require evaluation of a suitable scale factor s. One robust choice that has gained a measure of popularity is

$$s = \text{median } |e_i - \text{median } (e_i)|/0.6745, \qquad (25.2.12)$$

which would provide a roughly unbiased estimator of $\text{sd}(Y_i) = \sigma$ if n were large and the errors were normally distributed. (This suggestion of Hampel is given, with a misprinted denominator, in Andrews et al., 1972, p. 12. It is also sometimes shown elsewhere with the denominator already inverted as 1.4826.) Several alternative scale factors have been proposed and discussed; see Hill and Holland (1977, pp. 830–831), for example. The actual choice for s *does* have an effect on the results; that is, the robust regression parameter estimates are not invariant to choice of s. This is a third practical difficulty of robust estimation.

25.3. STEEL EMPLOYMENT EXAMPLE

Table 25.2 shows steel employment by country in thousands of people for the years 1974 and 1992. We propose to fit the model

$$Y = \beta_0 + \beta_1 X + \epsilon, \qquad (25.3.1)$$

where $Y = $ 1992 employment and $X = $ 1974 employment. The least squares fit is

$$\hat{Y} = -0.314 + 0.4004X, \qquad (25.3.2)$$

with $R^2 = 0.736$. Observations 1 and 4, with internally studentized residuals of 2.53 and -2.07, are designated "large." The corresponding externally studentized residuals are 5.34 and -2.83. We now make a robust fit using Huber's criterion with breakpoints at ± 2.

The residuals are 39.4, 11.9, -19.9, -36.4, -2.3, -0.3, 6.3, -0.9, 1.7, 0.5 rounded to one decimal place. The median of the residuals is $(0.5 - 0.3)/2 = 0.1$ and the $|e_i - \text{median }(e_i)|$ values are thus 39.3, 11.8, 20.0, 36.5, 2.4, 0.4, 6.2, 1.0, 1.6, 0.4. These have median $(6.2 + 2.4)/2 = 4.3$. This gives $s = 4.3/0.6745 = 6.375$ if we use (25.2.12). Dividing this into the residuals gives u values of

$$6.2, 1.9, -3.1, -5.7, -0.4, 0.0, 1.0, -0.1, 0.3, 0.1. \qquad (25.3.3)$$

Applying the weight formula of Table 25.1(b) with $a = 2$ results in initial weights of

$$0.323, 1, 0.645, 0.351, 1, 1, 1, 1, 1, 1. \qquad (25.3.4)$$

These weights are inserted in (25.2.10) and iterated upon, leading to the fit

$$\hat{Y} = 3.334 + 0.3205X \qquad (25.3.5)$$

with final weights of 0.208, 0.711, and 0.462 for observations 1, 2, and 4 and unity for

T A B L E 25.2. Steel Employment by Country in Europe, by Thousands, in 1974 and 1992

Country	1974	1992
Germany	232	132[a]
Italy	96	50
France	158	43
United Kingdom	194	41
Spain	89	33
Belgium	64	25
Netherlands	25	16
Luxembourg	23	8
Portugal	4	3
Denmark	2	1
Total	887	353

[a] Includes the former East Germany.

all others. The fitted least squares and robust fit lines are shown in Figure 25.2. The robust fit clearly down-weights heavily the outlying and influential first and fourth data points whose Cook's statistics are 2.54 and 0.85 in the original least squares fit. (The full listing is: 2.54, 0.02, 0.12, 0.85, 0, 0, 0.01, 0, 0, 0.)

Adjusting the First Observation

Prompted by the robust fit above, we now take note of the footnote in Table 25.2. Since the 1992 figure for Germany includes the former East Germany (whereas the 1974 figure does not), it may be too large. We can adjust it roughly by the factor $\frac{63}{80}$, approximately the ratio of West German and total German populations in 1992. This replaces 132 by $\frac{63}{80}$ (132) = 104. The new least squares fit is now

$$\hat{Y} = 2.803 + 0.3337X, \tag{25.3.6}$$

with $R^2 = 0.798$. Observations 1 and 4 have internally studentized residuals of 2.19 and -2.16. The externally studentized values are 3.22 and -3.11 and the ten Cook's statistics are, in observation order, 1.89, 0.07, 0.10, 0.92, 0, 0, 0.01, 0, 0, 0.01. Initial weights used for the Huber's criterion fit with breakpoints at ± 2 were calculated as in the previous fit, using (25.2.12) to get $s = 5.4855$, as

$$0.461, 0.721, 0.878, 0.414, 1, 1, 1, 1, 1, 1. \tag{25.3.7}$$

The robust fit that emerges is

$$\hat{Y} = 3.344 + 0.3205X \tag{25.3.8}$$

with final weights of 0.430, 0.711, and 0.462 for observations 1, 2, and 4 and unity for all others. Surprisingly, the robust fit is the same as before in (25.3.5), and the final weights of observations 2 and 4 are unchanged. The adjusted first observation is still heavily down-weighted but its weight has approximately doubled. The fitted least squares and robust fit lines are shown on Figure 25.3. On this scale they are not very distinguishable, in spite of the down-weighted observations. The adjustment of the largest Y appears to have been effective.

One might also question here whether the line should be forced to pass through the origin, since clearly if $X = 0$, $Y = 0$ necessarily. If this is done here, we get the no-intercept fitted model

$$\hat{Y} = 0.3516X, \tag{25.3.9}$$

the slope indicating the estimated reduction ratio of employment from 1974 to 1992

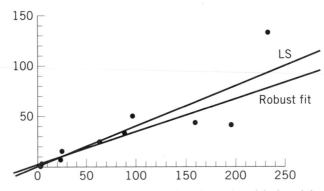

Figure 25.2. Least squares and robust fits to the original steel data.

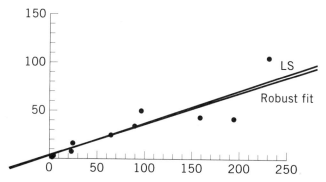

Figure 25.3. Least squares and robust fits to the adjusted steel data.

as about a third. This line is very close to the (two) lines in Figure 25.3 but passes through the origin and, at $X = 250$, lies merely 1.7 units higher than the least squares line and 4.4 units higher than the robust fit line.

Other Analyses Possible

There are several other robust analyses that could be made, using criteria other than Huber's from Table 25.1. Note that the weighting schemes used in both fits correspond well to the basic idea that the larger Y values might be subject to more error than the smaller Y values. This would not be surprising in data of this type. Our overall conclusion, however, is that any of the fits (25.3.6), (25.3.8), or (25.3.9) would be acceptable, all things considered. Perhaps (25.3.9) makes the most sense. The robust fits have greatly aided our view of how these data might be analyzed.

Standard Errors of Estimated Coefficients

Values for standard errors of the robust regression estimates can be obtained from the square roots of the diagonal entries in $(\mathbf{X}'\mathbf{W}_q\mathbf{X})^{-}s^2$, where q is set to its final value in the iterative procedure and s^2 is obtained from the final version of (25.2.12). A number of alternative calculations have also been proposed, however. For comments and additional references, see Hill and Holland (1977), Huber (1973), and Birch and Agard (1993).

25.4. TREES EXAMPLE

Table 25.3 shows tree data obtained by Tamra J. Burcar. The 25 observations (plotted in Figure 25.4) consist of values of

X = tree diameter in inches, measured at breast height (DBH),

Y = tree height in feet.

How well can we estimate height using a prediction equation based on tree diameter X?
We first carry out the least squares fit

$$\hat{Y} = 41.956 + 2.7836X, \tag{25.4.1}$$

with $R^2 = 0.633$. It is obvious that extrapolation of this equation below the data makes

T A B L E 25.3. Tree Data, from Tamra J. Burcar, on Diameters (X) and Heights (Y) of 24 Trees

Observation Number	Y	X	Observation Number	Y	X
1	58	5.5	14	75	10.1
2	60	5.7	15	72	10.2
3	42	5.8	16	78	10.4
4	64	6.5	17	65	10.6
5	60	6.6	18	80	10.6
6	65	6.7	19	82	10.8
7	56	6.9	20	70	11.3
8	57	7.0	21	74	11.3
9	70	7.3	22	68	11.6
10	68	8.3	23	68	11.6
11	65	8.6	24	82	13.0
12	70	9.5	25	88	18.0
13	63	10.0			

no sense, as it predicts heights of above 40 feet on trees with close to zero diameter. It is also clear from the plot and the computer output that the third observation and the 25th are inconsistent with the rest of the data:

Observation Number:	3	25
Residual	−16.10	−4.06
Internally studentized residual	−2.76	−0.87
Externally studentized residual	−3.29	−0.86
Cook's statistic	0.44[a]	0.28[a]

[a] The other Cook's statistics range from 0 to 0.07.

Both observations 3 and 25 are influential, but only 3 has a large residual. Both these data points had already attracted suspicion. The small tree was clearly stunted and the large tree protruded above the tree line and may have been damaged by winds. These trees also stand out on the plot of Figure 25.4. All in all, there are ample

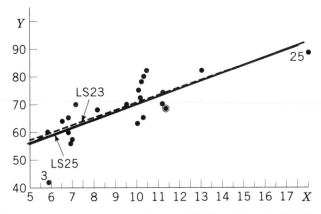

Figure 25.4. Tree data. Least squares fit to 25 points (LS25), to 23 points (LS 23), and a robust fit to 25 points (middle long-dashed line).

grounds for removing them entirely and fitting a straight line to the remaining data. However, we first procede with a robust analysis. We shall again use Huber's procedure with $a = 2$. Our estimate of scale, given via (25.2.12) using the least squares residuals, is

$$s = 4.1953/0.6745 = 6.22. \tag{25.4.2}$$

Only one observation, No. 3, is given a non-unit weight, of value 0.7726. Subsequently, iterations via (25.2.10) produce the robust fit

$$\hat{Y} = 42.872 + 2.7043X \tag{25.4.3}$$

with a weight 0.7377 for the third observation, and unity for all others. This line is compared with the least squares line (LS25) in Figure 25.4. It differs only slightly from it. The least squares line obtained by fitting to 23 data points, omitting the influential third and 25th observations, is

$$\hat{Y} = 44.353 + 2.6172X. \tag{25.4.4}$$

We see from Figure 25.4 that omission of the third point has allowed the intercept to rise very slightly, and that is the main change compared with the (quite similar) fits via least squares and robust estimation to all 25 observations. The robust fit tells us that observations 3 and 25 make only a slight difference compared to the least squares fit if all 25 observations are used. Thus the real decision is whether to use or discard observations 3 and 25. [Based on conversations with the scientist who provided these data, we would be inclined to discard them and use (25.4.4).] Again, making a robust fit to all 25 observations has helped our assessment of these data.

[If the least squares fit (25.4.4) is used as a "starter" for Huber's criterion with $a = 2$, the weights are all of size 1, so a robust fit to the 23 observations is not indicated.]

25.5. LEAST MEDIAN OF SQUARES (LMS) REGRESSION

Another robust regression method is the *least median of squares*, or LMS, method. This is given by minimizing, with respect to the elements of $\boldsymbol{\beta}$,

$$\underset{i}{\text{Median}} \, (Y_i - \mathbf{x}_i' \boldsymbol{\beta})^2.$$

For any given $\boldsymbol{\beta}$, we could find the median of the squared residuals; the solution is that $\boldsymbol{\beta}$ that produces the minimum such median. Geometrically, this method amounts to finding the narrowest strip (narrowest in the Y-axis direction) that covers "half" of the observations, where half means "the integer part of $n/2$, plus one." The LMS line lies at the center of this band. The computations are intricate and require a specially written program. See Rousseeuw and Leroy (1987), for example.

25.6. ROBUST REGRESSION WITH RANKED RESIDUALS (rreg)

There are a number of procedures of this type, which involve the minimization with respect to $\boldsymbol{\beta}$ of a weighted sum of the residuals. Suppose $a(\cdot)$ is a weight function (to be chosen) and $\mathbf{x}_i'$ is the ith row of the usual $\mathbf{X}$ matrix, $i = 1, 2, \ldots, n$. Then $Y_i -$

$\mathbf{x}_i'\boldsymbol{\beta}$ is the deviation of Y_i from the corresponding model function. Consider

$$D(\boldsymbol{\beta}) = \sum_{i=1}^{n} a\{R(Y_i - \mathbf{x}_i'\boldsymbol{\beta})\}\{Y_i - \mathbf{x}_i'\boldsymbol{\beta}\}, \qquad (25.6.1)$$

where $R(Y_i - \mathbf{x}_i'\boldsymbol{\beta})$ is the rank of the ith deviation. Minimization with respect to $\boldsymbol{\beta}$ of this dispersion function will give a robust rank estimator of $\boldsymbol{\beta}$. There are many possible choices for the function $a(\cdot)$. One possible choice is the Wilcoxon score function

$$a(i) = 12^{1/2}\left\{\frac{i}{n+1} - \frac{1}{2}\right\} \qquad (25.6.2)$$

for which

$$D_W(\boldsymbol{\beta}) = 12^{1/2}\sum_{i=1}^{n}\left\{\frac{i}{n+1} - \frac{1}{2}\right\}\{Y_i - \mathbf{x}_i'\boldsymbol{\beta}\}, \qquad (25.6.3)$$

where the notation i in (25.6.2) and (25.6.3) *now* implies that the residuals are in rank order. We first illustrate with a small constructed example. [The factor $12^{1/2}$ in (25.6.3) can be dropped in the fitting process.]

Example 1. To the five data points $(X, Y) = (2, 2), (3, 3), (5, 4), (9, 6), (12, 7)$, we fit a straight line. Minimizing (25.6.3), using the rreg option in MINITAB, for example, gives

$$\hat{Y}_r = 1.25 + 0.5X.$$

The least squares fit (which uses weights of 1 for all the observations) is

$$\hat{Y} = 1.3701 + 0.48870X.$$

The best horizontal straight line fit is $\hat{Y} = \overline{Y} = 4.4$. (This is not a good fit but provides another fit to compare.) The weights $(i/6 - \frac{1}{2})$ are $(-\frac{1}{3}, -\frac{1}{6}, 0, \frac{1}{3}, \frac{1}{6})$. Table 25.4 shows the three sets of residuals, how the weights would be allocated to these residuals, if rreg were applied, and the values of the statistic (25.6.3) for the three cases. We can see from this example that, since the more extreme residuals receive larger (Wilcoxon) weights in computing $D_W(\boldsymbol{\beta})$ and that the $\boldsymbol{\beta}$ is chosen to minimize $D_W(\boldsymbol{\beta})$, this procedure will tend to reduce some larger residuals to attain the fit. Since the least squares fit has the smallest residual sum of squares, other lower weighted residuals will necessarily increase in modulus in the rreg fit compared to the least squares fit.

T A B L E 25.4. Residuals and Wilcoxon Weights for Three Fits, Example 1

		rreg		Least Squares		$\hat{Y} = \overline{Y}$	
X	Y	Residuals	Weights	Residuals	Weights	Residuals	Weights
2	2	−0.25	−0.33	−0.35	−0.16	−2.4	−0.33
3	3	0.25	0.16	0.16	0.	−1.4	−0.16
5	4	0.25	0.	0.19	0.16	−0.4	0.
9	6	0.25	0.33	0.23	0.33	1.6	0.16
12	7	−0.25	−0.16	−0.23	−0.33	2.6	0.33
Cross-product/$12^{1/2}$		0.25		0.25942		2.167	
Residual SS		0.3125	—	0.29	—	17.2	—

Example 2. Steam data (Appendix 1A) with $Y = \beta_0 + \beta_8 X_8 + \epsilon$. The two fits are as follows.

Predictor	Coefficient		Standard Deviation of Coefficients	
	Rank	Least Squares	Rank	Least Squares
Constant	13.6194	13.6230	0.6929	0.5815
X_8	−0.07992	−0.07983	0.01254	0.01052
	Hodges–Lehmann estimate of tau = 1.061		Least squares s = 0.8901	

The "Hodges–Lehmann estimate of tau" provides an estimate of σ from the robust fit. The two fits are quite similar. An alternative "window estimate" of σ can also be used.

Example 3. Full set of steam data (Appendix 1A) with $Y = \beta_0 + \beta_2 X_2 + \cdots + \beta_{10} X_{10} + \epsilon$. The two fits are again quite similar.

Predictor	Coefficient		Standard Deviation of Coefficients	
	Rank	Least Squares	Rank	Least Squares
Constant	1.266	1.894	7.287	6.996
X_2	0.5544	0.7054	0.5884	0.5649
X_3	0.358	−1.894	4.319	4.146
X_4	1.4575	1.1342	0.7771	0.7461
X_5	0.1122	0.1188	0.2131	0.2046
X_6	0.14070	0.17935	0.08431	0.08095
X_7	−0.01300	−0.01818	0.02553	0.02451
X_8	−0.07430	−0.07742	0.01728	0.01659
X_9	−0.10952	−0.08585	0.05416	0.05200
X_{10}	−0.4227	−0.3450	0.2195	0.2107
	Hodges–Lehmann estimate of tau = 0.5934		Least squares s = 0.5697	

Other Weights

Other weights, for example, weights derived from the normal distribution, can be used in similar fashion. Quite similar results are obtained in practice. The weights are the normal deviates associated with cumulative normal levels of $100(i - \frac{1}{2})/n$ or, in MINITAB, $100(i - \frac{3}{8})/(n + \frac{1}{4})$, where n is the number of observations and $i = 1, 2, \ldots, n$.

Example 4. Hald data (Appendix 15A), fitting $Y = \beta_0 + \beta_1 X_1 + \beta_2 X_2 + \epsilon$. Again the two fits are quite similar.

Predictor	Coefficient		Standard Deviation of Coefficients	
	Rank	Least Squares	Rank	Least Squares
Constant	52.222	52.577	3.229	2.286
X_1	1.4631	1.4683	0.1713	0.1213
X_2	0.6678	0.6623	0.0648	0.0459
	Hodges–Lehmann estimate of tau = 3.399		Least squares s = 2.406	

We see throughout these Examples 2–4 that the standard deviation estimates are larger (more conservative) for the rreg fits.

As with many other alternatives to least squares, there is a degree of arbitrariness in the rreg fits in the choice of the weight functions used. In general, we recommend alternative fits mostly to make a comparison with the least squares fit. Large differences seen in the comparison often give useful clues about problems with the data set. Where differences are slight, proceeding with the least squares fit and its subsequent analyses is usually preferable.

Acknowledgment. We are grateful to Monica Hsieh who performed computations for Section 25.6 while pursuing "directed study" credits at the University of Wisconsin–Madison.

25.7. OTHER METHODS

Other robust estimation methods have been suggested. Some of these involve a replacement of one factor of each term in the sum of squares function by a score function, which could be a rank, for example. Currently, such other methods appear to be of limited practical value, and so we do not discuss them.

For a robust version of ridge regression, see Hogg (1979b) and Askin and Montgomery (1980).

25.8. COMMENTS AND OPINIONS

Is it sensible to use robust regression methods? The story here has similarities to that of ridge regression in the following sense. To use a robust estimator blindly is like using a ridge estimator blindly; it may or may not be appropriate. Any specific robust estimator is sensible when it provides (exactly or approximately) maximum likelihood estimation of the parameters under the alternative error assumptions believed to be true, when the assumption $\epsilon \sim N(\mathbf{0}, \mathbf{I}\sigma^2)$ is *not* true. The use of a robust regression method involves three practical difficulties, mentioned earlier:

1. What function $\rho(u)$ should we choose, if we do not know the distribution of the errors in the model? (This amounts to a choice of weight function type, for example, from Figure 25.1.)

2. How should we choose the tuning constants in the $\rho(u)$ we choose? (The values given in Table 25.1 are suggested by past studies. Different values would alter the shapes, but not the basic forms, of the weight functions in Figure 25.1.)

3. How should we choose a robust estimator of the scale factor s in (25.2.6) and (25.2.7)? (Different choices will give different numerical results for the regression coefficients.)

It is often not possible to make a choice of $\rho(u)$ based on factual knowledge about the errors. Usually too little is known. It might make sense simply to look at Figure 25.1 and pick a weight function. In our examples, we picked (*b*) with the breakpoint at $u = 2$. This would (typically) result in much of the data receiving weight 1, with a few observations receiving lesser weight. Selection (*f*) is a slight variation of (*b*). For other choices (*c*), (*d*), (*e*), and (*g*) only zero residuals receive full weight, and *all* others are down-weighted. This is more interesting but can also be confusing to assess.

There seems little harm in doing what we showed in our examples, namely, making both a least squares fit *and* a robust fit of some sort, reader's choice. A comparison between the two fits attracts attention to the observations down-weighted by the robust fit (or severely down-weighted if most or all observations receive weights less than 1) and attracts attention to outliers. This knowledge can be combined with, for example, examination of influence statistics like Cook's, and the data can be reassessed, and action taken to choose a good explanatory analysis. Where it is possible to justify a final least squares fit, that is often the best conclusion, simply because we know more about how to interpret the least squares solution.

25.9. REFERENCES

Books

Andrews, D. F., P. J. Bickel, F. R. Hampel, P. J. Huber, W. H. Rogers, and J. W. Tukey (1972). *Robust Estimates of Location.* Princeton, NJ: Princeton University Press.

Arthanari, T. S., and Y. Dodge (1981). *Mathematical Programming in Statistics.* New York: Wiley.

Birkes, D., and Y. Dodge (1993). *Alternative Methods of Regression.* New York: Wiley.

Bloomfield, P., and W. Steiger (1983). *Least Absolute Deviations: Theory, Applications, and Algorithms.* Boston: Birkhäuser.

Box, G. E. P., and N. R. Draper (1987). *Empirical Model-Building and Response Surfaces.* New York: Wiley.

Dodge, Y. (ed.) (1987a). *Statistical Data Analysis Based on the L_1-Norm and Related Methods.* New York: North-Holland.

Dodge, Y. (ed.) (1987b). Special Issue on Statistical Data Analysis Based on the L_1 Norm and Related Methods. *Computational Statistics & Data Analysis,* vol. 5, pp. 237–450.

Dodge, Y. (ed.) (1992). *L_1-Statistical Analysis and Related Methods.* Amsterdam: North-Holland.

Hampel, F. R., E. M. Ronchetti, P. J. Rousseeuw, and W. A. Stahel (1986). *Robust Statistics: The Approach Based on Influence Functions.* New York: Wiley.

Huber, P. (1981). *Robust Statistics.* New York: Wiley.

Rousseeuw, P. J., and A. M. Leroy (1987). *Robust Regression and Outlier Detection.* New York: Wiley.

Staudte, R. G., and S. J. Sheather (1990). *Robust Estimation and Testing.* New York: Wiley.

Articles

Adichie, J. N. (1967). Estimates of regression parameters based on rank test. *Annals of Mathematical Statistics,* **37**, 894–904.

Andrews, D. F. (1974). A robust method for multiple linear regression. *Technometrics,* **16**, 523–531.

Askin, R. G., and D. C. Montgomery (1980). Augmented robust estimators. *Technometrics,* **22**, 333–341.

Barrodale, I., and F. D. K. Roberts (1974). Algorithm 478: solution of an overdetermined system of equations in the l_1-norm. *Communications of the Association for Computing Machinery,* **17**, 319–320.

Beaton, A. E., and J. W. Tukey (1974). The fitting of power series, meaning polynomials, illustrated on band spectroscopic data. *Technometrics,* **16**, 147–185.

Birch, J. B., and D. B. Agard (1993). Robust inferences in regression: a comparative study. *Communications in Statistics, Simulation and Computation,* **22**(1), 217–244.

Davies, P. L. (1987). Asymptotic behaviour of S-estimates of multivariate location parameters and dispersion matrices. *Annals of Statistics,* **15**, 1269–1292.

Davies, L. (1990). The asymptotics of S-estimators in the linear regression model. *Annals of Statistics,* **18**, 1651–1675.

Davies, L. (1992). The asymptotics of Rousseeuw's minimum volume ellipsoid estimator. *Annals of Statistics,* **20**, 1828–1843.

Dielman, T., and R. Pfaffenberger (1982). LAV (least absolute value) estimation in linear regression: a review. In S. H. Zanakis and J. S. Rustagi (eds.). *Optimization in Statistics*. New York: North-Holland.

Dielman, T., and R. Pfaffenberger (1990). Tests of linear hypotheses and LAV estimation: a Monte Carlo comparison. *Communications in Statistics, Simulation and Computation*, **19**, 1179–1199.

Donoho, D. L., and P. J. Huber (1983). The notion of breakdown point. In P. J. Bickel, K. Doksum, and J. L. Hodges, Jr. (eds.). *A Festschrift for Erich Lehmann* pp. 157–184. Belmont, CA: Wadsworth.

Dutter, R. (1977). Numerical solution of robust regression problems: computational aspects, a comparison. *Journal of Statistical Computing and Simulation*, **5**, 207–238.

Farebrother, R. W. (1988). Algorithm AS 238: a simple recursive procedure for the L_1 norm fitting of a straight line. *Applied Statistics*, **37**, 457–465.

Forsythe, A. B. (1972). Robust estimation of straight line regression coefficients by minimizing pth power deviations. *Technometrics*, **14**, 159–166.

Gentle, J. E., S. C. Narula, and V. A. Sposito (1987). Algorithms for unconstrained L_1 linear regression. In Y. Dodge (ed.). *Statistical Data Analysis Based on the L_1-Norm and Related Methods*. New York: North-Holland.

Hampel, F. R. (1971). A general qualitative definition of robustness. *Annals of Mathematical Statistics*, **42**, 1887–1896.

Hampel, F. R. (1974). The influence curve and its role in robust estimation. *Journal of the American Statistical Association*, **69**, 383–393.

Hampel, F. R. (1975). Beyond location parameters: robust concepts and methods. *Proceedings of the 40th Session of the ISI*, **46**, 375–391.

Hampel, F. R. (1978). Optimally bounding the gross-error-sensitivity and the influence of position in factor space. *Proceedings of the Statistical Computing Section*, pp. 59–64. Washington, DC: American Statistical Association.

Hettmansperger, T. P., and S. J. Sheather (1992). A cautionary note on the method of least median squares. *The American Statistician*, **46**, 79–83.

Hill, R. W. (1979). On estimating the covariance matrix of robust regression M-estimates. *Communications in Statistics*, **A8**, 1183–1196.

Hill, R. W., and P. W. Holland (1977). Two robust alternatives to least squares regression. *Journal of the American Statistical Association*, **72**, 828–833.

Hogg, R. V. (1979a). Statistical robustness: one view of its use in applications today. *The American Statistician*, **33**, 108–115.

Hogg, R. V. (1979b). An introduction to robust estimation. In R. L. Launer and G. N. Wilkinson (eds.). *Robustness in Statistics*, pp. 1–18. New York: Academic Press.

Hogg, R. V., and R. H. Randles (1975). Adaptive distribution-free regression methods and their applications. *Technometrics*, **17**, 399–407.

Holland, P. W., and R. E. Welsch (1977). Robust regression using iteratively reweighted least-squares. *Communications in Statistics A*, **6**, 813–888.

Huber, P. (1964). Robust estimation of a location parameter. *Annals of Mathematical Statistics*, **35**, 73–101.

Huber, P. J. (1972). Robust statistics: a review. *Annals of Mathematical Statistics*, **43**, 1041–1067.

Huber, P. J. (1973). Robust regression: asymptotics, conjectures, and Monte Carlo. *Annals of Statistics*, **1**, 799–821.

Huber, P. J. (1975). Robustness and designs. *A Survey of Statistical Design and Linear Models*. Amsterdam: North-Holland.

Huber, P. J. (1983). Minimax aspects of bounded-influence regression (with discussion). *Journal of the American Statistical Association*, **78**, 66–80.

Jureckova, J., and S. Portnoy (1987). Asymptotics for one-step M-estimators with application to combining efficiency and high breakdown point. *Communications in Statistics, Theory & Methods*, **16**, 2187–2199.

Kelly, G. (1996). Adaptive choice of tuning constant for robust regression estimators. *The Statistician*, **45**, 35–40.

Koenker, R., and G. Bassett (1982). Tests of linear hypothesis and l_1 estimation. *Econometrica*, **50**, 1577–1583.

Krasker, W. S. (1980). Estimation in the linear regression model with disparate data points. *Econometrica*, **48**, 1333–1346.

Krasker, W. S., and R. E. Welsch (1982). Efficient bounded-influence regression estimation. *Journal of the American Statistical Association*, **77**, 595–604.

Lopuhaä, H. P. (1989). On the relation between S-estimators and M-estimators of multivariate location and covariance. *Annals of Statistics*, **17**, 1662–1683.

Lopuhaä, H. P., and Rousseeuw, P. J. (1991). Breakdown points of affine equivariant estimators of multivariate location and covariance matrices. *Annals of Statistics*, **19**, 229–248.

Marazzi, A. (1987). Solving bounded influence regression problems with ROBSYS. In: Y. Dodge (ed.). *Statistical Data Analysis Based on the L_1-Norm and Related Methods*. New York: North-Holland.

Maronna, R. A. (1976). Robust M-estimators of multivariate location and scatter. *Annals of Statistics*, **4**, 51–67.

Maronna, R. A., O. Bustos, and V. Yohai (1979). Bias- and efficiency-robustness of general M-estimators for regression with random carriers. In T. Gasser and M. Rosenblatt (eds.). *Smoothing Techniques for Curve Estimation*, pp. 91–116. New York: Springer.

Maronna, R. A., and V. J. Yohai (1981). Asymptotic behaviour of general M-estimates for regression and scale with random carriers. *Z. Wahrsch. Verw. Gebiete*, **58**, 7–20.

Maronna, R. A., and V. J. Yohai (1991). The breakdown point of simultaneous general M estimates of regression and scale. *Journal of the American Statistical Association*, **86**, 699–703.

McKean, J. W., and R. M. Schrader (1987). Least absolute errors analysis of variance. In Y. Dodge (ed.). *Statistical Data Analysis Based on the L_1-Norm and Related Methods*. New York: North-Holland.

McKean, J. W., and T. J. Vidmar (1994). A comparison of two rank-based methods for the analysis of linear models. *The American Statistician*, **48**, 220–229.

Morgenthaler, S. (1989). Comment on Yohai and Zamar. *Journal of the American Statistical Association*, **84**, 636.

Ramsay, J. O. (1977). A comparative study of several robust estimates of slope, intercept, and scale in linear regression. *Journal of the American Statistical Association*, **72**, 608–615.

Riedel, M. (1989). On the bias robustness in the location model. I. *Statistics*, **20**, 223–233. II. *Statistics*, **20**, 235–246.

Riedel, M. (1991). Bias-robustness in parametric models generated by groups. *Statistics*, **22**, 559–578.

Rocke, D. M., and D. F. Shanno (1986). The scale problem in robust regression M-estimates. *Journal of Statistical Computing and Simulation*, **24**, 47–69.

Ronchetti, E., and P. J. Rousseeuw (1985). Change-of-variance sensitivities in regression analysis. *Z. Wahrsch. Verw. Gebiete*, **68**, 503–519.

Rousseeuw, P. J. (1984). Least median of squares regression. *Journal of the American Statistical Association*, **79**, 871–880.

Rousseeuw, P. J. (1986). Multivariate estimation with high breakdown point. In W. Grossmann, G. Pflug, I. Vincze, and W. Wertz (eds.). *Mathematical Statistics and Application, Vol. B*. Dordrecht: Reidel.

Rousseeuw, P. J., and V. Yohai (1984). Robust regression by means of S-estimators. *Robust and Nonlinear Time Series Analysis. Lecture Notes in Statistics*, **26**, 256–272. New York: Springer.

Rousseeuw, P. J., and B. C. van Zomeren (1990). Unmasking multivariate outliers and leverage points. *Journal of the American Statistical Association*, **85**, 633–651.

Sadovski, A. N. (1974). Algorithm AS 74: L_1-norm fit of a straight line. *Applied Statistics*, **23**, 244–248.

SAS Institute Inc. (1990). *SAS/STAT User's Guide*, version 6, 4th ed., vol. 2. Cary, NC: SAS Institute.

Schrader, R. M., and T. P. Hettmansperger (1980). Robust analysis of variance based on a likelihood criterion. *Biometrika*, **67**, 93–101.

Schrader, R. M., and J. W. McKean (1987). Small sample properties of least absolute errors analysis of variance. In Y. Dodge (ed.). *Statistical Data Analysis Based on the L_1-Norm and Related Methods*. New York: North-Holland.

Shanno, D. F., and D. M. Rocke (1986). Numerical methods for robust regression: linear models. *SIAM Journal of Scientific and Statistical Computing*, **7**, 86–97.

Simpson, D. G., D. Ruppert, and R. J. Carroll (1992). On one-step GM-estimates and stability of inferences in linear regression. *Journal of the American Statistical Association*, **87**, 439–450.

Stahel, W. A. (1981). Robuste Schätzungen: Infinitesimale Optimalität und Schätzungen von Kovarianzmatrizen. Ph.D. dissertation, Dept. Statistics. ETH, Zürich.

Steele, J. M., and W. L. Steiger (1986). Algorithms and complexity for least median of squares regression. *Discrete Applied Mathematics*, **14**, 93–100.

Stefanski, L. A. (1991). A note on high breakdown estimators. *Statistics and Probability Letters*, **11**, 353–358.

Yohai, V. J. (1987). High breakdown point and high efficiency robust estimators for regression. *Annals of Statistics*, **15**, 642–656.

Yohai, V. J., and R. H. Zamar (1988). High breakdown point estimates of regression by means of the minimization of an efficient scale. *Journal of the American Statistical Association*, **83**, 406–413.

Welsch, R. E. (1975). Confidence regions for robust regression. *Statistical Computations Section Proceedings of the American Statistical Association*, Washington, DC.

EXERCISES FOR CHAPTER 25

A. Consider any of the data sets in this book and compare the least squares fit with a fit determined by any robust procedure computationally available to you. Do the fits differ much? If yes, what features, or what observations, seem to be causing the difference? Would you be happy using the least squares fit or not?

[Starter suggestions for data: (a) Table 2.1, (b) Appendix 1A, (c) Appendix 15A. One or more predictors can be used.]

B. For the data in Table 2.1, show that the least squares fit is $\hat{Y} = 1.43 + 0.316X$ while the rreg fit with Wilcoxon weights is $\hat{Y} = 1.64 + 0.25X$ and the rreg fit with normal weights is $\hat{Y} = 1.66 + 0.25X$. Plot all three lines. Why is the first fit so different from the other two?

CHAPTER 26

Resampling Procedures (Bootstrapping)

The ability to do a lot of computation extremely fast has led to the use of techniques that provide "new" sets of data by resampling numbers generated from a single data set. These methods are used in many different statistical situations. In this brief introduction, resampling methods are illustrated specifically in the context of linear regression. Via resampling methods, we can reexamine a regression analysis already made, by comparing it with a population of results that might have been obtained under certain assumed circumstances. Often, an important point of bootstrapping is not just to evaluate estimates of the parameters, but also to obtain good estimates of standard errors from the distributions generated by the parameter estimates in bootstrapped iterations. This would be especially valuable in fitting situations where standard errors were *not* directly derivable from theory, for example, in nonlinear estimation situations, or situations where least squares is inappropriate, such as for generalized linear models. In our linear model examples, we compare resampling parameter estimates and standard errors with the corresponding least squares values.

26.1. RESAMPLING PROCEDURES FOR REGRESSION MODELS

Two resampling procedures that can be used in the regression context are:

(a) Fit the linear model and obtain the n residuals. Choose a sample of size n from the residuals, generated with probability $1/n$ for each residual, and sampling with replacement. Attach these sampled values to the n predicted $\hat{Y}_i$ to give a resampled set of Y's. Thus if the model is $\mathbf{Y} = \mathbf{X}\boldsymbol{\beta} + \boldsymbol{\epsilon}$ and $\hat{\mathbf{Y}} = \mathbf{Xb}$, the new $\mathbf{Y}$-values are

$$\mathbf{Y}^* = \mathbf{Xb} + \mathbf{e}^*, \tag{26.1.1}$$

where $\mathbf{e}^*$ is a resampled set from the vector $\mathbf{e} = \mathbf{Y} - \hat{\mathbf{Y}}$. Least squares regression is now performed on the model

$$\mathbf{Y}^* = \mathbf{X}\boldsymbol{\beta} + \boldsymbol{\epsilon} \tag{26.1.2}$$

to obtain an estimate $\mathbf{b}^*$ (say). As many iterations as desired can be performed, and the usual sample mean and sample standard deviation of each of the elements of those vector estimates can be found. In our tables, these are called the "bootstrap averages" and the "bootstrap standard errors," respectively.

(b) The resampling could also be carried out on the pairs $(Y_i, \mathbf{x}_i')$, where Y_i is the ith observation and $\mathbf{x}_i'$ is the ith row of the $\mathbf{X}$ matrix. The resampling involves selecting a set of n of the $(Y_i, \mathbf{x}_i')$, each selected with probability $1/n$, and sampling with replacement, to obtain (say) $\mathbf{Y}^{**}$ and $\mathbf{X}^{**}$. The regression model

$$\mathbf{Y}^{**} = \mathbf{X}^{**}\boldsymbol{\beta} + \boldsymbol{\epsilon} \qquad (26.1.3)$$

is now fitted by least squares. Again, the properties of the corresponding $\mathbf{b}^{**}$ values can be examined after any desired number of iterations.

Note that, after either resampling (a) or (b) has been carried out, the model could also be fitted robustly, rather than by least squares. Numerous variations are possible.

Both procedures (a) and (b), and variations of them, are called *bootstrapping procedures*. In (a) we have "bootstrapped the residuals"; in (b) we have "bootstrapped pairs."

26.2. EXAMPLE: STRAIGHT LINE FIT

We make use of the steel employment data of Table 25.2, but with the 1992 Germany value of 132 adjusted to 104 (to remove the effect of the former East Germany; see Section 25.3). This gives the data X, Y of Table 26.1. The predicted values and the residuals are those from the fitted least squares equation $\hat{Y} = 2.803 + 0.3337X$ and are rounded to one decimal place.

(a) A MINITAB program (see Appendix 26A) was used to bootstrap residuals. Five example iterations gave the results of the upper portion of Table 26.2. We see that such a small data set can lead to some peculiarities in the samples, for example, iteration 2 used the same residual (No. 2, 15.2) five times out of 10. We see, however, that after 100 iterations, the averages of the estimates are only slightly different from the least squares values and the standard errors are of comparable size. Another anomaly that can arise is an occasional negative Y^* value, which is of course not physically possible. Such observations could be adjusted to zero, or discarded and replaced.

(b) A second MINITAB program (see Appendix 26B) was used to bootstrap pairs (Y, X) from the original data set. Table 26.3 shows results in a format similar to that of Table 26.2. The bootstrapped "averaged fits" provide lines quite close to the least squares values.

T A B L E 26.1. Steel Employment by Country in Europe in Thousands, in 1974 (X) and 1992 (Y) Together with Predicted Values and Residuals from the Least Squares Fit: Y_1 Has Been Adjusted to 104 from 132

Observation Number	1974 (X)	1992 (Y)	$\hat{Y}$	Residuals e_i
1	232	104	80.2	23.8
2	96	50	34.8	15.2
3	158	43	55.5	−12.5
4	194	41	67.5	−26.5
5	89	33	32.5	0.5
6	64	25	24.2	0.8
7	25	16	11.1	4.9
8	23	8	10.5	−2.5
9	4	3	4.1	−1.1
10	2	1	3.5	−2.5

T A B L E 26.2. Five Sample Iterations of Bootstrapping Residuals with Results of 100 Iterations

Iteration	Residual Numbers Used	b_0^*	b_1^*
1	9, 6, 8, 9, 1, 2, 2, 8, 2, 7	14.479	0.2784
2	1, 9, 8, 7, 2, 2, 5, 2, 2, 2	13.317	0.3294
3	10, 4, 8, 7, 8, 8, 8, 2, 7, 1	8.878	0.2762
4	4, 10, 8, 5, 7, 2, 7, 9, 6, 3	6.119	0.2750
5	7, 3, 10, 4, 4, 1, 7, 6, 5, 9	4.180	0.2794
Five iterations (residuals)	Bootstrap averages	9.395	0.2877
	Bootstrap standard errors	4.456	0.0234
Least squares values		2.803	0.3337
Least squares standard errors		6.992	0.0593
100 iterations (residuals)	Bootstrap averages	3.523	0.3296
	Bootstrap standard errors	6.032	0.0530

Using the Original Data

We reinstated the original first reading of 132 (which was replaced by 104 in the calculations above) and performed 100 simulations on bootstrapped residuals, and another 100 on bootstrapped pairs. The results were (standard errors in parentheses):

Least squares values $\quad b_0 = -0.314(9.997), \quad b_1 = 0.4004(0.0849)$
Bootstrapping residuals $\quad b_0^* = -0.469(8.626), \quad b_1^* = 0.4078(0.0712)$
Bootstrapping pairs $\quad b_0^{**} = -0.004(7.258), \quad b_1^{**} = 0.3927(0.1258)$

We see that the least squares values are approximately reproduced and that the first observation (which has the largest internally studentized residual of 2.53 in the least squares fit) is not "flagged" in any way by the bootstrap procedures. Bootstrapping procedures will not do anything for us in this regard because such residuals and/or the corresponding pairs $(Y_i, \mathbf{x}_i')$ have their share of appearances in the random choices of "new" data.

T A B L E 26.3. Five Sample Iterations of Bootstrapping Pairs and Results of 100 Iterations

Iteration	Numbers of (Y, X) Pairs Used	b_0^{**}	b_1^{**}
1	5, 7, 5, 9, 9, 8, 3, 10, 9, 5	3.378	0.2920
2	4, 5, 7, 4, 4, 1, 5, 6, 4, 1	1.910	0.3051
3	7, 3, 3, 7, 5, 3, 2, 8, 9, 1	2.718	0.3428
4	7, 3, 9, 10, 7, 1, 4, 1, 10, 6	0.918	0.3676
5	2, 3, 5, 6, 9, 7, 1, 5, 3, 9	1.664	0.3660
Five iterations (pairs)	Bootstrap averages	2.119	0.3347
	Bootstrap standard errors	0.953	0.0347
Least squares values		2.803	0.3337
Least squares standard errors		6.992	0.0593
100 iterations (pairs)	Bootstrap averages	3.810	0.3242
	Bootstrap standard errors	3.995	0.0841

T A B L E 26.4. Five Hundred Bootstrappings on the Steam Data (Y, X_5, X_6, X_8)

b_0	b_5	b_6	b_8	Method Used
−2.968	0.402	0.199	−0.074	Least
4.833	0.157	0.041	0.0072	squares
−2.724	0.393	0.200	−0.074	Bootstrapping
4.356	0.141	0.038	0.0068	residuals
−4.047	0.410	0.238	−0.074	Bootstrapping
6.880	0.234	0.107	0.0074	pairs

26.3. EXAMPLE: PLANAR FIT, THREE PREDICTORS

We next look briefly at a larger example, using the steam data (see Appendix 1A) to fit a plane

$$Y = \beta_0 + \beta_5 X_5 + \beta_6 X_6 + \beta_8 X_8 + \epsilon. \tag{26.3.1}$$

Table 26.4 gives results of 500 "connected" sets of sampled data. Averages of the 500 sets of bootstrapped parameter estimates are in the top row of each pair of lines and the corresponding bootstrapped standard errors are in the bottom row of each pair of lines. The two sets of iterations (a) on the residuals e_i and (b) on the $(Y_i, \mathbf{x}_i')$ pairs were done using the same choices of random numbers for the observations selected. That is, 500 sets of random choices were made, and each set was used on (a) and (b). This was done as a matter of convenience. We see that the results obtained by bootstrapping residuals are, on the whole, closer to the least squares results, and that bootstrapping pairs appears to provide larger variance estimates then does least squares.

26.4 REFERENCE BOOKS

Edgington, E. (1980). *Randomization Tests*, 2nd ed. New York: Marcel Dekker.

Efron, B. (1982). *The Jackknife, the Bootstrap, and Other Resampling Procedures*. Philadelphia: SIAM Publications.

Efron, B., and R. Tibshirani (1993). *Introduction to the Bootstrap*. London: Chapman and Hall.

Good, P. (1994). *Permutation Tests: A Practical Guide to Resampling Methods for Testing Hypotheses*. New York: Springer-Verlag.

Hall, P. (1995). *The Bootstrap and Edgeworth Expansion*. New York: Springer-Verlag.

Hjorth, J. S. U. (1994). *Computer Intensive Statistical Methods*. London: Chapman and Hall.

Manley, B. (1992). *Randomization and Monte Carlo Methods in Biology*. London: Chapman and Hall.

Maritz, J. S. (1994). *Distribution Free Statistical Methods*. London: Chapman and Hall.

Noreen, E. (1989). *Computer-Intensive Methods for Testing Hypotheses*. New York: Wiley.

Westfall, P., and S. Yound (1992). *Resampling-Based Multiple Testing*. New York: Wiley.

For other useful information, see the review by J. Albert and M. Berliner of the "Resampling Stats" computing package in *American Statistician*, **48**, 1994, 129–131 and "Bootstrap: more than a stab in the dark?" by G. A. Yound, *Statistical Science*, **9**, 1994, 382–415, which contains extensive discussion.

Acknowledgment. We are grateful to Ying Kuen (Ken) Cheung who provided computations and commentary while pursuing "directed study" credits at the University of Wisconsin–Madison.

APPENDIX 26A. SAMPLE MINITAB PROGRAMS TO BOOTSTRAP RESIDUALS FOR A SPECIFIC EXAMPLE

```
#ken.r   MAIN PROGRAM
oh=0
set cl
104 50 43 41 33 25 16 8 3 1
set c2
232 96 158 194 89 64 25 23 4 2
name cl 'Y'
name c2 'X'
name c4 'fitted'
name c5 'resids'
let k1=1
let k2=10                # number of data points
let k10=100              # number of iterations
let k20=1
let c9=0
let c10=0
name c7 'Y*'
noecho
print k2
print k10
regress cl 1 c2 c22 c4;
residuals c5.
name c5 'resids'
name c4 'fitted'
print c2 cl c4 c5
plot cl c2

exec 'kena.r' k10      #call subprogram 1
let c11=c9/k10
let c12=(c10-k10*c11**2)/(k10-1)
let c13=sqrt(c12)
print c11-c13
end
stop
```

```
#kena.r     SUBPROGRAM 1
random k2 c6;
integer k1 k2.
print k20
print c6
let k3=1

exec 'kenab.r' k2      #call subprogram 2
noecho
print 'Y*' 'X'
regress 'Y*' 1 'X';
coeff c8.
let c9=c9+c8
let c10=c10+c8**2
let k20=k20+1
end
```

```
#kenab.r    SUBPROGRAM 2
let k4=c6(k3)
let c7(k3)=c4(k3)+c5(k4)
let k3=k3+1
end
```

APPENDIX 26B. SAMPLE MINITAB PROGRAMS TO BOOTSTRAP PAIRS FOR A SPECIFIC EXAMPLE

```
#ken.p  MAIN PROGRAM
oh=0
set c1
104 50 43 41 33 25 16 8 3 1
set c2
232 96 158 194 89 64 25 23 4 2
name c2 'X'
let k1=1
let k2=10                  #number of data points
let k10=100                #number of iterations
let k20=1
let c9=0
let c10=0
name c4 'Y*'
name c5 'X*'
noecho
print k2
print k10
regress c1 1 c2
print c1 c2
plot c1 c2

exec 'kena.p' k10       #call subprogram 1
let c11=c9/k10
let c12=(c10-k10*c11**2)/(k10-1)
let c13=sqrt(c12)
print c11-c13
end
stop
```

```
#kena.p     SUBPROGRAM 1
random k2 c6;
integer k1 k2.
print k20
print c6
let k3=1

exec 'kenab.p' k2       #call subprogram 2
noecho
print 'Y*' 'X*'
regress 'Y*' 1 'X*';
coeff c8.
let c9=c9+c8
let c10=c10+c8**2
let k20=k20+1
end
```

```
#kenab.p     SUBPROGRAM 2
let k4=c6(k3)
let c4(k3)=c1(k4)
let c5(k3)=c2(k4)
let k3=k3+1
end
```

ADDITIONAL COMMENTS

Bootstrapping residuals can be thought of as working with a fixed $\mathbf{X}$ matrix; bootstrapping pairs corresponds to using a random $\mathbf{X}$. Both procedures above can be described as nonparametric procedures. If (as a third alternative) the residuals were independently sampled from a $N(0, s^2)$ distribution where s^2 was the least squares estimate of σ^2 from the original data, we would have a *parametric* bootstrap procedure. For discussion, see the references on page 588.

EXERCISES FOR CHAPTER 26

A. Consider any of the data sets in this book and use the least squares fit as a basis for bootstrapping the residuals as described in the text. Start by using 100 iterations and compare the least squares fit with the bootstrap results. Then do more iterations on the same data to see what effect that has on your conclusions.

B. Consider any of the data sets in this book and bootstrap pairs as described in the text. Begin with 100 iterations and then do more to see the effect on your results. Compare your results with the least squares fit to the original data and to your bootstrap results in Exercise A.

Bibliography

Adcock, R. J. (1878). A problem in least squares. *Analyst*, **5**, 53–54.

Adichie, J. N. (1967). Estimates of regression parameters based on rank test. *Annals of Mathematical Statistics*, **38**, 894–904.

Aia, M. A., Goldsmith, R. L., and Mooney, R. W. (1961). Predicting stoichiometric $CaHPO_4 \cdot 2H_2O$. *Industrial and Engineering Chemistry*, **53**, January, 55–57.

Aitken, M., D. Anderson, B. Francis, and J. Hinds (1989). *Statistical Modelling in GLIM*. Oxford, UK: Clarendon Press.

Albert, J., and M. Berliner (1994). Review of the "Resampling Stats" computer package. *The American Statistician*, **48**, 129–131.

Amer, F. A., and W. T. Williams (1957). Leaf-area growth in *Pelargonium zonale*. *Annals of Botany, New Series*, **21**, 339–342.

Anderson, D. A., and R. G. Scott (1974). The application of ridge regression analysis to a hydrologic target-control model. *Water Resources Bulletin*, **10**, 680–690.

Anderson-Sprecher, R. (1994). Model comparisons and R^2. *The American Statistician*, **48**, 113–117.

Andrews, D. F. (1974). A robust method for multiple linear regression. *Technometrics*, **16**, 523–531.

Andrews, D. F., P. J. Bickel, F. R. Hampel, P. J. Huber, W. H. Rogers, and J. W. Tukey (1972). *Robust Estimates of Location*. Princeton, NJ: Princeton University Press.

Andrews, D. F., and D. Pregibon (1978). Finding the outliers that matter. *Journal of the Royal Statistical Society, Series B*, **4**, 84–93.

Anscombe, F. J., and J. W. Tukey (1963). The examination and analysis of residuals. *Technometrics*, **5**, 141–160.

Arthanari, T. S., and Y. Dodge (1981). *Mathematical Programming in Statistics*. New York: Wiley.

Askin, R. G., and D. C. Montgomery (1980). Augmented robust estimators. *Technometrics*, **22**, 333–341.

Atkinson, A. C. (1985). *Plots, Transformations and Regression. Oxford, UK: Clarendon Press*.

Atkinson, A. C. (1994). Fast very robust methods for the detection of multiple outliers. *Journal of the American Statistical Association*, **89**, 1329–1339.

Barker, F., Y. C. Soh, and R. J. Evans (1988). Properties of the geometric mean functional relationship. *Biometrics*, **44**, 279–281.

Barnett, V. D. (1967). A note on linear structural relationships when both residual variances are known. *Biometrika*, **54**, 670–672.

Barnett, V. D. (1975). Probability plotting methods and order statistics. *Applied Statistics*, **24**, 95–108.

Barnett, V. D., and T. Lewis (1994). *Outliers in Statistical Data*, 3rd ed. New York: Wiley.

Barrodale, I., and F. D. K. Roberts (1974). Algorithm 478: solution of an overdetermined system of equations in the l_1-norm. *Communications of the Association for Computing Machinery*, **17**, 319–320.

Bartlett, M. S. (1947). The use of transformations. *Biometrics*, **3**, 39–52.

Bartlett, M. S. (1949). Fitting a straight line when both variables are subject to error. *Biometrics*, **5**, 207–212.

Bates, D. M., and D. G. Watts (1988). *Nonlinear Regression Analysis and Its Applications*. New York: Wiley.

Bauer, F. L. (1971). Elimination with weighted row combinations for solving linear equations and least squares problems. In J. H. Wilkinson and C. Reisch (Eds.). *Handbook for Automatic Computation Volume* II: *Linear Algebra*. New York: Springer Verlag.

Becker, W., and P. Kennedy (1992). A lesson in least squares and R squared. *The American Statistician*, **46**, 282–283.

Beaton, A. E., and J. W. Tukey (1974). The fitting of power series, meaning polynomials, illustrated on band spectroscopic data. *Technometrics*, **16**, 147–185.

Belsley, D. A. (1991). *Conditioning Diagnostics, Collinearity and Weak Data in Regression*, New York: Wiley. (This book is a revision of the following book.)

Belsley, D. A., E. Kuh, and R. E. Welsch (1980). *Regression Diagnostics*, New York: Wiley.

Berk, K. N., and D. E. Booth (1995). Seeing a curve in multiple regression. *Technometrics*, **37**, 385–398.

Berkson, J. (1950). Are there two regressions? *Journal of the American Statistical Association*, **45**, 164–180.

Bing, J. (1994). How to standardize regression coefficients. *The American Statistician*, **48**, 209–213.

Birch, J. B., and D. B. Agard (1993). Robust inferences in regression: a comparative study. *Communications in Statistics, Simulation and Computation*, **22**(1), 217–244.

Birkes, D., and Y. Dodge (1993). *Alternative Methods of Regression*. New York: Wiley.

Bisgaard, S., and H. T. Fuller (1994-95). Analysis of factorial experiments with defects or defectives as the response. *Quality Engineering*, **7**(2), 429–443.

Bloomfield, P., and W. Steiger (1983). *Least Absolute Deviations: Theory, Applications, and Algorithms*. Boston: Birkhäuser.

Box, G. E. P. (1966). Use and abuse of regression. *Technometrics*, **8**, 625–629.

Box, G. E. P., and G. A. Coutie (1956). Application of digital computers in the exploration of functional relationships. *Proceeding of the Institution of Electrical Engineers*, 103, Part B, Suppl. No. 1, 100–107.

Box, G. E. P., and D. R. Cox (1964). An analysis of transformations. *Journal of the Royal Statistical Society, Series B*, **26**, 211–243 (discussion pp. 244–252).

Box, G. E. P., and N. R. Draper (1965). The Bayesian estimation of common parameters from several responses. *Biometrika*, **52**, 355–365.

Box, G. E. P., and N. R. Draper (1987). *Empirical Model-Building and Response Surfaces*. New York: Wiley.

Box, G. E. P., and J. S. Hunter (1954). A confidence region for the solution of a set of simultaneous equations with an application to experimental design. *Biometrika*, **41**, 190–199.

Box, G. E. P., and W. G. Hunter (1962). A useful method for model building. *Technometrics*, **4**, 301–318.

Box, G. E. P., and W. G. Hunter (1965). Sequential design of experiment for nonlinear models. *Proceedings of the IBM Scientific Computing Symposium on Statistics*, October 21–23, 1963, pp. 113–137.

Box, G. E. P., W. G. Hunter, and J. S. Hunter (1978). *Statistics for Experimenters*. New York: Wiley.

Box, G. E. P., G. M. Jenkins, and G. C. Reinsel (1994). *Time Series Analysis, Forecasting and Control*, 3rd ed. Englewood Cliffs, NJ: Prentice Hall.

Box, G. E. P., and H. L. Lucas (1959). Design of experiments in non-linear situations. *Biometrika*, **46**, 77–90.

Box, G. E. P., and J. Wetz (1973). Criteria for judging adequacy of estimation by an approximating response function. University of Wisconsin Statistics Department Technical Report No. 9.

Box, J. F. (1978). *R. A. Fisher: The Life of a Scientist*. New York: Wiley.

Breiman, L. (1995). Better subset regression using the nonnegative garrote. *Technometrics*, **37**, 373–384.

Breiman, L. (1996). Heuristics of instability and stabilization in model selection. *Annals of Statistics*, **24**, 2350–2383.

Brieman, L., and J. H. Friedman (1997). Predicting multivariate responses in multiple linear regression. *Journal of the Royal Statistical Society, Series B*, **59**(1), 3–37 (discussion pp. 37–54).

Bright, J. W., and G. S. Dawkins (1965). Some aspects of curve fitting using orthogonal polynomials. *Industrial and Engineering Chemistry Fundamentals*, **4**, February, 93–97.

Bring, J. (1994). How to standardize regression coefficients. *The American Statistician*, **48**, 209–213.

Brown, P. J., and C. Payne (1975). Election night forecasting. *Journal of the Royal Statistical Society, Series A*, **138**, 463–483 (discussion pp. 483–498).

Brownlee, K. A. (1965). *Statistical Theory and Methodology in Science and Engineering*, 2nd ed. New York: Wiley.

Bunke, H., and O. Bunke (1989). *Nonlinear Regression, Functional Relations and Robust Methods*. New York: Wiley.

Bunke, O. (1975). Minimax linear, ridge and shrunken estimators for linear parameters. *Mathematische Operationsforchung und Statistiks*, **6**, 697–701.

Carroll, R. J., and D. Ruppert (1988). *Transformations and Weighting in Regression*. London: Chapman and Hall.

Carroll, R. J., and H. Schneider (1985). A note on Levene's tests for equality of variances. *Statistics and Probability Letters*, **3**, 191–194.

Chappell, R. (1994). Presenting the coefficients of the linear quadratic formula for clinical use. *International Journal of Radiation Oncology, Biology, Physics*, **29**(1), 191–193.

Chatterjee, S., and A. S. Hadi (1988). *Sensitivity Analysis in Linear Regression*. New York: Wiley.

Chaturvedi, A. (1993). Ridge regression estimators in the linear regression models with non-spherical errors. *Communications in Statistics, Theory and Methods*, **22**(8), 2275–2284.

Clayton, D. G. (1971). Gram–Schmidt orthogonalization. *Applied Statistics*, **20**, 335–338 (Fortran algorithm).

Cleveland, W. S., and B. Kleiner (1975). A graphical technique for enhancing scatter plots with moving statistics. *Technometrics*, **17**, 447–454.

Collett, D. (1991). *Modelling Binary Data*. London: Chapman and Hall.

Coniffe, D., and J. Stone (1973). A critical view of ridge regression. *The Statistician*, **22**, 181–187.

Cook, R. D. (1977). Detection of influential observations in linear regression. *Technometrics*, **19**, 15–18.

Cook, R. D., and S. Weisberg (1982). *Residuals and Influence in Regression. London: Chapman and Hall.*

Cook, R. D., and S. Weisberg (1994). Transforming a response variable for linearity. *Biometrika*, **81**, 731–737.

Cooper, B. E. (1971). The use of orthogonal polynomials with equal x-values. *Applied Statistics*, **20**, 209–213.

Copas, J. B. (1983). Regression, prediction and shrinkage. *Journal of the Royal Statistical Society, Series B*, **45**, 311–335 (discussion pp. 335–354).

Cornell, J. A. (1990). *Experiments with Mixtures*, 2nd ed. New York: Wiley.

Cox, D. R., and E. J. Snell (1989). *Analysis of Binary Data*. London: Chapman and Hall.

Creasy, M. A. (1956). Confidence limits for the gradient in the linear functional relationship. *Journal of the Royal Statistical Society, Series B*, **18**, 65–69.

Crouse, R. H., C. Jin, and R. C. Hanumara (1995). Unbiased ridge estimation with prior information and ridge trace. *Communications in Statistics, Theory and Methods*, **24**(9), 2341–2354.

Daniel, C., and F. S. Wood, assisted by J. W. Gorman. (1980) *Fitting Equations to Data*, 2nd ed. New York: Wiley

Davies, P. L. (1987). Asymptotic behaviour of S-estimates of multivariate location parameters and dispersion matrices. *Annals of Statistics*, **15**, 1269–1292.

Davies, L. (1990). The asymptotics of S-estimators in the linear regression model. *Annals of Statistics*, **18**, 1651–1675.

Davies, L. (1992). The asymptotics of Rousseeuw's minimum volume ellipsoid estimator. *Annals of Statistics*, **20**, 1828–1843.

De Lury, D. B. (1960). *Values and Integrals of the Orthogonal Polynomials up to $n = 26$*. Toronto: University of Toronto Press.

Derringer, G. C. (1974). An empirical model for viscosity of filled and plasticized elastomer products. *Journal of Applied Polymer Science*, **18**, 1083–1101.

Dielman, T., and R. Pfaffenberger (1982). LAV (least absolute value) estimation in linear regression: a review. In S. H. Zanakis and J. S. Rustagi (eds.). *Optimization in Statistics*. New York: North-Holland.

Dielman, T., and R. Pfaffenberger (1990). Tests of linear hypotheses and LAV estimation: a Monte Carlo comparison. *Communications in Statistics, Simulation and Computation*, **19**, 1179–1199.

Diggle, P. J. (1990). Time Series, A Biostatistical Introduction. Oxford, UK: Clarendon Press/Oxford Science Publications.

Ding, C. G., and R. E. Bargmann (1991). Statistical algorithms AS 260 and AS 261, concerning the distribution of R^2. *Applied Statistics*, **40**, 195–198.

Dixon, W. J. (chief ed.). *Biomedical Computer Programs, P-Series*. Berkeley: University of California Press. (Seek the most recent update.)

Dobson, A. J. (1990). *Introduction to Generalized Linear Models*. London: Chapman and Hall.

Dodge, Y. (ed.) (1987a). *Statistical Data Analysis Based on the L_1-Norm and Related Methods*. New York: North-Holland.

Dodge, Y. (ed.) (1987b). Special Issue on Statistical Data Analysis Based on the L_1 Norm and Related Methods. *Computational Statistics & Data Analysis*, **5**, 237–450.

Dodge, Y. (ed.) (1992). *L_1-Statistical Analysis and Related Methods*. Amsterdam: North-Holland.

Dodge, Y. (1996). The guinea pig of multiple regression. In H. Rieder (ed.). *Lecture Notes in Statistics,* No. 109, see pp. 91–118. New York: Springer.

Dolby, J. L. (1963). A quick method for choosing a transformation. *Technometrics*, **5**, 317–325.

Donoho, D. L., and P. J. Huber (1983). The notion of breakdown point. In P. J. Bickel, K. Doksum and J. L. Hodges, Jr. (eds.). *A Festschrift for Erich Lehmann*, pp. 157–184. Belmont, CA: Wadsworth.

Draper, N. R. (1984). The Box–Wetz criterion versus R^2. *Journal of the Royal Statistical Society, Series A*, **147**, 100–103. Also (1985), **148**, 357.

Draper, N. R. (1992). Straight line regression when both variables are subject to error. *Proceedings of the 1991 Kansas State University Conference on Applied Statistics in Agriculture*, pp. 1–18.

Draper, N. R., P. Prescott, S. M. Lewis, A. M. Dean, P. W. M. John, and M. G. Tuck (1993). Mixture designs for four components in orthogonal blocks. *Technometrics*, **35**, 268–276. Also see (1995), **37**, 131–132.

Draper, N. R., and I. Guttman (1995). Confidence intervals versus regions. *The Statistician*, **44**, 399–403.

Draper, N. R., and A. M. Herzberg (1987). A ridge-regression sidelight. *The American Statistician*, **41**, 282–283.

Draper, N. R., and W. G. Hunter (1969). Transformations: some examples revisited. *Technometrics*, **11**, 23–40.

Draper, N. R., and R. C. Van Nostrand (1979). Ridge regression and James–Stein estimation: review and comments. *Technometrics*, **21**, 451–466.

Draper, N. R., and Y. (Fred) Yang (1997). Generalization of the geometric mean functional relationship. *Computational Statistics and Data Analysis*, **23**, 355–372.

Dressler, A. (1984). International kinematics of galaxies in cluster, I. Velocity dispersions for elliptical galaxies in Coma and Virgo. *Astrophysical Journal*, **281**, 512–524.

Driscoll, M. F., and D. J. Anderson (1980). Point-of-expansion, structure and selection in multivariate polynomial regression. *Communications in Statistics, Theory and Methods*, **A9**, 821–836.

Durbin, J. (1969). Tests for serial correlation in regression analysis based on the periodogram of least squares residuals. *Biometrika*, **56**, 1–15.

Durbin, J. (1970). An alternative to the bounds test for testing for serial correlation in least squares regression. *Econometrica*, **38**, 422–429.

Durbin, J., and G. S. Watson (1950). Testing for serial correlation in least squares regression, I. *Biometrika*, **37**, 409–428.

Durbin, J., and G. S. Watson (1951). Testing for serial correlation in least squares regression, II. *Biometrika*, **38**, 159–178.

Durbin, J., and G. S. Watson (1971). Testing for serial correlation in least squares regression, III. *Biometrika*, **58**, 1–19.

Dutter, R. (1977). Numerical solution of robust regression problems: computational aspects, a comparison. *Journal of Statistical Computing and Simulation*, **5**, 207–238.

Edgington, E. (1980). *Randomization Tests*, 2nd ed. New York: Marcel Dekker.

Efron, B. (1982). *The Jackknife, the Bootstrap, and Other Resampling Procedures*. Philadelphia: SIAM Publications.

Efron, B., and R. Tibshirani (1993). *Introduction to the Bootstrap*. London: Chapman and Hall.

Egerton, G. M., and P. J. Laycock (1981). Some criticisms of stochastic shrinkage and ridge regression, with counterexamples. *Technometrics*, **23**, 155–159. (Also see letter, **25**, 1983, 303–304 and correction, 304.)

Eisenhart, C. (1964). The meaning of "least" in least squares. *Journal of the Washington Academy of Sciences*, **54**, 24–33.

Ellerton, R. R. W. (1978). Is the regression equation adequate—a generalization. *Technometrics*, **20**, 313–315.

Erjavec, J., G. E. P. Box, W. G. Hunter, and J. F. MacGregor (1973). Some problems associated with the analysis of multiresponse data. *Technometrics*, **15**, 33–51.

Fahrmeir, L., and G. Tutz (1994). *Multivariate Statistical Modelling Based on Generalized Linear Models*. New York: Springer-Verlag.

Farebrother, R. W. (1974). Gram–Schmidt regression. *Applied Statistics*, **23**, 470–476 (Algol 60 algorithm).

Farebrother, R. W. (1988). Algorithm AS 238: a simple recursive procedure for the L_1 norm fitting of a straight line. *Applied Statistics*, **37**, 457–465.

Feller, W. (1957). *An Introduction to Probability Theory and Its Applications, Volume 1*, 2nd ed. New York: Wiley.

Fisher, R. A., and F. Yates (1964). *Statistical Tables for Biological, Agricultural and Medical Research*, 6th ed. New York: Hafner Publishing.

Forsythe, A. B. (1972). Robust estimation of straight line regression coefficients by minimizing pth power deviations. *Technometrics*, **14**, 159–166.

Francis, B., M. Green, and C. Payne (1993). *The GLIM System: Release 4 Manual*. New York: Oxford University Press.

Frank, I., and J. Friedman (1993). A statistical view of some chemometrics regression tools. *Technometrics*, **35**, 109–135 (discussion pp. 136–148). See p. 124.

Freeman, M. F., and J. W. Tukey (1950). Transformations related to the angular and the square root. *Annals of Mathematical Statistics*, **21**(4), 607–611.

Freund, R. J. (1988). When is $R^2 > r_{yx_1}^2 + r_{yx_2}^2$? (Revisited). *The American Statistician*, **42**, 89–90.

Freund, R., and R. Littell (1991). *SAS System for Regression*, 2nd ed. Cary, NC: SAS Institute.

Füle, E. (1995). On ecological regression and ridge estimation. *Communications in Statistics, Simulation and Computation*, **24**(2), 385–398.

Furnival, G. M., and R. W. Wilson (1974). Regression by leaps and bounds. *Technometrics*, **16**, 499–511.

Gallant, A. R. (1987). *Nonlinear Statistical Models*. New York: Wiley.

Galton, F. (1877). Typical laws of heredity in man. Address to the Royal Institution, England.

Galton, F. (1885). Regression towards mediocrity in hereditary stature. *Journal of the Anthropological Institute*, **15**, 246–263.

Galton, F. (1885). Presidential address to Section H of the British Association, printed in *Nature*, September, 507–510.

Gentle, J. E., S. C. Narula, and V. A. Sposito (1987). Algorithms for unconstrained L_1 linear regression. In Y. Dodge (ed.). *Statistical Data Analysis Based on the L_1-Norm and Related Methods*. New York: North-Holland.

Gibbons, D. G. (1981). A simulation study of some ridge estimators. *Journal of the American Statistical Association*, **76**, 131–139.

Gibson, W. M., and G. H. Jowett (1957). Three-group regression analysis, Part I. *Applied Statistics*, **6**, 114–122.

Gilmour, S. G. (1996). The interpretation of Mallows C_p-statistic. *The Statistician*, **45**, 49–56.

Goldstein, M., and A. F. M. Smith (1974). Ridge-type estimators for regression analysis. *Journal of the Royal Statistical Society, Series B*, **36**, 284–291.

Good, P. (1994). *Permutation Tests: A Practical Guide to Resampling Methods for Testing Hypotheses*. New York: Springer-Verlag.

Gorman, J. W., and R. J. Toman (1966). Selection of variables for fitting equations to data. *Technometrics*, **8**, 27–51.

Gray, J. B., and W. H. Woodall (1994). The maximum size of standardized and internally studentized residuals in regression analysis. *The American Statistician*, **48**, 111–113.

Graybill, F. A. (1961). *An Introduction to Linear Statistical Models*. New York: McGraw-Hill.

Grechanovsky, E., and I. Pinsker (1995). Conditional p-values for the F-statistic in a forward selection procedure. *Computational Statistics and Data Analysis*, **20**, 239–263.

Green, P. J., and B. W. Silverman (1994). *Nonparametric Regression and Generalized Linear Models: A Roughness Penalty Approach*. London: Chapman and Hall.

Gregory, F. G. (1928). Studies in the energy relations of plants, II. *Annals of Botany*, **42**, 469–507.

Hadi, A. S. (1992). A new measure of overall potential influence in linear regression. *Computational Statistics and Data Analysis*, **14**, 1–27.

Hald, A. (1952). *Statistical Theory with Engineering Applications*. New York: Wiley.

Hall, P. (1995). *The Bootstrap and Edgeworth Expansion*. New York: Springer-Verlag.

Hamilton, D. (1987a). Sometimes $R^2 > r_{yx_1}^2 + r_{yx_2}^2$. *The American Statistician*, **41**, 129–132.

Hamilton, D. (1987b). Reply to Freund and Mitra. *The American Statistician*, **42**, 90–91.

Hampel, F. R. (1971). A general qualitative definition of robustness. *Annals of Mathematical Statistics*, **42**, 1887–1896.

Hampel, F. R. (1974). The influence curve and its role in robust estimation. *Journal of the American Statistical Association*, **69**, 383–393.

Hampel, F. R. (1975). Beyond location parameters: robust concepts and methods. *Proceedings of the 40th Session of the ISI*, **46**, 375–391.

Hampel, F. R. (1978). Optimally bounding the gross-error-sensitivity and the influence of position in factor space. *Proceedings of the Statistical Computing Section*, pp. 59–64. Washington, DC: American Statistical Association.

Hampel, F. R., E. M. Ronchetti, P. J. Rousseeuw, and W. A. Stahel (1986). *Robust Statistics: The Approach Based on Influence Functions*. New York: Wiley.

Hartley, H. O. (1961). The modified Gauss–Newton method for the fitting of non-linear regression functions by least squares. *Technometrics*, **3**, 269–280.

Harvey, P. H., and G. M. Mace (1982). Comparison between taxa and adaptive trends: problems of methodology. In *Current Problems in Sociobiology*. New York: Cambridge University Press.

Harville, D. A. (1997). Matrix Algebra From a Statistician's Perspective. New York: Springer-Verlag.

Hawkins, D. M. (1980). *Identification of Outliers*. London: Chapman and Hall.

Healy, M. J. R. (1984). The use of R^2 as a measure of goodness of fit. *Journal of the Royal Statistical Society, Series A*, **147**, 608–609.

Healy, M. J. R. (1988). *GLIM: An Introduction*. Oxford, UK: Oxford University Press.

Herschel, J. F. W. (1849). *Outliers of Astronomy*. Philadelphia: Lee & Blanchard.

Hettmansperger, T. P., and S. J. Sheather (1992). A cautionary note on the method of least median squares. *The American Statistician*, **46**, 79–83.

Hilbe, J. M. (1994). Generalized linear models. *The American Statistician*, **48**, 255–265. This article reviews seven software packages: GAIM (Version 1.1), GENSTAT 5 (Release 2), GLIM (Release 4.0), SAS 6.08 for Windows, S-Plus for Windows (Version 3.1), XLISP-Stat (Version 2.1R3), and XploRe (Version 3.1).

Hill, R. W. (1979). On estimating the covariance matrix of robust regression M-estimates. *Communications in Statistics*, **A8**, 1183–1196.

Hill, R. W., and P. W. Holland (1977). Two robust alternatives to least squares regression. *Journal of the American Statistical Association*, **72**, 828–833.

Hines, R. J. O., and W. G. S. Hines (1995). Exploring Cook's statistic graphically. *The American Statistician*, **49**, 389–394.

Hjorth, J. S. U. (1994). *Computer Intensive Statistical Methods*. London: Chapman and Hall.

Hocking, R. R. (1996). *Methods and Applications of Linear Models*. New York: Wiley.

Hoerl, A. E. (1954). Fitting curves to data. *Chemical Business Handbook*. New York: McGraw-Hill.

Hoerl, A. E., and R. W. Kennard (1970a). Ridge regression: biased estimation for non-orthogonal problems. *Technometrics*, **12**, 55–67.

Hoerl, A. E., and R. W. Kennard (1970b). Ridge regression: applications to non-orthogonal problems. *Technometrics*, **12**, 69–82 (correction, **12**, 723).

Hoerl, A. E., and R. W. Kennard (1975). A note on a power generalization of ridge regression. *Technometrics*, **17**, 269.

Hoerl, A. E., and R. W. Kennard (1976). Ridge regression: iterative estimation of the biasing parameter. *Communications in Statistics*, **A5**, 77–88.

Hoerl, A. E., and R. W. Kennard (1981). *Ridge Regression 1980, Advances, Algorithms, and Applications*. Columbus, OH: American Sciences Press.

Hoerl, A. E., R. W. Kennard, and K. F. Baldwin (1975). Ridge regression, some simulations. *Communications in Statistics*, **A4**, 105–123.

Hogg, R. V. (1979a). Statistical robustness: one view of its use in applications today. *The American Statistician*, **33**, 108–115.

Hogg, R. V. (1979b). An introduction to robust estimation. In R. L. Launer and G. N. Wilkinson (eds.). *Robustness in Statistics*, pp. 1–18. New York: Academic Press.

Hogg, R. V., and R. H. Randles (1975). Adaptive distribution-free regression methods and their applications. *Technometrics*, **17**, 399–407.

Holland, P. W., and R. E. Welsch (1977). Robust regression using iteratively reweighted least-squares. *Communications in Statistics*, **A6**, 813–888.

Hosmer, D. W. Jr., and S. Lemeshow (1989). *Applied Logistic Regression*. New York: Wiley.

Huber, P. (1964). Robust estimation of a location parameter. *Annals of Mathematical Statistics*, **35**, 73–101.

Huber, P. J. (1972). Robust statistics: a review. *Annals of Mathematical Statistics*, **43**, 1041–1067.

Huber, P. J. (1973). Robust regression: asymptotics, conjectures, and Monte Carlo. *Annals of Statistics*, **1**, 799–821.

Huber, P. J. (1975). Robustness and designs. *A Survey of Statistical Design and Linear Models*. Amsterdam: North-Holland.

Huber, P. (1981). *Robust Statistics*. New York: Wiley.

Huber, P. J. (1983). Minimax aspects of bounded-influence regression (with discussion). *Journal of the American Statistical Association*, **78**, 66–80.

Huff, D. (1954). *How to Lie with Statistics*. New York: W. W. Norton.

Hunter, W. G., and R. Mezaki (1964). A model-building technique for chemical engineering kinetics. *American Institute of Chemical Engineering Journal*, **10**, 315–322. [But note error in text below Table 4, p. 320; if the sixth residual were positive (it is just negative) the pattern would be obvious.]

Jackson, J. E., and W. H. Lawton (1967). Answer to Query 22. *Technometrics*, **9**(2), 339–340.

Jaske, D. R. (1994). Illustrating the Gauss–Markov theorem. *The American Statistician*, **48**, 237–238.

Jefferys, W. H. (1990). Robust estimation when more than one variable per equation of condition has error. *Biometrika*, **77**, 597–607.

John, J. A., and N. R. Draper (1980). An alternative family of transformations. *Applied Statistics*, **29**, 190–197.

Joiner, B. L., and J. R. Rosenblatt (1971). Some properties of the range in samples from Tukey's symmetric λ distributions. *Journal of the American Statistical Association*, **66**, 394–399.

Judge, G. G., and T. Takayama (1966). Equality restrictions in regression analysis. *Journal of the American Statistical Association*, **61**, 166–181.

Jureckova, J., and S. Portnoy (1987). Asymptotics for one-step M-estimators with application to combining efficiency and high breakdown point. *Communications in Statistics, Theory & Methods*, **16**, 2187–2199.

Kelly, G. (1984). The influence function in the errors in variables problem. *Annals of Statistics*, **12**, 87–100.

Kelly, G. (1996). Adaptive choice of tuning constant for robust regression analysis. *The Statistician*, **45**, 35–40.

Kendall, M. G., and A. Stuart (1961). *The Advanced Theory of Statistics*, Vol. 2. New York: Hafner Publishing.

Kennard, R. W. (1971). A note on the C_p statistic. *Technometrics*, **13**, 899–900.

Koenker, R., and G. Bassett (1982). Tests of linear hypothesis and l_1 estimation. *Econometrica*, **50**, 1577–1583.

Kozumi, H., and K. Ohtani (1994). The general expressions for the moments of Lawless and Wang's ordinary ridge regression estimator. *Communications in Statistics, Theory & Methods*, **23**(10), 2755–2774.

Krasker, W. S. (1980). Estimation in the linear regression model with disparate data points. *Econometrica*, **48**, 1333–1346.

Krasker, W. S., and R. E. Welsch (1982). Efficient bounded-influence regression estimation. *Journal of the American Statistical Association*, **77**, 595–604.

Kronmal, R. A. (1993). Spurious correlation and the fallacy of the ratio standard revisited. *Journal of the Royal Statistical Society, Series A*, **156**, 379–392.

Kurotori, I. S. (1966). Experiments with mixtures of components having lower bounds. *Industrial Quality Control*, **22**, 592–596.

Kvalseth, T. O. (1985). Cautionary note about R^2. *The American Statistician*, **39**, 279–285.

LaMotte, L. R. (1994). A note on the role of independence in t statistics constructed from linear statistics in regression models. *The American Statistician*, **48**, 238–240.

Largey, A., and J. E. Spencer (1996). F- and t-tests in multiple regression: the possibility of conflicting outcomes. *The Statistician*, **45**, 105–109.

Lawless, J. F., and P. Wang (1976). A simulation study of ridge and other regression estimators. *Communications in Statistics, Theory & Methods*, **A5**, 307–323.

le Cessie, S., and J. C. van Houwelingen (1992). Ridge estimators in logistic regression. *Applied Statistics*, **41**, 191–201.

Levenberg, K. (1944). A method for the solution of certain non-linear problems in least squares. *Quarterly of Applied Mathematics*, **11**, 164–168.

Lim, T.-S., and W.-Y. Loh (1996). A comparison of tests of equality of variances. *Computational Statistics and Data Analysis*, **22**, 287–301.

Lindley, D. V., and W. F. Scott (1984; 2nd ed., 1995). *New Cambridge Elementary Statistical Tables*. New York: Cambridge University Press.

Littell, R., R. Freund, and P. Spector (1991). *SAS System for Linear Models*, 3rd ed. Cary, NC: SAS Institute.

Lopuhaä, H. P. (1989). On the relation between S-estimators and M-estimators of multivariate location and covariance. *Annals of Statistics*, **17**, 1662–1683.

Lopuhaä, H. P., and P. J. Rousseeuw (1991). Breakdown points of affine equivariant estimators of multivariate location and covariance matrices. *Annals of Statistics*, **19**, 229–248.

Magee, L. (1990). R^2 measures based on Wald and likelihood ratio joint significance tests. *The American Statistician*, **44**, 250–253.

Mallows, C. L. (1973). Some comments on C_p. *Technometrics*, **15**, 661–675.

Mallows, C. L. (1995). More comments on C_p. *Technometrics*, **37**, 362–372. See also (1997), **39**, 115–116.

Manley, B. (1992). *Randomization and Monte Carlo Methods in Biology*. London: Chapman and Hall.

Marazzi, A. (1987). Solving bounded influence regression problems with ROBSYS, In: Y. Dodge (ed.), *Statistical Data Analysis Based on the L_1-Norm and Related Methods*. New York: North-Holland.

Maritz, J. S. (1994). *Distribution Free Statistical Methods*. London: Chapman and Hall.

Maronna, R. A. (1976). Robust M-estimators of multivariate location and scatter. *Annals of Statistics*, **4**, 51–67.

Maronna, R. A., O. Bustos, and V. Yohai (1979). Bias- and efficiency-robustness of general M-estimators for regression with random carriers. In T. Gasser and M. Rosenblatt (eds.). *Smoothing Techniques for Curve Estimation*, pp. 91–116. New York: Springer.

Maronna, R. A., and V. J. Yohai (1981). Asymptotic behaviour of general M-estimates for regression and scale with random carriers. *Z. Wahrsch. Verw. Gebiete*, **58**, 7–20.

Maronna, R. A., and V. J. Yohai (1991). The breakdown point of simultaneous general M estimates of regression and scale. *Journal of the American Statistical Association*, **86**, 699–703.

Marquardt, D. W. (1963). An algorithm for least squares estimation of nonlinear parameters. *Journal of the Society for Industrial and Applied Mathematics*, **11**, 431–441.

Marquardt, D. W. (1970). Generalized inverses, ridge regression, biased linear estimation, and nonlinear estimation. *Technometrics*, **12**, 591–612.

Marquardt, D. W., and R. D. Snee (1975). Ridge regression in practice. *The American Statistician*, **29**, 3–19.

McCullagh, P., and J. A. Nelder (1989). *Generalized Linear Models*, 2nd ed. London: Chapman and Hall.

McDonald, G. C. (1980). Some algebraic properties of ridge coefficients. *Journal of the Royal Statistical Society, Series B*, **42**, 31–34.

McKean, J. W., and R. M. Schrader (1987). Least absolute errors analysis of variance. In Y. Dodge (ed.). *Statistical Data Analysis Based on the L_1-Norm and Related Methods*. New York: North-Holland.

McKean, J. W., and T. J. Vidmar (1994). A comparison of two rank-based methods for the analysis of linear models. *The American Statistician*, **48**, 220–229.

Mead, R. (1970). Plant density and crop yield. *Applied Statistics*, **19**, 64–81.

Medawar, P. B. (1940). Growth, growth energy, and ageing of the chicken's heart. *Proceedings of the Royal Society of London*, **B-129**, 332–355.

Meeter, D. A. (1964). *Problems in the Analysis of Nonlinear Models by Least Squares*. University of Wisconsin Ph.D. Thesis.

Mickey, M. R., O. J. Dunn, and V. Clark (1967). Note on the use of stepwise regression in detecting outliers. *Computers and Biomedical Research*, **1**, 105–111.

Miller, A. J. (1990). *Subset Selection in Regression*. London: Chapman and Hall.

Miller, R. G. (1980). Kanamycin levels in premature babies. *Biostatistics Casebook*, Vol. III, 127–142. (Technical Report No. 57, Division of Biostatistics, Stanford University.)

Miller, R. G. (1981). *Simultaneous Statistical Inference*, 2nd ed. New York: Springer-Verlag.

Mitra, S. (1988). The relationship between the multiple and the zero-order correlation coefficients. *The American Statistician*, **42**, 89.

Morgenthaler, S. (1989). Comment on Yohai and Zamar. *Journal of the American Statistical Association*, **84**, 636.

Nagelkerke, N. J. D. (1991). A note on general definition of the coefficient of determination. *Biometrika*, **78**, 691–692.

Nelder, J. A. (1961). The fitting of a generalization of the logistic curve. *Biometrics*, **17**, 89–110.

Nelder, J. A., and R. W. M. Wedderburn (1972). Generalized linear models. *Journal of the Royal Statistical Society, Series A*, **135**, 370–384.

Nelson, L. S. (1980). The mean square successive difference test. *Journal of Quality Technology*, **12**, 174–175.

Nickerson, D. M. (1994). Construction of a conservative confidence region from projections of an exact confidence region in multiple linear regression. *The American Statistician*, **48**, 120–124.

Noreen, E. (1989). Computer-Intensive Methods for Testing Hypotheses. New York: Wiley.

Nyquist, H. (1988). Applications of the jackknife procedure in ridge regression. *Computational Statistics & Data Analysis*, **6**, 177–183.

Obenchain, R. L. (1978). Good and optimal ridge estimators. *Annals of Statistics*, **6**, 1111–1121.

Oman, S. D. (1981). A confidence bound approach to choosing the biasing parameter in ridge regression. *Journal of the American Statistical Association*, **76**, 452–461.

Panopoulos, P. (1989). Ridge regression: discussion and comparison of seven ridge estimators. *Statistica*, **49**, 265–275.

Park, S. H. (1981). Collinearity and optimal restrictions on regression parameters in ridge regression. *Technometrics*, **23**, 289–295.

Pearson, E. S., and H. O. Hartley (1958). *Biometrika Tables for Statisticians, Volume I.* New York: Cambridge University Press.

Peixoto, J. L. (1987). Hierarchical variable selection in polynomial regression models. *The American Statistician*, **41**, 311–313.

Peixoto, J. L. (1990). A property of well-formulated polynomial regression models. *The American Statistician*, **44**, 26–30. Also see (1991), **45**, 82.

Plackett, R. L. (1960). *Regression Analysis.* Oxford, UK: Clarendon Press.

Plackett, R. L. (1972). Studies in the history of probability and statistics. XXIX. The discovery of the method of least squares. *Biometrika*, **59**, 239–251.

Ralston, A., and H. S. Wilf (eds.) (1962). *Mathematical Methods for Digital Computers.* New York: Wiley. (See the article "Multiple regression analysis," by M. A. Efroymson.)

Ramsay, J. O. (1977). A comparative study of several robust estimates of slope, intercept, and scale in linear regression. *Journal of the American Statistical Association*, **72**, 608–615.

Rao, C. R. (1973). *Linear Statistical Inference and Its Applications*, 2nd ed. New York: Wiley.

Ratkowsky, D. A. (1983). *Nonlinear Regression Modeling.* New York: Marcel Dekker.

Ratkowsky, D. A. (1990). *Handbook of Nonlinear Regression Models.* New York: Marcel Dekker.

Rayner, R. K. (1994). The small-sample power of Durbin's h test revised. *Computational Statistics and Data Analysis*, **17**, 87–94.

Richards, F. J. (1959). A flexible growth function for empirical use. *Journal of Experimental Botany*, **10**, 290–300.

Riedel, M. (1989). On the bias robustness in the location model. I. *Statistics*, **20**, 223–233. II. *Statistics*, **20**, 235–246.

Riedel, M. (1991). Bias-robustness in parametric models generated by groups. *Statistics*, **22**, 559–578.

Riggs, D. A., J. A. Guarnieri, and S. Addelman (1978). Fitting straight lines when both variables are subject to error. *Life Sciences*, **22**, 1305–1360.

Robson, D. S. (1959). A simple method for constructing orthogonal polynomials when the independent variable is unequally spaced. *Biometrics*, **15**, 187–191.

Rocke, D. M., and D. F. Shanno (1986). The scale problem in robust regression M-estimates. *Journal of Statistical Computing and Simulation*, **24**, 47–69.

Ronchetti, E., and P. J. Rousseeuw (1985). Change-of-variance sensitivities in regression analysis. *Z. Wahrsch. Verw. Gebiete*, **68**, 503–519.

Ross, G. J. S. (1990). *Nonlinear Estimation.* New York: Springer-Verlag.

Rousseeuw, P. J. (1984). Least median of squares regression. *Journal of the American Statistical Association*, **79**, 871–880.

Rousseeuw, P. J. (1986). Multivariate estimation with high breakdown point. In W. Grossmann, G. Pflug, I. Vincze, and W. Wertz (eds.). *Mathematical Statistics and Application, Vol. B.* Dordrecht: Reidel.

Rousseeuw, P. J., and A. M. Leroy (1987). *Robust Regression and Outlier Detection.* New York: Wiley.

Rousseeuw, P. J., and B. C. van Zomeren (1990). Unmasking multivariate outliers and leverage points. *Journal of the American Statistical Association*, **85**, 633–651.

Rousseeuw, P. J., and V. Yohai (1984). Robust regression by means of S-estimators. *Robust and Nonlinear Time Series Analysis. Lecture Notes in Statistics*, **26**, 256–272. New York: Springer.

Roy, S. N., and A. E. Sarhan (1956). On inverting a class of patterned matrices. *Biometrika*, **43**, 227–231.

Royston, J. P. (1995). A remark on Algorithm AS181: the W-test for normality. *Applied Statistics*, **44**, 547–551.

Sadovski, A. N. (1974). Algorithm AS 74: L_1-norm fit of a straight line. *Applied Statistics*, **23**, 244–248.

Saleh, A. K. M. E., and B. M. G. Kibria (1993). Performance of some new preliminary test ridge regression estimators and their properties. *Communications in Statistics, Theory & Methods*, **22**(10), 2747–2764.

SAS Institute Inc. (1990). *SAS/STAT User's Guide*, Version 6, 4th ed., Vol. 2, Cary, NC: SAS Institute.

Saville, D. J., and G. R. Wood (1996). *Statistical Methods: A Geometric Primer*. New York: Springer-Verlag.

Savin, N. E., and K. J. White (1977). The Durbin–Watson test for serial correlation with extreme sample sizes or many regressors. *Econometrica*, **45**, 1989–1996.

Scheffé, H. (1958). Experiments with mixtures. *Journal of the Royal Statistical Society Series B*, **20**, 344–360.

Scheffé, H. (1963). The simplex-centroid design for experiments with mixtures. *Journal of the Royal Statistical Society Series B*, **25**, 235–263.

Schemper, M. (1990). The explained variation in proportional hazards regression. *Biometrika*, **77**, 216–218.

Schey, H. M. (1993). The relationship between the magnitudes of $SSR(x_2)$ and $SSR(x_2|x_1)$: a geometric description. *The American Statistician*, **47**, 26–30.

Schneider, H., and G. P. Barker (1973). *Matrices and Linear Algebra*, 2nd ed. New York: Holt, Reinhart and Winston. (Also reprinted by Dover Books.)

Schrader, R. M., and T. P. Hettmansperger (1980). Robust analysis of variance based on a likelihood criterion. *Biometrika*, **67**, 93–101.

Schrader, R. M., and J. W. McKean (1987). Small sample properties of least absolute errors analysis of variance. In Y. Dodge (ed.). *Statistical Data Analysis Based on the L_1-Norm and Related Methods*. New York: North-Holland.

Scott, A., and C. Wild (1991). Transformations and R^2. *The American Statistician*, **45**, 127–129.

Searle, S. R. (1971). *Linear Models*. New York: Wiley.

Searle, S. R. (1987). *Linear Models for Unbalanced Data*. New York: Wiley.

Searle, S. R. (1988). Parallel lines in residual plots. *The American Statistician*, **42**(3), 211.

Searle, S. R., G. Casella, and C. E. McCulloch (1992). *Variance Components*. New York: Wiley.

Seber, G. A. F. (1977). *Linear Regression Analysis*. New York: Wiley.

Seber, G. A. F., and C. J. Wild (1989). *Nonlinear Regression*. New York: Wiley.

Segerstedt, B. (1992). On ordinary ridge regression in generalized linear models. *Communications in Statistics, Theory & Methods*, **21**(8), 2227–2246.

Shah, A. K. (1991). Relationship between the coefficients of determination of algebraically related models. *The American Statistician*, **45**, 300–301.

Shanno, D. F., and D. M. Rocke (1986). Numerical methods for robust regression: linear models. *SIAM Journal of Scientific and Statistical Computing*, **7**, 86–97.

Simpson, D. G., D. Ruppert, and R. J. Carroll (1992). On one-step GM-estimates and stability of inferences in linear regression. *Journal of the American Statistical Association*, **87**, 439–450.

Smith, G. (1980). An example of ridge regression difficulties. *Canadian Journal of Statistics*, **8**, 217–225.

Smith, H. (1969). The analysis of data from a designed experiment. *Journal of Quality Technology*, **1**, 259–263.

Smith, H., and S. D. Dubey (1964). Some reliability problems in the chemical industry. *Industrial Quality Control*, **22**, 64–70.

Snee, R. D. (1973). Some aspects of non-orthogonal data analysis. Part I. Developing prediction equations. *Journal of Quality Technology*, **5**, 67–79.

Spang, H. A. (1962). A review of minimization techniques for non-linear functions. *Society for Industrial and Applied Mathematics Review*, **4**, 343–365.

Sprent, P., and G. R. Dolby (1980). The geometric mean functional relationship. *Biometrics*, **36**, 547–550. (see also, **38**, 859–860.)

Srivastava, V. K., and D. E. A. Giles (1991). Unbiased estimation of the mean squared error of the feasible generalised ridge regression estimator. *Communications in Statistics, Theory & Methods*, **20**, 2357–2386.

Stahel, W. A. (1981). Robuste Schätzungen: Infinitesimale Optimalität und Schätzungen von Kovarianzmatrizen. Ph.D. dissertation, Dept. Statistics. ETH, Zürich.

Stapleton, J. H. (1995). *Linear Statistical Models*. New York: Wiley.

Staudte, R. G., and S. J. Sheather (1990). *Robust Estimation and Testing*. New York: Wiley.

Steele, J. M., and W. L. Steiger (1986). Algorithms and complexity for least median of squares regression. *Discrete Applied Mathematics*, **14**, 93–100.

Stefanski, L. A. (1991). A note on high breakdown estimators. *Statistics and Probability Letters*, **11**, 353–358.

Stigler, S. M. (1974). Gergonne's 1815 paper on the design and analysis of polynomial regression experiments. *Historia Mathematica*, **1**, 431–447.

Stigler, S. M. (1986). *The History of Statistics*. Boston: Harvard University/Belknap Press.

Swed, F. S., and C. Eisenhart (1943). Tables for testing randomness of grouping in a sequence of alternatives. *Annals of Mathematical Statistics*, **14**, 66–87.

Swindel, B. F. (1981). Geometry of ridge regression. *The American Statistician*, **35**, 12–15. (See, also, pp. 268–269.)

Teissier, G. (1948). La relation d'allometrie sa signification statistique et biologiques. *Biometrics*, **4**, 14–48 (discussion, pp. 48–53).

Tibshirani, R. (1996). Regression shrinkage and selection via the lasso. *Journal of the Royal Statistical Society, Series B*, **58**, 267–288.

Tukey, J. W. (1957). On the comparative anatomy of transformations. *Annals of Mathematical Statistics*, **28**, 602–632.

von Bertalanffy, L. (1941). Stoffwechseltypen und Wachstumstypen. *Biologisches Zentralblatt*, **61**, 510–532.

von Bertalanffy, L. (1957). Quantitative laws in metabolism and growth. *Quarterly Review of Biology*, **32**, 218–231.

Wald, A. (1940). The fitting of straight lines if both variables are subject to error. *Annals of Mathematical Statistics*, **11**, 284–300.

Walker, E., and J. B. Birch (1988). Influence measures in ridge regression. *Technometrics*, **30**, 221–227.

Waterman, M. S. (1974). A restricted least squares problem. *Technometrics*, **16**, 135–136.

Wei, W. S. (1990). *Time Series Analysis, Univariate and Multivariate Methods*. Reading, MA: Addison-Wesley.

Weisberg, S. (1985). *Applied Linear Regression*. New York: Wiley.

Welsch, R. E. (1975). Confidence regions for robust regression. *Statistical Computations Section Proceedings of the American Statistical Association*, Washington, DC.

Westfall, P., and S. Yound (1992). *Resampling-Based Multiple Testing*. New York: Wiley.

Wilkie, D. (1965). Complete set of leading coefficients, $\lambda(r, n)$ for orthogonal polynomials up to $n = 26$. *Technometrics*, **7**, 644–648.

Wilkinson, J. H., and C. Reisch (eds.) (1971). *Handbook for Automatic Computation, Volume II: Linear Algebra*. New York: Springer Verlag. (For article by F. L. Bauer, see pp. 119–133.)

Willan, A. R., and D. G. Watts (1978). Meaningful multicollinearity measures. *Technometrics*, **20**, 407–412.

Willett, J. B., and J. D. Singer (1988). Another cautionary note about R^2: its use in weighted least-squares regression analysis. *The American Statistician*, **42**, 236–238.

Williams, E. J. (1959). *Regression Analysis*. New York: Wiley.

Wishart, J., and T. Metakides (1953). Orthogonal polynomial fitting. *Biometrika*, **40**, 361–369.

Wong, M. Y. (1989). Likelihood estimation of a simple linear regression model when both variables have error. *Biometrika*, **76**, 141–148.

Woods, H., H. H. Steinour, and H. R. Starke (1932). Effect of composition of Portland cement on heat evolved during hardening. *Industrial and Engineering Chemistry*, **24**, 1207–1214.

Yohai, V. J. (1987). High breakdown point and high efficiency robust estimators for regression. *Annals of Statistics*, **15**, 642–656.

Yohai, V. J., and R. H. Zamar (1988). High breakdown point estimates of regression by means of the minimization of an efficient scale. *Journal of the American Statistical Association*, **83**, 406–413.

Yound, G. A. (1994). Bootstrap: more than a stab in the dark? *Statistical Science*, **9**, 382–415.

True/False Questions

1. If data on (Y, X) are available at only two values of X, the models $Y = \beta_0 + \beta_1 X + \epsilon$ and $Y = \beta_1 X + \beta_2 X^2 + \epsilon$ will fit the data equally well.

2. A "$H_0 : \beta = 0$" t-test on one parameter can also be performed as an F-test, and $t^2 = F$ for the values of the test statistics.

3. If a $(1 - \alpha)$ confidence interval for slope β_1 contains zero, we will not reject $H_0 : \beta_1 = 0$ versus $H_1 : \beta_1 \neq 0$ at the α level.

4. If we fit the model $Y = \beta_0 + \epsilon$ to a set of data, we will always get $b_0 = \overline{Y}$ and $\hat{Y} = \overline{Y}$.

5. Choice of a test level α implies that we expect our test decision to be wrong $100\alpha\%$ of the time when we reject H_0.

6. When $R^2 = 1$, all the data lie on a line of positive, negative, or zero slope.

7. In regression work we are looking for an empirical relationship between a response and one or more predictor variables.

8. A straight line $\hat{Y} = b_0 + b_1 X$ fitted by least squares must contain the point $(\overline{X}, \overline{Y})$.

9. The assumption that the errors in the model are normally distributed is not needed for the construction of the analysis of variance table, which is simply an algebraic breakup of ΣY_i^2.

10. There are always exactly as many normal equations as there are parameters in the model.

11. $\Sigma(X_i - \overline{X})(Y_i - \overline{Y}) = \Sigma X_i Y_i - n\overline{XY}$.

12. If we fit the model $Y = \beta X + \epsilon$ to a set of data (X_i, Y_i), $i = 1, 2, \ldots, n$, the sum of the residuals is not necessarily zero.

13. In regression work, the assumption that the errors in Y are normally distributed is needed to validate use of F- and t-tests.

14. Even if we are using the wrong regression model, repeat runs can be used to provide an estimate of σ^2.

15. We obtain the "lack of fit" degrees of freedom by subtracting the "pure error" degrees of freedom from the "residual" degrees of freedom.

16. The R^2 statistic is not the average of the squared correlations of Y with X_1, Y with X_2, $\ldots$, Y with X_k, where $X_1, X_2, \ldots, X_k$ are the X's being fitted in the model.

17. A correctly calculated "joint confidence region" for β_0 and β_1 (say) has an elliptical shape.

18. The confidence interval statements we make about individual β's depend on the $\epsilon \sim N(\mathbf{0}, \mathbf{I}\sigma^2)$ assumption.

19. Even if we fit the wrong model, it would be possible for some or all of the parameter estimates we got to be unbiased (by the terms omitted).

20. The formula $\hat{V}(\mathbf{b}) = (\mathbf{X'X})^{-1}s^2$ provides the estimated variances and pairwise covariances of the b's we have fitted.

21. $H_0 : \beta_1 = \beta_2\beta_3$ is not a linear hypothesis.

22. The extra $SS(\mathbf{b}_2|\mathbf{b}_1)$, obtained from fitting the model $\mathbf{Y} = \mathbf{X}_1\boldsymbol{\beta}_1 + \mathbf{X}_2\boldsymbol{\beta}_2 + \epsilon$, is exactly the same as the regression sum of squares for the model $\mathbf{Y} = \mathbf{X}_2\boldsymbol{\beta}_2 + \epsilon$, provided that $\mathbf{X}_2'\mathbf{X}_2 = \mathbf{0}$.

23. *Both* of the following are true for a linear model: (a) $\Sigma e_i = 0$ when there is a β_0 in the model; (b) $\Sigma e_i \hat{Y}_i = 0$ *always*.

24. Even though we assume that the errors ϵ_i in a linear model are pairwise uncorrelated, this is not typically true of the residuals e_i, $i = 1, 2, \ldots, n$.

25. The plot of e_i versus Y_i always has a slope of size $1 - R^2$ in it.

26. The Durbin–Watson statistic always lies between 0 and 4, no matter what the (nonsingular) linear regression problem may be.

27. If we have ten residuals, five positive and five negative, there are only two ways that two runs can occur out of the 252 possibilities of rearrangements of signs. (Assume 252 is right; don't worry about *that* aspect.)

28. When we add a second predictor X_2 to a model $Y = \beta_0 + \beta_1 X_1 + \epsilon$, the value of s^2 may go down, or up, but the value of R^2 cannot go down.

29. Confidence bands for the true mean of Y given X can be evaluated for a straight line model fit, and similar bands exist when there are two or more predictors.

30. A test of $H_0 : \beta_2 = 0$ versus $H_1 : \beta_2 \neq 0$ in the model $Y = \beta_0 + \beta_1 X_1 + \beta_2 X_2 + \epsilon$ can be made either in (a) extra sum of squares F-test form, or (b) $t = b_2/(\text{Est. Var } (b_2))^{1/2}$ form, where $t = F^{1/2}$.

31. On a normal probability plot, a line should be drawn through the "middle bulk" of the residuals as a check on normality.

32. An observation can be both influential and an outlier.

33. An observation can be influential but not an outlier.

34. An observation can be an outlier but not influential.

35. If we fit a straight line $Y = \beta_0 + \beta_1 X + \epsilon$, and we find that $b_0 = 0$ exactly, then the residuals will still add to zero.

36. We can justify the use of least squares when $\epsilon \sim N(0, \mathbf{I}\sigma^2)$ via the application of maximum likelihood.

37. The extra sum of squares for b_1 and b_2 (say) given b_3, b_4, $\ldots$, b_q has two degrees of freedom, no matter how many other b's are "given."

38. $F(1, 22) = \{t(22)\}^2$.

39. The specific values of the (sequential) sums of squares for a series of input variables in a regression may be changed if we change the "order of entry" of those input X's.

40. $b_w = (\mathbf{X'V}^{-1}\mathbf{X})^{-1}\mathbf{X'V}^{-1}\mathbf{Y}$ is the "generalized" least squares solution where we assume $\epsilon \sim N(0, \mathbf{V}\sigma^2)$.

41. $(\mathbf{Y} - \hat{\mathbf{Y}})'\mathbf{1} = 0$ is true when the model contains a β_0 term.

42. The R^2 statistic is the square of the correlation between the $\mathbf{Y}$ and $\hat{\mathbf{Y}}$ columns.

43. The rectangle formed by individual confidence intervals for β_1 and β_2 (say) is not a correct "joint region" for the pair (β_1, β_2).

44. The formula $E(\mathbf{b}) = \boldsymbol{\beta} + (\mathbf{X'X})^{-1}\mathbf{X'X}_2\boldsymbol{\beta}_2$ tells us the bias effect on $\mathbf{b}$ of failing to include the terms $\mathbf{X}_2\boldsymbol{\beta}_2$ in the model.

45. If, in the model $\mathbf{Y} = \mathbf{X}\boldsymbol{\beta} + \epsilon$, $\epsilon \sim N(0, \mathbf{I}\sigma^2)$, then the elements of $\mathbf{b} = (\mathbf{X'X})^{-1}\mathbf{X'Y}$ are also normally distributed.

46. The vector of fitted values $\hat{\mathbf{Y}}$ is always orthogonal to the vector of residuals $\mathbf{e}$.

47. When we fit the model $Y = \beta_0 + \epsilon$ (i.e., no X's) to a set of data, $R^2 = 0$, always.

48. When $R^2 = 1$, all the residuals must be zero.

49. If there are n observations and e degrees of freedom for pure error, a unique fitted linear model can contain no more than $(n - e)$ parameters, including β_0.

50. The linear hypothesis $H_0 : \beta_1 - \beta_2 = \beta_2 - \beta_3 = \beta_3 - \beta_4 = 0$ will take up three degrees of freedom in a formal test.

51. The model $Y = \beta_0 + \beta_1 X_1 + \beta_2 X_2 + \beta_3 X_2 \ln X_1 + \epsilon$ is a linear model.

52. The model $Y = \beta_0 + \beta_1 X + \beta_2(\beta_3)^X + \epsilon$ is not a linear model.

53. If we "know" that σ_Y is proportional to the kth power of the response then we can use Y^{1-k} as a variance stabilizing transformation.

54. For binomial data, arcsin $(Y^{1/2})$ is a useful variance stabilizing transformation.

55. The model $Y = \theta + \alpha X_1 X_3 + \beta X_1 + \alpha \beta X_2 + \epsilon$ with parameters (θ, α, β) is nonlinear.

56. The model $Y = \beta_0 + \beta_1(X_1 - X_2) + \beta_2(X_1 - X_2)^2 + \epsilon$ is a linear model.

57. If a model $Y = \beta_0 + \beta_1 X + \beta_2 X^2$ is fitted and a t-test indicates we should keep the $b_2 X^2$ term in, we should also retain the $b_1 X$ term, significant t-test or not.

58. The transformation form $V = (Y^\lambda - 1)/(\lambda \dot{Y}^{\lambda-1})$ for $\lambda \neq 0$ is continuous at $\lambda = 0$, where $\dot{Y}$ is exp $\{n^{-1} \sum_{i=1}^{n} \ln Y_i\}$.

59. Even though we choose the best value of λ in the transformation form $V = (Y^\lambda - 1)/(\lambda \dot{Y}^{\lambda-1})$, we may still not get a good regression fit.

60. The $\sin^{-1}(Y^{1/2})$ transformation stabilizes the variance if the original data Y_i are binomial data.

61. After a set of regression data has been "centered and scaled," the coefficients in the normal equations are *all* correlations ρ, such that $-1 \leq \rho \leq 1$.

62. In "analysis of variance" problems, we should always work out the residuals and examine them after fitting the model.

63. "Analysis of variance" models are usually overparameterized, that is, have more parameters then we can estimate uniquely.

64. To take account of level differences between two groups of observations, we can make use of a dummy variable Z such that $Z = -1$ for the first group, and $Z = 1$ for the second group.

65. Fitting the model $Y = \beta_0 + \beta_1 X_1 + \beta_2 X_2 + \epsilon$ and fitting the model $Y = \beta_0' + \beta_1(X_1 - \bar{X}_1) + \beta_2(X_2 - \bar{X}_2) + \epsilon$, by least squares, both lead to the same fitted equation.

66. The correlation matrix of the estimated regression coefficients is a rescaled form of the $(\mathbf{X'X})^{-1}$ matrix.

67. If two columns of the $\mathbf{X}$ matrix are proportional to each other [e.g., column $A = 4$ (column B)], then the determinant of $\mathbf{X'X}$ will be zero.

68. Even if the determinant of $\mathbf{X'X}$ is zero, it is still possible to solve the least squares normal equations $\mathbf{X'Xb} = \mathbf{X'Y}$ but the solution will not be unique.

69. The backward elimination (selection) procedure can leave out of consideration a good combination of the X's that it "never gets to."

70. The "usual" F-statistic percentage point does not provide an accurate determination of the true percentage value at which entry or exit (removal) tests are made in selection procedures.

71. Ridge regression calculations for $\theta \neq 0$ provide residual sums of squares that are always greater than the least squares values.

72. In a certain metric, ridge regression can be regarded as "least squares subject to a restriction that the sum of squares of the parameters (except the intercept) is restricted to a spherical region."

73. If X is measured over a small range and the corresponding regression coefficient estimate b is *not* significant, X may, nevertheless, have an important effect on Y, which the data do not reveal.

74. The C_p statistic $[\text{RRS}_p/s^2 - (n - 2p)]$ will always take the value p exactly for at least one regression equation in the "all regressions" case.

75. Use of the backward elimination method may reveal ill-conditioning in the data at the first step.

76. $S_{YY} = (n - p)s^2/(1 - R^2)$ can be used to calculate S_{YY} from any regression printout, but it may be subject to some round-off error.

77. To fit a planar model uniquely in mixture ingredients X_1, X_2, X_3 such that $X_1 + X_2 + X_3 = 1$, we cannot use the model function $\beta_0 + \beta_1 X_1 + \beta_2 X_2 + \beta_3 X_3$ but must modify it.

78. $\hat{\mathbf{Y}} = \mathbf{Xb}$ lies in the estimation space and is always unique, even when $\mathbf{b}$ is not unique.

79. $\mathbf{e} = \mathbf{Y} - \hat{\mathbf{Y}}$ lies in the error space.

80. The estimation space is spanned (defined by) the column vectors of the $\mathbf{X}$ matrix.

81. Even if the columns of $\mathbf{X}$ are linearly dependent, they still define the estimation space.

82. If $\mathbf{M} = \mathbf{X}'\mathbf{X}$ is singular, we can still find a generalized inverse $\mathbf{M}^-$ such that $\mathbf{M}\mathbf{M}^-\mathbf{M} = \mathbf{M}$.

83. A generalized inverse is not unique.

84. If $\mathbf{Y}$ is a vector and $\mathbf{P}$ is a projection matrix, $\mathbf{P}\mathbf{Y} = \mathbf{P}^2\mathbf{Y} = \cdots = \mathbf{P}^m\mathbf{Y}$.

85. The fact that $\mathbf{P}\mathbf{Y} = \mathbf{P}^2\mathbf{Y} = \mathbf{P}(\mathbf{P}\mathbf{Y})$ means that, after the $\mathbf{Y}$ vector has been projected onto $R(\mathbf{P})$, projecting it again does not change it, as it is already in $R(\mathbf{P})$.

86. If $\mathbf{P} = \mathbf{X}(\mathbf{X}'\mathbf{X})^{-1}\mathbf{X}'$ is the projection matrix for a regression problem with model containing a β_0, the rows and columns of $\mathbf{P}$ all add to 1.

87. If $\mathbf{X}$ is not of full rank, we can still estimate some linear *combinations* of the β's uniquely, but not all of the individual β's.

88. A linear combination $\mathbf{c}'\boldsymbol{\beta}$ of the parameters can *always* be estimated provided $\mathbf{c}'$ is expressible as a linear combination of the rows of $\mathbf{X}$, and even if $\mathbf{X}$ is not of full rank.

89. Analysis of variance models are typically overparameterized, and some choice of suitable dummy variables must be made for a nonsingular regression treatment.

90. Residuals should be examined in analysis of variance type problems as well as in general regression problems.

91. "Analysis of variance" data analyses are really just variations of regression problems.

92. A one-way classification typically consists of several groups or "treatments" with multiple observations in each group.

93. The one-way classification model $Y_{ij} = \mu + t_i + \epsilon_{ij}$ must have a restriction attached to it and this restriction is not unique.

94. To begin doing a nonlinear estimation problem some "starting values" or "initial values" are needed.

95. Linear least squares approximations are used to get approximate confidence intervals for the parameters in nonlinear estimation problems.

96. $Y = \beta_1 \exp\{\beta_2 t\} + \epsilon$ is a nonlinear model.

97. The least squares method is valid for estimating the parameters of a nonlinear regression model.

98. Iterative procedures are needed to solve a general nonlinear regression problem.

99. The sum of squares surface for a nonlinear model is not an ellipsoidal "bowl" in general.

100. The model $Y = \exp(\theta_1 + \theta_2 t) + \epsilon$ is nonlinear in θ_1 and θ_2.

101. Fitting the model $\ln Y = \theta_1 + \theta_2 t + \epsilon$ would usually provide reasonable starting values for fitting the model $Y = \exp(\theta_1 + \theta_2 t) + \epsilon$.

Answers to Exercises

Chapters 1–3

A. 1. $b_1 = \frac{158}{110} = 1.44$; $b_0 = \frac{102}{11} = 9.27$; $\hat{Y} = 9.27 + 1.44X$.

2.

Analysis of Variance

Source of Variation	df	SS	MS	F
Total (corrected)	10	248.18		
Regression	1	$\dfrac{(158)^2}{110}$	226.95 ⌐	96.17*
Residual	9	21.23	2.36 ⌐	

*The hypothesis, $H_0: \beta_1 = 0$ is tested with $\alpha = 0.05$ by comparing the computed $F(1, 9)$ statistic with the critical $F(1, 9)$ for $\alpha = 0.05$. From the $\alpha = 0.05$ F-table, we find $F(1, 9, 0.95) = 5.12$. Since 96.17 is greater than 5.12, reject the hypothesis $\beta_1 = 0$.

3. The 95% confidence limits for β_1 are

$$1.11 \le \beta_1 \le 1.77.$$

4. The 95% confidence limits for the true average value of Y at $X_0 = 3$ are

$$12.15 \le \text{true average } Y \text{ at } X_0 = 3 \le 15.03.$$

5. The 95% confidence limits for the difference between the true average value of Y at $X_1 = 3$ and the true average value of Y at $X_2 = 2$. First determine the algebraic difference between $\hat{Y}_1$ and $\hat{Y}_2$:

$$\hat{Y}_1 = b_0 + b_1(3), \qquad \hat{Y}_2 = b_0 + b_1(-2).$$

Thus

$$\hat{Y}_1 - \hat{Y}_2 = b_1(3 + 2) = 5b_1 = 5(1.44) = 7.20,$$

$$s^2_{(\hat{Y}_1 - \hat{Y}_2)} = 25s^2_{b_1} = 25\left(\frac{2.36}{110}\right) = 0.53635,$$

$$s_{(\hat{Y}_1 - \hat{Y}_2)} = \sqrt{0.53635} = 0.732,$$

$$ts_{(\hat{Y}_1 - \hat{Y}_2)} = (2.262)(0.732) = 1.656.$$

Thus the 95% confidence band on the true difference is

$$7.20 - 1.66 \le \text{true difference} \le 7.20 + 1.66$$
$$5.54 \le \text{true difference} \le 8.86.$$

6. Calculate the residuals and look for patterns.

X	Y	$\hat{Y}$	$Y - \hat{Y}$
−5	1	2.07	−1.07
−4	5	3.51	1.49
−3	4	4.95	−0.95
−2	7	6.39	0.61
−1	10	7.83	2.17
0	8	9.27	−1.27
1	9	10.71	−1.71
2	13	12.15	0.85
3	14	13.59	0.41
4	13	15.03	−2.03
5	18	16.47	1.53

There is no obvious alternative to the model.

7. If the tentative assumption of a first-order model is reasonable, there is little point in using eleven different experimental levels. Of course, we need at least two levels to estimate the parameters in the model, and at least one more to detect curvature in the true model, if curvature exists. By taking repeat observations at some or all levels, we can obtain a pure error estimate of σ^2 to use in checking lack of fit. Thus for an experiment of about the same size, one possibility would be to choose three widely spaced levels—the extremes of the X-range and the center, for example—and to take four observations at each of these levels. This would lead to an analysis of variance table of the form below.

ANOVA ($n = 12$)

Source of Variation	df
Total (corrected)	11
Regression	1
Residual	10
Lack of fit	1
Pure error	9

Since we now have only 1 degree of freedom for lack of fit, this is not entirely satisfactory. Slightly better would be the choice of three runs at each of four levels:

ANOVA ($n = 12$)

Source of Variation	df
Total (corrected)	11
Regression	1
Residual	10
Lack of fit	2
Pure error	8

There are many other possibilities. See Section 1.8.

B. 1. Randomized order.

2a. $\hat{Y} = 0.5 + 0.5X$.

2b.

Source	df	SS	MS
Corrected total $\Sigma(Y_i - \bar{Y})^2$	19	83.2	
Due to regression $\dfrac{[\Sigma(X_i - \bar{X})(Y_i - \bar{Y})]^2}{\Sigma(X_i - \bar{Y})^2}$	1	40.0	40.0*
Residual	18	43.2	2.4

2c. (1) $\hat{Y} = 3.0 \pm 0.73 = 2.27$ to 3.73.

   (2) $\hat{Y} = 5.0 \pm 1.26 = 3.74$ to 6.26.

3a.

Residual	18	43.2	2.4
Lack of fit	3	1.2	0.4 ⌐ NS
Pure error	15	42.0	2.8 ⌐

No significant lack of fit.

3b. Yes.

4a. Same ANOVA and conclusion as in 3a.

4b. Confidence limits not applicable. Error variance is dependent on level of Y.

4c. First-order model suitable

5a.

Residual	18	43.2	2.4
Lack of fit	3	20.0	6.67 ⌐ *
Pure error	15	23.2	1.55 ⌐

Significant lack of fit indicates inadequacy of model.

5b. Since the model is incorrect, confidence intervals will be invalid.

5c. A second-order model is suggested.

C. Best fitting straight line is

$$\hat{Y} = b_0 + b_1 X = 129.7872 - 24.0199X.$$

ANOVA

Source	df	SS	MS	F
Total	12	3396.62		
Regression	1	3293.77	3293.77	
Residual	11	102.85	9.35	
Lack of fit	5	91.08	18.22	9.30*
Pure error	6	11.77	1.96	

The model is inadequate.

D. First note that the sum of squares function for the no-intercept model is

$$\Sigma(Y_i - \beta X_i)^2 = \Sigma\{Y_i - bX_i + bX_i - \beta X_i\}^2$$

$$= \Sigma(Y_i - bX_i)^2 + (b - \beta)^2 \Sigma X_i^2$$

$$+ 2\Sigma(Y_i - bX_i)(b - \beta)X_i.$$

The cross-product term can be rewritten as

$$2(b - \beta)\Sigma(X_i Y_i - bX_i^2) = 2(b - \beta)\{\Sigma X_i Y_i - b\Sigma X_i^2\}$$

and vanishes by definition of b. What is left is a function of β that is minimized when $\beta = b$. This proves the least squares result. For the rest of the question, we see first that $n + 1 = (1 + a)^2$. If we use (U, V) to denote the new data, then, for example,

$$\bar{U} = (\text{Sum of all new observations})/(n + 1)$$

$$= (n\bar{X} + n\bar{X}/a)/(n + 1)$$

$$= n\bar{X}/\{a(1 + a)\},$$

$$S_{UU} = \Sigma X_i^2 + (m\bar{X})^2 - (n + 1)\bar{U}^2$$

$$= \Sigma X_i^2 + n^2\bar{X}^2/a^2 - n^2\bar{X}^2(1 + a)^2/\{a^2(1 + a)^2\}$$

$$= \Sigma X_i^2.$$

Similarly $S_{UV} = \Sigma X_i Y_i$ and so $b_1 = S_{UV}/S_{UU} = b$.

   (The estimate $b_0 = \bar{V} - b\bar{U} = n(\bar{Y} - b\bar{X})/\{a(1 + a)\}$ is not zero in general. Ignore it! This trick gets the right slope, but not the right zero intercept in general!)

For our example, the equations obtained are $\hat{Y} = 0.22857X$ for the three data points, and $\hat{Y} = 0.1714 + 0.22857X$ for the four data points. Note that the standard error and t-statistic for $b = 0.22857$ remain the same. You might consider why this happens.

E. There are only seven observations, so one cannot hope for too much here. For response Y_2, we get:

1. $\hat{Y} = 38.067 - 358.2X$. We leave the plot to the reader.

2. Residuals 19.36, 2.11, 0.25, -15.99, -13.55, -0.11, 7.92. Their sum is -0.01, zero within rounding error.

3.

Source	df	SS	MS	F	
$b_1\|b_0$	1	891.5	891.5	5.06	$(p = 0.074)$
Residual	5	881.4	176.3	—	
Total, corrected		1772.8	—	—	

4. $se(b_0) = 9.146$ $se(b_1) = 159.3$

5. $se(\hat{Y}) = \left\{ \dfrac{1}{7} + \dfrac{(X_0 - 0.048)^2}{0.006946} \right\}^{1/2} (176.3)^{1/2}.$

X_0	$\hat{Y}$	$se(\hat{Y})$	95% Limits ($t = 2.571$)
0.01	34.49	7.864	14.27, 54.71
0.03	27.32	5.780	12.46, 42.18
0.05	20.16	5.029	7.23, 33.09
0.07	12.99	6.121	-2.98, 28.96
0.09	5.83	8.364	-15.67, 27.33
0.10	2.25	9.686	-22.65, 27.15

6. For F see (3). Note that $t = -358.2/159.3 = -2.249 = -F^{1/2}$, $R^2 = 891.5/1772.8 = 0.5029$.

7. The pure error SS $= 162.33 + 21.78 = 184.1$ with $2 + 1 = 3$ df.
The lack of fit SS $= 881.4 - 184.1 = 697.3$ with $5 - 3 = 2$ df.
$F = (697.3/2)/(184.1/3) = 5.68$ with $(2,3)$ df, close to the 10% upper-tail point of 5.46. Technically then, no lack of fit is shown. Of course, we are working with very few df here.

8. The straight line gives some basis ($p = 0.07$ for the F-test) for the idea that the trend is downward, that is, that higher costs for alcohol are associated with fewer deaths. Whether the deaths really depend on the costs is not known. It would be nice to think we *could* reduce the number of deaths by raising the price. The early data vary much more than the later data. This would seem to deny our assumption that the σ^2 is constant, but with so few points we cannot be sure. Obviously the practical conclusion is to raise prices in France, Italy, Germany, and Belgium and see what happens!

F. 1. $\hat{Y} = -21.33 + 5X$

2. $2.984 \le \beta_1 \le 7.016$

ANOVA

Source	df	SS	MS	F
Corrected total	11	69.67		
Due to regression	1	52.50	52.50	
Residual	10	17.17	1.72	
Lack of fit	4	5.50	1.375	0.706 (not significant
Pure error	6	11.67	1.945	at $\alpha = 0.05$)

The model seems to be adequate.

G. 1. $\hat{Y} = 323.628 + 131.717X$.

ANOVA

Source	df	SS	MS	F
Corrected total	16	2,305,042		
Due to regression	1	1,099,641.1	1,099,641.10 ⌐	13.68 significant
Residual	15	1,205,400.9	80,360.06 ⌐	at $\alpha = 0.05$

2.

ANOVA

Source	df	SS	MS	F
Corrected total	16	2,305,042		
Due to regression	1	1,099,641.1		
Residual	15	1,205,400.9		
Lack of fit	1	520,648.6	104,129.72	1.52 not significant
Pure error	10	684,752.3	68,475.23	at $\alpha = 0.05$

A straight line relationship seems reasonable.

H. Prediction equation: $\hat{Y} = 1.222 + 0.723\,X$

ANOVA

Source	df	SS	MS	F	$F_{0.95}$
Corrected total	13	2.777			
Due to regression	1	1.251	1.251 ⌐	9.850	4.75
Residual	12	1.526	0.127 ⌐		

$9.850 > F(1, 12, 0.95) = 4.75$; $\therefore$ reject $H_0 : \beta_1 = 0$ if no lack of fit.

ANOVA

Source	df	SS	MS	F	$F_{0.95}$
Corrected total	13	2.777			
Due to regression	1	1.251			
Residual	12	1.526			
Lack of fit	7	0.819	0.117 ⌐	0.830	4.88
Pure error	5	0.707	0.141 ⌐		

$0.830 < F(7, 5, 0.95) = 4.88$, $\therefore$ lack of fit is not significant.
Conclusion: Use the prediction equation

$$\text{Cup loss}(\%) = 1.222 + (0.723)[\text{bottle loss}(\%)].$$

I. Prediction equation: $\hat{Y} = 17.146 + 11.836X$.

ANOVA

Source	df	SS	MS	F	$F_{0.95}$
Corrected total	12	22,126.308			
Due to regression	1	6,034.379	6,034.379	4.125	4.840
Residual	11	16,091.929	1,462.903		

$4.125 < F(1, 11, 0.95) = 4.840$; $\therefore$ do not reject $H_0 : \beta_1 = 0$. The regression is not significant.

$$R^2 = \frac{\text{SS due to regression}}{\text{Corrected total SS}} = \frac{6034.379}{22,126.308} = 27.27\%.$$

Conclusions: (*i*) The model is not useful. (*ii*) Further investigation of alternative variables will be necessary. (*iii*) Check pure error.

J. 1. $\hat{Y} = 2.5372000 - 0.004718X$

2.

ANOVA

Source	df	SS	MS	F	$F_{0.95}$
Corrected total	14	0.209333			
Due to regression	1	0.110395	0.110395	14.50	4.67
Residual	13	0.098938	0.007611		

$14.50 > F(1, 13, 0.95) = 4.67$; ∴ reject $H_0: \beta_1 = 0$. The regression is significant, if there is no lack of fit.

ANOVA

Source	df	SS	MS	F	$F_{0.95}$
Residual	13	0.098938			
Lack of fit	5	0.018938	0.003788	0.38	3.69
Pure error	8	0.080000	0.010000		

$0.38 < F(5, 8, 0.95) = 3.69$; ∴ Lack of fit not significant.

3. The 95% confidence interval on the true mean value of Y, calculated at four points: $X = 0, X = \overline{X}, X = 400, X = 460$:

At $X = 0$ $\hat{Y} \pm (2.160)(0.527) = \hat{Y} \pm 1.138$
At $X = \overline{X}$ $\hat{Y} \pm (2.160)(0.022) = \hat{Y} \pm 0.048$
At $X = 400$ $\hat{Y} \pm (2.160)(0.039) = \hat{Y} \pm 0.084$
At $X = 460$ $\hat{Y} \pm (2.160)(0.048) = \hat{Y} \pm 0.104$

K. 1. Plot the data and draw line by eye. The line drawn by eye here may well be somewhat different from that fitted by least squares later.

2. $\Sigma X_u = 1244.5$ $\Sigma Y_u = 30.458$
     $\Sigma X_u^2 = 73{,}920.05$ $\Sigma Y_u^2 = 27.573638$
$\Sigma X_u Y_u = 1032.4865$

For later numerical work it is wise to keep all digits in the sums of squares calculations.

3. $b_1 = -0.00290351$
$b_0 = \overline{Y} - b_1 \overline{X} = 1.00210$

The line contains (for example) the points (0, 1.0021) and (100, 0.7117).

4. There do not appear to be any peculiarities severe enough to warrant corrective action. (There is a slight tendency for small residuals to be associated with small X's, however, and this might bear further investigation.)

5.

ANOVA

Source	df	SS	MS
Regression (b_0)	1	27.28500	
Regression ($b_1 \mid b_0$)	1	0.23914	0.23915
Residual	32	0.04950	$s^2 = 0.001547$
Total	34	27.57364	

6. $se(b_1) = s/\{\Sigma X_u^2 - (\Sigma X_u)^2/n\}^{1/2} = 0.00023$,
$se(b_0) = s/[\Sigma X_u^2/\{n\Sigma X_u^2 - (\Sigma X_u)^2\}]^{1/2} = 0.01089$.

7. $se(\hat{Y}_0) = s\{1/n + (X_0 - \overline{X})^2/(\Sigma X_u^2 - (\Sigma X_u)^2/n)\}^{1/2}$.

The formula for any particular X_0 is obtained by substituting for the known quantities (except X_0). Then the 95% confidence band for the true mean value of Y at X_0 is given by

$$\hat{Y} \pm t(32, 0.975)se(\hat{Y}_0).$$

For $t(32, 0.975)$ we can use $t(30, 0.975) = 2.042$ or interpolate in the table. The plot of confidence limits will look something like Figure 3.1.

8. The F-test statistic for overall regression is $0.23915/0.001547 = 154.6$, as compared to $F(1, 30, 0.95) = 4.17$. We could interpolate for 32 df, but it is clearly not necessary here. We therefore reject the null hypothesis that $\beta_1 = 0$. $R^2 = 0.83$, so that 83% of the variation about the mean $\overline{Y}$ is explained by our linear regression.

L. We calculate the approximate pure error SS as 0.01678 with 10 df. So we have the table:

ANOVA

Source	df	SS	MS
Total (corrected)	33	0.2886	
Regression ($b_1\|b_0$)	1	0.23915	0.23915
Residual	32	0.04950	$s^2 = 0.001547$
Lack of fit	22	0.03272	$MS_L = 0.001487$
Pure error	10	0.01678	$s_e^2 = 0.001678$

To test lack of fit we calculate an F-statistic of $MS_L/s_e^2 = 0.8862$; clearly no lack of fit is indicated since $F(22, 10, 0.95) = 2.75$. We thus pool the lack of fit SS with the pure error SS to calculate s^2. *Conclusion:* The data appear to be adequately described by a straight line regression of Y upon X. One could use the fitted relationship for predicting the true mean value of Y for any particular X_0, and the confidence bands drawn previously give an indication of what the accuracy of such predictions would be, assuming the model to be correct.

M. The conclusions are that (1) R^2 can equal 1 if there are no repeat runs in the data, but (2) R^2 cannot achieve 1 if nonidentical repeat runs exist, because the model cannot explain the pure error sum of squares. These conclusions are true in general regression situations as the following algebra shows.

Let the observations be

$Y_{11}, Y_{12}, \ldots, Y_{1n_1}$ at the first location in X-space

$Y_{21}, Y_{22}, \ldots, Y_{2n_2}$ at the second location in X-space

$\vdots$

$Y_{k1}, Y_{k2}, \ldots, Y_{kn_k}$ at the kth location in X-space

$$R^2 = 1 - \frac{\sum_{r=1}^{k} \sum_{u=1}^{n_r} (Y_{ru} - \hat{Y}_{ru})^2}{\sum_{r=1}^{k} \sum_{u=1}^{n_r} (Y_{ru} - \overline{Y}_{ru})^2} = 1 - \frac{\text{Residual SS}}{\text{Total corrected SS}}.$$

Now $\hat{Y}_{ru} = \hat{Y}_r$, the same for each u. Let $\overline{Y}_r = \sum_{u=1}^{n_r} Y_{ru}/n_r$ be the mean response at the rth location. Then

$$\sum_{u=1}^{n_r} (Y_{ru} - \hat{Y}_{ru})^2 = \sum_{u=1}^{n_r} \{(Y_{ru} - \overline{Y}_r) + (\overline{Y}_r - \hat{Y}_r)\}^2$$

$$= \sum_{u=1}^{n_r} (Y_{ru} - \overline{Y}_r)^2 + \sum_{u=1}^{n_r} (\overline{Y}_r - \hat{Y}_r)^2$$

$$+ 2(\overline{Y}_r - \hat{Y}_r) \sum_{u=1}^{n_r} (Y_{ru} - \overline{Y}_r)$$

$$= \sum_{u=1}^{n_r} (Y_{ru} - \overline{Y}_r)^2 + n_r(\overline{Y}_r - \hat{Y}_r)^2,$$

the last summation being zero.

Thus

$$R^2 = 1 - \frac{\sum_{r=1}^{k} \sum_{u=1}^{n_r} (Y_{ru} - \overline{Y}_r)^2 + \sum_{r=1}^{k} n_r (\overline{Y}_r - \hat{Y}_r)^2}{\sum_{r=1}^{k} \sum_{u=1}^{n_r} (Y_{ru} - \overline{Y})^2}.$$

It follows that R^2 can attain 1 if (i) each $Y_{ru} = \overline{Y}_r$ and (ii) $\overline{Y}_r = \hat{Y}_r$.

(i) is true, however, only if either there are no repeats, that is, all $n_r = 1$, or all repeats at a location are identical, for *all* locations.

(ii) implies the model fits all the means perfectly, which can happen, sometimes.

Thus, in general, when pure error exists, $R^2 < 1$.

N. Test for lack of fit before checking the regression. If there *is* lack of fit, the F-test for regression and the calculations for confidence intervals and so forth are not valid. We calculate pure error as follows:

X	Pure Error Contribution		df
10	$\frac{1}{2}[-2 - (-4)]^2$	$= 2$	1
20	$\frac{1}{2}[1 - 3]^2$	$= 2$	1
30	$\frac{1}{2}[2 - 5]^2$	$= 4.5$	1
40	$0^2 + 1^2 + 2^2 - 3^2/3$	$= 2$	2
50	$(-2)^2 + (-3)^2 + (-4)^2 - \frac{(-9)^2}{3}$	$= 2$	2
	Pure error	$= 12.5$	7

Split-up of Residual SS

Source	df	SS	MS	F
Lack of fit	3	73.177	24.392	$F(3, 7) = 13.66$, significant lack of fit shown
Pure error	7	12.5	1.786	
Residual	10	85.677		

The model suffers from lack of fit; the next step is to plot residuals and check patterns, to see if the model can be improved.

O. The fitted equation is $\hat{Y} = 2.0464 + 0.1705X$.

ANOVA

Source	df	SS	MS	F
Regression ($b_1|b_0$)	1	16.514	16.514	9.47, significant at 2.76% level
Residual	5	8.723	$s^2 = 1.745$	
Total, corrected	6	25.237		
SS(b_0)	1	272.813		
Total	7	298.050		

$$R^2 = \frac{16.514}{25.237} = 0.6544, \qquad \begin{aligned} \hat{Y}(0) &= 2.05, \\ \hat{Y}(100) &= 19.10, \end{aligned}$$

$$\text{Est } V\{\hat{Y}(X_0)\} = s^2 \left\{ \frac{1}{n} + \frac{(X_0 - \overline{X})^2}{\Sigma (X_i - \overline{X})^2} \right\}$$

$$= 1.745 \left\{ \frac{1}{7} + \frac{(X_0 - 24.614286)^2}{568.168571} \right\};$$

Est $V(\hat{Y}(0)) = 1.745\{0.142857 + 1.066344\} = 2.110056 = (1.452603)^2$;
Est $V(\hat{Y}(100)) = 1.745\{0.142857 + 10.002324\} = 17.70334 = (4.207534)^2$.

The 95% confidence limits for true mean value of Y are

$$\hat{Y} = t_\nu(1 - \tfrac{1}{2}\alpha)\sqrt{\text{Est } V(\hat{Y})} = \hat{Y} \pm 2.571\sqrt{\text{Est } V(\hat{Y})}.$$

$X = 0$: $2.05 \pm 3.73 = -1.68$ to 5.78;
$X = 100$: $19.10 \pm 10.82 = 8.28$ to 29.92.   (See plot below.)

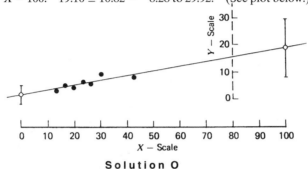

Solution O

These are wide (compared, e.g., with range of Y, 6 units) even if model is correct at 0 and 100, which is uncertain for two reasons:

1. 0 and 100 are *well* outside data range
2. There is no reason to believe a linear relationship will be preserved at crucial places like 0 and 100 from practical considerations.

Conclusion: The predictions at $X = 0$ and $X = 100$ must be regarded with caution.

P. 1. $r_{XY}^2 = \dfrac{\{\Sigma (X_i - \overline{X})(Y_i - \overline{Y})\}^2}{\{\Sigma (X_i - \overline{X})^2\}\{\Sigma (Y_i - \overline{Y})^2\}} = \dfrac{\text{SS}(R|b_0)}{\Sigma (Y_i - \overline{Y})^2} = R^2.$

2. $r_{Y\hat{Y}} = \dfrac{\Sigma (\hat{Y}_i - \overline{\hat{Y}}_i)(Y_i - \overline{Y})}{\{\Sigma (Y_i - \overline{\hat{Y}}_i)^2\}^{1/2}\{\Sigma (Y_i - \overline{Y})^2\}^{1/2}}.$

Now $\hat{Y}_i = b_0 + b_1 X_i$, $\overline{\hat{Y}}_i = b_0 + b_1\overline{X} = \overline{Y}$. Thus we can substitute for $\hat{Y}_i - \overline{\hat{Y}}_i = b_1(X_1 - \overline{X})$, cancel out b_1 top and bottom, and we are left with r_{XY}.

Q. Solution depends on the data collected.

R. (Partial solution) The fitted equation is $\hat{A} = 14.410649 + 0.130768T$.

ANOVA

Source	df	SS	MS	F
Total	43	39,554.000	—	
b_0	1	38,221.488		
$b_1\|b_0$	1	764.646	764.646	55.2, significant at 1%
Residual	41	567.866	$s^2 = 13.851$	
Lack of fit	20	243.866	12.19	<1, not significant
Pure error	21	324.000	15.43	

$R^2 = 0.574$.

There appear to be no peculiarities in the residuals. There is a significant fit. Of the variation about the mean, 57.4% is explained out of a possible maximum of $100(1332.512 - 324)/1332.512 = 75.7\%$. (The remaining variation is due to pure error.) Thus a reasonably satisfactory fit has been obtained although there is room for improvement.

Residuals in same order, rounded to the nearest unit:

-2				-3	
$-2, -1$				$0,\ 1$	
$-7, -2$				$2, -7$	
$0,\ 3$				$-7,\ 0$	
$-5, -1,$	5			-2	
-2				$2, -1$	
$3,\ 2,$	6			$6, -2$	
$7,\ 4$				$5, -5, 2$	
$1, -1,$	1			-4	
$2,\ 7,$	1			-2	
$6, -5, -2$				0	

S. **1.** $\hat{X}_0 = 45.38$, $(X_L, X_U) = (39.56, 50.05)$.

    2. $g = 0.074418$. For $g = 0$, $(X_L, X_U) = (39.83, 49.86)$, fairly close to the values in (1), but less conservative.

    3. $\hat{X}_0 = 45.38$ [same as in (1)]. An additional 1 is inserted within the right brackets in Eq. (3.2.8) to provide end points $(X_L, X_U) = (19.33, 70.28)$, which are very widely spaced.

    4. $g = 0.074418$ [same as in (2)]. For $g = 0$, $(X_L, X_U) = (21.23, 68.46)$. The large sizes of the two intervals concerned make these look reasonably close to those in (3), in spite of the large numerical differences.

T. The only way the confidence bands can look parallel is if the second term of $V(\hat{Y}_0)$ is small and, since $150 \le X_0 \le 170$, whatever $\bar{X}$ is, the top of the second term will be large for some X_0 values. So $S_{XX} = \Sigma (X_i - \bar{X})^2$ must be very large, and the answer is (3).

U. **1.** $b_1^* = b_1(1 - q)$. In the ANOVA table, $SS(b_1^*|b_0) = SS_{XY}^{*2}/S_{XX}^* = S_{XY}^2/S_{XX}$, so no change in the ANOVA, or F-tests. The fitted values and confidence calculations are changed, however. Replace X_i by $X_i/(1 - q)$ in all formulas.

    2. There would be complicated changes throughout due to the fact that $(1 - q_i)$ cannot be pulled through the summations. Replace X_i by $X_i/(1 - q_i)$ in all formulas.

    3. Replace $(1 - q_i)^2$ by $1 - 2q_i$ where it occurs. If $q_i = q$ pull all q factors through the summations.

V. We can fit the straight line $\hat{Y} = 9.930 - 0.010987X$. The analysis of variance table is as follows:

ANOVA

Source	df	SS	MS	F
b_0	1	4,230.1602		
$b_1\|b_0$	1	1.1777	1.1777	0.73
Residual	48	77.2321	1.6090	
Lack of fit	29	45.7771	1.5785	0.95
Pure error	19	31.4550	1.6555	
Total	50	4,308.5700		

$F(30, 18, 0.95) = 2.11$; $F(1, 48, 0.95) = 4.05$. There is no apparent lack of fit, and the regression slope is not significant.

Conclusion: These data do *not* confirm the idea that length of life is related to length of lifeline.

Notes: (i) The contributions to pure error at $X = 75$ and 82 are (comparatively) extremely large because of two extreme observations, 6.45 and 13.20, respectively. (ii) An improved analysis would take account of the covariate "body length" or some similar covariate and would adjust the observations for this possible source of variation. It is anticipated that this would not alter our main conclusion.

W. Parts 1 and 2 lead you to think in terms of using *age* as a predictor, and to consider the possibility of transforming price and/or age, perhaps as in (3).

3. $\mathbf{Y} = (3.91, 3.56, \ldots, 1.61)'$.

$\mathbf{Z} = (82, 72, \ldots, 12)'$.

$\hat{Y} = 1.143181 + 0.0346564Z$.

$100\mathbf{e} = (-8, -8, 27, -8, 25, 13, -31, 8, -32, 22, 3, -23, 6, 5)'$.

ANOVA

Source	df	SS	MS	F
b_0	1	80.6880	—	
$b_1\|b_0$	1	6.4075	6.4075	157.4
Residual	12	0.4884	$s^2 = 0.0407$	
Total	14	87.5859		

The regression is highly significant and $R^2 = 0.9292$.

4. Note the difficulty in answering apparently straightforward questions when transformations have been used. The per year rate of increase of price is b_1 (price) and depends not only on b_1 but also price. Thus it is simpler to say that $Y = \ln$ (price) increases at rate b_1 per year.

5. Again, some care is needed. If we forecast Y (on the basis of the fitted equation) at $Z = 38$, we would have $\hat{Y}_{38} = 2.46$. However, in 1975 the actual $Y = \ln(20) = 3.00$, so that the prediction falls short by 0.54, a comparatively large amount. If a straight line relationship holds in 1975, it does not look like the same one, but it is difficult to tell on the basis of one new point whether the whole line has risen so that b_0 alone has increased (possible) or if b_1 alone has increased (unlikely, since younger port would then be cheaper!) or both b_0 and b_1 have increased (possible). More data are needed to investigate further.

X. See the quoted source for a full discussion of this exercise. For a way of generating such data, see "Computer generation of data sets for homework exercises in simple regression," by S. R. Searle and P. A. Firey, *The American Statistician,* **34,** February 1980, 51–54.

Y. Answer is implicit in the question. It is true generally, as we now show in matrix algebra (see Chapters 5 and 8).

$$\mathbf{Y} = \mathbf{X}\boldsymbol{\beta} + \boldsymbol{\epsilon},$$
$$\hat{\mathbf{Y}} = \mathbf{Xb} = \mathbf{X(X'X)}^{-1}\mathbf{X'(X}\boldsymbol{\beta} + \boldsymbol{\epsilon}) = \mathbf{X}\boldsymbol{\beta} + \mathbf{H}\boldsymbol{\epsilon},$$

where $\mathbf{H} = \mathbf{X(X'X)}^{-1}\mathbf{X'}$. Thus $\mathbf{Y} - \hat{\mathbf{Y}} = (\mathbf{I} - \mathbf{H})\boldsymbol{\epsilon}$. This is a useful result to know.

Z. **1.** Not shown.

2. $\hat{Y} = -10.9 + 18.449X$.

3. The residuals range from -1340.81 to 539.35 and add to 0.000

4.

Source	df	SS/10^6	MS/10^6	F
Regression $\|b_0$	1	22.01	22.01	97.68
Residual	15	3.38	0.225	
Total, corrected	16	25.39		

5. $R^2 = 22.01/25.39 = 0.8669$.

6. $s_e^2 = 584291.745/4 = 146072.9363$;

$F = \{2795543.255/11\}/14607.9363 = 1.74$
with df (4, 11). No lack of fit.

7. Yes, $F(1,15) = 97.68$, $p < 0.001$.

8. $se(b_0) = 141.7$, $se(b_1) = 1.867$.

Note that the nonsignificant $b_0 = -10.9$ is not practically reasonable (implying a negative consumption when the population is zero) and that refitting a straight line through the origin would make sense here. The result is $\hat{Y} = 18.365X$.

9.

	Limits of 95% Bands	
X	Lower	Upper
25	193	707
50	665	1158
100	1503	2165
150	2269	3244
200	3012	4346
250	3746	5456

10. The Swiss data point would not have much effect as it is down close to other small results. The U.S. Data point is at an extreme and changing it would have an effect. The point is *influential*; see Cook's statistic elsewhere.

11. The e versus $\hat{Y}$ plot has funnel-shaped characteristics, which imply more variation in larger Y-values. Consideration should be given to transforming the Y's; see *transformations*.

12. $\hat{X} = 6.401 + 0.046989Y$ or $Y = 136.20 + 21.281\hat{X}$;

$c_1 = (18.449/0.046989)^{1/2} = 19.815$.

The third line is $\hat{Y} - \overline{Y} = c_1(X - \overline{X})$ or $\hat{Y} - 804.93 = 19.815(X - 44.224)$, namely, $\hat{Y} = 71.357 + 19.815X$. This lies between the other two lines; all three lines intersect at $(\overline{X}, \overline{Y})$.

AA. 1. $\hat{Y} = -2.679 + 9.5X$.

2.

ANOVA

Source of Variation	df	SS	MS	F
Total	6	651.714		
Regression: $b_1|b_0$	1	631.750	631.750	158.33
Residual	5	19.964	3.99	

Since $158.33 > 6.61$, the regression is significant, $\alpha = 0.05$.

3. No evidence to suggest a more complicated model needed.

BB. 1. $b_1 = 9.13$; $\hat{Y} = 9.13X$.

Obs. No.	X	Y	$\hat{Y}$	Residuals
1	3.5	24.4	31.955	−7.555
2	4.0	32.1	36.520	−4.420
3	4.5	37.1	41.085	−3.985
4	5.0	40.4	45.650	−5.250
5	5.5	43.3	50.215	−6.915
6	6.0	51.4	54.780	−3.380
7	6.5	61.9	59.345	2.555
8	7.0	66.1	63.910	2.190
9	7.5	77.2	68.475	8.725
10	8.0	79.2	73.040	6.160

3. The plot of the residuals against $\hat{Y}$ indicates the omission of the β_0 term in the model.

4. The model $Y = \beta_0 + \beta_1 X_1 + \epsilon$ is recommended, restricting its use to within the X_1 range, 3.5 to 8.0, shown by the data.

If the true model really has $\beta_0 = 0$, then data will have to be obtained nearer to the zero response before any insight into the right model for the range of X from 0 to a large value of X can be obtained.

CC. 1. $\hat{Y} = -252.298 + 8.529X.$

2.

ANOVA

Source of Variation	df	SS	MS	F
Total	9	219270.5000		
$b_1\|b_0$	1	200772.3188		
Residual	8	18498.1812		
Lack of fit	4	18454.6812	4613.6703	424.2455
Pure error	4	43.5000	10.8750	

The lack of fit test indicates that the model is inadequate.

The plot of the residuals against $\hat{Y}$ indicates a definite trend from negative residuals to positive residuals as the value of $\hat{Y}$ increases.

There is also evidence of an outlier; namely, $Y = 415$ when $X = 90$. This point should be investigated further.

DD. 1. $\hat{Y} = 7.950 - 0.0179T.$

2.

ANOVA

Source of Variation	df	SS	MS	Calc. F	$F_{0.95}$
Total	8	6.260			
Regression	1	1.452	1.452	2.114	5.59
Residual	7	4.808	0.687		
Lack of fit	3	1.763	0.588	<1	
Pure error	4	3.045	0.761		

The regression is nonsignificant, $R^2 = 23.2\%$.

3. $s_{b_1} = 0.01228. \ -0.04689 \le \beta_1 \le 0.01119.$

4.

Batch No.	Y_i	$\hat{Y}_i$	$Y_i - \hat{Y}_i$
1	2.10	2.95	-0.85
2	3.00	3.49	-0.49
3	3.20	2.59	0.61
4	1.40	2.24	-0.84
5	2.60	2.42	0.18
6	3.90	2.95	0.95
7	1.30	2.24	-0.94
8	3.40	2.59	0.81
9	2.80	2.24	0.56

No discernible pattern in the residuals.

5. $0.193 \le \hat{Y}_0 \le 4.453.$

6. No. The slope of the line fitted is not significant and, in any case, 360 is well beyond the temperature range, making the use of the fitted equation even more dangerous.

EE. 1. $\hat{Y} = 44.353 + 2.6172X.$

3. −0.75, 0.73, 2.63, −1.63, 3.11, −6.41, −5.67, 6.54, 1.92, −1.86, 0.78, −7.53, 4.21, 0.95, 6.43, −7.10, 7.90, 9.38, −3.93, 0.07, −6.71, −6.71, 3.62. Their sum = −0.03.

4.

Source	df	SS	MS
b_0	1	107,170.00	—
$b_1\|b_0$	1	728.09	728.09
Residual	21	580.34	27.64
Lack of fit	18	459.84	25.55
Pure error	3	120.50	40.17
Total	23	108,478.43	

5. $R^2 = 728.09/1308.43 = 0.5565$.

6. $F = 25.55/40.17 = 0.64$. This value does not cause us to suspect the model, at least as far as this test is concerned.

7. As far as we can tell, yes. $F = 728.09/27.64 = 26.34$, which exceeds $F(1, 21, 0.95) = 4.32$ handily. We reject the idea that $\beta_1 = 0$.

8. Use s^2. We get $se(b_0) = \{s^2 \Sigma X_i^2/(nS_{XX})\}^{1/2} = 4.785$ and $se(b_1) = \{s^2/S_{XX}\}^{1/2} = 0.510$. The corresponding t-ratios are 9.27 and 5.13. [Note $5.13^2 = 26.32$, the F-value in (7) apart from rounding error.]

9. $se(\hat{Y}_0) = s\{1/23 + (X_0 - \overline{X})^2/S_{XX}\}^{1/2}$.

X_0	$\hat{Y}_0$	95% Limits	
6	60.1	56.0	64.1
7	62.7	59.5	65.9
8	65.3	62.7	67.9
9	67.9	65.6	70.2
10	70.5	68.1	73.0
11	73.1	70.1	76.2
12	75.8	72.0	79.6

10. The new trees lie well away from the rest of the data. The small tree is clearly of low height compared with trees of similar diameter. The large tree would lie close to the fitted line, as recorded. If damaged, it would indicate that the model might have to be rethought to curve upward if larger diameter, nondamaged trees were included in future data collection.

FF. The two individual lines are

$$\hat{Y} = 44.353 + 2.6172X,$$

$$\hat{X} = -5.378 + 0.21261Y,$$

$$\text{or} \quad Y = 25.295 + 4.7034\hat{X};$$

$$c_1 = (2.6172/0.21261)^{1/2} = 3.5085.$$

Thus $c_0 = \overline{Y} - c_1\overline{X} = 68.261 - 3.5085(9.1348) = 36.211$.
The compromise line is thus

$$\hat{Y} = 36.211 + 3.5085X.$$

GG. Detroit has low turnover but also a low wage, so it seems likely that this is the (atypical) city with high unemployment. The fitted line is $\hat{Y} = 110.79 - 11.515X$. $R^2 = 0.825$, $s^2 = 65.50$ (2 df). When $X = 6$, $\hat{Y} = 41.7$ and $se(\hat{Y}) = 4.40$. The confidence interval is thus $41.7 \pm 4.303(4.40) = (22.8, 60.6)$. This is very wide, but not surprisingly so, since the fit is based on only four data points.

HH. $\hat{Y} = 1919.8 + 3.1376X$ is the fitted equation. If conditions remained the same, we could use this as a predictive equation *within the current limits of the X-data*. When $X = 0$, $\hat{Y} =$ 1920 and the requested limits are (1521, 2319). Does this mean that the Protestants will receive no money at all, or lose money (!), if Catholic attendance drops low enough? This seems most improbable! The model is unlikely to be valid below the data we have. For example, the true model could turn sharply downward, with X staying positive, below the present data. We are grateful to R. Peter Hypher for this interesting example.

II. The underlying idea is that there might be some relationship between how well the TA was rated and the course number (ranging from the elementary 201 to the most advanced 824). The fitted line $\hat{Y} = 3.6854 + 0.00140X$ explains only 16.3% of the variation even though the F is technically significant at about 2.2%. (See Chapter 11 for more discussion on this.) A plot of the data makes it obvious that observations 6, 7, and 11 are very influential in determining the slope of the line. This would show up in high values of Cook's statistic (see Chapter 8). These three observations are from courses that require highly trained teaching assistants. The omission of these three influential points actually improves the R^2 value to 21.0% with an F significant at about 1.2% from $\hat{Y} = 2.9895 + 0.003932X$. So overall we might claim some modest effect is apparent, probably because the higher the course number, the more experienced the teaching assistant needs to be. In the second fit, the two observations from course 424 now become influential. Omitting those gives $R^2 = 29.8\%$ and a significant F at about 0.3%, from $\hat{Y} = 2.4412 + 0.006071X$. So the effect seems fairly persistent. Note that spacing the data by an X-coordinate based on the course number may not be a good way to represent their relative difficulty or the requirements for their teaching assistants.

JJ. **1, 3.** Not provided.

 2. $\hat{Y} = 76.01 + 1.27667X$.

 4.

Source	df	SS	MS	F
Regression $|b_0$	1	6,895,054	6,895,054	289.52
Residual	46	1,095,495	23,815	
Total, corrected	47	7,990,548		

 5. $R^2 = 0.8629$.

 6.

Lack of fit SS	1,012,189.6	(29 df)	MS = 34,903.1
Pseudo pure error SS	83,305.4	(17 df)	MS = 4,900.3

 $F = 7.12$; $F(29, 17, 0.95) = 2.16$.
 Significant lack of fit.
 If only the exact repeats are used, the pure error MS $= 6893/2 = 3446.5$ (2 df).
 $F = \{1,092,048.5/44\}/3,446.5 = 7.20$, while $F(44, 2, 0.95) = 99.47$, so this test is very insensitive with only 2 df for pure error. One would be inclined to believe that there is lack of fit, relying on the pseudo test above.
 Examination of the data plot shows that the first observation is very influential, and that the impression of a curved plot comes mostly from it.

 10. Apart from two observations that stand out, a reasonable broad band is seen.

 11. The analysis is little altered by these changes. Lack of fit persists in alternative analyses. The fitted straight line gives a reasonable impression of the data ($R^2 = 0.8629$) but there is a lot of variation unexplained at some of the lower data points by it.

KK. The original fit has a slope of 0.979 and an $R^2 = 0.985$. Removal of observation 9, which has the smallest X-value, the largest residual, and is the most influential (via Cook's D, see Section 8.3) leads to a regression with slope 1.013 and $R^2 = 0.995$. Now the second largest observation becomes most influential. Although eight of the papers suffered circulation losses, papers 2 and 9, which did not suffer losses, slightly distort the effects of these losses in the regression. The original fit is probably good enough for prediction, although some analysts would argue the point. A fit through the origin is also not unreasonable in this problem; this fit has a slope of 0.990.

LL. The fitted equation is $\hat{Y} = 43.84 + 37.23X$.

ANOVA

Source	df	SS	MS	F
Total	13	857,500		
b_0	1	687,700		
$b_1\|b_0$	1	155,258	155,258	
Residual	11	14,542	1,322	
Lack of fit	6	13,075	2,179	7.44*
Pure error	5	1,467	293	

* Significant lack of fit at the 5% level.

Residuals in order are: -31, -8, -28, -6, -16, 24, -3, 43, 63, 36, 18, -36, -56.
Conclusions: A plot of the residuals in order, or a plot of the data, both show a clear "quadratic curve" pattern and, as we have already noted, there is significant lack of fit. We need to improve the model. One way would be to fit a quadratic curve in X to the data. If we do this we obtain the fitted equation

$$\hat{Y} = -49.05 + 83.18X - 4.07X_1^2.$$

Further analysis shows that no lack of fit is indicated and the overall regression is highly significant with $F(2, 10) = 237.4$. This model explains $100R^2 = 97.94\%$ of the variation about the mean. For an alternative analysis in which additional information is used, see Exercise H in "Exercises for Chapters 5 and 6."

MM.
$$b_0 = -1.752, \quad b_1 = 0.908$$

$$a_0 = -1.58, \quad a_1 = 1.097$$

$$Y = 89.19 + 0.910(X - 96.25)$$

All four sets of residuals $(Y_i - \hat{Y}_i, X_i - \hat{X}_i)$, and both set of residuals from a gmfr line sum to zero. All four plots of residuals versus fitted values suggest a nonconstant variance structure.

Chapter 4

A.
1. False. Matrices are of different sizes.
2. False. **A** is 3×2, **C** is 3×3.
3. True.
4. True.
5. False. Only square matrices have inverses.
6. False. The right-hand side should be 2×2 not 3×3.

B.
1. Impossible. **B** is 3×3 and **C** is 2×2.

2.
$$\mathbf{BB'} = \begin{bmatrix} -1 & 1 \\ 2 & 3 \\ 3 & 2 \end{bmatrix} \begin{bmatrix} -1 & 2 & 3 \\ 1 & 3 & 2 \end{bmatrix} = \begin{bmatrix} 2 & 1 & -1 \\ 1 & 13 & 12 \\ -1 & 12 & 13 \end{bmatrix}.$$

3. Impossible. **A** is 3×3. **B'B** is 2×2.

4.
$$\mathbf{BC} = \begin{bmatrix} -1 & 1 \\ 2 & 3 \\ 3 & 2 \end{bmatrix} \begin{bmatrix} 4 & 3 \\ 3 & 2 \end{bmatrix} = \begin{bmatrix} -1 & -1 \\ 17 & 12 \\ 18 & 13 \end{bmatrix}.$$

5. $\mathbf{AA^{-1}BC = IBC = BC}$, given above.

6. $\mathbf{CB}' = (\mathbf{BC}')' = (\mathbf{BC})' = \begin{bmatrix} -1 & 17 & 18 \\ -1 & 12 & 13 \end{bmatrix}$

(or multiply it out).

7. Impossible. $\mathbf{C}$ is 2×2. $\mathbf{A}$ is 3×3.

8. $\begin{bmatrix} -1 & 1 \\ 2 & 3 \\ 3 & 2 \end{bmatrix} \begin{bmatrix} -2 & 3 \\ 3 & -4 \end{bmatrix} = \begin{bmatrix} 5 & -7 \\ 5 & -6 \\ 0 & 1 \end{bmatrix}$.

9. Partitioned matrix.

Invert middle to get $\frac{1}{4}$. Corners are $\mathbf{C}$, invert to give $\mathbf{C}^{-1}$. Put these together to give

$$\begin{bmatrix} -2 & 0 & 3 \\ 0 & \frac{1}{4} & 0 \\ 3 & 0 & -4 \end{bmatrix}.$$

10. $\mathbf{A}$ is symmetric so $\mathbf{A}' = \mathbf{A}$. Thus $\mathbf{A}'\mathbf{A}(\mathbf{A}')^{-1}\mathbf{A}^{-1} = \mathbf{AAA}^{-1}\mathbf{A}^{-1} = \mathbf{AIA}^{-1} = \mathbf{I}_{3\times3}$.

C. $\mathbf{A}$ is 2×3, $\mathbf{b}$ is 2×1, $\mathbf{C}$ is 2×2, $\mathbf{D}$ is 3×3.

So (1), (5), and (6) are all false. The matrices are not the right size for the designated operations. (3) is also false:

$$\mathbf{AD} = \begin{bmatrix} 1 & 4 & 1 \\ 0 & 8 & 1 \end{bmatrix}.$$

The (2, 1) element is not 6, as given. The remaining equations, (2), (4), (7), (8), and (9), are all true. However, in (7) one must check that $\mathbf{C}^{-1}$ does, in fact, exist.

D.

$$(\mathbf{X}'\mathbf{X})^{-1} = \begin{bmatrix} 14 & 790 \\ 790 & 49300 \end{bmatrix}^{-1} = \frac{1}{66100}\begin{bmatrix} 49300 & -790 \\ -790 & 14 \end{bmatrix} \quad \mathbf{X}'\mathbf{Y} = \begin{bmatrix} 295 \\ 18030 \end{bmatrix}.$$

$$\mathbf{b} = \begin{bmatrix} 299800/66100 \\ 19370/66100 \end{bmatrix} = \begin{bmatrix} 4.535552 \\ 0.293041 \end{bmatrix}.$$

(Plot data, line, residuals.)

ANOVA

Source	df	SS	MS	F
Total	14	6641.00		
b_0	1	6216.07		
$b_1\|b_0$	1	405.45	405.45	250 significant
Residual	12	19.48	$s^2 = 1.62$	
Lack of fit	4	0.81	0.20	Not significant
Pure error	8	18.67	2.33	

$$\mathbf{V}(\mathbf{b}) = (\mathbf{X}'\mathbf{X})^{-1}s^2 = \begin{bmatrix} 1.208 & -0.019 \\ -0.019 & 0.000343 \end{bmatrix}$$

When $X = 65$,

$$V(\hat{Y}) = (1, 65)\begin{bmatrix} 1.208 & -0.019 \\ -0.019 & 0.000343 \end{bmatrix}\begin{bmatrix} 1 \\ 65 \end{bmatrix} = 0.1872;$$

$$\hat{Y}(65) = 23.583; \quad t(12, 0.975) = 2.179.$$

The 95% confidence interval for $E(Y|X = 65)$ is $23.583 \pm 2.179(0.1872)^{1/2} = 22.640$ to 24.526.

E. $\mathbf{b} = \begin{bmatrix} 10 & 36 \\ 36 & 160 \end{bmatrix}^{-1}\begin{bmatrix} 26.5 \\ 86.1 \end{bmatrix} = \frac{1}{304}\begin{bmatrix} 160 & -36 \\ -36 & 10 \end{bmatrix}\begin{bmatrix} 26.5 \\ 86.1 \end{bmatrix} = \begin{bmatrix} 3.75132 \\ -0.305921 \end{bmatrix}.$

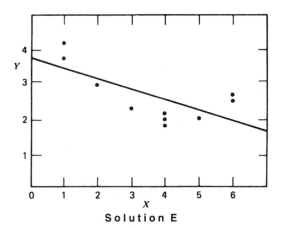

Solution E

Y_i	4.2	3.8	3.0	2.3	1.8	2.0	2.2	2.0	2.5	2.7
$\hat{Y}_i$	3.45	3.45	3.14	2.83	2.53	2.53	2.53	2.22	1.92	1.92
$e_i = Y_i - \hat{Y}_i$	0.75	0.35	−0.14	−0.53	−0.73	−0.53	−0.33	−0.22	0.58	0.78

$\Sigma \, e_i = -0.02$

ANOVA

Source	df	SS	MS	F
b_0	1	70.225		
$b_1\vert b_0$	1	2.845		
Lack of fit	4	2.740	$MS_L = 0.685$	15.222*
Pure error	4	0.180	$s_e^2 = 0.045$	
Total	10	75.990		

* We must test lack of fit first. If lack of fit exists, most other calculations (e.g., F-test for regression, confidence intervals, confidence bands) are not valid and should not be performed at all. Now, from the F-table, $F(4, 4, 0.95) = 6.39$. Thus there is significant lack of fit because $15.222 > 6.39$.

$R^2 = 2.845/5.765 = 0.4935$. The straight line explains only 49.35% of the variation about the mean, that is, not much.

We see, from the residuals, which have a clear positive–negative–positive pattern, that the data have a quadratic tendency, which a straight line model cannot follow. This is obvious from the plot, of course.

We conclude that the straight line model is inadequate and not usable and that a model $Y = \beta_0 + \beta_1 X + \beta_{11} X^2 + \epsilon$ should be tried next. (If we do this, we get $\hat{Y} = 5.462 - 1.6380X + 0.192840X^2$ with a good fit.)

F. 1. $\mathbf{b} = \begin{bmatrix} 10 & 60 \\ 60 & 482 \end{bmatrix}^{-1} \begin{bmatrix} 5 \\ 32 \end{bmatrix} = \begin{bmatrix} 49/122 \\ 1/61 \end{bmatrix}.$

$$\hat{Y} = \frac{49}{122} + \frac{1}{61} X.$$

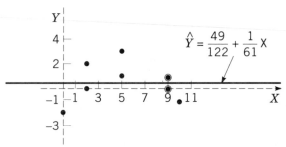

Solution F

3. Basic ANOVA:

$$\mathbf{b'X'Y} = \left[\frac{49}{122}, \frac{1}{61}\right]\left[\begin{array}{c} 5 \\ 32 \end{array}\right] = \frac{245 + 64}{122} = 2\tfrac{65}{122}.$$

ANOVA

Source	df	SS
b_0, b_1	2	$2\tfrac{65}{122}$
Residual	8	$18\tfrac{57}{122}$
Total	10	21

4.
$$SS(b_0) = (\Sigma\,Y)^2/n = 2\tfrac{1}{2}.$$

X	Pure Error (p.e.) Contribution at X	df
2	2	1
5	2	1
9	1	3
—	5 = Pure error SS	5 = p.e. df

ANOVA

Source	df	SS	MS
b_0	1	$2\tfrac{1}{2}$	
$b_1|b_0$	1	$\tfrac{2}{61} = .033$	
Lack of fit	3	$13\tfrac{57}{122} = 13.467$	$4\tfrac{179}{366}$
Pure error	5	5	1
Total	10	21	

5. $F = \dfrac{\text{MS}_L}{s_e^2} = 4\tfrac{179}{366} < F(3, 5, 0.95) = 5.41.$

Thus lack of fit is not significant by this test. We recombine lack of fit SS and residual SS to give

$$s^2 = 18\tfrac{57}{122}/8 = 2\tfrac{301}{976} = 2.308525.$$

6. $F = \dfrac{2/61}{2\frac{301}{976}} < 1.$ Obviously not significant. Yes, test is valid because lack of fit is not significant.

7. $V(\hat{Y}_0) = \left\{\dfrac{1}{n} + \dfrac{(X_0 - \overline{X})^2}{\Sigma\,(X_i - \overline{X})^2}\right\}\sigma^2 = \left\{\dfrac{1}{10} + \dfrac{(X_0 - 6)^2}{122}\right\}\sigma^2.$

When $X_0 = \sqrt{122} + 6$, this reduces to $1.1\sigma^2$, which is estimated by $1.1s^2 = (1.1)(2\tfrac{301}{906}) = (1.1)(2.308525) = 2.5393775;$

$$t(8, 0.975) = 2.306.$$

So the interval required is $\hat{Y}_0 \pm 2.306\sqrt{2.5393775}$
or $\hat{Y}_0 \pm 3.6747.$

Yes, it is valid because lack of fit is not significant. However, it is a long way outside the data and so is somewhat dangerous as we cannot guarantee the model out there.

8.

X_i	Y_i	$\hat{Y}_i$	$e_i = Y_i - \hat{Y}_i$
0	-2	0.40	-2.40
2	0	0.43	-0.43

X_i	Y_i	$\hat{Y}_i$	$e_i = Y_i - \hat{Y}_i$
2	2	0.43	1.57
5	1	0.48	0.52
5	3	0.48	2.52
9	1	0.54	0.46
9	0	0.54	−0.54
9	0	0.54	−0.54
9	1	0.54	0.46
10	−1	0.56	−1.56
			$0.06 = \Sigma\, e_i$

(Two decimal places is quite enough; even one decimal place would do.) The plot of e_i versus Y_i looks like the plot in (2) with different axes, of course. (Ask yourself why.)

9. $V(b) = (X'X)^{-1}\sigma^2 = \begin{bmatrix} 10 & 60 \\ 60 & 482 \end{bmatrix}^{-1}\sigma^2 = \dfrac{\sigma^2}{1220}\begin{bmatrix} 482 & -60 \\ -60 & 10 \end{bmatrix}$

$= \dfrac{\sigma^2}{610}\begin{bmatrix} 241 & -30 \\ -30 & 5 \end{bmatrix}.$

10. $R^2 = \dfrac{2/61}{18\frac{1}{2}} = \dfrac{2}{61}\cdot\dfrac{2}{37} = \dfrac{4}{2257} \approx 0.18\%.$

11. Identical to R^2 for a straight line model.

12. No significant regression, no lack of fit. Nevertheless, the residuals pattern indicates a systematic departure from randomness that needs exploring further in spite of the nonsignificance of the lack of fit test. The model fitted is not of much value.

Chapters 5 and 6

A. 1. $b_0 = 14$, $b_1 = -2$, $b_2 = -\frac{1}{2}$.

2.

ANOVA

Source of Variation	df	SS	MS	F
Total (corrected)	10	190		
Due to regression	2	122	61.0	7.17*
Residual	8	68	8.5	

3. Test of significance:

$$\text{Compare } F = \frac{\text{MS regression}}{\text{MS residual}} \text{ with } F(2, 8, 0.95) = 4.46.$$

Since 7.17 is greater than the critical F, we reject the hypothesis of no planar fit and use the fitted equation

$$\hat{Y} = 14 - 2X_1 - \tfrac{1}{2}X_2.$$

4. $R^2 = \frac{122}{190} = 64.21\%.$

5a. Estimated variance of $b_1 = 1.4365$.

b. Estimated variance of $b_2 = 0.3587$.

c. Estimated variance of $\hat{Y} = 1.95075$.

6.

ANOVA

Source of Variation	df	SS	MS	F
Total (corrected)	10	190.00		
Regression	2	122.00	61.00	7.18
Due to b_1	1	116.08	116.08	13.64*
b_2 given b_1	1	5.92	5.92	<1
Residual	8	68.00	8.50	

7.

ANOVA

Source of Variation	df	SS	MS	F
Total (corrected)	10	190.00		
Regression	2	122.00	61.00	7.18
Due to b_2	1	98.33	98.33	11.57*
b_1 given b_2	1	23.67	23.67	2.78
Residual	8	68.00	8.50	

8. *Conclusions:*
 (i) While the regression equation, $\hat{Y} = 14 - 2X_1 - \frac{1}{2}X_2$, is statistically significant, the mean square error is larger than that obtained when $\hat{Y} = 9.162 - 1.027X_1$ is used.
 (ii) Independent estimates of β_1 and β_2 are not obtainable from these data. If independent estimates of β_1 and β_2 are desired, a balanced experiment in X_1 and X_2 should be done.
 (iii) When problems arise as to a choice of a model, more experimental work of a balanced nature is usually necessary.

B.
$$\hat{X} = 1.0607 + 0.0056Y - 0.0013Z.$$

ANOVA

Source of Variation	df	SS	MS	F
Total (corrected)	11	0.294867		
Regression	2	0.236409	0.118204	18.20
Due to Y	1	0.236275	0.236275	36.38*
Due to $Z\|Y$	1	0.000134	0.000134	<1
Due to Z	1	0.236006	0.236006	36.34*
Due to $Y\|Z$	1	0.000403	0.000403	<1
Residual	9	0.058458	0.006495	

Conclusion: The inclusion of both Y and Z in the model is not useful. This is further demonstrated by the correlation coefficient, $r_{yz} = -0.9978$.

C.

ANOVA

Source of Variation	df	SS	MS	F
Total (corrected)	14	85386.0000		
Regression	2	49791.1751	24895.58755	8.39*
Due to X_1 (alone)	1	48186.1482	48186.1482	16.24*
Due to $X_2\|X_1$	1	1605.0249	1605.0269	<1
Residual	12	35594.8249	2966.2354	

Conclusions
(i) The predictive model

$$\hat{Y} = 124.063977 + 3.512038X_1 + 0.834632X_2$$

explains only 58.31% of the total corrected variability in G.C.E. examination scores. While it proves to be statistically significant for an α-risk of 0.011, the standard deviation of the residuals is 54.46 and expressed as a percentage of the mean exam score, 9.725%. This shows that there is a great deal of unexplained variation, and thus the equation will not be very useful for prediction.

(ii) The addition of X_2, the previous performance in S.C. English Language, adds little to the predictability of a candidate's total mark in the G.C.E. examination. A simple model $Y = \beta_0 + \beta_1 X_1 + \epsilon$ would do almost as well.

D. 1. $\hat{Y} = -94.552026 + 2.801551X_1 + 1.072683X_2$.

2.

ANOVA

Source of Variation	df	SS	MS	F
Total (corrected)	7	2662.14		
Regression	2	2618.98	1309.49	151.74*
Residual	5	43.16	8.63	

Since $F(2, 5, 0.95) = 5.79$, the overall regression is statistically significant, that is, $151.74 > 5.79$.

3. $R^2 = \dfrac{\text{SS Regression}}{\text{Corr. Tot.}} = \dfrac{2618.98}{2662.14} = 98.38\%$.

E. 1. $\hat{Y} = 67.234527 + 0.906089(X_1 - 164) - 0.064122(X_2 - 213)$.

2.

ANOVA

Source of Variation	df	SS	MS	F
Total (corrected)	15	8429.14444		
Regression	2	6796.77105	3398.385525	26.90*
Residual	13	1632.37339	126.336415	

Multiple $R^2 = 80.5\%$.
Standard deviation of residuals $= 11.239947$.
The fitted model is statistically significant. However, 20% of the variability remains unexplained; more work needs to be done on this problem.

3.

ANOVA

Source of Variation	df	SS	MS	F
Total (corrected)	15	8429.14444		
Regression	2	6796.77105	3398.385525	26.90*
X_1	1	6777.72877	6777.72877	53.65*
$X_2\|X_1$	1	19.04228	19.04228	NS
X_2	1	25.10057	25.10057	NS
$X_1\|X_2$	1	6771.67048	6771.67048	53.60*
Residual	13	1632.37339	126.33645	

X_1 is the more important variable.

4. *Conclusions*

(i) In the region of this machine's operation as indicated by the levels of Plate Clearance and Plate Temperature, the Plate Clearance has a pronounced effect on the percent properly sealed.

(ii) There is little evidence present that the Plate Temperature has an additive effect on the percent properly sealed.

There are some indications in the data that a different model should be fitted. It is helpful to examine the observations after they have been rearranged into the format below.

		Sealer Plate Temperature		
		176–208	210–220	225–240
Sealer Plate Clearance	130–148	35 42.5 43.5	34.5	56.7 51.7
	156–178	81.7 94.3	44.3 91.4	52.7
	186–194	82.0	98.3 83.3	95.4 84.4

One can see a definite interaction occurring between the clearance and the temperature. Thus a second-order model would be more appropriate.

F. 1. $\hat{Y} = 72.25 + 0.0286X_1 + 0.0487X_2$.

2.

Source	df	SS	MS	F
$b_1, b_2\|b_0$	2	35.0	17.5	<1
$b_1\|b_0$	1	1.7	1.7	<1
$b_2\|b_0, b_1$	1	33.3	33.3	<1
$b_2\|b_0$	1	34.8	34.8	<1
$b_1\|b_0, b_2$	1	0.6	0.6	<1
Residual	14	1509.1	$s^2 = 107.8$	
Lack of fit	5	898.6	179.7	2.65
Pure error	9	610.5	67.8	
Total, corrected	16			

There is no apparent lack of fit.

3. Residuals are 15, −4, 1, −11, −11, 6, 6, 3, −13, 5, −11, 0, 7, 21, 0, −9, −2. (Examination is left to the reader.)

4. Neither X_1 nor X_2 has any value in explaining Y on the basis of this data set. Together they explain only 100 $R^2 = 2.3\%$ of the variation about the mean.

G. For the model with both X_1 and X_2 in, we obtain: $\hat{Y} = 65.13 + 0.286X_1 + 0.487X_2$ with analysis of variance table as follows:

ANOVA

Source	df	SS	MS	F
b_0	1	79.973.88		
$b_1, b_2\|b_0$	2	35.01	17.51	0.16
Lack of fit	5	898.61	179.72	2.65
Pure error	9	610.50	67.83	
Total	17	81,518.00		

Because $F(5, 9, 0.95) = 3.48$, there is no reason to suspect lack of fit. Thus $s^2 = (898.61 + 610.50)/(5 + 9) = 107.79$. Test for overall regression: $F = 17.51/107.79 = 0.16$, not significant. The residuals show nothing worth remarking.

$$SS(b_1|b_0, b_2) = 0.21; \quad SS(b_2|b_0, b_1) = 33.32.$$

Both partial F-tests are nonsignificant. We conclude that the model $\hat{Y} = \overline{Y} = 68.588$ is as good as any.

Note: It is possible for the overall F-test for $H_0 : \beta_1 = \beta_2 = 0$ to be nonsignificant, but the partial F-test for one of the hypotheses $H_0 : \beta_j = 0, j = 1, 2$, to be significant. This would denote a weak relationship with the corresponding X_j.

H. The fitted equation is

$$\hat{Y} = -5.95 + 54.35X - 27.40Z.$$

Now $\mathbf{b'X'Y} = 21{,}205{,}018{,}600/24{,}780 = 855{,}731.$

This last figure is the $SS(b_0, b_1, b_2)$. Because of the pattern of Z's it will be found that the pure error sum of squares is exactly the same as before, 1467 with five degrees of freedom. (This is not true in general; usually the addition of a new variable, like Z here, will lead to fewer degrees of freedom for pure error, because responses Y_i with exactly the same X-values generally have different Z-values. In our data, such responses always have the same Z-values, which is uncommon.) The overall analysis of variance table is as follows:

ANOVA

Source	df	SS	MS	F
b_0	1	687,700		
$b_1, b_2\|b_0$	2	168,031	84,015	
Lack of fit	5	302	60	<1 NS*
Pure error	5	1,467	293	
Total	13	857,500		

* So $s^2 = (302 + 1{,}467)/(5 + 5) = 177.$

Test for overall regression: $F(2, 10) = 84{,}015/177 = 474.7$, which is very highly significant. The extra sum of squares test for $H_0 : \beta_2 = 0$ versus $H_1 : \beta_2 \neq 0$ is as follows:

$$SS(b_2|b_1, b_0) = SS(b_2, b_1|b_0) - SS(b_1|b_0)$$

$$= 168{,}031 - 155{,}258$$

$$= 12{,}773.$$

$F(1, 10) = MS(b_2|b_1, b_0)/s^2 = 12{,}773/177 = 72.2$, which is very highly significant. Thus addition of Z as a predictor is extremely worthwhile.

Conclusions: The "number of men working" is an important variable. Because $b_2 = -27.40$, it has a negative effect. Note that this does *not* mean that the men do negative work but it does mean that their presence tends to cause less work to be done than might otherwise be anticipated. It would probably be a little more informative to fit the equation in the form

$$Y = \beta_0 + \beta_1(X - Z) + \beta_2 Z + \epsilon \tag{1}$$

so that the two predictors $(X - Z)$ and Z are the numbers of women and men working, respectively. Our original fitted equations in X and Z given above can be rewritten as

$$\hat{Y} = -5.95 + 54.35(X - Z) + 26.95Z,$$

which is what we would have obtained by fitting Eq. (1) by least squares directly. We see that, for the type of data we have here, the women do about twice as much work as the men! The indicated practical conclusion is to use women employees instead of men (or perhaps just different men) in the future, and see how that turns out.

I. 1. Fit model with both X_1 and X_2:

$$\mathbf{b} = (\mathbf{X'X})^{-1}\mathbf{X'Y} = \begin{bmatrix} 7 & 0 & 0 \\ 0 & 68 & -67 \\ 0 & -67 & 68 \end{bmatrix}^{-1} \begin{bmatrix} 46 \\ -66 \\ 69 \end{bmatrix} = \begin{bmatrix} \frac{46}{7} \\ 1 \\ 2 \end{bmatrix}.$$

ANOVA

Source	df	SS	MS	F
b_0	1	302.29		
$b_1, b_2 \| b_0$	2	72.00	36.00	83.72
Residual	4	1.71	$s^2 = 0.43$	
Total	7	376.00		

F is significant at 1% level so we reject $H_0 : \beta_1 = \beta_2 = 0$.

2. Fit model with X_1 alone:

$$\hat{Y} = 46/7 - (66/68)X_1,$$

$$SS(b_1 | b_0) = 64.06.$$

Thus $SS(b_2 | b_1, b_0) = 72 - 64.06 = 7.94$.

Test for $\beta_2 = 0$ (β_1 in model): $F = 7.94/0.43 = 18.53$; significant at 5% level but not at 1% because $F(1, 4, 0.99) = 21.20$.

3. Fit model with X_2 alone:

$$\hat{Y} = 46/7 + (69/68)X_2,$$

$$SS(b_2 | b_0) = 70.01.$$

Thus $SS(b_1 | b_2, b_0) = 72 - 70.01 = 1.99$.

Test for $\beta_1 = 0$ (β_2 in model): $F = 1.99/0.43 = 4.64$ not significant at 5% level, because $F(1, 4, 0.95) = 7.71$.

4. Implications: If X_2 is in, we don't need X_1.

If X_1 is in, X_2 helps out significantly.

Thus X_2 is clearly the more useful variable and alone explains $R^2 = 70.01/73.71 = 0.9498$ of the variation about the mean, whereas X_1 alone explains 0.8691, and X_1 and X_2 together explain 0.9768. Note that X_1 and X_2 are highly correlated in this data set.

J.

$$\mathbf{b} = (\mathbf{X'X})^{-1}\mathbf{X'Y} = \begin{bmatrix} 10 & 0 & 0 \\ 0 & 8 & -4 \\ 0 & -4 & 8 \end{bmatrix}^{-1} \begin{bmatrix} 105 \\ -7 \\ 17 \end{bmatrix} = \begin{bmatrix} 10.50 \\ 0.25 \\ 2.25 \end{bmatrix}.$$

Source	df	SS	MS	F
b_0	1	1102.50		
$b_1, b_2 \| b_0$	2	36.50		
Lack of fit	2	6	3.0	$F = 2.50 \ (<5.79)$
Pure error	5	6	1.2	
Total	10	1151		

No lack of fit; $s^2 = 1.714$

$$\text{Extra SS } F = \{(36.50 - 6.125)/1\}/1.74 = 17.72 \ (\text{df } 1,7),$$

$$> 5.59, \text{reject } H_0.$$

This F is the square of the corresponding t_7-statistic since the df are 1, 7.

Note that the extra SS F for $H_0 : \beta_1 = 0$ versus $H_1 : \beta_1 \neq 0$ is $F = \{(36.50 - 36.125)/1\}/1.714 = 0.219$, not significant. So X_1 could be dropped to give the equation $\hat{Y} = 10.5 + 2.125X_2$.

K. Add a column of 1's to the X_1, X_2 columns to get $\mathbf{X}$.

1.
$$\begin{bmatrix} 5 & 0 & 0 \\ 0 & 4 & 2 \\ 0 & 2 & 2 \end{bmatrix} \begin{bmatrix} b_0 \\ b_1 \\ b_2 \end{bmatrix} = \begin{bmatrix} 50.3 \\ 9.9 \\ 5.7 \end{bmatrix}.$$

2.
$$\mathbf{b} = \begin{bmatrix} 0.2 & 0 & 0 \\ 0 & 0.5 & -0.5 \\ 0 & -0.5 & 1 \end{bmatrix} \begin{bmatrix} 50.3 \\ 9.9 \\ 5.7 \end{bmatrix} = \begin{bmatrix} 10.06 \\ 2.10 \\ 0.75 \end{bmatrix}.$$

3. $\mathbf{b'X'Y} = 531.083$.

4. $s^2 = (531.19 - 531.083)/(5 - 3) = 0.107/2 = 0.0535$.

5. $se(b_0) = (0.2 \times 0.0535)^{1/2} = 0.103$,
  $se(b_1) = (0.5 \times 0.0535)^{1/2} = 0.164$,
  $se(b_2) = (1 \times 0.0535)^{1/2} = 0.231$.

6. $\hat{Y}_0 = (1, 0.5, 0)\mathbf{b} = 11.11$.

7. $se(\hat{Y}_0) = \{\mathbf{X}_0'(\mathbf{X'X})^{-1}\mathbf{X}_0 s^2\}^{1/2} = 0.132$.

8. Set $\beta_2 = 0$, refit, get $SS(b_0, b_1) = 530.5205$; $SS(b_2|b_1, b_0) = 0.5625$.

9. The $\mathbf{V}$ matrix is diag$(1, 1, 0.25, 1, 1)$ with $\mathbf{V}^{-1}$ = weight matrix = diag$(1, 1, 4, 1, 1)$. The new estimates are $\mathbf{b}_W = (9.9625, 2.1, 0.75)$.
[*Note*: Here are some (X_1, X_2, Y) data for a similar exercise: $(-2, -4, 22)$, $(-1, -1, 19)$, $(0, 0, 19)$, $(1, 1, 24)$, $(2, 4, 26)$.]

L. 1. The negative residuals occur at low levels of concentration and the positive residuals occur at high levels of concentration.

2. $\hat{Y} = 2.693374 - 0.277361 X_1 + 0.365028 X_2$.

3.

ANOVA

Source of Variation	df	SS	MS	Calc. F	$F_{0.95}$
Total (corrected)	8	6.2600			
Regression	2	4.7381	2.3690	9.34	5.15
$b_1\|b_0$	1	1.4521	1.4521	5.73	5.99
$b_2\|b_0, b_1$	1	3.2860	3.2860	12.96	5.99
Residual	6	1.5219	0.2536		

a. Since there are no replicates, the lack of fit test cannot be done.

b. A model $\hat{Y} = \bar{Y}$ explains $62.41/68.67 = 90.88\%$ of the crude variation in the data measured from $Y = 0$. Of the remaining variation, the model $\hat{Y} = b_0 + b_1 X_1 + b_2 X_2$ explains 75.69% of it, or a total of 97.78% of the crude variation.

c. The addition of β_2 to the model improves the fit as shown by R^2 going from 23.30% to 75.69%.

4. $R^2 = 75.69\%$.

5. $se (\tilde{b}_1) = 0.00795$, se $(\tilde{b}_2) = 0.10141$.

6.

Batch	Y	$\hat{Y}$	$Y - \hat{Y}$
1	2.100	2.518	-0.418
2	3.000	3.350	-0.350
3	3.200	2.693	0.507
4	1.400	1.774	-0.374
5	2.600	2.781	-0.181
6	3.900	3.248	0.652
7	1.300	1.044	0.256
8	3.400	3.058	0.342
9	2.800	3.234	-0.434

7. Var $\hat{Y}$(coded) $= 0.044871$.

M. 1. The estimated point (b_0, b_1) is

$$(b_0, b_1) = (13.623005, -0.079829).$$

2. The 90% confidence contour for (β_0, β_1) is given by $(\boldsymbol{\beta} - \mathbf{b})'\mathbf{X}'\mathbf{X}(\boldsymbol{\beta} - \mathbf{b}) \leq ps^2F(p, \nu, 1 - \alpha)$. Let

$$\begin{pmatrix} \gamma_0 \\ \gamma_1 \end{pmatrix} = \boldsymbol{\gamma} = \boldsymbol{\beta} - \mathbf{b} = \begin{pmatrix} \beta_0 - b_0 \\ \beta_1 - b_1 \end{pmatrix}.$$

This is equivalent to making a transformation to a new origin at $\mathbf{b}$. So now

$$\boldsymbol{\gamma}'\mathbf{X}'\mathbf{X}\boldsymbol{\gamma} \leq ps^2F(p, \nu, 1 - \alpha). \tag{1}$$

$$\text{LHS} = \boldsymbol{\gamma}'\mathbf{X}'\mathbf{X}\boldsymbol{\gamma} = (\gamma_0, \gamma_1) \begin{bmatrix} n & \Sigma_{i=1}^n X_i \\ \Sigma_{i=1}^n X_i & \Sigma_{i=1}^n X_i^2 \end{bmatrix} \begin{pmatrix} \gamma_0 \\ \gamma_1 \end{pmatrix}$$

$$= [n\gamma_0 + \gamma_1 \Sigma X_i, \gamma_0 \Sigma X_i + \gamma_1 \Sigma X_i^2] \begin{bmatrix} \gamma_0 \\ \gamma_1 \end{bmatrix}$$

$$= (n\gamma_0 + \gamma_1 \Sigma X_i)\gamma_0 + (\gamma_0 \Sigma X_i + \gamma_1 \Sigma X_i^2)\gamma_1$$

$$= n\gamma_0^2 + 2\gamma_0\gamma_1 \Sigma X_i + \gamma_1^2 \Sigma X_i^2.$$

Then Eq. (1) is

$$n\gamma_0^2 + 2\gamma_0\gamma_1 \Sigma X_i + \gamma_1 \Sigma X_i^2 \leq ps^2F(p, \nu, 1 - \alpha) = c, \text{ say.}$$

The confidence region thus consists of the interior points of the ellipse whose boundary is

$$n\gamma_0^2 + 2\gamma_0\gamma_1 \Sigma X_i + \gamma_1^2 \Sigma X_i^2 = c. \tag{2}$$

in the (γ_0, γ_1) space.

To find points on the boundary, we select some value of β_0 and thus of γ_0, and solve the quadratic Eq. (2) for γ_1. There will be two roots (providing an upper and a lower point on the ellipse):

$$\gamma_{11} = \{-\gamma_0 \Sigma X_i - \sqrt{(\gamma_0 \Sigma X_i)^2 - (\Sigma X_i^2)(n\gamma_0^2 - c)}\}/\Sigma X_i^2,$$

$$\gamma_{12} = \{-\gamma_0 \Sigma X_i + \sqrt{(\gamma_0 \Sigma X_i)^2 - (\Sigma X_i^2)(n\gamma_0^2 - c)}\}/\Sigma X_i^2.$$

(Imaginary roots mean we have chosen a γ_0 value that lies outside the ellipse.) We then put $\beta_0 = \gamma_0 + b_0$, $\beta_{1j} = \gamma_{1j} + b_1$ to give rise to two points (β_0, β_{11}), (β_0, β_{12}) in the original (β_0, β_1) space. These are plotted, a new β_0 is chosen, and the whole cycle is repeated until the shape of the ellipse is clear and can be sketched in. We know $\Sigma X_i = 1315$, $\Sigma X_i^2 = 76{,}323.42$, $c = ps^2F(p, \nu, 1 - \alpha) = 2(0.7926)(2.55) = 4.04225$. Via a short Fortran program, we obtain the following output.

β_0	β_{11}	β_{12}
12.4	-0.0614	-0.0561
12.5	-0.0643	-0.0567
12.6	-0.0668	-0.0576
12.7	-0.0691	-0.0587
12.8	-0.0713	-0.0600
12.9	-0.0734	-0.0613
13.0	-0.0755	-0.0627
13.1	-0.0775	-0.0641
13.2	-0.0794	-0.0657

β_0	β_{11}	β_{12}
13.3	−0.0813	−0.0672
13.4	−0.0832	−0.0688
13.5	−0.0850	−0.0705
13.6	−0.0867	−0.0722
13.7	−0.0884	−0.0739
13.8	−0.0901	−0.0757
13.9	−0.0917	−0.0775
14.0	−0.0933	−0.0794
14.1	−0.0948	−0.0813
14.2	−0.0963	−0.0832
14.3	−0.0977	−0.0853
14.4	−0.0991	−0.0873
14.5	−0.1004	−0.0895
14.6	−0.1015	−0.0918
14.7	−0.1025	−0.0942
14.8	−0.1033	−0.0969
14.9	−0.1035	−0.1001

3. The 95% confidence limits for β_1 are

$$b_1 \pm t(23, 0.975)s/\{\Sigma (X_i - \overline{X})^2\}^{1/2}$$

or $-0.0798 \pm (2.069)(0.0105)$, providing the interval $-0.1015 \le \beta_1 \le -0.0581$.
The 95% confidence limits for β_0 are

$$b_0 \pm t(23, 0.975)s/\{\Sigma X_i^2/n \Sigma (X_i - \overline{X})^2\}^{1/2},$$

or $13.623 \pm (2.069)(0.5814)$, providing the interval $12.420 \le \beta_0 \le 14.826$.
The rectangle produced by these two intervals is shown in the figure (Solution M).

Comments
(i) The joint 90% confidence region for the parameters β_0 and β_1, is shown as the long thin ellipse and encloses values (β_0, β_1), which the data regard as jointly reasonable for the parameters.
(ii) If we interpret the 95% confidence intervals of β_0 and β_1 simultaneously (and wrongly) as a "joint 90.25% confidence region" (rectangle), it is clear that we shall be led astray to some extent in this example. See Section 5.5.

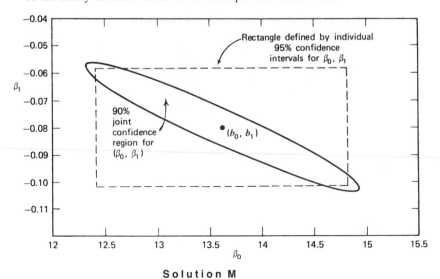

Solution M

N. We have that $\hat{Y}_i - \overline{\hat{Y}}_i = \hat{Y}_i - \overline{Y}$

$$Y_i - \overline{Y} = (\hat{Y}_i - \overline{Y}) + e_i.$$

The numerator of $r_{Y\hat{Y}}$ is the sum of cross-products of these two and it reduces to $\Sigma(\hat{Y}_i - \overline{Y})^2$, the other terms vanishing due to the facts that (a) the residuals are orthogonal to the $\hat{Y}_i$'s and (b) $\Sigma e_i = 0$. [To prove (a): $\hat{Y}'e = (HY)'(I - H)Y = Y'H'(I - H)Y = 0$ because H is symmetric ($H' = H$) and idempotent ($H = H^2$).] The square root of the numerator now cancels out part of the denominator, and what is left is H.

O. The vector on the right-hand side of the appropriate normal equations consists of the three elements $\Sigma e_i = 0$, $\Sigma e_i \hat{Y}_i = 0$, and $T_{12} = \Sigma e_i \hat{Y}_i^2$. Thus all three estimated coefficients depend on T_{12}, which is thus a measure of the amount of quadratic trend in the e_i versus $\hat{Y}_i$ plot.

P. $SS(\text{regression}) = b'X'Y = ((X'X)^{-1}X'Y)'X'Y = Y'X(X'X)^{-1}X'Y$
$$= Y'HY.$$

Remember $H = H^2 = H^3 = \cdots = H^m \cdots$. Also,

$$\hat{Y} = Xb = X(X'X)^{-1}X'Y = HY.$$

So $\hat{Y}'\hat{Y} = \hat{Y}H'HY = \hat{Y}HHY = \hat{Y}HY = SS(\text{regression})$.
$\hat{Y}'H^3Y = \hat{Y}'HY = \hat{Y}'\hat{Y}$.

Q. $X'e = X'(I - X(X'X)^{-1}X')Y$
$= (X' - X')Y = 0$

R. Write X_i' for the ith row of X. Then

$$\sum_{i=1}^{n} V(\hat{Y}_i) = \sum_{i=1}^{n} X_i'(X'X)^{-1}X_i\sigma^2$$

$$= \text{trace}\{X(X'X)^{-1}X')\}\sigma^2 \tag{1}$$

$$= \text{trace}\{(X'X)^{-1}X'X\}\sigma^2$$

$$= p\sigma^2$$

and we divide both sides by n. For step (1) see Appendix 6A.

S. This can be proved algebraically or demonstrated on the computer using one set of errors and different β's.

T. Follows directly from the hint.

U. Follows directly as indicated.

V. We get $\hat{Y} = 6.059 - 30.22U + 34.39U^2$, where $U = (X - 0.048)/0.048$.
Note that, in fitting this model, we have coded the X's to U's. Some programs will not carry out the calculation properly if some sort of coding is not done, because the correlation between X and X^2 for these data is high, 0.990, something that typically happens when polynomial terms are combined in a model and when orthogonal polynomials are not used. The minimum $\hat{Y} = -0.58$ occurs when $U = -(-30.22)/\{2(34.39)\} = 0.4394$, that is, when $X = 0.048 + 0.048(0.4394) = 0.069$. Of course, the value of the minimum is senseless, being negative, but we might hope that, in fact, a positive or zero minimum might really lie around $X = 0.069$. The data set is too small. Nevertheless, it provides a nice example of what might be achievable if one had a lot of good quality data.

W. Solutions in Appendix 7B.

X.   **1.** $\Sigma\,(e_i - \bar{e})(Y_i - \bar{Y}) = \Sigma\,e_i(Y_i - \bar{Y})$ (because $\bar{e} = 0$ if a β_0 term is in the model)

$$= \Sigma\,e_i Y_i \qquad\qquad (\bar{e} = 0)$$

$$= \mathbf{e'Y}$$

$$= \mathbf{e'e} \qquad\qquad \text{[because}$$

$$\mathbf{e'e} = \mathbf{Y'(I - H)'(I - H)Y}$$

$$= \mathbf{Y'(I - H)Y}$$

$$= \mathbf{Y'e}$$

$$= \text{Residual SS} \qquad = \mathbf{e'Y},\text{ where}$$

$$\mathbf{H = X(X'X)^{-1}X'}$$

$$\Sigma\,(e_i - \bar{e})^2 = \Sigma\,e_i^2 = \mathbf{e'e}$$

$$\Sigma\,(Y_i - \bar{Y})^2 = \text{Total corrected SS}$$

$$r_{eY} = \frac{\mathbf{e'e}}{\{(\mathbf{e'e})\,\Sigma\,(Y_i - \bar{Y})^2\}^{1/2}} = \left\{\frac{\text{Residual SS}}{\text{Total corrected SS}}\right\}^{1/2} = \{1 - R^2\}^{1/2}.$$

2. Slope $= \mathbf{e'e}/S_{YY} = (1 - R^2).$

3. $\Sigma\,(e_i - \bar{e})(\hat{Y}_i - \bar{\hat{Y}}) = \Sigma\,e_i\hat{Y}_i$ (by similar reduction to that above)

$$= \mathbf{e'\hat{Y}}$$

$$= \mathbf{Y'(I - H)'HY} \qquad \text{[because } \hat{\mathbf{Y}} = \mathbf{Xb} =$$
$$\mathbf{X(X'X)^{-1}X'Y = HY]}$$

$$= \mathbf{Y'(H - H^2)Y} = 0 \qquad \text{so that } r_{e\hat{Y}} = 0 \text{ also.}$$

Reference: Jackson and Lawton (1967).

Y. The basic reason for the computer's failure to provide estimates is that the experimental design and model used provide a singular $(\mathbf{X'X})$ matrix, which cannot be inverted. Here, $X_1^4 = 10X_1^2 - 9$ and $X_2^4 = 10X_2^2 - 9$, for each row of $\mathbf{X}$.

Both models have the same singularity problems.

Z. 1. $\hat{Y} = 22.561235 + 1.668017X - 0.067958X^2.$

2.

ANOVA

Source of Variation	df	SS	MS	F
Total (corrected)	18	204.481053		
Regression	2	201.994394	100.997197	649.8*
Residual	16	2.486659	0.155416	

The regression is statistically significant.

3. Test for lack of fit (breakdown of residual term in the above ANOVA table)

Source	df	SS	MS	F
Residual	16	2.486659		
Lack of fit	8	1.733325	0.216666	2.30
Pure error	8	0.753334	0.094168	

The lack of fit is nonsignificant since $2.30 < F(8, 8, 0.95) = 3.44$. Thus the quadratic model is sufficient for predictive purposes.

4. $\hat{Y} = 23.346374 + 1.045463X$.

ANOVA

Source of Variation	df	SS	MS	F
Total	18	204.481053		
Regression	1	195.242967	195.242967	359.29*
Residual	17	9.238086	0.543417	
Lack of fit	9	8.484752	0.942750	10.01*
Pure error	8	0.753334	0.094168	

The lack of fit is statistically significant since $10.01 > F(9, 8, 0.95) = 3.39$.

Residuals from the fitted equation also demonstrate the inadequacy of the model.

By plotting the residuals against the values of X, the predictor variable, one sees a definite curvature in the residuals which indicates the need for a second-order term in X.

5. *Conclusions:* The cloud point can be predicted by a second-order function of the percent $I - 8$ in the base stock,

$$\hat{Y} = 22.561235 + 1.668017X - 0.067958X^2.$$

There is no indication that a more complicated model is needed.

AA. First note that the total sum of squares is $4^2 + 2^2 + 1^2 + 5^2 = 46$ and this is the maximum sum of squares for all sets of linear functions. $SS(L_1, L_2) = \mathbf{z'CY}$, where

$$\begin{bmatrix} 10 & -2 \\ -1 & 4 \end{bmatrix} \mathbf{z} = \begin{bmatrix} 15 \\ 6 \end{bmatrix} \quad \text{or} \quad \mathbf{z} = \begin{bmatrix} 2 \\ 2.5 \end{bmatrix} \quad \text{while } \mathbf{CY} = \begin{bmatrix} 15 \\ 6 \end{bmatrix}.$$

So $SS(L_1, L_2) = 30 + 15 = 45$.

$SS(L_1, L_2, L_3)$ can be done similarly, but we can also note that the vector for L_3, $(-3, 0, 0, 3)$, is orthogonal to the vectors for L_1 and L_2, so we can work this out separately as $(\mathbf{c_3'Y})^2/\mathbf{c_3'c_3} = 3^2/18 = 0.5$ and add it on to get $SS(L_1, L_2, L_3) = 45.5$. Any fourth linearly independent linear function L_4 will add exactly 0.5 to this subtotal. Pick any L_4 and try it. An easy (orthogonal) choice is $L_4 = Y_2 - Y_3$.

Chapter 7

A. 1. The plots are not given here.

2. The histogram of residuals is somewhat skewed toward negative values, and a similar characteristic appears in the curved normal plot. The "residuals versus order" plot shows one comparatively low residual value (the 11th). There is also a drop (representing overprediction) in the last four residual values; this may have resulted from better control over the steam plant in the later part of the observed period. The lowest five negative residuals give rise to the skewness mentioned above. The "residuals versus $\hat{Y}$" plot is also affected by these five negative residuals. For that reason, the impression of a widening scatter is probably illusory.

3. $d = 1.39$, $k = 3$, $n = 25$. The d-value is inconclusive at the (two-tailed) 10%, 5%, and 2% levels and is close to being nonsignificant at the 2% level. Thus there is weak evidence of positive serial correlation, perhaps.

4. $n_1 = 10$, $n_2 = 15$, $r = 8$.
$\mu = 13$, $\sigma^2 = 5.5$.
$z = (8 - 13 + \tfrac{1}{2})/(5.5)^{1/2} = -1.919$. The two-tailed p-value is 0.055. As in (3), weak evidence (for a 5% person) of positive serial correlation is indicated.

B. For plots, see the source reference, especially p. 372–373 in which Figures 3, 4, and 5 of Watts and Bacon correspond, respectively, to the plots for our data sets 1, 2, and 3. The runs test results look like this:

Data Set:	1	2	3
r	9	22	13
n_1	19	29	16
n_2	28	17	30
n	47	46	46
z(lower)	−4.33	—	−2.76
z(upper)	—	—	—

Only the two z-values shown are relevant, the others being nowhere near their corresponding tail areas. There are pronounced indications of positive serial correlation in data sets 1 and 3. When this test is applied to unequally spaced data, inaccurate results can arise. Here, however, the long stretches of equally spaced data are reassuring.

C. See the source reference.

D. $\mu = 28.857$, $\sigma^2 = (3.688)^2$, $r = 38$.
$z = (38 - 28.857 - 0.5)/3.688 = 2.3435$. Yes.

E. $z = (5 - 13.48 + 0.5)/2.443 = -3.266$. Yes.

F. $d = 2.33$ is in the upper tail, so use $4 - d = 1.67$. The 2.5% lower tail bounds are $(d_L, d_U) = (1.51, 1.65)$. So $4 - d$ is not significant.

G. The Durbin–Watson statistic cannot exceed 1, a significant value for $n = 51$, $k = 5$. Advice: Conclude evidence of positive serial correlation on the basis of this test.

H. The Durbin–Watson statistic for these residuals is $d = 2225/834 = 2.67$, so $4 - d = 1.33$. This is at the upper end of the range $(d_L, d_U) = (1.16, 1.33)$ for $n = 24$, $k = 1$, in the 2.5% table. So it is not significant, and the answer is no.

Chapter 8

A. The matrix H should behave as described in Section 8.1. When outliers are in the data, the residuals plots may be slightly different.

B. The last point $(6, 4.5)$ is very influential. With it, the line has positive slope; without it, negative. Some observations between $X = 3$ and $X = 6$ would be useful here.

C. A majority of Cook's statistics in a regression are typically small. You seek ones that are large *compared with a majority*. Some computer printouts indicate influential observations and you can see by experience how big the Cook's statistics needs to be to trigger that indicator. There is, however, no firm rule on how big the Cook's statistics should be for an observation to be influential, and the choice of a critical level is arbitrary. When you find an influential observation, try to see where it is in the X-space. Is it an isolated point? Does it have a large residual too? (An influential point *could* have a small or zero residual if it pulls the model to it, remember.) If the influential point is isolated can we "fill in" new data to bridge the gap to the main pattern of the data?

D. See source reference.

Chapter 9

A. H_0 is $\beta_1 = \beta_3$ and $\beta_2 = \beta_4$, so that a suitable (nonunique)

$$C = \begin{bmatrix} 0 & 1 & 0 & -1 & 0 \\ 0 & 0 & 1 & 0 & -1 \end{bmatrix}.$$

B. The reduced model is $E(Y) = \beta_0 + 2\beta_2 X_1 + \beta_2 X_2 = \beta_0 + (2X_1 + X_2)\beta_2$. The fitted equation is $\hat{Y} = 10.06 + 0.980769(2X_1 + X_2)$ with residual sum of squares 0.162 (3 df). Thus $F = \{(0.162 - 0.107)/1\}/\{0.107/2\} = 1.03$. So the hypothesis is reasonable (not rejected).

C. For the full fit, the residual sum of squares is 47.8635 (8 df). The reduced model is $Y = \beta_0 +$

$\beta_1(X_1 + X_3) + \beta_2(X_2 + X_4) + \epsilon$ with fit $\hat{Y} = 274.1 - 1.106(X_1 + X_3) - 2.104(X_2 + X_4)$ and residual sum of squares 2510.1 (10 df). It follows that $F = \{(2510.1 - 47.8635)/(10 - 8)\}/ \{47.8635/8\} = 205.77$ (2, 8 df). H_0 is rejected. The hypothesis pairs the two aluminates $(X_1 + X_3)$ and the two silicates $(X_2 + X_4)$. Both pairs are highly negatively correlated so H_0 looks reasonable at first sight, at least to a layperson.

D. $\hat{Y} = 1.7213 + 0.22434X$. Note that the slope has been much reduced compared with the ordinary least squares fit $\hat{Y} = 1.4364 + 0.3379X$ because of the smaller weight attached to the (possibly faulty) last observation.

E. $\hat{Y} = 1.8080 + 0.19750$. With even less weight on the largest observation, the slope is reduced even further. Down-weighting observations does not always change the fit. For example, an observation with a zero residual can be omitted (weighted zero) without affecting the fitted line.

F. It is obvious that this must be true, but it can be formally proved by applying Eq. (9.5.2) with $\mathbf{C} = (0, 1, 0)$, $\boldsymbol{\beta} = (\beta_0, \beta_1, \beta_2)'$, and $d = 1$. It then emerges, after reduction, that

$$\hat{\beta}_0 = b_0 + C_{01}(1 - b_1)/C_{11},$$

$$\hat{\beta}_1 = 1,$$

$$\hat{\beta}_2 = b_2 + C_{21}(1 - b_1)/C_{11},$$

where b_0, b_1, b_2 are the least squares estimates when there is no restriction on β_1 and $(\mathbf{X'X})^{-1} = \{C_{ij}\}$ for $i, j = 0, 1, 2$. It can then be shown that the same estimates arise from the alternative fit. This requires some tedious algebra. It is easier to work a numerical example both ways on the computer.

G. See F solution first. Again obviously true, and again provable in a similar manner. Forcing a model through specific points can be unwise. (The comments that follow arise from a conversation with J. K. Little.)

Suppose we decide to fit a straight line $Y = \beta X + \epsilon$ through the origin when there are data points at $X = 0$. Because

$$b = \Sigma X_i Y_i / \Sigma X_i^2,$$

the points at $X = 0$ will not contribute to the regression slope estimate at all. Thus the decision to fit a line through the origin means we effectively *ignore the data at the origin,* an unwise move.

Another way of seeing this is that the sum of squares function

$$S(\beta) = \Sigma(Y_i - \beta X_i)^2$$

contains a constant portion $Y_1^2 + \cdots + Y_q^2$, say, for the q data values at $X = 0$, and so these data do not contribute to the fitting process.

A similar point arises when we fit a nonlinear model function that takes zero value at $\xi = \mathbf{0}$ for any value of the parameter vector $\boldsymbol{\theta}$.

The overall moral is that, if we have data at the predictor variables origin, we should be *especially* cautious about fitting models forced *through* the origin, because the fitting process ignores important information. (In general, we advise that, for linear models at least, β_0 should *never* be omitted. One can always test whether β_0 is zero after the fit has been made.)

By extension of the idea above, any criterion such as "minimize

$$\Sigma W_i \{Y_i - f(\mathbf{X}_i, \boldsymbol{\theta})\}^\gamma,$$

where W_i are given weights, independent of $\boldsymbol{\theta}$, and where γ is a chosen constant" applied to data such that $f(\mathbf{X}_i, \boldsymbol{\theta}) =$ constant for some i, independent of $\boldsymbol{\theta}$, will exhibit similar behavior. Forcing a regression equation through any point can thus create potential problems.

Chapter 10

A. We can go directly to part 2 and evaluate $E(\mathbf{b}) = \boldsymbol{\beta} + \mathbf{A}\boldsymbol{\beta}_2$ for both designs. We get

$$A = \frac{8}{8 + n_0} \begin{bmatrix} 1 & 1 & c \\ 0 & 0 & 0 \\ 0 & 0 & 0 \end{bmatrix},$$

where $c = -0.5$ for design A and $c = 0$ for design B. So

$$E(b_0) = \beta_0 + \frac{8}{8 + n_0}(\beta_{11} + \beta_{22} + c\beta_{12})$$

while b_1 and b_2 are unbiased. The symmetric design B is slightly better from this viewpoint. Adding center points reduces the bias directly, through the factor $8/(8 + n_0)$. Design B is also better in that all estimates are uncorrelated and have the same or smaller variances compared with design A. Evaluate $(\mathbf{X}'\mathbf{X})^{-1}\sigma^2$ for both designs to see this. For design B, $V(b_0) = (8 + n_0)^{-1}\sigma^2$, $V(b_1) = \sigma^2/8 = V(b_2)$, whereas for design A, $V(b_0) = (8 + n_0)^{-1}\sigma^2$, $V(b_1) = \sigma^2/6 = V(b_2)$ and $\text{cov}(b_1, b_2) = \sigma^2/12$.

B. For design A,

$$A = \begin{bmatrix} 8 + n_0 & 0 \\ 0 & 8 \end{bmatrix}^{-1} \begin{bmatrix} 0 \\ -4 \end{bmatrix} = \begin{bmatrix} 0 \\ -4/(8 + n_0) \end{bmatrix}.$$

So $E(b_0) = \beta_0$, (unbiased), $E(b_2) = \beta_2 - \dfrac{4\beta_1}{8 + n_0}$.

For design B, $\mathbf{X}'\mathbf{X}_2 = \mathbf{0}$ so both b_0 and b_2 are unbiased. Again, the orthogonal design B is better from this viewpoint.

C.

$$A = (\mathbf{X}'\mathbf{X})^{-1}\mathbf{X}'\mathbf{X}_2 = \begin{bmatrix} 6 & 0 \\ 0 & 70 \end{bmatrix}^{-1} \begin{bmatrix} 70 \\ 0 \end{bmatrix} = \begin{bmatrix} 35/3 \\ 0 \end{bmatrix}.$$

$$E(b_0) = \beta_0 + 35\beta_2/3,$$

$$E(b_1) = \beta_1 \qquad \text{(unbiased)}.$$

D. We need

$$\mathbf{X}' = \begin{bmatrix} 1 & 1 & 1 & 1 & 1 & 1 \\ -5 & -3 & -1 & 1 & 3 & 5 \end{bmatrix}, \qquad \boldsymbol{\beta} = \begin{bmatrix} \beta_0 \\ \beta_1 \end{bmatrix},$$

$$\mathbf{X}_2' = [25, 9, 1, 1, 9, 25], \qquad \beta_2 = \beta_2,$$

$$n = 6, \; p = 2,$$

$$A = \begin{bmatrix} 35/3 \\ 0 \end{bmatrix}.$$

We then find that

$$E\,(\text{MS due to } b_0) = \sigma^2 + (6\beta_0 + 70\beta_2)^2/6,$$

$$E\,(\text{MS due to } b_1|b_0) = \sigma^2 + 70(\beta_1^2 + 7\beta_2^2),$$

$$E\,(\text{MS due to lack of fit}) = \sigma^2 + 1792\beta_2^2/3.$$

Note that, if $\beta_2 = 0$, a test for $H_0: \beta_1 = 0$ becomes feasible as described in the text. Note also that only the first expectation contains β_0.

E. The key difference between this quadratic term and the previous one in Exercise C is that the current vector $\mathbf{X}_2$, which is $(5, -1, -4, -4, -1, 5)'$, is orthogonal to both columns of $\mathbf{X}$ so that $\mathbf{X}'\mathbf{X}_2 = 0$. This means that $\mathbf{A} = \mathbf{0}$ and so no biases exist. Note that if we rewrite the quadratic model function of Exercise C as

$$\{\beta_0 + 35\beta_2/3\} + \beta_1 X + (8\beta_2/3)\{0.375(X^2 - 35/3)\}$$

with all estimates now orthogonal since $\mathbf{X}'\mathbf{X}$ is diagonal, it becomes apparent that the estimate of the first coefficient is an estimate of $\beta_0 + 35\beta_2/3$, as given in Exercise C.

F. 1. Either express the lack of fit mean square as a quadratic form and apply the result in Section 10.3 directly, or evaluate

$$E(\mathbf{Y}'\mathbf{Y}) = E(\mathbf{Y}'\mathbf{I}\mathbf{Y})$$

$$= E(\mathbf{Y}')E(\mathbf{Y}) + \text{trace } \mathbf{I}\sigma^2$$

$$= (\mathbf{X}\boldsymbol{\beta} + \mathbf{X}_2\boldsymbol{\beta}_2)'(\mathbf{X}\boldsymbol{\beta} + \mathbf{X}_2\boldsymbol{\beta}_2) + n\sigma^2$$

and then obtain E(lack of fit sum of squares) by difference, assuming the other results in the table, and remembering to multiply by the degrees of freedom where necessary. Remember that $(\mathbf{I} - \mathbf{H}) = (\mathbf{I} - \mathbf{H})^2$, where $\mathbf{H} = \mathbf{X}(\mathbf{X}'\mathbf{X})^{-1}\mathbf{X}'$ and that $(\mathbf{I} - \mathbf{H})\mathbf{X} = \mathbf{0}$ and

$$(\mathbf{I} - \mathbf{H})\mathbf{X}_2\boldsymbol{\beta}_2 = (\mathbf{X}_2 - \mathbf{X}\mathbf{A})\boldsymbol{\beta}_2.$$

2. E(residual mean square$|\beta_2 = 0) = (f\sigma^2 + e\sigma^2)/(f + e) = \sigma^2$.

G.

$$\mathbf{A} = \begin{bmatrix} 5 & 83 \\ 83 & 2189 \end{bmatrix}^{-1} \begin{bmatrix} 0.5 \\ 5.0 \end{bmatrix} = \begin{bmatrix} 0.167530 \\ 0.004068 \end{bmatrix}.$$

b_0 is biased by 0.168β; b_1 is biased by 0.004β.

Chapter 11

A. Results depend on α and the degrees of freedom used. For example, if a plane $Y = \beta_0 + \beta_1 X_1 + \beta_2 X_2 + \epsilon$ is fitted to 20 observations, $\nu_1 = 2$, $\nu_2 = 17$ and $F(\nu_1, \nu_2, 1 - 0.05) = 3.59$. Using Eq. (11E) gives $R^2 = 2(3.59)/(2(3.59) + 17) = 0.2969$. So a "just significant" regression explains only about 30% of the variation about the mean $\overline{Y}$.

B. Here $\nu_1 = 2$, $\nu_2 = 30$, $\alpha = 0.05$. $F(2, 30, 0.95) = 3.32$. So $R^2 = 2(3.32)/(6.64 + 30) = 0.1812$, quite low.

C. Because $n = 46$ and the number of parameters fitted is 6, then $\nu_1 = 5$, $\nu_2 = 40$ and we need an F such that, at minimum, $0.90 = 5F/(5F + 40)$, that is, $F = 72$. Note that, since there are 25 degrees of freedom for pure error, there will be a maximum possible value for R^2 in such a set of data. It may well lie below the desired 90%!

D. 1. There are $n = 50$ observation and $\nu_1 = 5$, $\nu_2 = 44$. The observed $F = 10F(5, 44, 0.95) = 24.3$ and $R^2 = 5(24.3)/\{5(24.3) + 44\} = 0.7341$.

2. There are $5(7) + 9 = 44$ degrees of freedom for pure error and these comprise the entire residual degrees of freedom. Thus no further increase in R^2 is possible.

Chapter 12

A. 1. The model contains ten parameters. Examination of the data reveals only eight different data sites. Thus it is impossible to fit the model as stated.

2. $s^2 = 4.325$ with 10 degrees of freedom, calculated from pure error.

B.

$$a_0 = 12.565663, \qquad b_0 = 0.038437, \qquad c_1 = -0.032454,$$

$$a_1 = -0.006327, \qquad b_1 = -0.013571, \qquad c_2 = 0.001248,$$

$$a_2 = -0.090698, \qquad b_2 = 0.001376, \qquad c_3 = 0.000198.$$

ANOVA

Source	df	SS	MS	F
Regression$\mid b_0$	8	44.40797	5.55100	
Residual	11	0.03989	0.00363	
Lack of fit	6	0.03461	0.00577	5.46, significant at $\alpha = 0.05$ level
Pure error	5	0.00528	0.001056	
Total (corrected)	19	44.44785		

$R^2 = 44.40797/44.44785 = 0.9991$. This is an interesting set of data. Practically all the variation is explained by the model, but the lack of fit test is significant. This would lead one to first question whether the repeat runs are really *that* precise. If they are, the analysis says that there is still variation that is in excess of natural variation, which could be explained. (From a practical point of view, however, most of the variation *is* explained, so why not use the equation? The author did.) A plot of the residuals in the X-space shows up a possible $X_1 X_2$ interaction and suggests the possible addition of terms $(\delta_0 + \delta_1 Z + \delta_{11} Z^2) X_1 X_2$ to the model. If this is done, δ_{11} cannot be estimated separately because of column dependence in the $\mathbf{X}$ matrix and the least squares estimates are

$$a_0 = 11.918982, \qquad b_0 = 0.049215,$$

$$a_1 = 0.057033, \qquad b_1 = -0.014627,$$

$$a_2 = -0.090698, \qquad b_{11} = 0.001376,$$

$$c_0 = -0.001660, \qquad d_0 = 0.000513,$$

$$c_1 = -0.001769, \qquad d_1 = -0.000050.$$

$$c_{11} = 0.000198,$$

The extra SS for d_0 and d_1 given the other estimates is 0.02183, which leads to an F of $5.43 > F(1, 9, 0.95) = 5.12$, just significant. The new lack of fit F is $3.03 < F(4, 5, 0.95) = 5.19$, not significant, and R^2 goes up very slightly to 0.9996. The regression F-value (given a_0) is 2214, highly significant.

C. $\widehat{\log Y} = -4.927 + 16.85\{1000/(T + 460)\}$.

ANOVA

Source	df	SS	MS	
b_0	1	238.731		
$b_1 \mid b_0$	1	1.785	1.785	
Lack of fit	2	0.064	0.032	1.24, not significant
Pure error	20	0.515	0.026	
Total	24	241.095		

Test for regression $F = 1.785/(0.579/22) = 67.87 > F(1, 22, 0.95) = 4.30$, very significant. $R^2 = 0.7552$.

The residuals show a declining scatter as T increases, an indication that the log Y transformation has not adequately removed the inhomogeneity of variance in the original set of data.

Further (or different) transformation and use of weighted least squares are two possibilities to consider.

The increasing slope of residuals when plotted against fitted values within each temperature is a spurious "effect" that is caused by the grouping. This essentially (for equal groups) forces the straight line to attempt to fit the group means, so inevitably the smaller observations *within* the group will here have lower residuals, and so on. The overall (ungrouped) plot shows only the funnel effect noted above.

D. The design is *not* rotatable. If it were, the axial points would be at distance 1.682 from the origin, not 1.2154.

Response Y_1: A second-order model is not needed. The plane $\hat{Y}_1 = 76.220 - 7.013X_1 - 3.324X_2 - 4.305X_3$ explains 76.8% of the variation about the mean. $\hat{Y}$ increases when all three X's are reduced.

Response Y_2: Again, a plane $\hat{Y}_2 = 63.573 - 10.133X_1 - 5.381X_2 - 6.009X_3$ explains a lot of the variation about the mean, 82.1%, and the quadratic curvature is nonsignificant. Reducing X's again increases the predicted response.

Response Y_3: There is not much variation in these data. A plane explains only 46.6% of the variation around the mean. Adding second-order terms increases this to 83.7% even though the second-order terms as a group are not significant, and nor are any of them marginally. The best second-order term is the one in X_1^2. Adding this we find that the equation $\hat{Y} = 96.626 + 0.1598X_1 - 0.3091X_2 - 0.3803X_3 + 0.5036X_1^2$ explains 64.3% of the variation about the mean. This may be a reasonable compromise using the original response variable. There is very little variation in the response data so a transformation would do little to help.

E. Use $X_1 = $ (fat -12)/4, $X_2 = $ (flour -20)/10, $X_3 = $ (water -50)/4, and $X_4 = $ (rpm -130)/40 to translate the first point into $(-1, -1, -1, -1)$, and so on. The axial points are at a distance 1 and the design is not rotatable because a distance 2 is needed to achieve that. A plane explains 64.5% ($R^2 = 0.645$) and the full quadratic 82.9%. Only b_2 and b_4 show large marginal t-values in the planar fit. Adding the quadratic terms weakens the first-order coefficients but draws attention to the $b_{24}X_2X_4$ term. A possible compromise is to fit $\hat{Y} = 552.44 - 37.78X_2 - 84.39X_4 + 2.30X_2^2 + 32.80X_4^2 + 34.63X_2X_4$, which explains $R^2 = 0.718$.

F. 1. The fitted surface is

$$\hat{Y} = 6.087 - 0.240X_1 + 0.340X_2 - 0.495X_3$$

$$-0.036X_1^2 + 0.021X_2^2 + 0.118X_3^2$$

$$+ 0.040X_1X_2 - 0.070X_1X_3 + 0.045X_2X_3.$$

2. The analysis of variance table takes the form:

ANOVA

Source	df	SS	MS	F
Mean (b_0)	1	758.173		
First order	3	5.671	1.890	
Second order$\mid b_0$	6	0.299	0.050	
Lack of fit	5	0.839	0.168	2.37
Pure error	5	0.354	0.071	
Total	20	765.336		

3. No lack of fit is shown and the second-order terms do not make a significant contribution to explaining the variation in the data when the mean square due to second order is compared to the residual mean square $s^2 = (0.839 + 0.354)/(5 + 5) = 0.119$, which has ten degrees of freedom.

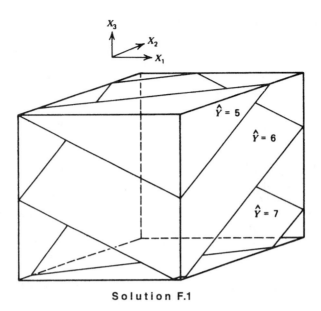

Solution F.1

4. When a (reduced) first-order model is fitted to the data, the fitted equation takes the form

$$\hat{Y} = 6.157 - 0.240X_1 + 0.340X_2 - 0.495X_3,$$

there is no significant lack of fit, and the new estimate of σ^2 is $s^2 = 0.093$ based on 16 degrees of freedom.

Note: The fitted (planar) contours take the form illustrated in Solution F1. The cube shown has vertices (±2, ±2, ±2).

An alternative representation is shown in Solution F.2. The three parts of this figure show straight line contours for X_1 and X_2 in the three planes $X_3 = -1, 0$, and 1. The contours must be imagined as continuous between these three planes. Design points that fall on the planes are shown as dots. Two axial points that do not lie on any of the planes are not shown. Note that these contours cover a smaller region than that shown by the other figure.

A series of cross-sectional diagrams of this type is especially useful when second-order contours must be examined. For larger numbers of predictors than three, the methods of canonical reduction are especially useful.

We leave the examination of the residuals to the reader.

G. Yes.

H. The model chosen is

$$\hat{Y} = 120.627 + 490.412X_2 - 5.716X_3 - 1107.847X_2^2.$$

A plot of the residuals reveals runs of + and − signs indicating the presence of unconsidered X-variables. Adding second-order terms in X_2 and X_3 is of only marginal help. This equation has $R^2 = 90.27\%$, with a standard deviation of 6.2233.

The model in Anderson and Bancroft is

$$\hat{Y} - 84.204 = 2.463(X_1 - 1.86) - 75.369(X_2 - 0.188)$$
$$+ 1.584(X_3 - 7.64)$$
$$- 1.380(X_1X_2 - 0.3507).$$

This model is not as good a fit. The residuals have a definite pattern and $R^2 = 75.49\%$.

Warning: The example in Anderson and Bancroft was used to illustrate regression calculations; there was no intention of building a best model.

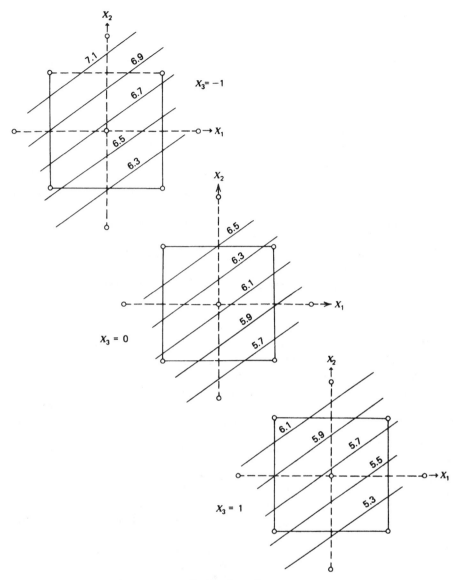

Solution F.2

Chapter 13

A.

X	Y	$Z = \log_{10} Y$	$V = \log_{10} \log_{10} Y$
1830	30	1.4771	0.1694
1905	130	2.1139	0.3251
1930	400	2.6021	0.4153
1947	760	2.8808	0.4595
1952	1500	3.1761	0.5019
1969	25000	4.3979	0.6432

1. The plot of Y versus X shows all the "action" as having been in the most recent years and the plot is not particularly informative. The need for a transformation is apparent. In such a situation, proportional increases, determined by Z, are often more informative.

2. The plot of Z versus X is much easier to assess and is preferable to the plot of Y versus X.

3. $U \equiv V$ is not bad, although you may find a better one.

4. $\hat{V} = -5.4874 + 0.00307284X$.

The residuals plots overall and against time are somewhat unsatisfactory but with only six residuals and four degrees of freedom between them, we cannot hope to say much more.

5.

		ANOVA		
Source	df	SS	MS	F
$b_1 \mid b_0$	1	0.1185	0.11850	41.58
Residual	4	0.0114	0.00285	
Total (corrected)	5	0.1299		

$41.58 > F(1, 4, 0.99) = 21.20$, significant at 1%. $R^2 = 0.9122$. Most of the variation in the data is explained by the model.

Note that covariance $(b_0, b_1) = -0.999717$, very high. This is caused by the remoteness of the origin, and a reparameterization by taking a new origin closer to $\bar{X} = 1922.167$ would help. (See the closely related comments in Section 24.4.)

6. Use inverse interpolation by setting

$$\log\{\log(186,000 \times 3600)\} = -5.4874 + 0.00307284\hat{X}$$

$$\hat{X} = 2094.$$

7. This is a risky extrapolation of a small, difficult set of data under a tentative transformation that may well not hold up for new data. It depends on the trend being maintained for another 100 years or so. Don't rely on it!

B. In both solutions, values of $\lambda = -1(0.1)1$ were used. A finer grid would give greater accuracy. We leave detailed examination of the residuals to the reader.

First response (N990), $n = 23$. The best λ is around -0.6, where $S(\hat{\lambda}, \mathbf{V}) = 23.87$ is the residual sum of squares. A conservative (large) 95% confidence band based on $S(\lambda, \mathbf{V}) = 23.87 \exp(3.84/20) = 28.92$ stretches from -1.1 to -0.2, very roughly. The inverse square root $\lambda = -0.5$ might be a good compromise here, explaining 0.975 of the variation about the mean.

Second response (Silica B), $n = 24$. The best λ is around -0.3, where $S(\hat{\lambda}, \mathbf{V}) = 46.1$. A 95% conservative confidence band based on $S(\lambda, \mathbf{V}) = 46.1 \exp(3.84/21) = 67.06$ stretches from roughly -0.7 to 0.1 so that both -0.5 and 0 are possible choices. The corresponding R^2 values range from about 0.974 to 0.971 with better values in between. For $\hat{\lambda} = 0.3$, a very significant regression explains $R^2 = 0.981$ of the variation about the mean. The more appealing inverse square root value $\lambda = -0.5$ comes close with $R^2 = 0.979$.

If the same transformation were to be used for both responses, the inverse square root would be a sensible choice.

C. (Y_1 data)

λ	$L_{\max}(\lambda)$
-1.0	-22.9
-0.5	-11.0
-0.4	-9.5
-0.3	-8.0
-0.2	-6.5
-0.1	-5.3
0.0	-4.8
0.1	-5.3
0.2	-6.5
0.3	-8.1
0.4	-9.8
0.5	-11.5
1.0	-24.8

An appropriate 95% confidence interval is $-0.35 \leq \lambda \leq 0.32$. We select $\lambda = 0$, that is, use the transformation $W = \ln Y_1$. The fitted equation is $\hat{W} = 4.234 + 0.204X_1 + 0.098X_2 - 0.139X_3 - 0.070X_4$.

$R^2 = 0.9963$. All partial F-values for individual coefficients are highly significant. With only six degrees of freedom in eleven residuals, residual examination cannot be expected to be very revealing and it isn't.

(Y_2 data: solution not provided.)

D. The best value of λ is at about 2.1. The widest confidence interval based on $S(\lambda, \mathbf{V}) = 86.50$ $\exp(3.84/24) = 114$ is roughly $1.4 \leq \lambda \leq 3.0$. (Substituting the residual degrees of freedom 14 for $n = 24$ changes this to about 1.5 to 2.8, still quite wide.) Using $\lambda = 1$ permits a second-order fit that explains $R^2 = 0.906$ and using $\lambda = 2.1$ raises this to 0.940. This fact, combined with the wide confidence band for λ makes it clear that using a transformation is probably not worthwhile.

E. $\hat{\lambda} = 0.11$. The 95% confidence band is about $-0.16 \leq \lambda \leq 0.39$, so we can take $\lambda = 0$, that is, use $W = \ln M$. Then,

$$\hat{W} = -5.728 + 2.031 \ln T,$$
$$R^2 = 0.8056,$$
$$SS(b_1|b_0) = 16.240 \text{ (1 df)},$$
$$\text{Residual SS} = 3.919 \text{ (41 df)},$$
$$\text{Total, corrected} = 20.159 \text{ (42 df)},$$
$$F = 169.9 > F(1, 41, 0.99) = 4.08, \text{ very significant regression.}$$

Plots of the residuals versus T and $\hat{Y}$ show nothing unusual. The Durbin–Watson $d = 1.94$ obtained from the listed order is not significant.

F. $\widehat{\log Y} = 1.9929 + 0.5428 \log X_1 + 0.2740 \log X_2$.

Only 37.52% of the variation is explained by this model. The regression is not significant, the overall F-value being $(6.49/2)/(12.03/13) = 3.51$, which is $<F(2, 13, 0.95) = 3.81$.

The sixth residual is extremely large and negative (-2.626), indicating that the corresponding Y seems far too low.

Plots of residual versus $\widehat{\log Y}$ exhibit curvature.

It seems doubtful that the errors are additive and it would be sensible to try the nonlinear model $Y = \alpha X_1^\beta X_2^\gamma + \epsilon$ for which the methods of Chapter 24 are needed. Initial estimates for the parameters (needed for nonlinear estimation) would be taken from the model fitted above, namely, $\alpha_0 = 10^{1.9929} \div 100$, $\beta_0 = 0.5428$, and $\gamma_0 = 0.2740$.

G. Model $Y = \alpha X_1^\beta X_2^\gamma X_3^\delta \cdot \epsilon$.

By taking logarithms to the base e we can convert the model into the linear form.

$$\ln Y = \ln \alpha + \beta \ln X_1 + \gamma \ln X_2 + \delta \ln X_3 + \ln \epsilon$$

or

$$\widehat{\ln Y} = 8.5495297 + 0.1684244 \ln X_1$$
$$-0.537137 \ln X_2 - 0.0144135 \ln X_3.$$

X_2 or peripheral wheel velocity.

All except $X_3 =$ feed viscosity, which provides an F-ratio of

$$2.15 < F(1, 31, 0.95) = 4.16.$$

With 95.52% variation explained and a small standard deviation of 1.563% of the response mean, this looks like a good prediction equation. Plots of residuals reveal no peculiarities.

H. $\hat{Y} = 0.4875 + 0.13X_1 - 0.08X_2 + 0.125X_3 - 0.0975X_4$ explains $R^2 = 0.907$ of the variation about the mean; $s = 0.115$.

Chapter 14

A. Set up a dummy variable Z with values 1, 3, 4, 6, 10, 13, 14, 15, 16, 17, 19 corresponding to November 17, 19, . . . , December 5. The model $Y = \beta_0 + \beta_1 X_1 + \beta_2 X_2 + \alpha_1 Z + \alpha_2 Z^2 + \epsilon$ is then fitted. (Note that a Z term as well as Z^2 term is needed to properly represent the quadratic trend. We also would need an α_0 term if we did not already have a β_0 in the model.) Alternatively, we could use dummies $(Z - 1)$ or $(Z - \bar{Z})$; these would change the interpretation of the constant term only.

B. To account for the tillers, we employ two dummy variables Z_1 and Z_2, assigning $Z_1 = 1$ for tiller 1 and zero elsewhere, and $Z_2 = 1$ for tiller 2 and zero elsewhere. A third dummy, Z_3, is assigned the value one for the Waldron variety and zero for the Ciano variety. The N rate variable is called N. We can then fit by least squares the equation

$$\hat{Y} = -333.5 + 185.01X - 7.801X^2 - 0.1294N$$

$$+ 106.0Z_1 + 33.7Z_2 + 19.6Z_3.$$

(We quote sufficient significant figures to produce predictive accuracy of the same order as that of the original data, that is, to the nearest unit.) The analysis of variance table is as follows:

ANOVA

Source	df	SS	MS	F
$b_1 \mid b_0$	1	1,112,031	1,112,031	176.71
$b_{11} \mid b_0, b_1$	1	33,968	33,968	5.40
$b_2 \mid b_0, b_1, b_{11}$	1	20,392	20,392	3.24
$a_1, a_2, a_3 \mid$ others	3	14,994	4,998	0.79
Residual	11	69,225	6,293	
Total (corrected)	17	1,250,610		

$F(1, 11, 0.95) = 4.84$; $F(1, 11, 0.99) = 9.65$.

We see that the total contribution of the dummy variables is not significant. Omission of the dummies leads to an $R^2 = 0.933$ from the fitted equation $\hat{Y} = -338.3 + 200.3X - 7.594X^2 - 0.3696N$, whereas retention of the dummies provides $R^2 = 0.945$, an unimportant increase. Note, however, that if *only* the three dummy variables are used, they account for $R^2 = 0.571$ of the variation about the mean. This is due to correlation between the dummies and X; note, for example, the low tiller 3 responses.

The N rate is, not surprisingly, highly correlated ($r = 0.672$) with X and ($r = 0.629$) with X^2. Its nonsignificant contribution could be omitted from the equation, as could those of the dummies. If this is done, we have $\hat{Y} = -337 + 196.0X - 8.13X^2$ with $R^2 = 0.916$. A plot of the data confirms what the analysis of variance table shows: that the plot has only a slight quadratic bend in it and that the first-order term takes up most of the variation. The values of Y, $\hat{Y}$, $se(\hat{Y})$, e, and the standardized residuals are tabulated below for further examination by the reader.

Y	$\hat{Y}$	$se(\hat{Y})$	e	$e/se(e)$
370	343	25	27	0.34
659	652	32	7	0.09
935	803	50	132	1.97
390	263	24	127	1.59
753	683	32	70	0.91
733	819	62	-86	-1.53
182	192	26	-10	-0.13
417	462	31	-45	-0.58
686	697	31	-11	-0.14
188	148	30	40	0.51
632	546	33	86	1.12
538	690	31	-152	-1.96

Y	$\hat{Y}$	$se(\hat{Y})$	e	$e/se(e)$
27	22	46	5	0.07
141	148	30	-7	-0.09
262	263	24	-1	-0.01
34	133	31	-99	-1.28
222	192	26	30	0.37
242	355	26	-113	-1.43

C. 1.

$$X = \begin{array}{cccc} X_0 & X_1 & X_2 & X_3 \\ \begin{bmatrix} 1 & 1 & 0 & -1 \\ 1 & 1 & 0 & 0 \\ 1 & 1 & 0 & 1 \\ 1 & 0 & 1 & -1 \\ 1 & 0 & 1 & 0 \\ 1 & 0 & 1 & 1 \\ 1 & -1 & -1 & -1 \\ 1 & -1 & -1 & 0 \\ 1 & -1 & -1 & 1 \end{bmatrix} \end{array}$$

$$\hat{Y} = 248 + 2X_1 - 10X_2 - 7.33X_3.$$

2.

ANOVA

Source of Variation	df	SS	MS	F
Total (corrected)	8	1466.0		
Regression	3	826.7	257.57	NS
$b_1 \mid b_0$	1	54.0⎫	252.00	NS
$b_2 \mid b_0, b_1$	1	450.0⎭504.0		
$b_3 \mid b_0, b_1, b_2$	1	322.7	322.70	NS
Residual	5	639.3	127.9	

Operator differences are not statistically significant; that is, $\dfrac{252.0}{127.9} = 1.97$ is less than $F(2, 5, 0.95) = 5.79$.

Operator No. 1: $\hat{Y} = 248 + 2(1) = 250$;

Operator No. 2: $\hat{Y} = 248 - 10(1) = 238$;

Operator No. 3: $\hat{Y} = 248 + 2(-1) - 10(-1) = 256.$

3. There is not sufficient evidence to say line speed affects bar appearance with an α risk of 0.05.

4. Residual plots indicate a second-order model with line speed a better choice.

D. Such problems can often be done several ways, for example:

1. Add the Z_0 vector to all the others. The resulting six vectors are clearly linearly independent, so the system will work.

2. More tediously, solve

$$aZ_0 + bZ_1 + cZ_2 + dZ_3 + eZ_4 + fZ_5 = 0. \tag{1}$$

If $a = b = c = d = e = f = 0$, then columns are linearly independent and the system works. If any of $a, b, \ldots, f$ are *nonzero*, it does not work.

Here we get

$$a + b - c - d - e - f = 0,$$
$$a - b + 2c - d - e - f = 0,$$
$$a - b - c + 3d - e - f = 0,$$
$$a - b - c - d + 4e - f = 0,$$
$$a - b - c - d - e + 5f = 0,$$
$$a - b - c - d - e - f = 0.$$

Subtraction of the last equation from the others in succession shows that $b = c = d = e = f = 0$, whereupon $a = 0$ from the last equation. It works.
   3. Find the determinant of the 6×6 matrix. If it is zero, the system fails; if nonzero, it works. Here the determinant has value -720 and so the system works. To get the determinant in a computer, request the eigenvalues. The determinant is the product of these.
E. Remember to put in the Z_0 column of 1's. Refer to the D solution for methods. It works.
F. 1. Leave in u and v and write out the six equations that would hold ($a\mathbf{Z}_0 + \cdots + f\mathbf{Z}_5 = 0$) for linear dependence. Solve them to obtain $a = b = c = d = e = f = 0$, for *any* values of u and v. So the system works if $u = v = 0$ in particular.
   2. Since it works for any u and v, the answer is no.
   3. Not relevant.

G. System B is given in the text as a suitable one.
   System A also is ok. Putting subscripts on the columns we can see that $X_{0A} = X_{0B}$, $X_{1A} = X_{1B} - 5X_{3B}$, $X_{2A} = X_{2B} - X_{3B}$ and $X_{3A} = X_{3B} - X_{0B}$. Thus the columns of A are linearly independent columns derivable from linear combinations of the columns of B. This can also be done by showing that $\hat{Y}_A$ and $\hat{Y}_B$ are identical.
H.

$$
\mathbf{X} =
\begin{array}{c}
\begin{array}{ccccc} X_0 & X_1 & X_1^2 & X_2 & X_3 \end{array} \\
\begin{bmatrix}
1 & 1 & 1 & 0 & 0 \\
1 & 2 & 4 & 0 & 0 \\
1 & 3 & 9 & 0 & 0 \\
1 & 4 & 16 & 0 & 0 \\
1 & 5 & 25 & 1 & 1 \\
1 & 5 & 25 & 2 & 1 \\
1 & 5 & 25 & 3 & 1
\end{bmatrix}
\end{array}
$$

$$Y = \beta_0 + \beta_1 X_1 + \beta_{11} X_1^2 + \beta_2 X_2 + \beta_3 X_3 + \epsilon.$$

As in all dummy variable situations, many other valid answers are possible.
I. Using the dummy variable suggested in the similar example in the text, we get $\hat{Y} = -0.5 + 2X_1 + X_2 + 0.2X_3$ with separate equations $\hat{Y}_1 = -0.5 + 2X_1$ and $\hat{Y}_2 = 9.7 + X_2$ for the two lines that intersect at $X_1 = 5.2$ (or $X_2 = 0.2$).
J. We can use dummies with values

$$X_1 = 1, 2, 3, 4, 5, 6, 7, 8, 8, 8, 8, \ldots, 8,$$
$$X_2 = 0, 0, 0, 0, 0, 0, 0, 0, 1, 2, 3, \ldots, 64,$$
$$\mathbf{X}_3 = 0, 0, 0, 0, 0, 0, 0, 1, 1, 1, 1, \ldots, 1.$$

to get

$$\hat{Y} = 0.421429 + 0.039643X_1 + 0.004062X_2 + 0.016366X_3.$$

$SS(b_1, b_2, b_3 \mid b_0) = 1.00910$ (3 df).
Residual SS $= 0.03829$ (68 df).
$F = 597.3 > F(3, 68, 0.99) = 4.10$ (approximately). Significant regression. $R^2 = 0.9634$.
Overall an excellent fit appears to have been achieved by these criteria (but see below).
   The lines intersect at an age of 7.96 ($X_1 = 8.46$).
   The Durbin–Watson statistic is $d = 2.514$; $4 - d = 1.486$ is significant at the 5% level indicating negative serial correlation in the residuals, which needs further investigation. The corresponding upper-tail runs test is also significant ($r = 46$, $n_1 = 32$, $n_2 = 40$, $z = 2.27$, upper-tail probability, 0.0116). The presence of negative serial correlation affects the validity of the regression tests above to some extent. A possible next step would be to estimate the value of ρ and use generalized least squares.
   Note: In the original paper, the data were fitted by a quadratic curve and a line rather than two straight lines as here, and no assumption was made as to which points were on which portions of the model. For this more general fit, see the quoted reference.

K. We can use the same dummy layout as in the previous exercise except that the values run only to 32 observations here. Now

$$\hat{Y} = 0.421429 + 0.039643X_1 + 0.004277X_2 + 0.012905X_3.$$

$$SS(b_1, b_2, b_3 \mid b_0) = 0.33925 \text{ (3 df).}$$

$$\text{Residual SS} = 0.02152 \text{ (28 df).}$$

$$F = 147.13 > F(3, 28, 0.99) = 4.57; \quad \text{significant regression.}$$

$R^2 = 0.9403$. Again an excellent fit by these criteria, but see below.
   The lines intersect at an age of 7.86 ($X_1 = 8.36$).
   The Durbin–Watson statistic is $d = 2.722$; $4 - d = 1.278$ is almost but not quite significant at the 5% level (where $d_L = 1.24$). The corresponding upper-tail runs test is significant at the 2.5% level (one tail), however. ($r = 23$, $n_1 = 18$, $n_2 = 14$, $z = 2.10$, upper-tail probability, 0.0179). The possibility of negative serial correlation affects the validity of the regression tests above to some extent. A possible next step would be to estimate the values of ρ_s and use generalized least squares.
   (Read the note in the previous solution.)

L. If we attach a dummy variable Z to distinguish the two groups, we can look at all four possibilities at once. For example, with $Z = 0$ for set A and $Z = 1$ for set B, we can fit the model $Y = \beta_0 + \beta_1 X + \alpha_0 Z + \alpha_1 XZ + \epsilon$. The fitted equation is $\hat{Y} = 1.142 + 0.506X - 0.0418Z - 0.0360XZ$. The last two coefficients have large standard errors. We can test if a single line is sufficient by ignoring the Z and ZX terms to fit $\hat{Y} = 1.075 + 0.492X$. The extra sum of squares $F = (0.1818/2)/(0.3272/4) = 1.11$. So a single straight line seems appropriate.

M. For the model, see Exercise N. For testing for two "parallel" quadratics, H_0: $\alpha_1 = \alpha_{11} = 0$ is appropriate.

N. 1. $\alpha_0 = \alpha_1 = \alpha_{11} = 0$.
   2. $\beta_{11} = \alpha_0 = \alpha_1 = \alpha_{11} = 0$.
   3. By fitting the model as given and setting first $Z = -1$ and then $Z = 1$.
   4. No, their Z values would be different.

O. See Section 14.2.

P. Solution is implicit in the question.

Q. Models: $Y_{iu} - \overline{Y}_i = \beta_i(X_{iu} - \overline{X}_i)$, $i = 1, 2, \ldots, m$
$u = 1, 2, \ldots, n$

1.

X_1	Y_1	$(X_{iu} - \overline{X}_i)$	$(Y_{iu} - \overline{Y}_i)$
3.5	24	−1.529	−17.286
4.1	32	−0.929	−9.286
4.4	37	−0.629	−4.286
5.0	40	−0.029	−1.286
5.5	43	0.471	1.714
6.1	51	1.071	9.714
6.6	62	1.571	20.714

$$\overline{X}_1 = 5.029, \quad \overline{Y}_1 = 41.286, \quad n_1 = 7,$$

$$b_1 = \left\{ \sum_{u=1}^{7} (X_{iu} - \overline{X}_i)(Y_{iu} - \overline{Y}_i) \right\} \bigg/ \left\{ \sum_{u=1}^{7} (X_{iu} - \overline{X}_i)^2 \right\}$$

$$= 81.542858/7.434 = 10.969.$$

$$\text{SS}(b_1) = b_1^2 \left\{ \sum_{u=1}^{7} (\overline{X}_{iu} - \overline{X}_i)^2 \right\}$$

$$= (120.318961)(7.434) = 894.451.$$

2.

X_2	Y_2		$(X_{iu} - \overline{X}_i)$	$(Y_{iu} - \overline{Y}_i)$
3.2	22	$\overline{X}_2 = 5.533$ $\overline{Y}_2 = 41.333$	−2.333	−19.333
3.9	33		−1.633	−8.333
4.9	39	$n_2 = 6$	−0.633	−2.333
6.1	44		0.567	2.667
7.0	53		1.467	11.667
8.1	57		2.567	15.667

$$b_2 = \left\{ \sum_{u=1}^{6} (X_{iu} - \overline{X}_i)(Y_{iu} - \overline{Y}_i) \right\} \bigg/ \left\{ \sum_{u=1}^{6} (X_{iu} - \overline{X}_i)^2 \right\}$$

$$= 119.033334/17.573334 = 6.773520.$$

$$\text{SS}(b_2) = b_2^2 \left\{ \sum_{u=1}^{6} (X_{iu} - \overline{X}_i)^2 \right\}$$

$$= (45.880573)(17.573334) = 806.274633.$$

3.

X_3	Y_3	$(X_{iu} - \overline{X}_i)$	$(Y_{iu} - \overline{Y}_i)$
3.0	32	−2.775	−18.750
4.0	36	−1.775	−14.750
5.0	47	−0.775	−3.750
6.0	49	0.225	−1.750
6.5	55	0.725	4.250
7.0	59	1.225	8.250
7.3	64	1.525	13.250
7.4	64	1.625	13.250

$$\overline{X}_3 = 5.775, \quad \overline{Y}_3 = 50.750, \quad n_3 = 8.$$

$$b_3 = \left\{ \sum_{u=1}^{8} (X_{iu} - \overline{X}_i)(Y_{iu} - \overline{Y}_i) \right\} \bigg/ \left\{ \sum_{u=1}^{8} (X_{iu} - \overline{X}_i)^2 \right\}$$

$$= (135.650000)/(18.495) = 7.334415.$$

$$\text{SS}(b_3) = (53.793643)(18.495) = 994.913427.$$

$$b = \left\{ \sum_{i=1}^{m=3} \sum_{u=1}^{n_i} (X_{iu} - \overline{X}_i)(Y_{iu} - \overline{Y}_i) \right\} \Bigg/ \left\{ \sum_{i=1}^{m=3} \sum_{u=1}^{n_i} (X_{iu} - \overline{X}_i)^2 \right\}$$

$$= (81.542858 + 119.033334 + 135.650)$$

$$\div (7.434 + 17.573334 + 18.495)$$

$$= 336.226192/43.502334 = 7.728923.$$

$$SS(b) = b^2 \left\{ \sum_{i=1}^{3} \sum_{u=1}^{n_i} (X_{iu} - \overline{X}_i)^2 \right\}$$

$$= (59.736251)(43.502334) = 2598.666299.$$

$$\text{SS due to all } b_i|b = \sum_{i=1}^{3} SS(b_i) - SS(b)$$

$$= 894.451000 + 806.274633 + 994.913427$$

$$- 2598.666343 = 96.972717.$$

$$\text{Residual} = \text{Total SS} - SS(b) - \text{SS due to all } b_i|b$$

$$= 2792.261906 - 2598.666343 - 96.972717 = 96.622846$$

ANOVA

Source	df	SS	MS	F
b	1	2598.666343	2598.666343	403.42
All $b_i|b$	2	96.972717	48.486359	7.53
Residual	15	96.622846	6.441523	
Total	18	2792.261906		

$H_0: \beta_i = \beta$ $F_2 = 7.53 > F(2, 15, 0.95) = 3.68$, $\therefore H_0$ is rejected.

R. If we code boot A as $X_1 = -1$ and boot B as $X_1 = 1$ and enter the temperature (as listed) as X_2, the model $\hat{Y} = 10.6 + 0.9208X_1 - 0.15186X_2$ explains 45.1% of the variation. While this is not very good, it seems difficult to do better with the variables available. Lower responses are better and the coefficient of X_1 is positive. So boot A is better, and significantly so ($p = 0.027$). Fitting five dummy variables orthogonally to separate the subjects leaves the fitted coefficients of X_1 and X_2 unchanged and raises R^2 to 57.6%. Boot A is still better, of course.

S. We fit $E(Y) = \beta_0 + \beta_1 X + Z(\alpha_0 + \alpha_1 X)$, where $Z = -1, 1$, because this choice makes the (Z and XZ) columns orthogonal to the (1 and X) columns. Note that the columns *within* the parentheses are *not* orthogonal to each other, however.

Line 1 is then $\hat{Y} = 7.461 + 1.016X$
Line 2 is then $\hat{Y} = 9.670 + 0.320X$.

We can express H_0 as $H_0: \alpha_1 = 0$, and fit the reduced model to get the extra sum of squares for a_1 given b_0, b_1, a_0. The results are:

Source	df	SS	MS	F
$a_0, b_1|b_0$	2	4.771	2.386	170.4
$a_1|a_0, b_1, b_0$	1	1.147	1.147	81.9
Residual	4	0.056	0.014	
Total, corrected	7	5.974		

Clearly, H_0 is rejected and the two separate lines are needed.

T. (a) We can fit $\hat{Y} = 0.878 + 0.0670X + 3.641Z - 0.0147XZ$ ($R^2 = 0.945$), where Z defines the dummy variable column $\mathbf{Z} = (1, 1, 1, 1, 1, 1, 1, 1, 1, 0, 0, 0, 0, 0, 0)'$. The XZ term can be dropped, leading to the pair of parallel lines $\hat{Y} = 1.329 + 0.0589X + 2.972Z$ (for $Z = 1$ and $Z = 0$). This reduced equation explains $R^2 = 0.938$ of the variation about the mean $\overline{Y}$.

(b) Now, $\hat{Y} = -10.216 + 3.718U + 9.820Z - 1.780UZ$, where $U = \ln X$ (with $R^2 = 0.972$), but no terms can be dropped, so that two different lines are needed. Thus, with these data, little advantage is gained by using $\ln X$ rather than X. We remind the reader that our example data were interpolated from a published graph and that results with the actual data, not given in the paper, may well differ.

U. Using $Z_1 = 1$ for cherry and -1 for yellow oval, and using $(Z_2, Z_3) = (1, 0)$, $(0, 1)$, and $(0, 0)$ for the early, middle, and late planting times, respectively, leads to the fitted equation

$$\hat{Y} = 9.214 - 1.5\, Z_1 + 33.786\, Z_2 + 21.29\, Z_3.$$

$$(5.63) \quad (4.5) \qquad (7.2) \qquad (10.6)$$

The standard errors (shown in parentheses) and the analysis of variance table (not shown) reveal "no differences" between the two tomato types as far as production numbers are concerned. The dummies Z_2 and Z_3 should be regarded as a unit and not as separate variables. By choosing Z_2 and Z_3 in various ways we can make one look more important than the other at our choice. Thus dropping one of these variables is usually pointless. As a unit they show significant differences caused by different planting times.

The experimental layout is not well-planned. It would have been better to increase the direct comparisons between cherry and yellow oval tomatoes by planting some yellow ovals in the middle period and some cherries in the late period.

Chapter 15

A. 1. $\hat{Y} = 6360.3385 + 13.868864X_1 + 0.211703X_2$

$$- 126.690360X_3 - 21.817974X_4.$$

2. The above least squares equation shows an R^2 of 76.702710, which is the highest one in the given regression information. The overall $F = 9.8770256$ is statistically significant. The partial F-values are also all statistically significant. None of the 95% confidence limits on the β coefficients includes zero. The standard deviation as percent of response mean $= 7.536\%$, which is lower than any other in the given information.

3. The random vector X_5 does not contribute significantly to the explanation of variation. Actually, it contributes less than 1% and it increases the standard deviation as percent of the response from 7.536 to 7.759. The partial F-test also shows that this variable is not statistically significant.

4. This part is left to the reader.

B. 1. The stepwise procedure enters X_1 ($F = 12.60$), enters X_2 ($F = 2.04$), enters X_3 ($F = 3.62$). Now, however, X_1 has weakened with partial $F = 0.06$. X_1 is rejected and both X_2 and X_3 remain. The final equation is

$$\hat{Y} = 63.021 + 11.517X_2 - 0.816X_3.$$

2.

Source of Variation	df	SS	MS	F	Partial F
Total (corrected)	8	1279.20010			
Regression	2	1079.12600	539.56300	16.181	
$b_2 \mid b_0$	1	754.40445	754.40445	22.624	17.197
$b_3 \mid b_0, b_2$	1	324.72155	324.72155	9.738	9.738
Residual	6	200.07410	33.34568		

Residuals plots reveal no problems.

C. A study of the correlation matrix shows right away that X_1 and X_5 are perfectly correlated, caused by the fact that workers 1 and 5 are either *both* on duty or *both* absent in every run. Thus their effects cannot be separately assessed and, for regression purposes, we can drop X_5 immediately (or X_1, it makes no difference which). In these data, there are only eight distinct runs, there being twelve degrees of freedom for pure error. The sum of squares for pure error is 131.929 (12 df), so that $s_e^2 = 10.994$. The accompanying table shows the residual sums of squares for the various models in variables X_1, X_2, X_3, and X_4.

Variables in Model[a]	Residual df	Residual SS[b]	$100R^2$
—	19	42,644.00	—
1	18	8,352.28	80.14
2	18	36,253.69	14.99
3	18	36,606.19	14.16
4	18	27,254.91	36.09
12	17	7,713.10	81.91
13	17	762.55	98.21
14	17	6,071.56	85.76
23	17	32,700.17	23.32
24	17	24,102.10	43.48
34	17	16,276.60	61.83
123	16	761.41	98.21
124	16	5,614.59	86.83
134	16	163.93	99.62
234	16	15,619.01	63.37
1234	15	163.10	99.62

[a] In addition to β_0, which is in all models.
[b] Before removal of pure error.

If we adopt a stepwise procedure, we are led through this sequence:
(a) Enter X_1. $F_1 = (42,644 - 8,352.28)/(8352.28/18) = 73.90.$ Retain X_1.
(b) Add X_3. $F_3 = (8352.28 - 762.55)/(762.55/17) = 169.20.$ Retain X_3.
           $F_1 = (36,606.19 - 762.55)/(762.55/17) = 799.08.$ Retain X_1.
(c) Add X_4. $F_4 = (762.55 - 163.93)/(163.93/16) = 58.43.$ Retain X_4.
           $F_3 = (6071.56 - 163.93)/(163.93/16) = 576.60$ Retain X_3.
           $F_1 = (16,276.60 - 163.93)/(163.93/16) = 1572.64$ Retain X_1.

D. We give here the results of applying backward elimination, forward selection, and stepwise regression. For an analysis that uses the C_p statistic, see either of the source references.
   1. *Backward Elimination*
      For the full equation (123456), SS(residual) = 0.307 with $31 - 7 = 24$ df. The residual sum of squares for each five-variable equation is:

Variables	SS(residual)
12345	0.412
12346	0.311 ←— This is the best, so test variable 5.
12356	0.364 Partial F-test for 5, given 12346:
12456	0.313
13456	0.545 $F_{1,24} = \dfrac{0.311 - 0.307}{0.307/24} = 0.31.$
23456	0.939

The partial F-test indicates that variable 5 is not necessary for a good fit. Drop variable 5.

Our model now includes just the variables (12346), so SS(residual) = 0.311 with $31 - 6 = 25$ df. This model has the following four-variable submodels:

Variables	SS(residual)
1234	0.441
1236	0.365
1246	0.323 ←— Test variable 3.
1346	0.555 Partial F-test for 3, given 1246:
2346	0.995

$$F_{1,25} = \frac{0.323 - 0.311}{0.311/25} = 0.96.$$

Variable 3 appears unnecessary. Drop variable 3.

Our model now is (1246). SS(Residual) = 0.323 with $31 - 5 = 26$ df. This model has the following three-variable submodels:

Variables	SS(residual)
124	0.450
126	0.367 ←— Test variable 4.
146	0.558 Partial F-test for 4, given 126:
246	1.192

$$F_{1,26} = \frac{0.367 - 0.323}{0.323/26} = 3.54.$$

This F-value is significant at the 0.10 level, but not at the 0.05 level. This is a borderline case. If the equation is to be used just to summarize the data, it would probably be wise to keep variable 4 in the equation. The backward elimination procedure then stops and the final model is (1246). If the equation is to be used for prediction, however, the fact that variable 4 (a dummy variable that expresses an overall difference in response between the two sets of runs) is not reproducible would probably cause it to be dropped.

If variable 4 is dropped, the resulting model is (126), and SS(residual) = 0.367, with $31 - 4 = 27$ df.

This model has the following two-variable submodels:

Variables	SS(residual)
12	0.499 ←— Test variable 6.
16	0.576 Partial F-test for 6, given 12:
26	9.192

$$F_{1,27} = \frac{0.499 - 0.367}{0.367/27} = 9.71.$$

This F-value is significant at the 0.01 level. Conclude variable 6 is necessary and cannot be dropped. Procedure ends and final model is (126).

2. *Forward Selection.*

To find the variable most correlated with the response, we consider all six one-variable equations:

Variable	SS(residual)	Variable	SS(residual)
1	0.607	4	1.522
2	10.795	5	9.922
3	10.663	6	9.196

The model with variable 1 is the best. The *F*-statistic for variable 1 is:

$$F_{1,29} = \frac{11.058 - 0.607}{0.607/29}$$

$$= 499.31$$

which is highly significant. We now try adding each of the 5 remaining variables:

Variables	SS(residual)	
12	0.499	← Test variable 2.
13	0.600	Partial *F*-test for 2, given 1:
14	0.582	
15	0.597	$F_{1,28} = \dfrac{0.607 - 0.499}{0.499/28} = 6.06.$
16	0.576	

This *F*-value is significant at the 0.05 level. Add variable 2 to the model. Our model now includes the variables (12). Try adding each of the four remaining variables:

Variables	SS(residual)	
123	0.498	
124	0.450	
125	0.447	
126	0.367	← Test variable 6.
		Partial *F*-test for 6, given 12:

$$F_{1,27} = \frac{0.499 - 0.367}{0.367/27} = 9.71.$$

This *F*-value is significant at the 0.01 level. Add variable 6 to the model. Our model is now (126). Try adding each of the three remaining variables:

Variables	SS(residual)	
1236	0.365	
1246	0.323	← Test variable 4.
1256	0.364	Partial *F*-test for 4, given 126:

$$F_{1,26} = \frac{0.367 - 0.323}{0.323/26} = 3.54.$$

This is a borderline case. (See discussion above where the same decision had to be made in the backward elimination procedure.) If we decide not to include variable 4, the procedure stops and the final model is (126). If we include variable 4, the procedure continues, and we try adding each of the two remaining variables to the model (1246):

Variables	SS(residual)	
12346	0.311	← Test variable 3.
12456	0.313	Partial *F*-test for 3, given 1246:

$$F_{1,25} = \frac{0.323 - 0.311}{0.311/25} = 0.96.$$

This *F*-value is not significant. Do not add variable 3. Stop procedure. Final model is (1246).

(*Note*: At several stages, the best-fitting equation turned out to be the same under the forward selection procedure as under the backward elimination procedure. This will not necessarily be the case with other sets of data.)

3. Stepwise Regression.

This procedure starts out like the forward selection procedure described above, with the inclusion of variable 1, then variable 2. At this point we do a partial F-test for 1, given 2. Since SS(residual) for 2 is 10.795, and SS(residual) for (12) is 0.499,

$$F_{1,28} = \frac{10.795 - 0.499}{0.499/28} = 577.73.$$

This F-value is highly significant, so we cannot remove variable 1. We shall assume that the critical F-value for deletion does not exceed the critical F-value for entry (as recommended in Section 15.4) so we need not test variable 2, the last variable entered, for possible deletion. Resuming the forward selection procedure, we enter variable 6 as above, then test to see whether either of the previously entered variables (1 or 2) can be removed.

$$\text{Partial } F\text{-test for 1 given 26: } F_{1,27} = \frac{9.192 - 0.367}{0.367/27} = 649.25.$$

$$\text{Partial } F\text{-test for 2 given 16: } F_{1,27} = \frac{0.576 - 0.367}{0.367/27} = 15.38$$

Both of these F-values are significant at the 0.01 level, so we cannot remove either of them. Resume the forward selection procedure, choosing variable 4 as the next candidate for entry. If we decide that the partial F-value (3.51) does not warrant the addition of variable 4, the stepwise regression procedure stops and the final model is (126).

If we decide to keep variable 4, continue the stepwise regression procedure by testing to see if any of the previously entered variables (1, 2, or 6) can be dropped.

$$\text{Partial } F\text{-test for 1 given 246: } F_{1,26} = \frac{1.192 - 0.323}{0.323/26} = 69.95.$$

$$\text{Partial } F\text{-test for 2 given 146: } F_{1,26} = \frac{0.558 - 0.323}{0.323/26} = 18.92.$$

$$\text{Partial } F\text{-test for 6 given 124: } F_{1,26} = \frac{0.450 - 0.323}{0.323/26} = 10.22.$$

These F-values are all significant at the 0.01 level. None of them can be dropped. Continuing, we try to add variable 3, but find that it does not significantly improve the fit. The final model is thus (1246).

E. Both stepwise and backward elimination procedures produce the equation

$$\hat{Y} = 3.068360 + 0.0007259X_1 + 0.0446022X_4 \tag{1}$$

with an $R^2 = 0.7874$. However, there is lack of fit indicated, the appropriate F-statistic for the lack of fit test being $F = \{0.27351/73\}/\{0.02154/14\} = 2.435$, compared to $F(73, 14, 0.95) = 2.21$, approximately. The residuals plots show a quadratic tendency in X_1, and the Durbin–Watson statistic $d = 0.8480$ indicates evidence of positive serial correlation.

When the term $\beta_{11}X_1^2$ is added to the model, other terms now enter, also. The stepwise procedure, for example, produces the following sequence of entries:

Predictor	R^2	Change in R^2	Significance Level (α)
X_1 in	0.5999	0.5999	0.000
X_4 in	0.7874	0.1875	0.000
X_1^2 in	0.8917	0.1043	0.000
X_6 in	0.9000	0.0083	0.009
X_7 in	0.9056	0.0056	0.028

The final fitted model is

$$\hat{Y} = 2.997526 + 0.0019418X_1 - 0.00000289X_1^2 \qquad (2)$$
$$+ 0.0222738X_4 + 0.0607877X_6 - 0.0224359X_7.$$

There is now no lack of fit shown by the F-test, the value of the statistic being $F = \{0.10944/70\}/\{0.02154/14\} = 1.016 < F(70, 14, 0.95) = 2.21$, approximately. The Durbin–Watson statistic is $d = 1.60$, an inconclusive result by the usual test. The table above shows that X_6 and X_7 contribute little to R^2 even though their entry into the model is "statistically significant," where we use the standard (but wrong, see p. 343) test. If we omit X_6 and X_7 and refit, we obtain the fitted equation

$$\hat{Y} = 3.0016115 + 0.0018344X_1 - 0.000002641X_1^2 \qquad (3)$$
$$+ 0.0390777X_4.$$

The lack of fit test now provides the statistic $F = \{0.12877/72\}/\{0.02154/14\} = 1.162$, which is not significant compared to $F(72, 14, 0.95) = 2.21$ approximately. The Durbin–Watson statistic is $d = 1.64$, an inconclusive result by the usual test.

Thus if we wish to use the model indicated by the formal stepwise procedure, we should use Eq. (2), but Eq. (3) seems a sensible practical compromise. Examination of the corresponding printouts shows that the predictions from Eq. (3) are close to those from Eq. (2).

Additional Comments:

(*i*) An X_1^3 term can also be considered. If it is, the stepwise entry sequence is X_1 ($R^2 = 0.5999$), X_4 (incremental change in $R^2 = 0.1875$), X_1^2 (0.1043), X_6 (0.0083), X_1^3 (0.0079), X_7 (0.0052).

(*ii*) If X_4^2 and X_4^3 are considered in addition (as well as X_1^2 and X_1^3), the stepwise entry sequence (with incremental R^2 values) is exactly the same as in (*i*) but with two extra steps, namely X_4^3 (0.0052) comes in, but now X_4 (-0.0010) goes out.

In both cases (*i*) and (*ii*), by arguing as above, we would probably use just X_1, X_4, and X_1^2, since all the other significant variables add less than 0.01 each to R^2.

F. $$\hat{Y} = 0.4368012 + 0.0001139X_1 - 0.0051897X_3$$

$$- 0.0018887X_4 + 0.0044263X_5.$$

The plot of the residuals versus $\hat{Y}$ indicates that the variance is not homogeneous. One should try weighted least squares, or perhaps a transformation on the Y_i.

This model explains only 76.9% of the total variation, and the confidence limits on β_1 and β_4 include zero. The standard deviation of the residuals is 3.3% of the mean response. Thus the model predicts well but is not as good as one would like. If one can get rid of the large variance for large Y's, the model will be much better.

G. Model: $\hat{Y} = b_0 + b_2X_2 + b_8X_8$

or $\hat{Y} = 9.4742224 + 0.7616482X_2 - 0.0797608X_8$

$R^2 = 86.0\%$

Standard deviation as percent of response mean = 6.761%.

H. 1. $\hat{Y} = 87.158859 + 0.8519104X_1 + 0.5988662X_2$
$\qquad + 2.3613018X_6 - 0.9755309X_9.$

where X_1 = year,
$\qquad X_2$ = preseason precipitation in inches,
$\qquad X_6$ = rainfall in July in inches,
$\qquad X_9$ = August temperature.

2. The most important variable is X_1, which accounts for the upward trend in corn yield. Of all the other variables, only preseason precipitation, July rainfall, and August temperature contribute significantly to the regression.

3. With an R^2 of 72.06% and standard deviation as percent of response mean of 14.903%, this prediction equation needs to be improved. New variables should be found to bring R^2 up, and to decrease the standard deviation of residuals. Investigation of the residuals may yield some insight into this problem. (See "Fast very robust methods for the detection of multiple outliers," by A. C. Atkinson, *Journal of the American Statistical Association*, **89**, 1994, 1329–1339. Observations 7, 8, and 11 "may be outliers" on p. 1336.)

I. 1. There is a lot of replication in the data. Thus an independent estimate of pure error can be obtained. The analysis of variance can be written down as:

ANOVA

Source of Variation	df
Total	47
Regression	5
Residuals	42
Lack of fit	2
Pure error	40

2. $\hat{Y} = 134.258 + 0.050X_1 - 0.012X_2$
$\qquad + 0.834X_3 - 0.154X_4 - 3.804X_5.$

3. The model is not adequate since the lack of fit is statistically significant at $\alpha = 0.05$.

Source of Variation	df	SS	MS	F
Total	47	2850.3107		
Regression	5	1817.1055		
Residual	42	1033.2052		
Lack of fit	2	383.7052	191.8526⌐	11.82*
Pure error	40	649.5000	16.2375⌐	

4. This model explains only 63.75% of the variation, and it is not a good one. The residuals show definite nonrandom patterns.

5. This experiment is poorly designed; there are too many replicates and not enough different design points.

J. The *prediction* equation obtained by the stepwise procedure using a critical F of 2.00 for acceptance and rejection is

$$\hat{Y} = 250.1875 - 2.3124998 \left(\frac{X_1 - 146}{3} \right)$$

$$- 14.687499 \left(\frac{X_2 - 69.5}{3.5} \right) - 2.8124997 \left(\frac{X_6 - 289.5}{93.5} \right).$$

The optimum rate using this prediction equation will be at the point $\hat{Y} = 270$; $X_1 = 143$, $X_2 = 66$, $X_6 = 196$; and the other variables held at their mean levels, namely, $X_3 = -10$, $X_4 = 132.5$, $X_5 = 91.5$.

K. $\hat{Y}_1 = -2.80512 + 0.15176X_1 + 3.60191X_3,$
$\hat{Y}_2 = -2.84492 + 0.11344X_1 + 3.67343X_3.$

L. 1. Both the stepwise and the backward elimination procedures (with F-to-enter and F-to-remove values set to 4) produce the equation

$$\hat{Y} = -1.2471 + 0.510X_2 + 0.768X_4,$$

with an $R^2 = 0.9770$, and $s^2 = 0.01$. So an excellent fit is obtained.

2. High correlations (in excess of 0.8) are observed between the following variables (indicated by the subscripts):

$$r_{23} = 0.901, \qquad r_{34} = 0.850, \qquad r_{37} = 0.835, \qquad r_{47} = 0.879.$$

The main hint from these is that question 3 may be redundant and could be dropped.

3. If the relationship *were* causal, the regression would be saying this: Teach a well organized course and answer questions in a helpful manner and you will be assured of a good overall grade as an instructor. The conclusion appears to be a very reasonable one!

M. 1. Look at the equations 145 (RSS = 0.569), 245 (1.030), 345 (1.383), 456 (1.352). The combination 145 is best (lowest RSS) and the consequent

$$F_{1|45} = \{(1.397 - 0.569)/1\}/\{0.569/27\} = 39.26,$$

which is significant, exceeding 4.21.

2. The RSS for 12345 is 0.412. The least increase for any four of the five variables is attained by 1245 (0.413). So $F_{3|1245} = \{(0.413 - 0.412)/1\}/\{0.412/25\} = 0.06$. Variable 3 would be removed.

3. $S_{YY} = 11.058$. $R^2 = (11.058 - 0.441)/11.058 = 0.9601$.

4. Using 3456 (1.342) and 123456 (0.307) we get

$$C_p = \frac{1.342}{0.307/24} - (31 - 10) = 83.91.$$

This is not a useful regression.

5. For 1246, RSS = 0.323. Dividing the latter by $31 - 5 = 26$ gives $s^2 = 0.0124$.

N. 1. $\hat{Y} = -50.359 + 0.6711545X_1 + 1.295351X_2.$
No lack of fit, $F = 1.42$ with 12, 6 df. $R^2 = 0.8506$.

2. The residuals show that the fitted equation is least satisfactory at $(X_1, X_2, X_3) = (70, 20, 91)$, that is, at run 21; thus one would be reluctant to use the equation in that neighborhood. Future runs should be chosen to provide more balanced coverage of the X-space.

Additional Notes: For a more detailed and extensive analysis of these data, see Chapters 5 and 7 (p. 138) of Daniel and Wood (1980). This particular set of data is one of the most analyzed regression problems in the literature! It has provided wonderful ammunition for those critical of least squares, because many authors maintained that there were four questionable observations, mentioned below, not all detected immediately via a least squares fit. In a thoroughly researched and amusing article, Dodge (1996) has investigated the history of these data. Among other things, he points out that various methods of analyzing them have provided at least 26 *distinct sets of detected outliers*, the most cited set being observations 1, 3, 4, and 21. Many references are provided by Dodge. See also Atkinson (1994, pp. 1330–1331).

O. For the types of detailed calculations needed, see the fully worked solution to Exercise 15D. We summarize the steps needed for this example, only.

1. *Backward Elimination.* The RSS for 123456 is 12.508. If we look at the RSS values for all sets of five, we see the smallest increase is for 12356 with RSS = 12.542. The $F(1,19)$ value is $\{(12.542 - 12.508)/1\}/\{12.508/19\} = 0.05$, not significant; so drop 4.

Proceeding similarly from 12356 we are led to drop 1. Then at the $\alpha = 0.10$ level, we quit at 2356. At the $\alpha = 0.05$ level we proceed to 25, where we quit.

2. *Stepwise Regression.* We first select 2, then 5. Variable 3 is next up. At the $\alpha = 0.05$ level we quit at 25. At the $\alpha = 0.10$ level we proceed from 25 to 235 to 2356.

3. C_p *Statistic.* The candidate sets of predictors with low C_p values are the following:

$p = 3$	$p = 4$	$p = 5$	$p = 6$
25 ($C_p = 5.44$)	125 (6.28)	1235 (5.39)	12345 (6.85)
	235 (4.31)	2345 (5.09)	12356 (5.05)
	245 (6.43)	2356 (3.28)	23456 (5.12)

Clearly, the equations 25, 235, and 2356 attract attention, with a final choice depending on how may predictors it is decided to include to achieve the indicated C_p reductions as p increases.

P. The selection procedure identifies x_1, x_2, and x_1x_2 as the most important terms. Residual No. 12 seems far too remote from the others. The four observations (Nos. 4, 8, 12, and 16) at $(x_1, x_2) = (1, 1)$ are 139.7, 141.4, 48.6, and 172.6, leading to the suspicion that 48.6 perhaps should have been 148.6. With this replacement, the fit improves to an $R^2 = 0.978$ from the previous 0.753. The largest residual is now the 16th, but the overall fit is excellent, and the two factors x_1 and x_2 and their interaction provide an excellent explanation of the data. (In the original paper, the author replaced 48.6 by the rounded average, 151.2, of the other three numbers mentioned above. This is also clearly a sensible way to deal with the matter.)

Chapter 16

A. No, because the columns are related via $-2X_1 + X_2 + X_3 = 0$. Dropping any one of the three related columns allows a unique fit. If we form the X matrix by adding a column $X_0 = 1$, and apply the Gram–Schmidt procedure, we get the matrix

$$
Z = \begin{bmatrix}
1 & -\dfrac{17}{5} & \dfrac{215}{186} & 0 \\[2mm]
1 & -\dfrac{12}{5} & -\dfrac{822}{186} & 0 \\[2mm]
1 & -\dfrac{2}{5} & \dfrac{824}{186} & 0 \\[2mm]
1 & \dfrac{13}{5} & \dfrac{131}{186} & 0 \\[2mm]
1 & \dfrac{18}{5} & -\dfrac{348}{186} & 0
\end{bmatrix}
$$

B. No. We first add an $X_0 = 1$ column to get X. The second column is already orthogonal to the first. We find

$$Z_{3T} = \tfrac{1}{2}(-5, -3, -1, 1, 3, 5)'$$
$$Z_{4T} = (0, 0, 0, 0, 0, 0)', \quad \text{so column dependence exists.}$$

In fact, the sum of the second, third, and fourth columns of X is zero.

C. No. We add an $X_0 = 1$ column and columns generated by X_1^2, X_2^2 and X_1X_2 to form X. Note that $X_1^2 = X_2^2$ always, so the X matrix is singular. We can fit the model $Y = \beta_0 + \beta_1 X_1 + \beta_2 X_2 + \beta(X_1^2 + X_2^2) + \beta_{12}X_1X_2 + \epsilon$ but cannot estimate β_{11} and β_{22} individually.

D. We first note that X_1 and X_2 are centered already and we center Y about $\overline{Y} = 12$. We next see that $S_{11} = 10$, $S_{22} = 34$, and $S_{YY} = 38$ and the square roots of these will be the scale factors. After centering and scaling, the data become:

Z_1	Z_2	Y
−0.632	−0.343	0
−0.316	−0.171	−0.487
0	0	−0.487
0.316	0.171	0.324
0.632	0.343	0.649

The normal equations are

$$a_1 + 0.976\ a_2 = 0.667,$$
$$0.976a_1 + a_2 = 0.584,$$

with solution $a_1 = 2.055$, $a_2 = 1.421$. The determinant of the correlation matrix is $1 - (0.976)^2 = 0.047$. The square root of this is 0.218. (See Section 5.5.)

Chapter 17

A. No, although the ridge estimators can be expressed in terms of the least squares estimators. See Appendix 17A.

B. Note that $\overline{X} = 0$ and $S_{XX}^{1/2} = 4$ is the scaling factor for X. We thus get a new column

$$\mathbf{Z} = (-0.5, -0.25, -0.25, -0.25, 0, 0.25, 0.5, 0.5)'$$

and the ridge equation is simply $b_1(\theta) = (1 + \theta)^{-1}\Sigma Z_i Y_i = 8.75/(1 + \theta)$. Then $b_0(\theta) = \overline{Y} - b_1(\theta)\overline{X} = \overline{Y} = 40$. The fitted ridge equation is thus $\hat{Y} = 40 + 8.75Z/(1 + \theta) = 40 + 2.1875X/(1 + \theta)$. The least squares fit appears when $\theta = 0$. When $\theta = 0.4$ we get $\hat{Y} = 40 + 6.25Z = 40 + 1.5625X$.

Chapter 18

A. In all cases we might consider fitting a linear model function $\eta = \beta_0 + \beta_1 X_1 + \beta_2 X_2 + \cdots + \beta_k X_k$ but there are several ways this could be done.
1. Fit $f(Y_i) = \eta_i + \epsilon_i$ using a generalized linear model fit with, for example, $f(Y_i) = \ln\{Y_i/(1 - Y_i)\}$ (Chapter 18).
2. Fit the responses $f(Y_i) = \ln\{Y_i/(1 - Y_i)\}$ by generalized (weighted) least squares with weights inversely proportional to the variances $V\{f(Y_i)\} = \{Y_i(1 - Y_i)m\}^{-1}$ (Chapter 9, also see Chapter 13).
3. Transform $U_i = \sin^{-1}(Y_i^{1/2})$ and use least squares (Chapter 13).

Chapter 19

A. $\hat{Y} = 6.333 + 0.770z_1 + 0.666z_2 - z_1^2 - 0.333z_2^2 + 0.385z_1z_2$.

B.

$$\mathbf{T} = \begin{bmatrix} -a & -a & 0 & 0 \\ -b & -b & 2b & 0 \\ -c & -c & -c & 3c \\ 1 & 1 & 1 & 1 \end{bmatrix}.$$

New coordinates are, respectively, $(a, -b, -c)$, $(0, -b, -c)$, $(0, 0, -c)$, $(0, 0, 0)$, where $a = (2/3)^{1/2} = 0.816$, $b = 2^{1/2}/3 = 0.471$, $c = 0.333$.

C. Look at Figure 19.6b and think of it (ignoring the corner markings) as a slice of a fourth variable. Imagine two such slices, one above the other, one smaller than the other (see also Figure 19.4c). It is clear that there will be, at most, 12 extreme vertices. In fact there are fewer here mainly because the 0.80 (80%) cutoff plane for x_1 is so high that no vertices appear on it. A mechanical way to generate the extreme vertices for q dimensions is to choose any $(q - 1)$ pairs of limits temporarily ignoring the remaining pair and writing points with all possible combinations of the $(q - 1)$ pairs of levels. For example, if we ignore x_4 first, we write, converting 0.80 to 80% and so on, to get rid of decimals:

x_1	x_2	x_3	x_4	Feasible x_4?	Vertex
10	25	20	45	Yes	1
80	25	20	—		
10	45	20	25	Yes	2
80	45	20	—		
10	25	40	25	Yes	3
80	25	40	—		
10	45	40	5	No	
80	45	40	—		

The x_1, x_2, x_3 levels are all possible combinations of the restriction limits; x_4 values are added, which bring $x_1 + x_2 = x_3 + x_4 = 100$. For the point to be a vertex, we need $0.15 \le x_4 \le 55$. Repeating this by next ignoring x_3, then x_2, then x_1 gives further vertices as follows:

4. $(10, 25, 50, 15)$;   5. $(10, 45, 30, 15)$;   6. $(10, 35, 40, 15)$;   7. $(40, 25, 20, 15)$;

8. $(20, 45, 20, 15)$;   9. $(20, 25, 40, 15)$.

To get face centroids we must identify faces on which an x_i is constant at a boundary level. For example:

$x_1 = 10$: Vertices 1, 2, 3, 4, 5, 6, essentially similar to Figure 19.6b.
$x_2 = 25$: Vertices 1, 3, 4, 7, 9.
$x_2 = 45$: Vertices 2, 5, 8.
$x_3 = 20$: Vertices 1, 2, 7, 8.
$x_3 = 40$: Vertices 3, 6, 9.
$x_4 = 15$: Vertices 4, 5, 6, 7, 8, 9.

The centroid of the $x_3 = 20$ face, for example, is at

$$\{(10 + 10 + 40 + 20)/4, (25 + 45 + 25 + 45)/4,$$
$$(20 + 20 + 20 + 20)/4, (45 + 25 + 15 + 15)/4\},$$

namely, at $(20, 35, 20, 25)$. The overall centroid, the average of all vertex points, is at $(15.6, 32.8, 31.1, 20.6)$ whose ingredients add to 100.1 due to rounding error.

D. $\hat{Y} = 40.91x_1 + 25.432x_2 + 28.61x_3 - 8.214x_1x_2 - 39.339x_1x_3 + 8.002x_2x_3$. At the centroid, $\hat{Y} = 27.26$. The contours are curves with highest response value at $(1, 0, 0)$, declining as x_1 decreases. (They have not been plotted here.)

E. The points are on the vertices of the triangular mixture space, on the midpoints of the sides, and the final point is right in the middle.

F. For $m = 2$, $q = 3$, only points with coordinates 0 or $\frac{1}{2}$ or 1 whose coordinates add to 1 are allowed. Of the 27 possible combinations, six are valid, namely, the first six points mentioned in Exercise E. The number of valid points in the general case is

$$\binom{q + m - 1}{m}.$$

For $m = 2$, $q = 3$ this is

$$\binom{4}{2} = 6.$$

G. The vertices are $(1, 0, 0)$, $(0, 1, 0)$, and $(0, 0, 1)$. Averaging of pairs gives $(\frac{1}{2}, \frac{1}{2}, 0)$, $(\frac{1}{2}, 0, \frac{1}{2})$, and $(0, \frac{1}{2}, \frac{1}{2})$. The average of all three is $(\frac{1}{3}, \frac{1}{3}, \frac{1}{3})$.

H. Six, namely, β_1, β_2, β_3, β_{12}, β_{13}, β_{23}. Thus data from (a) will provide an exact fit while (b) will provide only one degree of freedom for error.

Chapter 20

A.   **1.** $S(\beta_0, \beta_1) = (1.4 - \beta_0 - \beta_1)^2 + (2.2 - \beta_0 - 2\beta_1)^2 + (2.3 - \beta_0 - 3\beta_1)^2 + (3.1 - \beta_0 - 4\beta_1)^2.$

    2. $b_0 = 0.95$, $b_1 = 0.52$.

    3.

$$\hat{\mathbf{Y}} = \begin{bmatrix} 1.47 \\ 1.99 \\ 2.51 \\ 3.03 \end{bmatrix}, \qquad \mathbf{e} = \mathbf{Y} - \hat{\mathbf{Y}} = \begin{bmatrix} -0.07 \\ 0.21 \\ -0.21 \\ 0.07 \end{bmatrix}.$$

    4.

$$\mathbf{P} = \frac{1}{10} \begin{bmatrix} 7 & 4 & 1 & -2 \\ 4 & 3 & 2 & 1 \\ 1 & 2 & 3 & 4 \\ -2 & 1 & 4 & 7 \end{bmatrix}.$$

    5.

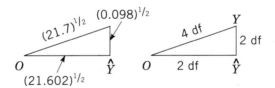

Solution A5

    6.

Source	SS	df	MS	$F(2, 2)$
Regression	$O\hat{Y}^2 = 21.602$	2	10.801	220.43
Residual	$Y\hat{Y}^2 = 0.098$	2	0.049	
Total	$OY^2 = 21.7$	4		

$F = (O\hat{Y}^2/2)/(Y\hat{Y}^2/2)$ in diagram.

    7. $\boldsymbol{\beta}_0 = (1, 0.5)'$, $\mathbf{X}\boldsymbol{\beta}_0 = (1.5, 2.0, 2.5, 3.0)'$ and this point replaces O in figures similar to those in 5, but with different lengths except for $Y\hat{Y}$. The new ANOVA is based on

$$\mathbf{Y} - \mathbf{X}\boldsymbol{\beta}_0 = (\hat{\mathbf{Y}} - \mathbf{X}\boldsymbol{\beta}_0) + (\mathbf{Y} - \hat{\mathbf{Y}})$$

or, for the squared lengths (sum of squares of the elements),

$$0.1 = 0.002 + 0.098.$$

$F = 0.001/0.049 = 0.02$, not significant.

8.

$$\mathbf{X} = \begin{bmatrix} 1 & 1 \\ 1 & 2 \\ 1 & 3 \\ 1 & 4 \end{bmatrix} \quad \text{and} \quad \mathbf{P} = \mathbf{X}(\mathbf{X'X})^{-1}\mathbf{X'} = \mathbf{X}\alpha, \quad \text{say.}$$

So we need

$$\alpha = (\mathbf{X'X})^{-1}\mathbf{X'} = \begin{bmatrix} 1 & 0.5 & 0 & -0.5 \\ -0.3 & -0.1 & 0.1 & 0.3 \end{bmatrix}.$$

9. $\mathbf{PX} = \mathbf{X}$, so the two columns of $\mathbf{X}$ define the required two combinations.
10. $\mathbf{X}_{1.0} = (-1.5, -0.5, 0.5, 1.5)$,
  $\hat{\mathbf{Y}} = 2.25\mathbf{X}_0 + 0.52\mathbf{X}_{1.0}$.
11. From 10, we get an orthogonal SS split of the regression SS as

$$21.602 = \hat{\mathbf{Y}}'\hat{\mathbf{Y}} = (2.25\mathbf{X}_0)'(2.25\mathbf{X}_0) + (0.52\mathbf{X}_{1.0})'(0.52\mathbf{X}_{1.0})$$

$$21.602 = 20.25 + 1.352.$$

Thus $F = (1.352/1)/0.049 = 27.59 > F(1, 2, 0.95) = 18.51$.

B. 1.

$$\begin{bmatrix} 4 & 0 & 40 \\ 0 & 20 & 60 \\ 40 & 60 & 580 \end{bmatrix} \begin{bmatrix} b_0 \\ b_1 \\ b_2 \end{bmatrix} = \begin{bmatrix} 16 \\ -8 \\ 136 \end{bmatrix}.$$

These are dependent equations: $E_3 = 10E_1 + 3E_2$.
2. Let $b_2 = b$ (anything). Then $b_0 = 4 - 10b$, $b_1 = -0.4 - 3b$.
3.

$$\hat{\mathbf{Y}} = \mathbf{Xb} = \begin{bmatrix} 1 & -3 & 1 \\ 1 & -1 & 7 \\ 1 & 1 & 13 \\ 1 & 3 & 19 \end{bmatrix} \begin{bmatrix} 4 - 10b \\ -0.4 - 3b \\ 6 \end{bmatrix} = \begin{bmatrix} 5.2 \\ 4.4 \\ 3.6 \\ 2.8 \end{bmatrix}.$$

Note that $\hat{\mathbf{Y}}$ is unique; the b has disappeared!
4. Because the normal equations are dependent, $\mathbf{b}$ is not unique. Thus $\hat{\mathbf{Y}} = \mathbf{Xb}$ can be *described* in many ways. There is exactly one $\hat{\mathbf{Y}}$, however, the foot of the perpendicular from $\mathbf{Y}$ onto $R(\mathbf{X})$. This simple example illustrates regression when $\mathbf{X}$ is not of full rank.
C. 1. $\mathbf{X'X}$ is singular, so any generalized inverse is needed. The one that puts zeros in the last row and column and the inverse of the first 3 rows and columns of $\mathbf{X'X}$ in the corresponding positions is

$$(\mathbf{X'X})^- = \frac{1}{8} \begin{bmatrix} 2 & 0 & -2 & 0 \\ 0 & 1 & 0 & 0 \\ -2 & 0 & 3 & 0 \\ 0 & 0 & 0 & 0 \end{bmatrix}.$$

This leads to

$$\mathbf{P} = \frac{1}{4}\begin{bmatrix} \mathbf{J} & \mathbf{0} & \mathbf{0} \\ \mathbf{0} & \mathbf{J} & \mathbf{0} \\ \mathbf{0} & \mathbf{0} & \mathbf{J} \end{bmatrix},$$

where each $\mathbf{J}$ is a 4 by 4 block of 1's and each $\mathbf{0}$ is a 4 by 4 block of 0's.

2. $\mathbf{P}$ is unique, although $(\mathbf{X'X})^-$ is not unique. $\mathbf{PY}$, the vector from O to the foot of the perpendicular from $\mathbf{Y}$ to $R(\mathbf{P}) = R(\mathbf{X})$, is also unique.

3. For $\mathbf{c}'\boldsymbol{\beta}$ to be estimable, the $\mathbf{c}'$ vector must be expressible as a linear combination of the rows of $\mathbf{X}$. There are 12 rows but only three are distinct, namely,

$$\mathbf{d}_1' = (1, -1, 1, -1),$$
$$\mathbf{d}_2' = (1, 0, 0, 0),$$
$$\mathbf{d}_3' = (1, 1, 1, 1).$$

Checking the five $\mathbf{c}'$ vectors, we see that

$$\mathbf{c}_1' = (1, 0, 1, 0) = (\mathbf{d}_1' + \mathbf{d}_3')/2; \quad \text{estimable.}$$

$$\mathbf{c}_2' = (0, 1, 0, 1); \quad \text{not estimable.}$$

$$\mathbf{c}_3' = (1, -1, 0, 0); \quad \text{not estimable.}$$

$$\mathbf{c}_4' = (0, 1, 0, 1) = (\mathbf{d}_3' - \mathbf{d}_1')/2; \quad \text{estimable.}$$

$$\mathbf{c}_5' = (1, 1, 1, 1) = \mathbf{d}_3; \quad \text{estimable.}$$

D. 1. The basic triangle has sides of lengths $OY = (786)^{1/2}$, $O\hat{Y} = (530)^{1/2}$, $Y\hat{Y} = 16$ in spaces of dimensions 3, 2, 1, respectively.

2. $786 = 530 + 256$, for SS.
$\quad\ 3 = \quad 2 + 1, \quad$ for df.

3. The point Y lies high above the plane of the estimation space. In fact, the F for regression is $(530/2)/256 = 1.04$, because the error MS is large compared with the regression MS.

4. Yes, it is $U_3 = 0$. That is why $Y(19, 13, 16)$ projects to $\hat{Y}(19, 13, 0)$.

5. $\mathbf{X}_{2.1}' = (1.2, -0.6, 0)$. The axes are now perpendicular and $530 = 405 + 125$ provides an orthogonal breakup of the regression SS in the rectangle formed by O, $\hat{Y}$, and the projections of $\mathbf{Y}$ on to the new axes.

6. If Y is moved to (19, 13, 2) from (19, 13, 16), the perpendicular to $\hat{Y}$ is the same but Y is much closer, that is, the error vector is shorter.

7. The new $F = (530/2)/4 = 66.25$
Moral: A good regression is one where Y lies close to the estimation space.

E. 1.

$$\mathbf{P} = \mathbf{A}(\mathbf{A'A})^{-1}\mathbf{A}' = \frac{1}{20}\begin{bmatrix} 19 & 3 & -3 & 1 \\ 3 & 11 & 9 & -3 \\ -3 & 9 & 11 & 3 \\ 1 & -3 & 3 & 19 \end{bmatrix}.$$

2. In four-dimensional space there is only one basis vector $\mathbf{v} = (a, b, c, d)'$, say, orthogonal to the columns of **A** and it must satisfy the orthogonality conditions

$$a + b + c + d = 0,$$

$$-3a - b + c + 3d = 0,$$

$$a - b - c + d = 0.$$

So **v** = $(1, -3, 3, -1)'$ or any multiple of this.

3. Obviously **Pv** = **0** so **v** projects to the origin and **P** and **A** span the same space, that is, $R(\mathbf{P}) = R(\mathbf{A})$.

F. All are! We need $\mathbf{MM^-M} = \mathbf{M}$ to hold.
Suppose we call

$$\mathbf{M}^- = \begin{bmatrix} a & b \\ c & d \end{bmatrix}.$$

Then

$$\mathbf{MM^-M} = (a + b + c + d) \begin{bmatrix} 1 & 1 \\ 1 & 1 \end{bmatrix}.$$

It follows that any such matrix with $a + b + c + d = 1$ will work, and those given all satisfy this condition.

G. The 10 df = 2 (model) + 3 (lack of fit) + 5 (pure error). The $(\mathbf{1}, \mathbf{X})$ vectors span the estimation space, five vectors $(-1, 1, 0, 0, \ldots, 0, 0), \ldots, (0, 0, \ldots, 0, 0, -1, 1)$ span the pure error space, and we need three more vectors to span the lack of fit space. Because of the form of the **X** vector (see below), we can use the orthogonal polynomials for $n = 5$ from the table in Section 22.2 and double them up. For example, from $\psi_2' = (2, -1, -2, -1, 2)$ we make the 10-dimensional vector $\phi_2' = (2, 2, -1, -1, -2, -2, -1, -1, 2, 2)$ and similarly for ψ_3' and ψ_4'. The **X** vector is a doubled up form of ψ_1', of course.

H. The row sums of **P** are given by **P1**. The vector **1** is already in the estimation space because β_0 is in the model, so $\mathbf{P1} = \mathbf{1}$, because when we project a vector already in the estimation space it remains as is. The column sums are $\mathbf{1'P} = (\mathbf{P1})'$, since $\mathbf{P} = \mathbf{X(X'X)^{-1}X'}$ is symmetric, and so $\mathbf{1'P} = \mathbf{1'}$.

Chapter 21

A. 1.

$$\frac{1}{20}\begin{bmatrix} 19 & 3 & -3 & 1 \\ 3 & 11 & 9 & -3 \\ -3 & 9 & 11 & 3 \\ 1 & -3 & 3 & 19 \end{bmatrix} - \frac{1}{20}\begin{bmatrix} 14 & 8 & 2 & -4 \\ 8 & 6 & 4 & 2 \\ 2 & 4 & 6 & 8 \\ -4 & 2 & 8 & 14 \end{bmatrix} = 0.25(\mathbf{X}_2, -\mathbf{X}_2, -\mathbf{X}_2, \mathbf{X}_2).$$

The dimensions are $3 - 2 = 1$, so that, respectively, only 3, 2, and 1 column(s) are linearly independent.

2. Obviously $\mathbf{X}_2$ is $\omega^\perp \cap \Omega$ by definition, since it is the only vector in Ω that is orthogonal to the vectors in ω.

3. $\mathbf{X}_2$.

4. $\mathbf{P}_\Omega - \mathbf{P}_\omega$ has dimension (rank) $d = 1$ and is idempotent. A theorem says that idempotent matrices of rank d have eigenvalues 1 (d times over) and all other eigenvalues are zero. So the answer is 1, 0, 0, 0.
More directly, we solve $\det(\mathbf{P}_\Omega - \mathbf{P}_\omega - \lambda\mathbf{I}) = 0$ to get $\lambda^3(\lambda - 1) = 0$.

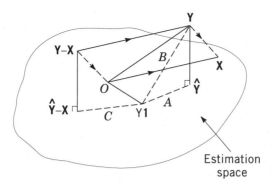

Solution C1. The geometrical effect of changing the response from **Y** to **Y** − **X** in a regression on 1 and **X**. Note that the residual sum of squares is unchanged.

B. The least squares fit is **b** = (20. 3̇, 6.25, 2.41̇6̇)′. **A** = (1, −2, 0), **c** = 4.
   Z = **A**(**X′X**)$^{-1}$**A′** = $\frac{4}{3}$, so **Z**$^{-1}$ = 0.75.
   c − **Ab** = −3.83̇.
   b$_H$ = (19.375, 7.6875, 3.375).
   Note that $b_{0H} - 2b_{1H} = 4$ as required.
   (This exercise can also be done by substituting for the restriction into the model.)

C. Look first at the vector parallelogram in Solution C1. We see that the **Y** − **X** vector is the same (in length and orientation) as the vector joining the tips of the **Y** and **X** vectors. It follows that the perpendiculars to the estimation space must also be equivalent vectors with the same lengths. The squares of these lengths are the residual sums of squares for regressions of (a) **Y** on 1 and **X** and (b) **Y** − **X** on 1 and **X**.

   We now consider the R^2 statistics for the two regressions, shown in Solution C2. Obviously they can be different and R_a^2 could be larger or smaller than R_b^2 depending on the original geometry. Similar sorts of comments apply when there are k regression vectors $\mathbf{X}_1, \mathbf{X}_2, \ldots, \mathbf{X}_k$ and $\mathbf{Y} - \Sigma \theta_i \mathbf{X}_i$ is used in the (b) regression.

   As an extreme case of this moving of the response vector, consider **Y** − **Xb**, where we reduce **Y** by the particular combination of columns of **X** designated by the elements of the least squares vector **b**. Of course, the resulting response vector is the residual vector **e**; this is orthogonal to the estimation space and consequently $R^2 = 0$.

   The author of the reference quoted suggests that use of R^2 is unsatisfactory because of this feature. We disagree. R^2, which measures the correlation between the selected response values and the predicted response values, accurately reflects the geometry of each regression situation shown. Looking at the geometry clarifies what the data manipulation has done.

G. The hypothesis $H_0 : \beta_1 = 7, \beta_2 = 4$ cannot be tested because β_1 and β_2 are not individually estimable; **X** has dependent columns. There are 6 df available, 3 of which carry pure error. The remaining 3 df of the estimation space can be associated with the three averages obtained

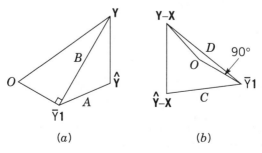

(a)   (b)

Solution C2. The R^2 statistics for the two regressions: (a) $R_a^2 = A^2/B^2$, (b) $R_b^2 = C^2/D^2$.

from (a) the three estimates of $\beta_0 + \beta_1$ (namely, 101, 105, 94), (b) the two estimates of $\beta_0 + \beta_2$ (84, 88), and (c) the one estimate of $\beta_0 + \beta_3$ (32). Obviously, the comparison $\hat{\beta}_1 - \hat{\beta}_2 = \frac{1}{3}(101 + 105 + 94) - \frac{1}{2}(84 + 88) = 14$ provides an estimate of $\beta_1 - \beta_2$ and this has a standard error of $\{\sigma^2(\frac{1}{3} + \frac{1}{2})\}^{1/2} = \{5\hat{\sigma}^2/6\}^{1/2}$. The pure error estimate of σ^2 is $\hat{\sigma}^2 = \{(1^2 + 5^2 + 6^2) + (2^2 + 2^2)\}/(2 + 1) = 70/3 = 23.333$, so that the standard error is 4.410. We can thus test whether $\beta_1 - \beta_2 = 7 - 4 = 3$. We get a t-statistic of $t = \{14 - (7 - 4)\}/4.410 = 2.494$. The corresponding $F(1, 3)$ statistic is $(2.494)^2 = 6.22$. The result is that we cannot reject $\beta_1 - \beta_2 = 3$ at the $\alpha = 0.05$ level.

This simple example (in which the n-dimensional space is divided up only into the estimation and pure error spaces) makes it clear that we can estimate only quantities that are linear combinations of the rows of $\mathbf{X}\boldsymbol{\beta}$. For some intricate algebra in the general case of nontestable hypotheses and some possible alternatives, see the papers mentioned in the source references.

H. 1.

$$70\mathbf{P} = \begin{bmatrix} 62 & 18 & -6 & -10 & 6 \\ & 26 & 24 & 12 & -10 \\ & & 34 & 24 & -6 \\ & & & 26 & 18 \\ \text{symmetric} & & & & 62 \end{bmatrix}.$$

2.

$$70\mathbf{P}_\omega = \begin{bmatrix} 42 & 28 & 14 & 0 & -14 \\ & 21 & 14 & 7 & 0 \\ & & 14 & 14 & 14 \\ & & & 21 & 28 \\ \text{symmetric} & & & & 42 \end{bmatrix}.$$

3. $\mathbf{P}_1 = \mathbf{P} - \mathbf{P}_\omega$, so

$$14\mathbf{P}_1 = \begin{bmatrix} 4 & -2 & -4 & -2 & 4 \\ & 1 & 2 & 1 & -2 \\ & & 4 & 2 & -4 \\ & & & 1 & -2 \\ \text{symmetric} & & & & 4 \end{bmatrix}.$$

4. All are true.

5. $\mathbf{P}_1\mathbf{x} = \mathbf{0}$. This is obvious because $\mathbf{x}$ is a vector in ω and so is orthogonal to the space $R(\mathbf{P}_1)$, which is $\Omega - \omega$.

Chapter 22

A. Note that Company A data can be fitted in original Y or in $\ln Y$ form by orthogonal polynomials, whereas Company B data cannot be logged, because of the negative values, which are losses. Both sets of data given show severe cyclical trends, so that more than a straight line fit is needed. Data like these are notoriously difficult to predict into the future, and the polynomial, at best, explains what has happened. Prediction in other more stable cases is likely to be more useful.

B. $\hat{Y} = 11.2 + 0.9143(2X) + 1.5833\{1.5(X^2 - 35/12)\} + 0.006\{(5/3)(X^3 - 101X/20)\}$, with successive regression sums of squares 627.2, 58.51, 210.58, 0.006; and residual sum of squares 207.70 (2 df). The cubic term can be dropped, and the remaining terms rearranged in the form

$$\hat{Y} = 4.273 + 1.8286X + 2.3750X^2.$$

The sums of squares mentioned in the question are identical to the corresponding sums of squares of the same order given above.

C. Let Z = week number. The fitted equation is

$$\hat{Y} = 136.227 + 2.687Z + 0.167Z^2.$$

Source of Variation	df	SS	MS	F
Total	20	74,628.00		
a_0	1	48,609.80		
a_1	1	25,438.75	25,438.75	4,558.92*
a_2	1	489.00	489.00	87.63*
a_3	1	1.15	1.15	0.21
Residual	16	89.30	5.58	

D. 1. $\hat{Y} = -0.0037 - 2.8008t + 0.2314t^2.$
   2. Residual analysis: There is no evidence for increasing the degree of the polynomial in t.
E. The analysis of variance is shown below.

Value of b_j	Source	df	SS
15.111	b_0	1	2055.11
1.866667	b_1	1	209.07
0.165945	b_2	1	76.33
0.072727	b_3	1	5.24
—	Residual	5	4.36
	Total	9	2350.00

The cubic term is not significant. A suitable model is

$$\hat{Y} = b_0 + b_1X + b_2(3X^2 - 20)$$

$$= 11.792 + 1.8667X + 0.4978X^2,$$

where the dyestuff levels are coded $X = -4, -3, -2, -1, 0, 1, 2, 3, 4$. This model accounts for $R^2 = 285.4/294.89 = 0.9678$ of the variation about the mean; $s^2 = 9.6/6 = 1.6$.

F. Partial solution.

i	Estimated Coefficient a_i	Before Compression df	Before Compression SS(a_i)	After Compression df	After Compression SS(a_i)
0	66.25	1	52,668.75	1	52,669
1	−0.02273	1	0.30		
2	−0.00774	1	0.72		
3	1.96018	1	19,780.16	1	19,780
4	0.01349	1	1.46		
5	−0.02137	1	7.26		
6	−0.06551	1	19.26		
Residual		5	13.09	10	42
Total		12	72,491.00	12	72,491

All the a_i eliminated in the compression are not significant at the $\alpha = 0.05$ level except a_6. However, use of a_6 increases R^2 only from 0.9979 to 0.9988, so that we rate a_6 as statistically significant, but numerically unimportant. The corresponding F-value of $19.36/\{13.09/5\} = 7.39$ only slightly exceeds $F(1, 5, 0.95) = 6.61$.

The selected model is $\hat{Y} = 66.25 + 1.96\{\frac{2}{3}(X^2 - \frac{85}{4})\}X$.

Durbin–Watson $d = 2.33$, not significant.

G. First read Section 22.3. Enter 322 observations as described there. Let Z_1, Z_2 be dummy variables such that $Z_1 = 1$ for category A, 0 otherwise, and $Z_2 = 1$ for category B, 0 otherwise. We fit the model $Y = \beta_0 + \alpha_1 Z_1 + \alpha_2 Z_2 + \beta_1 (\text{age}) + \epsilon$ to obtain $\hat{Y} = 80.58 + 5.73Z_1 + 7.54Z_2 + 1.08 (\text{age})$. The adjusted ANOVA table is:

ANOVA

Source	df	SS	MS	F
$a_1, a_2\|b_0$	2	4,398.97	2,199.49	
$b_1\|b_0, a_1, a_2$	1	1,000.18	1,000.18	4.52
Residual	318	70,441.50	$s^2 = 221.51$	
Lack of fit	19	2,379.08	125.21	0.55
Pure error	299	68,062.42	227.63	
Total, corrected	321	75,840.64		

Lack of fit is not significant and we reject $\beta_1 = 0$ at the $\alpha = 0.05$ level, because $4.52 >$ (value interpolated between 3.92 and 3.84 in F-table). $R^2 = (4{,}398.97 + 1{,}000.18)/75{,}840.64 = 0.0712$. This value (only 7% of the variation about the mean is explained), and the fact that the F-value 4.52 is only slightly bigger than the percentage point with which it is compared, indicate that the fitted equation is not a very useful one. Tests of $\alpha_1 = 0$, $\alpha_2 = 0$, and $\alpha_1 - \alpha_2 = 0$ confirm the fact that the two groups of men weigh about the same and are heavier than the group of women. (The appropriate t-values are 2.55, 3.82, and 0.84.) We conclude then that, although weight increases with age in these data, age is not a good predictor of it and that, although there *are* weight differences in the diet categories, they seem to be confounded with gender and may well be due to that, rather than diet.

Chapter 23

A. 1. Both methods of analysis will yield the following analysis of variance table:

ANOVA

Source of Variation	df	SS	MS	F	$F_{0.95}$
Total SS	18	24,403.750			
Mean	1	22,352.027			
Corrected total	17	2,051.723			
Steam pressure	2	963.721	481.861	11.728*	4.26
Blowing time	2	37.481	18.741	0.456	4.26
Interaction	4	680.756	170.189	4.142*	3.63
Pure error	9	369.765	41.085		

Thus both steam pressure and the interaction of steam pressure and blowing time are statistically significant.

The regression model obtained for this problem is

$$\hat{Y} = 35.239 + 4.794X_1 - 10.339X_2 - 0.206X_3 + 1.861X_4$$
$$+ 5.773X_1X_3 - 2.394X_1X_4 + 6.506X_2X_3 - 3.361X_2X_4,$$

where X_1, X_2 are dummy variables for steam pressure defined as follows:

$$
\begin{array}{cc}
X_1 & X_2 \\
1 & 0 = 10 \text{ pounds steam pressure} \\
0 & 1 = 20 \text{ pounds steam pressure} \\
-1 & -1 = 30 \text{ pounds steam pressure}
\end{array}
$$

and X_3, X_4 are dummy variables for blowing time defined as follows:

$$
\begin{array}{cc}
X_3 & X_4 \\
1 & 0 = \text{blowing time 1 hour} \\
0 & 1 = \text{blowing time 2 hours} \\
-1 & -1 = \text{blowing time 3 hours}
\end{array}
$$

Residual analysis indicates that the experiments have a much smaller variance at the low level of steam pressure. Since only two repeat runs are available at each set of conditions, the analysis is not necessarily invalid, but further investigation is clearly indicated. To appreciate the interaction effect, a table of mean values could be examined, or the mean responses could be plotted against blowing time for each level of steam pressure.

2.

ANOVA

Source	df	SS	MS	F	$F_{0.95}$
Total	18	417.000			
Mean	1	34.722			
Corrected total	17	382.278			
Premix speed	2	24.111	12.055	1.080	4.26
Finished mix speed	2	7.444	3.722	0.333	4.26
Interaction	4	250.223	62.556	5.602*	3.63
Pure error	9	100.500	11.167		

Thus only the interaction term is statistically significant in this experiment.
    The fitted regression model obtained for this problem is

$$
\hat{Y} = 1.39 - 1.39X_1 + 1.44X_2 - 0.89X_3 + 0.61X_4 + 1.39X_1X_2 \\
+ 1.89X_1X_4 + 4.56X_2X_3 - 1.44X_2X_4,
$$

where X_1, X_2 are dummy variables for premix speeds, and X_3, X_4 are dummy variables for finished mix speeds as follows:

$$
\begin{array}{cc}
X_1 & X_2 \\
1 & 0 = \text{premix speed 1} \\
0 & 1 = \text{premix speed 2} \\
-1 & -1 = \text{premix speed 3}
\end{array}
\qquad
\begin{array}{cc}
X_3 & X_4 \\
1 & 0 = \text{finished mix speed 1} \\
0 & 1 = \text{finished mix speed 2} \\
-1 & -1 = \text{finished mix speed 3}
\end{array}
$$

Residual analysis indicates a much larger variability in the observations at the lowest level of finished mix speed. This should be investigated.
    The significant interaction is most easily seen in the following table:

Table of Mean Responses

		X_2		
		-1	0	$+1$
	-1	0.5	2.5	-3
X_1	0	6.5	2.0	0
	$+1$	-5.5	1.5	8

B. 1a.

		ANOVA		
Source	df	SS	MS	F
Total	16	921.0000		
Regression	4	881.2500	220.3125	
b_0	1	798.0625	798.0625	
b_1	1	18.0625	18.0625	5.45*
b_2	1	60.0625	60.0625	18.13*
b_{12}	1	5.0625	5.0625	1.53 NS
Residual	12	39.7500	3.3125	

b. (1) Regression equation is significant. $F(3, 12) = 8.37*$.
    (2) All except b_{12}.
c. $R^2 = 67.67\%$.

2a.

$$b_0 = \frac{798.0625}{113} = 7.0625,$$

$$b_1 = \frac{18.0625}{17} = 1.0625,$$

$$b_2 = \frac{60.0625}{31} = 1.9375,$$

$$b_{12} = \frac{5.0625}{-9} = -0.5625.$$

$\therefore \hat{Y} = 7.0625 + 1.0625X_1 + 1.9375X_2 - 0.5625X_1X_2$.

b. $s^2(\mathbf{X'CX}) = 0.6875$

$(3.3125)(\mathbf{X'CX}) = 0.6875$

$$\mathbf{X'CX} = \frac{0.6875}{3.3125} = 0.207547.$$

Variance of a single observation $= s^2(1 + \mathbf{X'CX}) = (3.3125)(1 + 0.207547)$
                                             $= 4.0000$.

c. $\hat{Y}$ is 54 at $X_1 = 70$ and $X_2 = 150$.
    $V(\hat{Y}) = 0.6875$.
    $\therefore$ Confidence limits for the true mean value of Y are

$$\hat{Y} \pm t(12, 0.95)\text{se}(\hat{Y}) = 54 \pm (2.179)\sqrt{0.6875}$$

$$= 54 \pm (2.179)(0.8292)$$

$$= 54 \pm 1.8068.$$

3a. The prediction equation determined by this analysis is

$$\hat{Y} = 7.0625 + 1.0625X_1 + 1.9375X_2 - 0.5625X_1X_2.$$

b. The interaction term, X_1X_2, is not statistically significant at an α level of 0.05. Thus there is some doubt as to the validity of the assumed model. However, this doubt is based on a small number of observations, $n = 16$, and the original model was based on the knowledge of the chemist. Before considering dropping the X_1X_2 term, more experimental work should be done. This is an example of the statement: "Even though a variable is nonsignificant statistically, it should *not* be considered to have a zero effect on the result of the experiment."

C. Solution not provided.

D. Our solution uses Section 23.5. Transformed data:

Sample No.	1	2	3	4	5
$Y_{ij} - 1430$	260	120	195	295	100
	150	15	20	120	115
	315	215	80	0	135
	255	115		15	90
$J_i \overline{Y}_i$	980	465	295	430	440
J_i	4	4	3	4	4
$\overline{Y}_i = b_i$	245	116.25	98.33	107.5	110

From Eqs. (23.5.5) and (23.5.6), residual SS is $524{,}450 - 417{,}788.6 = 106{,}661.4$. To test equality of groups we obtain, from Eqs. (23.5.10) and (23.5.6), the SS due to H_0 as $(524{,}450 - 358{,}531.6) - 106{,}661.4 = 59{,}257$. $F = (59{,}257/4)/(106{,}661.4/14) = 1.94 < F(4, 14, 0.95) = 3.11$. Do not reject H_0. We conclude that the data do not indicate that the strength of the concrete is significantly affected by varying the amount of coal dust added to the sand.

Working with $Y_{ij} - 1430$ affects only the total SS and the correction factor. The between group and within group SS values are unaffected.

E. 1. We recode the factor combination bcd as $X_1 = -1$, $X_2 = X_3 = X_4 = 1$, putting -1 when a letter does *not* appear in the combination and 1 when it *does* appear. We can fit a model with terms in β_0, $\beta_i X_i$, and $\beta_{ij} X_i X_j$, $i \neq j = 1, 2, 3, 4$. Only X_2 and X_3 are worth retaining and the equation of choice is $\hat{Y} = 39.68 - 0.6688 X_2 + 0.8937 X_3$ with $R^2 = 0.7229$.

2. The data come from a four-way classification, each classification having two levels, with one observation per cell. The model of Section 23.6 would have to be extended from two-way to four-way to treat this in an ANOVA framework.

F. 1. The original setup produces a singular $\mathbf{X}'\mathbf{X}$.

2. Any valid dummy setup can be used. Suppose we choose two dummies $(Z_1, Z_2) = (-1, 1)$, $(0, -2)$, and $(1, 1)$ for groups 1, 2, and 3, respectively. This produces orthogonal $\mathbf{X}$-columns because there are the same numbers of observations in each group. (These dummies are orthogonal polynomials; see Section 22.1.) Then $\mathbf{b} = (6, 2, 0)$ and the orthogonal SS breakup of $\mathbf{b}'\mathbf{X}'\mathbf{Y}$ is $324 + 24 + 0 = 348$ (3 df). The residual SS $= 366 - 348 = 18$ (5 df). The F-statistic for testing equality of groups is thus $F(2, 5) = \{(24 + 0)/2\}/(18/5) = 3.33 < 5.79$, not significant. The F-value does not depend on the dummy system chosen, as the reader can confirm by trying the problem another way.

G. For both (1) and (2), many choices of dummies can be made, for example:

$$(X_1, X_2) = (1, 0) \quad \text{for tube 1,}$$
$$= (0, 1) \quad \text{for tube 2,}$$
$$= (0, 0) \quad \text{for tube 3,}$$
$$(X_3, X_4, X_5, X_6) = (1, 0, 0, 0) \quad \text{for level 1,}$$
$$(0, 1, 0, 0) \quad \text{for level 2,}$$
$$(0, 0, 1, 0) \quad \text{for level 3,}$$
$$(0, 0, 0, 1) \quad \text{for level 4,}$$
$$(0, 0, 0, 0) \quad \text{for level 5.}$$

Thus, if we copy the observations row by row into the $\mathbf{Y}$ vector, we get for corresponding rows of the **X** matrix and the original **Y** vector, the following:

| \multicolumn{7}{c}{Subscript of X} | | | | | | | |
0	1	2	3	4	5	6	Y
1	1	0	1	0	0	0	78.4
1	0	1	1	0	0	0	84.5
1	0	0	1	0	0	0	62.1
1	1	0	0	1	0	0	10.6
1	0	1	0	1	0	0	77.5
1	0	0	0	1	0	0	94.4
1	1	0	0	0	1	0	62.7
1	0	1	0	0	1	0	4.4
1	0	0	0	0	1	0	24.8
1	1	0	0	0	0	1	92.5
1	0	1	0	0	0	1	17.5
1	0	0	0	0	0	1	0.9
1	1	0	0	0	0	0	11.6
1	0	1	0	0	0	0	0.9
1	0	0	0	0	0	0	20.4

1. We fit three models:

$$\hat{Y} = 40.52 + 10.64X_1 - 3.56X_2,$$

$$\hat{Y} = 10.97 + 64.03X_3 + 49.87X_4 + 19.67X_5 + 26.00X_6,$$

$$\hat{Y} = 8.61 + 10.64X_1 - 3.56X_2 + 64.03X_3 + 49.87X_4 + 19.67X_5 + 26.00X_6.$$

The respective regression sums of squares (given b_0) are

$$546 \ (2 \ \text{df}),$$

$$7672 \ (4 \ \text{df}),$$

$$8218 \ (6 \ \text{df}),$$

and the residual mean square is $s^2 = 1294(8 \ \text{df})$. It follows that the "between tubes" $F_{2,8} = \{(8218 - 7672)/2\}/1294 = 0.21$ and the "between levels" $F_{4,8} = \{(8218 - 546)/4\}/1294 = 1.48$.

2. The details are exactly parallel. Now, however, $F_{2,8} = \{(17300 - 17016)/2\}/159.2 = 0.89$ and $F_{4,8} = \{(17300 - 284)/4\}/159.2 = 26.72$. So now there are differences between levels, in this (fake) data set.

H. The four regressions explain (1) 98.7% (2) 99.1% (3) 98.5%, and (4) 99.1% of the variation about the mean $\overline{Y}$. Essentially the ranges of Y and X_1 are such that the transformations make little difference in explanatory value, at least in this set of data. All equations appear plausible ones, and the residuals look most satisfactory for the (1) fit. Equation (1) is

$$\hat{Y} = 0.3688 + 0.0340X_1 - 0.1852X_2 + 0.0998X_3 + 0.002313X_2X_3.$$

The dummy variable system suggested makes the X_2, X_3, and X_{23} columns, which carry the 3 df among the four columns DD, DS, SD, and SS, orthogonal to the 1 and X_1 columns. The b_2 and b_3 coefficients are very highly significant, the b_{23} is not significant. Additive main effects are thus indicated. Hatchabilities are best for the DD combination followed by DS, SD, and SS, and the differences are significant ones.

Chapter 24

A. $\hat{\theta} = 0.20345$, $S(\hat{\theta}) = 0.00030$; $0.179 \le \theta \le 0.231$.

B. $\hat{\theta} = 0.20691$, $S(\hat{\theta}) = 0.01202$; $0.190 \leq \theta \leq 0.225$.

C. $(\hat{\alpha}, \hat{\beta}) = (0.38073, 0.07949)$, $S(\hat{\alpha}, \hat{\beta}) = 0.00005$;
$S(\alpha, \beta) = 0.001$ (or 0.0009 to one more decimal place)

D. $(\hat{\alpha}, \hat{\beta}, \hat{\rho}) = (72.4326, 28.2519, 0.5968)$, $S(\hat{\alpha}, \hat{\beta}, \hat{\rho}) = 3.5688$; $S(\alpha, \beta, \rho) = 106.14$.

E. $(\hat{\alpha}, \hat{\beta}, \hat{\gamma}) = (5.2673, 8.5651, 294.9931)$, $S(\hat{\alpha}, \hat{\beta}, \hat{\gamma}) = 1718.2108$; $S(\alpha, \beta, \gamma) = 3400$.

F. Write the model as

$$Y = \theta + \alpha(X_1 X_3 + \gamma X_1) + \beta(X_2 X_3 + \gamma X_2) + \epsilon.$$

Fix γ, solve for $\hat{\theta}$, $\hat{\alpha}$, $\hat{\beta}$. Repeat for other values of γ, iterating on γ until minimum $S(\hat{\theta}, \hat{\alpha}, \hat{\beta}, \hat{\gamma})$ is obtained.

G. $(115.2, 2.310, -22.022)$, $S(\hat{\theta}) = 7.0133$. $S(\theta) = 209.0$.

H. $(0.00376, 27,539)$, $S(\hat{\theta}) = 0.00429326$. $S(\theta) = 0.00507559$.

I. $(0.00366, 27,627)$, $\hat{S}(\hat{\theta}) = 0.000754$. $S(\theta) = 0.00204$.

J. $(3.57, 12.77, 0.63)$, $S(\hat{\theta}) = 0.00788$. $S(\theta) = 0.01665$.

K. $(0.480, 1.603)$, $S(\hat{\theta}) = 7.301$. $S(\theta) = 53.8$.
The long thin contour indicates that a large number of pairs of values (θ_1, θ_2) are almost as suitable as the actual least squares values.

L.

No.	$\hat{\theta}_1$	$\hat{\theta}_2$	$S(\hat{\theta})$	$S(\theta)$
1	205.25	0.431	252	835.8
2	2.498	0.202	0.0262	0.0712
3	892.67	0.245	3,376.5	15,093
4	25.475	0.323	17.004	76.007
5	13.809	0.398	0.866	3.871
6	19.903	0.441	3.716	16.61
7	213.82	0.547	1,168	5,221
8	19.142	0.531	25.99	116.18
9	10,525	0.569	68,349	1,366,980

M. $\hat{\alpha} = 0.009229$, with $se(\hat{\alpha}) = 0.010725$.
$\hat{\beta} = 1.825450$, with $se(\hat{\beta}) = 0.234714$.
$s = 19.3049$, based on 41 df.
The estimates are highly correlated, with $r = 0.999$.
(Note that, if we now take natural logarithms, we obtain $\ln \hat{M} = \ln \hat{\alpha} + \hat{\beta} \ln T = -4.6854 + 1.82545 \ln T$. This compares with the linear least squares fit with coefficients $a = -5.728$, $b = 2.031$ in Exercise 13E. One could next examine the residuals from each fitted equation, the nonlinear as fitted and the linear, and make an assessment of which fit seemed to be preferable.)

N. (Partial solution)
Tree No. 1

$(24.7.2)$ $\hat{\alpha} = 268.3$, $\hat{\beta} = 0.9478$, $\hat{k} \times 10^4 = 4.740$,

$(24.7.5)$ $\hat{\alpha} = 154.1$, $\hat{\beta} = 5.643$, $\hat{k} \times 10^3 = 2.759$,

$(24.7.8)$ $\hat{\delta} = 5.032$, $\hat{\beta} = 5.792$, $\hat{k} \times 10^3 = 2.814$,

$(24.7.10)$ $\hat{\alpha} = 172.2$, $\hat{\beta} = 2.813$, $\hat{k} \times 10^3 = 1.626$.

Visually, (24.7.8) looks best although this may be due to the reduced response range induced by the ln transformation. The first model (24.7.2) does not pick up well the "S-shaped" tendency in the data. With only seven observations, however, the data set is too small to permit too definite conclusions.

Tree No. 2

$$(24.7.2) \quad \hat{\alpha} = 519.3, \quad \hat{\beta} = 0.9820, \quad \hat{k} \times 10^4 = 3.208,$$

$$(24.7.5) \quad \hat{\alpha} = 218.9, \quad \hat{\beta} = 8.225, \quad \hat{k} \times 10^3 = 3.010,$$

$$(24.7.8) \quad \hat{\delta} = 5.398, \quad \hat{\beta} = 8.228, \quad \hat{k} \times 10^3 = 2.962,$$

$$(24.7.10) \quad \hat{\alpha} = 248.4, \quad \hat{\beta} = 2.645, \quad \hat{k} \times 10^3 = 1.703.$$

O.

β	2.00	2.01	2.02	2.03
α	−1.03	−1.06	−1.09	−1.12
SS	0.0275	0.0134	0.0126	0.0255

Solution is $\hat{\alpha} = -1.09$, $\hat{\beta} = 2.02$, correct to the second decimal place in β.

P. The range of γ values $0.68 \le \gamma \le 0.70$ is best, with $\hat{\alpha} = 43.2$, $\hat{\beta} = 123.2$ at $\gamma = 0.69$ and minor variations nearby.

For ϕ, $0.93 \le \phi \le 0.96$ is best with minor variations about values $\hat{\delta} = 34.0$, $\hat{\theta} = 106.0$ at $\phi = 0.94$.

Q.

$$\hat{\alpha} = 0.6588, \quad se(\hat{\alpha}) = 0.250.$$

$$\hat{\beta} = 1.0272, \quad se(\hat{\beta}) = 0.040.$$

$$\hat{\gamma} = 0.5184, \quad se(\hat{\gamma}) = 0.011.$$

$$\text{corr}\,(\hat{\alpha}, \hat{\beta}) = -0.963, \quad \text{corr}\,(\hat{\alpha}, \hat{\gamma}) = -0.268, \quad \text{corr}\,(\hat{\beta}, \hat{\gamma}) = 0.$$

$$s = 9176.93 \ (13 \text{ df}).$$

R.

Group	Equation $\hat{Y} =$
185 kg B	$0.8822 + 2.2290 \,(0.7512)^X$
275 kg B	$0.8444 + 1.7083 \,(0.796)^X$
275 kg H × B	$0.8079 + 1.7701 \,(0.7483)^X$
450 kg $\frac{3}{4}$ B	$0.7066 + 0.5700 \,(0.7905)^X$
450 kg $\frac{3}{4}$ H	$0.6699 + 0.2448 \,(0.8458)^X$

S. (Main details.) The parameter estimates are:

Sample	$\hat{\delta}$	$\hat{\beta}$	$\hat{k}$
8	4.96	78.76	0.15
14	5.08	222.36	0.21
16	4.88	103.70	0.15
22	4.79	128.24	0.17
24	5.71	23.84	0.07
32	4.88	217.93	0.20
52	5.18	50.66	0.12
54	4.95	49.17	0.13
All	4.96	89.32	0.15

Although samples 14, 24, and 32 do catch the eye, the various plots reveal an overall consistency in these data sets.

Individual readings would provide a pure error estimate of σ^2 and enable us to check its (assumed) constancy; because the *same* number of observations determines each mean, the actual parameter estimates would not change.

T. The main points are as follows.

$$\hat{\boldsymbol{\theta}} = (0.10579, 1.7007)'.$$

$$\text{Pure error SS} = 1.998 \times 10^{-4} \text{ (6 df)}.$$

$$\text{Lack of fit SS} = 1.08 \times 10^{-6} \text{ (4 df)}.$$

The model seems satisfactory. The best single run is at the largest X value practically feasible; the best pair are both at that same value.

U. We can use the initial estimates $\theta_{10} = 0.11$, $\theta_{20} = 1.7$ from the results of Exercise T. Substituting these values in the model function and (for example) putting in (X, Y) values $(2, 0.0571)$ and $(0.2, 0.0108)$ derived from the first and last data pairs, gives two equations from which θ_4 can be eliminated to give $0.06834 = 0.0529(2^{\theta_3}) - 0.0992(0.2^{\theta_3})$. An iterated solution is $\theta_{30} = 0.93$, whereupon either equation can be solved, or both solutions can be averaged, to give $\theta_{40} = -0.075$ approximately. The various fits give the following results.

Model with These θ's	$S(\hat{\boldsymbol{\theta}})$	df	$s = \text{se}(Y)$
$\theta_1, \theta_2, \theta_3, \theta_4$	0.0002	8	0.0050
$\theta_1, \theta_2, \theta_3(\theta_4 = 0)$	0.0002	9	0.0047
$\theta_1, \theta_2(\theta_3 = \theta_4 = 0)$	0.0002	10	0.0045

We recall from T that the pure error SS $= 0.0002$ (6 df). All the $S(\hat{\boldsymbol{\theta}})$ and the pure error sum of squares are about the same size, 0.0002, so the extra SS are very tiny, their small differences being reflected in the minor changes in the last column. Obviously the smallest model is satisfactory. For this, $\hat{\theta}_1 = 0.105643$ (se $= 0.018$), $\hat{\theta}_2 = 1.70269$ (se $= 0.476$).

V. Data Set 1 has the characteristics of the data in Figure 24.2a. The approximate 95% confidence contour is very large and $\hat{\theta}_1$ and $\hat{\theta}_2$ are highly correlated. Data Set 2 is like the additional data in Figure 24.2b. It provides much better estimation and a very small approximate 95% confidence contour. As an additional exercise, use all nine observations, omitting one of the duplicates at $t = 1$, and compare the three analyses.

W. By ignoring the error and forcing the curve through any of the data points, we can solve $Y = \theta_{10}\{1 - \exp(-t/4)\}$ for θ_{10}. Substitution of each data point in turn gives $\theta_{10} = 2.13, 2.22, 2.25, 2.45$. Any of these could be used; so could the average 2.26 or 2.3.

X. *Variety No. 1 Data.* If we initially ignore the error and just work with the model function we see that

$$\omega^{-\theta_1} = \theta_2 + \theta_3 X^{\theta_4}.$$

Thus, if we knew θ_1 and θ_4, the plot of $\omega^{-\theta_1}$ versus X^{θ_4} would be a straight line; see the discussion in point 4 of the last subsection of Section 24.2.

In fact, $\theta_1 = \theta_4 = 1$ gives an excellent line with intercept 0.0055 and slope 0.0016, approximately, fitted by eye. So we can take $\boldsymbol{\theta}_0 = (1, 0.0055, 0.0016, 1)$. Nonlinear estimation from there gives $\hat{\boldsymbol{\theta}} = (1.1528, 1.7793 \times 10^{-3}, 1.0189 \times 10^{-3}, 1.0046)$ with $S(\hat{\boldsymbol{\theta}}) = 0.0225461$. The approximate linearized individual confidence bands are

$$\theta_1: \quad (-2.9614, 5.2671),$$

$$\theta_2: \quad (-3.2602 \times 10^{-2}, 3.6161 \times 10^{-2}),$$

$$\theta_3: \quad (-2.1540 \times 10^{-2}, 2.3578 \times 10^{-2}),$$

$$\theta_4: \quad (-2.0443, 4.0535).$$

It is clear that the parameter estimation is poor and possible overparameterization is indicated by the off-diagonal elements of the linearized parameter estimates correlation matrix, all of which are close to ± 1. The $S(\boldsymbol{\theta})$ contour is elongated and many different final sets of estimates for the parameters, all with more or less the same $S(\hat{\boldsymbol{\theta}})$, are possible. Initial values that differ only slightly from those above will give rise to alternative $\hat{\boldsymbol{\theta}}$ values. The plots of residuals do not reveal any abnormalities. In view of the plot that showed that values $\theta_1 = \theta_4 = 1$ seemed perfectly reasonable and the fact that the model seems overparameterized anyway, we could try to fit the simpler model

$$\ln W_i = -\ln(\alpha + \beta X_i) + \epsilon_i$$

using the same initial values (in revised notation), $\alpha_0 = 0.0055$ and $\beta_0 = 0.0016$. This leads to the solution $(\hat{\alpha}, \hat{\beta}) = (5.309 \times 10^{-3}, 1.587 \times 10^{-3})$ with $S(\hat{\alpha}, \hat{\beta}) = 0.026239$. Previously we had $S(\hat{\theta}) = 0.0225461$ so that the increase is small considering that two parameters have been dropped. The approximate linearized confidence intervals are now

$$\alpha: \quad (4.345 \times 10^{-3}, 6.272 \times 10^{-3}),$$
$$\beta: \quad (1.587 \times 10^{-3}, 1.705 \times 10^{-3}),$$

neither of which includes zero. Overall, the fit seems adequate.

Y. To get initial values, one can do something like the following, for example. Substitute into the model function for $(X_1, X_2, Y) = (0.25,0,0.310)$ and $(5,0,1.075)$ where the Y's are roughly guessed from the data. This gives $\theta_1 = 0.310(4\theta_2 + 1) = 1.075(0.2\theta_2 + 1)$ which can be solved for $\theta_2 = 0.692$, $\theta_1 = 1.168$. Now set $(X_1, X_2, Y, \theta_1, \theta_2) = (5,0.5,0.280,1.168,0.692)$ into the model function and solve for $\theta_3 = 0.0228$. We thus obtain $\theta_0 = (1.17,0.69,0.023)'$.

| | Parameter Estimates and Linearized 95% Limits | | |
	Lower	$\hat{\theta}$	Upper
θ_1	1.235	1.260	1.284
θ_2	0.792	0.847	0.901
θ_3	0.0254	0.0272	0.0290

$S(\hat{\theta}) = 0.01370.$

Correlations of parameter estimates in order 12, 13, 23: 0.866, 0.466, 0.673.

Source	SS	df	MS
Lack of fit	0.00761	15	0.000507
Pure error	0.00609	28	0.000218
Residual, $S(\hat{\theta})$	0.01370	43	

The "approximate $F(15, 28)$" ratio is 2.33 which exceeds $F(15, 28, 0.95) = 2.04$, indicating lack of fit. However, the three groups of data for which $X_2 = 0, 0.2$, and 0.5 contribute pure error sums of squares of 0.005490 (16 df, 0.000343 per df), 0.000093 (6 df, 0.000016 per df), and 0.000508 (6 df, 0.000085 per df), respectively. The pairs of observations are considerably less variable than the groups of three and four observations, casting some doubt on what is only an approximate F-test in the best of circumstances.

Z. 1. $\hat{Q} = -2.08302$, $\hat{p} = 0.194833$, $\hat{N} = 76.8884$ with se$(\hat{Q}) = 6.113$, se$(\hat{p}) = 0.1858$, se$(\hat{N}) = 93.8601$. The correlations are 0.965 $(\hat{Q}, \hat{p})$, -0.990 $(\hat{Q}, \hat{N})$, and -0.984 $(\hat{p}, \hat{N})$. The residual se $= 2.42384$ (7 df). $S(\hat{\theta}) = 41.1252$. The negative value for $\hat{Q}$ is meaningless and the correlations are large, so it makes sense to set $Q = 0$ and refit.

2. Set $Q = 0$, then $\hat{p} = 0.278715$, $\hat{N} = 48.9678$ with se$(\hat{p}) = 0.0765$ and se$(\hat{N}) = 9.134$. The correlation between estimates is -0.733. The residual se $= 2.30168$ (8 df). $S(\hat{\theta}) = 42.3819$. In this fit the residual sum of squares is only slightly higher and the residual se is lower. The $\hat{N}$ of about 49 makes sense, the se values of the coefficients are relatively small and the prediction equation $\hat{Y} = (0.278715)(48.9678) \exp\{-0.278715t\}$ produces estimates of 12, 9, 7, 5, 4, 3, 2, 2, 1, 1 for the times given—not bad in the circumstances.

Chapter 25

A. Various solutions are obtained depending on the robust method used and the tuning constants selected. Data sets with outliers tend to be better fitted by robust methods compared with least squares and the differences in the fits provide important clues to the possibility of adjusting the data.

B. The largest observation in the data set is the cause of the difference seen, because it "pulls the least squares line up toward itself." The robust fits accord the largest observation less control.

Chapter 26

A. No solution provided. Results will vary.

B. No solution provided. Results will vary.

Solutions to True/False Questions

1–101. All statements are true. They can be amended to become false if used in quizzes. (It was felt unwise to present false statements here and risk a misunderstanding.)

Tables

NORMAL DISTRIBUTION

Normal Distribution (Single-Sided): Proportion (A) of Whole Area Lying to Right of Ordinate Through $x = \mu + z\sigma$ [$z = (x - \mu)/\sigma$]

Deviate (z)	Prefix	0.00	0.01	0.02	0.03	0.04	0.05	0.06	0.07	0.08	0.09	Prefix	Deviate (z)
0.0	0.5	000	960	920	880	840	801	761	721	681	641	0.4	0.0
0.1	0.4	602	562	522	483	443	404	364	325	286	247	0.4	0.1
0.2	0.4	207	168	129	090	052	013	974	936	897	859	0.3	0.2
0.3	0.3	821	783	745	707	669	632	594	557	520	483		0.3
0.4		446	409	372	336	300	264	228	192	156	121	0.3	0.4
0.5	0.3	085	050	015	981	946	912	877	843	810	776	0.2	0.5
0.6	0.2	743	709	676	643	611	578	546	514	483	451		0.6
0.7		420	389	358	327	296	266	236	206	177	148	0.2	0.7
0.8	0.2	119	090	061	033	005	977	949	922	894	867	0.1	0.8
0.9	0.1	841	814	788	762	736	711	685	660	635	611		0.9
1.0		587	562	539	515	492	469	446	423	401	379		1.0
1.1		357	335	314	292	271	251	230	210	190	170	0.1	1.1
1.2	0.1	151	131	112	093	075	056	038	020	003	985	0.0	1.2
1.3	0.0	968	951	934	918	901	885	869	853	838	823		1.3
1.4		808	793	778	764	749	735	721	708	694	681		1.4
1.5		668	655	643	630	618	606	594	582	571	559		1.5
1.6		548	537	526	516	505	495	485	475	465	455		1.6
1.7		446	436	427	418	409	401	392	384	375	367		1.7
1.8		359	351	344	336	329	322	314	307	301	294		1.8
1.9		287	281	274	268	262	256	250	244	239	233		1.9
2.0		228	222	217	212	207	202	197	192	188	183		2.0
2.1		179	174	170	166	162	158	154	150	146	143		2.1
2.2		139	136	132	129	125	122	119	116	113	110	0.0	2.2
2.3	0.0	107	104	102	990	964	939	914	889	866	842	0.00	2.3
2.4	0.00	820	798	776	755	734	714	695	676	657	639		2.4
2.5		621	604	587	570	554	539	523	508	494	480		2.5
2.6		466	453	440	427	415	402	391	379	368	357		2.6
2.7		347	336	326	317	307	298	289	280	272	264		2.7
2.8		256	248	240	233	226	219	212	205	199	193		2.8
2.9	0.00	1.87	181	175	169	164	159	154	149	144	139	0.00	2.9

Source: Adapted with permission from O. L. Davies (ed.), *The Design and Analysis of Industrial Experiments*, 2nd ed., Oliver and Boyd, Edinburgh, 1956, condensed and adapted with permission from E. S. Pearson and H. O. Hartley, *Biometrika Tables for Statisticians*, Vol. 1, Cambridge University Press, New York, 1954.

Extension for Higher Values of the Deviate

Deviate (z)	Proportion of Whole Area (A)	Deviate (z)	Proportion of Whole Area (A)	Deviate (z)	Proportion of Whole Area (A)	Deviate (z)	Proportion of Whole Area (A)
3.0	0.00135	3.5	0.000233	4.0	0.0^4317	4.5	0.0^5340
3.1	0.000968	3.6	0.000159	4.1	0.0^4207	4.6	0.0^5211
3.2	0.000687	3.7	0.000108	4.2	0.0^4133	4.7	0.0^5130
3.3	0.000483	3.8	0.0^4723	4.3	0.0^5854	4.8	0.0^6793
3.4	0.000337	3.9	0.0^4481	4.4	0.0^5541	4.9	0.0^6479
						5.0	0.0^6287

Source: Adapted with permission from O. L. Davies (ed.), *The Design and Analysis of Industrial Experiments,* 2nd ed., Oliver and Boyd, Edinburgh, 1956, condensed and adapted with permission from E. S. Pearson and H. O. Hartley, *Biometrika Tables for Statisticians,* Vol. 1, Cambridge University Press, New York, 1954.

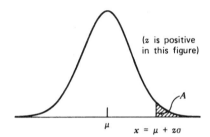

(z is positive in this figure)

μ

$x = \mu + z\sigma$

The illustration shows a normal curve. The scales are such that the total area under the curve is unity. The shaded portion is the area A given in the table above. The entries refer to positive values of the argument z. For negative values of z write down the complements of the entries.

Examples. Let $z = +1.96$. The prefix $= 0.0$ and the entry $= 250$, together $0.0250 =$ area to right. Area to left $= 1 - 0.0250 = 0.9750$.

Let $z = -3.00$. The tabulated value $= 0.00135$. Since z is negative, this represents the area to the *left*. Area to right $= 1 - 0.00135 = 0.99865$.

Let $z = +4.50$. Tabulated value $= 0.00000340$. Area to left $= 0.99999660$.

To find the value of z corresponding to a given A, we can use the table in reverse, thus:

Let area to right (i.e., A) $= 0.10$. The two adjacent tabulated values are $A = 0.1003$ for $z = 1.28$, and $A = 0.0985$ for $z = 1.29$. We interpolate linearly to obtain the required value of z. Thus $z = 1.28 + (3)(0.01)/18 = 1.2817$.

PERCENTAGE POINTS OF THE *t*-DISTRIBUTION

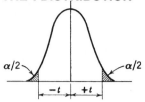

Distribution of *t*

"Probability = Area in Two Tails of Distribution Outside ± *t*-Value in Table"

Degrees of Freedom	\ Probability 0.9	0.7	0.5	0.3	0.2	0.1	0.05	0.02	0.01	0.001
1	0.158	0.510	1.000	1.963	3.078	6.314	12.706	31.821	63.657	636.619
2	0.142	0.445	0.816	1.386	1.886	2.920	4.303	6.965	9.925	31.598
3	0.137	0.424	0.765	1.250	1.638	2.353	3.182	4.541	5.841	12.924
4	0.134	0.414	0.741	1.190	1.533	2.132	2.776	3.747	4.604	8.610
5	0.132	0.408	0.727	1.156	1.476	2.015	2.571	3.365	4.032	6.869
6	0.131	0.404	0.718	1.134	1.440	1.943	2.447	3.143	3.707	5.959
7	0.130	0.402	0.711	1.119	1.415	1.895	2.365	2.998	3.499	5.408
8	0.130	0.399	0.706	1.108	1.397	1.860	2.306	2.896	3.355	5.041
9	0.129	0.398	0.703	1.100	1.383	1.833	2.262	2.821	3.250	4.781
10	0.129	0.397	0.700	1.093	1.372	1.812	2.228	2.764	3.169	4.587
11	0.129	0.396	0.697	1.088	1.363	1.796	2.201	2.718	3.106	4.437
12	0.128	0.395	0.695	1.083	1.356	1.782	2.179	2.681	3.055	4.318
13	0.128	0.394	0.694	1.079	1.350	1.771	2.160	2.650	3.012	4.221
14	0.128	0.393	0.692	1.076	1.345	1.761	2.145	2.624	2.977	4.140
15	0.128	0.393	0.691	1.074	1.341	1.753	2.131	2.602	2.947	4.073
16	0.128	0.392	0.690	1.071	1.337	1.746	2.120	2.583	2.921	4.015
17	0.128	0.392	0.689	1.069	1.333	1.740	2.110	2.567	2.898	3.965
18	0.127	0.392	0.688	1.067	1.330	1.734	2.101	2.552	2.878	3.922
19	0.127	0.391	0.688	1.066	1.328	1.729	2.093	2.539	2.861	3.883
20	0.127	0.391	0.687	1.064	1.325	1.725	2.086	2.528	2.845	3.850
21	0.127	0.391	0.686	1.063	1.323	1.721	2.080	2.518	2.831	3.819
22	0.127	0.390	0.686	1.061	1.321	1.717	2.074	2.508	2.819	3.792
23	0.127	0.390	0.685	1.060	1.319	1.714	2.069	2.500	2.807	3.767
24	0.127	0.390	0.685	1.059	1.318	1.711	2.064	2.492	2.797	3.745
25	0.127	0.390	0.684	1.058	1.316	1.708	2.060	2.485	2.787	3.725
26	0.127	0.390	0.684	1.058	1.315	1.706	2.056	2.479	2.779	3.707
27	0.127	0.389	0.684	1.057	1.314	1.703	2.052	2.473	2.771	3.690
28	0.127	0.389	0.683	1.056	1.313	1.701	2.048	2.467	2.763	3.674
29	0.127	0.389	0.683	1.055	1.311	1.699	2.045	2.462	2.756	3.659
30	0.127	0.389	0.683	1.055	1.310	1.697	2.042	2.457	2.750	3.646
40	0.126	0.388	0.681	1.050	1.303	1.684	2.021	2.423	2.704	3.551
60	0.126	0.387	0.679	1.046	1.296	1.671	2.000	2.390	2.660	3.460
120	0.126	0.386	0.677	1.041	1.289	1.658	1.980	2.358	2.617	3.373
∞	0.126	0.385	0.674	1.036	1.282	1.645	1.960	2.326	2.576	3.291

Source: Abridged from Table III of R. A. Fisher and F. Yates, *Statistical Tables for Biological, Agricultural and Medical Research* (6th ed.) published by Oliver and Boyd, Ltd., Edinburgh, 1964, by permission of the authors and publishers.

PERCENTAGE POINTS OF THE χ^2-DISTRIBUTION

χ^2-Distribution, Selected Upper Percentage Points for ν df, 10% (0.1) to 0.1% (0.001)

ν	\multicolumn{6}{c}{Upper-Tail Area Probability}					
	0.10	0.05	0.025	0.01	0.005	0.001
1	2.71	3.84	5.02	6.63	7.88	10.8
2	4.61	5.99	7.38	9.21	10.6	13.8
3	6.25	7.81	9.35	11.3	12.8	16.3
4	7.78	9.49	11.1	13.3	14.9	18.5
5	9.24	11.1	12.8	15.1	16.7	20.5
6	10.6	12.6	14.4	16.8	18.5	22.5
7	12.0	14.1	16.0	18.5	20.3	24.3
8	13.4	15.5	17.5	20.1	22.0	26.1
9	14.7	16.9	19.0	21.7	23.6	27.9
10	16.0	18.3	20.5	23.2	25.2	29.6
11	17.3	19.7	21.9	24.7	26.8	31.3
12	18.5	21.0	23.3	26.2	28.3	32.9
13	19.8	22.4	24.7	27.7	29.8	34.5
14	21.1	23.7	26.1	29.1	31.3	36.1
15	22.3	25.0	27.5	30.6	32.8	37.7
16	23.5	26.3	28.8	32.0	34.3	39.3
17	24.8	27.6	30.2	33.4	35.7	40.8
18	26.0	28.9	31.5	34.8	37.2	42.3
19	27.2	30.1	32.9	36.2	38.6	43.8
20	28.4	31.4	34.2	37.6	40.0	45.3
21	29.6	32.7	35.5	38.9	41.4	46.8
22	30.8	33.9	36.8	40.3	42.8	48.3
23	32.0	35.2	38.1	41.6	44.2	49.7
24	33.2	36.4	39.4	43.0	45.6	51.2
25	34.4	37.7	40.6	44.3	46.9	52.6
26	35.6	38.9	41.9	45.6	48.3	54.1
27	36.7	40.1	43.2	47.0	49.6	55.5
28	37.9	41.3	44.5	48.3	51.0	56.9
29	39.1	42.6	45.7	49.6	52.3	58.3
30	40.3	43.8	47.0	50.9	53.7	59.7
40	51.8	55.8	59.3	63.7	66.8	73.4
50	63.2	67.5	71.4	76.2	79.5	86.7
60	74.4	79.1	83.3	88.4	92.0	99.6
70	85.5	90.5	95.0	100.4	104.2	112.3
80	96.6	101.9	106.6	112.3	116.3	124.8
90	107.6	113.1	118.1	124.1	128.3	137.2
100	118.5	124.3	129.6	135.8	140.2	149.4

For $\nu > 100$, the "Wilson–Hilferty formula"

$$\chi^2 = \nu\left\{1 - \frac{2}{9\nu} + z\left(\frac{2}{9\nu}\right)^{1/2}\right\}^3$$

provides an approximation, where z is the appropriate normal deviate for the required tail area, namely:

Tail	0.1	0.05	0.025	0.01	0.005	0.001
z	1.2816	1.6449	1.9600	2.3263	2.5758	3.0902

PERCENTAGE POINTS OF THE *F*-DISTRIBUTION
F-Distribution, Upper 10% Points [$F(\nu_1, \nu_2, 0.90)$]

ν_2 \ ν_1	1	2	3	4	5	6	7	8	9	10
1	39.86	49.50	53.59	55.83	57.24	58.20	58.91	59.44	59.86	60.19
2	8.53	9.00	9.16	9.24	9.29	9.33	9.35	9.37	9.38	9.39
3	5.54	5.46	5.39	5.34	5.31	5.28	5.27	5.25	5.24	5.23
4	4.54	4.32	4.19	4.11	4.05	4.01	3.98	3.95	3.94	3.92
5	4.06	3.78	3.62	3.52	3.45	3.40	3.37	3.34	3.32	3.30
6	3.78	3.46	3.29	3.18	3.11	3.05	3.01	2.98	2.96	2.94
7	3.59	3.26	3.07	2.96	2.88	2.83	2.78	2.75	2.72	2.70
8	3.46	3.11	2.92	2.81	2.73	2.67	2.62	2.59	2.56	2.54
9	3.36	3.01	2.81	2.69	2.61	2.55	2.51	2.47	2.44	2.42
10	3.29	2.92	2.73	2.61	2.52	2.46	2.41	2.38	2.35	2.32
11	3.23	2.86	2.66	2.54	2.45	2.39	2.34	2.30	2.27	2.25
12	3.18	2.81	2.61	2.48	2.39	2.33	2.28	2.24	2.21	2.19
13	3.14	2.76	2.56	2.43	2.35	2.28	2.23	2.20	2.16	2.14
14	3.10	2.73	2.52	2.39	2.31	2.24	2.19	2.15	2.12	2.10
15	3.07	2.70	2.49	2.36	2.27	2.21	2.16	2.12	2.09	2.06
16	3.05	2.67	2.46	2.33	2.24	2.18	2.13	2.09	2.06	2.03
17	3.03	2.64	2.44	2.31	2.22	2.15	2.10	2.06	2.03	2.00
18	3.01	2.62	2.42	2.29	2.20	2.13	2.08	2.04	2.00	1.98
19	2.99	2.61	2.40	2.27	2.17	2.11	2.06	2.02	1.98	1.96
20	2.97	2.59	2.38	2.25	2.16	2.09	2.04	2.00	1.96	1.94
21	2.96	2.57	2.36	2.23	2.14	2.08	2.02	1.98	1.95	1.92
22	2.95	2.56	2.35	2.22	2.13	2.06	2.01	1.97	1.93	1.90
23	2.94	2.55	2.34	2.21	2.11	2.05	1.99	1.95	1.92	1.89
24	2.93	2.54	2.33	2.19	2.10	2.04	1.98	1.94	1.91	1.88
25	2.92	2.53	2.32	2.18	2.09	2.02	1.97	1.93	1.89	1.87
26	2.91	2.52	2.31	2.17	2.08	2.01	1.96	1.92	1.88	1.86
27	2.90	2.51	2.30	2.17	2.07	2.00	1.95	1.91	1.87	1.85
28	2.89	2.50	2.29	2.16	2.06	2.00	1.94	1.90	1.87	1.84
29	2.89	2.50	2.28	2.15	2.06	1.99	1.93	1.89	1.86	1.83
30	2.88	2.49	2.28	2.14	2.05	1.98	1.93	1.88	1.85	1.82
40	2.84	2.44	2.23	2.09	2.00	1.93	1.87	1.83	1.79	1.76
60	2.79	2.39	2.18	2.04	1.95	1.87	1.82	1.77	1.74	1.71
120	2.75	2.35	2.13	1.99	1.90	1.82	1.77	1.72	1.68	1.65
∞	2.71	2.30	2.08	1.94	1.85	1.77	1.72	1.67	1.63	1.60

Degrees of Freedom for Numerator

Source: Reproduced with permission from E. S. Pearson and H. O. Hartley, *Biometrika Tables for Statisticians*, Vol. 1, Cambridge University Press, New York, 1954.

10%

Degrees of Freedom for Numerator								
12	15	20	24	30	40	60	120	∞
60.71	61.22	61.74	62.00	62.26	62.53	62.79	63.06	63.33
9.41	9.42	9.44	9.45	9.46	9.47	9.47	9.48	9.49
5.22	5.20	5.18	5.18	5.17	5.16	5.15	5.14	5.13
3.90	3.87	3.84	3.83	3.82	3.80	3.79	3.78	3.76
3.27	3.24	3.21	3.19	3.17	3.16	3.14	3.12	3.10
2.90	2.87	2.84	2.82	2.80	2.78	2.76	2.74	2.72
2.67	2.63	2.59	2.58	2.56	2.54	2.51	2.49	2.47
2.50	2.46	2.42	2.40	2.38	2.36	2.34	2.32	2.29
2.38	2.34	2.30	2.28	2.25	2.23	2.21	2.18	2.16
2.28	2.24	2.20	2.18	2.16	2.13	2.11	2.08	2.06
2.21	2.17	2.12	2.10	2.08	2.05	2.03	2.00	1.97
2.15	2.10	2.06	2.04	2.01	1.99	1.96	1.93	1.90
2.10	2.05	2.01	1.98	1.96	1.93	1.90	1.88	1.85
2.05	2.01	1.96	1.94	1.91	1.89	1.86	1.83	1.80
2.02	1.97	1.92	1.90	1.87	1.85	1.82	1.79	1.76
1.99	1.94	1.89	1.87	1.84	1.81	1.78	1.75	1.72
1.96	1.91	1.86	1.84	1.81	1.78	1.75	1.72	1.69
1.93	1.89	1.84	1.81	1.78	1.75	1.72	1.69	1.66
1.91	1.86	1.81	1.79	1.76	1.73	1.70	1.67	1.63
1.89	1.84	1.79	1.77	1.74	1.71	1.68	1.64	1.61
1.87	1.83	1.78	1.75	1.72	1.69	1.66	1.62	1.59
1.86	1.81	1.76	1.73	1.70	1.67	1.64	1.60	1.57
1.84	1.80	1.74	1.72	1.69	1.66	1.62	1.59	1.55
1.83	1.78	1.73	1.70	1.67	1.64	1.61	1.57	1.53
1.82	1.77	1.72	1.69	1.66	1.63	1.59	1.56	1.52
1.81	1.76	1.71	1.68	1.65	1.61	1.58	1.54	1.50
1.80	1.75	1.70	1.67	1.64	1.60	1.57	1.53	1.49
1.79	1.74	1.69	1.66	1.63	1.59	1.56	1.52	1.48
1.78	1.73	1.68	1.65	1.62	1.58	1.55	1.51	1.47
1.77	1.72	1.67	1.64	1.61	1.57	1.54	1.50	1.46
1.71	1.66	1.61	1.57	1.54	1.51	1.47	1.42	1.38
1.66	1.60	1.54	1.51	1.48	1.44	1.40	1.35	1.29
1.60	1.55	1.48	1.45	1.41	1.37	1.32	1.26	1.19
1.55	1.49	1.42	1.38	1.34	1.30	1.24	1.17	1.00

F-Distribution, Upper 5% Points $[F(\nu_1, \nu_2, 0.95)]$

ν_2 \ ν_1	Degrees of Freedom for Numerator									
	1	2	3	4	5	6	7	8	9	10
1	161.4	199.5	215.7	224.6	230.2	234.0	236.8	238.9	240.5	241.9
2	18.51	19.00	19.16	19.25	19.30	19.33	19.35	19.37	19.38	19.40
3	10.13	9.55	9.28	9.12	9.01	8.94	8.89	8.85	8.81	8.79
4	7.71	6.94	6.59	6.39	6.26	6.16	6.09	6.04	6.00	5.96
5	6.61	5.79	5.41	5.19	5.05	4.95	4.88	4.82	4.77	4.74
6	5.99	5.14	4.76	4.53	4.39	4.28	4.21	4.15	4.10	4.06
7	5.59	4.74	4.35	4.12	3.97	3.87	3.79	3.73	3.68	3.64
8	5.32	4.46	4.07	3.84	3.69	3.58	3.50	3.44	3.39	3.35
9	5.12	4.26	3.86	3.63	3.48	3.37	3.29	3.23	3.18	3.14
10	4.96	4.10	3.71	3.48	3.33	3.22	3.14	3.07	3.02	2.98
11	4.84	3.98	3.59	3.36	3.20	3.09	3.01	2.95	2.90	2.85
12	4.75	3.89	3.49	3.26	3.11	3.00	2.91	2.85	2.80	2.75
13	4.67	3.81	3.41	3.18	3.03	2.92	2.83	2.77	2.71	2.67
14	4.60	3.74	3.34	3.11	2.96	2.85	2.76	2.70	2.65	2.60
15	4.54	3.68	3.29	3.06	2.90	2.79	2.71	2.64	2.59	2.54
16	4.49	3.63	3.24	3.01	2.85	2.74	2.66	2.59	2.54	2.49
17	4.45	3.59	3.20	2.96	2.81	2.70	2.61	2.55	2.49	2.45
18	4.41	3.55	3.16	2.93	2.77	2.66	2.58	2.51	2.46	2.41
19	4.38	3.52	3.13	2.90	2.74	2.63	2.54	2.48	2.42	2.38
20	4.35	3.49	3.10	2.87	2.71	2.60	2.51	2.45	2.39	2.35
21	4.32	3.47	3.07	2.84	2.68	2.57	2.49	2.42	2.37	2.32
22	4.30	3.44	3.05	2.82	2.66	2.55	2.46	2.40	2.34	2.30
23	4.28	3.42	3.03	2.80	2.64	2.53	2.44	2.37	2.32	2.27
24	4.26	3.40	3.01	2.78	2.62	2.51	2.42	2.36	2.30	2.25
25	4.24	3.39	2.99	2.76	2.60	2.49	2.40	2.34	2.28	2.24
26	4.23	3.37	2.98	2.74	2.59	2.47	2.39	2.32	2.27	2.22
27	4.21	3.35	2.96	2.73	2.57	2.46	2.37	2.31	2.25	2.20
28	4.20	3.34	2.95	2.71	2.56	2.45	2.36	2.29	2.24	2.19
29	4.18	3.33	2.93	2.70	2.55	2.43	2.35	2.28	2.22	2.18
30	4.17	3.32	2.92	2.69	2.53	2.42	2.33	2.27	2.21	2.16
40	4.08	3.23	2.84	2.61	2.45	2.34	2.25	2.18	2.12	2.08
60	4.00	3.15	2.76	2.53	2.37	2.25	2.17	2.10	2.04	1.99
120	3.92	3.07	2.68	2.45	2.29	2.17	2.09	2.02	1.96	1.91
∞	3.84	3.00	2.60	2.37	2.21	2.10	2.01	1.94	1.88	1.83

Source: Reproduced with permission from E. S. Pearson and H. O. Hartley, *Biometrika Tables for Statisticians,* Vol. 1, Cambridge University Press, New York, 1954.

5%

Degrees of Freedom for Numerator								
12	15	20	24	30	40	60	120	∞
243.9	245.9	248.0	249.1	250.1	251.1	252.2	253.3	254.3
19.41	19.43	19.45	19.45	19.46	19.47	19.48	19.49	19.50
8.74	8.70	8.66	8.64	8.62	8.59	8.57	8.55	8.53
5.91	5.86	5.80	5.77	5.75	5.72	5.69	5.66	5.63
4.68	4.62	4.56	4.53	4.50	4.46	4.43	4.40	4.36
4.00	3.94	3.87	3.84	3.81	3.77	3.74	3.70	3.67
3.57	3.51	3.44	3.41	3.38	3.34	3.30	3.27	3.23
3.28	3.22	3.15	3.12	3.08	3.04	3.01	2.97	2.93
3.07	3.01	2.94	2.90	2.86	2.83	2.79	2.75	2.71
2.91	2.85	2.77	2.74	2.70	2.66	2.62	2.58	2.54
2.79	2.72	2.65	2.61	2.57	2.53	2.49	2.45	2.40
2.69	2.62	2.54	2.51	2.47	2.43	2.38	2.34	2.30
2.60	2.53	2.46	2.42	2.38	2.34	2.30	2.25	2.21
2.53	2.46	2.39	2.35	2.31	2.27	2.22	2.18	2.13
2.48	2.40	2.33	2.29	2.25	2.20	2.16	2.11	2.07
2.42	2.35	2.28	2.24	2.19	2.15	2.11	2.06	2.01
2.38	2.31	2.23	2.19	2.15	2.10	2.06	2.01	1.96
2.34	2.27	2.19	2.15	2.11	2.06	2.02	1.97	1.92
2.31	2.23	2.16	2.11	2.07	2.03	1.98	1.93	1.88
2.28	2.20	2.12	2.08	2.04	1.99	1.95	1.90	1.84
2.25	2.18	2.10	2.05	2.01	1.96	1.92	1.87	1.81
2.23	2.15	2.07	2.03	1.98	1.94	1.89	1.84	1.78
2.20	2.13	2.05	2.01	1.96	1.91	1.86	1.81	1.76
2.18	2.11	2.03	1.98	1.94	1.89	1.84	1.79	1.73
2.16	2.09	2.01	1.96	1.92	1.87	1.82	1.77	1.71
2.15	2.07	1.99	1.95	1.90	1.85	1.80	1.75	1.69
2.13	2.06	1.97	1.93	1.88	1.84	1.79	1.73	1.67
2.12	2.04	1.96	1.91	1.87	1.82	1.77	1.71	1.65
2.10	2.03	1.94	1.90	1.85	1.81	1.75	1.70	1.64
2.09	2.01	1.93	1.89	1.84	1.79	1.74	1.68	1.62
2.00	1.92	1.84	1.79	1.74	1.69	1.64	1.58	1.51
1.92	1.84	1.75	1.70	1.65	1.59	1.53	1.47	1.39
1.83	1.75	1.66	1.61	1.55	1.50	1.43	1.35	1.25
1.75	1.67	1.57	1.52	1.46	1.39	1.32	1.22	1.00

F-Distribution, Upper 1% Points [$F(\nu_1, \nu_2, 0.99)$]

ν_2 \ ν_1	Degrees of Freedom for Numerator									
	1	2	3	4	5	6	7	8	9	10
1	4052	4999.5	5403	5625	5764	5859	5928	5982	6022	6056
2	98.50	99.00	99.17	99.25	99.30	99.33	99.36	99.37	99.39	99.40
3	34.12	30.82	29.46	28.71	28.24	27.91	27.67	27.49	27.35	27.23
4	21.20	18.00	16.69	15.98	15.52	15.21	14.98	14.80	14.66	14.55
5	16.26	13.27	12.06	11.39	10.97	10.67	10.46	10.29	10.16	10.05
6	13.75	10.92	9.78	9.15	8.75	8.47	8.26	8.10	7.98	7.87
7	12.25	9.55	8.45	7.85	7.46	7.19	6.99	6.84	6.72	6.62
8	11.26	8.65	7.59	7.01	6.63	6.37	6.18	6.03	5.91	5.81
9	10.56	8.02	6.99	6.42	6.06	5.80	5.61	5.47	5.35	5.26
10	10.04	7.56	6.55	5.99	5.64	5.39	5.20	5.06	4.94	4.85
11	9.65	7.21	6.22	5.67	5.32	5.07	4.89	4.74	4.63	4.54
12	9.33	6.93	5.95	5.41	5.06	4.82	4.64	4.50	4.39	4.30
13	9.07	6.70	5.74	5.21	4.86	4.62	4.44	4.30	4.19	4.10
14	8.86	6.51	5.56	5.04	4.69	4.46	4.28	4.14	4.03	3.94
15	8.68	6.36	5.42	4.89	4.56	4.32	4.14	4.00	3.89	3.80
16	8.53	6.23	5.29	4.77	4.44	4.20	4.03	3.89	3.78	3.69
17	8.40	6.11	5.18	4.67	4.34	4.10	3.93	3.79	3.68	3.59
18	8.29	6.01	5.09	4.58	4.25	4.01	3.84	3.71	3.60	3.51
19	8.18	5.93	5.01	4.50	4.17	3.94	3.77	3.63	3.52	3.43
20	8.10	5.85	4.94	4.43	4.10	3.87	3.70	3.56	3.46	3.37
21	8.02	5.78	4.87	4.37	4.04	3.81	3.64	3.51	3.40	3.31
22	7.95	5.72	4.82	4.31	3.99	3.76	3.59	3.45	3.35	3.26
23	7.88	5.66	4.76	4.26	3.94	3.71	3.54	3.41	3.30	3.21
24	7.82	5.61	4.72	4.22	3.90	3.67	3.50	3.36	3.26	3.17
25	7.77	5.57	4.68	4.18	3.85	3.63	3.46	3.32	3.22	3.13
26	7.72	5.53	4.64	4.14	3.82	3.59	3.42	3.29	3.18	3.09
27	7.68	5.49	4.60	4.11	3.78	3.56	3.39	3.26	3.15	3.06
28	7.64	5.45	4.57	4.07	3.75	3.53	3.36	3.23	3.12	3.03
29	7.60	5.42	4.54	4.04	3.73	3.50	3.33	3.20	3.09	3.00
30	7.56	5.39	4.51	4.02	3.70	3.47	3.30	3.17	3.07	2.98
40	7.31	5.18	4.31	3.83	3.51	3.29	3.12	2.99	2.89	2.80
60	7.08	4.98	4.13	3.65	3.34	3.12	2.95	2.82	2.72	2.63
120	6.85	4.79	3.95	3.48	3.17	2.96	2.79	2.66	2.56	2.47
∞	6.63	4.61	3.78	3.32	3.02	2.80	2.64	2.51	2.41	2.32

Source: Reproduced with permission from E. S. Pearson and H. O. Hartley, *Biometrika Tables for Statisticians,* Vol. 1, Cambridge University Press, New York, 1954.

1%

			Degrees of Freedom for Numerator					
12	15	20	24	30	40	60	120	∞
6106	6157	6209	6235	6261	6287	6313	6339	6366
99.42	99.43	99.45	99.46	99.47	99.47	99.48	99.49	99.50
27.05	26.87	26.69	26.60	26.50	26.41	26.32	26.22	26.13
14.37	14.20	14.02	13.93	13.84	13.75	13.65	13.56	13.46
9.89	9.72	9.55	9.47	9.38	9.29	9.20	9.11	9.02
7.72	7.56	7.40	7.31	7.23	7.14	7.06	6.97	6.88
6.47	6.31	6.16	6.07	5.99	5.91	5.82	5.74	5.65
5.67	5.52	5.36	5.28	5.20	5.12	5.03	4.95	4.86
5.11	4.96	4.81	4.73	4.65	4.57	4.48	4.40	4.31
4.71	4.56	4.41	4.33	4.25	4.17	4.08	4.00	3.91
4.40	4.25	4.10	4.02	3.94	3.86	3.78	3.69	3.60
4.16	4.01	3.86	3.78	3.70	3.62	3.54	3.45	3.36
3.96	3.82	3.66	3.59	3.51	3.43	3.34	3.25	3.17
3.80	3.66	3.51	3.43	3.35	3.27	3.18	3.09	3.00
3.67	3.52	3.37	3.29	3.21	3.13	3.05	2.96	2.87
3.55	3.41	3.26	3.18	3.10	3.02	2.93	2.84	2.75
3.46	3.31	3.16	3.08	3.00	2.92	2.83	2.75	2.65
3.37	3.23	3.08	3.00	2.92	2.84	2.75	2.66	2.57
3.30	3.15	3.00	2.92	2.84	2.76	2.67	2.58	2.49
3.23	3.09	2.94	2.86	2.78	2.69	2.61	2.52	2.42
3.17	3.03	2.88	2.80	2.72	2.64	2.55	2.46	2.36
3.12	2.98	2.83	2.75	2.67	2.58	2.50	2.40	2.31
3.07	2.93	2.78	2.70	2.62	2.54	2.45	2.35	2.26
3.03	2.89	2.74	2.66	2.58	2.49	2.40	2.31	2.21
2.99	2.85	2.70	2.62	2.54	2.45	2.36	2.27	2.17
2.96	2.81	2.66	2.58	2.50	2.42	2.33	2.23	2.13
2.93	2.78	2.63	2.55	2.47	2.38	2.29	2.20	2.10
2.90	2.75	2.60	2.52	2.44	2.35	2.26	2.17	2.06
2.87	2.73	2.57	2.49	2.41	2.33	2.23	2.14	2.03
2.84	2.70	2.55	2.47	2.39	2.30	2.21	2.11	2.01
2.66	2.52	2.37	2.29	2.20	2.11	2.02	1.92	1.80
2.50	2.35	2.20	2.12	2.03	1.94	1.84	1.73	1.60
2.34	2.19	2.03	1.95	1.86	1.76	1.66	1.53	1.38
2.18	2.04	1.88	1.79	1.70	1.59	1.47	1.32	1.00

Index of Authors Associated with Exercises

(To facilitate agreed copyright acknowledgments, any related references are quoted in the exercises indicated. Names may be cross-checked against the sequential listing by exercise number which follows, in order to find associated coauthors.)

Listing by Exercise Number, Authors, Year

Index

(Note: Authors of books and articles listed in Sections 17.7 (p. 396), 18.5 (p. 408), 25.9 (pp. 581–584), 26.4 (p. 588) and in the main Bibliography (pp. 593–603) are not individually referenced in this index unless their names appear otherwise in the text.)

WILEY SERIES IN PROBABILITY AND STATISTICS
ESTABLISHED BY WALTER A. SHEWHART AND SAMUEL S. WILKS

Editors
Vic Barnett, Ralph A. Bradley, Noel A. C. Cressie, Nicholas I. Fisher,
Iain M. Johnstone, J. B. Kadane, David G. Kendall, David W. Scott,
Bernard W. Silverman, Adrian F. M. Smith, Jozef L. Teugels,
Geoffrey S. Watson; J. Stuart Hunter, Emeritus

Probability and Statistics Section

*ANDERSON · The Statistical Analysis of Time Series
ARNOLD, BALAKRISHNAN, and NAGARAJA · A First Course in Order Statistics
BACCELLI, COHEN, OLSDER, and QUADRAT · Synchronization and Linearity:
 An Algebra for Discrete Event Systems
BASILEVSKY · Statistical Factor Analysis and Related Methods: Theory and
 Applications
BERNARDO and SMITH · Bayesian Statistical Concepts and Theory
BILLINGSLEY · Convergence of Probability Measures
BOROVKOV · Asymptotic Methods in Queuing Theory
BRANDT, FRANKEN, and LISEK · Stationary Stochastic Models
CAINES · Linear Stochastic Systems
CAIROLI and DALANG · Sequential Stochastic Optimization
CONSTANTINE · Combinatorial Theory and Statistical Design
COVER and THOMAS · Elements of Information Theory
CSÖRGŐ and HORVÁTH · Weighted Approximations in Probability Statistics
CSÖRGŐ and HORVÁTH · Limit Theorems in Change Point Analysis
DETTE and STUDDEN · The Theory of Canonical Moments with Applications in
 Statistics, Probability, and Analysis
*DOOB · Stochastic Processes
DRYDEN and MARDIA · Statistical Analysis of Shape
DUPUIS and ELLIS · A Weak Convergence Approach to the Theory of Large Deviations
ETHIER and KURTZ · Markov Processes: Characterization and Convergence
FELLER · An Introduction to Probability Theory and Its Applications, Volume 1,
 Third Edition, Revised; Volume II, *Second Edition*
FULLER · Introduction to Statistical Time Series, *Second Edition*
FULLER · Measurement Error Models
GELFAND and SMITH · Bayesian Computation
GHOSH, MUKHOPADHYAY, and SEN · Sequential Estimation
GIFI · Nonlinear Multivariate Analysis
GUTTORP · Statistical Inference for Branching Processes
HALL · Introduction to the Theory of Coverage Processes
HAMPEL · Robust Statistics: The Approach Based on Influence Functions
HANNAN and DEISTLER · The Statistical Theory of Linear Systems
HUBER · Robust Statistics
IMAN and CONOVER · A Modern Approach to Statistics
JUREK and MASON · Operator-Limit Distributions in Probability Theory
KASS and VOS · Geometrical Foundations of Asymptotic Inference
KAUFMAN and ROUSSEEUW · Finding Groups in Data: An Introduction to Cluster
 Analysis

*Now available in a lower priced paperback edition in the Wiley Classics Library.

*Now available in a lower priced paperback edition in the Wiley Classics Library.

*Now available in a lower priced paperback edition in the Wiley Classics Library.

*Now available in a lower priced paperback edition in the Wiley Classics Library.

*Now available in a lower priced paperback edition in the Wiley Classics Library.

Texts and References Section

AGRESTI · An Introduction to Categorical Data Analysis

ANDERSON · An Introduction to Multivariate Statistical Analysis, *Second Edition*

ANDERSON and LOYNES · The Teaching of Practical Statistics

ARMITAGE and COLTON · Encyclopedia of Biostatistics: Volumes 1 to 6 with Index

BARTOSZYNSKI and NIEWIADOMSKA-BUGAJ · Probability and Statistical Inference

BERRY, CHALONER, and GEWEKE · Bayesian Analysis in Statistics and Econometrics: Essays in Honor of Arnold Zellner

BHATTACHARYA and JOHNSON · Statistical Concepts and Methods

BILLINGSLEY · Probability and Measure, *Second Edition*

BOX · R. A. Fisher, the Life of a Scientist

BOX, HUNTER, and HUNTER · Statistics for Experimenters: An Introduction to Design, Data Analysis, and Model Building

BOX and LUCEÑO · Statistical Control by Monitoring and Feedback Adjustment

BROWN and HOLLANDER · Statistics: A Biomedical Introduction

CHATTERJEE and PRICE · Regression Analysis by Example, *Second Edition*

COOK and WEISBERG · An Introduction to Regression Graphics

COX · A Handbook of Introductory Statistical Methods

DILLON and GOLDSTEIN · Multivariate Analysis: Methods and Applications

DODGE and ROMIG · Sampling Inspection Tables, *Second Edition*

DRAPER and SMITH · Applied Regression Analysis, *Third Edition*

DUDEWICZ and MISHRA · Modern Mathematical Statistics

DUNN · Basic Statistics: A Primer for the Biomedical Sciences, *Second Edition*

FISHER and VAN BELLE · Biostatistics: A Methodology for the Health Sciences

FREEMAN and SMITH · Aspects of Uncertainty: A Tribute to D. V. Lindley

GROSS and HARRIS · Fundamentals of Queueing Theory, *Third Edition*

HALD · A History of Probability and Statistics and their Applications Before 1750

HALD · A History of Mathematical Statistics from 1750 to 1930

HELLER · MACSYMA for Statisticians

HOEL · Introduction to Mathematical Statistics, *Fifth Edition*

JOHNSON and BALAKRISHNAN · Advances in the Theory and Practice of Statistics: A Volume in Honor of Samuel Kotz

JOHNSON and KOTZ (editors) · Leading Personalities in Statistical Sciences: From the Seventeenth Century to the Present

JUDGE, GRIFFITHS, HILL, LÜTKEPOHL, and LEE · The Theory and Practice of Econometrics, *Second Edition*

KHURI · Advanced Calculus with Applications in Statistics

KOTZ and JOHNSON (editors) · Encyclopedia of Statistical Sciences: Volumes 1 to 9 wtih Index

KOTZ and JOHNSON (editors) · Encyclopedia of Statistical Sciences: Supplement Volume

KOTZ, REED, and BANKS (editors) · Encyclopedia of Statistical Sciences: Update Volume 1

KOTZ, REED, and BANKS (editors) · Encyclopedia of Statistical Sciences: Update Volume 2

LAMPERTI · Probability: A Survey of the Mathematical Theory, *Second Edition*

LARSON · Introduction to Probability Theory and Statistical Inference, *Third Edition*

LE · Applied Survival Analysis

MALLOWS · Design, Data, and Analysis by Some Friends of Cuthbert Daniel

MARDIA · The Art of Statistical Science: A Tribute to G. S. Watson

MASON, GUNST, and HESS · Statistical Design and Analysis of Experiments with Applications to Engineering and Science

MURRAY · X-STAT 2.0 Statistical Experimentation, Design Data Analysis, and Nonlinear Optimization

*Now available in a lower priced paperback edition in the Wiley Classics Library.

Texts and References (Continued)

PURI, VILAPLANA, and WERTZ · New Perspectives in Theoretical and Applied
 Statistics
RENCHER · Methods of Multivariate Analysis
RENCHER · Multivariate Statistical Inference with Applications
ROSS · Introduction to Probability and Statistics for Engineers and Scientists
ROHATGI · An Introduction to Probability Theory and Mathematical Statistics
RYAN · Modern Regression Methods
SCHOTT · Matrix Analysis for Statistics
SEARLE · Matrix Algebra Useful for Statistics
STYAN · The Collected Papers of T. W. Anderson: 1943–1985
TIERNEY · LISP-STAT: An Object-Oriented Environment for Statistical Computing
 and Dynamic Graphics
WONNACOTT and WONNACOTT · Econometrics, *Second Edition*

WILEY SERIES IN PROBABILITY AND STATISTICS

ESTABLISHED BY WALTER A. SHEWHART AND SAMUEL S. WILKS

Editors
*Robert M. Groves, Graham Kalton, J. N. K. Rao, Norbert Schwarz,
Christopher Skinner*

Survey Methodology Section

BIEMER, GROVES, LYBERG, MATHIOWETZ, and SUDMAN · Measurement
 Errors in Surveys
COCHRAN · Sampling Techniques, *Third Edition*
COX, BINDER, CHINNAPPA, CHRISTIANSON, COLLEDGE, and KOTT (editors) ·
 Business Survey Methods
*DEMING · Sample Design in Business Research
DILLMAN · Mail and Telephone Surveys: The Total Design Method
GROVES and COUPER · Nonresponse in Household Interview Surveys
GROVES · Survey Errors and Survey Costs
GROVES, BIEMER, LYBERG, MASSEY, NICHOLLS, and WAKSBERG ·
 Telephone Survey Methodology
*HANSEN, HURWITZ, and MADOW · Sample Survey Methods and Theory,
 Volume 1: Methods and Applications
*HANSEN, HURWITZ, and MADOW · Sample Survey Methods and Theory,
 Volume II: Theory
KASPRZYK, DUNCAN, KALTON, and SINGH · Panel Surveys
KISH · Statistical Design for Research
*KISH · Survey Sampling
LESSLER and KALSBEEK · Nonsampling Error in Surveys
LEVY and LEMESHOW · Sampling of Populations: Methods and Applications
LYBERG, BIEMER, COLLINS, de LEEUW, DIPPO, SCHWARZ, TREWIN (editors) ·
 Survey Measurement and Process Quality
SKINNER, HOLT, and SMITH · Analysis of Complex Surveys

*Now available in a lower priced paperback edition in the Wiley Classics Library.